AF861502

NONLINEAR SURFACE ELECTROMAGNETIC PHENOMENA

MODERN PROBLEMS IN CONDENSED MATTER SCIENCES

Volume 29

Series editors

V.M. AGRANOVICH

Moscow, USSR

A.A. MARADUDIN

Irvine, California, USA

Advisory editorial board

F. Abelès, Paris, France
F. Bassani, Pisa, Italy
N. Bloembergen, Cambridge, MA, USA
E. Burstein, Philadelphia, PA, USA
I.L. Fabelinskii, Moscow, USSR
P. Fulde, Stuttgart, FRG
M.D. Galanin, Moscow, USSR
V.L. Ginzburg, Moscow, USSR
H. Haken, Stuttgart, FRG
R.M. Hochstrasser, Philadelphia, PA, USA
I.P. Ipatova, Leningrad, USSR
A.A. Kaplyanskii, Leningrad, USSR
L.V. Keldysh, Moscow, USSR
R. Kubo, Tokyo, Japan
R. Loudon, Colchester, UK
Yu.A. Ossipyan, Moscow, USSR
L.P. Pitaevskii, Moscow, USSR
A.M. Prokhorov, Moscow, USSR
K.K. Rebane, Tallinn, USSR
J.M. Rowell, Red Bank, NJ, USA

N·H

NORTH-HOLLAND
AMSTERDAM · LONDON · NEW YORK · TOKYO

NONLINEAR SURFACE ELECTROMAGNETIC PHENOMENA

Volume editors

H.-E. PONATH

Friedrich-Schiller Universität
Jena, Germany

G.I. STEGEMAN

CREOL
University of Central Florida
Orlando, FL, USA

1991
NORTH-HOLLAND
AMSTERDAM · LONDON · NEW YORK · TOKYO

© Elsevier Science Publishers B.V., 1991

All rights reserved. No part of this publication may be reproduced, stored in a retrieval system, or transmitted, in any form or by any means, electronic, mechanical, photocopying, recording or otherwise, without the written permission of the publisher, Elsevier Science Publishers B.V., P.O. Box 211, 1000 AE Amsterdam, The Netherlands.
Special regulations for readers in the USA: This publication has been registered with the Copyright Clearance Center Inc. (CCC), Salem, Massachusetts. Information can be obtained from the CCC about conditions under which photocopies of parts of this publication may be made in the USA. All other copyright questions, including photocopying outside of the USA, should be referred to the publisher.

No responsibility is assumed by the Publisher for any injury and/or damage to persons or property as a matter of products liability, negligence or otherwise, or from any use or operation of any methods, products, instructions or ideas contained in the material herein.

ISBN: 0 444 88359 2

North-Holland

Elsevier Science Publishers B.V.
P.O. Box 211
1000 AE Amsterdam
The Netherlands

Sole distributors for the USA and Canada:

Elsevier Science Publishing Company, Inc.
655 Avenue of the Americas
New York, NY 10010
USA

Library of Congress Cataloging-in-Publication Data

Nonlinear surface electromagnetic phenomena / volume editors,
H.-E. Ponath, G.I. Stegeman.
p. cm. – (Modern problems in condensed matter sciences, vol. 29)
Includes bibliographical references and index.
ISBN 0-444-88359-2
1. Surfaces (Physics) 2. Electromagnetic surface waves. 3. Nonlinear theories. I. Ponath, H.-E. II. Stegeman, G.I. III. Series.
QC173.4.S94N66 1991
530.4'27–dc20 91-23834
CIP

Printed on acid-free paper
Printed in The Netherlands

MODERN PROBLEMS IN CONDENSED MATTER SCIENCES

Vol. 1. SURFACE POLARITONS
V.M. Agranovich and D.L. Mills, *editors*

Vol. 2. EXCITONS
E.I. Rashba and M.D. Sturge, *editors*

Vol. 3. ELECTRONIC EXCITATION ENERGY TRANSFER IN CONDENSED MATTER
V.M. Agranovich and M.D. Galanin

Vol. 4. SPECTROSCOPY AND EXCITATION DYNAMICS OF CONDENSED MOLECULAR SYSTEMS
V.M. Agranovich and R.M. Hochstrasser, *editors*

Vol. 5. LIGHT SCATTERING NEAR PHASE TRANSITIONS
H.Z. Cummins and A.P. Levanyuk, *editors*

Vol. 6. ELECTRON–HOLE DROPLETS IN SEMICONDUCTORS
C.D. Jeffries and L.V. Keldysh, *editors*

Vol. 7. THE DYNAMICAL JAHN–TELLER EFFECT IN LOCALIZED SYSTEMS
Yu.E. Perlin and M. Wagner, *editors*

Vol. 8. OPTICAL ORIENTATION
F. Meier and B.P. Zakharchenya, *editors*

Vol. 9. SURFACE EXCITATIONS
V.M. Agranovich and R. Loudon, *editors*

Vol. 10. ELECTRON–ELECTRON INTERACTIONS IN DISORDERED SYSTEMS
A.L. Efros and M. Pollak, *editors*

Vol. 11. MEDIUM-ENERGY ION REFLECTION FROM SOLIDS
E.S. Mashkova and V.A. Molchanov

Vol. 12. NONEQUILIBRIUM SUPERCONDUCTIVITY
D.N. Langenberg and A.I. Larkin, *editors*

MODERN PROBLEMS IN CONDENSED MATTER SCIENCES

Vol. 13. PHYSICS OF RADIATION EFFECTS IN CRYSTALS
R.A. Johnson and A.N. Orlov, *editors*

Vol. 14. INCOMMENSURATE PHASES IN DIELECTRICS
(Two volumes)
R. Blinc and A.P. Levanyuk, *editors*

Vol. 15. UNITARY TRANSFORMATIONS IN SOLID STATE PHYSICS
M. Wagner

Vol. 16. NONEQUILIBRIUM PHONONS IN NONMETALLIC CRYSTALS
W. Eisenmenger and A.A. Kaplyanskii, *editors*

Vol. 17. SOLITONS
S.E. Trullinger, V.L. Pokrovskii and V.E. Zakharov, *editors*

Vol. 18. TRANSPORT IN PHONON SYSTEMS
V.L. Gurevich

Vol. 19. CARRIER SCATTERING IN METALS AND SEMICONDUCTORS
V.F. Gantmakher and I.B. Levinson

Vol. 20. SEMIMETALS – 1. GRAPHITE AND ITS COMPOUNDS
N.B. Brandt, S.M. Chudinov and Ya.G. Ponomarev

Vol. 21. SPECTROSCOPY OF SOLIDS CONTAINING RARE EARTH IONS
A.A. Kaplyanskii and R.M. Macfarlane, *editors*

Vol. 22. SPIN WAVES AND MAGNETIC EXCITATIONS
(Two volumes)
A.S. Borovik-Romanov and S.K. Sinha, *editors*

Vol. 23. OPTICAL PROPERTIES OF MIXED CRYSTALS
R.J. Elliott and I.P. Ipatova, *editors*

MODERN PROBLEMS IN CONDENSED MATTER SCIENCES

Vol. 24. THE DIELECTRIC FUNCTION OF CONDENSED SYSTEMS
L.V. Keldysh, D.A. Kirzhnitz and A.A. Maradudin, *editors*

Vol. 25. CHARGE DENSITY WAVES IN SOLIDS
L.P. Gor'kov and G. Grüner, *editors*

Vol. 26. HELIUM THREE
W.P. Halperin and L.P. Pitaevskii, *editors*

Vol. 27. LANDAU LEVEL SPECTROSCOPY
(Two volumes)
G. Landwehr and E.I. Rashba, *editors*

Vol. 28. HOPPING TRANSPORT IN SOLIDS
M. Pollak and B. Shklovskii, *editors*

Vol. 29. NONLINEAR SURFACE ELECTROMAGNETIC PHENOMENA
H.E. Ponath and G. Stegeman, *editors*

Vol. 30. MESOSCOPIC PHENOMENA IN SOLIDS
B.L. Altshuler, P.A. Lee and R.A. Webb, *editors*

In preparation

ELECTRIC STRAIN FIELDS AND DISLOCATION MOBILITY
V.L. Indenbom and J. Lothe, *editors*

ELECTRONIC PHASE TRANSITIONS
W. Hanke and Yu. Kopaev, *editors*

NONRADIATIVE RECOMBINATION IN SEMICONDUCTORS
V.N. Abakumov, V.I. Perel and I.N. Yassievich

Oh, how many of them there
are in the fields!
But each flowers in its
own way –
In this is the highest achievement
of a flower!

Matsuo Bashó
1644–1694

PREFACE TO THE SERIES

Our understanding of condensed matter is developing rapidly at the present time, and the numerous new insights gained in this field define to a significant degree the face of contemporary science. Furthermore, discoveries made in this area are shaping present and future technology. This being so, it is clear that the most important results and directions for future developments can only be covered by an international group of authors working in cooperation.

"Modern Problems in Condensed Matter Sciences" is a series of contributed volumes and monographs on condensed matter science that is published by Elsevier Science Publishers under the imprint of North-Holland. With the support of a distinguished Advisory Editorial Board, areas of current interest that have reached a maturity to be reviewed, are selected for the series. Both Soviet and Western scholars are contributing to the series, and each contributed volume has, accordingly, two editors. Monographs, written by either Western or Soviet authors, are also included. The complete series will provide the most comprehensive coverage available of condensed matter science.

Another important outcome of the foundation of this series is the emergence of a rather interesting and fruitful form of collaboration among scholars from different countries. We are deeply convinced that such international collaboration in the spheres of science and art, as well as other socially useful spheres of human activity, will assist in the establishment of a climate of confidence and peace.

The publishing house 'Nauka' publishes the volumes in the Russian language. This way the broadest possible readership is ensured.

The General Editors of the Series,

V.M. Agranovich A.A. Maradudin

PREFACE

In recent years the physics of electromagnetic surface phenomena has developed rapidly, evolving in some recent cases into technologies for communications and industry, for example, fiber and integrated optics. The variety of phenomena based on electromagnetism at surfaces is rich and the need for research in this field is clear, even to the non-specialist. For many years research on guided waves was concentrated on understanding the fundamentals of linear propagation at surfaces and in waveguides. This included research on surface polaritons of various types, and integrated and fiber optics. This work was a necessary first step to the investigation of nonlinear interactions at surfaces and in layered ordered systems. Most recently, the multiplicity and richness of novel phenomena which occurs when nonlinearity is introduced into surfaces and waveguides has attracted research interest to this field. For example, many fundamental questions have been raised concerning the dynamic nonlinear behaviour of structured systems and surfaces interacting with ultrafast electromagnetic excitations. All of these nonlinear phenomena have forced researchers to refine their understanding of supposedly well-known linear relationships, as well as to devise novel methods for performing theoretical and experimental studies. Such nonlinear processes require a wide range of physics and optics in order to deal with electromagnetic surface phenomena. They also offer novel possibilities for applications based on a deep understanding of the physics. The present book was written with the goal of summarizing the available knowledge in selected areas of nonlinear electromagnetic surface phenomena.

The first three chapters deal with similar problems. Nonlinear waves based on second- and third-order nonlinearities are investigated with respect to propagation and interaction for both weakly and strongly nonlinear waveguides. Second-order effects which give rise to important practical devices, such as second-harmonic generators and optical parametric amplifiers, are discussed in chapter 1. Chapter 2 contains a complete summary of the current knowledge about steady-state wave propagation in geometries containing multiple layers some of which exhibit strong nonlinearities. The effects considered here cannot be described by first-order perturbation theory (i.e. coupled mode theory). Detailed solutions to nonlinear wave equations are needed. The problem of the stability of nonlinear waves guided by surfaces and waveguides of various symmetry is highlighted in chapter 3, as the principal means of exciting such strongly nonlinear surface waves. Non-steady-state surface and guided waves in materials with large nonlinearities are not treated in the present book. The material contained in these first three chapters is particularly relevant to optoelectronics and photonics.

In chapter 4 the quasi-stationary optical switching at nonlinear interfaces is explained theoretically in terms of plane-wave-concepts. There is an acute need for experimental verification of the multiplicity of possible effects arising from the present theoretical considerations. Hopefully the reader will be inspired and provoked to think of new experiments and theoretical approaches to these problems.

Experiments in surface physics, their conception, analysis and applicability to problems of practical relevance require, as a minimum, detailed theoretical knowledge of the phenomenological and microscopic treatment of excitations characteristic of surfaces, in particular of their interaction with electromagnetic waves. For example, the breaking of inversion symmetry at surfaces and interfaces leads to novel features, many of which are treated in chapters 5 to 8. As stated in the introductions of the respective chapters, many interesting and stimulating questions are posed in addition to, or extensions of those found in present reviews and monographs. In chapter 5 second-order nonlinear optical effects at interfaces of centrosymmetric media are investigated. Chapter 6 describes the electromagnetic surface modes of semiconductors with free charge carriers as well as the influence of external magnetic fields (gyrotropy induced by magnetic fields). In chapter 7 scattering effects involving acoustic and optical phonons, phonon-polaritons as well as spin waves in thin films and structures of thin layers are analyzed and reported. This part of the book also contains diverse phenomena such as phase transitions at surfaces and interfaces, and their interactions with electromagnetic waves. Unfortunately, due to space restrictions the present book can deal with this complicated and broad field of the physics of surfaces only in a very limited way. More detailed discussions will be presented in a further edition of this series. For instance, the phase transitions of a hot laser-induced charge carrier plasma in semiconductors, the dynamics of the formation of electron–hole liquids in layered systems and on surfaces and the interaction with electromagnetic guided and bulk waves are all very relevant to modern optoelectronics. In chapter 8 are presented magnetic order–disorder and orientational phase transitions based on the Landau–Ginzburg–Levanyuk theory. Experimental investigations of electromagnetic nonlinear phenomena in such systems should be encouraged.

Chapters 9 and 10 deal with the problems of laser-induced surface gratings and the surface excitations caused by local energy deposition through interaction with surface excitations, for example, surface polaritons. These effects exhibit several universal characteristics and can be observed in metals, semiconductors and dielectrics over a broad spectrum of incident light frequencies.

Therefore, they are of fundamental interest for basic investigations as well as for possible applications.

This book contains many new ideas. The editors hope that it will prove a stimulation for further theoretical and experimental research in this very rich area of science.

H.-E. Ponath
G.I. Stegeman

CONTENTS

CHAPTER 1

Second-Order Nonlinear Guided Wave Interactions

W. SOHLER

Angewandte Physik
Universität-GH-Paderborn
D-4790 Paderborn, Fed. Rep. Germany

© *Elsevier Science Publishers B.V.*, 1991

Nonlinear Surface Electromagnetic Phenomena
Edited by
H.-E. Ponath and G.I. Stegeman

Contents

1. *Introduction*

Among the many fascinating new fields, the invention of the laser in 1960 originated, nonlinear optics with its broad scope certainly is one of the most interesting (Bloembergen 1965, Akhmanov and Khokhlov 1972, Shen 1984). Its inception was the experimental work of Franken et al. (1961) on optical second-harmonic generation (SHG) in 1961 and the theoretical work of Bloembergen and co-workers (Armstrong et al. 1962) on optical wave mixing in 1962. Since that time, the field has grown rapidly; today it includes besides the coherent generation of new frequencies a host of nonlinear spectroscopies (Demtröder 1982) and a variety of nonlinear phenomena, such as phase conjugation (Fisher 1982), optical bistability (Gibbs 1985) and optical switching (see, e.g., Caglioti et al. 1988) with applications to all optical signal processing and computing.

In general, nonlinear optical interactions occur whenever the (coherent) optical fields of one or more laser beams are large enough to produce in a medium – additionally to a linear response – nonlinear polarization fields. They can be understood as the second- and higher-order terms of a series expansion of the polarization proportional to the product of two or more of the incident fields. A second-order term can only arise in certain classes of crystals which lack an inversion center (Nye 1957), whereas third-order terms occur in all materials. The nonlinear polarization fields are the sources of optical fields at new frequencies; if phase matched they grow linearly with the propagation distance.

Optical waveguides can be ideal media for performing efficient nonlinear interactions. Due to the concentration of optical fields in a waveguide with cross-section dimensions of the order of a wavelength, large power densities can be produced with relatively small absolute power levels. They can be maintained along the whole waveguide length, if the propagation losses are small. Beam spreading due to diffraction, as in bulk optics, is avoided. Furthermore, there are additional possibilities to achieve phase matching, e.g., by taking advantage of the waveguide dispersion.

All these attractive prospects of nonlinear guided-wave optics stimulated many groups throughout the world to investigate a host of different nonlinear interactions in optical waveguides. These interactions are classified according to the order of the nonlinear polarization term exploited. Third-order interactions comprise processes like degenerate four-wave mixing, optical bistability, intensity-dependent coupling, optical switching, etc.; they are discussed in several other contributions to this book.

It is the aim of this chapter to review second-order nonlinear guided wave interactions mainly in integrated optical waveguides. They comprise second-harmonic generation (section 6), sum and difference frequency generation and parametric amplification (section 7), spontaneous parametric fluorescence (section 8) and parametric oscillation (section 9). The coupled-mode theory is generally used to describe these effects (section 2). Nonlinear materials and waveguides are presented in section 3, whereas phase matching and field overlap problems are discussed in sections 4 and 5. Emphasis is laid on recent developments as several review papers described the field in different depth some years ago (Sohler 1984, 1985, Sohler and Suche 1983, Stegeman et al. 1986a,b, Stegeman and Seaton 1985, Ostrowsky 1985).

2. *Coupled-mode analysis of second-order nonlinear interactions*

In this section we just give an outline of the coupled-mode theory (Yariv 1973, 1975a, 1979, Bava et al. 1987, Sipe and Stegeman 1979, Kogelnik 1975, 1979, Marcuse 1974) which is usually applied to describe second-order nonlinear interactions. More detailed theoretical models and descriptions can be found in the references cited above and in the following sections.

As in most nonlinear effects to be described high photon densities are involved (spontaneous parametric fluorescence is an exception; see section 8), we are allowed to use classical electromagnetic fields. In second-order processes, in general three optical fields of different frequencies ω_1, ω_2 and ω_3 interact. We confine the discussion to guided modes, as only the concepts are to be emphasized; the extension of the theory to take radiating (substrate) modes into account, can be found in the literature.

We start to consider the fields of the guided modes, bound in x- and y-directions, propagating in the z-direction; they can be written as

$$\begin{aligned} \boldsymbol{E}_1(x, y, z, t) &= A_1(z)\mathscr{E}_1(x, y)\cos(\omega_1 t - \beta_1 z) \\ &= \tfrac{1}{2}A_1(z)\mathscr{E}_1(x, y)\,\mathrm{e}^{\mathrm{i}(\omega_1 t - \beta_1 z)} + \text{c.c.}, \end{aligned} \tag{1}$$

with analogous expressions for $\boldsymbol{E}_2$ and $\boldsymbol{E}_3$.

$\mathscr{E}_1(x, y)\cos(\omega_1 t - \beta_1 z)$ and the corresponding fields of frequencies ω_2 and ω_3 are eigenmodes of the guide obeying the unperturbed wave equation

$$\nabla^2 \boldsymbol{E}(x, y, z, t) = \frac{n_{\text{eff}}^2}{c^2}\frac{\partial^2}{\partial t^2}\boldsymbol{E}(x, y, z, t). \tag{2}$$

The $A_{1,2,3}(z)$ are the (slowly varying) amplitudes describing the power transfer from one wave to the other caused by a nonlinear coupling, the $\beta_{1,2,3} = (\omega_{1,2,3}/c)n_{\text{eff},1,2,3}$ are the propagation constants, c is the velocity of light and n_{eff} an effective index of refraction. The driving forces of second-order interactions are nonlinear polarization fields, which arise at high optical power densities in crystals without an inversion center (Nye 1957). Under these conditions, the response of the material is no longer a linear function of the optical fields; nonlinear polarization fields are produced proportional to the product of two optical fields of (in general) different frequencies. A second-order nonlinear susceptibility $\chi_{ijk}^{(2)}$ (a third-rank tensor) has been defined such that (Kaminow 1974, Boyd and Kleinman 1968)

$$P_i^{\mathrm{nl}}(x, y, z, t) = \varepsilon_0 \chi_{ijk}^{(2)} E_j(x, y, z, t)\, E_k(x, y, z, t) \quad \text{with} \quad i = x, y, z. \tag{3}$$

A summation over the same indices from $1 = x$ to $3 = z$ is understood. E_j and E_k represent the total fields, polarized along j and k. Both driving optical fields $\boldsymbol{E}_1$ and $\boldsymbol{E}_2$ may have a component along j and k resulting in

$$\begin{aligned} P_i^{\mathrm{nl}}(x, y, z, t) &= \varepsilon_0 \chi_{ijk}^{(2)} (E_{1j} + E_{2j})(E_{1k} + E_{2k}) \\ &= \varepsilon_0 \chi_{ijk}^{(2)} (E_{1j}E_{1k} + E_{1j}E_{2k} + E_{2j}E_{1k} + E_{2j}E_{2k}) \end{aligned} \tag{4}$$

with

$$E_{1j} = E_{1j}(x, y, z, t) = \tfrac{1}{2} A_1(z)\, \mathscr{E}_{1j}(x, y)\, \mathrm{e}^{\mathrm{i}(\omega_1 t - \beta_1 z)} + \text{c.c.},$$

and analogue expressions for E_{2j}, E_{1k} and E_{2k}.

If we evaluate the 4 field products in eq. (4), we get

$$\begin{aligned} E_{1j}E_{1k} &= \tfrac{1}{4} A_1^2 \mathscr{E}_{1j} \mathscr{E}_{1k} \mathrm{e}^{\mathrm{i}(2\omega_1 t - 2\beta_1 z)} + \tfrac{1}{4} A_1 A_1^* \mathscr{E}_{1j} \mathscr{E}_{1k} + \text{c.c.}, \\ E_{1j}E_{2k} + E_{2j}E_{1k} &= \tfrac{1}{2} A_1 A_2 \mathscr{E}_{1j} \mathscr{E}_{2k} \mathrm{e}^{\mathrm{i}[(\omega_1 + \omega_2)t - (\beta_1 + \beta_2)z]} \\ &\quad + \tfrac{1}{2} A_1 A_2^* \mathscr{E}_{1j} \mathscr{E}_{2k} \mathrm{e}^{\mathrm{i}[(\omega_1 - \omega_2)t - (\beta_1 - \beta_2)z]} + \text{c.c.}, \\ E_{2j}E_{2k} &= \tfrac{1}{4} A_2^2 \mathscr{E}_{2j} \mathscr{E}_{2k} \mathrm{e}^{\mathrm{i}(2\omega_2 t - 2\beta_2 z)} + \tfrac{1}{4} A_2 A_2^* \mathscr{E}_{2j} \mathscr{E}_{2k} + \text{c.c.} \end{aligned} \tag{5}$$

The essential feature of this result is that the optical waves of frequencies ω_1 and ω_2 produce nonlinear polarization fields of frequencies $2\omega_1$, $2\omega_2$, $\omega_1 + \omega_2$, $\omega_1 - \omega_2$ and zero. They give rise to second-harmonic generation (SHG), sum and difference frequency generation and optical rectification.

To describe these effects mathematically, the nonlinear fields enter the wave equation as a perturbation, yielding

$$\nabla^2 \boldsymbol{E}(x, y, z, t) = \frac{n_{\mathrm{eff}}^2}{c^2} \frac{\partial^2}{\partial t^2} \boldsymbol{E}(x, y, z, t) + \mu_0 \frac{\partial^2}{\partial t^2} \boldsymbol{P}^{\mathrm{nl}}. \tag{6}$$

Fortunately, to find an optical field $\boldsymbol{E}$ obeying the perturbed wave equation, it can be split into a set of equations for the different frequency components. Then the orthogonality of waveguide modes allows to express the field $\boldsymbol{E}$ as a superposition of guided modes of different order and of radiating modes. As we are interested in the coupling of guided modes only, we neglect the radiating modes and obtain

$$\boldsymbol{E} = \boldsymbol{E}_3(x, y, z, t) = \sum_m \tfrac{1}{2} A_3^{(m)}(z)\, \mathscr{E}_3^{(m)}(x, y)\, \mathrm{e}^{\mathrm{i}(\omega_3 t - \beta_3^{(m)} z)} + \text{c.c.}, \tag{7}$$

ω_3 is the frequency of interest, e.g., $\omega_3 = \omega_1 - \omega_2$ or $\omega_3 = 2\omega_1$, etc., as determined by one of the nonlinear polarization fields.

Substituting eq. (7) into the wave equation (6), assuming 'slow' varying amplitudes with

$$\frac{d^2}{dz^2} A_3^{(m)} \ll \beta_3^{(m)} \frac{d}{dz} A_3^{(m)}, \tag{8}$$

and recalling that $\mathscr{E}_3^{(m)} \exp[i\omega_3 t - \beta_3^{(m)} z)]$ obeys the unperturbed wave equation (2), gives for one cartesian component i,

$$-i \sum_m \left\{ \beta_3^{(m)} \frac{dA_3^{(m)}}{dz} \mathscr{E}_{3i}^{(m)}(x, y) e^{i(\omega_3 t - \beta_3^{(m)} z)} + \text{c.c.} \right\} = \mu_0 \frac{\partial^2}{\partial t^2} P_i^{\text{nl}}(\omega_3). \tag{9}$$

To facilitate the discussion, we now assume quasi-transversal modes with one dominant component in the y-direction ($i = y$; TE-modes) and suppress this index in the following. Multiplying eq. (9) with $\mathscr{E}_3^{(n)}(x, y)$, integrating the resulting equation and making use of the orthogonality relations of guided modes, we obtain (only modes traveling the same direction are considered)

$$\frac{dA_3}{dz} = -i \frac{\omega_3}{4} \varepsilon_0 \chi_{\text{eff}}^{(2)} \int \mathscr{E}_1^{(l)}(x, y) \mathscr{E}_2^{(m)}(x, y) \mathscr{E}_3^{(n)}(x, y) \, dx \, dy \, A_1 A_2 e^{i\Delta\beta z}, \tag{10}$$

with $\Delta\beta = \beta_3 - \beta_2 - \beta_1$.

A coupling coefficient κ is defined which is essentially determined by the overlap integral of the interacting modes of orders l, m and n and frequencies ω_1, ω_2 and ω_3,

$$\kappa = \frac{\varepsilon_0}{4} \chi_{\text{eff}}^{(2)} \int \mathscr{E}_1^{(l)}(x, y) \mathscr{E}_2^{(m)}(x, y) \mathscr{E}_3^{(n)}(x, y) \, dx \, dy. \tag{11}$$

Of course, the new field of frequency ω_3 can interact with the fields of ω_1 and ω_2, respectively, to produce a nonlinear polarization field of frequency ω_2 and ω_1, respectively. The corresponding equations complete the coupled differential equation system to describe second-order nonlinear interactions

$$\frac{dA_1}{dz} = -i\omega_1 \kappa A_3 A_2^* e^{-i\Delta\beta z},$$

$$\frac{dA_2}{dz} = -i\omega_2 \kappa A_3 A_1^* e^{-i\Delta\beta z},$$

$$\frac{dA_3}{dz} = -i\omega_3 \kappa A_1 A_2 e^{i\Delta\beta z}. \tag{12}$$

These are the essential results. We recognize that it is absolutely necessary to adjust phase matching of the coupled modes, i.e., $\Delta\beta = 0$, to get an efficient nonlinear interaction along the whole waveguide length. How to realize phase matching in integrated optics is discussed in section 4. Furthermore, the coupling coefficient should be as large as possible not only by an appropriate choice of the waveguide material (section 3), but also by an optimization of the overlap of the interacting modes (section 5). With both means very efficient nonlinear guided-wave interactions can be achieved, up to several orders of magnitude more efficient than in bulk optics.

The equation system (12) can be specialized to describe SHG, sum and difference frequency generation, parametric amplification and oscillation; all these effects are discussed in subsequent sections.

3. Nonlinear materials and waveguides

Second-order nonlinear optical materials play an important role in many applications of modern optical technology. In particular, they are used to convert the frequency of available laser systems by parametric processes such as second-harmonic generation (SHG). Mainly inorganic materials are employed for this purpose; among them are ammonium dihydrogen phosphate (ADP; $NH_4H_2PO_4$), potassium dihydrogen phosphate (KDP; KH_2PO_4), deuterated KDP (KD*P; KD_2PO_4), lithium niobate ($LiNbO_3$) and others, all well-known for a long time and available as large crystals of excellent quality. More recent developments of nonlinear materials include beta barium borate (BBB; β-BaB_2O_4), potassium niobate ($KNbO_3$), potassium titanate phosphate (KTP; $KTiOPO_4$) and L-arginine phosphate (LAP; $(NH_2)_2+CNH(CH_2)_3CH(NH_3)+COO^-\cdot H_2PO_4^-\cdot H_2O$). Furthermore, a whole new class of organic and polymeric nonlinear optical materials of great potential has been developed in the last years. Also in the field of semiconductor materials a considerable progress was reported, primarily for higher-order, resonant nonlinear processes allowing applications such as all-optical switching or optical storage in bistable devices. (An interesting assessment of the current status of research in the U.S. on the different classes of nonlinear optical materials has been given by a large group of American scientists (Auston et al. 1987)).

All the different nonlinear optical materials mentioned above, most of them with well-documented properties (Kurtz 1972, see, e.g., table 16.2 in the book of Yariv (1975b)), are candidates to be a substrate or waveguide material for second-order guided-wave interactions. However, only few of them have been used in nonlinear integrated optics up to now. It turned out that a fabrication of waveguides of the required high quality failed in most cases or has not yet been tried with the necessary intensity.

There are several different waveguide configurations possible to allow nonlinear interactions. As the optical field of a guided mode extends into core and cladding(s) materials, not necessarily all of them must be nonlinear. Some of the possible configurations are illustrated in fig. 1. Either the guide (fig. 1a, d, e) or the substrate (fig. 1b) or the cover (layer) (fig. 1c) may be nonlinear. The

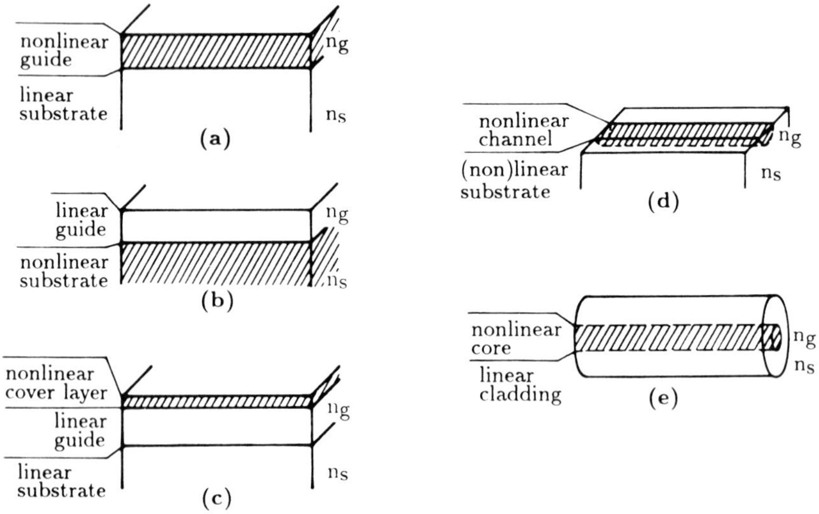

Fig. 1. Possible waveguide configurations for nonlinear optical interactions. (After Zyss 1985.)

optimum structure is a channel guide or even a fiber built up by nonlinear materials only (not shown in fig. 1).

To prepare the optical waveguides, several techniques have been developed; among them are diffusion ($LiNbO_3$), ion exchange ($LiNbO_3$, $KTiOPO_4$), ion implantation ($KNbO_3$, $LiNbO_3$), fiber growth ($LiNbO_3$), different epitaxial techniques (semiconductors), spinning and dipping (organic materials) and monolayer deposition with the Langmuir–Blodgett technique (organic materials); they are discussed in more detail in context with the different materials. Up to now the indiffusion technique (of Ti in $LiNbO_3$) has proven to be the most successful method to prepare not only low-loss, but also extremely homogeneous waveguides. Such properties are absolutely necessary to allow phase matching in a long waveguide structure and in this way efficient, second-order nonlinear interactions. As an example, Ti:$LiNbO_3$ channel guides could be fabricated with an optical attenuation down to 0.03 dB/cm (Regener and Sohler 1988); their homogeneity was sufficient to get an effective interaction length of phase-matched processes approaching the physical waveguide length of some cm. This is proved by the SHG response presented in fig. 12 (section 4); its small width results from a phase-matched interaction over nearly the full length (of 4.7 cm) of the waveguide.

The quality of the substrate material (in particular its homogeneity) essentially determines the waveguide quality. This is especially true for waveguides embedded into the substrate, e.g., by diffusion techniques, but also for waveguides fabricated on top of the substrate, e.g., by epitaxial methods. In both geometries the homogeneity of the optical waveguide and thus the effective interaction length of nonlinear processes are more or less influenced by the homogeneity of the substrate. Therefore, sensitive methods are necessary to test the substrate quality before the waveguide fabrication. Of special interest are nonlinear optical methods, such as a Maker fringe analysis of $LiNbO_3$ substrates (Lunt et al. 1985) to investigate the homogeneity of the substrate. Even more sensitive is the analysis of the nonlinear response, if the laser beam traverses the substrate in longitudinal direction parallel to the projected waveguide (Ivanova and Kholodnykh 1980, Itskhoki et al. 1982). Figure 2 presents as an example the SHG response, i.e., the SH-power versus the substrate temperature around the phase-match condition (Seibert, unpublished results).

By fitting a calculated response to the measured one, the effective (phase-matched) interaction length can be evaluated. It was – within the errors of this procedure – identical to the substrate length of 4.7 cm, characterizing a $LiNbO_3$ crystal of excellent longitudinal homogeneity. It is obvious that by scanning the probing laser beam the whole substrate can be investigated.

By comparing the guided-wave result of fig. 12 with the (nearly) plane-wave results obtained by SHG in the waveguide substrate (fig. 2), the waveguide fabrication technology (here the Ti-indiffusion) can be assessed. It induces only slight further inhomogeneities, as both measured results can be fitted by a theoretical response assuming nearly the same effective interaction length. Not all techniques yield such homogeneous guides.

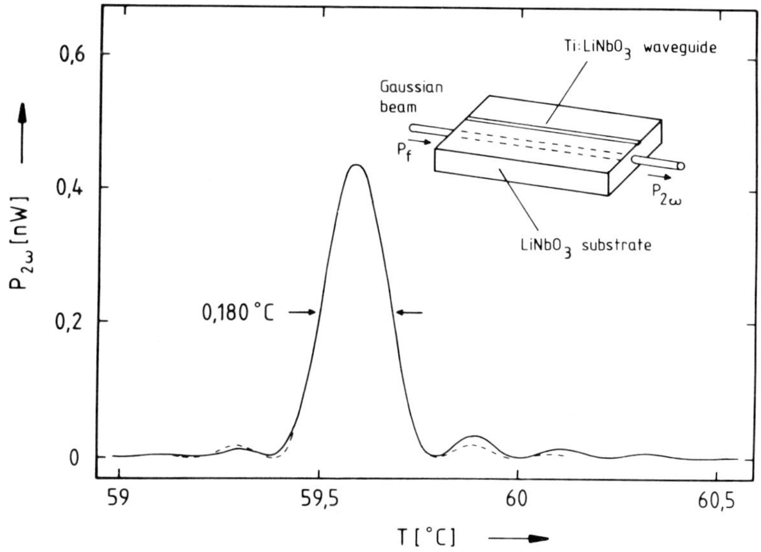

Fig. 2. SHG ($\lambda_f = 1.09$ μm) in a bulk $LiNbO_3$ substrate of 4.7 cm length: phase matching by temperature tuning of the birefringence. (After Seibert, unpublished results.)

In the following, nonlinear materials and waveguide structures of actual interest are discussed more specifically with emphasis on recent results. The earlier work up to 1985 is summarized and tabulated in an excellent review paper by Stegeman and Seaton (1985) (see also Sohler 1984, Sohler et al. 1986).

3.1. *Lithium niobate ($LiNbO_3$)*

$LiNbO_3$ is a very attractive, well-known material of unique electro-optic, photoelastic, piezoelectric and nonlinear properties. They are exploited in a variety of (bulk) devices. By the invention of the indiffusion technique to fabricate high-quality waveguides (Schmidt and Kaminov 1974), $LiNbO_3$ became one of the most important materials of integrated optics. Ti-indiffused waveguides in $LiNbO_3$ (Ti:$LiNbO_3$) are not only used for electro-optic modulators, switches, polarization transformers, etc. and for acousto-optic beam deflectors, frequency shifters, tunable wavelength filters, etc., all of outstanding performance (Sohler 1989); they are also excellent candidates for efficient nonlinear second-order guided-wave interactions. They have low losses down to 0.03 dB/cm and can be very homogeneous (as discussed above), so that phase matching also in long-waveguide structures is feasible; the high birefringence allows phase-matched interactions of low-order TE- and TM-modes with reasonable field overlap. A large variety of second-order interactions such as second-harmonic and sum frequency generation, difference frequency generation and parametric amplification, spontaneous parametric fluorescence and parametric oscillation have been investigated in Ti:$LiNbO_3$ waveguides with very encouraging results; they are discussed in more detail below.

The indiffusion of titanium in $LiNbO_3$ yields a concentration profile which follows a nearly Gaussian function in the depth [as confirmed by secondary ion mass spectroscopy (Lüdtke et al. 1980) or electron microprobe analysis (Minakata et al. 1978, 1979, Steffen et al. 1989) (see fig. 3)] and a superposition of two errorfunctions in the width (Bava et al. 1987).

However, the extraordinary and ordinary indices of refraction are changed in different manners; only the increase of the extraordinary index is proportional to the concentration of the Ti^{4+}-ions (see figs. 3 and 4) (Lüdtke et al. 1980).

This special property allows by an adjustment of the fabrication parameters to tailor the profiles in such a way that the field overlap of the interacting modes is optimized. There is a unique solution for the optimized parameters at relatively low Ti concentrations (Bava et al. 1987).

An example of the (ordinary) index profile (calculated from diffusion theory) of such an optimized monomode Ti:$LiNbO_3$ channel guide is presented in fig. 5 together with the (measured) intensity profile of the fundamental mode (Brucksch et al. 1986).

A further possibility to tailor index and mode profiles was demonstrated by indiffusing Mg(O) in a second step; this lowers both the extraordinary and the ordinary indices of refraction of $LiNbO_3$ (Bulmer 1984). By an appropriate

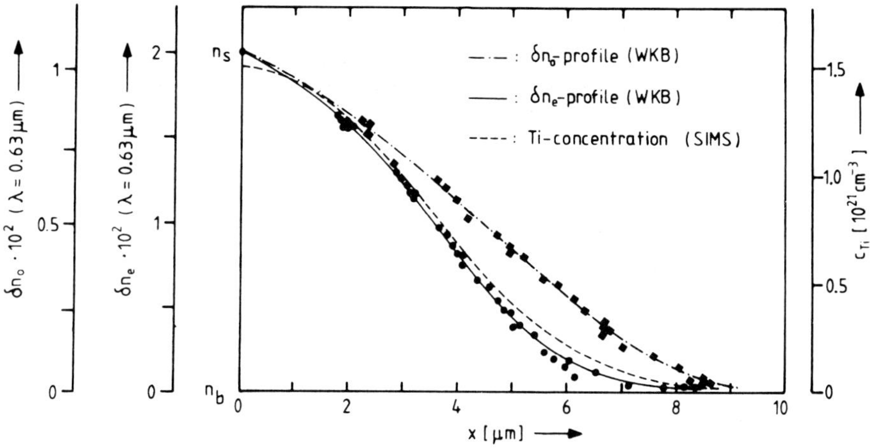

Fig. 3. Ti-concentration profile (right ordinate) and profiles of the extraordinary and ordinary indices of refraction (left ordinate) of a Ti:$LiNbO_3$-slab waveguide. (After Lüdtke et al. 1980.)

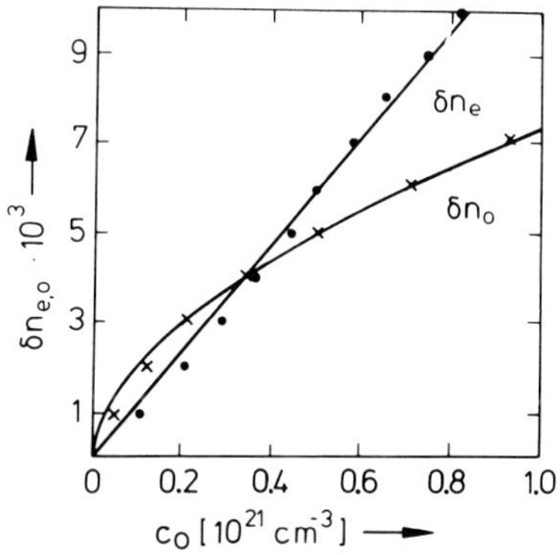

Fig. 4. Measured local increase of extraordinary and ordinary indices by Ti-indiffusion in $LiNbO_3$ as function of Ti concentration (dots and crosses) fitted by analytical functions. (After Lüdtke et al. 1980 and Bava et al. 1987.)

shallow diffusion, for example, nearly buried waveguides can be fabricated (Komatsu et al. 1986). Furthermore, Mg-doped $LiNbO_3$ has a lower susceptibility to optical damage than (nominally) pure $LiNbO_3$ (see below) (Bryan et al. 1985).

Another technique to prepare (nonlinear) waveguides in $LiNbO_3$ is the 'proton exchange', invented by Jackel et al. (1982). By immersing a (masked) $LiNbO_3$ substrate in a molten acid (e.g., benzoic acid) used as proton source, a substantial part of the Li^+-ions near the surface is replaced by protons. This procedure leads to a large increase of the extraordinary index of refraction (up to 0.12), but to a decrease of the ordinary index (see fig. 6).

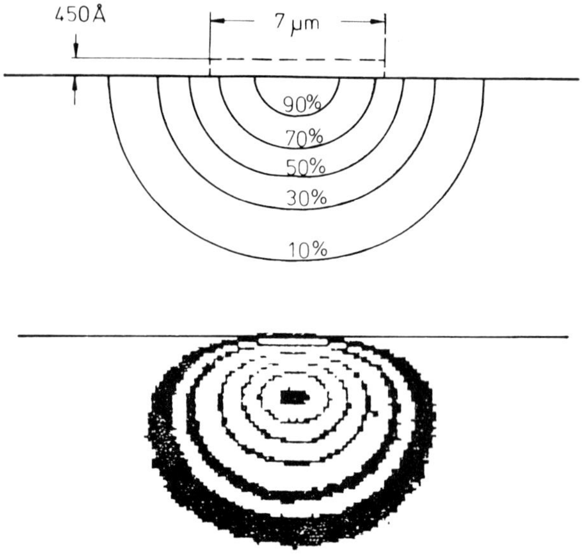

Fig. 5. Calculated extraordinary index profile of a Ti:$LiNbO_3$ strip guide (in percent of maximum index increase) together with the measured intensity distribution of the fundamental TM_{00}-mode at $\lambda = 1.15$ μm (lines of constant intensity: 100%, 80%, 60%, 40%, 20%, 10%). (After Brucksch et al. 1986.)

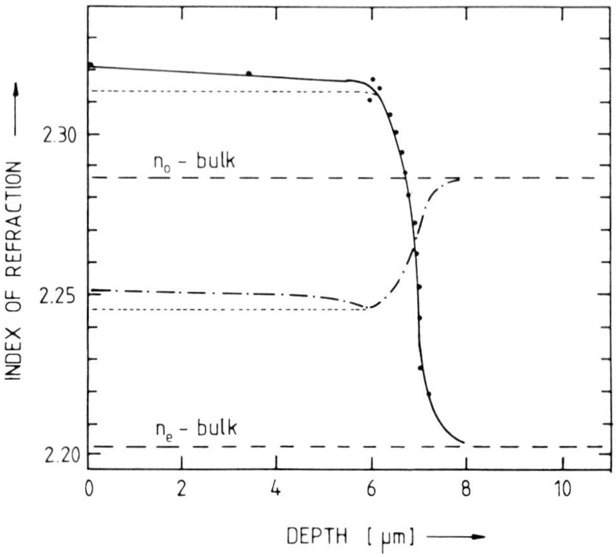

Fig. 6. Profiles of the extraordinary (measured) and ordinary (estimated) indices of refraction of a ('proton exchanged') $H_xLi_{1-x}NbO_3$ surface layer on $LiNbO_3$ (here with superimposed Ti-doping). (After Hinkov and Ise 1986.)

Therefore, waveguides are formed for the extraordinary polarization only; phase-matched nonlinear interactions of TM- and TE-guided modes are impossible. However, proton-exchanged guides have been successfully used for SHG in the form of Cerenkov radiation (see section 6) by a coupling of guided to substrate modes of equal polarization. This process can be very efficient, especially in the strongly guiding $H_xLi_{1-x}NbO_3$ channels, taking advantage of the largest nonlinear coefficient d_{33} of $LiNbO_3$.

An increased flexibility of waveguide design in $LiNbO_3$ is obtained by combining both the Ti-indiffusion and the proton exchange techniques. De Micheli and co-workers (De Micheli et al. 1982, De Micheli 1983) introduced so-called TIPE-guides (Ti-Indiffused/Proton-Exchanged) of different type, fabricated by controlling the depth of the proton exchange performed after the Ti-indiffusion (see fig. 7). They can be used to extend the phase-matching range of nonlinear second-order interactions considerably (see section 4).

Hinkov and Ise (1986) have shown how to control the birefringence of a TIPE-guide of type II (see fig. 7) by annealing the $H_xLi_{1-x}NbO_3$ layer or stripe embedding the Ti-diffused waveguide (see fig. 8). These structures can not only be applied in nonlinear integrated optics, but also in integrated acousto-optics, where phase matching is required as well (Hinkov et al. 1988).

Not only integrated optical waveguides can be fabricated in $LiNbO_3$. Fejer et al. (1982) reported a successful growth of single crystalline $LiNbO_3$ fibers for applications in nonlinear optics. Although encouraging results have been reported (Fejer et al. 1984) it seems, however, to be very difficult to grow long (> 5 cm), low-loss, homogeneous, if possible monomode fibers of the appropriate crystal orientation to allow phase matching of guided modes.

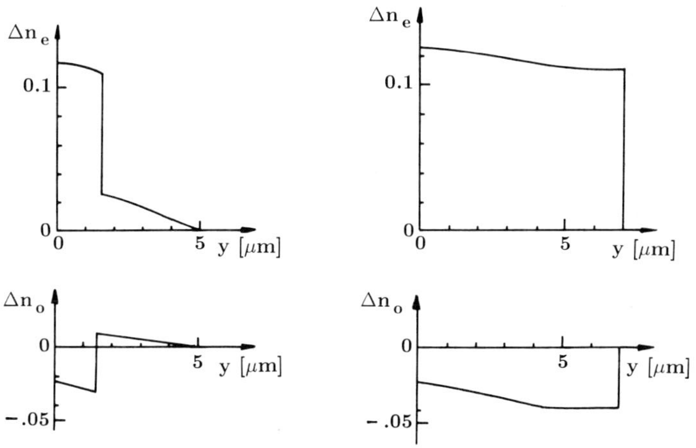

Fig. 7. Schematical sketch of TIPE (Ti-Indiffused/Proton Exchanged) waveguides of type I (left) and type II (right). (After De Micheli 1983.)

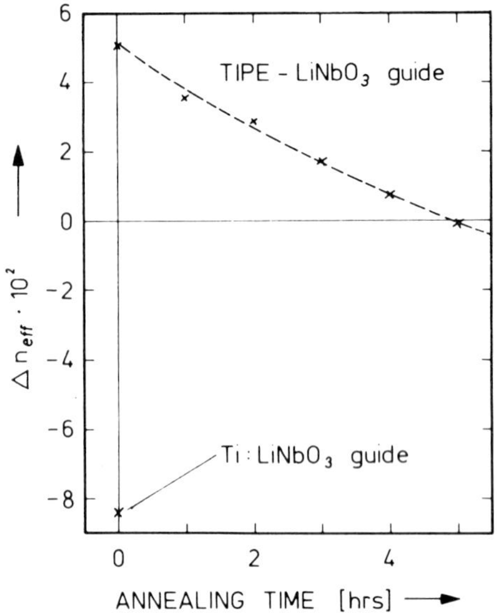

Fig. 8. Control of birefringence of TIPE guides of type II (see fig. 7) by annealing. (After Hinkov and Ise 1986.)

Severe limitations for the application of all $LiNbO_3$ optical waveguides in phase-sensitive (nonlinear) devices are set by the susceptibility of this material to optically induced changes of the (effective) indices of refraction ('optical damage'). This photorefractive effect is caused by the photoionization of impurities, especially Fe^{2+}-ions; the electrons generated by this process migrate along the optical axis of the crystal out of the waveguide core into the surrounding dark regions where they are trapped by Fe^{3+}-centers. In this way, a space charge is built up with an electrical field across the waveguide channel, which changes the indices of refraction via the electro-optic effect (Glass 1978). This optical damage has a strong wavelength dependence determined by the photoionization processes involved. Therefore, mainly visible radiation causes a problem (e.g., optically induced index changes prevent a cw-operation of integrated optical parametric oscillators up to now; see section 9), whereas the optical damage induced by long-waveguide radiation ($\lambda > 1$ µm) usually can be neglected. On the other hand, the magnitude of optical damage effects is determined by the conductivity of the material. Therefore, proton-exchanged waveguides with their enhanced conductivity are orders of magnitude less sensitive than Ti-indiffused structures (Jackel et al. 1984); moreover, the Ti-doping increases the ratio of Fe^{2+}/Fe^{3+}-impurities yielding in this way an enlarged photorefractive sensitivity even larger than in (nominally pure) undoped $LiNbO_3$ (Gericke et al. 1984).

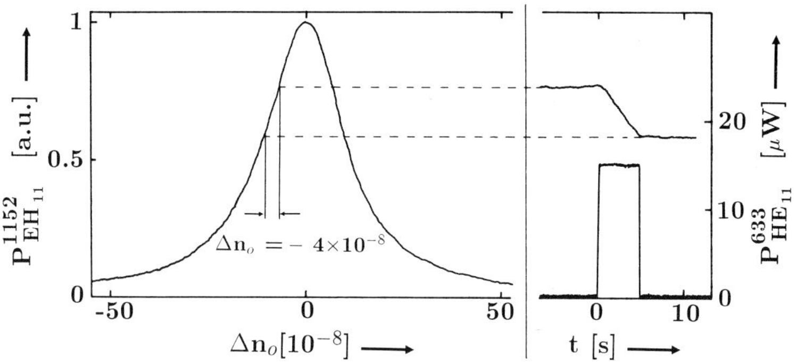

Fig. 9. 'Excite and probe' technique to measure optically-induced changes of the ordinary index of refraction Δn_0 in a Ti:$LiNbO_3$ channel-guide resonator: Resonator characteristic ($\lambda = 1.15$ μm) on the left; transmitted power of probe and excite ($\lambda = 0.63$ μm) radiation versus time on the right. (After Volk and Sohler 1988.)

Recently, an extremely sensitive 'excite and probe' technique was presented to measure optically induced refractive index changes in Ti:$LiNbO_3$ channel-guide resonators (Volk and Sohler 1988) as used for resonant second-order nonlinear interactions. To measure damage effects, the resonator is temperature-stabilized in a resonance of the long-wavelength 'probe' radiation (which does not induce index changes) at the point of steepest slope (the circle at the resonator characteristic in fig. 9). If then a second, nonresonant, short-wavelength 'excite' wave induces a change of the index of refraction, the resonator is detuned. This can be easily monitored by the 'probe' beam transmission as function of time (right part of fig. 9). An example of such a measurement obtained with a high-finesse resonator ($F = 50$ at $\lambda_{probe} = 1.15$ μm) is given in fig. 9. The influence of the visible ($\lambda = 0.63$ μm) 'excite' radiation switched on for only 5 s can be studied with an enormous sensitivity (of $\Delta n_0 \approx 5 \times 10^{-9}$).

The new method allows to study the intensity-, wavelength-, polarization- and temperature-dependence as well as the dynamical properties of optically induced index changes in great detail. It could help to get a more quantitative understanding of the damage effects in optical waveguides and to stimulate in this way a further improvement of the purity of $LiNbO_3$ substrates.

3.2. Potassium niobate ($KNbO_3$)

Another material of very favorable nonlinear optical properties is $KNbO_3$; it is a perovskite ferroelectric; between -50°C and $+223$°C the material is orthorhombic with point group symmetry mm2 (Flückiger and Arend 1978). It allows efficient noncritical type-I phase-matched SHG of semiconductor laser radiation for the wavelength $\lambda = 860$ nm at room temperature (Baumert et al. 1983). The largest nonlinear optical coefficient d_{32}, which was exploited, is about three times as large as the corresponding coefficient in $LiNbO_3$. Up to now, the material withstood all attempts to fabricate optical waveguides by the indiffusion and proton exchange techniques. Only Günter and co-workers (Baumert et al. 1985) demonstrated waveguiding in electro-optically induced channels. Furthermore, exciting new results have been achieved very recently; using ion implantation (He^{4+}), it was possible to fabricate planar and even channel guides (Bremer et al. 1988). It is hoped that they confirm the great potential of this material also in applications of nonlinear integrated optics.

3.3. Potassium titanate phosphate ($KTiOPO_4$)

KTP ($KTiOPO_4$) is a nonlinear optical material of very attractive properties including an electro-optic 'figure of merit' about twice that of $LiNbO_3$, a high optical damage threshold and a good thermal stability (Zumsteg et al. 1976). It is phase matchable in a broad wavelength range and has large second-order nonlinear coefficients; type-II phase-matched SHG has been demonstrated in a bulk crystal with an energy conversion efficiency of up to 59% ($\lambda = 1.064$ μm; 380 MW/cm^2) (Fan et al. 1987). KTP seems to become a standard doubler material in many commercial laser systems.

Recently, planar and channel optical waveguides have been fabricated in KTP using ion exchange processes (Bierlein et al. 1987). Increases in surface refractive index Δn of 0.23 have been observed by a Tl–K-exchange; a Rb–K-exchange resulted in a Δn of 0.02. With the latter process using an evaporated metal mask on top of the KTP substrate, also channel guides have been fabricated; their optical attenuation was less than 0.4 dB/cm. This technology was also applied to develop the first integrated electro-optical devices in that material: phase modulator and Mach–Zehnder interferometer as intensity modulator (Chouinard et al. 1988). However, though the nonlinear properties of KTP are excellent, no parametric guided-wave interactions have been reported up to now.

3.4. Organic materials

Recently, there is a large interest in the attractive nonlinear optical properties of organic materials, which are reviewed in several excellent papers and books (Williams 1983, 1984, Chemla and Zyss 1987, Carter and Zyss 1987). Acentric organic molecules with conjugated double bonds, such as dyes, have a very high second-order polarizability. Such molecules are characterized by their delocalized π-electron system, which is responsible for their extraordinary nonlinear optical properties. As an example, the molecular structure of 2-methyl-4-nitroaniline (MNA) is given in fig. 10.

If such molecules crystallize in a non-centrosymmetric structure, also macroscopic second-order susceptibilities arise yielding a nonlinear figure of merit up to two orders of magnitude higher than in $LiNbO_3$. In table 1, taken from the

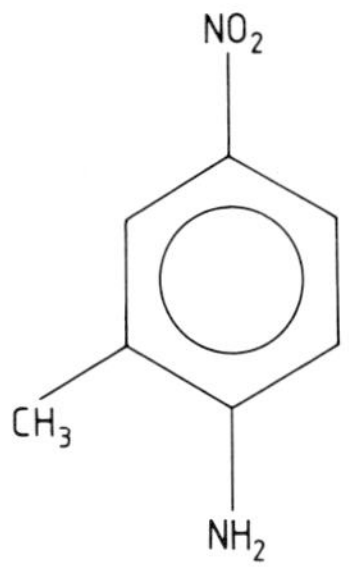

Fig. 10. Molecular structure of 2-methyl-4-nitroaniline (MNA).

Table 1

Highest phase-matchable coefficient d and nonlinear parametric figure of merit d^2n^{-3} for some organic molecular and polymeric crystals relative to the corresponding quantities of $LiNbO_3$. POM denotes 3-methyl-4-nitropyridine-1-oxide; MAP, methyl-(2,4-dinitrophenyl)-aminopropanoate; mNA, meta-nitroaniline; MNA, 2-methyl-4-nitroaniline; and NPP, N-(4-nitrophenyl)-(L)-prolinol.

	POM [1]	MAP [2]	mNA [3]	MNA [4]	NPP [5]
d	1.6	2.8	3.2	4.8	14.1
d^2n^{-3}	3.3	15	18	45	45

References: [1] Zyss et al. (1981); [2] Oudar and Hierle (1977); [3] Kato (1980); Carenco et al. (1977); [4] Levine et al. (1979); [5] Ledoux et al. (1986).

article by Zyss (1985) and supplemented by the data of NPP the highest phase-matchable coefficient d and the nonlinear parametric figure of merit d^2n^{-3} of some organic molecular and polymeric crystals are given relative to the corresponding quantities of $LiNbO_3$ ($d_{31} = 14.2 \times 10^{-9}$ esu).

These high quadratic nonlinearities together with the wide transparency range over the visible and near-infrared spectral range explain the current interest in these materials. Some of them can be grown as large crystals of good optical and structural quality; as an example, crystals of (−) 2-(α-methylbenzylamino)-5-nitropyridine (MBA-NP) have been prepared up to a maximum size of $7 \times 7 \times 5\ cm^3$ and analyzed by SHG (Bailey et al. 1988).

However, to use these attractive organic materials in integrated optics applications, they have to be prepared as thin films or stripes. Several (review) papers discuss the potential of such organic waveguiding structures, but also the problems and difficulties of preparing them (Zyss 1985, Kowel et al. 1987, Lytel et al. 1989). There are several ways to grow, e.g., the materials mentioned above as single crystalline films. This has been successfully done with MNA (Sasaki et al. 1984, Itoh et al. 1986), which indeed is a very good candidate for nonlinear integrated optics (Stegeman and Liao 1983). Even phase-matched SHG could be demonstrated in these films (Sasaki et al. 1984, Itoh et al. 1986). (The first observation of efficient SHG in a highly nonlinear organic film was reported by Hewig and Jain, using a polycrystalline PCPU (parachlorophenylurea) waveguide (Hewig and Jain 1983).)

However, the more favorable single crystalline waveguide would be a strip guide or even a fiber. The latter approach is especially attractive, as it potentially allows long nonlinear interaction lengths. In 1983, Nayar (1982, 1983, 1986) presented a novel technique to produce waveguiding structures by the growth of organic crystals in glass capillaries. He could even demonstrate SHG in a guided-radiating mode interaction. In the following years, the preparation of such organic crystal cored fibers and their nonlinear properties have been studied by several groups; an excellent paper by Vidakovic et al. lists all the experiments and materials investigated till 1987 (Vidakovic et al. 1987). Also the possible crystal orientations necessary to achieve efficient nonlinear interactions were analyzed (Vidakovic et al. 1987). It seems that most of the nonlinear materials crystallize in capillaries in very inefficient orientations; therefore, the growth methods have to be adapted to allow the optimum crystal orientation or organic materials belonging to lower-symmetry classes have to be used (Vidakovic et al. 1987). In any case, much work remains to be done to obtain low-loss crystal cored fibers which have the desired crystal orientations and furthermore allow phase-matched nonlinear interactions.

Another way of preparing nonlinear waveguides is the Langmuir–Blodgett technique (Pitt and Walpita 1980). It offers a perfect controllability of the film thickness by depositing one mono-molecular layer after another on top of the preceding ones. Non-centrosymmetric films require depositing different materials using an alternating layer approach (Cross et al. 1987) or making so-called Z-type structures. Z-type structures are head–tail–head–tail– ... arrangements of successive layers of the deposited material. The head refers to the hydrophilic, the tail to the hydrophobic portion of the organic molecule. Enhanced SHG from such a multilayer structure of a mixture of a hemicyanine dye and PMMA have recently been reported (Hayden et al. 1986). However, good waveguide quality together with the required nonlinearity apparently has not yet been achieved.

Perhaps the most promising way to fabricate organic nonlinear waveguides is to use either nonlinear molecules dissolved in a polymer host (Hill et al. 1987, Wang 1986) or attached as side groups to a molecular (polymer) backbone (side-chain polymers) (Lytel et al. 1988, 1989, Möhlmann et al. 1989). The latter approach allows a higher concentration of the nonlinear molecules. Both kinds of polymers can easily be prepared by spinning, dipping or casting. Poling in an electrical field yields an alignment of the nonlinear molecules and removes the initially present centrosymmetry (Singer et al. 1987); by a selective poling procedure even single-mode channel waveguides can be created (Thackara et al. 1988) with electro-optic coefficients similar to those of $LiNbO_3$ (Möhlmann et al. 1989). This exciting progress allowed to fabricate channel-guide phase and intensity (Mach–Zehnder interferometer and directional coupler) modulators (Lytel et al. 1989, Möhlmann et al. 1989). However, three-wave frequency converting interactions have not yet been reported, although a considerable birefringence (Lytel et al. 1989) should allow phase matching even in monomode channel guides.

Finally, nonlinear molecules can also be deposited on top of a thin-film (e.g., polymeric) waveguide. This situation has been theoretically analyzed by Sipe et al. (1983) in comparison with SHG and parametric mixing in nonlinear molecules adsorbed at a surface of transparent or metallic media and illuminated by plane waves from above the surface (Chen et al. 1981, Heinz et al. 1982), a reduction of the incident laser power requirements by orders of magnitude can be expected in the guided-wave geometry (Sipe et al. 1983).

4. *Phase matching in waveguide structures*

Second-order parametric nonlinear effects require phase matching of the interacting optical waves ($\Delta\beta = 0$) to allow a steady flow of energy from the pump wave(s) to the generated new field(s). If phase matching is adjusted, the amplitude of the new field(s) (linearly) grows with the propagation length in the waveguide; the power conversion efficiency grows quadratically. However, if there is a phase mismatch ($\Delta\beta \neq 0$), the amplitude oscillates along the propagation length like $[\sin(\Delta\beta z/2)]/(\Delta\beta z/2)$ and remains very small (see also section 2).

In bulk optics only birefringent nonlinear materials allow to achieve phase matching by a nonlinear coupling of waves of orthogonal polarization (a material of suitable symmetry with a corresponding nonzero off-diagonal element of the second-order susceptibility tensor is assumed). In integrated optics, however, several methods have been demonstrated to achieve phase matching. Besides the technique analogous to bulk optics, the waveguide mode dispersion can be used to compensate the material dispersion and to get phase matching even in guides of isotropic materials. Furthermore, waveguides of artificial birefringence or with a suitable grating structure also allow to get the necessary synchronism of nonlinear polarization and parametrically generated waves. Another specialty of integrated optics is the efficient, automatically phase-matched generation of the second-harmonic as Cerenkov radiation. Finally, the electro-optical fine-tuning of phase matching is by far more attractive in integrated optics than in bulk optics. All these methods are discussed in the following subsections. As most methods have been demonstrated in SHG experiments, we restrict the discussion to that case. Now, phase matching means that the propagation constant of the nonlinear polarization wave $2\beta(\omega_f)$ equals the propagation constant of the harmonic field $\beta(\omega_{SH} = 2\omega_f)$.

Before going into details, a general remark is necessary. Experimentally, a position-dependent phase mismatch $\Delta\beta = \Delta\beta(z)$ often occurs due to non-ideal fabrication processes or inhomogeneities of the substrate material or of the waveguide temperature. Indeed, for a long time it seemed that the extremely stringent requirements for the waveguide homogeneity would hinder a phase-matched interaction length longer than a few hundred micrometers (Tien 1971). It is mainly due to improved fabrication technologies that at least in the weakly guiding Ti:$LiNbO_3$ guides effective interaction lengths up to 4.7 cm could be achieved (Seibert, unpublished results) (see also fig. 12). However, even in that material system extremely small waveguide inhomogeneities producing a locally varying phase mismatch $\Delta\beta(z)$ are sufficient to reduce the efficiency of a parametric conversion process drastically. This is demonstrated by calculations of the second-harmonic intensity (which is proportional to the conversion efficiency) as function of the waveguide temperature taking a parabolic phase mismatch along the guide into account (Laurell and Arvidsson 1988) (see fig. 11). A maximum mismatch of $\Delta\beta_{max} = 1 \times 10^{-4}\ \mu m^{-1}$ in the 15 mm long guide reduces the conversion efficiency of the unperturbed, homogeneous sample with $\Delta\beta = 0$ to about 40%. The ideal sinc^2-phase-matching curve is perturbed; it obtains some further structures characteristic of the form of the waveguide disturbance. They result in the examples of fig. 11 from the interference of harmonic waves generated in the first and second halfs of the sample around the loci of $\Delta\beta = 0$; a change of the waveguide temperature symmetrically shifts these two positions and in that way the relative phase of the harmonic waves. Constructive or destructive interference is the result leading to the numerically calculated temperature dependent response of fig. 11.

In the example cited above the maximum phase mismatch is $\Delta\beta_{max} = 1 \times 10^{-4}\ \mu m^{-1}$. To illustrate this figure: it corresponds to a perturbation of the width of the 10 μm wide channel of only 0.04 μm or to a temperature deviation of 0.12°C (Laurell and Arvidsson 1988). More general, the allowed phase mismatch $\Delta\beta(z)$ should be very much smaller than $2\pi/L$ in the whole waveguide [i.e., $n_{eff}(\lambda_f) - n_{eff}(\lambda_{SH}) < \lambda_f/2L$ in the case of second-harmonic generation] to guarantee an effective interaction length approaching the actual waveguide length. This is a very stringent condition which we should have in mind when we are going to discuss the different ways of integrated optics to get a phase-matched interaction. Only perfect phase matching along the whole waveguide length allows to get the high efficiency of nonlinear interactions guided-wave optics promises.

4.1. *Natural birefringence*

In a nonlinear waveguide an optical mode of frequency ω_f and propagation constant $\beta(\omega_f)$ generates a (second-order) polarization field which propagates with $2\beta(\omega_f)$ through the guide. This nonlinear polarization is the source of a second-harmonic field propagating with $\beta(\omega_{SH})$ as an optical mode of frequency

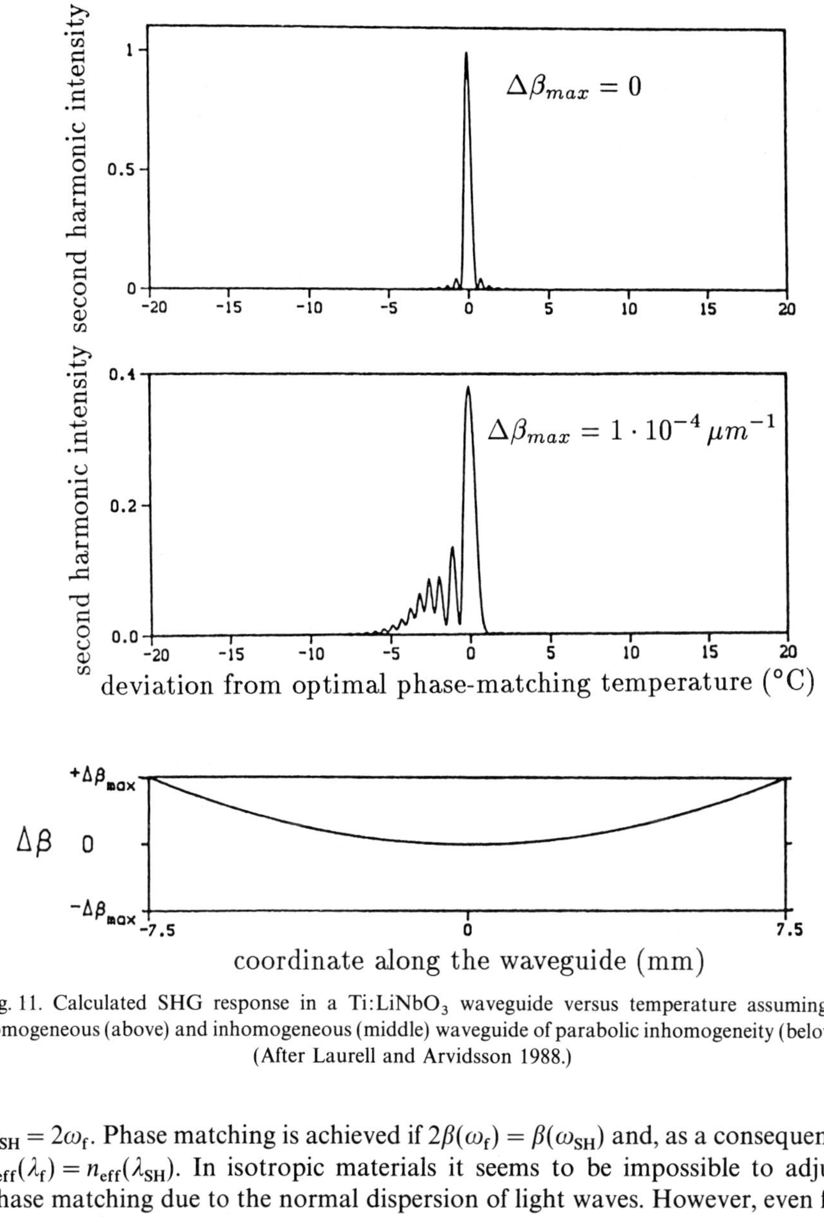

Fig. 11. Calculated SHG response in a Ti:$LiNbO_3$ waveguide versus temperature assuming a homogeneous (above) and inhomogeneous (middle) waveguide of parabolic inhomogeneity (below). (After Laurell and Arvidsson 1988.)

$\omega_{SH} = 2\omega_f$. Phase matching is achieved if $2\beta(\omega_f) = \beta(\omega_{SH})$ and, as a consequence $n_{eff}(\lambda_f) = n_{eff}(\lambda_{SH})$. In isotropic materials it seems to be impossible to adjust phase matching due to the normal dispersion of light waves. However, even for that case there are several integrated optics approaches to compensate the material dispersion (see following subsections). But the more conventional way is – similar to bulk optics – to use optical waveguides in birefringent, e.g., optically negative nonlinear crystals. Here we have the possibility to get $n^o_{eff}(\lambda_f) = n^e_{eff}(\lambda_{SH})$. It implies a fundamental mode of 'ordinary' polarization, generating not only an 'ordinary' harmonic nonlinear polarization field but also an 'extraordinary' one which can be phase matched to become an efficient source of a second-harmonic wave.

$LiNbO_3$ is an excellent example of such a nonlinear material; it is an optically negative, uniaxial crystal which allows to get phase-matched second-harmonic generation in Ti:$LiNbO_3$ waveguides with an 'ordinary' fundamental and an 'extraordinary' harmonic mode. To be specific, in a Y-cut crystal with X-propagating optical modes as in the example of fig. 12, the TM-like fundamental mode excites a TE-like second-harmonic mode (Seibert unpublished result). The waveguide is monomode in the near-infrared and multimode in the visible region; therefore a coupling of the fundamental TM_{00} mode to harmonic TE_{mn}-modes of different orders is possible by adjusting the appropriate phase-matching temperature. (It is mainly the extraordinary index of refraction that shifts by a change of the temperature; in this way $n_{TM_{00}}(\lambda_f) = n_{TE_{mn}}(\lambda_{SH})$ can be adjusted for a special pair of interacting modes.) As the indiffusion of titanium only slightly increases the index of refraction in the waveguide, the effective mode indices differ not very much from the bulk indices $[n_{TM_{00}}(\lambda_f) \approx n^o(\lambda_f)$ and $n_{TE_{mn}}(\lambda_{SH}) \approx n^e(\lambda_{SH})]$. Therefore it is mainly the *natural birefringence* of $LiNbO_3$ which determines for that weakly guiding channel the phase-match wavelength (at a given waveguide temperature) or the phase-match temperature (for a given fundamental wavelength in a small band in the near-infrared).

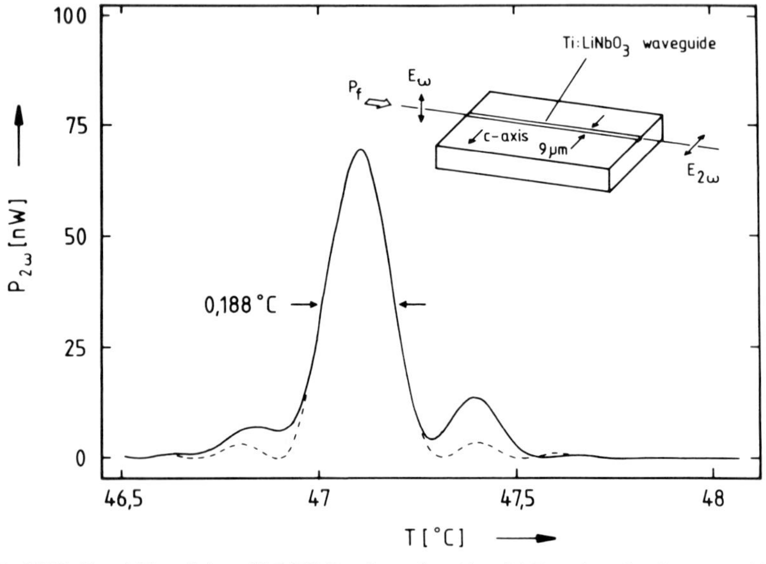

Fig. 12. SHG ($\lambda_f = 1.09$ µm) in a Ti:$LiNbO_3$ channel guide of 4.7 cm length: phase matching by temperature tuning of the birefringence: $\beta_{\omega}^{TM_{00}} + \beta_{\omega}^{TM_{00}} = \beta_{2\omega}^{TE_{00}}$. (After Seibert, unpublished results.)

The waveguide sketched in fig. 12 was designed to optimize the TM_{00}–TE_{00} coupling; with 0.5 mW coupled to the fundamental mode a conversion efficiency of 1.4×10^{-4} has been achieved at the phase-match temperature of 47.1°C ($\lambda_f = 1.092$ μm). Remarkable is the small response curve, indicating a waveguide of excellent homogeneity. The effective interaction length determined by fitting a theoretical $\mathrm{sinc}^2(\Delta\beta L_{\mathrm{eff}}/2)$ phase-match curve (dashed line in fig. 12) to the experimental result, is nearly equal to the actual waveguide length. This means an almost perfect phase matching along the whole 4.7 cm long channel.

As mentioned above, the phase-match wavelength in weakly guiding channels like Ti:$LiNbO_3$ waveguides is essentially determined by the natural birefringence of the bulk crystal. In $LiNbO_3$ it can be very easily changed within a surface layer by an exchange of Li^+-ions versus protons ('proton exchange') to get TIPE-waveguides of type II (De Micheli et al. 1982) (see fig. 7). This changes the sign of the birefringence; its magnitude can be controlled by annealing (Hinkov and Ise 1986) (see fig. 8), yielding a possibility to adjust the waveguide for a phase-match wavelength shifted further into the infrared.

This shift of the phase-match wavelength has been predicted and experimentally verified by De Micheli and co-workers (De Micheli 1983, De Micheli et al. 1983) (see fig. 13). Furthermore, these authors have analyzed the more complicated waveguide structure if a shallow proton exchange is superimposed to

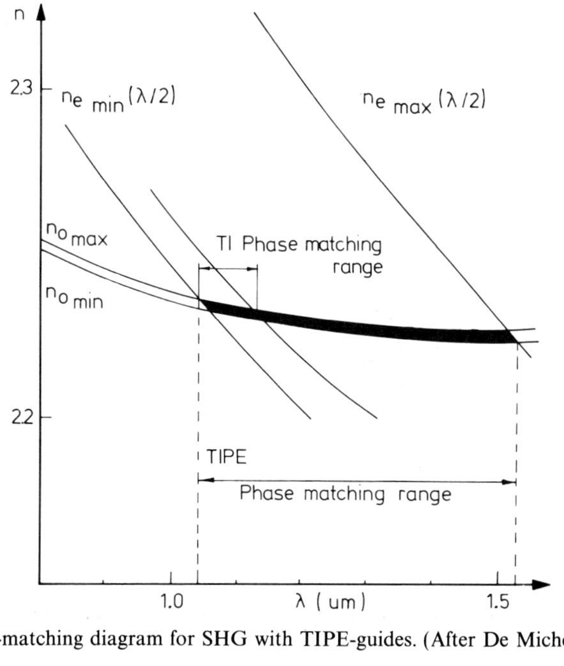

Fig. 13. Phase-matching diagram for SHG with TIPE-guides. (After De Micheli et al. 1983.)

the Ti:$LiNbO_3$ guide ('TIPE I'-guide (De Micheli 1983), see fig. 7). The consequence is an extension of the phase-match range to longer wavelengths, too, but the conversion efficiency of second-harmonic generation is strongly reduced due to a drastically smaller overlap of the interacting mode field distributions.

4.2. *Artificial birefringence*

Another very interesting approach to achieve phase matching is to exploit the artificial anisotropy of a periodic multilayer waveguide structure. Sotin and Shevtsov (1984) have shown that a periodic structure of isotropic layers of alternatively large and small index of refraction (e.g., Ta_2O_5 and SiO_2) with a layer thickness much smaller than the wavelength behaves as a homogeneous medium having the anisotropic properties of a negative uniaxial crystal with the optic axis directed normal to the plane of the layers. If a nonlinear material like ZnO is incorporated as in the example of fig. 14 (Deryugin et al. 1986), a phase-matched nonlinear interaction can be achieved by a coupling of orthogonally polarized modes as in the case of $LiNbO_3$ guides. Now the phase-match wavelength is determined by the indices of refraction and the thicknesses of the layers of the waveguide structure.

Using the multilayer optical waveguide as sketched in fig. 14, and taking advantage of its *artificial birefringence*, Deryugin et al. (1986) succeeded to achieve phase-matched second-harmonic generation of a $TM_0(2\omega)$-mode by a $TE_0(\omega)$ fundamental mode of wavelength $\lambda = 1.064$ µm. The effective interaction length was about 3 mm, measured by scanning the scattered second-

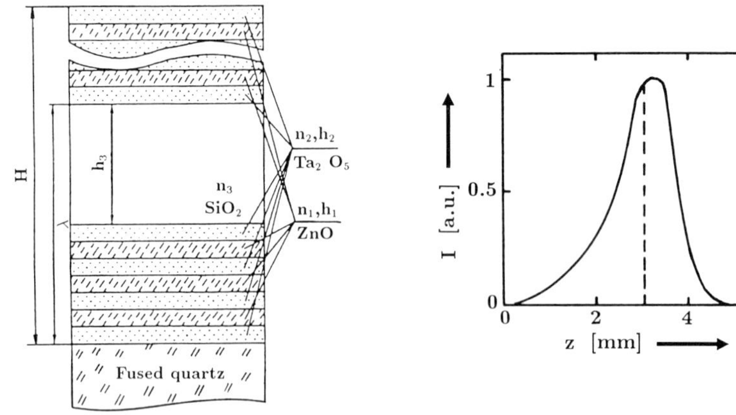

Fig. 14. SHG ($\lambda_f = 1.06$ µm) in a guide of artificial birefringence (left): $\beta_{\omega}^{TE_0} + \beta_{\omega}^{TE_0} = \beta_{2\omega}^{TM_0}$. Distribution of the SH emission along the waveguide in propagation direction (right). (After Deryugin et al. 1986.)

harmonic light intensity along the propagation direction (see the right-hand side of fig. 14). A conversion efficiency of about 1×10^{-4} was achieved with 0.85 W coupled to the fundamental TE_0-mode.

4.3. Modal dispersion

In multimode optical waveguides a nonlinear coupling of modes of different order is possible. If weakly guiding structures are considered, a natural or artificial birefringence of the waveguide is necessary to allow phase matching of modes of orthogonal polarizations (as discussed in the preceding subsections). In strongly guiding structures, however, a phase-matched nonlinear coupling even of modes of equal polarization becomes possible if a low-order fundamental mode interacts with a high-order harmonic mode. In the following we present examples for both, weakly and strongly guiding multimode structures.

Ti:$LiNbO_3$ waveguides are prominent examples of weakly guiding structures. If a multimode guide is used for second-harmonic generation (at least at the harmonic frequency every guide is a multimode one), a distinct fundamental mode can excite one of several harmonic modes. Which one is excited depends on the fundamental wavelength, which has to be properly adjusted to achieve phase matching of a certain pair of optical modes. This is illustrated by figs. 15 and 16 (Sohler and Suche 1983).

For a specific, planar, multimode Ti:$LiNbO_3$ guide the measured dispersion relations of optical modes of both polarizations are drawn in fig. 15 as $2\omega_{TM}$ and ω_{TE} versus the effective index of refraction. At each point of intersection phase

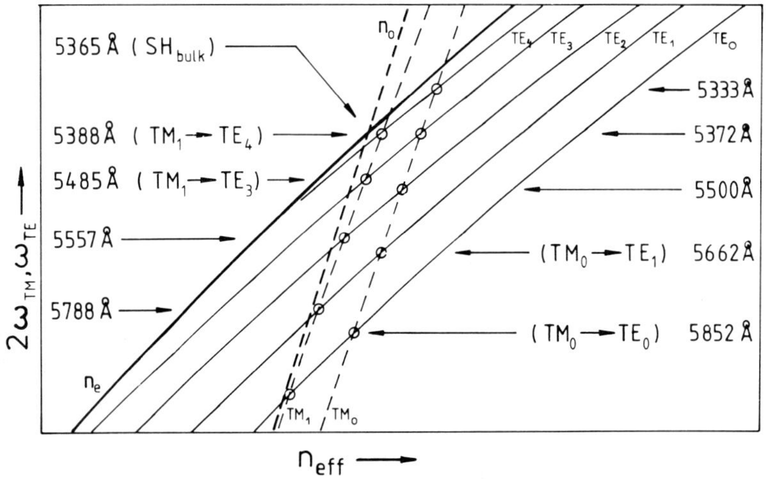

Fig. 15. Dispersion relation of optical modes of both polarizations in a Ti:$LiNbO_3$ guide. Measured phase-matching wavelengths for $\beta^{TM_j}(\omega) + \beta^{TM_j}(\omega) = \beta^{TE_k}(2\omega)$ ($j = 0, 1$; $k = 0, 1, \ldots, 4$) are given. (After Sohler and Suche 1983.)

matching of a fundamental TM- and a harmonic TE-mode is possible. Nearly all of these combinations could be experimentally observed (see the figures of the harmonic wavelengths given in fig. 15) by exciting the TM_0 or TM_1 fundamental mode via prism coupling at a specific wavelength in the near infrared. An example of the results of such a measurement is given in fig. 16, where the harmonic power is plotted versus the fundamental wavelength. The corresponding far-field patterns of the harmonic modes are presented on the right of fig. 16.

An example of second-harmonic generation in a strongly guiding structure was presented by Ito and Inaba (1975). They achieved phase matching of a fundamental $TE_0(\omega)$- and a harmonic $TE_2(2\omega)$-mode in a planar ZnS-film on $LiNbO_3$ using an optical parametric oscillator as the fundamental light source of $\lambda_f \approx 1.1$ μm. Figure 17 presents their experimental results together with a calculated phase-match curve assuming an interaction length of 40 μm.

Figure 18 gives the corresponding qualitative dispersion of the effective index of refraction of the modes of one polarization with the two phase-matched modes marked. By coupling a low-order fundamental and a higher-order harmonic mode, integrated optics allows to use even guides of isotropic material for phase-matched nonlinear interactions which is impossible in bulk optics. This is of great importance as strongly nonlinear, but optically isotropic materials like GaAs (Anderson and Boyd 1971) or the organic PCPU (Hewig and Jain 1983) could be used as waveguide materials for efficient nonlinear interactions. However, by coupling modes of different order the efficiency of the nonlinear process is often drastically reduced (see also section 5).

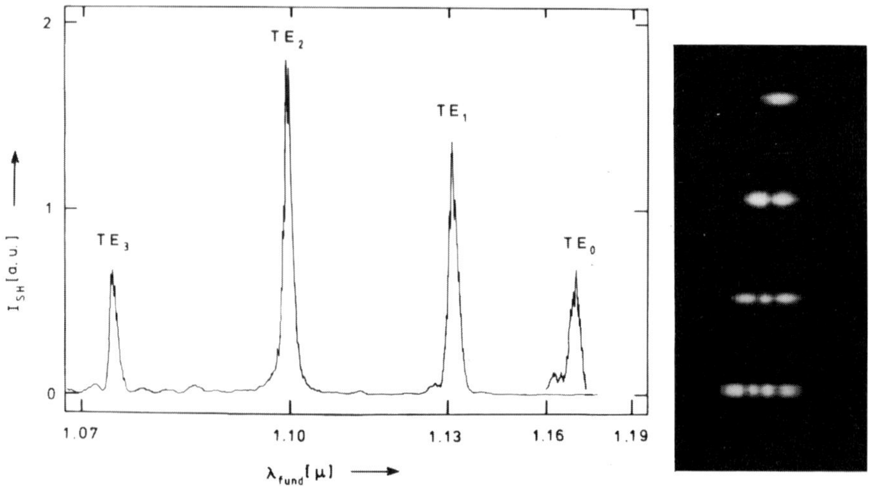

Fig. 16. SH power versus fundamental wavelength of the TM_0-mode (left) together with the corresponding far-field patterns (right). (After Sohler and Suche 1983.)

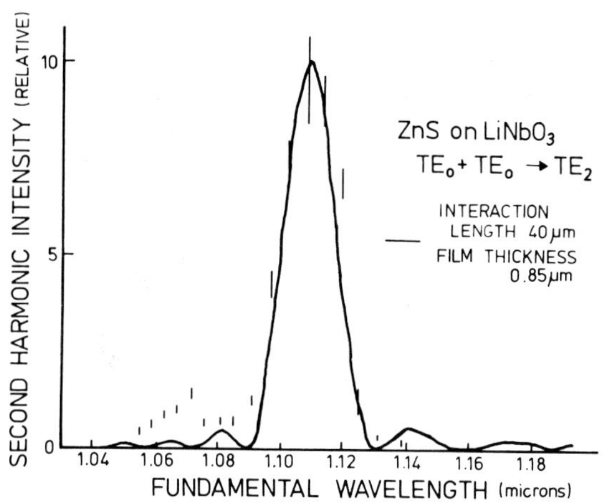

Fig. 17. SHG in a ZnS-waveguide on $LiNbO_3$: phase matching by modal dispersion: $\beta^{TE_0}(\omega) + \beta^{TE_0}(\omega) = \beta^{TE_2}(2\omega)$. (After Ito and Inaba 1975.)

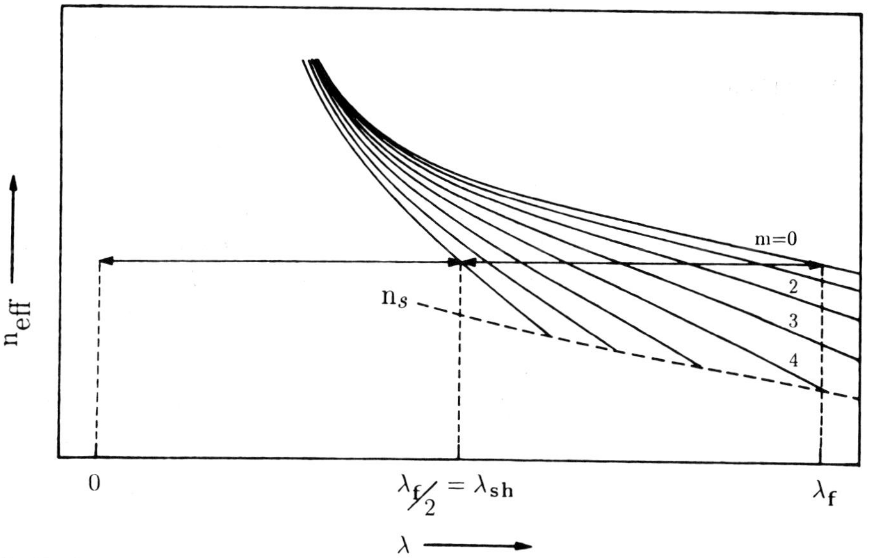

Fig. 18. Qualitative dispersion of optical modes of equal polarization in a high-index slab waveguide. Phase-matched SHG with $m = 0$ fundamental and $m = 7$ harmonic modes is possible.

This reduction of the efficiency can be avoided if a more complex waveguide structure is used as proposed by Duguay and Weiner (1985). They suggested to use a five-layer structure designed in such a way that it creates two coupled asymmetric waveguides at the harmonic, but only one guide at the fundamental frequency (see fig. 19). By a proper choice of the composition (determining the

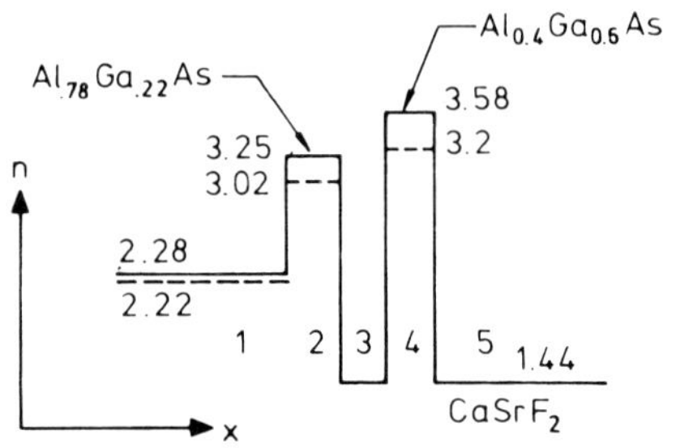

Fig. 19. Proposed waveguide structure to achieve efficient SHG: $\beta^{TE_0}(\omega) + \beta^{TE_0}(\omega) = \beta^{TM_1}(2\omega)$. (After Duguay and Weiner 1985.)

index of refraction) and the thickness of the AlGaAs layers phase matching can be achieved for the fundamental $TE_0(\omega)$- and the harmonic $TM_1(2\omega)$-modes at a certain spacing. This approach is similar to the tunable coupler filter (Taylor 1973), which is essentially a directional coupler consisting of two nonidentical strip waveguides having intersecting dispersion characteristics. Moreover, the waveguide parameters of the structure sketched in fig. 19 are chosen so that most of the field of the $TM_1(2\omega)$-mode resides in one of the coupled waveguides. As a result of this asymmetry, the overlap integral (see section 5) of harmonic and squared fundamental modes, which mainly determines the conversion efficiency, yields a relatively high value (though a coupling of zero- and first-order modes is considered). The authors calculated an efficiency of second-harmonic conversion of about 12% ($\lambda = 1.3$ μm), assuming a phase-matched interaction length of 2 mm and an input power of 10 mW per micrometer of guide width in the lateral direction. However, these data have not yet been confirmed experimentally.

A similar approach has been analyzed by Maier (1980) and experimentally demonstrated by Bozhevol'nyi et al. (1981). They used two identical, coupled, parallel Ti:$LiNbO_3$ channel waveguides to compensate the material dispersion partially by taking advantage of the different wave vectors of symmetric and antisymmetric modes in a TM_{11}–TE_{13}-mode coupling. However, the overlap of the fundamental field with the harmonic field is reduced relative to the single-channel case.

4.4. *Angle-tuning*

Similar to bulk optics angle-tuning is possible in planar waveguides to achieve phase matching. This has been experimentally demonstrated by Zolotov et al. (1978) using Ti:$LiNbO_3$ guides. However, the range of allowed angles is small in this material system; otherwise coupling to radiation modes sets in yielding increased radiation losses (Burns et al. 1979).

4.5. Electro-optic fine-tuning

In an electro-optic material, such as $LiNbO_3$, a fine-tuning of the phase-match condition can be achieved by applying an electric field. This method has been experimentally demonstrated by two groups using Ti:$LiNbO_3$ (channel) guides (Uesugi et al. 1979, Zolotov et al. 1981). However, due to the small index change, which can be electro-optically induced, only a small shift of the phase-match wavelength can be achieved.

4.6. Grating structure

Another way to achieve phase matching is the coupling of fundamental and harmonic modes by a periodic grating structure on top of the guide or (more generally) by a periodic modulation of the (linear or/and nonlinear) waveguide properties (van der Ziel et al. 1976, Tang and Bey 1973, Somekh and Yariv 1972). This approach is especially useful if the other methods fail to achieve phase matching.

By a proper choice of the grating periodicity a nonlinear (harmonic) polarization wave can be excited as the source of a harmonic field propagating in phase with the driving polarization. The phase condition is:

$$2\beta(\omega) + mK = \beta(2\omega), \tag{13}$$

with $K = 2\pi/\Lambda$; Λ is the grating periodicity and $m = (0), \pm 1, \pm 2, \ldots$ (Yariv 1973). Modes of equal or orthogonal polarization can be coupled in materials of appropriate symmetry to take advantage of the largest nonlinear coefficient. Unfortunately, the efficiency of a grating-induced nonlinear effect is in general drastically reduced.

A very illustrative example was presented by Van der Ziel et al. (1976): these authors investigated second-harmonic generation in a short GaAs waveguide with a surface grating to get a phase-matched interaction of fundamental $TE_0(\omega)$- and harmonic $TM_0(2\omega)$-modes (see fig. 20).

In such an experiment nonlinear and linear waveguide properties are modulated both exciting nonlinear harmonic polarization waves, which can partially cancel each other, depending on the sign of the nonlinearity (Jaskorzynska et al. 1986). Therefore, the modulation of the nonlinear or linear properties alone is preferable. In a recent paper Jaskorzynska et al. (1986) theoretically analyzed the grating-induced SHG in $LiNbO_3$ waveguides and presented results of experiments with planar guides of periodically varying titanium concentration. In this way the dominant mechanism was the modulation of the (linear) index of refraction alone. Gratings of 6.4, 6.5 and 6.6 μm periodicity allowed the coupling of the $TM_0(\omega)$- and $TM_0(2\omega)$- or $TM_1(2\omega)$-modes in Z-cut (Y-propagating) $LiNbO_3$ via the largest (d_{33}) nonlinear coefficient. To achieve phase-matched second-harmonic generation with a fundamental wavelength $\lambda_f = 1.064$ μm, a fine-tuning of the effective indices of

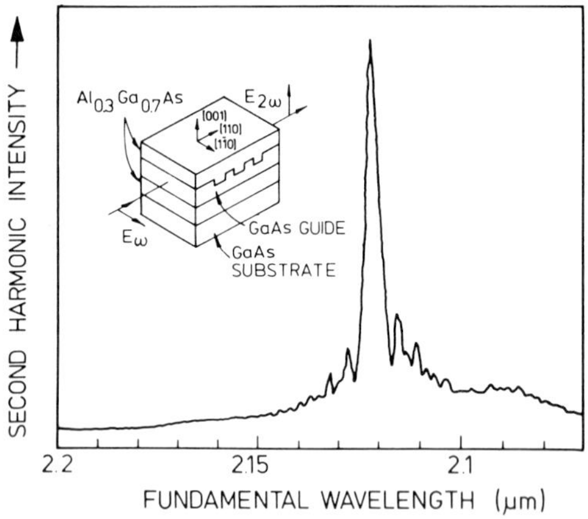

Fig. 20. SHG in a GaAs waveguide: phase matching by a grating structure of periodicity Λ: $\beta^{TE_0}(\omega) + \beta^{TE_0}(\omega) + 2\pi/\Lambda = \beta^{TM_0}(2\omega)$. (After van der Ziel et al. 1976.)

the waveguide modes was necessary by an adjustment of the crystal temperature (see fig. 21). However, the resulting conversion efficiency was orders of magnitude lower than in the case of conventional phase matching (using natural birefringence). The authors believe that the efficiency cannot be increased so far to be of practical interest by a stronger modulation of the linear properties. Instead of this they propose a strong modulation of the nonlinearity, a periodic domain structure, for example, as obtained in a bulk crystal earlier (Feisst and Koidl 1985). This has been very recently achieved by two groups, who succeeded in periodical poling $LiNbO_3$ waveguides for SHG (Webjörn et al. 1989, Lim et al. 1989).

4.7. Coupling to substrate modes (Cerenkov radiation)

All the methods discussed above lead to a second-harmonic radiation in form of a guided wave. However, also a coupling to radiating substrate modes can occur, which is of special interest if guided harmonic waves cannot be phase matched. Moreover, the continuous spectrum of substrate modes allows an 'automatic' phase matching:

$$2\beta_{\text{guided}}(\omega) = \beta_{\text{substrat}}(2\omega).$$

The nonlinear polarization wave produced by the guided fundamental mode travels along the waveguide faster than a free wave of the same frequency in the substrate. Consequently, the nonlinear polarization is a source of Cerenkov radiation, which is the second-harmonic wave in the substrate.

The first experiment of SHG in form of coherent Cerenkov radiation was reported by Tien et al. (1970) (see fig. 22). They used a polycrystalline ZnS-film

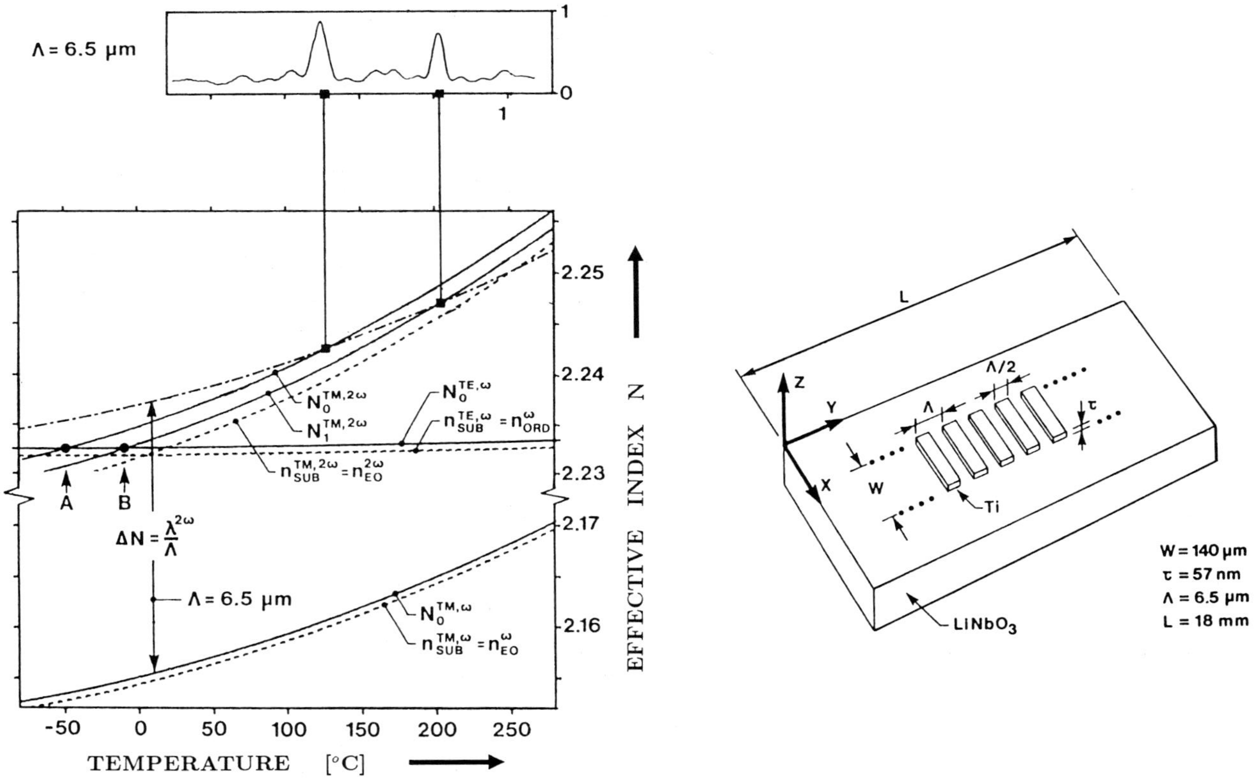

Fig. 21. SHG in a Ti:$LiNbO_3$ guide of periodically varying Ti concentration (patterned sample before indiffusion on the right): $\beta^{TM_0}(\omega) + \beta^{TM_0}(\omega) + 2\pi/\Lambda = \beta^{TM_{0,1}}(2\omega)$. Measured SH power (upper inset) and effective indices versus temperature. (After Jaskorzynska et al. 1986.)

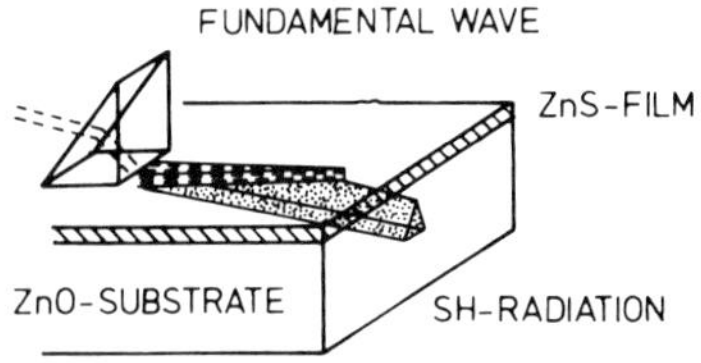

Fig. 22. Experimental arrangement of first guided-wave generation of SH-Cerenkov radiation. (After Tien et al. 1970.)

on a crystalline (nonlinear) ZnO substrate so that only the evanescent wave of the guided fundamental mode excited a nonlinear polarization wave in the substrate as the source of the second-harmonic radiation. Recently, Taniuchi and Yamamoto (1987) demonstrated that, by using proton-exchanged $H_xLi_{1-x}NbO_3$ guides, a surprisingly high conversion efficiency can be achieved for SHG in the form of Cerenkov radiation (see also section 6).

5. *Overlap of modal fields*

The efficiency of nonlinear processes with guided optical waves is essentially determined by the overlap of the interacting fields. This is described by the coupling coefficient κ, which is just the overlap integral of the (normalized) transverse modal field distributions (see section 2). The optimization of the overlap of the modal fields is one of the key problems of nonlinear integrated optics. To illustrate this problem, fig. 23 presents (calculated) field distributions for the case of second-harmonic generation in a Ti:$LiNbO_3$ waveguide. As expected, the fundamental (near-infrared) field is much larger than the harmonic one, yielding a modest overlap. However, this can be improved by optimizing the waveguide structure which determines the field distributions and in this way the overlap of the interacting fields. For a given material system and frequency set, an appropriate choice of thickness, width, (profile of) index of refraction of substrate, waveguide and cover yields the most efficient nonlinear interaction.

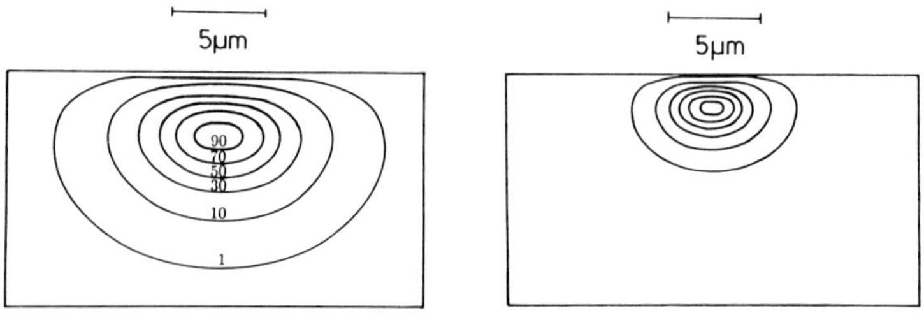

Fig. 23. Calculated near-field distributions of the fundamental ($\lambda_f = 1.15$ μm) and harmonic ($\lambda = 0.576$ μm) modes in a Ti:$LiNbO_3$ strip guide. (After Strake, unpublished result.)

There are several examples of such an optimization reported in the literature. For instance, the nonlinear dependence of the ordinary index of refraction on the Ti concentration in $LiNbO_3$ leads to optimized guides which have a relatively low Ti concentration. This was studied in particular for integrated optical parametric oscillators by Bava et al. (1987). Also a Russian group has studied the dependence of the overlap or the coupling coefficient, respectively, on various fabrication parameters to find the best waveguide design (Zolotov et al. 1979). Another way to improve the overlap of the interacting fields in $Ti:LiNbO_3$ waveguides was demonstrated by De Micheli (1984): he used a Nb_2O_5 overlayer to modify the field distributions and to improve the conversion efficiency of second-harmonic generation.

In waveguides with a gradient index profile (like the $Ti:LiNbO_3$ guides) the nonlinear coupling of zero- and first-order modes can be even stronger than that of zero-order modes (Sohler and Suche 1978); the relative coupling strength is determined by the waveguide structure.

On the other hand, in step-index guides, especially in symmetrical ones, the spatial overlap of modes of different orders is much smaller, as positive and negative contributions to the overlap integral tend to cancel each other. However, Ito and Inaba (1978) demonstrated how a layered waveguide structure allows an efficient nonlinear coupling of zero- and first-order modes. They used a sandwich structure composed of two high-index layers, one of a linear (TiO_2), the other of a nonlinear (ZnS) material (see fig. 24). Now the field overlap in the nonlinear part only determines the coupling coefficient. The improvement of the second-harmonic conversion efficiency achieved in this way for a $TM_0(\omega)$–$TE_1(2\omega)$ coupling was considerable (Ito and Inaba 1978).

An even more attractive sandwich structure was discussed by Akhmediev and Novak (1985); they proposed to use a combination of layers of nonlinear susceptibilities which change sign with a sign change of the product of the distributions of the interacting fields. It is evident that in this way a substantial enhancement of the nonlinear conversion efficiency can be expected. Another

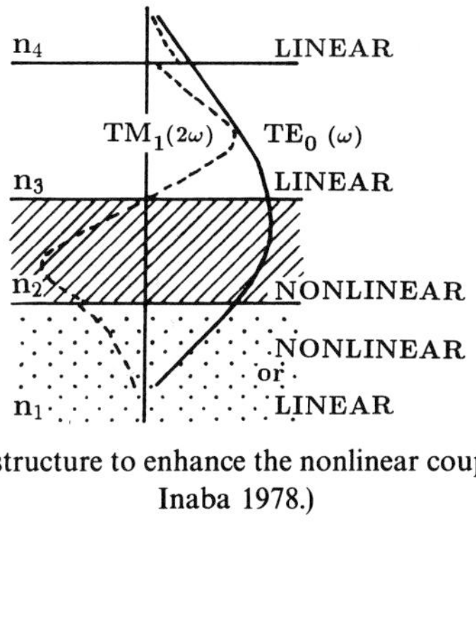

Fig. 24. Proposed waveguide structure to enhance the nonlinear coupling coefficient. (After Ito and Inaba 1978.)

method was proposed by Duguay and Weiner (1985). They suggested to use not only a four-layer, but even a five-layer structure of coupled waveguides (see fig. 19) to achieve a good nonlinear coupling together with the possibility of adjusting phase matching (see section 4.3).

All the examples cited above consider the nonlinear interaction of guided modes. However, also the coupling of guided to substrate modes is critical and should be optimized by a proper waveguide design. Only in that case a strong nonlinear interaction can be achieved, as already experimentally demonstrated by Taniuchi and Yamamoto (1987) for SHG in the form of Cerenkov radiation.

6. Second-harmonic generation (SHG)

In the last 18 years many experiments were reported from a lot of laboratories throughout the world demonstrating SHG in planar, channel and even fiber waveguides. Numerous different material combinations have been investigated for waveguide, substrate and cover; one example is ZnS as waveguide material on a $LiNbO_3$, glass, ZnO or TiO_2 substrate with air as cover. Besides 'classical' birefringent nonlinear materials like $LiNbO_3$, also isotropic semiconductors like GaAs on AlGaAs and GaP on CaF_2 have been studied. More recently, even organic materials as MNA (see section 3) on glass have been used. A good survey of all these experiments up to 1984 can be found in an excellent review paper of Stegeman and Seaton (1985) (see also the papers by Sohler (1984, 1985), Sohler and Suche (1983), Stegeman et al. (1986a, b)). Therefore, we shall concentrate our discussion mainly on more recent results.

6.1. SHG in planar waveguides

To illustrate SHG in planar waveguides, one early experiment of our own group from 1978 is presented (Sohler and Suche 1978); it allows to point again to the essentials of nonlinear integrated optics. Figure 25 presents the schematical experimental arrangement together with the results of measurements. A coherent light beam from an optical parametric oscillator as fundamental source was coupled into and out of the Ti:$LiNbO_3$ waveguide by rutile prisms. The internal reflectivity of the input prism versus the angle of incidence is shown in the upper left drawing. From the reflectivity dip the coupling efficiency of the input beam to the waveguide can be calculated (=37% in this example). The harmonic power versus the wavelength is found in the upper right inset; the lower right inset gives the transmission of the fundamental power versus the fundamental wavelength. At the phase-match wavelength efficient SHG sets in and can be seen as a dip in the transmission curve corresponding to a depletion of the fundamental power. From this depletion a SH conversion efficiency of 22% is evaluated for this TM_0–TE_1 nonlinear coupling with 45 W fundamental power at the beginning of the waveguide (pulsed-mode operation). Although this con-

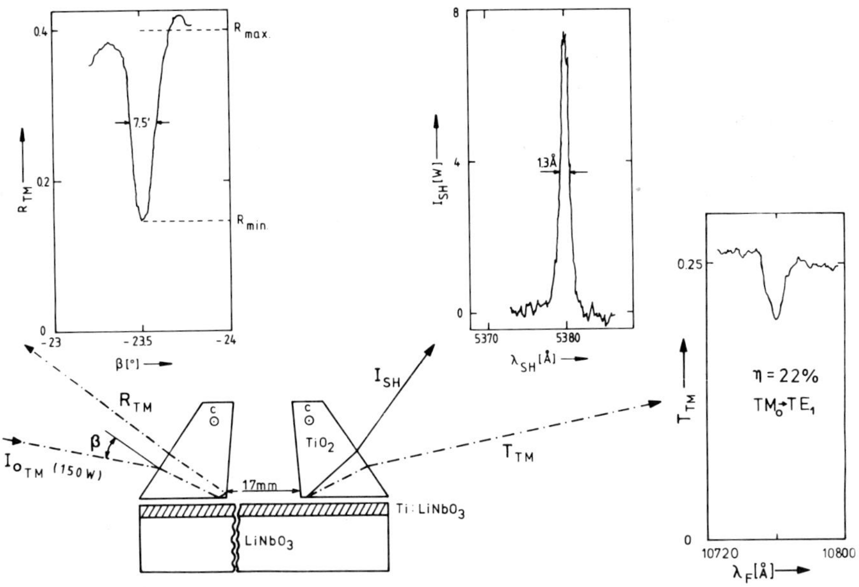

Fig. 25. SHG in a Ti:$LiNbO_3$ optical waveguide: internal reflectivity of the coupling prism versus the angle of incidence (left inset); transmitted fundamental (lower right inset) and generated harmonic (upper right inset) powers versus wavelength around phase matching. (After Sohler and Suche 1978.)

version efficiency is approximately two times better than that achieved with bulk optics (using $LiNbO_3$ crystals of the same length), the potential of nonlinear integrated optics has not yet been fully exploited. However, since 1978, considerable progress has been achieved to improve the efficiency of second-order nonlinear guided-wave interactions. Especially for the Ti:$LiNbO_3$ system we learned how to optimize the field overlap (see section 5) and how to fabricate extremely homogeneous low-loss waveguides to maintain a phase-matched interaction over the whole waveguide length (see section 4). Furthermore, the damage susceptibility of $LiNbO_3$ could be reduced by doping the material with MgO (Bryan et al. 1985). All these improvements were used to optimize nonlinear interactions, in particular in strip waveguides (see next subsection). Using Ti-diffused planar waveguides in a MgO-doped crystal, Fejer et al. (1986) succeeded to generate up to 22 mW average power of 532 nm radiation by frequency doubling.

6.2. SHG in strip waveguides

The two-dimensional confinement of fundamental and harmonic waves in an optical strip guide can improve the conversion efficiency of SHG further. In contrast to planar waveguides or bulk crystals, diffraction spreading of the optical modes can be avoided in both dimensions normal to the propagation direction. Thus a high optical power density can be maintained in the whole channel, low losses assumed.

The conversion efficiency η of SHG is given by the following equation (assuming negligible pump depletion) (Regener and Sohler 1988)

$$\eta = K\kappa^2 P_\omega(z=0)\, h(\alpha_\omega, \alpha_{2\omega}, l), \tag{14}$$

with

$$h(\alpha_\omega, \alpha_{2\omega}, l) = \left[\frac{\exp(-\alpha_\omega l) - \exp(-\alpha_{2\omega} l/2)}{(\alpha_\omega - \alpha_{2\omega}/2)}\right]^2,$$

where h reduces to l^2 for vanishing losses. $K = \omega^2 K'$ with a material-dependent parameter K', κ is the nonlinear coupling coefficient (see section 2), α_ω and $\alpha_{2\omega}$ are the loss coefficients of the waveguide at fundamental and harmonic frequencies, l is the waveguide length and P_ω the fundamental power.

In fig. 26, $h(\alpha_\omega, \alpha_{2\omega}, l)$ is plotted for three sets of loss parameters versus the guide length l. Furthermore, assuming a typical Ti:$LiNbO_3$ channel guide, a fundamental wavelength $\lambda = 1.09$ μm and a fundamental power (at $z = 0$) of 1 mW, absolute figures of the conversion efficiency are given (on the right ordinate). It can be seen that low-loss waveguides are absolutely necessary to take full advantage of the integrated optics approach of SHG.

The advantage of using waveguides instead of bulk crystals for harmonic generation is emphasized by a comparison of the conversion efficiency of both methods. In bulk optics the maximum efficiency is obtained by focusing a Gaussian beam into a nonlinear crystal in such a way that the Rayleigh range equals the crystal length (Yariv 1975). This length-dependent optimum focusing results in a linearly growing conversion efficiency (assuming perfect phase

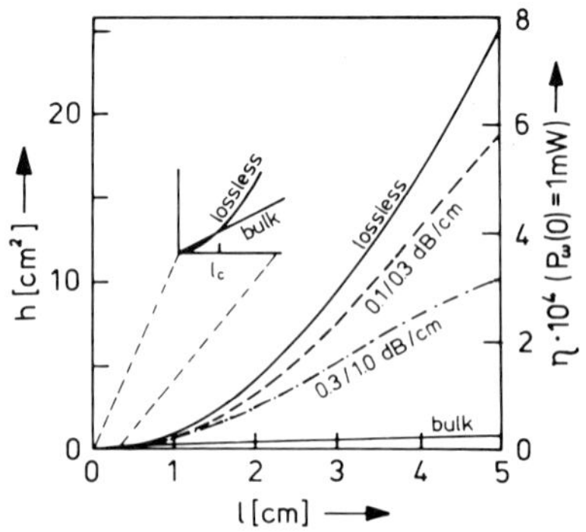

Fig. 26. The function h (see eq. (14), left ordinate) versus length l of a Ti:$LiNbO_3$ channel guide for several loss coefficients α_ω and $\alpha_{2\omega}$ as parameters. h is proportional to the conversion efficiency of SHG (right ordinate).

matching and a lossless crystal). Therefore, the conversion efficiency in a waveguide exceeds that in the bulk above a critical length l_c (the crossover of both efficiency curves in fig. 26),

$$l_c = F_{\text{eff}} 2n/\lambda. \tag{15}$$

The better the nonlinear coupling (or the smaller the effective interaction area $F_{\text{eff}} := \kappa^{-2}$), the shorter is l_c and the greater is the improvement that can be achieved in an optical waveguide. With optimized Ti:$LiNbO_3$ channels ($l_c \approx 0.5$ mm) of 5 cm length an improvement of the conversion efficiency up to 2 orders of magnitude should be possible, low waveguide losses assumed.

Up to now, the best experimental results for SHG have been achieved in Ti:$LiNbO_3$ strip guides. As an example, fig. 12 presents the results obtained in an extremely homogenous channel of 4.7 cm length with a fundamental cw-power of 0.5 mW coupled to the TM_{00}-mode ($\lambda_f = 1.092$ μm). The conversion efficiency of this TM_{00}–TE_{00} coupling is 1.4×10^{-4}, corresponding to a figure of 2.8×10^{-4}/mW, which is somewhat lower than expected from the calculated response curves (see fig. 26, but $\alpha_\omega = 0.24$ dB/cm and $\alpha_{2\omega} = 0.5$ dB/cm).

Even better results obtained with a pulsed Nd-laser ($\lambda_f = 1.064$ μm) were reported by Arvidsson and Laurell (1986). They used a 2.7 cm long Ti:$LiNbO_3$ channel guide and measured a conversion efficiency of 4.2% with a fundamental power of 120 mW corresponding to a figure of 3.5×10^{-4}/mW. Comparing again this value with the calculated response curves (fig. 26) allows the conclusion that a guide of smaller effective interaction area has been investigated than assumed for our calculations.

The same authors investigated SHG in Ti:MgO:$LiNbO_3$ channel guides (Laurell and Arvidsson 1988). However, in this material the measured conversion efficiency (2.4×10^{-5}/mW in a 16 mm long guide) was about 4 times lower than expected from theory. The authors attributed this result to waveguide inhomogeneities; it seems that the quality of the MgO-doped $LiNbO_3$ is not yet as good as that of undoped material.

Figure 27 shows a calculated diagram of the effective indices of the optical modes in a 10 μm wide Ti:MgO:$LiNbO_3$ strip waveguide (Laurell and Arvidsson 1988). Phase matching of the fundamental TM_{00}- and various TE_{mn}-modes is achieved at different temperatures, when $N_{00} = N_{mn}$ (open circles around the intersections). The observed near-field patterns of the harmonic modes are presented as well, together with the measured phase-match temperature. It is remarkable that, due to the enlarged birefringence of the MgO-doped material, phase matching was observed at considerably higher (up to nearly 100°C) temperatures than in the MgO-free $LiNbO_3$. Therefore, it was not possible to decide whether the observed increased power-handling capability for visible cw-light in Ti:MgO:$LiNbO_3$ guides was due to the MgO-doping or due to the elevated temperatures which also reduce the optical damage susceptibility.

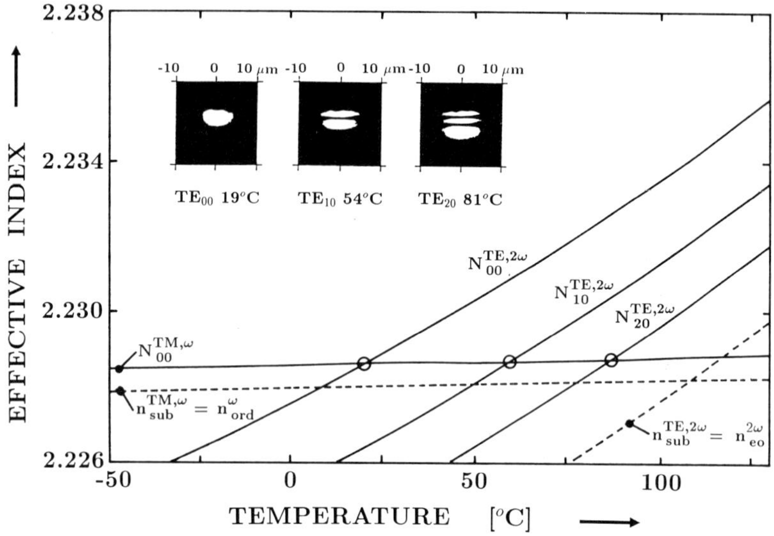

Fig. 27. Calculated diagram of effective indices of refraction of modes in a 10 μm wide Ti:MgO:$LiNbO_3$ strip guide versus the temperature. SHG ($\lambda_f = 1.06$ μm) with several mode combinations is possible. Near fields of harmonic modes and phase-match temperatures are given as inset. (After Laurell and Arvidsson 1988.)

6.3. SHG in waveguide resonators

Although promising results have been reported on the performance of integrated optical harmonic generators using Ti:$LiNbO_3$ and Ti:MgO:$LiNbO_3$ strip waveguides (see preceding subsection) as well as proton-exchanged $H_xLi_{1-x}NbO_3$ guides (see section 6.4), the conversion efficiency achieved so far is too low to permit efficient frequency doubling with low-power fundamental sources like semiconductor lasers. (A conversion efficiency of 10% achieved with a fundamental power of 100 mW has been proclaimed a challenging aim to stimulate new applications (Ostrowsky 1985).) Therefore a resonator approach was proposed to enlarge the conversion efficiency by a resonant enhancement of the fundamental power inside an integrated optical cavity (Regener et al. 1981). The fabrication of high-quality resonators requires nonlinear strip waveguides of very low losses. It seems that at present only Ti:$LiNbO_3$ guides are suitable for such an application; loss figures down to 0.03 dB/cm have already been demonstrated with near-infrared ($\lambda = 1.15$ μm) monomode channel guides (Regener and Sohler 1988).

Indeed, the internal power enhancement f in a resonator can be considerable. It depends on the reflectivities of the cavity mirrors and the internal (single-pass) loss $\alpha_\omega l$ (Regener and Sohler 1988),

$$f = \frac{1 - R_f}{[1 - (R_f R_r)^{1/2} \exp(-\alpha_\omega l)]^2}. \tag{16}$$

R_f is the reflectivity of the front mirror, R_r that of the rear mirror. Figure 28 presents the dependence of this power enhancement factor on the internal loss expressed in decibels. Two different types of resonators are considered: a symmetrical (f_s) and a matched resonator (f_m) with a front-mirror reflectivity given by $R_{fm} = R_r \exp(-2\alpha_\omega l)$.

Although an appreciable power enhancement can be achieved in a symmetric low-loss resonator, an even further improvement is possible with a matched device (see fig. 28). It is designed to maximize the power enhancement in a cavity of given internal loss αl and coupling loss $(1 - R_r)$. With a dichroitic, dielectric mirror of $R_r \approx 1$ for the fundamental wavelength (and R_r small for the harmonic wave), the coupling loss becomes negligible. Simultaneously, the fundamental power in the nonlinear waveguide is maximized yielding the best conversion efficiency for SHG possible with external (lossy) resonators; it grows with f^2.

A comparison of the calculated conversion efficiency of SHG in a nonresonant channel guide and in a symmetric and a matched resonator is presented in fig. 29 as a function of the interaction length in these devices. Absolute figures are given for the generated harmonic power as well, assuming in all three cases a fundamental power of 1 mW. Furthermore, perfect phase matching of TM_{00}- ($\lambda_f = 1.09$ μm) and TE_{00}- ($\lambda_{SH} = 0.546$ μm) modes is assumed and waveguide parameters typical for good, nonlinear Ti:$LiNbO_3$ strip guides: $\alpha_\omega = 0.1$ dB/cm; $\alpha_{2\omega} = 0.3$ dB/cm; $F_{eff} = 260$ μm^2. For the nonresonant waveguide the dependence on the interaction length is nearly quadratic, as already shown in more detail in fig. 26.

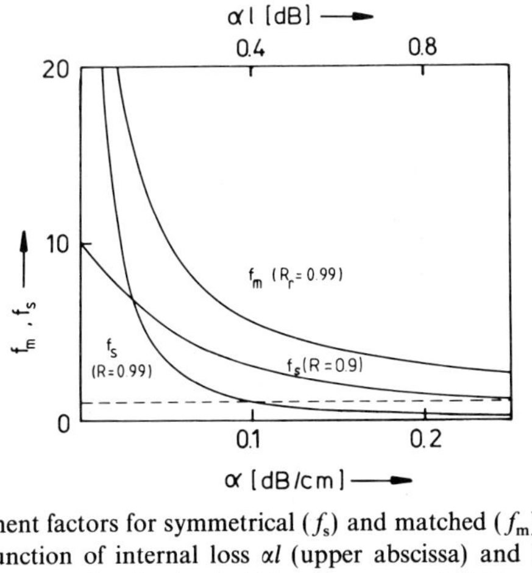

Fig. 28. Power enhancement factors for symmetrical (f_s) and matched (f_m) resonators for different mirror reflectivities as function of internal loss αl (upper abscissa) and loss coefficient α (lower abscissa, assuming $l = 4$ cm). (After Regener and Sohler 1988.)

With a symmetric resonator, a conversion efficiency can be achieved that surpasses that of the simple channel by as great as 2 orders of magnitude. However, by far the best device for efficient SHG (note the interrupted scale of the ordinate in fig. 29) is the matched resonator. Its conversion efficiency exceeds that of the symmetrical resonator by a factor of at least 4 for the special parameters chosen. The more the rear-mirror reflectivity R_r approaches 1, the higher the conversion efficiency becomes and the smaller is its dependence on the resonator length. This feature is attractive as it should allow the construction of relatively short, nevertheless efficient, harmonic generators.

The calculated conversion efficiency of the matched resonator of fig. 29 is higher than 1%/mW fundamental power, but could be further improved by more than 1 order of magnitude in resonators of waveguides with lower losses. Assuming $\alpha_\omega = 0.03$ dB/cm and $\alpha_{2\omega} = 0.09$ dB/cm, as it is experimentally feasible, the conversion efficiency surpasses 10%/mW (Regener and Sohler 1988).

This concept of resonant SHG in symmetric and matched resonators as a means of enhancing the conversion efficiency was also experimentally studied by Regener and Sohler (1988). They were able to demonstrate a large improvement of the conversion efficiency from 2.4×10^{-4}/mW in a nonresonant Ti:$LiNbO_3$ strip guide to 1×10^{-2}/mW in a (nearly) matched waveguide resonator (normalized figures). A single-frequency Argon laser was used as the fundamental source ($\lambda_f = 1.092$ μm). Figure 30 presents the second-harmonic power

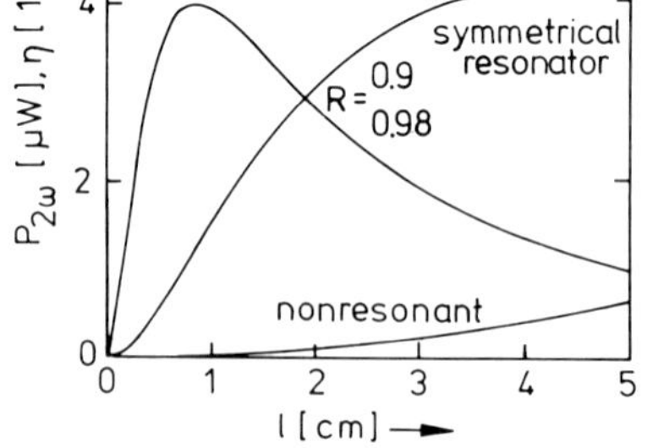

Fig. 29. Calculated SHG conversion efficiencies η and generated SH power $P_{2\omega}$, respectively, in a nonresonant Ti:$LiNbO_3$ guide and in symmetric as well matched waveguide resonators as function of the device length l, assuming $P_\omega = 1$ mW ($\lambda_f = 1.09$ μm) and typical waveguide parameters. (After Regener and Sohler 1988.)

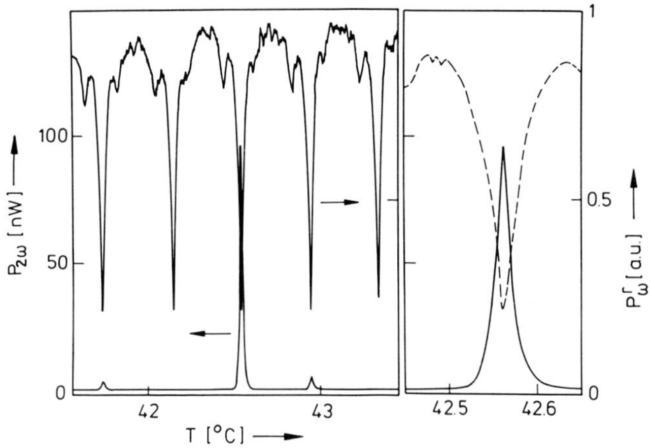

Fig. 30. SHG in a matched Ti:$LiNbO_3$ waveguide resonator: in forward direction generated SH power (left ordinate) and reflected fundamental power (right ordinate) versus the device temperature with $P_\omega = 100$ μW ($\lambda_f = 1.09$ μm). The right part of the figure is an enlarged view of the central resonances on the left. (After Regener and Sohler 1988.)

generated in the device together with the reflected fundamental power as function of the waveguide temperature. With only 100 μW of fundamental power measured in front of the waveguide resonator a harmonic power of 100 nW could be generated at the phase-matching temperature of 42.5°C in forward direction. (A similar harmonic emission was produced in backward direction, but not analyzed.) However, at harmonic power levels exceeding 200 nW in cw-operation, the onset of optically induced changes of the indices of refraction ('optical damage') of $LiNbO_3$ prevented higher conversion efficiencies and caused instabilities. Therefore, a further improvement of the material purity of $LiNbO_3$ is necessary to reduce its damage susceptibility (see also section 3) and to allow in this way to take advantage of the attractive resonator concept also at higher power levels.

6.4. SHG as Cerenkov radiation

Ti:$LiNbO_3$ optical waveguides suffer from a susceptibility to optical damage at higher (cw-) power levels; furthermore, phase matching utilizing the material birefringence cannot be achieved at fundamental wavelengths smaller than about 1.0 μm. Both limitations can be overcome by using proton-exchanged waveguides in $LiNbO_3$ (see section 3). They withstand considerably higher (cw-) power densities and yield an 'automatic' phase matching of guided fundamental and radiating harmonic modes (Cerenkov radiation). The Cerenkov scheme of SHG further benefits from the largest nonlinear coefficient d_{33} in $LiNbO_3$ and from the strong confinement of the fundamental field in the waveguide.

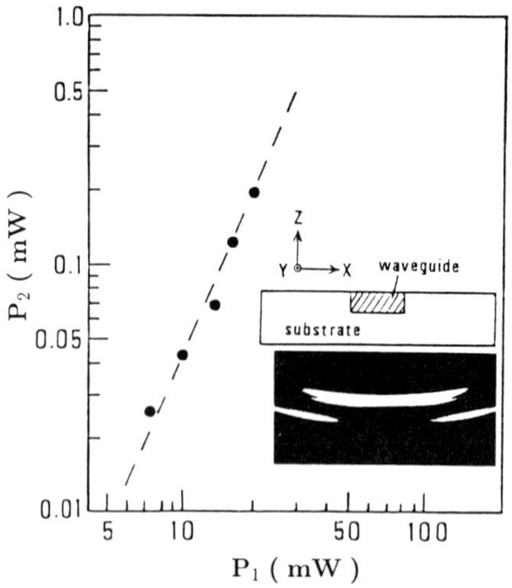

Fig. 31. SH power P_2 as function of fundamental power P_1. The inset shows the far-field pattern of the SH-Cerenkov radiation. (After Taniuchi and Yamamoto 1986a.)

Very promising results have been reported for SHG in $H_xLi_{1-x}NbO_3$ strip waveguides: using semiconductor lasers at 0.84 μm wavelength as fundamental source, Taniuchi and Yamamoto (1986a, b, 1987a, b) succeeded in generating up to 0.2 mW harmonic power with 20 mW fundamental power (see fig. 31).

Also Sanford et al. (1987) reported SHG in proton-exchanged waveguides together with sum frequency generation by mixing $\lambda = 1.32$ μm ($\lambda = 0.84$ μm) and $\lambda = 1.06$ μm radiations. They also utilized Cerenkov SH radiation as a tool for the contact-free measurement of the effective indices of the guided modes (Sanford and Robinson 1987). (A determination of the emerging angle of the harmonic radiation allows to evaluate the effective index of the guided fundamental mode due to phase matching.)

In a quite recent work (Arvidsson et al. 1989), the influence of annealing on the proton-exchanged guide has been studied yielding a complicated dependence of the conversion efficiency on the waveguide parameters. This could be explained by the parameter-dependent overlap of the interacting modes. Modeling Cerenkov SHG using a step-index waveguide model (Li et al. 1989) allowed to compare calculated (η_c) and measured (η_m) conversion efficiencies; a ratio η_c/η_m of about 3.5 was found. However, it was not yet possible to determine the origin of this factor.

6.5. SHG for signal processing

SHG in optical waveguides can also be used for signal processing applications by mixing no longer codirectional waves, as discussed above, but counterpropagating waves. This has been theoretically and experimentally studied in detail by

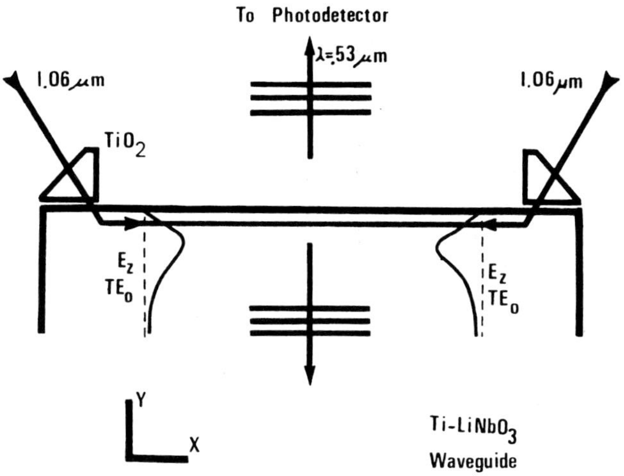

Fig. 32. Experimental setup with $LiNbO_3$ waveguide for SHG with counterpropagating waves, used in ps-signal processing applications. (After Normandin and Stegeman 1979.)

Stegeman and colleagues (Normandin et al. 1981, Normandin and Stegeman 1979, 1980a, 1982, Liao et al. 1983, Stegeman 1983). Also the counterpropagating waves produce a nonlinear polarization as source of an optical field at the harmonic frequency. However, this polarization field can only radiate normal to the waveguide surface because it has no wave-vector component parallel to the surface. Plane waves are radiated into both the substrate and the cladding (see fig. 32). Their intensity (many orders of magnitude lower than in the codirectional, phase-matched situation) is proportional to the square convolution of the envelopes of the input pulses of fundamental frequency. This temporal convolution (with a time compression of a factor of two) could also be measured as a spatial convolution by imaging the radiating surface on an integrating detector array (Normandin and Stegeman 1982). If then one of the input pulses has a δ-function-like envelope, the temporal profile of (the envelope of) the counterpropagating pulse is displayed on the detector array. Hence this device can act as an optical transient digitizer with psec resolution (Normandin and Stegeman 1980b).

7. *Sum and difference frequency generation (DGF); parametric amplification*

A nonlinear polarization wave with sum and difference frequency components is generated in a second-order nonlinear optical waveguide, if two modes of different frequencies (and sufficient power density of at least one mode) are simultaneously excited. Phase matching determines if the sum or difference frequency optical field, radiated by the nonlinear polarization, linearly grows with the interaction length.

Sum frequency generation would be especially interesting to produce tunable coherent radiation in the visible and ultraviolet. However, there are only two experiments reported in the literature demonstrating sum frequency generation in integrated optical waveguides: Reutov and Tarashchenko (1975) mixed the radiation of a Nd:YAG laser ($\lambda = 1.06$ µm) with that of a cw He–Ne laser ($\lambda = 1.15$ µm) in an extremely thin $LiNbO_3$ plate. Uesugi et al. (1978) on the other hand, used a tunable (pulsed) optical parametric oscillator to mix its radiation with that of a He–Ne laser in a Ti:$LiNbO_3$ strip guide. They achieved a tuning range of the sum frequency wave of $0.543\ \mu m < \lambda < 0.55\ \mu m$.

Difference frequency generation (DFG) has been investigated by some more groups as a means to produce tunable near-, mid- and far-infrared, coherent radiation. Again, integrated optics promises higher conversion efficiencies than bulk optics. Using the nomenclature for parametric processes, DFG requires a strong 'pump' laser beam of frequency ω_p (wavelength λ_p) and a weaker 'signal' beam ($\omega_s < \omega_p$; λ_s) simultaneously coupled to the nonlinear waveguide. If phase matching is adjusted, the difference frequency 'idler' field ($\omega_i = \omega_p - \omega_s$; λ_i) is generated (we assume $\omega_i < \omega_s < \omega_p$ throughout this paper). By using a laser beam of frequency ω_i, generation of a difference frequency 'signal' field ($\omega_s = \omega_p - \omega_i$) with the same 'pump' source is possible as well.

DFG is always accompanied by parametric amplification of the signal (or idler) wave. This becomes especially clear in the language of quantum mechanics: one pump photon is annihilated creating one idler and one signal photon simultaneously (energy conservation!).

Most (if not all) experiments to achieve DFG and parametric amplification have been performed with $LiNbO_3$ waveguides. Therefore, it is of interest to analyze the spectral ranges accessible by DFG in $LiNbO_3$.

Figure 33 presents the tuning characteristics for parametric three-wave interactions (including SHG, sum frequency generation, DFG and parametric amplification and oscillation) as signal and idler (vacuum) wavelengths versus the pump (vacuum) wavelength. These characteristics include energy ($1/\lambda_p = 1/\lambda_s + 1/\lambda_i$) and wave vector conservation ($n_p/\lambda_p = n_s/\lambda_s + n_i/\lambda_i$); parameter is the crystal (waveguide) temperature, which changes the indices of refraction and in that way the phase-match wavelengths. In fig. 33 only the spectral window between the (electronic) interband absorption edge ($\lambda \approx 0.4$ µm) and the onset of absorption by phonon excitation ($\lambda \approx 4$ µm) is considered; further windows exist in the far-infrared (see below).

The tunable generation of near-infrared radiation by DFG in Ti:$LiNbO_3$ channel waveguides was studied by two groups. They used a tunable, pulsed (Hampel et al., unpublished result, Sohler 1985) or cw (Uesugi 1980) dye laser as pump source, operated in the wavelength range $0.58\ \mu m < \lambda_p < 0.67\ \mu m$

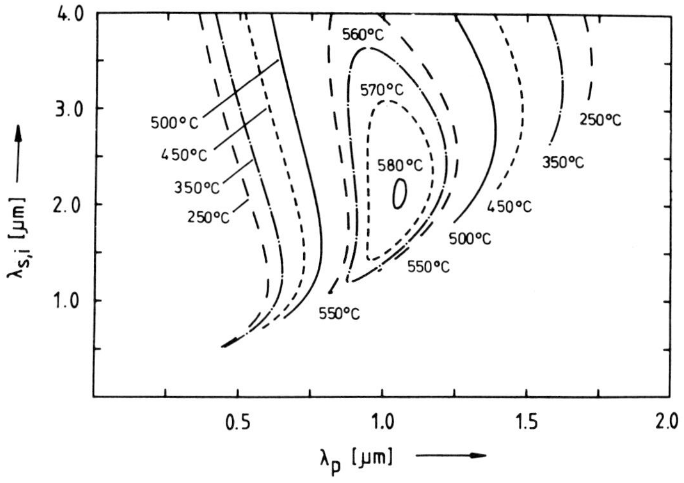

Fig. 33. Calculated tuning characteristic for three-wave, nonlinear parametric interaction in bulk $LiNbO_3$ as signal and idler wavelengths versus pump wavelength (parameter: crystal temperature). (After Herrmann and Sohler 1988.)

(Hampel et al., unpublished result) or 0.58 μm $< \lambda_p <$ 0.63 μm (Uesugi 1980), respectively. A cw He–Ne ($\lambda = 1.15$ μm) (Hampel et al., unpublished result) or a cw Nd:YAG laser ($\lambda = 1.06$ μm (Hampel et al., unpublished result); $\lambda = 1.32$ μm (Uesugi 1980)) was used as the signal source. Both beams were fed by endfire coupling into the nonlinear waveguide. Tuning was achieved by changing the pump wavelength together with the crystal temperature to maintain phase matching. Experimentally, tuning ranges 1.17 μm $< \lambda_i <$ 1.85 μm (Hampel et al., unpublished result) (see fig. 34) and 1.035 μm $< \lambda_i <$ 1.19 μm (Uesugi 1980) could be demonstrated. An external conversion efficiency defined as the ratio of idler (DFG) to signal power of 0.14% was quoted in one experiment (Uesugi 1980).

The mid-infrared range 2 μm $< \lambda <$ 4 μm is also attainable by a near-infrared pump source with 1.4 μm $< \lambda_p <$ 1.7 μm at not too high crystal temperatures (see fig. 35 which is mainly a magnified section of fig. 33). Such an approach has several advantages (Herrmann and Sohler 1988); in particular, the selective excitation of the lowest-order pump mode by end-fire coupling is facilitated, if the wavelengths of the three interacting waves are not too different (assuming a monomode guide for the longest wavelength involved).

Using a tunable $Tl^0(1)$:KCl color center laser (1.45 μm $< \lambda_p <$ 1.6 μm) as pump source and a He–Ne laser ($\lambda_i = 3.39$ μm) as idler source, tunable DFG in the ranges from $\lambda_s = 2.5$ μm to $\lambda_s = 3.0$ μm could be recently demonstrated (Herrmann and Sohler 1988) (see also the dotted line in fig. 35). Besides cw-operation, also ps-pulse generation was achieved by pumping the waveguide with pulses of the mode-locked color center laser.

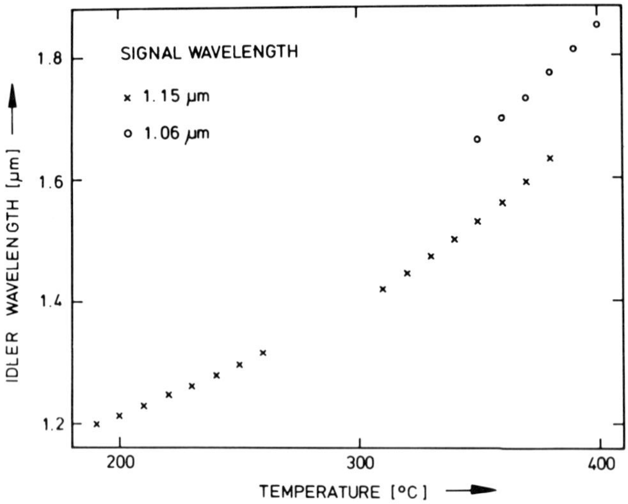

Fig. 34. DFG with He–Ne ($\lambda = 1.15$ µm) and Nd–YAG ($\lambda = 1.06$ µm) laser radiation, respectively, mixed in a Ti:$LiNbO_3$ channel guide with a dye laser beam as pump: (difference frequency) idler wavelength versus device temperature (pump wavelength is simultaneously tuned). (After Hampel et al., unpublished results.)

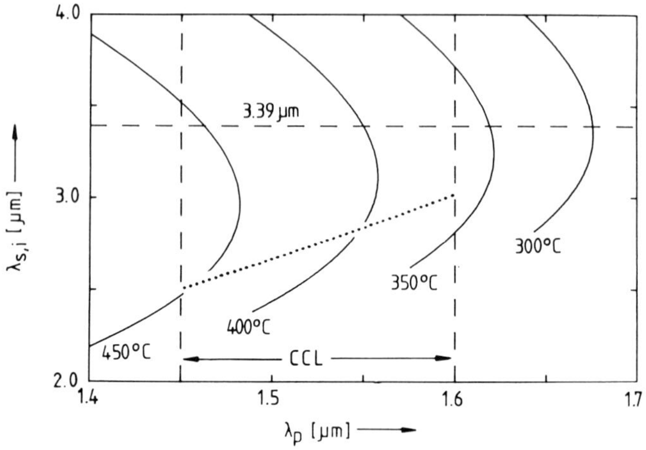

Fig. 35. Enlarged section of fig. 33: tuning characteristic for pump wavelength in the tuning range of a color center laser (CCL). Dots correspond to tunable DFG using a He–Ne laser ($\lambda_i = 3.39$ µm) and a CCL. (After Herrmann and Sohler 1988.)

Figure 36 shows a measured curve as an example of the experimental results (Herrmann and Sohler 1988); the generated signal power is plotted versus the pump wavelength. The temperature of the 30 µm wide, 39.4 mm long waveguide, fabricated by indiffusion (30 h, 1060°C) of a 1400 Å thick Ti stripe into Z-cut

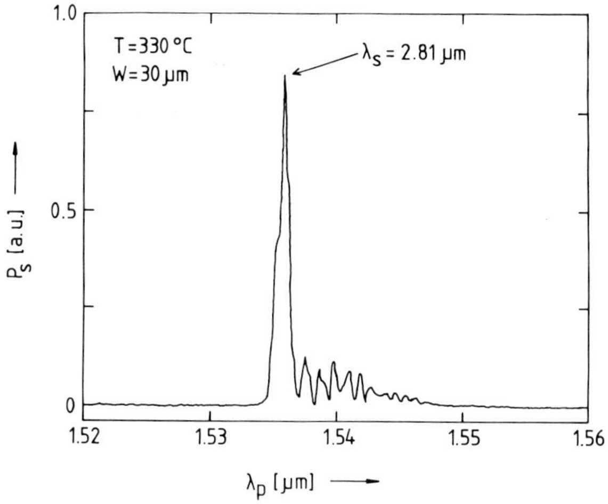

Fig. 36. Experimentally generated signal power (DFG with $\lambda_i = 3.39$ μm) versus pump wavelength. (After Herrmann and Sohler 1988.)

$LiNbO_3$, was set to 330°C. Under these conditions, phase matching occurred at a pump wavelength of 1.536 μm resulting in a signal wavelength of 2.81 μm. On the long-wavelength side a periodic structure appeared due to the slightly inhomogeneous temperature profile of the crystal oven; this results in phase-matched interactions along short sections near the waveguide endfaces with phase-dependent constructive and destructive interference (see also the discussion in section 4).

The efficiency of DFG in cw-operation was 1×10^{-4} achieved with 50 mW pump power (transmitted by the waveguide); with 100 μW idler power a signal power of approximately 10 nW was generated. This efficiency is roughly one order of magnitude better than possible with bulk optics using a nonlinear crystal of equivalent length. In the mode-lock operation of the pump laser, with pulse peak powers up to 50 W, an efficiency increase up to 1×10^{-1} was estimated.

To generate tunable far-infrared radiation, Hägele et al. (1981) used a tunable optical parametric oscillator and a Nd:YAG laser to mix not too different frequencies in a thin $LiNbO_3$ platelet as the nonlinear waveguide. Figure 37 presents the schematical experimental setup.

By tuning the parametric oscillator and restoring the necessary phase matching by adjusting the thickness t (see fig. 37) of the far-infrared waveguide, the wavelength of the far-infrared radiation could be shifted from 280 to 410 μm in another window of high transparency in $LiNbO_3$. Despite of overlap problems of near- and far-infrared modes, Hägele et al. succeeded to produce far-infrared pulses of a peak power of up to 1.0 μW.

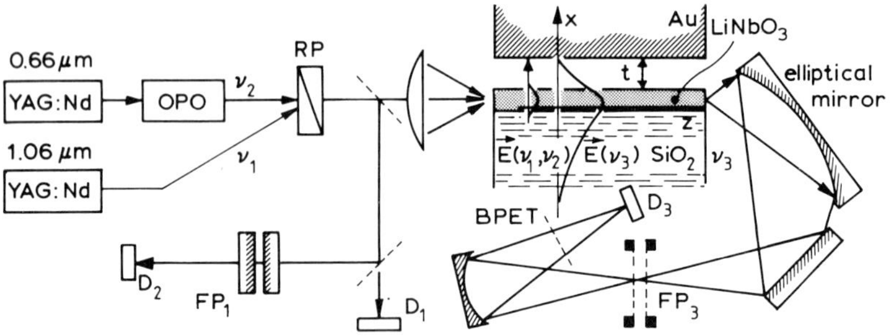

Fig. 37. Experimental setup for far-infrared generation as DFG in an optical $LiNbO_3$ waveguide. (After Hägele et al. 1981.)

In DFG experiments the simultaneous amplification of signal (or idler) wave can be observed. Figure 38 presents an example (Sohler and Suche 1980): the two laser beams can excite distinct waveguide modes to investigate a nonlinear interaction of various mode combinations. By inserting an appropriate filter in front of the detector, either DFG or parametric amplification of the He–Ne laser beam can be studied separately. The inset shows as an example of a measured result the sum of signal and idler powers as function of time; when a pump pulse passed the waveguide, amplification and DFG could be observed; as the

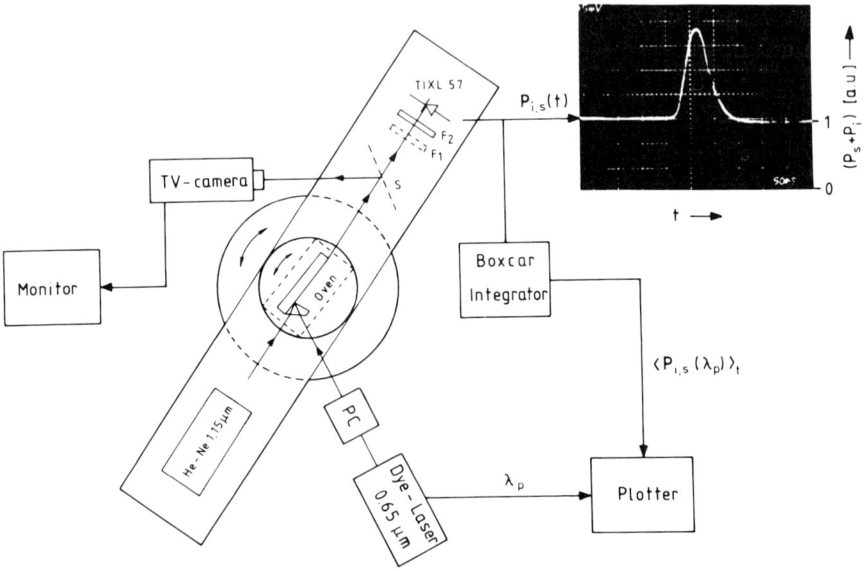

Fig. 38. Experimental setup to study parametric amplification and DFG in optical waveguides. F_1, F_2: bandpass and cuton wavelength filters, respectively; PC: Pockels cell. Inset: sum of transmitted signal and idler powers versus time (50 ns/div.). (After Sohler and Suche 1980.)

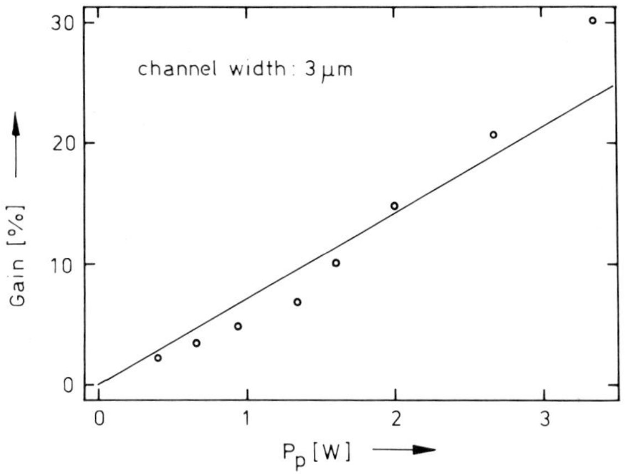

Fig. 39. Parametric gain in a Ti:LiNbO$_3$ channel guide versus the pump power ($\lambda_s = 1.15$ μm, $\lambda_p \approx 0.6$ μm). (After Seibert, unpublished results.)

detector sensitivity was similar at both, signal and idler wavelengths, half of the peak is due to parametric amplification. The maximum parametric gain achieved in the 50 μm wide, 32 mm long Ti:LiNbO$_3$ stripe guide, was 16 dB; however, the necessary pump power level was 150 W (in a pulse-mode operation), intolerable in cw-operation.

In smaller, 43 mm long Ti:LiNbO$_3$ channels of modified index profile to improve the field overlap (see section 5), parametric amplification up to 1.14 dB was obtained with a pulsed pump power level of 3 W (see fig. 39) (Seibert, unpublished results).

Both results demonstrate that high-gain cw amplifiers, as needed for fiber optical communication, cannot be designed on the basis of nonlinear Ti:LiNbO$_3$ optical waveguides for parametric amplification. For such an application new materials of larger nonlinearity have to be found or developed (see section 3). Nevertheless, there is one important application also of small parametric amplification, if the nonlinear waveguide is positioned inside an optical resonator: if then the parametric gain exceeds the roundtrip losses of an optical wave, parametric oscillation at signal and idler frequencies starts from noise (from spontaneous parametric fluorescence). This is discussed in more detail in section 9.

8. Spontaneous parametric fluorescence

All the mixing experiments described above can be understood classically by the interaction of two electromagnetic (optical) waves in a nonlinear dielectric material; both generate a nonlinear polarization wave as source of a third optical wave at the difference or sum frequency. If, however, only one of both input waves is present, the classical description fails to describe the situation adequately. In such a case a spontaneous 'decay' of very few of the input (pump) photons can be observed in the nonlinear material. This decay is determined by energy and momentum (phase matching) conservation like all the parametric effects discussed above. It can be understood semiclassically by the mixing of the input field of frequency ω_p with the quantum-mechanical zero-point fluctuations of the electromagnetic field at frequency ω_i (and ω_s); it leads via a nonlinear polarization of frequency $\omega_s = \omega_p - \omega_i$ (and $\omega_i = \omega_p - \omega_s$) to a fluorescence radiation at ω_s (and ω_i) (Lang 1975). This 'parametric scattering' (a term used as synonym for 'spontaneous parametric fluorescence') has first been observed independently by Harris et al. (1967) and Magde and Mahr (1967) in 1967 using ADP bulk crystals as nonlinear medium. Later on, this effect has been successfully used to measure the nonlinear susceptibility of the most interesting nonlinear materials and to determine the tuning characteristics of optical parametric effects.

Up to now, there exists only one investigation of this fundamental process of spontaneous parametric fluorescence in optical waveguides (Hampel and Sohler 1986). Using an experimental setup with careful polarization, wavelength and spatial filtering as sketched in fig. 40, it was possible to measure the guided fluorescence generated in $Ti:LiNbO_3$ channel guides. The dependence on the channel width, the pump wavelength, the pump power and the waveguide temperature was investigated.

In contrast to bulk optics three different 'decay channels' for pump photons have been found: either both signal and idler photons belong to guided modes or both to radiating substrate modes or one to a guided mode and the other to a substrate mode. The first and the latter processes are sketched in fig. 41 with wave vector diagrams on the left and schematical signal and idler spectra of the guided parametric fluorescence on the right. The spectra of the fluorescence emitted as substrate modes are schematically drawn in the middle.

The essential difference of parametric fluorescence in bulk crystals and in optical waveguides arises from the different structure of the optical fields involved. In (homogeneous) bulk crystals plane waves are parametrically coupled, whereas in waveguide structures guided and/or radiating modes

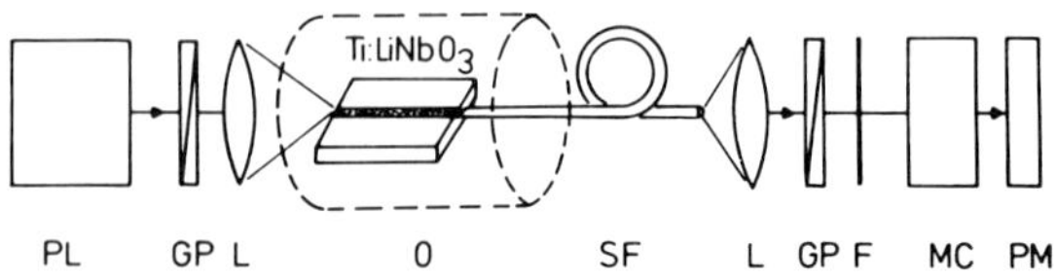

Fig. 40. Experimental setup to study spontaneous parametric fluorescence. PL: pump laser; GP: Glan–Thomson polarizer; O: oven; SF: single mode fiber; L: lens; F: filter; MC: monochromator; PM: photomultiplier. (After Hampel and Sohler 1986.)

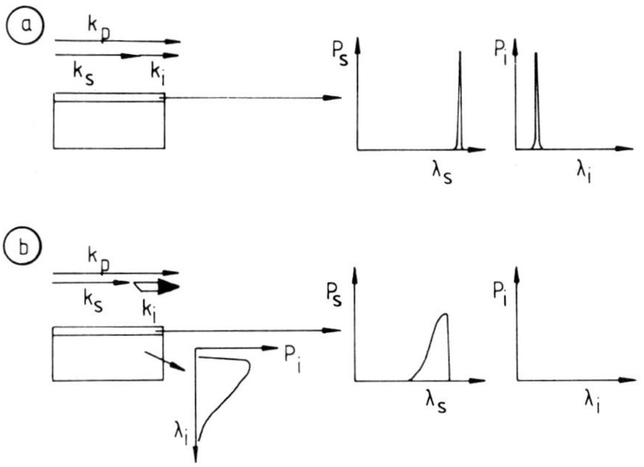

Fig. 41. Schematical wave vector diagrams (left) and corresponding guided (right) and unguided (middle) fluorescence spectra of two different processes of spontaneous parametric fluorescence in an optical waveguide. Only guided modes (a) or guided and substrate modes (b) are involved. Small (broad) arrows refer to guided (substrate) modes, respectively. (After Hampel and Sohler 1986.)

interact, which have field distributions depending on the transverse profile of the index of refraction. The overlap of these field distributions determines in all cases the coupling coefficient κ (see section 5).

If the pump, signal and idler waves are all guided optical modes, energy and momentum conservation definitely determine the narrow frequency bands of guided signal and idler fluorescence, whereas κ determines its intensity. The spectral width of the (e.g., signal) emission is given by the allowed mismatch of the propagation constants of the optical modes involved, yielding (Lang 1975)

$$\Delta\omega_s = \frac{2\pi c}{l\left[n_i - n_s + \left(\omega_i \frac{\partial n_i}{\partial \omega} - \omega_s \frac{\partial n_s}{\partial \omega}\right)\right]}, \tag{17}$$

$n_{s,i}$ is the effective index of refraction of guided signal (idler) at frequency $\omega_{s,i}$, c the velocity of light and l the waveguide length. The described situation is schematically depicted in fig. 41a.

If the pump and signal (or idler, respectively) waves only are guided optical modes, the spectrum of the guided fluorescence broadens. Due to the continuum of substrate modes with a corresponding continuum of different wave vectors a band of emission frequencies is allowed. It is separated from the frequency of the 'all-guided' fluorescence by a small gap which depends on the waveguide structure. The overlap of the interacting modes determines the wavelength dependent intensity (schematically drawn in fig. 41b).

If only the pump wave is guided and both the signal and idler waves are substrate modes, an even broader spectrum arises, shifted further to shorter wavelengths. As in the bulk case, the emission wavelength changes with the emission angle; the intensity of the fluorescence is again determined by the [angle-(wave vector) dependent] overlap of the optical modes.

Most of these considerations have been impressively confirmed by the experimental investigation of parametric fluorescence (Hampel and Sohler 1986). Figure 42 shows the measured spectra of the guided part of the signal fluorescence in 5 μm and 10 μm wide Ti:$LiNbO_3$ channel guides, respectively. The pump wavelength was 514 nm, the waveguide temperature 220°C.

The smaller channel is beyond cutoff at the idler wavelength ($\lambda_i \approx 2.1$ μm). Therefore, an 'all-guided' parametric fluorescence as described before (see fig. 41a) was not allowed; the measured spectrum (fig. 42a) must be the signal (short wavelength) part of a fluorescence with idler waves as substrate modes (see fig. 41b).

The wider channel is above cutoff. Therefore, an 'all-guided' fluorescence became possible leading to a sharp emission beyond the long-wavelength edge of the broad fluorescence with substrate modes involved (fig. 42b). The spectral structure of the latter is astonishingly well reproduced by numerical results based on a planar waveguide model. It should be separated by a small wavelength gap from the 'all-guided' emission as a consequence of wave-vector conservation; however, only a pronounced minimum was experimentally seen, probably due to waveguide inhomogeneities and due to the limited spectral resolution of the spectrometer used.

A tuning of the spontaneous parametric fluorescence was easily achieved by changing the waveguide temperature and/or the pump wavelength. With both means another pair of signal and idler frequencies is selected as a result of the momentum conserving decay of pump photons. Figure 43 presents as an

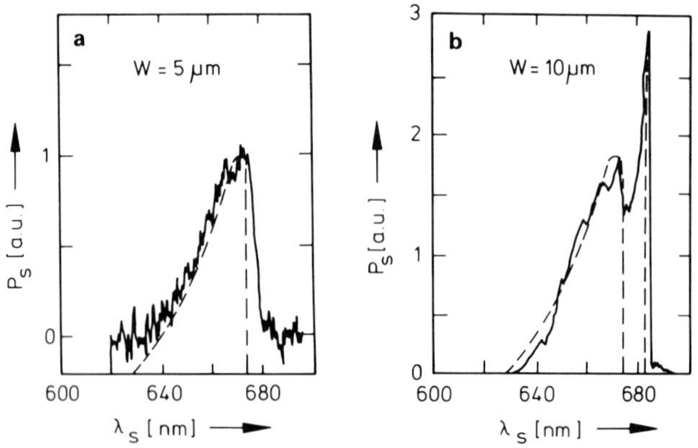

Fig. 42. Measured signal spectrum of spontaneous parametric fluorescence in a 5 μm (left) and 10 μm (right) wide Ti:$LiNbO_3$ channel guide. (After Hampel and Sohler 1986.)

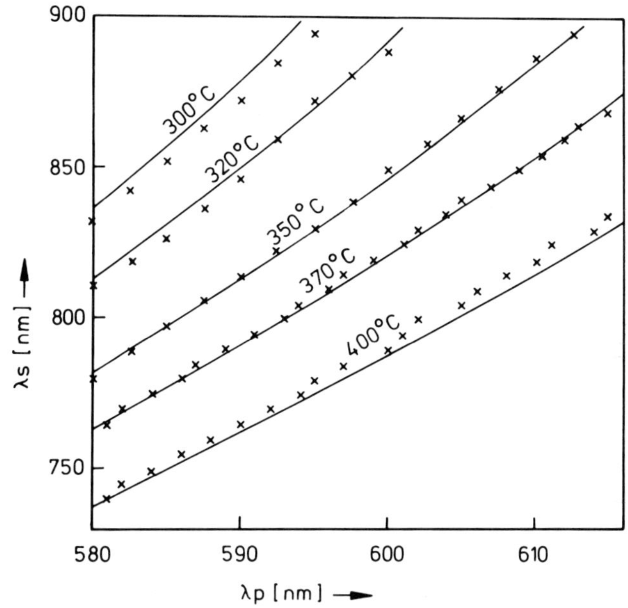

Fig. 43. Signal wavelength of the peak spontaneous fluorescence versus pump wavelength. Parameter of the set of graphs is the waveguide temperature. (After Hampel and Sohler 1986.)

example the signal wavelength of 'all-guided' parametric fluorescence generated in a 15 μm wide Ti:$LiNbO_3$ channel guide by the radiation of a dye laser. Parameter is the waveguide temperature.

The wide tunability of the parametric fluorescence is mainly exploited in the parametric oscillator (see section 9) for spectroscopic applications.

Spontaneous parametric fluorescence is a very weak process; in the experiment described above only a fraction of about 1×10^{-13} of the pump photons decayed spontaneously (Hampel and Sohler 1986). However, this decay can be stimulated by increasing the pump power density (or the interaction length) leading to a parametric amplification of signal and idler waves. The power of this stimulated fluorescence almost exponentially grows with increasing pump power in accordance with experimental results obtained with Ti:$LiNbO_3$ waveguides (Hampel and Sohler 1986) (see fig. 44).

In bulk optics stimulated parametric fluorescence is a well-known method to generate tunable near-infrared radiation [see, e.g., T.G. Giallorenzi and C.L. Tang (1969)]. Using a pulsed high-power laser as pump source a frequency conversion efficiency exceeding 10% has been achieved (Seilmeier et al. 1978). In integrated optics, however, analogous experiments have not yet been reported although guided-wave techniques would allow to reduce the necessary pump power level drastically. On the other hand, the stimulation in a resonator

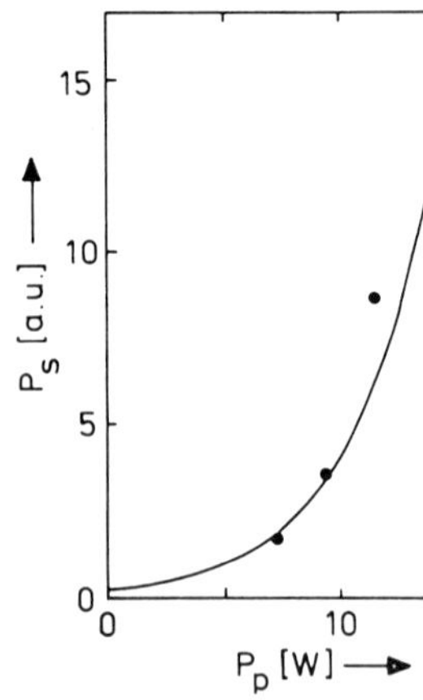

Fig. 44. Signal part of the fluorescence power as function of the incident pump power (pump wavelength 590 nm; temperature 370°C). (After Hampel and Sohler 1986.)

structure has been intensively studied; it can lead to parametric oscillation, which is discussed in detail in the next section.

9. Parametric oscillation

As discussed above, spontaneous parametric fluorescence (see section 8) can be further amplified in the nonlinear waveguide by parametric amplification (see section 7). At sufficiently high pump power densities the parametric gain surpasses the inherent (absorption and scattering) losses an optical mode suffers during propagation. In such a case, optical feedback, as provided by a resonator structure, can give rise to optical parametric oscillation. This means a further growth of signal and idler waves in the cavity until depletion of the pump wave and a corresponding gain saturation set in. In this section, the threshold of parametric oscillation, the output power and conversion efficiency, and the spectral properties of the emission are discussed.

9.1. Oscillation threshold

If a nonlinear optical waveguide is placed inside a resonator structure and pumped by coherent radiation of sufficient power, appropriate wavelength and polarization, parametric oscillation sets in. It starts from spontaneous parametric fluorescence, which grows due to parametric amplification. A steady state is achieved if gain and (roundtrip) losses just compensate each other. In that case, the amplitudes of signal and idler waves reproduce themselves after each roundtrip in the cavity.

By equating gain and losses the threshold pump power can be calculated using coupled-mode theory. For a three-wave interaction of pump, signal and idler the threshold power is given by (Bava et al. 1987)

$$P_{\mathrm{th}} = \frac{1}{\omega_s \omega_i \kappa^2}\left[\frac{\alpha_p}{1-\exp(-\alpha_p L)}\ln(Q+\sqrt{Q^2-1})\right]^2, \tag{18}$$

with

$$Q = \frac{1+R_s R_i \exp(-4\alpha L)}{(R_s+R_i)\exp(-2\alpha L)},$$

κ is the nonlinear coupling coefficient as given by eq. (11); α_p is the pump attenuation coefficient and L the cavity length, Q is a function of the cavity reflectivities R_s and R_i, the cavity length and the attenuation coefficient α at signal and idler wavelengths ($\alpha_s = \alpha_i = \alpha$ is assumed).

To achieve a low oscillation threshold, optical waveguides of large nonlinearity and well-overlapping interacting modes must be used. Furthermore, high-quality resonators of low waveguide and mirror transmission losses are necessary.

As the field overlap depends on the waveguide structure, it can be optimized by tailoring the index profile of the guide appropriately. This problem has been theoretically studied by Bava et al. (1987) for the Ti:$LiNbO_3$ system. Assuming a certain resonator structure with given losses they varied the Ti-concentration, width and depth of the waveguide to find the minimum threshold pump power for parametric oscillation. Due to the nonlinear dependence of the ordinary index of refraction on the Ti-concentration there is a unique solution as optimized waveguide structure at relatively low Ti-concentrations. Fortunately, the dependence of the threshold pump power on the waveguide parameters is weak. Figure 45 gives contours of constant threshold pump power in a (D, c_0)-

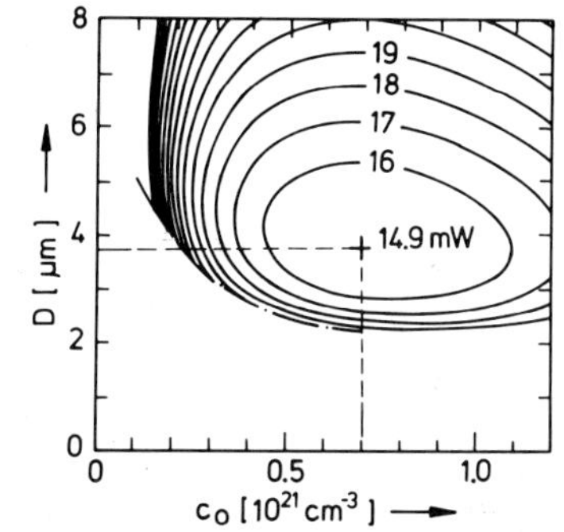

Fig. 45. Contours of constant threshold pump power in a (D, c_0)-plane. D: 1/e-depth of a Gaussian Ti-concentration profile normal to the waveguide surface; c_0: surface Ti-concentration of an equivalent slab waveguide with a Ti-layer thickness equal to that of the actual strip guide. (After Bava et al. 1987.)

plane, where D is the waveguide depth due to diffusion and c_0 is the surface Ti-concentration of an equivalent slab waveguide (Bava et al. 1987).

Also, the resonator configuration (i.e., wavelength-dependent mirror reflectivities and waveguide losses) strongly determines the threshold level. Singly- and doubly-resonant devices should be discussed separately as they differ not only with respect to the threshold pump power but also to the spectral properties (see section 9.3).

Singly-resonant oscillators have high-reflectivity mirrors for only one of both emission wavelengths. For the other one zero reflectivity is usually assumed in calculations although such a 'pure' situation is never obtained in an experiment. Assuming an optimized Ti:$LiNbO_3$ waveguide profile, Bava et al. (1987) calculated the threshold pump power of a singly-resonant device as function of waveguide losses and mirror reflectivity (see fig. 46a). For parameters as adjusted in early experiments ($L = 4.8$ cm, $R = 0.98$, and $\alpha = 0.1$ dB/cm) a threshold pump power of 0.25 W is expected. This figure is more than one order of magnitude smaller than the experimentally determined result (Sohler and Suche 1981). However, the calculated threshold indicates the power level inside the cavity and results from assuming a single-frequency pump. The experimental result corresponds to the pump power measured in front of the cavity and was achieved with a multimode laser of 3 cm^{-1} linewidth as pump source. Linewidth reduction to 0.1 cm^{-1} reduced the threshold by a factor of 4 (Suche et al. 1987). The estimated internal pump power then reasonably agreed with the predicted figure.

Doubly-resonant oscillators have high-reflectivity mirrors at signal *and* idler wavelengths. Due to the resonant enhancement of both waves, a lower oscillation threshold is expected. Assuming identical reflectivities for signal and idler waves, Bava et al. (1987) also calculated the threshold pump power of a

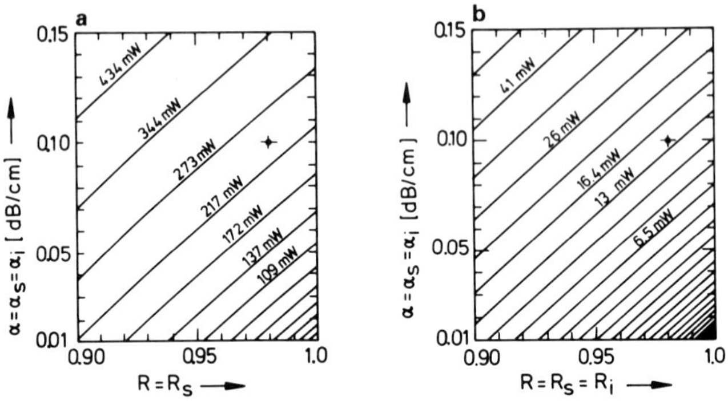

Fig. 46. Contours (in steps of 1 dB) of constant threshold pump power in an (α, R)-plane. $\alpha = \alpha_s = \alpha_i$ is the attenuation coefficient of signal and idler waves, R the mirror reflectivity. Left: results for a singly-resonant oscillator ($R = R_s$, $R_i = 0$). Right: results for a doubly-resonant oscillator ($R = R_s = R_i$). (After Bava et al. 1987.)

doubly-resonant device with the same optimized Ti:LiNbO$_3$ waveguide as in the singly-resonant oscillator discussed before. The results are presented in fig. 46b. They clearly demonstrate that a considerable reduction of the threshold pump power of more than one order of magnitude should be possible. Indeed, experimentally a threshold figure of only 26.5 mW was achieved using a single-frequency pump source (Suche and Sohler 1988). This figure is about four times larger than expected from the calculations. However, theory requires a further refinement by taking only resonator eigenfrequency pairs (of a certain linewidth) into account to model the oscillator emission (Bava et al. 1989a).

The measured oscillation threshold of the doubly-resonant device of 26.5 mW is already compatible with power levels available from single-mode diode lasers. Therefore, a combination of a semiconductor pump laser with an integrated optical parametric oscillator as tunable frequency converter is a very attractive goal. However, even short-wavelength diode lasers still have an emission wavelength too long to allow pumping a Ti:LiNbO$_3$ parametric oscillator at low device temperature.

There is another possibility to reduce the threshold pump power of singly- and doubly-resonant devices further. Using a rear mirror of additional high pump reflectivity results in a double pass of the pump wave through the oscillator. In this way, the effective interaction length is increased and a lower oscillation threshold can be expected. Moreover, with this configuration the power conversion efficiency can be significantly improved.

9.2. Output power and conversion efficiency

If the pump power to operate an (integrated) optical parametric oscillator is increased above the threshold level, the output of signal and idler waves grows. More and more pump photons are annihilated creating signal and idler photons simultaneously. Therefore, the pump wave depletes until gain saturation is achieved. All these processes strongly depend on the pump power level and the resonator structure of the oscillator. Therefore, a simultaneous, time-resolved measurement of the different power levels helps to better understand parametric oscillation.

The experimental setup, sketched in fig. 47, allows the simultaneous observation of four power levels. By inserting appropriate filters, incident, reflected and transmitted pump powers can be monitored as well as signal or (and) idler powers generated in forward or (and) backward directions. As an example, the pump power transmitted by a doubly-resonant oscillator and the sum of signal and idler powers generated in forward direction are presented as function of time in fig. 48a (Suche and Sohler 1988). The pump pulse of 140 mW peak power (in front of the oscillator) was 500 ns long. Within the first 100 ns, the transmitted power is strictly proportional to the input power. Parametric fluorescence is generated and amplified in the cavity; however, the corresponding

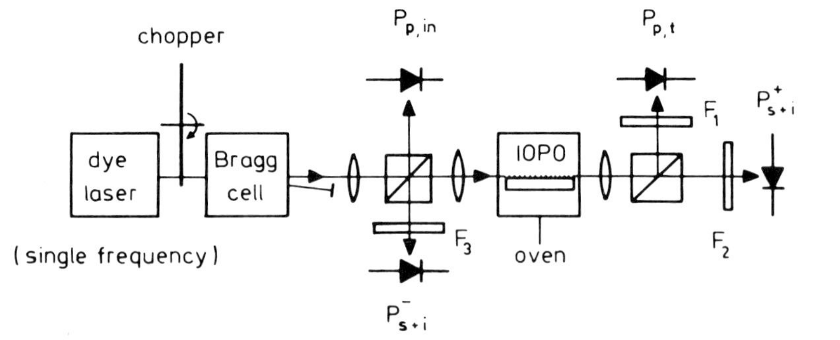

Fig. 47. Experimental setup to study doubly-resonant optical parametric oscillation. $F_{1,2,3}$: filters or spectrometers. (After Suche et al. 1989.)

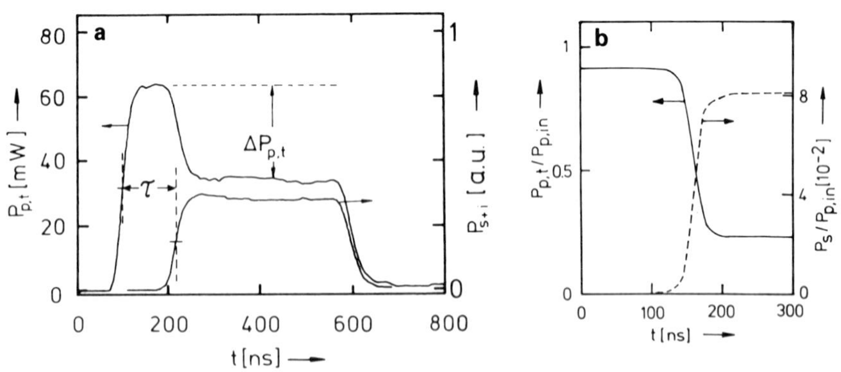

Fig. 48. Left: measured transmitted pump power $P_{p,t}$ and generated signal and idler powers versus time. $\Delta P_{p,t}$: pump depletion. (After Suche and Sohler 1988.) Right: calculated transmitted pump and generated signal powers normalized to the input pump power ($4P_{th}$) versus time. (After Bava et al. 1989b.)

power level is too low to be seen on the linear scale of the ordinate. After the build-up time τ, the output power almost grows to the same order of magnitude as the transmitted pump and saturates 120 ns after switch-on of the pump pulse. This is accompanied by a depletion of the pump of about 45% in this case. Obviously, the build-up time goes to infinity and the depletion to zero as the pump power level approaches the cw-threshold.

The amount of depletion is a measure of the power transfer from the pump to the forward travelling signal and idler waves. In the doubly-resonant device, however, the pump power is partly regenerated due to sum frequency mixing of the backward travelling signal and idler waves. This nonlinear optical reflection of pump power can degrade the conversion efficiency of the doubly-resonant device considerably. Thus, up to 80% depletion of the transmitted pump power could be observed in a doubly-resonant integrated optical parametric oscillator, but the totally emitted infrared power (in forward and backward directions) was less than 10% of the input pump power (also due to losses of the resonant waves) (Suche et al. 1989). Fortunately, a nonlinear optical reflection of the pump power can be prevented using a singly-resonant oscillator or a double-pass pumping scheme for the doubly-resonant device.

The main feature of the build-up of the output power, accompanied by the depletion of the pump, can be satisfactorily evaluated using a numerical integration of the coupled nonlinear differential equation system describing the parametric interactions in the oscillator (Bava et al. 1989b). An example is given in fig. 48b. However, the theoretical model has to consider further waveguide and resonator parameters to allow a quantitative description of the experimental results.

The power characteristic of a parametric oscillator resembles that of a laser: beyond the threshold the output power grows nearly linearly with increasing pump power. An example is presented in fig. 49 obtained with a doubly-resonant device (Suche and Sohler 1988): here the depletion of the transmitted pump power (see fig. 48) is plotted as a measure of the generated infrared power (1.07 μm $< \lambda <$ 1.28 μm) versus the incident pump power. Up to about 6 times the threshold power of 26.5 mW the emission consists of a single line as shown for the idler on the right of fig. 49. The resolution of the spectrometer of 0.1 nm did not allow to determine the longitudinal mode structure of the oscillator emission (mode spacing of 1.57 GHz).

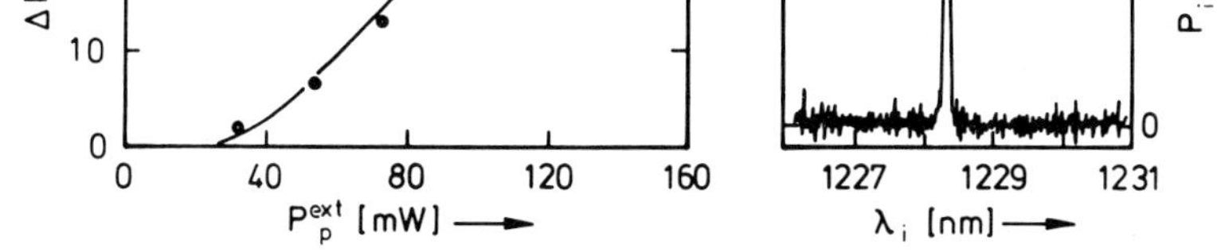

Fig. 49. Left: depletion of transmitted pump power $\Delta P_{p,t}$ versus incident pump power. Right: signal power (at $6P_{th}$) versus wavelength, measured with $f = 0.6$ m monochromator. (After Suche and Sohler 1988.)

9.3. Spectral properties

The emission wavelengths of a parametric oscillator are mainly determined by energy and wave-vector (phase-matching) conservation, as spontaneous parametric fluorescence (see section 8) is the basic process. Therefore, the tuning characteristic of a parametric oscillator as function of temperature and/or pump wavelength at least on the average follows the peak fluorescence gain. However, the fine-tuning properties are influenced by the resonator structure. We therefore have to discuss singly- and doubly-resonant devices separately.

In a singly-resonant device the range of emission wavelengths of the resonant wave is determined by the spectral width of the high-reflectivity band of the oscillator mirrors. The bandwidth of the nonresonant wave adjusts according to energy conservation. The two emission bands are normally separated, but they can join one another by a proper choice of the spectral position of the mirror-reflectivity bands and the device temperature. In this way a very broad continuous tuning range can be achieved.

An example of a tuning characteristic of a singly-resonant Ti:$LiNbO_3$ integrated optical parametric oscillator is given in fig. 50; this device had mirrors of reflectivity $R > 0.9$ in the wavelength range $0.9\ \mu m < \lambda < 1.2\ \mu m$. The solid lines are calculated graphs assuming Sellmeier fits for the indices of refraction [Hobden and Warner (1986), more recent Sellmeier fits for the refractive indices of $LiNbO_3$ can be found in the article by Edwards and Lawrence (1984)]. The dots and circles correspond to measured results. The dashed line indicates degeneracy and separates the upper branches of the idler wavelengths from those of the signal. With increasing temperature, degeneracy shifts to longer wavelengths out of the range of high mirror-reflectivity yielding a separation of idler and signal emission bands. At lower temperatures a continuous tuning from $\lambda = 0.96\ \mu m$ up to $\lambda = 1.6\ \mu m$ was possible (Suche and Sohler 1983). The linewidth was well below 1 nm with the exception near degeneracy (Suche 1981).

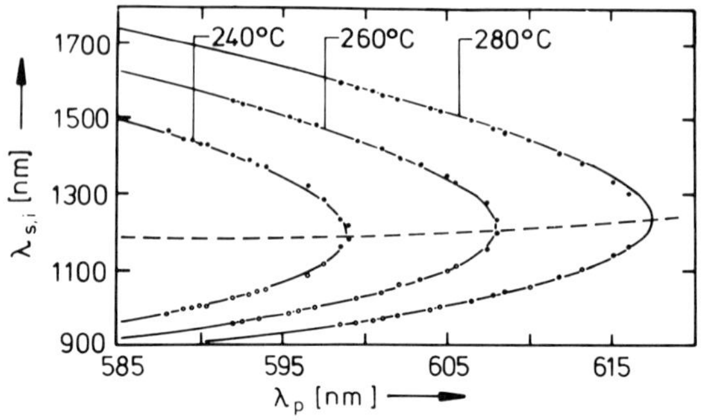

Fig. 50. Tuning characteristic of an almost singly-resonant, integrated optical (Ti:$LiNbO_3$) parametric oscillator: measured signal (circles) and idler (dots) wavelengths versus pump wavelength. Parameter of the set of graphs is the device temperature. (After Suche and Sohler 1983.)

In a doubly-resonant device, tuning is restricted to the range of high mirror-reflectivity usually centered around the wavelength of degeneracy of signal and idler. Figure 51 presents as an example the tuning characteristics of a special device with Ti:$LiNbO_3$ channel-guide resonator (Suche and Sohler 1988).

However, the spectral structure of the emission and the fine-tuning properties of a doubly-resonant oscillator are more complicated than those of a singly-resonant device. Not only energy conservation and phase matching determine the oscillator output, but also the simultaneous resonance condition for signal and idler waves. Only those pairs of cavity eigenfrequencies are allowed which add up to the pump frequency (energy conservation). Due to dispersion only few of such pairs are within the bandwidth of the parametric gain yielding a phase-matched interaction. Typically they are separated by several hundred axial-mode spacings (see fig. 52). (Due to the finite linewidth of the axial modes, not only the energy conserving eigenfrequency pairs are expected in the emission spectrum, but 'clusters' of frequencies belonging to the neighboring modes.) Two of these 'cluster' spacings correspond exactly to the spectral halfwidth of the parametric fluorescence. Therefore, just one, two or even three clusters can be emitted, depending on the cluster position with respect to the phase-match (gain) curve and on the excess gain above threshold. In figs. 52 and 49 experimental examples of a three- and one-cluster emission are presented, respectively.

If the pump frequency or the device temperature is tuned, the cluster positions and the gain curve shift in opposite directions with different velocities (the cluster positions shift much faster). The result is a tuning behavior as schematically sketched in the inset of fig. 51 and discussed in the following. At pump power levels just above threshold only discrete emission wavelengths exist; they belong to the pump frequencies yielding maximum gain for an 'energy conserving' eigenfrequency pair. If the pump power is increased, small discrete tuning ranges arise. Their center wavelengths follow the continuous theoretical

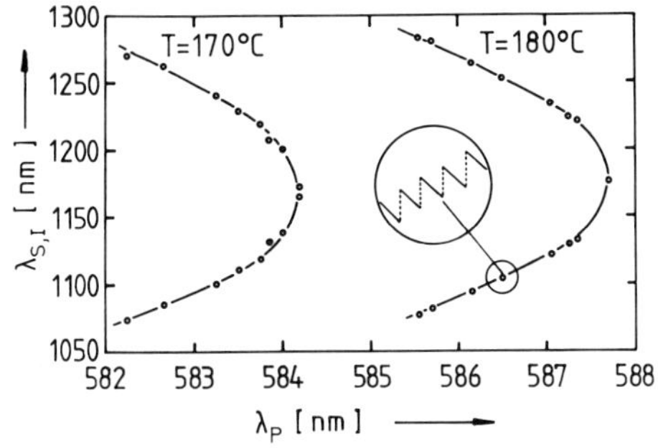

Fig. 51. Tuning characteristic of a doubly-resonant integrated optical (Ti:$LiNbO_3$) parametric oscillator: measured signal and idler wavelengths versus pump wavelength. Parameter of the graphs is the temperature. Inset: schematical fine-tuning characteristic. (After Suche and Sohler 1988.)

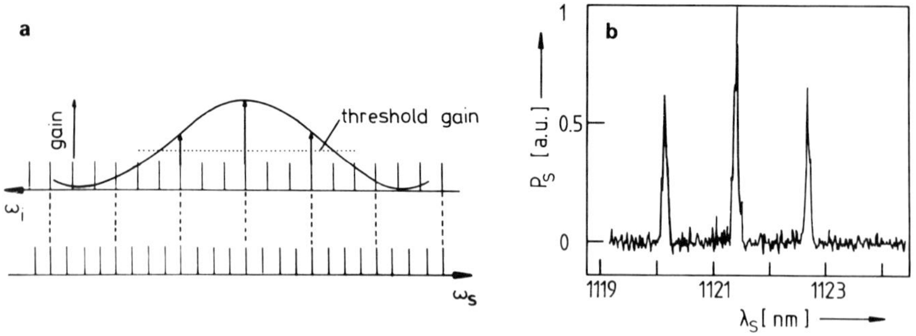

Fig. 52. Left: illustration of influence of dispersion on signal and idler passive cavity eigenfrequencies. Only few pairs give $\omega_i + \omega_s = \omega_p$, necessary for onset of parametric oscillation. Indicated threshold gain level would allow emission of three cluster lines. Right: measured three-line spectrum. (After Suche and Sohler 1988.)

tuning curve, which results by only taking energy and wave-vector conservation into account. At low pump power levels a continuous tuning of the pump wavelength therefore results in a periodic switch-off and -on of the oscillator emission accompanied by jumps of the emission wavelength and a strong modulation of the output power in the on-condition. Increasing the pump power yields a broadening of the discrete tuning ranges until they combine with a continuous saw-tooth tuning characteristic. At this pump power level there is still a one-cluster emission. If, e.g., the pump wavelength is increased, the signal wavelength decreases (accompanied by a corresponding increase of the idler wavelength) until the emission jumps to the neighboring cluster at longer signal and shorter idler wavelengths. This tuning is accompanied by a strong modulation of the emission power. At higher pump power levels the emission changes from a one- (or even three-) cluster operation near the gain maximum to a two-cluster operation and vice versa by tuning the pump wavelength. Furthermore, the higher the pump power level the smaller is the amplitude of the modulation of the emission power.

All these features of the tuning behavior of an integrated optical doubly-resonant parametric oscillator have been observed experimentally and could be modeled theoretically (Bava et al. 1989b). Figure 53 presents as an example the modulation of the output power (sum of signal and idler powers) as function of a slight tuning of the pump frequency. The periodic switch-off and -on at low pump power levels is observed as well. The amplitude of the saw-tooth tuning characteristic was about 1 nm. As a consequence, the single-frequency pumped doubly-resonant oscillator is not an ideal source for high-resolution spectroscopy, if continuous tuning over a broader wavelength range is required. For such an application the singly-resonant device should be preferred.

On the other hand, the linewidth of the doubly-resonant oscillator in the single cluster regime is extremely small. (The measured linewidth shown in

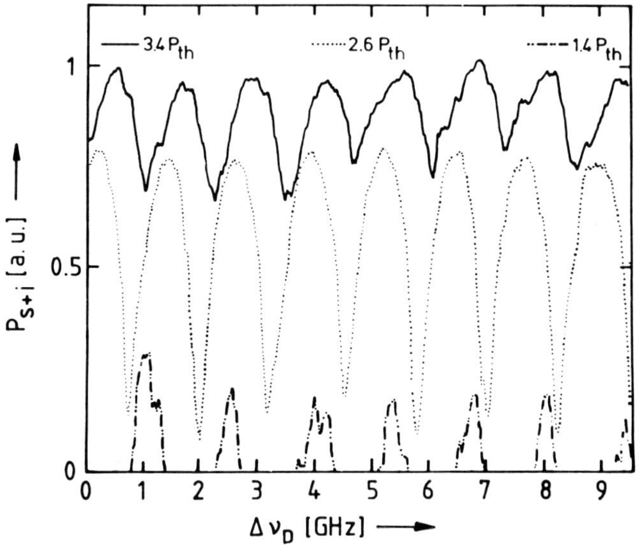

Fig. 53. Sum of signal and idler output powers of a doubly-resonant integrated optical ($Ti:LiNbO_3$) parametric oscillator versus pump frequency change at 1.4, 2.6 and 3.4 P_{th}. (After Bava et al. 1989b.)

fig. 49 was determined by the resolution of the spectrometer.) Using a fast integrated optical spectrum analyzer (Suche and Wersig 1990) it was possible to demonstrate clearly a single-frequency emission up to six times the threshold pump power (Suche et al. 1988); fig. 54 presents a corresponding result. The resolution of the spectrum analyzer of 146 MHz (9.8 GHz free spectral range) determined the experimentally observed linewidth. As the FWHM of the passive

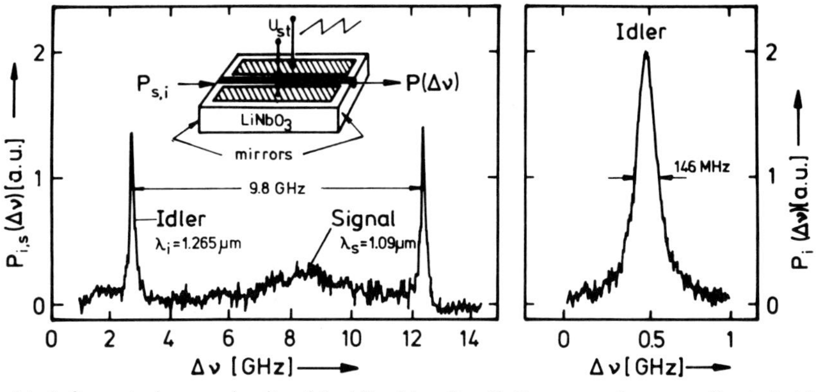

Fig. 54. Left: spectral power density of the idler (signal) radiation versus frequency. Inset: sketch of the fast electro-optical guided-wave spectrum analyzer. Right: high-resolution scan of the idler power density. (After Suche et al. 1988.)

cavity modes of the special oscillator investigated is only 22 MHz, a true emission bandwidth well below that figure is expected.

The temporal development of the emission could be experimentally observed and theoretically described (Bava et al. 1989b). Just after switching on the pump power, a couple of modes (a cluster) start to oscillate. However, the axial mode of largest gain grows with time at the expense of the neighboring modes leading to a single-frequency emission after some hundred nanoseconds (see fig. 55). At four times the threshold pump power level, the extinction of the neighboring modes was calculated to be at least −15 dB after 1.3 μs (Bava et al. 1989b).

With the fast integrated optical spectrum analyzer the development from a three axial-mode to a single axial-mode emission could be clearly observed (see fig. 56). The saturation of the surviving axial mode of largest gain takes much

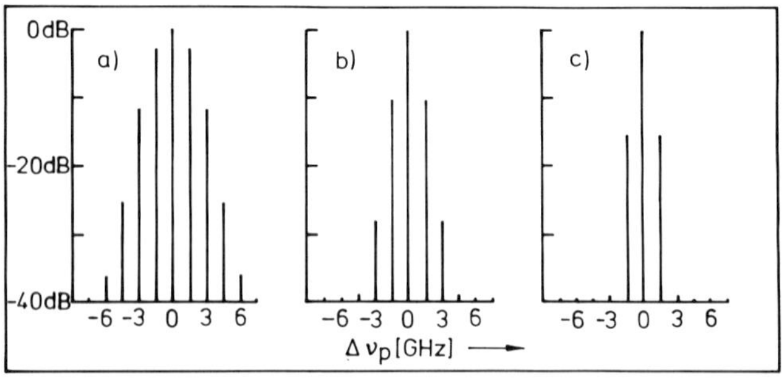

Fig. 55. Calculated relative power in dB of the signal axial modes versus signal frequency deviation $\Delta\nu$ from the peak parametric gain: (a) after 325 ns (500 cavity round trips); (b) after 650 ns; (c) after 1300 ns. (After Bava et al. 1989b.)

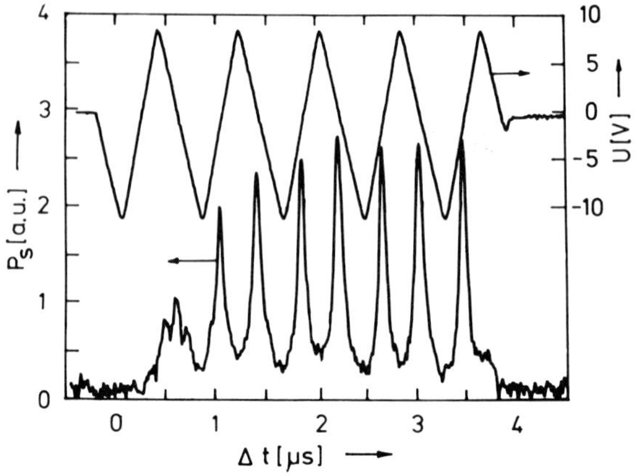

Fig. 56. Upper trace (right scale): drive voltage of integrated optical spectrum analyzer versus time. Lower trace (left scale) spectral power density of the signal radiation as function of time. (After Bava et al. 1989b.)

longer than the build-up of a constant total oscillator output power. Both features qualitatively agree with the theoretical predictions.

Also in a singly-resonant bulk parametric oscillator a single-frequency emission, obtained with the help of an intracavity etalon, was reported (Kreuzer 1969). It should be emphasized that in the integrated optical, doubly-resonant device a single-frequency emission was achieved without further mode-selecting means.

10. Conclusions

Second-order nonlinear guided-wave interactions in integrated optical waveguides have been reviewed with emphasis on more recent developments. Impressive results could be reported, in particular for SHG, DFG, and parametric oscillation. We expect future progress in this field; e.g., stimulated single-pass parametric fluorescence has not yet been demonstrated with integrated optical waveguides, although it is an efficient means for frequency conversion in bulk optics. Furthermore, parametric oscillation could be achieved in resonant waveguide structures not only in the near-, but also in the mid-infrared, if the improvement of waveguide fabrication technology goes on.

Up to now, $LiNbO_3$ is the dominant material in the field of second-order nonlinear integrated optics. By applying different techniques to achieve phase matching a very broad spectral range from the visible into the mid-infrared is accessible. The high quality (with respect to waveguide losses and homogeneity) of $Ti{:}LiNbO_3$ waveguides even allows to fabricate resonant structures for nonlinear interactions. However, the susceptibility of $LiNbO_3$ to optically induced changes of the index of refraction limits, mainly in the visible, the (average) allowed optical power density in the waveguide. This is a severe drawback especially for phase-sensitive processes like second-order nonlinear interactions, which require, of course, high optical fields to be efficient. Therefore, a reduction of this 'damage' susceptibility still is an important goal.

Besides $LiNbO_3$, a lot of further excellent nonlinear materials from well-known inorganics (like $KNbO_3$) to recently developed organics (like MNA) is successfully used in bulk optics. They all have an even greater potential in nonlinear integrated optics, if it is possible to fabricate optical waveguides of the required quality. This means in particular low-loss waveguides of very good homogeneity. To achieve this goal, the appropriate waveguide fabrication technology (indiffusion, ion implantation, ion exchange, etc.) must be found and optimized or even developed for the different materials; here a lot of work remains to be done.

Furthermore, greater efforts should be directed towards the applications of semiconductor lasers as pump for second-order nonlinear interactions. This is possible, as the required (pump) power levels in integrated optical waveguides can be orders of magnitude lower than in bulk devices. Such a development was successfully demonstrated by the combination of a ('proton-exchanged') $H_xLi_{1-x}NbO_3$ strip guide and an AlGaAs laser ($\lambda = 0.84$ μm) for efficient SHG (Taniuchi and Yamamoto 1987a). It is, hopefully, just the first example of many new miniaturized (tunable), coherent, all solid state light sources to be developed.

Acknowledgments

I am indebted to many colleagues, who gave the permission to use figures of their results in this paper. Thanks are also due to Mrs. I. Zimmermann, who typed the whole manuscript, and to O. Schiller and U. Nolte, who prepared many of the figures.

References

Akhmanov, S.A., and R.V. Khokhlov, 1972, Problems of Nonlinear Optics (Gordon and Breach, New York).

Akhmediev, N.N., and V.R. Novak, 1985, Opt. Spektrosk. (USSR) **58**, 558.

Anderson, D.B., and J.T. Boyd, 1971, Appl. Phys. Lett. **19**, 266.

Armstrong, J.A., N. Bloembergen, J. Ducuing and P.S. Pershan, 1962, Phys. Rev. **127**, 1918.

Arvidsson, G., and F. Laurell, 1986, Thin Solid Films **136**, 29.

Arvidsson, G., F. Laurell, B. Jaskorzynska and J. Webjörn, 1989, in: Topical Meeting on Nonlinear Guided-Wave Phenomena: Physics and Applications, 1989 Technical Digest Series, Vol. 2 (Optical Society of America, Washington, DC) p. 10.

Auston, D.A., et al., 1987, Collaboration of 48 authors, Appl. Opt. **26**, 211.

Bailey, R.T., F.R. Cruickshank, S.M.G. Guthrie, B.J. McArdle, H. Morrison, D. Pugh, E.A. Shepherd, J.N. Sherwood, C.S. Yoon, R. Kashyap, B.K. Nayar and K.I. White, 1988, Opt. Commun. **65**, 229.

Baumert, J.C., P. Günter and H. Melchior, 1983, Opt. Commun. **48**, 215.

Baumert, J.C., C. Walther, P. Buchmann, H. Melchior and P. Günter, 1985, in: Integrated Optics, Springer Series in Optical Sciences, Vol. 48, eds H.P. Nolting and R. Ulrich (Springer, Berlin) p. 44.

Bava, G.P., I. Montrosset, W. Sohler and H. Suche, 1987, IEEE J. Quantum Electron. **QE-23**, 42.

Bava, G.P., I. Montrosset and R. Pelissero, 1989a, Topical Meeting on Nonlinear Guided-Wave Phenomena: Physics and Applications, Houston, USA, Technical Digest Series, Vol. 2 (Optical Society of America, Washington, DC) paper THB18, p. 84.

Bava, G.P., I. Montrosset, R. Pelissero, W. Sohler and H. Suche, 1989b, Proc. 5th Europ. Conf. on Integrated Optics, ECIO 1989, Paris, Proc. SPIE **1141**, 26.

Bierlein, J.D., A. Ferretti, L.H. Brixner and W.Y. Hsu, 1987, Appl. Phys. Lett. **50**, 1216.

Bloembergen, N., 1965, Nonlinear Optics (Benjamin, New York).

Boyd, G.D., and D.A. Kleinman, 1968, J. Appl. Phys. **39**, 3597.

Bozhevol'nyi, S.I., K.S. Buritskii, E.M. Zolotov and V.A. Chernykh, 1981, Sov. Tech. Phys. Lett. **7**, 278.

Bremer, T., W. Heiland, B. Hellermann, P. Hertel, E. Krätzig and D. Kollewe, 1988, Ferroelectrics Lett. **9**, 11.

Brucksch, M., M. Fröhlich, W. Sohler, E. Strake, R. Volk and H. Ziegler, 1986, Proc. SPIE **651**, 246.

Bryan, D.A., R.R. Rice, R. Gerson, H.E. Tomaschke, K.L. Sweeney and L.E. Halliburton, 1985, Opt. Eng. **24**, 138.

Bulmer, C.H., 1984, Electron. Lett. **20**, 902.

Burns, W.K., S.K. Sheen and A.F. Milton, 1979, IEEE J. Quantum Electron. **QE-15**, 1282.

Caglioti, E., S. Trillo, S. Wabnitz and G.I. Stegeman, 1988, J. Opt. Soc. Am. B **5**, 472.

Carenco, A., J. Jerphagnon and A. Perigaud, 1977, J. Chem. Phys. **66**, 3806.

Carter, G.M., and J. Zyss, feature eds, 1987, Nonlinear Organic Materials, special issue of J. Opt. Soc. Am. B **4**(6).

Chemla, D.S., and J. Zyss, eds, 1987, Nonlinear Optical Properties of Organic Molecules and Crystals (Academic Press, New York).

Chen, C.K., T.F. Heinz, D. Ricard and Y.R. Shen, 1981, Phys. Rev. Lett. **46**, 1010.

Chouinard, M.P., E.P. Gargiulo, V. Guerra, J.L. Hohman, D.B. Laubacher, J.Y. Liou, M. Oren and P.H. Wyant, 1988, in: Proc. 14th European Conf. on Optical Communication ECOC 1988, Brighton, IEE Conf. Publ. no. 292, part 1, p. 45.

Cross, G.H., I.R. Girling, I.R. Peterson, N.A. Cade and J.D. Earls, 1987, J. Opt. Soc. Am. B **4**, 962.

De Micheli, M., 1983, J. Opt. Commun. **4**, 25.

De Micheli, M., 1984, in: New Directions in Guided Wave and Coherent Optics, Vol. II, eds D.B. Ostrowsky and E. Spitz, NATO ASI Series E, No. 79 (The Hague) p. 495.

De Micheli, M., J. Botineau, P. Sibillot, D.B. Ostrowsky and M. Papuchon, 1982, Opt. Commun. **42**, 101.

De Micheli, M., J. Botineau, S. Neveu, P. Sibillot and D.B. Ostrowsky, 1983, Opt. Lett. **8**, 116.

Demtröder, W., 1982, Laser Spectroscopy, 2nd Ed. (Springer, Berlin).

Deryugin, L.N., V.E. Sotin and V.M. Shevtsov, 1986, Sov. Tech. Phys. Lett. **12**, 33.

Duguay, M.A., and J.S. Weiner, 1985, Appl. Phys. Lett. **47**, 547.

Edwards, G.J., and M. Lawrence, 1984, Opt. & Quantum Electron. **16**, 373.

Fan, T.Y., C.E. Huang, B.Q. Hu, R.C. Eckardt, Y.X. Fan, R.L. Byer and R.S. Feigelson, 1987, Appl. Opt. **26**, 2390.

Feisst, A., and P. Koidl, 1985, Appl. Phys. Lett. **47**, 1125.

Fejer, M.M., R.L. Byer, R. Feigelson and W. Kway, 1982, Proc. SPIE **320**, 50.

Fejer, M.M., J.L. Nightingale, G.A. Magel and R.L. Byer, 1984, Rev. Sci. Instrum. **55**, 1791.

Fejer, M.M., M.J.F. Digonnet and R.L. Byer, 1986, Opt. Lett. **11**, 230.

Fisher, R.A., ed., 1982, Optical Phase Conjugation (Academic Press, New York).

Flückiger, U., and H. Arend, 1978, J. Crystal Growth **43**, 406.

Franken, P.A., A.E. Hill, C.W. Peters and G. Weinreich, 1961, Phys. Rev. Lett. **7**, 118.

Gericke, V., P. Hertel, E. Krätzig, J.P. Nisius and R. Sommerfeldt, 1984, Appl. Phys. B **44**, 155.

Giallorenzi, T.G., and C.L. Tang, 1969, Phys. Rev. **184**, 353.

Gibbs, H.M., 1985, Optical Bistability: Controlling Light with Light (Academic Press, New York).

Glass, A.M., 1978, Opt. Eng. **17**, 470.

Hägele, K.H., E. Kisela and R. Ulrich, April 1981, Technical Digest of the 3rd Int. Conf. on Integrated Optics and Optical Fiber Communication, San Francisco, paper WB 2, p. 90.

Hampel, B., and W. Sohler, 1986, Proc. Integrated Optical Circuit Engineering III, SPIE **651**, 229.

Hampel, B., W. Sohler and H. Suche, 1987, University of Paderborn, unpublished results.

Harris, S.E., M.K. Oshman and R.L. Byer, 1967, Phys. Rev. Lett. **18**, 732.

Hayden, L.M., St.T. Kowel and M.P. Srinivasan, 1986, Paper THQ 6, in: Conf. on Lasers and Electro-Optics (CLEO 1986) (Optical Society of America, Washington, DC) p. 324.

Heinz, T.F., C.K. Chen, D. Ricard and Y.R. Shen, 1982, Phys. Rev. Lett. **48**, 478.

Herrmann, H., and W. Sohler, 1988, J. Opt. Soc. Am. B **5**, 278.

Hewig, G.H., and K. Jain, 1983, Opt. Commun. **47**, 347.

Hill, J.R., P.L. Dunn, G.J. Davies, S.N. Oliver, P. Pantelis and J.D. Rush, 1987, Electron. Lett. **23**, 700.

Hinkov, V., and E. Ise, 1986, IEEE J. Lightwave Techn. **LT-4**, 444.
Hinkov, V.P., R. Opitz and W. Sohler, 1988, IEEE J. Lightwave Techn. **LT-6**, 903.
Hobden, M.V., and J. Warner, 1966, Phys. Lett. **22**, 243.
Ito, H., and H. Inaba, 1975, Opt. Commun. **15**, 104.
Ito, H., and H. Inaba, 1978, Opt. Lett. **2**, 139.
Itoh, H., H. Takara, K. Hotta and K. Sasaki, 1986, Proc. Conf. on Lasers and Electro-Optics (CLEO 1986) paper WR4.
Itskhoki, I.Ya., M.A. Kashintsev, B.G. Lysoi and A.A. Solov'ev, 1982, Sov. J. Quantum Electron. **12**, 747.
Ivanova, Z.I., and A.I. Kholodnykh, 1980, Sov. J. Quantum Electron. **10**, 342.
Jackel, J.L., C.E. Rice and J.J. Veselka, 1982, Appl. Phys. Lett. **41**, 607.
Jackel, J.L., A.M. Glass, G.E. Peterson, C.E. Rice, D.H. Olson and J.J. Veselka, 1984, J. Appl. Phys. **55**, 269.
Jaskorzynska, B., G. Arvidsson and F. Laurell, 1986, Proc. SPIE **651**, 221.
Kaminow, I.P., 1974, An Introduction to Electrooptic Devices (Academic Press, New York).
Kato, K., 1980, IEEE J. Quantum Electron. **QE-16**, 1288.
Kogelnik, H., 1975, in: Integrated Optics, Topics in Applied Physics, Vol. 7, ed. T. Tamir (Springer, Berlin) p. 66.
Kogelnik, H., 1979, in: Fiber and Integrated Optics, NATO Advanced Study Institutes Series B: Physics, Vol. 41, ed. D.B. Ostrowsky (Plenum Press, New York) p. 281.
Komatsu, K., M. Kondo and Y. Ohta, 1986, Electron. Lett. **22**, 882.
Kowel, St.T., L.Ye.Y. Zhang and L.M. Hayden, 1987, Opt. Eng. **26**, 107.
Kreuzer, L.B., 1969, Appl. Phys. Lett. **15**, 263.
Kurtz, S.K., 1972, Nonlinear Optical Materials, in: Laser Handbook, Vol. 1, eds F.T. Arecchi and E.O. Schulz-Dubois (North-Holland, Amsterdam) p. 923.
Lang, C.L., 1975, in: Quantum Electronics, Vol. I, part A, eds H. Rabin and C.L. Tang (Academic Press, New York) p. 419.
Laurell, F., and G. Arvidsson, 1988, J. Opt. Soc. Am. B **5**, 292.
Ledoux, I., D. Josse, P. Vidakovic and J. Zyss, 1986, Opt. Eng. **25**, 202.
Levine, B.F., C.G. Bethea, C.D. Thurmond, R.T. Lynch and J.L. Bernstein, 1979, J. Appl. Phys. **50**, 2523.
Li, M.J., M.P. De Micheli and D.B. Ostrowsky, 1989, in: Topical Meeting on Nonlinear Guided-Wave Phenomena: Physics and Applications, 1989 Technical Digest Series, Vol. 2 (Optical Society of America, Washington, DC) p. 74.
Liao, C., P. Bundman and G.I. Stegeman, 1983, J. Appl. Phys. **54**, 6213.
Lim, E.J., M.M. Fejer and R.L. Byer, 1989, in: Nonlinear Guided-Wave Phenomena: Physics and Applications, 1989 Technical Digest Series, Vol. 2 (Optical Society of America, Washington, DC) post-deadline paper PD3-1.
Lüdtke, H., W. Sohler and H. Suche, 1980, in: Digest Workshop Integrated Optics, eds R.Th. Kersten and R. Ulrich (Technical University Berlin, Germany) p. 122.
Lunt, S.R., G.E. Peterson, R.J. Holmes and Y.S. Kim, 1985, Proc. SPIE **578**, 22.
Lytel, R., G.F. Lipscomb and J.L. Thackara, 1988, in: Nonlinear Optical Properties of Polymers, eds A.J. Heeger, J. Orenstein and D.R. Ulrich, Proc. Mater. Res. Soc. **109**, 19.
Lytel, R., G.F. Lipscomb, M. Stiller, J.I. Thackara and A.J Ticknor, 1989, in: Integrated and Guided-Wave Optics, 1989 Technical Digest Series, Vol. 4 (Optical Society of America, Washington, DC) p. 126.
Madge, D., and H. Mahr, 1967, Phys. Rev. Lett. **18**, 905.
Maier, A.A., 1980, Sov. J. Quantum. Electron. **10**, 925.
Marcuse, D., 1974, Theory of Dielectric Optical Waveguides (Academic Press, New York).
Minakata, M., S. Saito, M. Shlbata and S. Miyazawa, 1978, J. Appl. Phys. **49**, 4677.
Minakata, M., S. Saito and M. Shibata, 1979, J. Appl. Phys. **50**, 3063.

Möhlmann, G.R., W.H.G. Horsthuis, M.B.J. Diemeer, F.M.M. Suyten, E.S. Trommel, A.Mc. Donach and N.Mc. Fadyen, 1989, Topical Meeting on Nonlinear Guided-Wave Phenomena: Physics and Applications, 1989 Technical Digest Series, Vol. 2 (Optical Society of America, Washington, DC) p. 171.
Nayar, B.K., 1982, in: Technical Digest, Topical Meeting on Integrated and Guided-Wave Optics (Optical Society of America, Washington, DC) paper ThA2.
Nayar, B.K., 1983, in: Nonlinear Optical Properties of Organic and Polymer Materials, ed. D.J. Williams, ACS Symposium Series, No. 233 (American Chemical Society, Washington, DC) p. 154.
Nayar, B.K., 1986, in: Proc. Conf. on Lasers and Electro-Optics (CLEO 1986) paper THQ1, p. 322.
Normandin, R., and G.I. Stegeman, 1979, Opt. Lett. **4**, 58.
Normandin, R., and G.I. Stegeman, 1980a, in: Proc. Int. Conf. on Solid State Devices, Tokyo, Jpn. J. Appl. Phys. **19**, 465.
Normandin, R., and G.I. Stegeman, 1980b, Appl. Phys. Lett. **36**, 253.
Normandin, R., and G.I. Stegeman, 1982, Appl. Phys. Lett. **40**, 759.
Normandin, R., P.J. Vella and G.I. Stegeman, 1981, Appl. Phys. Lett. **38**, 759.
Nye, J.F., 1957, Physical Properties of Crystals (Oxford, New York).
Ostrowsky, D.B., 1985, in: Integrated Optics, Springer Series in Optical Sciences, Vol. 48, eds H.P. Nolting and R. Ulrich (Springer, Berlin) p. 146.
Oudar, J.L., and R. Hierle, 1977, J. Appl. Phys. **48**, 2699.
Pitt, C.W., and L.M. Walpita, 1980, Thin Solid Films **68**, 101.
Regener, R., and W. Sohler, 1988, J. Opt. Soc. Am. B **5**, 267.
Regener, R., W. Sohler and H. Suche, 1981, Proc. 1st Europ. Conf. on Integrated Optics, London, UK, September 1981, IEEE Conference publishing no. 201, Post deadline papers, p. 19.
Reutov, A.T., and P.P. Tarashchenko, 1975, Sov. J. Quantum Electron. **4**, 910.
Sanford, N.A., and W.C. Robinson, 1987, Opt. Lett. **12**, 445.
Sanford, N.A., W.A. Dyes, J.M. Connors and A. Mahapatra, 1987, J. Opt. Soc. Am. **4**, 77.
Sasaki, K., T. Kinoshita and N. Karasawa, 1984, Appl. Phys. Lett. **45**, 333.
Schmidt, R.V., and I.P. Kaminov, 1974, Appl. Phys. Lett. **25**, 458.
Seibert, H., 1985, University of Paderborn, unpublished results.
Seibert, H., 1988, University of Paderborn, unpublished results.
Seibert, H., 1989, University of Paderborn, unpublished results.
Seilmeier, A., K. Spanner, A. Laubereau and W. Kaiser, 1978, Opt. Commun. **24**, 237.
Shen, Y.R., 1984, The Principles of Nonlinear Optics (Wiley, New York).
Singer, K.D., M.G. Kuzyk and J.E. Sohn, 1987, J. Opt. Soc. Am. B **4**, 968.
Sipe, J.E., and G.I. Stegeman, 1979, J. Opt. Soc. Am. **69**, 1676.
Sipe, J.E., G.I. Stegeman, C. Karaguleff, R. Fortenberry, R. Moshrefzadeh, W.M. Hetherington III and N.E. Van Wyck, 1983, Opt. Lett. **8**, 461.
Sohler, W., 1984, in: New Directions in Guided-Wave and Coherent Optics, Vol. II, eds D.B. Ostrowsky and E. Spitz, NATO ASI Series E, no. 79 (Martinus Nyhoff, The Hague) p. 449.
Sohler, W., 1985, Proc. SPIE **492**, 346.
Sohler, W., 1989, Thin Solid Films, to appear and references therein.
Sohler, W., and H. Suche, 1978, Appl. Phys. Lett. **33**, 518.
Sohler, W., and H. Suche, 1980, Appl. Phys. Lett. **37**, 255.
Sohler, W., and H. Suche, 1981, Technical Digest of the 3rd Int. Conf. on Integrated Optics and Fiber Communication, IOOC 1981, San Francisco, paper WB1, p. 89.
Sohler, W., and H. Suche, 1983, Proc. SPIE **408**, 163.
Sohler, W., B. Hampel, R. Regener, R. Ricken, H. Suche and R. Volk, 1986, IEEE J. Lightwave Techn. **LT-4**, 772.
Somekh, S., and Y. Yariv, 1972, Appl. Phys. Lett. **21**, 140.
Sotin, V.E., and V.M. Shevtsov, 1984, Sov. Tech. Phys. Lett. **10**, 200.

Steffen, J., A. Neyer and E. Voges, 1989, in: Integrated and Guided-Wave Optics, 1989 Technical Digest Series, Vol. 4 (Optical Society of America, Washington, DC) p. 121.
Stegeman, G.I., 1983, J. Opt. Commun. **4**, 20.
Stegeman, G.I., and C. Liao, 1983, (Letter to the editor) Appl. Opt. **22**, 2518.
Stegeman, G.I., and C.T. Seaton, 1985, J. Appl. Phys. **58**, R57.
Stegeman, G.I., C.T. Seaton, W.M. Hetherington III, A.D. Boardman and P. Egan, 1986, in: Electromagnetic Surface Excitations, Springer Series on Wave Phenomena, Vol. 3, eds R.F. Wallis and G.I. Stegeman (Springer, Berlin) p. 261.
Stegeman, G.I., C.T. Seaton, W.M. Hetherington III, A.D. Boardman and P. Egan, 1986, Nonlinear Optics: Materials and Devices, Springer Proceedings in Physics, Vol. 7, eds C. Flytzanis and J.L. Oudar (Springer, Berlin) p. 31.
Strake, E., 1986, University of Paderborn, unpublished results.
Suche, H., 1981, Ph.D. (Thesis University of Dortmund, FRG).
Suche, H., and W. Sohler, 1983, Proc. 2nd Europ. Conf. on Integrated Optics, ECIO 1983, IEE Conference Publishing No. 227, p. 98.
Suche, H., and W. Sohler, 1988, Topical Meeting on Itegrated and Guided-Wave Optics, IGWO 1988, Santa Fe, USA, Technical Digest Series, Vol. 5 (Optical Society of America, Washington, DC) paper MF15, p. 176.
Suche, H., and C. Wersig, 1990, to be published.
Suche, H., R. Ricken and W. Sohler, 1987, Proc. 4th Europ. Conf. on Integrated Optics, ECIO 1987, Glasgow, Great Britain, p. 202.
Suche, H., H. Teichmann and W. Sohler, 1988, 1st Europ. Quantum Electronics Conf., EQEC 1988, Hannover, FRG, Digest of Technical Papers, paper WeBB3.
Suche, H., B. Hampel-Vogedes, R. Ricken, W. Sohler and H. Teichmann, 1989, Topical Meeting on Nonlinear Guided-Wave Phenomena: Physics and Applications, Houston, USA, Technical Digest Series, Vol. 2 (Optical Society of America, Washington, DC) paper THA1, p. 2.
Tang, C.L., and P.P. Bey, 1973, IEEE J. Quantum Electron. **QE-9**, 9.
Taniuchi, T., and K. Yamamoto, 1986a, Technical Digest of the 12th Europ. Conf. on Optical Communication, Madrid, 1986, p. 171.
Taniuchi, T., and K. Yamamoto, 1986b, Technical Digest of the Conference on Lasers and Electro-Optics (CLEO 1986), San Francisco, 1986, p. 230 (paper WR3).
Taniuchi, T., and K. Yamamoto, 1987a, Optoelectronics – Devices and Technologies **2**, 53.
Taniuchi, T., and K. Yamamoto, 1987b, Advanced Optoelectronic Technology, Cannes, Proc. SPIE 864-09.
Taylor, H.F., 1973, Opt. Commun. **8**, 421.
Thackara, J.I., M. Stiller, G.F. Lipscomb, A.J. Ticknor and R. Lytel, 1988, Appl. Phys. Lett. **52**, 1031.
Tien, P.K., 1971, Appl. Opt. **10**, 2410.
Tien, P.K., R. Ulrich and R.J. Martin, 1970, Appl. Phys. Lett. **17**, 447.
Uesugi, N., 1980, Appl. Phys. Lett. **36**, 178.
Uesugi, N., K. Daikoku and M. Fukuma, 1978, J. Appl. Phys. **49**, 4945.
Uesugi, N., K. Daikoku and K. Kubota, 1979, Appl. Phys. Lett. **34**, 60.
van der Ziel, J.P., M. Ilegems, P.W. Foy and R.M. Mikulyak, 1976, Appl. Phys. Lett. **29**, 775.
Vidakovic, P.V., M. Coquillay and F. Salin, 1987, J. Opt. Soc. Am. B **4**, 998.
Volk, R., and W. Sohler, 1988, in: Technical Digest Series, Vol. 5, Topical Meeting of Integrated and Guided-Wave Optics, Santa Fe/New Mexico, 1988 (Optical Society of America, Washington, DC) paper MF6.
Wang, Y., 1986, Chem. Phys. Lett. **126**, 209.
Webjörn, J., F. Laurell and G. Arvidsson, 1989, in: Nonlinear Guided-Wave Phenomena: Physics and Applications, 1989 Technical Digest Series, Vol. 2 (Optical Society of America, Washington, DC) p. 6.

Williams, D.J., ed., 1983, Nonlinear Optical Properties of Organic and Polymeric Materials, ACS Symposium Series, no. 233 (American Chemical Society, Washington, DC).
Williams, D.J., 1984, Angew. Chem. Int. Ed. Engl. **23**, 690.
Yariv, A., 1973, IEEE J. Quantum Electron. **QE-9**, 919.
Yariv, A., 1975a, Quantum Electronics, 2nd Ed. (Wiley, New York).
Yariv, A., 1975b, Quantum Electronics, 2nd Ed. (Wiley, New York) table 16.2.
Yariv, A., 1979, in: Fiber and Integrated Optics, NATO Advanced Study Institutes Series B: Physics, Vol. 41, ed. D.B. Ostrowsky (Plenum Press, New York). p. 223.
Zolotov, E.M., V.G. Mikhalevich, V.M. Pelekhatyi, A.M. Prokhorov, V.A. Chernykh and E.A. Shcherbakov, 1978, Sov. Tech. Phys. Lett. **4**, 89.
Zolotov, E.M., V.M. Pelekhatyi, A.M. Prokhorov and V.A. Chernykh, 1979, Sov. Phys. JETP **49**, 603.
Zolotov, E.M., A.M. Prokhorov and V.A. Chernykh, 1981, Sov. Tech. Phys. Lett. **7**, 129.
Zumsteg, F.C., J.D. Bierlein and T.E. Gier, 1976, J. Appl. Phys. **47**, 4980.
Zyss, J., 1985, J. Mol. Electron. **1**, 25.
Zyss, J., D.S. Chemla and J.F. Nicoud, 1981, J. Chem. Phys. **74**, 4800.

CHAPTER 2

Third-Order Nonlinear Electromagnetic TE and TM Guided Waves

A.D. BOARDMAN and P. EGAN

Department of Physics
University of Salford
Salford M5 4WT, UK

F. LEDERER and U. LANGBEIN

Department of Physics
University of Jena
Jena 6900, Germany

and

D. MIHALACHE

Central Institute of Physics
Department of Fundamental Physics
P.O. Box MG-6, Bucharest, Romania

Nonlinear Surface Electromagnetic Phenomena
Edited by
H.-E. Ponath and G.I. Stegeman

© *Elsevier Science Publishers B.V., 1991*

Contents

1. Introduction

In 1962 Askaryan made an important contribution to nonlinear optics. In a study of microwaves, he predicted the way a nonlinear refractive index can oppose the linear tendency to geometrical and diffraction-induced laser beam spreading, through a new phenomenon called self-focussing. This mechanism was subsequently investigated for optical frequencies by Chiao et al. (1964), who assumed that the refractive index of an isotropic dielectric, in which the light beam is propagating, increases from its linear value by a third-order nonlinear term proportional to its intensity. This is an example of a self-trapping state in which the electromagnetic wave essentially forms its own waveguide. During this process a laser beam will preserve its profile so that the term 'focussing' is meant in an extended sense, rather than the ordinary definition of focussing at a point. Short contributions were then made by Alanakyan (1967), on nonlinear TM (p-polarised) surface waves on a semi-infinite plasma, and Litvak and Mironov (1968). The latter provided exact analytical solutions of the nonlinear field equations for TE (s-polarised) surface waves on the interface between two semi-infinite positive dielectric media, together with a very brief discussion of TM surface waves. These authors were the first to address the problem of finding exact solutions for planar nonlinear surface-guided waves, to which should also be added the contribution by Haus (1966) who computed exact solutions to the TE-type solutions to a nonlinear field equation for azimuthally symmetric modes describing a light beam of finite radial extent. These solutions implied rings of energy about the central beam, and were independently predicted, many years later, for nonlinear optical fibres by Boardman et al. (1986d). The initial developments were followed by predictions of solitary waves and stationary waves in bounded and unbounded self-focussing media carrying waves with TE polarisation (Miyagi and Nishida 1972, 1973, 1975a, b). The treatments of TE and TM nonlinear polarisations present different scales of difficulty, however, as will be discussed in this article, and this is especially so when guided waves are considered. Nevertheless, detailed TM treatments for bulk media, and limited exercises for a single interface, did emerge quite a long time ago (Alanakyan 1967, Litvak and Mironov 1968, Berkhoer and Zakharov 1970, Miyagi and Nishida 1974, 1975a, b, Bhat and Tripathi 1975). The theoretical techniques included phase-plane and equivalent potential models for bulk media, but the rather limited discussions for guided waves fell a long way short of the exact solutions needed to cope with the various types of nonlinearities that can be identified in condensed matter physics. One of the results that was established for bulk waves, for example, is that TM solitons of the type that arise from transverse modulational instability (beam solitons) exist (Miyagi and Nishida 1975a, b). In general, however, the literature listed above does not offer much help in solving boundary value problems. Indeed, nonlinear TM waves in planar guided wave structures have only recently received an exact numerical treatment (Boardman et al. 1987a, b).

Although an exact formula for the nonlinear wave power flow carried by TE-polarised waves propagating along the interface between a nonlinear semi-infinite medium and a linear semi-infinite medium was produced by Litvak and Mironov (1968), the work remained unnoticed. This is probably due to the fact that much of the effort in solid state physics was at that time being directed towards the linear problems involved in surface polariton theory. The present day activity in nonlinear waves has, instead, emerged from the pioneering investigations of Agranovich and Mills (1982), Maradudin (1981) and Tomlinson (1980), who independently established exact theoretical results for TE waves and TM waves (within an unusual approximation) propagating along a single interface. The subject also received a great impetus from the seminal experiment of Vach et al. (1984) who showed that some of the new features of these strong nonlinear waves, that were beginning to emerge theoretically, could be observed with existing materials, such as a liquid crystal bounding a planar glass waveguide.

An important feature of nonlinear waves, strongly emphasised by Maradudin (1981), is that nonlinearity may be expected to promote the existence of surface-guided waves that cannot exist in the linear limit. This is a general rule that is exemplified by the existence of TE nonlinear surface waves at the interface between two materials that have positive dielectric constants. Since, in the linear (zero power) limit, such waves cannot exist, they must be associated with a threshold power that must be reached before propagation can occur. This threshold power was predicted by Litvak and Mironov but the surface polariton context was greatly clarified by Maradudin. For strong nonlinear waves that do have a linear limit, such as surface plasmon-polaritons, the wave number acquires a term that is an arbitrary function of the wave intensity that vanishes as the power tends to zero. It is easy to see, with hindsight, that the latter is an entirely expected result and that even the existence of nonlinear s-polarised surface polaritons, which cannot have a linear limit, must involve the establishment of a self-focussed channel in the vicinity of the interface. The shape of this channel can be found only from the exact solution of the field equations.

The modern work (Boardman et al. 1987b) on third-order nonlinear waves is driven by the anticipated demand for a fast and powerful all-optical signal processing capability. This need is now directing the attention of many laboratories to the creation of new nonlinear materials that have accessible self-focussing and other nonlinear properties. The optical nonlinearities that materials can sustain depends, among other things, upon their crystal symmetry, and attention is usually limited to terms in the polarisation that are second and/or third order in the electric field components of any electromagnetic wave propagating in them. It is currently anticipated that such new materials may well be organic (Zyss 1985), and the new optical processing that will be achieved with them is one of the most exciting developments in nonlinear optics. Of the nonlinearities that are immediately accessible, it is the third-order one (usually characterised for isotropic materials by a single scalar parameter if the nonlinear mechanism is known) that has attracted recent attention. The emphasis on the third-order nonlinearity arises technically because its use permits all-optical signal processing with equal input and output frequencies (Stegeman et al. 1984a–c), if possible 3ω processes are negligible due to lack of phase matching. The effects and associated functions that can be achieved (Seaton et al. 1985a–d, Zyss 1985) with third-order nonlinearities include four-wave mixing (coherent Raman spectroscopy), phase gratings (real-time holography), the Kerr effect (ultra high-speed optical gates) and optical bistability (amplifiers, logic gates).

The recent theoretical work on third-order nonlinear waves guided by surfaces or thin films that is reviewed in this chapter concerns the CW propagation of surface-guided waves in materials that have optical nonlinearities that are local in time. The nonlinearity will also usually be spatially local, although a brief discussion of diffusive nonlinearity will also be given. In the main, therefore, the third-order polarisation response of the nonlinear medium to the electric field of the waves will be instantaneous and will depend only on the local field at any point in the system. Before the form of the third-order polarisation, needed for this article, is discussed in detail it is useful to point out the difference between the weak and strong nonlinear regimes.

First, if a waveguide can support stable linear guided waves then a weakly nonlinear regime can always be found in which the nonlinearity will only slightly disturb the linear system. A mathematical treatment of this situation that retains only the lowest order of perturbation results in modal fields that are unchanged in shape and position from their linear values, but with wavenumbers that have acquired a nonlinear shift. This shift will be linearly proportional to the power of the wave and, as the wave propagates it will change its phase in proportion to the distance that it travels. The general theoretical treatment of this regime involves coupled-mode theory and is presented in section 2. The nonlinearity discussed in the remaining sections of this review is not weak. The exact solutions, derived later on, reveal that the field distributions are strongly dependent upon the nonlinearity and that they readily distort from their linear value to reach a self-focussing condition. This strongly nonlinear regime is entered when the refractive index change due to the nonlinearity becomes the same order of magnitude as the difference between the linear parts of the refractive indices of any two adjacent media. This is very interesting because it implies that it may not be necessary to seek large absolute values of nonlinearity for use within a waveguide format, provided that precise matching of materials is possible.

The theory presented in the remaining sections concerns materials with a centre of symmetry, or waveguiding situations that do not allow second harmonic generation so that only the third-order polarisation vector needs to be considered here. This can be written, after neglecting spatial dispersion and suppressing the $\boldsymbol{r}$ dependence, as

$$P_i^{(3)}(t) = \varepsilon_0 \int_{0^+}^{\tau_j} \int_{0^+}^{\tau_k} \int_{0^+}^{\tau_l} \chi_{ijkl}(\boldsymbol{r}, t_j, t_k, t_l) E_j(t - t_j) E_k(t - t_k) \times E_l(t - t_l)\, \mathrm{d}t_j\, \mathrm{d}t_k\, \mathrm{d}t_l, \tag{1.1}$$

where $P_i^{(3)}(\boldsymbol{r}, t)$ is a real polarisation vector, $\chi_{ijkl}(\boldsymbol{r}, t_j, t_k, t_l)$ is the real, dielectric tensor, $\boldsymbol{E}(t - t_{j,k,l})$ is the electric field and $\tau_{j,k,l}$ are response times of the nonlinearity. The latter can be defined vaguely here, since it is only necessary to assert that the field $\boldsymbol{E}$ does not vary very much on the τ time scale. If that is the case, then the $E_{j,k,l}(t - t_{j,k,l})$ can be Taylor expanded into a slowly varying term and a sequence of time derivatives. It is usually sufficient to neglect the derivatives, as will be done here, and then to remove the slowly varying part of the fields $\boldsymbol{E}$ from under the integral signs in eq. (1.1). More precisely, the field components can be written, after suppressing the $\boldsymbol{r}$ dependence, as

$$E_j(t) = \tfrac{1}{2}(\xi_j(t)\mathrm{e}^{-\mathrm{i}\omega t} + \xi_j^*(t)\mathrm{e}^{\mathrm{i}\omega t}), \tag{1.2}$$

where $\xi(t)$ are the slowly varying parts. Usually, they are considered to be so slowly varying as to be constants in time. This means that no derivatives of $\xi_j(t)$, with respect to time, will emerge from eq. (1.1) and $\xi(t)$ may, with confidence, be taken outside any integral sign. For this case we can use a somewhat less satisfactory, but common, notation and write $\xi(t) \equiv E(\omega)$ so that eq. (1.2) then takes on a conventional appearance. After substituting into eq. (1.1) we obtain

$$P_i^{(3)}(t) = \tfrac{1}{8}\varepsilon_0\{\mathrm{e}^{-3\mathrm{i}\omega t}\chi_{ijkl}^{(3)}(\omega, \omega, \omega)E_j(\omega)E_k(\omega)E_k(\omega) + 3\mathrm{e}^{-\mathrm{i}\omega t}\chi_{ijkl}^{(3)}(\omega, -\omega, \omega)E_j(\omega)E_k^*(\omega)E_k(\omega)\}, \tag{1.3}$$

where

$$\chi_{ijkl}^{(3)}(\omega', \omega'', \omega''') = \iiint \chi_{ijkl}(t', t'', t''')\mathrm{e}^{\mathrm{i}[\omega' + \omega'' + \omega''']}\, \mathrm{d}t'\, \mathrm{d}t''\, \mathrm{d}t''', \tag{1.4}$$

is the nonlinear susceptibility tensor in the frequency domain.

The vector $P_i^{(3)}(t)$ can now be written as

$$P_i^{(3)}(t) = P_i^{(3)}(3\omega)\mathrm{e}^{-3\mathrm{i}\omega t} + P_i^{(3)}(\omega)\mathrm{e}^{-\mathrm{i}\omega t} + \text{c.c.} \tag{1.5}$$

Hence, after the application of crystal symmetry requirements, an expression for $P_i^{(3)}(\omega)$ and $P_i^{(3)}(3\omega)$ can be found.

The general nonlinear electromagnetic wave equation, even for an isotropic material, shows that all the field components are coupled to each other. They may be uncoupled into a TE or a TM set only by *assuming* a TE or TM state of polarisation. If this is done then Maxwell's equations do not permit TE or TM polarisations to degrade into each other. In order to apply the results of such a single polarisation theory it is necessary to maintain one principal polarisation in a given experiment. The more complex problem of TE–TM coupling, that can be generated even in isotropic media, will be addressed much later in the review. For nonlinear s-polarized (TE) waves propagating down the x-axis only the E_y field component will appear in the field equations. For p-polarized (TM) waves both E_x and E_z will appear, however, and the phase relationship of E_x to E_z creates additional serious difficulties when an exact theory of nonlinear p-polarized waves is being formulated.

For an isotropic nonlinear medium the third-order susceptibility tensor, where the superscript (3) and frequency arguments are now regarded as redundant, has elements (Butcher 1965)

$$\chi_{yyzz} = \chi_{zzyy} = \chi_{zzxx} = \chi_{xxzz} = \chi_{xxyy} = \chi_{yyxx}, \tag{1.6a}$$

$$\chi_{yzyz} = \chi_{zyzy} = \chi_{zxzx} = \chi_{xzxz} = \chi_{xyxy} = \chi_{yxyx}, \tag{1.6b}$$

$$\chi_{yzzy} = \chi_{zyyz} = \chi_{zxxz} = \chi_{xzzx} = \chi_{xyyx} = \chi_{yxxy}, \tag{1.6c}$$

$$\chi_{xxxx} = \chi_{yyyy} = \chi_{xxyy} + \chi_{xyxy} + \chi_{xyyx}. \tag{1.6d}$$

For surface-guided waves propagating along the x-axis, with no variation along the y-axis, the electric and magnetic fields, $\boldsymbol{E}$ and $\boldsymbol{H}$, can be divided into two groups of field components, (i) E_y, E_z, H_y and H_z and (ii) E_x and H_x, where the field components within a group lie on the same axis of the complex phase plane and the two groups are orthogonal to each other. The introduction of the substitutions $E_x \to \mathrm{i}E_x$, $E_y \to E_y$, $E_z \to E_z$, $H_x \to \mathrm{i}H_x$, $H_y \to H_y$ and $H_z \to H_z$, where E_i and H_i are now real, and eqs (1.6), (1.3) and (1.5) for the fundamental leads to

$$P_x^{(3)}(\omega) = 2\varepsilon_0 \chi_{xxyy}[E_x^2 + E_y^2 + E_z^2]E_x - \varepsilon_0 \chi_{xyyx}[-E_x^2 + E_y^2 + E_z^2]E_x, \tag{1.7a}$$

$$P_y^{(3)}(\omega) = 2\varepsilon_0 \chi_{xxyy}[E_x^2 + E_y^2 + E_z^2]E_y + \varepsilon_0 \chi_{xyyx}[-E_x^2 + E_y^2 + E_z^2]E_y, \tag{1.7b}$$

$$P_z^{(3)}(\omega) = 2\varepsilon_0 \chi_{xxyy}[E_x^2 + E_y^2 + E_z^2]E_z + \varepsilon_0 \chi_{xyyx}[-E_x^2 + E_y^2 + E_z^2]E_z, \tag{1.7c}$$

where the electric field vector for TM waves is $\boldsymbol{E} = (E_x, 0, E_z)$, for TE waves is $\boldsymbol{E} = (0, E_y, 0)$, the frequency dependence in E_i has been dropped and, from permutation symmetry, $\chi_{xyxy} = \chi_{xxyy}$. Note here that χ_{xxyy} describes an 'isotropic' part of the nonlinear polarisation and would by itself give rise to a nonlinear refractive index change that is the same in all directions. The term involving the coefficient χ_{xyyx} describes an 'anisotropic' part of the nonlinear polarisation and gives rise to a nonlinear birefringence. The substitutions (Boardman et al. 1987a)

$$\gamma = \left[\frac{1 - \dfrac{\chi_{xyyx}}{2\chi_{xxyy}}}{1 + \dfrac{\chi_{xyyx}}{2\chi_{xxyy}}}\right], \tag{1.8}$$

and

$$a = 2\chi_{xxyy}\left(1 + \frac{\chi_{xyyx}}{2\chi_{xxyy}}\right), \tag{1.9}$$

results in the following compact form for the nonlinear polarisations

TE: $$P_y^{(3)} = \varepsilon_0 a E_y^3, \tag{1.10a}$$

TM: $$P_x^{(3)} = \varepsilon_0 a(E_x^2 + \gamma E_z^2)E_x, \tag{1.10b}$$

$$P_z^{(3)} = \varepsilon_0 a(E_z^2 + \gamma E_x^2)E_z, \tag{1.10c}$$

where $\gamma = \frac{1}{3}$, $-\frac{1}{2}$ or 1 for electronic distortion, molecular orientational or thermal nonlinear mechanisms, respectively (Svelto 1974). The dielectric functions used for TE and TM guided wave propagation are therefore

TE: $$\varepsilon = \bar{\varepsilon} + aE_y^3, \tag{1.11a}$$

TM: $$\varepsilon = \begin{pmatrix} \varepsilon + aE_x^2 + \gamma a E_z^2 & 0 \\ 0 & \varepsilon + aE_z^2 + \gamma a E_x^2 \end{pmatrix}. \tag{1.11b}$$

Section 2 contains a brief review of the weakly nonlinear regime that can be described in terms of coupled-mode theory. This is followed in sections 3 and 4 by a very detailed theory of strong nonlinear waves on thin film and single surface structures, for both Kerr and non-Kerr media. Sections 5 and 6 are devoted to numerical solutions, targetted, in section 5, on the variation of the effective guide index with the total power flow for stationary states, and on nonstationary propagation effects in section 6. The final section also addresses questions concerning nonlinear wave stability.

2. *Nonlinear coupling of linear waveguide modes*

2.1. *Introduction*

Electromagnetic waves can be guided by a variety of optically linear configurations, e.g., the interface between two semi-infinite media, by single or multiple films bounded by two semi-infinite media, or by refractive index channels embedded near the interface of some semi-infinite medium.

The guiding mechanism usually results from an interplay between the electromagnetic field and some resonance. The resonance can be of geometric origin, as occurs in thin film and channel waveguides, or can be related to the carrier dynamics of the optical material, and leads to surface-polaritons or -plasmons (Stegeman et al. 1986b).

The key feature of guided waves is that the fields penetrate typically only a fraction of their wavelength into the bounding media of the waveguide. This diffraction-less propagation of optical fields, confined up to the order of their wavelength perpendicular to the guiding interfaces, leads to maximum power densities that can be achieved with small total powers over distances in the centimeter range.

The propagation wave vectors of guided waves are determined by a configuration-specific dispersion relation which offers new opportunities for field phase control, in particular phase matching between interacting fields.

Both basic features have attractive implications for *nonlinear* guided wave optics. In this field a variety of new developments have centered on phenomena that are of potential interest both for all-optical signal processing and for spectroscopy. In particular, those wave interactions that are automatically phase-matched, e.g., degenerate four-wave mixing and power-dependent refractive index effects, have been extensively studied in recent years. The tuning of the difference frequency of two guided wave fields through characteristic molecular vibrational frequencies leads to resonant enhancements of the signals and hence can be used for spectroscopy.

The interplay of the material nonlinearity with the evanescent nature of the guided waves, their particular dispersion properties and their distributed coupling to plane waves respectively, gives rise to a new class of nonlinear optical phenomena that have no analog in nonlinear optics with plane waves. Last, but not least, both advanced thin-film fabrication technology and the discovery of new highly nonlinear optical materials have recently stimulated the investigation of third-order nonlinear effects within guided wave geometries.

There are several excellent review articles (e.g., Seaton et al. 1985b, c, Stegeman et al. 1985a, 1986b) which give a detailed description of the state-of-art achieved in recent years. For that reason the present section will be devoted specifically to the following aspects.

(i) Introduction of the two basic concepts for the analysis of nonlinear guided wave phenomena, i.e., the so-called 'weak' and 'strong' nonlinear regime. The 'coupled-mode' formalism, related to the 'weak' nonlinear regime, will be summarized here in a simple, unified notation, oriented at TE-polarised fields in planar geometries. The structure of the formulae will be essential, not their detailed notation.

(ii) Discussion of principal effects that will arise when linear waveguide modes propagate in configurations with intensity dependent (complex-valued) refractive indices. These results should be compared with the predictions derived within the framework of the 'strong' nonlinear regime later in this section.

(iii) Review of recent experimental progress, including the theoretical simulations, which have been achieved for phenomena in the 'weak' nonlinear regime, i.e., which can be interpreted as nonlinear coupling effects of linear guided waves. Particularly, power-dependent coupling phenomena, either between two guided modes (directional coupler) or between a guided mode and the spectrum of radiation modes (distributed coupler) will be studied. Both schemes serve as promising candidates for ultrafast all-optical switching applications, and have attracted by far the most recent interest. Due to their very recent origin, the outcome of these investigations is not yet covered by other existing reviews.

2.2. The two operational regimes in nonlinear waveguide optics

The electromagnetic theory states that the interaction of optical fields propagating in appropriate media can be described by means of some characteristic polarisation fields. These polarisation fields play the role of driving terms or sources in the relevant field equations (see, e.g., Jackson 1975, Shen 1982).

The proper decomposition of a given configuration into its basic, i.e., 'unperturbed' part, which is governed by a homogeneous field equation, and some additional contributions that perturb the basic configuration via appropriate polarisation fields, is, by no means, a straightforward task. In nonlinear waveguide optics, a widely adopted criterion is the effect imposed by the nonlinear polarisation sources on the waveguiding mechanism relevant for a given geometry.

In the 'weakly nonlinear regime', the waveguiding effect is completely determined by the linear optical properties of the waveguide. In this case, the nonlinear polarisation sources can be regarded as perturbational terms that lead to a mutual coupling of the normal modes of the linear guide. The normal modes themselves, which manifest the properties of the linear 'unperturbed' waveguide, remain unaffected.

On the contrary, the 'strong nonlinear regime' characterizes a situation where the nonlinear polarisation sources are strong enough to compete with the linear waveguide mechanism. In this case the guided-mode concept that is essentially of linear origin no longer applies. Now, even the 'unperturbed' guide shows nonlinear features and its basic guided field structures are the so-called 'nonlinear guided waves'.

The remainder of this section deals exclusively with phenomena that can be described within the framework of the weakly nonlinear regime.

2.3. The basic theoretical model

2.3.1. The coupled-mode equations

The theoretical calculus appropriate to the weak nonlinear regime is the coupled-mode theory. This calculus is well documented (see, e.g., Marcuse 1974, Kogelnik 1975, Sipe and Stegeman 1979, Langbein et al. 1984a–c) and we recall only those arguments that lie within the scope of this section as described above. Accordingly, all formulae will refer to planar isotropic geometries and colinear propagation of TE-polarised fields. This situation still covers approximately that range of phenomena we are interested in. Extensions of the present model are often straightforward and can be found elsewhere.

Our basic, 'unperturbed' waveguide will be described by a real-valued, scalar dielectric function $\varepsilon'(z, \omega)$, where z is the coordinate perpendicular to the planar guiding interfaces, and ω the frequency of some monochromatic, TE-polarised field component $\boldsymbol{E}(\boldsymbol{r}, t) = (0, E_y, 0)$, and

$$E_y(\boldsymbol{r}, t) = \tfrac{1}{2}E(\boldsymbol{r})\mathrm{e}^{-\mathrm{i}\omega t} + \mathrm{c.c.} \tag{2.1}$$

$E(\boldsymbol{r})$ is subject to the wave equation

$$[\partial^2/\partial x^2 + \partial^2/\partial z^2]E(\boldsymbol{r}) + k_0^2\varepsilon'(z, \omega)E(\boldsymbol{r}) = 0, \quad k_0 = \omega/c, \tag{2.2}$$

to the continuity requirements at every interface, and to some appropriate boundary conditions at infinity (Jackson 1975).

Equation (2.2), together with these conditions, defines the complete set of orthogonal *normal modes* the spectrum of which usually consists of both a discrete part (*guided modes*) and a continuous part (*radiation modes*). Radiation modes characterize the optical response of the waveguide for incident plane waves. These modes consist of certain standing wave patterns that range to infinity. The guided modes show a periodic field pattern within the waveguiding film only while the field tails in the bounding media are of an evanescent nature. The propagation wave vectors of the guided modes are determined by a *dispersion relation* (see, e.g., Marcuse 1974). Every solution $E(\boldsymbol{r})$ to eq. (2.2) can be decomposed into the normal modes of the waveguide

$$E(\boldsymbol{r}) = \sum_{\mu} A^{(\mu)}E^{(\mu)}(z)\exp(\mathrm{i}k_0 n^{(\mu)}x), \tag{2.3}$$

where the contribution of the radiation modes has been formally included into the μ-sum. Both the propagation constant $n^{(\mu)}$ and the modal field pattern $E^{(\mu)}(z)$ are determined by the eigenvalue equation

$$[\mathrm{d}^2/\mathrm{d}z^2 + k_0^2\varepsilon'(z, \omega)]E^{(\mu)}(z) = (k_0 n^{(\mu)})^2E^{(\mu)}(z). \tag{2.4}$$

Since the $E^{(\mu)}(z)$ are normal modes, they satisfy the orthogonality relation

$$\int_{-\infty}^{\infty} E^{(\mu)}(z)E^{(\mu')}(z)\,\mathrm{d}z = 2N\omega/(\varepsilon_0 c^2 k_0 n^{(\mu)})\delta_{\mu\mu'}, \tag{2.5}$$

where N fixes the guided power flow carried by a wavefront stripe of unit width.

The essential feature is that the mode amplitudes are completely determined by the boundary conditions as long as field propagation along the 'unperturbed' linear guide [characterized here by $\varepsilon'(z, \omega)$] is considered. The excited normal modes travel simultaneously and without any mutual coupling down the 'unperturbed' reference waveguide.

Every perturbation of the reference waveguide, however, will give rise to a mutual coupling of the normal modes. Their amplitudes will not then remain constant but will become solutions of appropriate equations of motion that contain the perturbational terms in the form of 'driving forces'.

The basic field equation, replacing eq. (2.2), is now

$$[\partial^2/\partial x^2 + \partial^2/\partial z^2]E(\boldsymbol{r}) + k_0^2\varepsilon'(z,\omega)E(\boldsymbol{r}) = -k_0^2/(\varepsilon_0)P(\boldsymbol{r}), \tag{2.6}$$

where the (monochromatic) polarisation $P(\boldsymbol{r})$ stands for any perturbation of the reference waveguide geometry, P comprises both linear and nonlinear contributions to be specified later in this section.

Now the total wavefield $E(\boldsymbol{r})$ will be expanded in terms of normal modes with *variable* modal amplitudes: $a^{(\mu)}(x)$,

$$E(\boldsymbol{r}) = \sum_{\mu} a^{(\mu)}(x)E^{(\mu)}(z)\exp(\mathrm{i}k_0 n^{(\mu)}x). \tag{2.7}$$

The modal amplitudes $a^{(\mu)}(x)$ are assumed to play the role of *slowly varying envelopes* with respect to the rapidly oscillating exponential term, i.e., the spatial variations of $a^{(\mu)}(x)$ are slow in a $(k_0 n^{(\mu)})^{-1}$ scale. The final equations of motion for the $a^{(\mu)}(x)$ can be derived by inserting eq. (2.7) into eqs (2.5) and (2.6) thereby making access of eqs (2.4) and (2.5). Eventually, one retains the first-order derivatives of the envelope functions $a^{(\mu)}(x)$ only and arrives at

$$[\mathrm{d}/\mathrm{d}x]a^{(\mu)}(x) = (\mathrm{i}\omega/4N)\mathrm{e}^{-\mathrm{i}k_0 n^{(\mu)}x}\int_{-\infty}^{\infty} \mathrm{d}z'\,(P(x,z')E^{(\mu)}(z')). \tag{2.8}$$

Equation (2.8) represents essentially a system of coupled differential equations since the total field, eq. (2.7), usually enters the polarization term P.

In contrast to plane-wave optics, only the 'overlap' of P with the specific mode $E^{(\mu)}(z)$ forms the relevant 'driving' term in eq. (2.8). This 'overlap' manifests itself as an averaging procedure which can be exploited to promote the efficiency of a specific interaction process. Coupled-mode theory is not restricted to spatial mode coupling but can be extended easily to transient mode coupling processes. Now the appropriate decomposition into coupled normal modes looks like

$$E_y(\boldsymbol{r},t) = \tfrac{1}{2}\sum_{\mu} a^{(\mu)}(x,t)E^{(\mu)}(z,\omega_\mu)\mathrm{e}^{\mathrm{i}\omega_\mu(n^{(\mu)}x - ct)/c} + \text{c.c.}, \tag{2.9}$$

where the mean frequency ω_μ of the μth normal mode determines both the relevant propagation constant $k_0 n^{(\mu)}$ and the mode profile $E^{(\mu)}(z)$ according to eq. (2.4). Again the temporal evolution of the mode amplitude $a^{(\mu)}(x,t)$ is assumed to be slow in a ω_μ^{-1} scale, i.e., $a^{(\mu)}(x,t)$ plays the role of a pulse envelope of the μth normal mode.

Equation (2.9) can be interpreted as a continuous sum over all monochromatic contributions of the type of eq. (2.7), i.e.,

$$
\begin{aligned}
\boldsymbol{E}(\boldsymbol{r}, t) &= \int_0^{\infty} \mathrm{d}\omega \, \tfrac{1}{2} \sum_{\mu} a^{(\mu)}(x, \omega) E^{(\mu)}(z, \omega) \, \mathrm{e}^{\mathrm{i}\omega(n^{(\mu)}(\omega)x - ct)/c} + \text{c.c.} \\
&= \tfrac{1}{2} \sum_{\mu} \mathrm{e}^{\mathrm{i}\omega_\mu(n^{(\mu)}x - ct)/c} \\
&\quad \times \int_0^{\infty} \mathrm{d}\omega \, E^{(\mu)}(z, \omega) a^{(\mu)}(x, \omega) \\
&\quad \times \mathrm{e}^{\mathrm{i}[(\omega n^{(\mu)}(\omega) - \omega_\mu n^{(\mu)})x/c - (\omega - \omega_\mu)t]} + \text{c.c.}
\end{aligned}
$$

For the sake of clarity, the ω-dependence of the appropriate functions has been explicitely indicated, and the $n^{(\mu)} = n^{(\mu)}(\omega_\mu)$ are the mean propagation constants for the corresponding modes. For some reasonable spectral pulse width $\Delta\omega \ll \omega_\mu$, the carrier frequency ω_μ determines the transverse mode profile

$$E^{(\mu)}(z, \omega) \simeq E^{(\mu)}(z, \omega_\mu) \equiv E^{(\mu)}(z). \tag{2.11a}$$

While only phase terms that vary linearly with $(\omega - \omega_\mu)$,

$$
\begin{aligned}
\omega n^{(\mu)}(\omega)/c &\simeq \omega_{\mu n}^{(\mu)}(\omega_\mu)/c + [\mathrm{d}/\mathrm{d}\omega](\omega n^{(\mu)}(\omega)/c)_{\omega\mu}(\omega - \omega_\mu) \\
&\equiv \omega_\mu n^{(\mu)}/c + v_\mu^{-1}(\omega - \omega_\mu),
\end{aligned} \tag{2.11b}
$$

will be considered, $v_\mu = \{[\mathrm{d}/\mathrm{d}\omega](\omega n^{(\mu)}(\omega)/c)_{\omega\mu}\}^{-1}$ is the group velocity of the guided wave pulse formed by the μth mode. Note that v_μ is affected both by material dispersion $\varepsilon'(z, \omega)$ and by the modal dispersion which can be controlled by the configuration parameters.

Obviously, approximations (2.11a, b) are consistent with the slowly varying envelope assumption introduced above. The application of both approximations in eq. (2.10) leads to

$$
\begin{aligned}
E(x, z, t) &\simeq \tfrac{1}{2} \sum_{\mu} \mathrm{e}^{\mathrm{i}\omega_\mu(n^{(\mu)}x - ct)/c} E^{(\mu)}(z) \\
&\quad \times \int \mathrm{d}\omega \, a^{(\mu)}(x, \omega) \, \mathrm{e}^{\mathrm{i}(\omega - \omega_\mu)(v_\mu^{-1}x - t)} + \text{c.c.},
\end{aligned}
$$

which implies

$$a^{(\mu)}(x, t) = \int \mathrm{d}\omega \, a^{(\mu)}(x, \omega) \, \mathrm{e}^{\mathrm{i}(\omega - \omega_\mu)(v_\mu^{-1}x - t)}. \tag{2.12}$$

Now it is easy to verify, that our mode amplitudes $a^{(\mu)}(x, t)$ obey the following equations of motion

$$[\partial/\partial x + v_\mu^{-1}\partial/\partial t]a^{(\mu)}(x,t) = \frac{\mathrm{i}\omega_\mu}{4N}\mathrm{e}^{-\mathrm{i}\omega_\mu(n^{(\mu)}x-ct)/c} \times \int_{-\infty}^{\infty} \mathrm{d}z'\, P(x,z',t)E^{(\mu)}(z'). \tag{2.13}$$

The verification of eq. (2.13) takes advantage of eq. (2.8) to replace $[\partial/\partial x]a^{(\mu)}(x,\omega)$, and $P(x,z,t)$ is related to $P(x,z,\omega)$ of eq. (2.8) by a Fourier transform.

Equations (2.8) and (2.13) are the fundamental equations of the coupled-mode theory. Together with appropriate material relations, which describe the generation of the polarisation field, these equations can be integrated analytically in many cases. Thus the coupled-mode calculus permits a flexible description of fairly general nonstationary propagation phenomena with respect to both the propagation coordinate and time. This flexibility, essentially based on the superposition principle, is not available in the strong nonlinear regime. The price paid by the coupled-mode model is that the waveguiding mechanism itself, i.e., the transverse mode profiles, is held fixed. This assumption has to be verified by the concrete experimental situation. The generalization of the coupled-mode equations (2.8) and (2.13) to guided wave fields with components along the propagation wave vector, i.e., surface plasmon-polaritons, TM integrated optics and channel guided waves (Sipe and Stegeman 1979) or to non-colinear mode propagation (Langbein et al. 1985e) needs some care but retains the structure of both equations.

The inclusion of higher-order dispersion effects would lead to higher-order derivatives with respect to time (Crosignani et al. 1982, Langbein et al. 1984b). While important for pulse propagation in fibers, these effects could not yet be observed over propagation lengths of a few millimeters that are typical for integrated optical schemes.

2.3.2. The Kerr-law nonlinearity and beyond

On a macroscopic level, polarisation fields describe the response of some given configuration to an applied electromagnetic wave. Evidently, a macroscopic material description can reflect only some essential aspects of the underlying complex microscopic dynamics. The emphasis of this chapter lies on phenomena that can be understood by means of intensity dependent susceptibilities, and the first section gave an introduction on how to deal with such susceptibilities.

In our present simplified (TE-oriented) notation, the polarisation field, central to the considerations of this chapter, obeys the Kerr law

$$P = \varepsilon_0 a^{\mathrm{NL}}(z)|E|^2 E, \tag{2.14}$$

with E being the electric field component of some TE-polarised wavefield. The sign of the nonlinear coefficient characterizes self-focusing ($a^{NL} > 0$) and self-defocusing media ($a^{NL} < 0$) respectively. The z-coordinate $a^{NL}(z)$ indicates a layered configuration.

For moderate applied intensities, this simple non-resonant nonlinearity is a reasonable approximation to many materials and has been used to interpret many new fundamental phenomena in nonlinear waveguide optics (Stegeman et al. 1985a, Seaton et al. 1985a). However, the Kerr-nonlinearity is of purely dispersive origin and describes essentially the steady-state material response only. With regard to the actual experimental situation, predictions based on Kerr-law nonlinearities should be complemented by estimating the effect of the following material properties:

(i) nonlinear refraction index saturation;
(ii) linear and nonlinear attentuation;
(iii) memory controlled material response.

To this end, several phenomenological extensions of the Kerr-law nonlinearity have been studied and the outcome will be reported at the appropriate places of this section.

However, from the theoretical point-of-view, there is a lack of models which comprise all these effects and their potential interplay in a concise manner. On the one hand, such models call for a more subtle description of the light–matter interaction taking place in a specific medium. On the other hand, such models should fit with the phenomenological language of nonlinear optics.

A classic example is the 'two-level-system' model, where the material is assumed to consist of noninteracting two-level atomic systems. In this case, the dynamical properties are described by the Bloch-equations (see, e.g., Flytzanis 1984, Langbein et al. 1984a). Here the dispersive/absorptive material response is related to the resonance detuning, and the time response is determined by two relaxation times responsible for phase and energy relaxation respectively.

For resonant nonlinearities in semiconductor materials, the light-induced carrier dynamics has to be taken into account. Phenomenological models that have already been applied in nonlinear optics include the effect of carrier plasma screening (Banyai and Koch 1986), and the thermal shift of the band edge (Lambsdorff et al. 1986).

The application of such models is still at an initial state but will disclose essential limitations of the basic Kerr-law model. A detailed description of these models lies beyond the scope of this chapter.

2.4. Power-controlled propagation of linear guided modes

The modification of optical properties, specifically the refractive index of a medium, in the presence of high-power optical fields has been well known since the early days of nonlinear optics (Chiao et al. 1964). In bulk optics, this leads to self-(de-)focussing effects (Kelly 1965) which may provoke a subtle transverse pattern formation when several beams are interacting (see, e.g., Moloney 1985, 1987). The introduction of a power-dependent refractive index material in appropriate optical resonator structures leads to a power-controlled phase shift between the interfering waves which may result in a multistable response characteristic (Flytzanis 1984). Such bistable schemes have far-reaching implications in all-optical switching and logic (Smith and Tomlinson 1984).

The consideration of power-dependent refractive index materials in integrated optical schemes began only recently but with increasing interest. At the present time the full range of associated phenomena, observed experimentally or simulated numerically, is not completely understood. In this subsection, we intend to illustrate basic phenomena of guided mode propagation in the presence of intensity-dependent nonlinear materials.

To this end we consider the steady-state propagation of a TE-polarised guided wave field in a planar symmetric configuration characterized by the refractive index profile $n(z)$ where $n^2(z) = \varepsilon'(z)$. A homogeneously distributed Kerr-law nonlinearity is assumed, giving rise to a polarisation field according to eq. (2.14) with a^{NL} now a constant (Stegeman et al. 1985a). Starting with the propagation of a single mode,

$$E(x, z) = a^{(\mu)}(x)\, E^{(\mu)}(z)\, \mathrm{e}^{\mathrm{i}k_0 n^{(\mu)} x}. \tag{2.15}$$

we are specifically interested in the interaction of the lowest order symmetric and antisymmetric guided modes supported by our configuration,

$$E(x, z) = \sum_{\mu=0,1} a^{(\mu)}(x)\, E^{(\mu)}(z)\, \mathrm{e}^{\mathrm{i}k_0 n^{(\mu)} x}. \tag{2.16}$$

This scheme covers two limiting cases, namely the directional coupler limit, where two parallel identical waveguides are interacting, and the single multimode film (see also insets to fig. 2.2).

2.4.1. Single-mode propagation

For the single-mode case, eq. (2.8) takes the form

$$[\mathrm{d}/\mathrm{d}x]\, a^{(\mu)}(x) = \mathrm{i}Q^{(\mu)} |a^{(\mu)}(x)|^2 a^{(\mu)}(x), \tag{2.17}$$

where

$$Q^{(\mu)} = \omega\varepsilon_0 a^{\mathrm{NL}}/(4N) \int (E^{(\mu)}(z))^4 \,\mathrm{d}z$$

is a real-valued constant. Equation (2.17) has the first integral $|a^{(\mu)}(x)|^2 = |a^{(\mu)}(0)|^2 = J^{(\mu)}/N = \text{const.}$, stating the conversation of the integral power flow $J^{(\mu)}$. The solution to eq. (2.17),

$$a^{(\mu)}(x) = a^{(\mu)}(0)\, \mathrm{e}^{\mathrm{i}k_0 \Delta n^{(\mu)} x}, \tag{2.18}$$

displays a power-dependent correction to the propagation constant

$$k_0 \Delta n^{(\mu)} = Q^{(\mu)} |a^{(\mu)}(0)|^2 = Q^{(\mu)} J^{(\mu)}/N \tag{2.19}$$

(Stegeman 1982b).

The correction is of importance when the excitation of single guided modes through distributed couplers, i.e., prisms or gratings, is to be described (see section 2.6). Note, however, that the weakly nonlinear regime implies guided mode powers, ensuring

$$\Delta n^{(\mu)} \ll \max_z [n(z)] - \min_z [n(z)]. \tag{2.20}$$

For larger powers the transverse field profile of the guided wave will be seriously affected.

The power-controlled shift of the propagation constant is the only essential effect on a single guided mode provoked by a Kerr-nonlinearity. In multi-mode guides, the excitation of a single mode may generate some other guided modes by the nonlinear polarization field. In many cases the power transfer to this mode is not significant because of phase mismatch reasons or small coupling constants (Stegeman 1982b).

The consideration of absorption effects on guided mode propagation is usually performed in the linear limit, by adding an imaginary part $\varepsilon''(z)$ to the dielectric function. This 'perturbation' can be included into the polarisation term of eq. (2.8) giving rise to an attenuation coefficient

$$\alpha^{(\mu)} = \varepsilon_0 c/(4N) \int_{-\infty}^{\infty} \varepsilon''(z)(E^{(\mu)}(z))^2 \, dz, \tag{2.21}$$

that can be included into a complex valued propagation constant $\tilde{n}^{(\mu)} = n^{(\mu)} + i\alpha^{(\mu)}$ (Marcuse 1974, Stegeman 1982b). The net absorption is determined by the mutual 'overlap' of both field- and $\varepsilon''(z)$-profile.

The effect of saturable absorption has been discussed by Langbein et al. (1984c), in the framework of a two-level-system (TLS) model. Particularly, the reversed situation, i.e., the amplification of the guided mode resonantly interacting with inverted TLSs was of interest (Langbein et al. 1984c). The relevant coupled-mode equation for the power flow $J(x) = N|a^{(0)}(x)|^2$ of a TE_0-mode is here

$$[d/dx]J(x) + \alpha J(x) = KJ(x) \int_{-\infty}^{\infty} dz \frac{D_0(E^{(0)}(z))^2}{1 + 8T_1 KJ(x)(E^{(0)}(z))^2/\hbar\omega}, \tag{2.22}$$

where K contains characteristic parameters of the TLS, T_1 is the energy relaxation time, $\alpha \equiv 2k_0^{\alpha(0)}$, and D_0 is the initial density of the excited TLS. $D_0 > 0$ permits amplification, $D_0 < 0$ saturable absorption. The competition between linear absorption and saturable amplification leads to the introduction of a net gain coefficient, g, that is a function of the configuration parameters α and D_0. $g > 0$ implies amplification (see fig. 2.1).

Both α and g determine also the maximum attainable (normalized) power flow $\tilde{J}_s$,

$$\ln(1 + \tilde{J}_s) = \tilde{J}_s\alpha/(g + \alpha), \tag{2.23}$$

where all TLS have been depleted. Since $\tilde{J}_s$ is independent of the input guided mode power, the amplification ratio is larger for small input signals than for larger ones (see fig. 2.1).

The shortening of a guided wave pulse due the resonant interaction with an array of TSLs has been considered by Langbein et al. (1984b). The TLSs extract energy from the leading edge of the pulse thus acquiring the excited state. For pulse-lengths much larger than the transverse TLS-relaxation time this extracted energy fraction gets lost by phase relaxation processes. While for TLS the maximum pulse shortening is $\frac{1}{2}$, realistic dye molecules representing multi-level systems should permit a more effective pulse shortening by resonant two- and multi-photon processes.

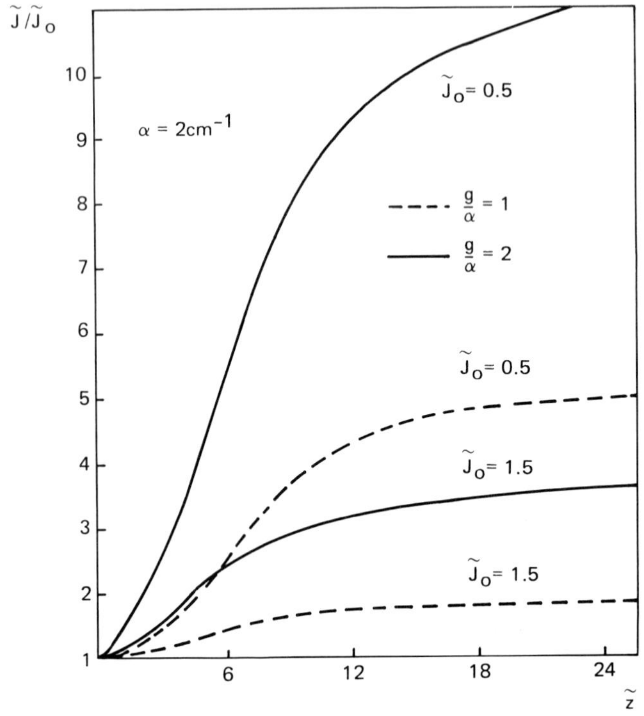

Fig. 2.1. Amplification of a single TE_0-mode by an externally pumped cladding for different initial (normalised) power flows J_0 and gain to intrinsic absorption ratios g/χ. (After Langbein et al. 1984a.)

Nonlinear optical switching and optical bistability due to induced absorption in a ZnSe single-crystal waveguide have been reported by Kim et al. (1987). The incident guided wave pulse leads to a local heating of the sample which in its turn causes a shift of the absorption edge to lower energies. Switching times of 10 μs and contrast ratios of 16/1 with power thresholds of the order of 1 mW were obtained with a sample length of 830 μm ($\lambda = 488$ nm). More subtle switching and bistability features have been observed in GaAlAs waveguide Fabry–Pérot resonator configurations (Aitchison et al. 1987, Walker et al. 1986). The reported effects are due to an interesting interplay of nonlinear cavity tuning and increasing absorption bistability.

2.4.2. Coherent interaction of two waveguide modes

Next we will discuss the steady-state nonlinear interaction of two normal modes according to eq. (2.16) thereby following essentially the instructive paper of Silberberg and Stegeman (1987). The emerging phenomena will result from the interplay of both power-controlled mode interference and power transfer effects between both modes.

According to eq. (2.16), the relevant nonlinear polarisation field consists of six terms. Their appearance in the corresponding coupled-mode equations (2.8) is determined by the relevant overlap integrals. In detail, one obtains

$$-\mathrm{i}[\mathrm{d}/\mathrm{d}x]a^{(0)} = C_0|a^{(0)}|^2 a^{(0)} + 2C_2|a^{(1)}|^2 a^{(0)} + C_2(a^{(1)})^2(a^{(0)})\mathrm{e}^{-2\mathrm{i}\Delta\beta x}, \tag{2.24}$$

$$-\mathrm{i}[\mathrm{d}/\mathrm{d}x]a^{(1)} = C_1|a^{(1)}|^2 a^{(1)} + 2C_2|a^{(0)}|^2 a^{(1)} + C_2(a^{(0)})^2(a^{(1)})\mathrm{e}^{2\mathrm{i}\Delta\beta x}, \tag{2.25}$$

where

$$\Delta\beta = k_0 n^{(0)} - k_0 n^{(1)}.$$

The real-valued constants

$$C_\mu = \omega\varepsilon_0 a^{\mathrm{NL}}/(4N)\int_{-\infty}^{\infty}(E^{(\mu)}(z))^4\,\mathrm{d}z, \quad \mu = 0, 1,$$

$$C_2 = \omega\varepsilon_0 a^{\mathrm{NL}}/(4N)\int_{-\infty}^{\infty}(E^{(0)}(z))^2(E^{(1)}(z))^2\,\mathrm{d}z, \tag{2.26}$$

are sensitive to the mode shapes. They coincide only in the weakly coupled directional coupler limit, but can be quite different for a single waveguide (negative waveguide separation, see fig. 2.2).

The complex mode amplitudes $a^{(\mu)} = (J^{(\mu)})^{1/2}\mathrm{e}^{\mathrm{i}\phi(\mu)}$, $\mu = 1, 2$, are controlled by two integrals of eqs (2.18), namely

$$J(x) = J^{(0)} + J^{(1)} = J(0) \equiv J, \tag{2.27a}$$

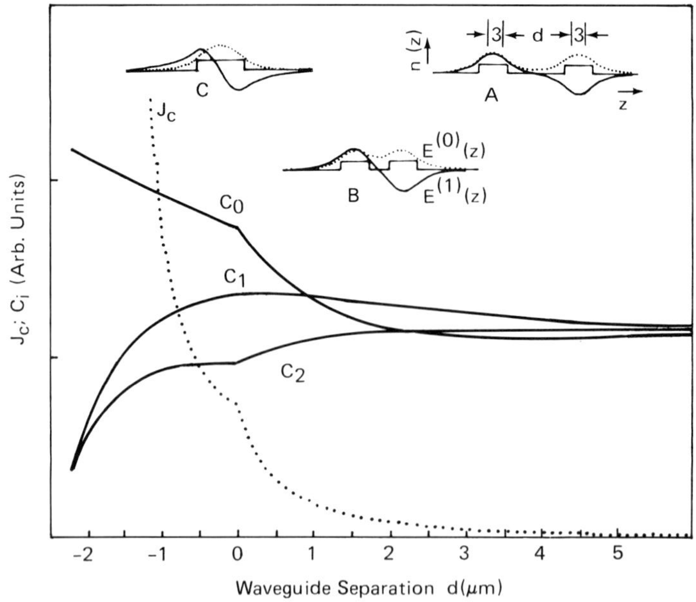

Fig. 2.2. Overlap integrals C_0, C_1 and C_2 in a slab structure composed of two equal guides, as a function of the separation between both guides. The insets illustrate the different waveguide structures and their lowest modes $E^0(z)$, $E^1(z)$: (A) directional coupler configuration, (C) single multimode film. The dotted line shows the critical power J_0 for the same structure. (After Silberberg and Stegeman 1987.)

and, with $V(x) = J^{(0)} - J(1)$, $\phi(x) = \phi^{(0)} - \phi^{(1)} + \Delta\beta x$,

$$G(x) = C_2(J^2 - V^2)\cos(2\phi) + [\tfrac{1}{2}(C_0 + C_1) - 2C_2]V^2 + [2\Delta\beta + (C_0 - C_1)J]V = G(0) \equiv G. \tag{2.27b}$$

From eqs (2.24), these integrals permit a single equation for V to be derived which is solvable in terms of an elliptic integral (Silberberg and Stegeman 1987).

The case of equal excitation of the two modes at $z = 0$, i.e., $V = 0$ and $\phi = 0$ will be discussed in more detail. For very small input powers the guide runs in the linear regime. In this case the interference of both modes leads to the familiar oscillational motion of the collective field maximum down the guide. For the coherent coupler this implies a periodic power transfer between both guides (Marcuse 1974). The coupling length is $\pi/\Delta\beta$.

For increasing input powers a periodic power exchange between the two modes sets in (see fig. 2.3). Simultaneously, a nonlinear phase shift between both modes appears for $C_0 \neq C_1$, i.e., essentially for the single waveguide case. Increasing power pulls the phases together, increases the power transfer, and lengthens the oscillation period (fig. 2.3).

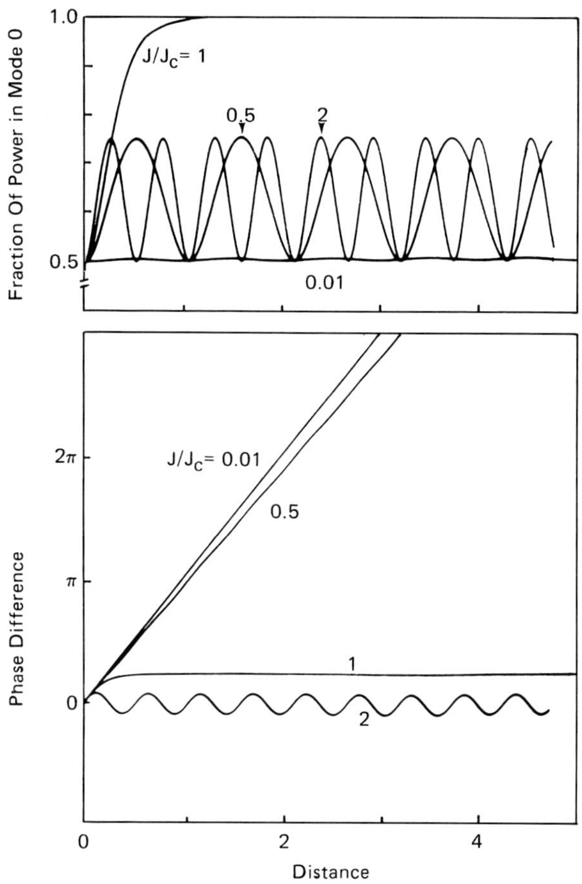

Fig. 2.3. Fraction of power guided in the TE_0-mode and the phase difference between the TE_0- and the TE_1-mode as a function of distance along the guide, for the directional coupler limit $C_0 = C_1 = C_2$. Distance is shown in units of coupling length. (After Silberberg and Stegeman 1987.)

There is a critical power J_c given for $2C_2 - C_0 - C_1 \gg C_2$ by

$$J_0 = 2\Delta\beta\{2C_2[3C_2 - \tfrac{1}{2}(C_0 + C_1)] - (C_0 - C_1)^2)^{-1/2}, \tag{2.28}$$

at which one mode grows asymptotically. The phase difference approaches $\frac{1}{4}\pi$ and the power is gradually transferred to the symmetric (antisymmetric) mode for a self-focussing (self-defocussing) nonlinearity (see fig. 2.3).

For $J > J_0$ the strong coupling produces a periodic but incomplete power transfer between both modes, while the phase difference remains nearly fixed (see fig. 2.3). In the coherent coupler limit, this leads effectively to a decoupling of both parallel guides (Jensen 1982).

Note, that generally both the initial power distribution and relative phase between the modes enter the integration constants (2.27a, b) and hence determine the detailed coupling process.

The transition between oscillatory and non-oscillatory behaviour at J_0 has been identified as the source of instabilities in the coherent coupler limit (Wabnitz et al. 1986). Based on these characteristics, several new schemes for all-optical signal processing have been proposed, such as a 'bifurcation optique active switch' (BOA) and a nonlinear phase shifter (Silberberg and Stegeman 1987). Other applications, relying on coherent coupler schemes, will be discussed in the next subsection.

The central section of a BOA switch consists of a nonlinear two-mode guide complemented by two cleaved off-branching output ports. The input ports are attached similarly at the opposite endface. Light incident on one input port excites equally the two guided modes in the central region. The distribution of the light in the output ports depends on the phase difference between the two modes at the output branching, which can be controlled by changing the input intensity (see also fig. 2.4a).

The validity of the coupled-mode calculus under conditions approaching or exceeding the critical power J_0 has been addressed by Finlayson et al. (1987). They solved the full wave equation [eq. (2.6)] numerically imposing only a paraxial approximation. Assuming again equal excitation of the two modes at the entrance face of the guide, they were able to simulate the resulting field evolution down the waveguide. The numerical technique, called 'beam propagation method' (BPM) is explained later in this chapter. Specifically it can allow for variations in the field shape that may go beyond the limits of the weakly nonlinear regime.

At low input powers ($J = \frac{1}{16}J_0$) the general features of the intensity pattern are in general accordance with the coupled-mode predictions. However, the mode coupling is not periodic, even at relatively low input powers (see fig. 2.4a).

At the coupled-mode value of the critical power, the numerically generated field of fig. 2.4b displays the gradual onset of the depletion of the asymmetric mode, but the phase curvature introduced by self-focussing causes the field to sharpen and broaden alternatively, a feature which is not predicted by the coupled-mode approach.

Figure 2.4c indicates that the formation of some TE_0-like, or still symmetric, field pattern increases more slowly than predicted. The difference between the percentage of power contained in all symmetric field components and the TE_0 overlap curve indicate that the TE_0-field shape has changed so that the coupled mode approach is no longer valid. This conclusion corresponds with the result that numerically the existence of a well-defined power instability could not be found for this configuration.

In summary, the coupled-mode approach proved to be a useful analytical model for describing, at least qualitatively, nonlinear guided mode coupling and interference effects at powers in the range of the order of the coupled-mode critical power J_0 and below. Device concepts that are based on the coupled-

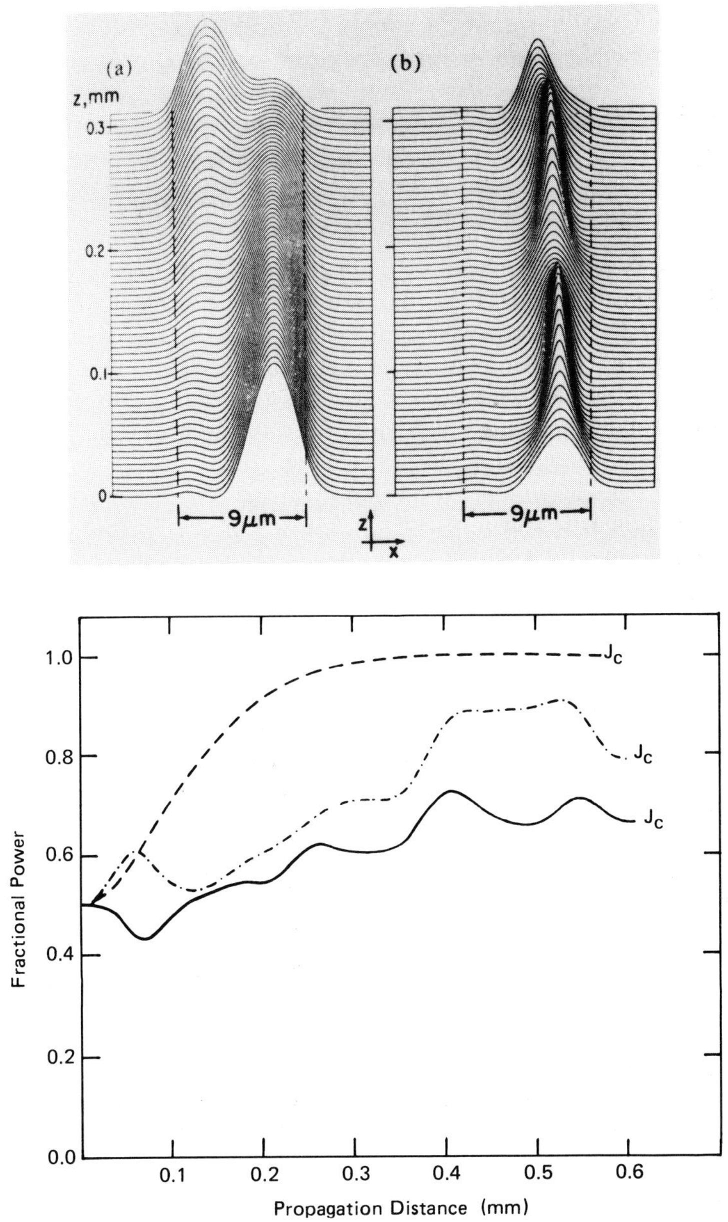

Fig. 2.4. Evolution of the intensity profiles in a symmetric single slab waveguide, capable of supporting two TE-modes in the linear regime, as a function of propagation distance: (a) with initial power $J = \frac{1}{16}J_0$; (b) with $J = J_0$. (c) Fractional power in the TE_0-mode as a function of propagation distance with initial power $J = J_0$. The broken line is the coupled-mode prediction, the solid line is the BPM result. The fractional power in all symmetric components of the field as the dot-dashed curve. (After Finlayson et al. 1987.)

mode power instability around J_0 deserve additional consideration by alternate models associated with the strong nonlinear regime.

2.5. *Nonlinear directional couplers*

2.5.1. *Operational characteristics and trajectories*

Many device concepts using nonlinear mode coupling rely on the directional coupler configuration. For that reason both the relevant features and related device concepts will be detailed for this particular configuration.

The directional coupler (DC) consists of two mono-mode waveguides placed in such proximity that the evanescent field tails of the guided modes of each waveguide overlap. In contrast to the preceding section, where this configuration has been regarded as a *single* symmetric three-slab guide, it is sometimes convenient to interpret the total field evolution in terms of the (TE_0-) modes $E(z)$, $E'(z)$ of *two* identical single waveguides. This is useful as long as both guides can be regarded as a perturbation which does not significantly distort the mode of the primary guide.

The coupled-mode equations

$$-\mathrm{i}[\mathrm{d}/\mathrm{d}x]a = Q_1 a + Q_2 a' + (Q_3|a|^2 + 2Q_4|a'|^2)a,$$
$$-\mathrm{i}[\mathrm{d}/\mathrm{d}x]a' = Q_1 a' + Q_2 a + (Q_3|a'|^2 + 2Q_4|a|^2)a', \tag{2.29}$$

have a slightly different structure compared with eq. (2.24) (Jensen 1982). $a(x)$ and $a'(x)$ are now the mode amplitudes in both guides, while the overlap integrals have to be correspondingly redefined. The linear coupling terms containing the coefficients Q_1 and Q_2 replace the phase-mismatch term in eq. (2.24). The coefficients Q_i, $i = 1$–4, are

$$Q_1 = \frac{\omega\varepsilon_0}{4N}\int_{\infty}^{\infty} \mathrm{d}z\ \varepsilon_\delta(z)(E)^2, \qquad Q_2 = \frac{\omega\varepsilon_0}{4N}\int_{\infty}^{\infty} \mathrm{d}z\ [\varepsilon_s(z) + \varepsilon_\delta(z)]E(z)E'(z),$$
$$Q_3 = \frac{\omega\varepsilon_0 a^{\mathrm{NL}}}{4N}\int_{-\infty}^{\infty} \mathrm{d}z\ |E(x)|^4, \qquad Q_4 = \frac{\omega\varepsilon_0 a^{\mathrm{NL}}}{4N}\int_{-\infty}^{\infty} \mathrm{d}z\ |E(z)|^2|E'(z)|^2, \tag{2.30}$$

where $\varepsilon_s(z)$ is the unperturbed dielectric coefficient of one guide, and $\varepsilon_\delta(z)$ is the linear perturbing contribution to that coefficient. Obviously, $\varepsilon'(z) = \varepsilon_s(z) + \varepsilon_\delta(z)$ holds with $\varepsilon'(z)$ being the dielectric profile of the global guide of the preceding section.

In the linear limit Q_2 now controls the periodic power transfer between both guides which was formerly a pure interference phenomenon. $\pi/2Q_2$ is the linear power transfer length. The nonlinear terms containing Q_3 describe self-phase modulation. The nonlinear coupling term (containing Q_4) is omitted in some models (Maier 1982, 1984).

Diano et al. (1985), derived an elegant solution to eq. (2.29) in terms of the generalized Stokes parameters,

$$S_{0/1} = |a|^2 \pm |a'|^2, \qquad S_2 = a[a']^* + a^* a', \qquad S_3 = \mathrm{i}(a^* a' - a[a']^*). \quad (2.31)$$

The corresponding evolution equations admit the invariants

$$R^2 = S_0^2 = S_1^2 + S_2^2 + S_3^2, \quad (2.32a)$$

$$\Gamma = S_2 + [(Q_3 - 2Q_4)/4Q_2]S_1^2, \quad (2.32b)$$

which are determined by the initial conditions. The invariants of eqs (2.32a, b) suggest a description of the field evolution in the three-dimensional space of S_1, S_2, S_3. R and r determine a set of closed curves, which are the trajectories of the spherical surfaces S^R with the parabolic cylinder C^r (see fig. 2.5) where a fixed R and $(Q_3 - 2Q_4)/Q_2 > 0$ has been assumed without loss of generality.

On S^R there are either two or four singular points: $F_1 = \{0, -R, 0\}$ which is always stable (see fig. 2.5a), and $F_2 = \{0, R, 0\}$ which is only stable for

$$R < 2Q_2/(Q_3 - 2Q_4), \quad (2.33)$$

(see fig. 2.5b). F_1 and F_2 describe a symmetric excitation of both modes travelling with equal amplitudes, but with the phase difference π and 0, respectively. Both excitations correspond to the lowest order normal modes of the (linear) global guide [eq. (2.16)].

Whenever relation (2.33) is not satisfied, the system of eq. (2.29) has a bifurcation, in the sense that F_2 is an unstable saddle point and additionally two

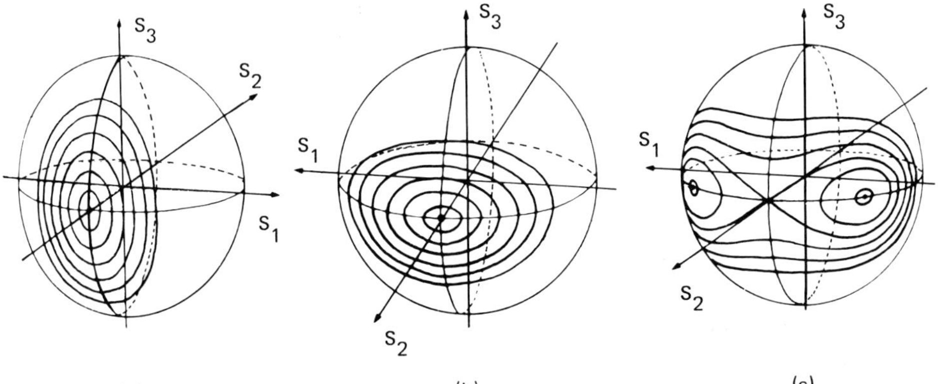

Fig. 2.5. Trajectories on the sphere S characterising different operational DC-regimes in terms of the Stokes parameters S_1, S_2 and S_3 (see text). (a) Orbits round the stable centre F_1. (b) Orbits round the stable centre F_2 [condition (2.36) holds]. (c) Characteristic orbits when eq. (2.36) does not hold. The orbit F_2 is the separatrix. (After Diano et al. 1985.)

stable critical points (centers) occur

$$F_{3,4} = \{\pm(R^2 - 4Q_2^2/(Q_3 - 2Q_4)^2)^{1/2}, 2Q_2/(Q_3 - 2Q_4), 0\},$$

(see fig. 2.5c).

There is a particular trajectory determined by eqs (2.32a, b) with

$$\Gamma = R \equiv J_c/N. \tag{2.34a}$$

This trajectory passes through F_2 and is called the separatrix. It determines three domains of different periodicity on the sphere, one for each stable singular point (Diano et al. 1985).

The separatrix describes the monotonic evolution into the (unstable) mixed state (equal power distribution) which corresponds to the depletion of the asymmetric global mode in the global guide picture discussed in the previous subsection. Physically, this particular evolution reflects an exact balance between the power-dependent mismatch, introduced by the nonlinear index change, and the linear coupling between both guides.

Any deviation from the separatrix leads to an abrupt change of the coupler behaviour either towards a periodic symmetric power exchange between both channels or towards some short period oscillations around a state of asymmetric power distribution between both channels.

Figures 2.5b and 2.5c give a good survey on the different operational regimes which can be understood in terms of the stability analysis summarized above. A definite regime will be selected by the initial conditions which are sensitive to both the power distribution and phase difference between the modes in the two channels.

Figure 2.6 displays the response curve for DCs of different length under initial conditions where all of the power is initially launched into one guide, i.e., $N|a(0)|^2 = J_0$, $|a'(0)|^2 = 0$. These initial conditions imply

$$J_c = 4NQ_2/(Q_3 - 2Q_4). \tag{2.34b}$$

The analytical solution to eq. (2.29), relevant to fig. 2.6, is

$$N|a(z)|^2 = J(z) = \tfrac{1}{2}J_0[1 + \mathrm{cn}(2Q_2 z|m)], \tag{2.35}$$

where $J(z)$ is the normalized straight-through output power and $\mathrm{cn}(\phi|m)$ is a Jacobi elliptic function with the modulus $m = J_0^2/J_c^2$ (Abramowitz and Stegun 1965). J_c is the critical incident power where the separatrix is crossed. In the limit of small input intensities $m \simeq 0$, and eq. (2.35) becomes

$$J(z) = \tfrac{1}{2}J_0[1 + \cos(2Q_2 z)]. \tag{2.36}$$

This is the well-known solution for a linear DC.

For the sake of brevity, we do not add here more general analytical solutions to eq. (2.29), but refer the interested reader to the original papers, e.g., Jensen (1982), Maier (1984), Diano et al. (1985), Trillo and Wabnitz (1986).

2.5.2. Device concepts and experiments

The different power- and phase-controlled features of nonlinear DCs suggest a varicty of possible applications. Table 2.1 gives a survey on some recently proposed schemes. The most common schemes rely on the response characteristic according to fig. 2.6 where only one input port is excited and the input *power* controls the branching ratio of the output power between both arms. In particular the steep slope of the response curve near the critical power J is well suited for optical switching, logic gates (Jensen 1982), power discrimination and demultiplexing (Stegeman et al. 1986a), and modulation amplification (Maier 1982, 1984, Trillo and Wabnitz 1986, Gibbons and Sarid 1987), respectively. The latter authors modified the primary scheme by feeding a holding beam with $J_0 \simeq J_c$ into one arm and the small input signal $J'_0 \ll J_c$ into the other, thus exploiting the response discontinuity near J_c.

Maier (1982) introduces different feedback loops into the DC-scheme in order to establish *bistable* response curves that are of interest for switches, logic and memory applications, Hoffe and Chrostowski (1986b) use an optical fiber scheme in order to realize a feedback loop. The resulting configuration forms an optical ring resonator coupled to an adjacent waveguide by a suitable DC.

Lattes et al. (1983) and Haus et al. (1984, 1985) implement the nonlinear DC into a Mach–Zehnder interferometer where some bias phase shift between both arms can be tuned electro-optically. In this way, different logic gates can be realized.

There is a recent suggestion to control the DC switching between the bar state and the mixed state by the *phase shift* of some small auxiliary signal sent into the second arm (Wabnitz et al. 1986). In this way also a crossing of the separatrix can be achieved, as discussed in the previous subsection.

Optical pulse shaping, particularly pulse compression, has been discussed by several authors (Kitayama and Wang 1983, Gusovskii et al. 1987, Maier et al. 1987, Hoffe and Chrostowski 1986a). In terms of the DC-response characteristic discussed above, the shaping mechanism manifests itselves as a partial power transfer under the pulse profile.

The DC length is adjusted such that both the leading and trailing pulse edge, revealing a low power, are coupled into the other DC-arm while the central light power part of the pulse appears as the straight-through output. Also, under suitable conditions, pulse breaking into several fractions should be feasible (Hoffe and Chrostowski 1986a).

To date, there are many more device concepts than successful experimental verifications. A survey on some recent experimental efforts is given in table 2.2. In only three cases has power-induced switching been observed (Li Kam Wa 1985a, b, Gusovskii et al. 1985, Maier et al. 1987), where the last two experiments refer to fiber optic implementations. Specifically, by means of a double core glass-fiber coupler, an interaction length of up to 1 m was achieved. In this way, the extremely small but very fast nonlinearity of the glass waveguide could be compensated, see table 2.2.

Table 2.1
Proposed nonlinear directional coupler schemes.

Reference	Scheme	Projected purpose	Control parameter	Comments
Jensen (1982); Stegeman (1986a)	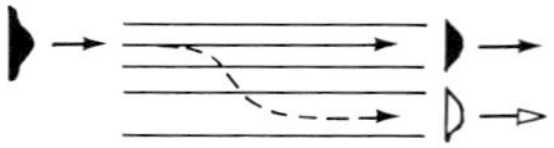	Power-controlled logic gate, optical switch, demultiplexer	Input power Input wave-length	Intensity and wave-length multiplexing (λ-dependent non-linearity)
Maier (1982, 1984)	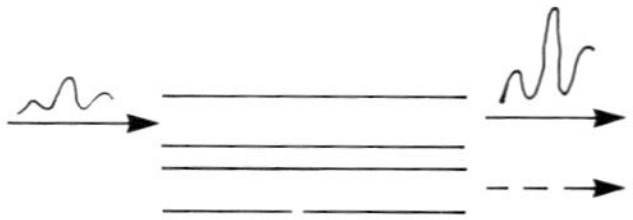	Optical transistor	Input power	Predicted amplifica-tion ratio: $\simeq 500$
Trillo (1986); Gibbons (1987)	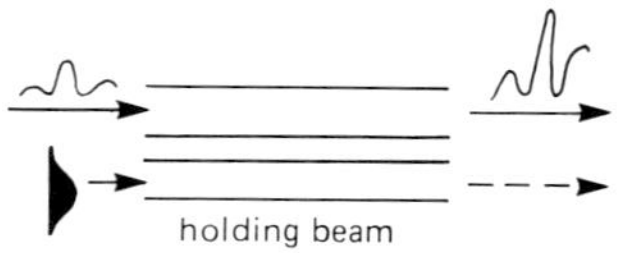	Modulation amplifier (optical transistor)	Input power in both channels	
Maier (1982); Hoffe (1986b)	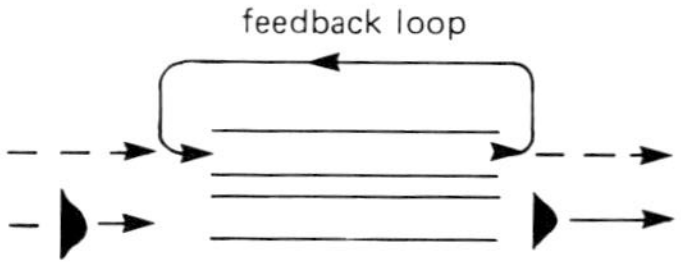	Bistable switching, logic	Input power, input phase	Different feedback loops discussed, fiber optic ring resonator
Wabnitz et al. (1986)	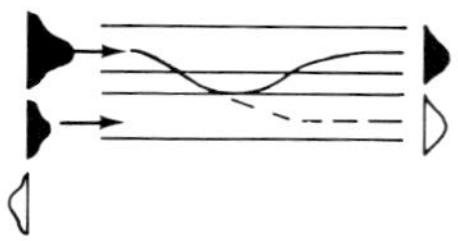	Phase-induced switching	Initial phase-shift between both guides	Also self-defocussing discussed

Lattes (1983); Haus (1983, 1985)	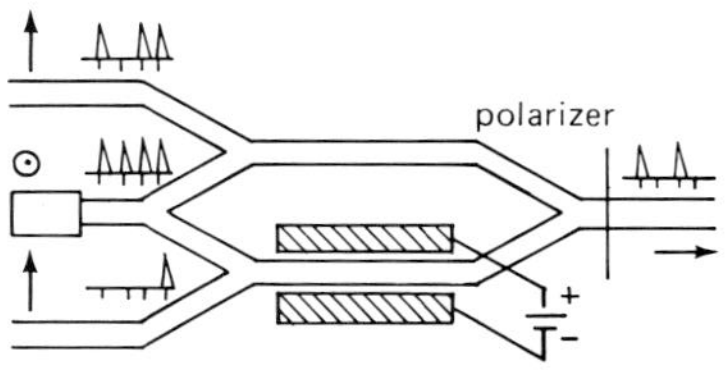	Logic gate	Input power	Optically modulated Mach–Zehnder waveguide interferometer
Kitayama (1983); Gusovskii (1987); Maier (1987); Hoffe (1986a)	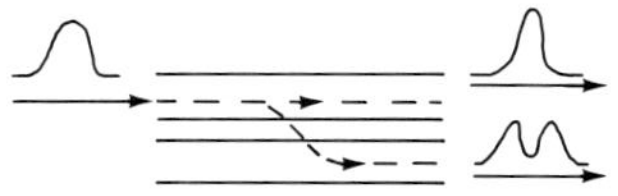	Pulse shaping	Input power	Fiber optic implementation, successful experiments
Silberberg (1987)	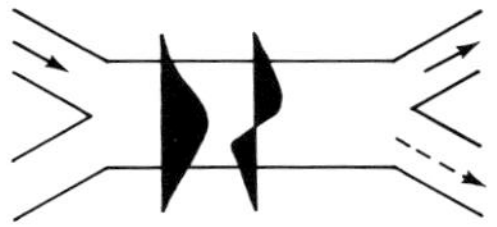	Bifurcation optic active coupler, power-dependent branching, switching	Input power	
	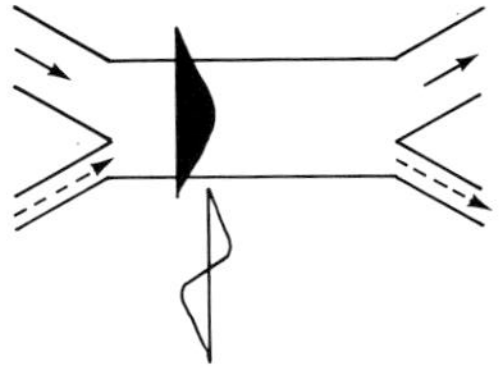	Power controlled phase shift between both arms		

Table 2.2

Experimental efforts towards nonlinear directional couplers.

Reference	Scheme	Waveguide material	n_2* [m²/W]	NL-medium	λ [μm]	Device length [mm]	Switch power [W]	Input (peak) power [W]	Pulse length [ps]	Comments
Lattes (1983)	Mach–Zehnder interferometer	$LiNbO_3$ (channel guides)	3×10^{-19}	$LiNbO_3$	0.85	20	150	2	5	Only pulse modulation, no switching
Haus (1985)	Mach–Zehnder interferometer	GaAlAs/GaAs-channel guides	5×10^{-17}	GaAs	1.06	5		No successful operation		Too large loss, slow opto-thermal index change, side effects
		InP-rib		InP	1.06	5				
Li Kam Wa (1985a, b)	DC-switch	GaAlAs-MQW stripes (strain-induced)	10^{-11}	GaAlAs-MQW (self-defocussing)	0.85	1.7	5×10^{-4}	10^{-3}	10^5	Cooling (180 K), resonant NL, uncomplete switching (large absorption)
Ironside (1985) Cullen (1986)	DC-switch	K^+/Na^+ ion-exchanged stripes in CdS_xSe_{1-x}-doped glass	10^{-14}	CdS_xSe_{1-x}-doped glass	0.61	1.13	5.4			Fast relaxation (16 ps), switching under investigation
Gusovskii (1985)	DC-switch	Double-core glass fiber	10^{-20}	SiO_2	0.53	180	95		10^5	Power controlled branching observed
Maier (1987)	DC-switch	Double-core glass fiber	10^{-20}	SiO_2	0.53	1000	$\simeq 60$		25–50	Fast relaxation (10 fs), full switching, pulse compression observed

$*a^{NL} = \varepsilon\alpha\varepsilon' c\eta_2$.

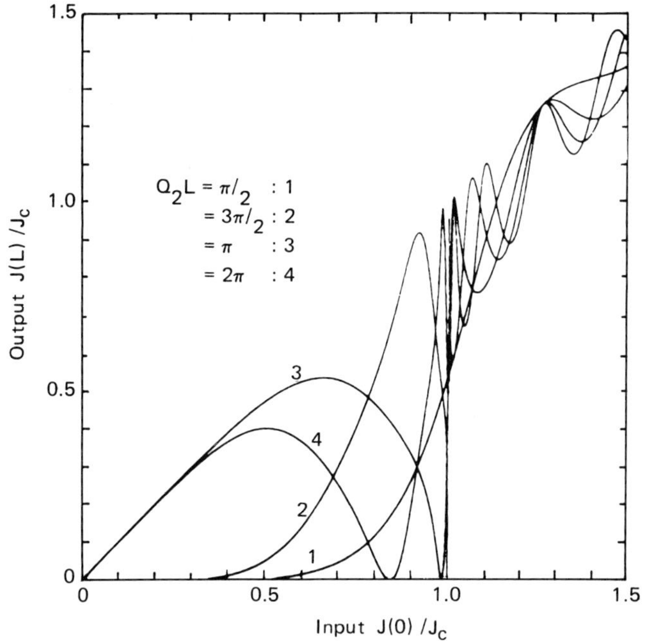

Fig. 2.6. Input/output characteristic for DCs of different length L. Curves 1 and 2 correspond to DCs that are in a crossed state at low input power $J(0)$, curves 3 and 4 illustrate a bar-state under the same conditions. (After Jensen 1982.)

A salient point responsible for the slow experimental progress in recent years is the lack of nonlinear materials with a sufficiently large nonlinear refractive index. Since in integrated optical geometries the available interaction length hardly exceeds a few mm, one is forced to drive the available nonlinear materials under conditions where their nonlinear properties deviate a significant amount from the Kerr-law. Depending on the concrete mechanism, such deviations may cover refractive index saturation as well as absorption which may be intensity-dependent, at least under near-resonance conditions. Both effects counteract nonlinear switching.

Another critical point is that the existing device concepts are based on stationary arguments, i.e., imply an instantaneous material response. Realistic nonlinear media reveal finite response times and the resulting memory effect has to be taken into account, particularly when the final pulse shape is of interest. In many cases, different competing nonlinear mechanisms have to be considered.

Eventually, we must recall that the coupled-mode approach used so far loses ground when the input power approaches the critical power of the device. The arising mode shape distortions may deteriorate the coupled-mode switching characteristic considerably, as has been considered in subsection 2.4.2. Also, the assumed strict symmetry between both waveguides, implying an exact synchronous linear coupling, is difficult to achieve in practice.

One convenient way to check the validity of the coupled-mode predictions based on eqs (2.29) and (2.32) is the numerical simulation of the field evolution using the BPM-method. This has been done by Wabnitz et al. (1986), in order to test the phase-induced switching mode of operation mentioned in subsection 2.5.1. It turned out that, for a positive Kerr-nonlinearity, the change in the field distribution due to self-focussing lowers the critical power with respect to the estimate furnished by the coupled-mode analysis. The related instability picture changes only quantitatively. Analoguous behaviour has been found for the self-defocussing case.

Love et al. (1986) studied numerically the power-induced branching of a signal launched into one input arm of the DC. The authors found good agreement with the coupled-mode results of Jensen (1982).

2.5.3. Generalized directional coupler models and transient effects

A look at the current experimental situation reveals that theoretical models based on the Kerr-nonlinearity are, in several respects, too poor to meet the actual experimental conditions. There are various initiatives towards generalized DC-modelling, i.e., to take important material properties, such as refractive index saturation (Stegeman et al. 1987a, c, Caglioti et al. 1987), intensity-dependent absorption (Stegeman et al. 1987b, c, Gibbons and Sarid 1987), into account.

The most general models considered so far can be described by

$$\begin{aligned} -\mathrm{i}[\partial/\partial x + v_0^{-1}\partial/\partial t]a(x,t) &= Q_1 a(x,t) + Q_2 a'(x,t)\mathrm{e}^{\mathrm{i}2\delta x} \\ &\quad + k_0(\Delta n(x,t) + \mathrm{i}\alpha(x,t))a(x,t), \\ -\mathrm{i}[\partial/\partial x + v_0^{-1}\partial/\partial t]a'(x,t) &= Q_1 a'(x,t) + Q_2 a(x,t)\mathrm{e}^{-\mathrm{i}2\delta x} \\ &\quad + k_0(\Delta n'(x,t) + \mathrm{i}\alpha'(x,t))a'(x,t), \end{aligned} \tag{2.37}$$

v_0 is the common group velocity of both guided modes [eq. (2.11b)]. The mode amplitudes $a(x, t)$, $a'(x, t)$ are now assumed slowly varying with respect to both space and time [eq. (2.9)]. In this approach, any dispersion of the linear coupling terms has been neglected and the nonlinear cross-coupling terms have been suppressed.

$\Delta n(x, t)$ and $\alpha(x, t)$ designate the nonlinear refractive index shift and nonlinear absorption respectively. Both terms may be subject to some dynamic equations determined by the specific material response, e.g., two-level-system (Stegeman et al. 1987c, d), charge-carrier-density-dependent refractive index shift and absorption (Gibbons and Sarid 1987), or Debye relaxation model (Stegeman et al. 1987b).

Also a deviation from the underlying configuration symmetry is allowed which leads to both dissimilar coupling coefficients and the additional phase-mismatch term $\exp(\pm i2\delta x)$. Note that the coordinate transform $\chi = x$, $\psi = x - v_0 t$ removes the second partial derivative in eqs (2.37) and their formal steady-state character is retained.

In what follows, we will give a survey on the predicted changes in the operational characteristic of a DC related to the modifications of the symmetric Kerr-model, i.e., to specific versions of eqs (2.37). At present, the complex consideration of different non-Kerr-like phenomena, resulting in important parameter trade-off estimations, is at an initial stage only.

The effect of a configuration asymmetry has been studied in a recent paper by Trillo and Wabnitz (1986). The authors analyze the coupled-mode equations in the following form

$$-i[d/dx]a = Q_2 a' e^{i2\delta x} + Q_3 |a|^2 a,$$
$$i[d/dx]a' = Q_2 a e^{-i2\delta x} + Q_3' |a'|^2 a', \tag{2.38}$$

which allows for an overall mismatch $\delta = k_0 n'^{(0)} - k_0 n^{(0)} > 0$ of the two propagation constants $n'^{(0)}$ and $n^{(0)}$ respectively. Now the modes in both guides $E^{(0)}(z)\exp[ik_0 n^{(0)} x]$ ('fast guide'), $E'^{(0)}(z)\exp[ik_0' n'^{(0)} x]$ ('slow guide') differ also with respect to their profiles $E(z)$, $E'(z)$. Together with a potential nonuniform material nonlinearity, this gives rise to non-equal coefficients Q_3, Q_3' in the self-phase modulation terms.

The outcome of the elegant analysis in terms of generalized Stokes parameters is that the linear mismatch, δ, breaks the pitchfork bifurcation picture which is characteristic for the symmetric configuration (see fig. 2.5). As a result, the DC becomes a highly nonreciprocal device. For input light into the slow guide, the nonlinear phase shift increases the primary mismatch. Therefore, this guide is virtually uncoupled at any powers.

Conversely, for input light into the fast guide the reasoning reverses and nearly complete switching is possible. There is also a critical input power (which is an increasing function of δ) where the power transfer drops to low values. This indicates the crossing of the separatrix also in this case.

Optical transistor operation according to scheme (c) in table 2.1 requires the pump beam to be launched into the fast guide while the weak signal is to be launched into the slow guide. In the case $Q_3 = -Q_3$, the device acts as a mismatched *linear* coupler (Trillo and Wabnitz 1986).

Another important step to meet the present experimental conditions was done by allowing for nonlinear index saturation. Assuming an exponential saturation law according to

$$\Delta n|a|^2 = \Delta n_{\mathrm{SAT}}[1 - e^{(-Q_3|a|^2/(k_0 \Delta n_{\mathrm{SAT}}))}], \tag{2.39}$$

valid both for a and a', and $\alpha = \alpha' = 0$, Caglioti et al. (1987) succeeded in solving eqs (2.37) analytically. They used the Hamiltonian associated to eqs (2.37) in order to reduce their solution through a quadrature and to perform a stability analysis as well. This analysis revealed that the switching characteristics can be excitingly modified in the presence of saturation as follows.

(i) There is a critical value of saturation $\Delta\tilde{n}_{SAT}$ below which effective switching cannot be achieved.

(ii) For $\Delta n_{SAT} > \Delta\tilde{n}_{SAT}$ there are always *two* discrete power levels at which switching occurs in a one beat length $L_B = \pi/Q_2$ long coupler.

The sharp '*on*'-switching of the bar-state, that occurs in the absence of saturation ($\Delta n_{SAT} \to \infty$) when the input power is just raised across the critical power $J_c = N\Delta Q_2/Q_3$ (see fig. 2.6), is now accompanied by a pronounced '*off*'-switching at a second (higher) power level, as shown in fig. 2.7a. Physically, this occurs because with saturation the unstable state, where the power-dependent mismatch exactly balances the linear coupling, is reached a second time at sufficiently high input powers.

Both power levels J_s where switching occurs have been plotted in fig. 2.7b for different saturation values Δn_{SAT}. The critical saturation Δn_{SAT} is determined by the instructive relation

$$\Delta n_{SAT} = 2455(2Q_2)/k_0 = 2455\lambda_0/L_B, \tag{2.40}$$

where the numerical factor is somewhat model-dependent (Caglioti et al. 1987). Although, according to eq. (2.40), Δn_{SAT} could be decreased by increasing the coupler length, absorption and scattering losses effectively limit this parameter.

The interplay of refractive index saturation and linear losses $\alpha(x, t) = \alpha^{(0)} = \alpha'^{(0)}$ has been checked by Stegeman et al. (1987a), having a specific material system, namely semiconductor-doped glass, in mind. The authors found that, except for a net loss in the total throughput due to attenuation, the input power level required for switching increases. Furthermore, the fraction of maximum output power switched decreases with increasing absorption. All-optical switching requires $\Delta n_{SAT} > \Delta\tilde{n}_{SAT}$ and $\alpha^0 L_B < 1$ as estimated by Stegeman et al. (1987a).

A two-level saturable absorptive medium has been studied by Stegeman et al. (1987c). It turned out that the critical saturation $\Delta\tilde{n}_{SAT}$ does not significantly depend upon the detuning of the two-level system. Conversely, the throughput is very sensitive to the detuning.

Very recently, Gibbons and Sarid (1987) combined the coupled-wave approach with a plasma model that describes the nonlinear response of certain bulk semiconductors such as GaAs, CdS_xSe_{1-x}-doped glasses, and CdS. The central physical parameter is the charge carrier density which generates the local changes of both refractive index and absorption coefficient. The authors gave only a very restricted account of the implications caused by their complex material model. For the considered choice of parameters, switching could be achieved only when the holding input power (appr. 12 mW) has been distributed

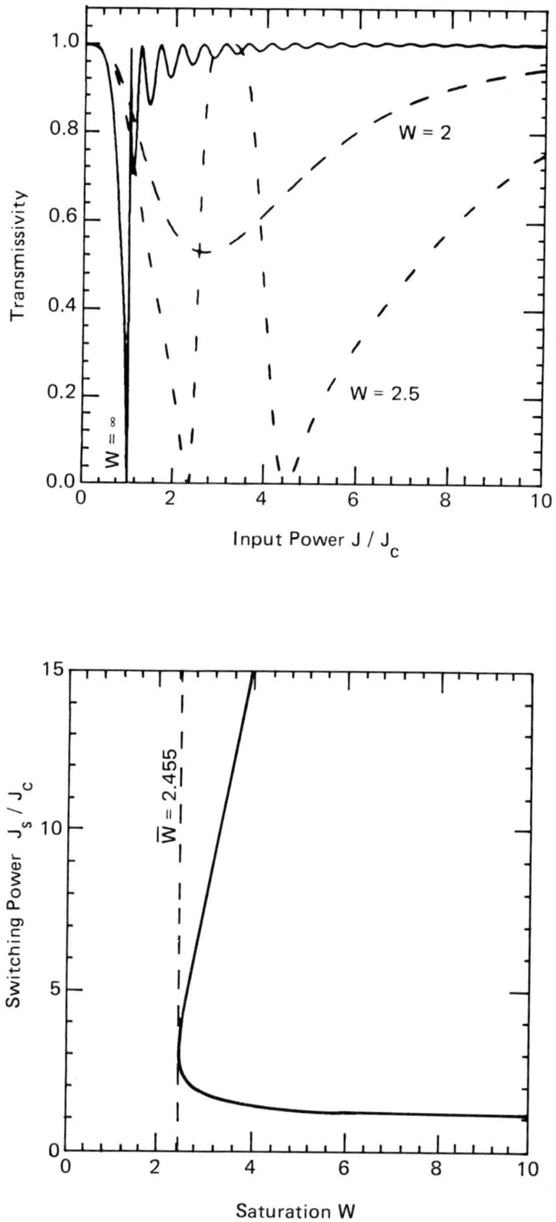

Fig. 2.7. Power-dependent switching of a DC in the presence of saturation. (a) Straight-through transmissivity vs. normalised power for a one beat length long DC when all the power J is launched into one guide for different values of the saturation parameter $L = \Delta n_{SAT} L_B/\lambda$. (b) Variation of the normalised power J/J_0 vs. saturation parameter ω. $\bar{\omega}$ is the critical saturation parameter value [eq. (2.43)]. (After Caglioti et al. 1987.)

nearly symmetrically into both DC channels. In this case, switching could be provoked by a 1 mW (additional) input into one guide. The device showed effectively a gain by a factor 4.

All theoretical models considered so far rely on media in which the nonlinearity responds instantaneously to the applied optical field. With regard to the available nonlinear media, this assumption is by no means self-evident, when just ultrafast all-optical switching is envisaged. Due to their travelling wave character, the switching time of nonlinear DCs is essentially determined by the relaxation times of the nonlinear materials involved. These times range from a few picoseconds to the order of milliseconds and will give rise to memory effects when the duration of the exciting light pulses becomes comparable or shorter than these relaxation times.

Stegeman et al. (1987b) took material memory into account by assuming

$$\Delta n(x, t) = Q_3/(k_0 \tau) \int_{-\infty}^{t} |a(x, t')|^2 \mathrm{e}^{-(t-t')/\tau} \, \mathrm{d}t', \tag{2.41}$$

for both amplitudes ($a' > \Delta n'(x, t)$). In the limit $\tau > 0$, the (instantaneous) Kerr-law nonlinearity is restored.

Specifically, the authors used eqs (2.37) and (2.41) including linear absorption, to examine the time response of a nonlinear DC in the limit $\tau \to \infty$, i.e., for optical pulses whose temporal width T_P is much shorter than τ. The material parameters correspond to semiconductor-doped glasses ($n_2/\tau = 10\ \mathrm{cm}^2/\mathrm{Ws}$). A train of successive pulses has been sent into one DC channel and the output pulse shape has been monitored after one (linear) coupling length $\frac{1}{2}L_\mathrm{B}$. For low pulse energies (10^{-2} Ws) the power transfer back onto the straight-through channel increases with increasing pulse number. This is because the index change accumulates in time at every point along the coupler. The pulse shapes remain symmetric, but show some slight retardation against one another. Higher pulse energies (10^{-11} Ws) produce large phase differences and lead to serious pulse shape distortions. The resulting asymmetric pulse shape is clearly due to the integrating nonlinearity and contrasts to the instantaneous case $T_\mathrm{P} \gg \tau$ (Kitayama and Wang 1983, Hoffe and Chrostowski 1986b).

The consideration of nonlinear index shift saturation in the form

$$\Delta \tilde{n}(x, t) = \Delta n_\mathrm{SAT}[1 - \mathrm{e}^{-\Delta n(x,t)/\Delta n_\mathrm{SAT}}], \tag{2.42}$$

with $\Delta n_\mathrm{SAT} = 3 \times 10^{-4}$ (Stegeman et al. 1987b), retained the general features. Some of the temporal pulse distortions got smoothed out. For smaller saturation values ($\Delta n_\mathrm{SAT} = 10^{-4}$) the nonlinear DC essentially behaves in a linear fashion.

Numerical calculations were made also for nonlinearities due to two-photon absorption having DCs in GaAs in mind. For 3×10^{-9} Ws-pulses ($T_\mathrm{P} = 30$ ps) the nonlinear attenuation proved sufficiently strong to produce an intensity minimum at the center of the pulse. The integrating nonlinearity, which changes the inter-channel transfer efficiency with time, causes asymmetric pulse shape distortions that evolve, for larger pulse energies, to a mutual pulse displacement. Pulse envelope modulation indicates a large nonlinear phase shift.

2.6. Nonlinear distributed couplers

2.6.1. Theoretical background

The excitation of a guided wave by an external radiation field is very frequently achieved by distributed couplers such as gratings or prisms. Both coupler types have many common properties. In what follows, we will refer to prism couplers since gratings are subject to a separate chapter of this volume.

A clear picture of the excitation process arises when the incident field is a plane wave. In this limiting case, the excitation process can be understood as a resonance phenomenon, i.e., the guided mode manifests itself as a resonance of the transfer function of the underlying geometry: substrate, film, gap, prism. Due to the assumed infinite lateral extent of the incident wave, the emerging total field represents an asymptotic state characterized by a completely balanced interplay between both the external radiation fields and the guided mode. Thus, the resulting field distribution parallel to the interfaces is completely homogeneous.

'Plane wave models' are of practical relevance for linear thin-film interferometry (see, e.g., Born and Wolf 1968), and attenuated total reflection (ATR) spectroscopy (see, e.g., the preceding volumes of this series).

'Plane wave models' have been developed also for thin-film and interface configurations where at least one of the involved media has an intensity-dependent refractive index (Reinisch et al. 1985, 1986, Montemayor and Deck 1985, Arlot et al. 1986, Hickernell and Sarid 1986, Gupta and Agarwal 1986b). The optically induced refractive index changes modify the field distribution *transverse* to the interfaces ('transverse' nonlinear effects), and control the resonance condition as well. Important predictions are an intensity-dependent detuning of the optimum coupling angle, and a distortion of the angular response characteristic that may lead to optical bistability with respect to both input intensity and angle of incidence. This is not surprising for the necessary feedback is implied in the resonant nature of the spatially stationary excitation process.

The practical relevance of the predictions based on plane wave arguments is still under discussion (Liao et al. 1984, Assanto et al. 1986). In practical situations, the finite lateral extent of any realistic incident radiation field breaks the assumed homogeneity inherent in plane wave models. Under these conditions the distributed cumulative nature of the excitation process comes into play, and field evolution effects *along* the interfaces arise ('longitudinal' effects).

For optically linear configurations the latter effects have been studied in the classical paper of Ulrich (1970), where an evolution equation for the guided mode amplitude under excitation has been derived. At least in a moderate power range this approach should be applicable also for configurations where some waveguide materials show an intensity-dependent refractive index. The resulting models (Chen and Carter 1983, Liao et al. 1984, 1985) disregard essentially a 'transverse' nonlinear effect, and belong to the coupled-mode approaches to be considered in this section.

In terms of the coupled-mode theory, the excitation process of a single waveguide mode can be understood as follows: the coupling efficiency of the prism coupler is essentially determined by the precision of the phase matching that is achieved between the incident radiation field and the excited guided wave. According to eqs (2.15), (2.18) and (2.19), the presence of an intensity-dependent refractive index gives rise to a guided wave propagation wave vector $n^{(\mu)}(J^{(\mu)}) = n^{(\mu)} + Q^{(\mu)}J^{(\mu)}/(Nk_0)$ that depends on the guided power flow $J^{(\mu)}$. Therefore, in contrast to the linear case, the phase-matching condition changes as the guided mode power $J^{(\mu)}$ grows. Hence, for a fixed angle of the input beam, the coupling efficiency becomes power dependent. This process can be described by the following evolution equation for the amplitude $a^{(\mu)}(x)$ of the guided mode under excitation (Liao et al. 1984, 1985)

$$[\mathrm{d}/\mathrm{d}x]a^{(\mu)}(x) = ta_{\mathrm{INC}}(x)\mathrm{e}^{\mathrm{i}\Delta\beta^{(\mu)}(J^{(\mu)})x} - ((l^{(\mu)})^{-1} + k_0\alpha^{(\mu)})a^{(\mu)}(x), \qquad (2.43)$$

$$\Delta\beta^{(\mu)}(J^{(\mu)}) = k_0[\eta_p \sin\theta - \eta^{(\mu)} - Q^{(\mu)}J^{(\mu)}(x)/(Nk_0)]. \qquad (2.44)$$

Here a_{INC} is the input field distribution on the prism base and θ is the incidence angle of the input beam axis in the prism with the refractive index 'n_{P}'. t is the linear plane wave transfer function (Ulrich 1970).

The second term on the right-hand side of eq. (2.44) includes guided wave losses through reradiation back into the prism (with a characteristic coupling distance l^{μ}) and waveguide loss [linear attenuation coefficient α^{μ}, see eq. (2.21)]. The phase-mismatch effect becomes clear from the exponential term. When $\Delta\beta^{\mu} \neq 0$, the mode amplitude no longer grows linearly with propagation distance and coupling efficiency is reduced.

The critical power level where nonlinear coupling sets in is reached when the maximum phase shift

$$\Delta\phi^{\mu} = Q^{\mu}/(N)\int_{-\infty}^{\infty} J^{\mu}(x)\,\mathrm{d}x \simeq 2\omega_0\gamma^{(\mu)}Q^{(\mu)}J_{\mathrm{IN}}/N, \quad (n_{\mathrm{P}}\sin\theta = n^{(\mu)}), \qquad (2.45)$$

is of the order of $\frac{1}{2}\pi$ or more. Here J_{IN} is the incident power per unit length (along the y-axis) in the prism, γ^{μ} is the optimized low-power coupling efficiency, and $2\omega_0$ is the Gaussian input beam half-width (Stegeman et al. 1985a). (We recall that the maximum power level of coupled-mode models is fixed by the onset of self-(de)focussing effects transverse to the film boundaries.)

According to eqs (2.43) and (2.44), the formation process of the guided mode manifests itself as a travelling wave interaction between the participating fields.

The dynamics of the system is now completely confined to the x-coordinate and the emerging phenomena will be labelled 'longitudinal' effects.

In the asymptotic spatially stationary limit, i.e., for $a_{\mathrm{INC}}(x) = a_{\mathrm{INC}} = \mathrm{const.}$, $|a^{(\mu)}(x)| = (J^{(\mu)}/N)^{1/2} = \mathrm{const.}$, $x \gg (1/l^{(\mu)} + k_0\alpha^{(\mu)})^{-1}$. Equations (2.43) and (2.44) have the implicit solution

$$J^{(\mu)}/N = |ta_{\mathrm{INC}}|^2/([k_0(n_{\mathrm{P}} \sin\theta - n^{(\mu)}) - Q^{(\mu)}J^{(\mu)}/N])^2 + [1/l^{(\mu)} + k_0\alpha^{(\mu)}]^2 \tag{2.46}$$

(Carter and Chen 1983a).

This is essentially the transfer function characteristic for plane wave models (Reinisch et al. 1985, Hickernell and Sarid 1986, and in the approximate treatment mentioned by Montemayor and Deck 1985). Here, the nonlinear medium effects the propagation vector, but not the transverse field profile that remains an evanescent plane wave. It is well known that equations of the type of eq. (2.46) may have multi-valued solutions provided the initial detuning $n_{\mathrm{P}} \sin\theta - n^{(\mu)}$ is large enough.

Several generalizations of eqs (2.43) and (2.44) towards more complex nonlinearities or transient regimes will be considered at the end of this section.

2.6.2. Nonlinear coupling phenomena: theoretical predictions and experimental results

2.6.2.1. Power-dependent optimum coupling conditions. Equations (2.43) and (2.44) have been integrated numerically for an incident Gaussian beam $a_{\mathrm{INC}}(x) - \exp(-x^2/w^2)$ and the initial condition $a(-w) = 0$* (Liao et al. 1984, 1985). Figure 2.8 illustrates the evolution of the coupling efficiency, $\gamma^{(\mu)}(x) \equiv |a^{(0)}(x)|^2/|a_{\mathrm{INC}}(x)|^2$, for zero low power mismatch; $n_{\mathrm{P}} \sin\theta = n^{(0)}$.

The higher the intensity of the input beam, the narrower the spatial region where phase synchronism is ensured. The peak coupling efficiency always decreases with increasing power, the rate of decrease depending on the coupling conditions. Note also that the position relative to the center of the Gaussian beam at which the prism should be cut off varies with the coupling conditions and the input intensity. These results illustrate that the realization of optimum coupling conditions calls for a judicious variation of parameters such as angular detuning, spot size, and coupling length. Also, a suitable contouring of the film or gap can decrease the deleterious effects of the nonlinear coupling process. In this case, a $n^{(\mu)}$-variation along the prism base has to be ensured, that compensates the nonlinear phase variation determined by eq. (2.44). Linearly chirped surface gratings show a similar effect (Liao et al. 1985).

*Note, that this boundary condition is not compatible with those inherent in the stationary limit [eq. (2.46)].

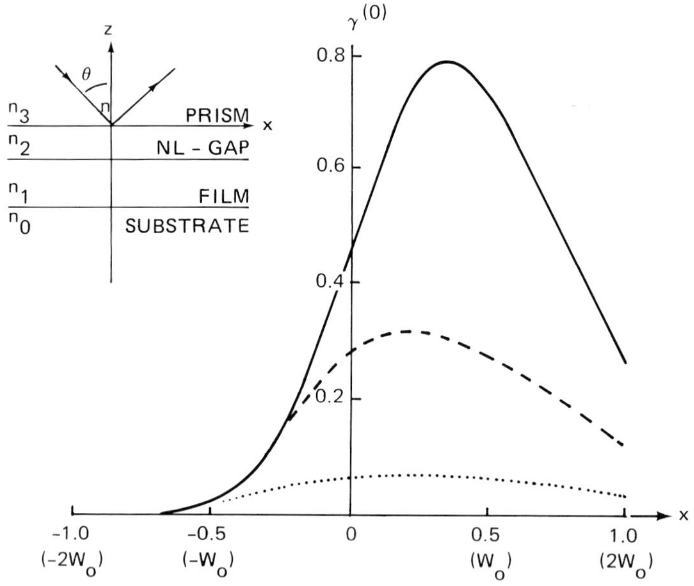

Fig. 2.8. Coupling efficiency $\gamma^{(0)}$ vs. propagation distance for a TE_0-mode for three different input powers. Solid line, 10^{-3} mW/mm; dashed line, 10 mW/mm; dotted line 100 mW/mm. The other parameters are $\omega_0 = 0.5$ mm; $l^{(0)} = 0.75$ mm; $n_0 = 1.52$; $n_1 = 1.57$; $n_2 = 1.55$ ($a_2^{NL} \simeq 5 \times 10^{-12}$ m^2/V^2 at $\lambda = 0.515$ μm; liquid MBBA). $x = 0$ coincides with the incident beam centre. (After Liao et al. 1985.)

2.6.2.2. Power-controlled distortions of the angular response characteristic. The incidence angle θ of the input beam controls directly the phase-matching conditions of the excitation process. Particularly, a suitable retuning of θ with respect to the low power case can recover phase synchronism to a certain extent (Carter and Chen 1983a, Liao et al. 1984, 1985, Sibilia and Bertolotti 1986). According to eq. (2.44), for self-focussing nonlinearities, the optimum high power coupling angle is further from the surface normal than for the low power case, and vice versa for self-defocussing nonlinearities.

Figure 2.9 displays experimental curves of the angle-tuned coupling efficiency $\gamma = f(\theta)$ for TM_0-mode excitation in a ZnS waveguide. These curves agreed well with model calculations according to eqs (2.43) and (2.44) (Assanto et al. 1986).

An important result is that the shift of the optimum coupling angle is accompanied by progressively larger asymmetries in the angular response curve. Particularly, there will be a steepening of the high angle edge (in the self-focussing case) that evolves into a sharp switching edge for angular shifts in the peak coupling angle greater than the angular width of the low power coupling peak (see fig. 2.9) (Carter et al. 1983a, b, Assanto et al. 1986, Liao et al. 1984, 1985).

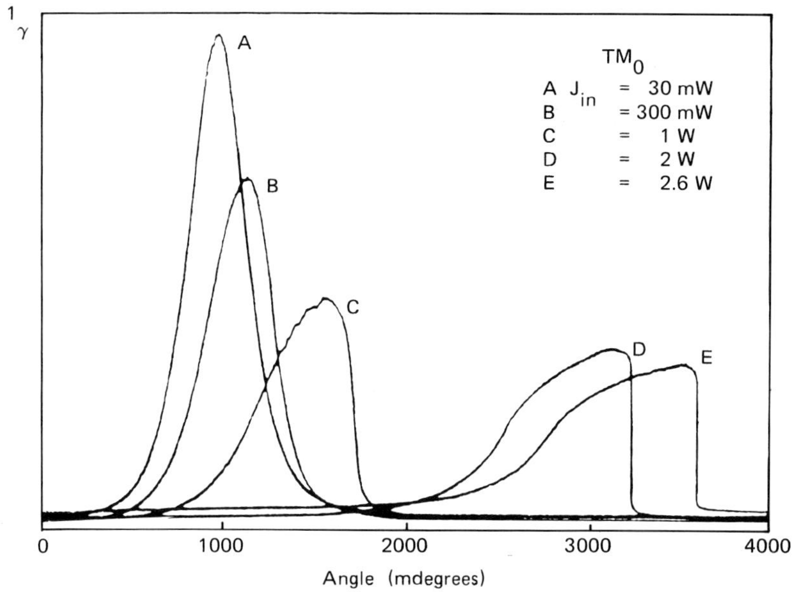

Fig. 2.9. Measured angular variation in coupling efficiency with increasing incident power J_{IN} (excitation of the TM_0-mode in a ZnS film waveguide, thickness 1.3 mm; $n = 2.36$; $a_1 = 5 \times 10^{-15}$ m^2/V^2; $n_0 = 1.51$; $\lambda = 0.515$ μm. (After Assanto et al. 1986.)

These nonlinear coupling phenomena have been demonstrated, respectively, for prism coupling to surface plasmons that propagate along a thin metal film in contact with a liquid crystal as the nonlinear medium (Arakelian et al. 1986), for grating coupling to surface plasmons guided by the interface between a silver substrate and a semiconductor material (Si, GaAs) (Chen and Carter 1982, 1984), and for grating coupling both into polydiacetylene (PDA) films (Carter et al. 1983b), as well as into GaAs/AlGaAs planar waveguides (Chen et al. 1986). These experiments were performed by measuring angular scans of the uncoupled reflected light, that is, via a standard attenuated total reflection geometry (ATR). Both position and shape of the reflexivity dip permit the determination of both the sign and the magnitude of the nonlinear coefficient a^{NL} even when the nonlinear material is absorbing (Chen and Carter 1982, 1984, Chen et al. 1986, Patela et al. 1986). The absorption coefficient has been estimated by the angular half-width of the coupling efficiency peak (Carter et al. 1983b), and time resolved measurements disclosed the nonlinearity response time (Chen and Carter 1984, Chen et al. 1986). The formation of asymmetric distortions of the angular response curve leading eventually to a sharp switching edge have been observed by Chen et al. (1986) and Assanto et al. (1986) (see fig. 2.8).

2.6.2.3. Optical limiting and switching. It is quite obvious that the power response curve, i.e., the coupled power versus incident power characteristic, shows an optical limiting behaviour. This has been predicted and experimentally verified by several groups (Valera et al. 1984, Jerominek et al. 1986, Patela et al. 1986). Representative experimental results are shown in fig. 2.10, which illustrates prism coupling into a thin glass film deposited on a CdS_xSe_{1-x}-doped glass substrate.

Time resolved measurements for 1 ns input light pulses revealed also a reduction of the maximal output power accompanied by changes in the pulse shape which disclosed the existence of a sub-nanosecond response time (Jerominek et al. 1986).

Switching within the temporal envelope of the pulses has been reported by Pardo et al. (1986) (ns-pulses with $\lambda = 1.06$ μm, grating coupled into a silicon film sandwiched between a sapphire substrate and a silver cladding) as well as by Fortenberry et al. (1986). The latter authors prism couple 10 ns-pulses into a low loss ZnO thin-film waveguide, see fig. 2.11.

In spite of the 1 μs nonlinearity relaxation time, 1 ns switching times has been identified within the nanosecond pulse envelopes.

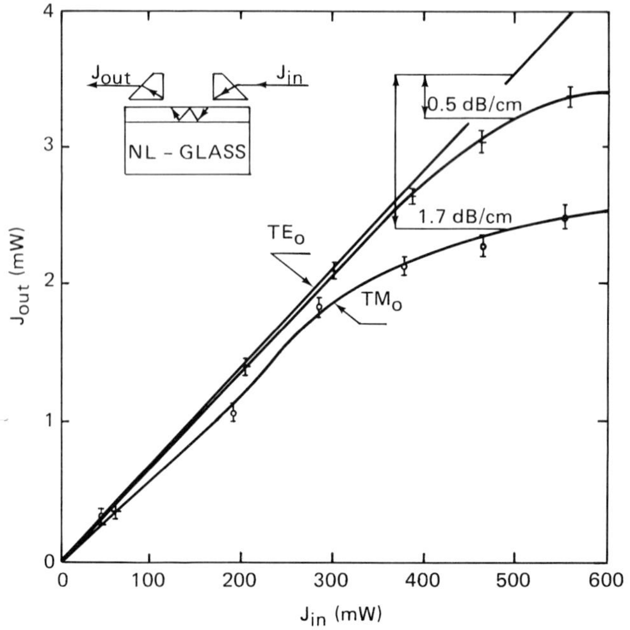

Fig. 2.10. Decoupled vs. incident light power for the TE_0- and TM_0-modes prism coupled into a monomode glass waveguide of 1.2 μm thickness. $n_{TE}^{(0)} = 1.593$; $n_{TM}^{(0)} = 1.595$; cw argon laser radiation, incident beam optimised at a low input power (below 0.1 mW), where 10% efficiency was achieved. The straight line represents an extrapolated low-power linear coupling. (After Patela et al. 1986.)

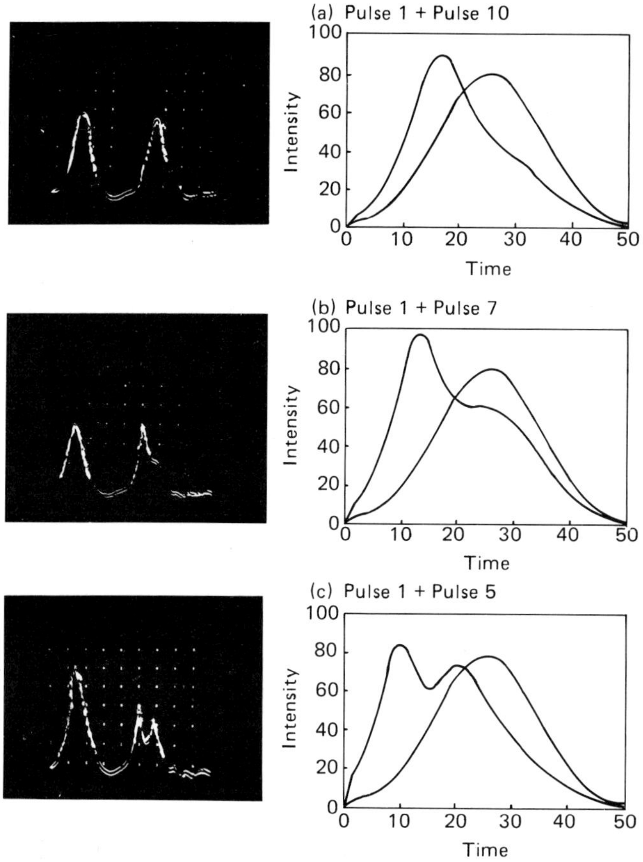

Fig. 2.11. All optical switching in the temporal envelope of a 10 ns laser pulse, coupled into a 0.6 mm thick ZnO film ($n_f \simeq 2.0$, $a^{NL} \simeq 5 \times 10^{-19}$ m^2/V^2) sputtered onto glass ($n_s = 1.47$), low power prism coupling efficiency 12%, $\lambda = 0.57$ μm (mode-locked dye laser). LHS: measured temporal profiles of output (first) and input (second) 10 ns single longitudinal mode pulse, scale 10 ns per division. RHS: corresponding calculated profiles. (a) Low pulse energy ($\simeq 1$ μJ): identical profiles. (b) Medium pulse energy ($\simeq 300$ μJ), fivefold reduced coupling efficiency: single switching. (c) High energy ($\simeq 5$ mJ): multiple switching. (After Fortenberry et al. 1986.)

The theoretical interpretation of these results has been given in terms of the evolution equations (2.43) and (2.44). Particularly, the phase-mismatch term of eq. (2.44) has been modified to

$$[\partial/\partial t + 1/\tau]\Delta\beta^{(\mu)}(x, t) = -Q^{(\mu)}/(\tau) J^{(\mu)}(x, t)/(Nk_0), \tag{2.47}$$

where a one-photon absorption mechanism (thermal nonlinearity) and low-power synchronous coupling ($n_p \sin\theta = n^{(\mu)}$) have been assumed.

The switching can be understood as a direct consequence of the nonlinear phase shift between the incident and the guided wave fields. Due to its cumulative (in space and time) nature, this phase shift may achieve multiple values of π which implies destructive interference and hence net decreases in the coupling efficiency. Particularly, the larger the incident power (and hence the faster the nonlinear phase shift change in time), the faster the switching. The switching time appears unrelated to the nonlinearity relaxation time (Fortenberry et al. 1986).

2.6.3. Optical bistability

By definition, the evolution equations (2.46) and (2.47) describe a travelling-wave interaction between the incident and the guided mode field. For the initial condition $a^{\mu}(x) = 0$ at the beam edge where the guided wave formation starts, all response curves are single-valued functions of their tuning parameters, e.g., input intensity, or angle of incidence. This is not surprising since there is no obvious feedback* mechanism implemented in eqs (2.46) and (2.47) as long as local instantaneous nonlinearities are assumed.

On the other hand, there is a clear evidence for bistable optical response in distributed coupling experiments (Martinot 1983, Martinot et al. 1985, Arakelian et al. 1986, Pardo et al. 1986, Lukosz et al. 1986, 1987, Lukosz and Briguet 1986, Assanto et al. 1986, Dannberg and Broese 1987, Vitrant and Ariot 1987), and also self-pulsing has been reported (Lukosz et al. 1986a, Briguet et al. 1987).

In some of these cases, the proper interpretation of the experimental results in terms of the coupled-mode approach, proved to be a delicate task. This applies particularly to the identification of the feedback mechanism responsible for the observed optical bistability (Assanto et al. 1986). For instance, feedback comes into the play when there are one or more 'global' variables which control the excitation process. 'Global' variables do not depend on the longitudinal (x-) coordinate thus providing a 'rigid' link between different points along the prism base, i.e., realizing a feedback link**.

Vitrant and Ariot (1987) observed hysteretic optical limiting characteristics when prism coupling through a thin liquid crystal (LC) film. They assumed that the nonlinear refractive index takes part in the coupling process in a longitudinally averaged form only thus forming a 'global' variable.

Lukosz et al. (1987) identified the gap width under the prism base as the 'global' variable when coupling into indium–tin oxide waveguides (see figs 2.12 and 2.13). The gap width is modified by a photothermal displacement of the waveguide surface, caused by partial absorption of the input power that heats the waveguide. For prism coupling of argon laser light into both ZnS and ZnSe waveguides, the interplay of photothermal displacement and nonlinear dispersion has been found responsible for both optical bistability and self-pulsing that have been observed (Lukosz et al. 1986, 1987) (see fig. 2.14).

*Since the system dynamics is confined to the longitudinal (x-) coordinate, any potential feedback has to effect different points *along* the interface, i.e., will be of longitudinal character.

**In the stationary limit, all variables are of global nature, thus realizing an overall feedback mechanism.

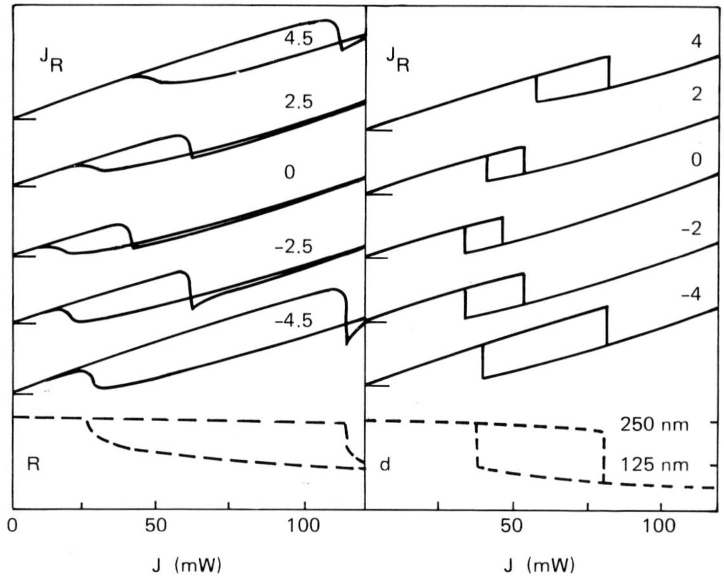

Fig. 2.12. Power-controlled optical bistability due to photothermal displacement of the waveguide (ITO-layer on glass) surface. Output power J_θ vs. input power J for different angular detunings (in units of 10^{-3}). LEFT: experimental results, scan time 120 sec. RIGHT: model calculations, d-gap width variation vs. J. (After Lukosz et al. 1987.)

Argon laser light coupled into SiO_2–TiO_2 waveguides experiences different types of optical bistability and self-pulsing, depending on the angle of incidence and the initial gap width. Again, two effects with different time constants have been identified to provide both nonlinear response and feedback, namely photothermal expansion of the waveguide, and heat induced desorption of H_2O molecules, from the SiO_2 film, accompanied by their lateral diffusion out of the coupling region (Lukosz and Briguet 1986, Briguet et al. 1987).

Dannberg and Broese (1987) relied on plane wave arguments (Montemayor and Deck 1985, 1986) when interpreting optical bistability observed for prism excitation of a film-guided wave under cw conditions. The nonlinear substrate was the nematic liquid crystal MBBA. The assumption of an incident plane wave, at least in the beam center, implies the existence of a 'global' incident field amplitude which provides the essential longitudinal feedback.

Note that the majority of plane wave models apply also for the strong nonlinear regime where self-(de-) focussing effects with respect to the transverse field profile have to be taken into account (Montemayor and Deck 1985, Reinisch et al. 1985, Agarwal and Gupta 1986a, b, Gupta and Agarwal 1986, Hickernell and Sarid 1986). The underlying resonance mechanism is now provided by a 'nonlinear guided wave' rather than a linear waveguide mode. Nonlinear guided waves show both power-dependent multiple branched disper-

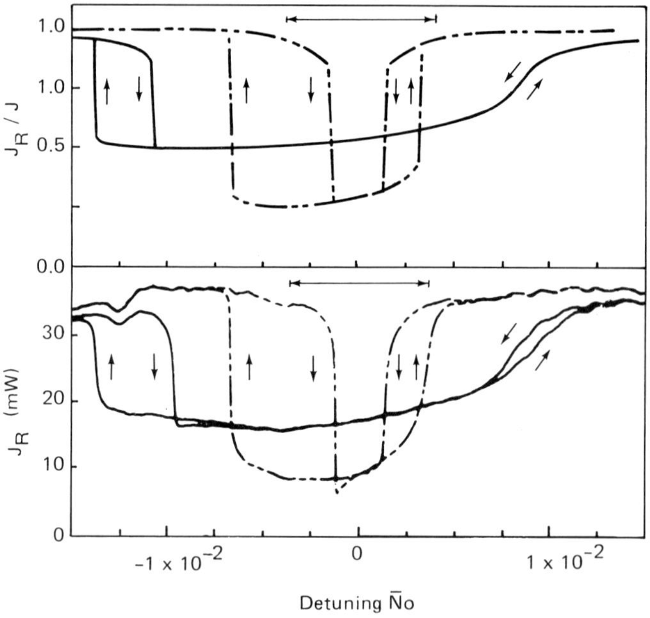

Fig. 2.13. Angle controlled optical bistability, same mechanism as in fig. 2.12. Output power J_R vs. angular detuning $n_0 = n^{(0)} - n_p \sin\theta$ for constant input power $J = 50$ mW. ----- half-width of lower power coupling efficiency at large gap width. TOP: theoretical results for $d_0 = 250$ nm (dashed line) and $d_0 \simeq 180$ nm (solid line), d_0 – initial low power gap width. BOTTOM: experimental results for $d_0 \simeq 220$ nm (dashed line) and $d_0 \simeq 150$ nm (solid line), scan time 10 min. (After Lukosz et al. 1987.)

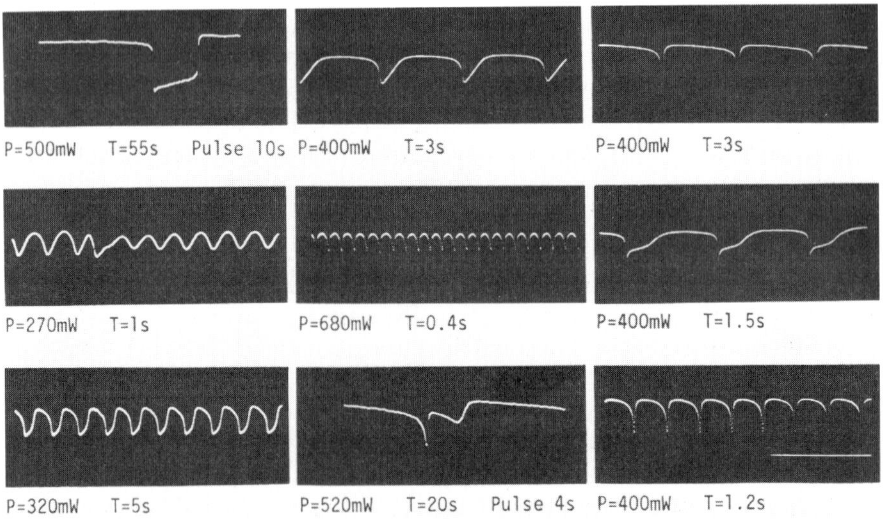

Fig. 2.14. Self-pulsing for prism coupling into a ZnS/SiO_2 waveguide for different input powers J, gap width and initial detunings, TE_0-mode, $\lambda = 488$ nm. Overview of pulse shapes and periods T. (After Lukosz et al. 1986a.)

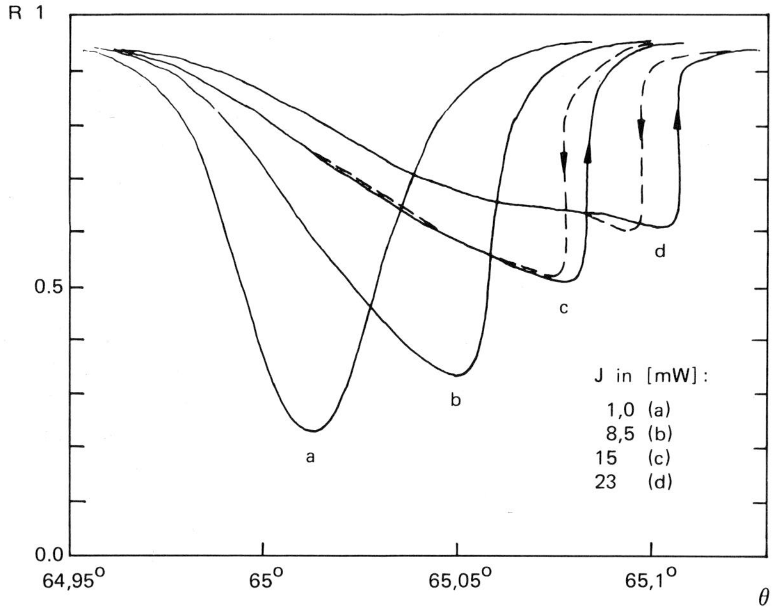

Fig. 2.15. Angular bistability for prism coupling into a SiO_2–TaO_2 waveguide ($n = 1.635$, thickness $= 1.6$ μm) with a liquid crystal (MBBA) substrate ($n_s = 1.57$; $a^{NL} \simeq 5 \times 10^{12}$ m^2/V^2). Reflexivity R vs. coupling angle θ for different input powers J. (After Dannberg and Boese 1987.)

sion curves; and power-controlled transverse field profiles as well. These waves will be studied in the remainder of this chapter.

Unfortunately, at present no formalism is known that permits the *combined* description of (longitudinal) travelling wave interactions along the waveguide in the strong nonlinear regime. Numerical simulations by means of the BPM predict nonlinear beating phenomena and refractive index channel formation partly reminiscent of the nonlinear optical interface (see, e.g., Smith and Tomlinson 1984, Akhmediev et al. 1985).

3. Nonlinear TE and TM waves at planar interfaces: Kerr regime

3.1. TE waves

If the nonlinearity is strong, the shape of the modal fields of nonlinear guided waves changes significantly with power (see, e.g., Boardman and Egan 1985, 1986a, b, and references therein). For this regime, the first-order coupled-mode approach of section 2 is quite inadequate. This is the situation that will be considered in this section, and will require exact solutions of the nonlinear equations. As was shown in the introduction, the dielectric tensor of a centrosymmetric, unsaturated, nonlinear material has third-order polarisation terms that have a quadratic dependence upon the electric field components

(Boardman et al. 1987a, Maker and Terhune 1965). These create nonlinear waves at the fundamental frequency ω and at the harmonic frequency 3ω. The generation of third-harmonic fields will be assumed to be small and exist only as a minor attenuation. There can be exceptions to this when the third harmonic and the fundamental are phase matched (synchronous), in which case the third harmonic will grow in a resonant manner. This can be avoided here since it is a reasonable assumption that the degree of phase matching required is generally absent in the waveguiding problems to be considered. The possibility that the achievable nonlinearity may saturate at some fairly modest value will be accounted for in the more general formulation of section 4 and considered quantitatively in section 5. Also, the role of absorption and the combined role of absorption (Gubbels et al. 1987) and saturation (Stegeman et al. 1987c) will be addressed in other parts of this review. The type of nonlinear polarisation to be considered here is unsaturated, and may arise from electronic distortion, molecular alignment, or electrostriction (thermal) effects. It lies in what is referred to as the Kerr regime. It may be assumed that most of the general arguments apply equally to self-focussing and self-defocussing nonlinearities but, for the most part, self-focussing, with a positive Kerr coefficient, will be used to sustain the particular examples.

Although, in principle, it is possible to discuss both TE- and TM-polarised nonlinear waves within the same all-embracing theoretical framework this will not be done here. The reason is that, whereas exact analytical solutions to the TE nonlinear equations can be found, only numerical solutions to the corresponding TM equations exist. Also, because of this fact, a very elegant and detailed formulation is possible in the general TE case but in the TM case a satisfactory infrastructure of the theory can often only be presented in numerical terms (Boardman et al. 1987b, Boardman and Twardowski 1988, 1989).

3.1.1. The nonlinear TE wave equation

The equation to be derived here has solutions that are nonlinear continuous TE waves polarised along the y-direction, have with a wave number k and an angular frequency ω. The electromagnetic field of the wave is, therefore,

$$\boldsymbol{E} = [0, E_y(k, \omega, z), 0]\mathrm{e}^{\mathrm{i}(kx-\omega t)},$$

$$\boldsymbol{H} = [H_x(k, \omega, z), 0, H_z(k, \omega, z)]\mathrm{e}^{\mathrm{i}(-\omega t)}, \tag{3.1}$$

and the operative component of the nonlinear dielectric function is simply

$$\varepsilon(\omega, |\boldsymbol{E}|^2) = \bar{\varepsilon}(\omega) + a(\omega)|E_y(k, \omega, z)|^2 = \bar{\varepsilon}(\omega) + \varepsilon^{\mathrm{NL}}(|\boldsymbol{E}|), \tag{3.2}$$

where $\bar{\varepsilon}(\omega)$ is the linear part of the dielectric function and $\varepsilon^{\mathrm{NL}}(|\boldsymbol{E}|)$ is the nonlinear part.

$\bar{\varepsilon}(\omega, |\boldsymbol{E}|^2)$ is generally a function of ω but, since $\bar{\varepsilon}(\omega)$ is the linear part of the dielectric function, it makes no difference to the generality of the formulation if it is assumed to be constant throughout this discussion. $a(\omega)$ is a nonlinear coefficient that is also assumed, without loss of generality, to be a constant a. The nonlinearity is assumed to be local in space and time, again without significant loss. Actually, nonlocality can arise through such processes as diffusion and this will be discussed in section 5. Its premature introduction, at this stage, would obscure the main thrust of the development.

The components of Maxwell's equations are, using eq. (3.1),

$$\frac{\mathrm{d}}{\mathrm{d}z} E_y(k, \omega, z) = -\mathrm{i}\omega\mu_0 H_x(k, \omega, z), \tag{3.3}$$

$$kE_y(k, \omega, z) = \omega\mu_0 H_z(k, \omega, z), \tag{3.4}$$

$$\frac{\mathrm{d}}{\mathrm{d}z} H_z(k, \omega, z) = \mathrm{i}kH_z(k, \omega, z) + \mathrm{i}\omega\varepsilon_0\varepsilon(\omega, |E_y|^2)\, E(k, \omega, z), \tag{3.5}$$

where ε_0 is the permittivity and μ_0 is the permeability of free space.

In general $E_y(\omega, z)$ is complex and has the form

$$E_y(k, \omega, z) = E(k, \omega, z)\mathrm{e}^{\mathrm{i}\phi(k,\omega,z)}, \tag{3.6}$$

where $E(k, \omega, z)$ and $\phi(k, \omega, z)$ are real functions. Equations (3.3)–(3.5), after equating the real and imaginary parts to zero, and defining $E \equiv E(k, \omega, z)$, $\phi \equiv \phi(k, \omega, z)$, $\bar{\varepsilon}(\omega) \equiv \bar{\varepsilon}$ give (Boardman and Egan 1986a)

$$\frac{\mathrm{d}^2 E}{\mathrm{d}z^2} - E\left(\frac{\mathrm{d}\phi}{\mathrm{d}z}\right)^2 + \left(\frac{\omega^2}{c^2}\bar{\varepsilon} - k^2\right)E + \frac{\omega^2}{c^2} aE^3 = 0, \tag{3.7}$$

$$\frac{\mathrm{d}\phi}{\mathrm{d}z} = \frac{K}{E^2}. \tag{3.8}$$

The last equation expresses the conservation of energy flux in the nonlinear medium and K is a constant of integration. The complete equation for the real quantity E is, therefore,

$$\frac{\mathrm{d}^2 E}{\mathrm{d}z^2} + \left(\frac{\omega^2}{c^2}\bar{\varepsilon} - k^2\right)E + \frac{\omega^2}{c^2} aE^3 - \frac{K^2}{E^3} = 0. \tag{3.9}$$

For guided or surface waves there is no flux of energy either to or from the guiding surfaces so that $E \to \pm\infty$, as $z \to \pm\infty$ and therefore $K = 0$.

This interesting feature can be expressed more succinctly through the z-component of the time-averaged Poynting vector, i.e.,

$$\langle S\rangle_z = -\tfrac{1}{2}\mathrm{Re}(E_y^* H_x) = \frac{c}{2\omega^2}\varepsilon_0 E^2 \frac{\mathrm{d}\phi}{\mathrm{d}z} = \frac{c^2}{2\omega^2}\varepsilon_0 K, \tag{3.10}$$

where c is the velocity of light in vacuo. From eq. (3.10) it is clearly seen that the condition $\langle S\rangle_z = 0$ implies that $K = 0$.

The scalar nonlinear wave equation that is left when $K=0$ can be integrated once to give the first integral (Boardman and Egan 1984b, Litvak and Mironov 1968, Maradudin 1983)

$$\left(\frac{\mathrm{d}E}{\mathrm{d}z}\right)^2+\left(\frac{\omega^2}{c^2}\bar{\varepsilon}-k^2\right)E^2+\frac{a\,\omega^2}{2\,c^2}E^4=C, \tag{3.11}$$

where C is another integration constant that will be expressed in terms of the fields evaluated at any boundaries when specific guiding situations are encountered in this discussion. In fact the constant C is non-zero, or zero, according to the application that is required. For example, in a nonlinear thin film C is non-zero, but for nonlinear waves propagating along an interface between two semi-infinite media, or along a linear dielectric layer bounded on one or both sides by semi-infinite nonlinear media, $C=0$. This is because in the latter cases both E and $\mathrm{d}E/\mathrm{d}z\rightarrow 0$ as $z\rightarrow\pm\infty$.

The nonlinear film will be considered later but for semi-infinite nonlinear media, for which $C=0$ eq. (3.11) factorises to (Boardman and Egan 1986b)

$$\left[\frac{\mathrm{d}E}{\mathrm{d}z}-E\sqrt{k^2-\frac{\omega^2}{c^2}\bar{\varepsilon}-\frac{a\,\omega^2}{2\,c^2}E^2}\right]\left[\frac{\mathrm{d}E}{\mathrm{d}z}+E\sqrt{k^2-\frac{\omega^2}{c^2}\bar{\varepsilon}-\frac{a\,\omega^2}{2\,c^2}E^2}\right]=0. \tag{3.12}$$

If $a>0$, for example, the solution to eq. (3.11) is

$$\int\left(\frac{\mathrm{d}E}{E\sqrt{k^2-\frac{\omega^2}{c^2}\bar{\varepsilon}-\frac{a\,\omega^2}{2\,c^2}E^2}}\right)=z\pm z_0, \tag{3.13}$$

where z_0 is a positive integration constant. This integration is elementary and the functional form of E is

$$E=\frac{c}{\omega}\sqrt{k^2-\frac{\omega^2}{c^2}\bar{\varepsilon}}\sqrt{\frac{2}{a}}\,\mathrm{sech}\left[\sqrt{k^2-\frac{\omega^2}{c^2}\bar{\varepsilon}}(z\pm z_0)\right], \tag{3.14}$$

where the sign adopted in eq. (3.14) depends upon the waveguide structure. For instance, nonlinear waves propagating along the interface between two semi-infinite nonlinear media are characterised by $\bar{\varepsilon}_{1,2}$ and $a_{1,2}$, so that the field solutions are

$$E_2=\frac{c}{\omega}\sqrt{\frac{2}{a_2}}\sqrt{k^2-\frac{\omega^2}{c^2}\bar{\varepsilon}_2}\;\mathrm{sech}\left[\sqrt{k^2-\frac{\omega^2}{c^2}\bar{\varepsilon}_2}\,(z-z_{02})\right],\quad z>0, \tag{3.15a}$$

$$E_1=\frac{c}{\omega}\sqrt{\frac{2}{a_1}}\sqrt{k^2-\frac{\omega^2}{c^2}\bar{\varepsilon}_1}\;\mathrm{sech}\left[\sqrt{k^2-\frac{\omega^2}{c^2}\bar{\varepsilon}_1}\,(z+z_{01})\right],\quad z<0. \tag{3.15b}$$

For much of the discussion that is to follow, however, it will be seen that considerable progress can be made through boundary field relationships alone. This means that it is often the case that a specific knowledge of field solutions is not actually required (Boardman and Egan 1984a, b, 1985, 1986b).

3.1.2. Boundary field relationships

The most general waveguiding structure that will be considered here consists of a nonlinear planar layer (film) bounded by dissimilar nonlinear media, as shown in fig. 3.1. In such a structure the relationship between the square of the field amplitude at the lower boundary, E_0^2, and the square of the field amplitude at the upper boundary, E_d^2, is of fundamental interest. It provides an elegant, yet simple, conceptual path to an appreciation of the various nonlinear waves that can propagate. This can be done without resorting to a knowledge of the nonlinear field solutions and can greatly simplify the numerical analysis. This (E_0^2, E_d^2) relationship can also be used to great effect in the nonlinear generalisation (Boardman and Egan 1984a, 1985, 1986a, b, Langbein et al. 1984d, 1985a, d) of the dispersion relationships of asymmetric and symmetric waveguides that are so familiar in linear solid state optics.

The first integrals associated with the substrate (s) the film (f) and the cladding (c) of the waveguide structure of fig. 3.1 are

$$E_s'^2 - k_0^2(q_2^2 - \tfrac{1}{2}a_s E_s^2)E_s^2 = 0, \quad -\infty \leqslant z \leqslant 0, \tag{3.16}$$

$$E_f'^2 - k_0^2(q_f^2 - \tfrac{1}{2}a_f E_f^2)E_f^2 = C_f, \quad 0 \leqslant z \leqslant d, \tag{3.17}$$

$$E_c'^2 - k_0^2(q_c^2 - \tfrac{1}{2}a_c E_c^2)E_c^2 = 0, \quad d \leqslant z \leqslant \infty, \tag{3.18}$$

where $k_0^2 = \omega^2/c^2$, $q_m^2 = n_x^2 - \bar{\varepsilon}_m$ $(m = c, f, s)$ and $n_x = ck_x/\omega$.

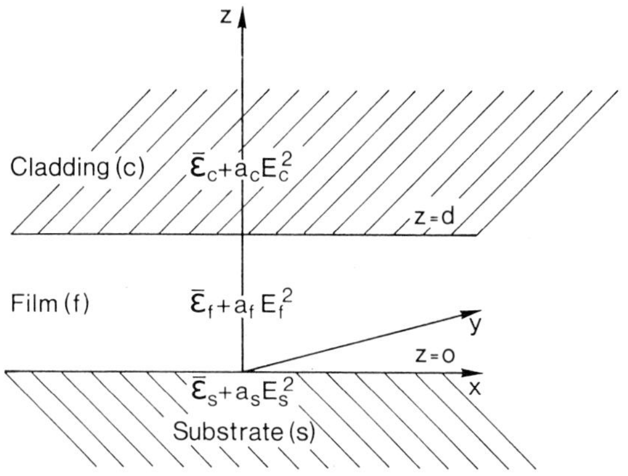

Fig. 3.1. A nonlinear optical waveguide consisting of a thin planar linear non-magnetic dielectric film, thickness d, with dielectric constant $\bar{\varepsilon}_f$, sandwiched between semi-infinite nonlinear substrate and cladding dielectrics. The substrate has, for TE waves, an effective dielectric function $\bar{\varepsilon}_s + a_s E_s^2$ and that of the cladding is $\bar{\varepsilon}_c + a_c E_c^2$.

The integration constants, introduced in eq. (3.11), are $C_s = 0$, $C_c = 0$, $C_f \neq 0$, and the notation $E'_m = dE_m/dz$ has been adopted.

The continuity conditions for the tangential electric fields E and their derivatives give at $z = d$:

$$E_c = E_f \equiv E_d, \tag{3.19}$$

$$E'_c = E'_f, \tag{3.20}$$

and at $z = 0$:

$$E_s = E_f \equiv E_0, \tag{3.21}$$

$$E'_s = E'_f. \tag{3.22}$$

The expression for the integration constant C_f, obtained from these boundary conditions, is (Boardman and Egan 1984a, 1985)

$$C_f = k_0^2 E_0^2(\bar{\varepsilon}_f - \bar{\varepsilon}_s + \tfrac{1}{2}(a_f - a_s)E_0^2) = k_0^2 E_d^2(\bar{\varepsilon}_f - \bar{\varepsilon}_c + \tfrac{1}{2}(a_f - a_c)E_d^2), \tag{3.23}$$

and the boundary field relationship is

$$E_0^2(\bar{\varepsilon}_f - \bar{\varepsilon}_s + \tfrac{1}{2}(a_f - a_s)E_0^2) = E_d^2(\bar{\varepsilon}_f - \bar{\varepsilon}_c + \tfrac{1}{2}(a_f - a_c)E_d^2). \tag{3.24}$$

This equation is very important because it is independent of frequency, wave number and thickness of the central guiding film. It depends only on the intrinsic parameters of the three nonlinear dielectric materials. As has already been shown, the field amplitudes must also be real. Hence eq. (3.24), in addition to showing the ranges over which solutions of the nonlinear wave equation are admissible, will also lessen the burden of the theoretical analysis of a nonlinear waveguide supporting waves with TE symmetry.

It is interesting that eq. (3.24) is in general a conic section, with coordinates (E_0^2, E_d^2), where one branch passes through the point $E_0^2 = 0$, $E_d^2 = 0$. After setting $\eta_i = \bar{\varepsilon}_f - \bar{\varepsilon}_i$ and $\delta_i = a_f - a_i$, $i = c, s$, this conic section is (Boardmann and Egan 1984a, 1985)

$$\frac{(\delta_s E_0^2 + \eta_s)^2}{\left(\eta_s^2 - \dfrac{\delta_s}{\delta_c}\eta_c^2\right)} - \frac{(\delta_c E_d^2 + \eta_c)^2}{\left(\dfrac{\delta_c}{\delta_s}\eta_s^2 - \eta_c^2\right)^2} = 1, \tag{3.25}$$

the centre of which is

$$\boldsymbol{R} = \left(\frac{-\eta_s}{\delta_s}, \frac{-\eta_c}{\delta_c}\right). \tag{3.26}$$

Now, since E_0 and E_d are real and one branch passes through the origin, only the part lying in the first quadrant is physically significant. Hence, for a given E_0^2, there may be two, one, or zero corresponding values of E_d^2 depending upon the shape of the conic section and the position of its centre.

The four types of conics (I, II, III and IV) that can occur are sketched in fig. 3.2 together with a specific example of a type II conic. The example actually refers to a linear centre layer for which $a_f = 0$. In this case, the general conditions for the hyperbola type I and II curves to occur are, allowing for the possibility of negative a_i ($i =$ s, f, c),

$$|a_c|\eta_s^2 > |a_s|\eta_c^2 \quad \text{(type I)}, \tag{3.27}$$

$$|a_c|\eta_s^2 < |a_s|\eta_c^2 \quad \text{(type II)}. \tag{3.28}$$

For a given system, the parameters a_i and $\bar{\varepsilon}_i$ will satisfy certain conditions so that the type of curve given by eq. (3.25) can be predicted. For example, if

$$|a_s| \leqslant |a_c| \quad \text{and} \quad (\bar{\varepsilon}_f > \bar{\varepsilon}_c > \bar{\varepsilon}_s \quad \text{or} \quad \bar{\varepsilon}_f < \bar{\varepsilon}_c < \bar{\varepsilon}_s), \tag{3.29}$$

then eq. (3.27) is satisfied and a type I conic occurs. If

$$|a_s| \geqslant |a_c| \quad \text{and} \quad (\bar{\varepsilon}_f > \bar{\varepsilon}_s > \bar{\varepsilon}_c \quad \text{or} \quad \bar{\varepsilon}_f < \bar{\varepsilon}_s < \bar{\varepsilon}_c), \tag{3.30}$$

then eq. (3.28) is satisfied and a type II conic curve occurs. Figures 3.3 and 3.4 show the real regions of E_0 and E_d covered by hyperbolic and elliptic types of curve.

The nonexistence of surface-guided TE waves in some systems can be easily shown by using these curves. For example, consider a nonlinear film, with an air cladding, deposited on a nonlinear substrate that has a smaller nonlinear coefficient (both coefficients positive) bounded by air. This cannot support TE waves at all if the linear part of the dielectric function of the film is greater than that of the substrate. This can be shown as follows. Since $a_s < a_f$ ($a_c = 0$ for air)

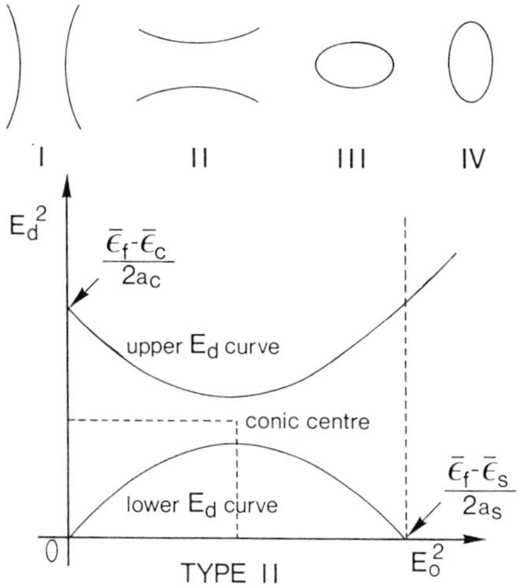

Fig. 3.2. Types of conic section produced by the boundary value field relationship, with a specific example of a type II conic (Boardman and Egan 1985).

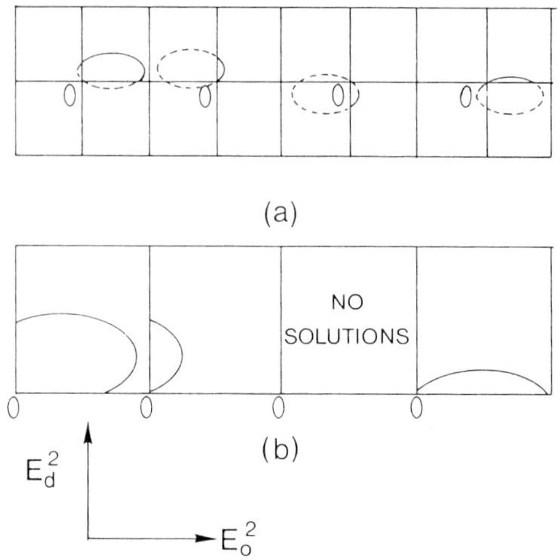

Fig. 3.3. Type III conic sections. These arise when a_s and a_c have opposite signs. (a) solid lines show the physically acceptable part of the conic section; (b) schematic drawing of the first quadrant. If the centre of the ellipse is in the third quadrant, no physical solution exists (Boardman and Egan 1985).

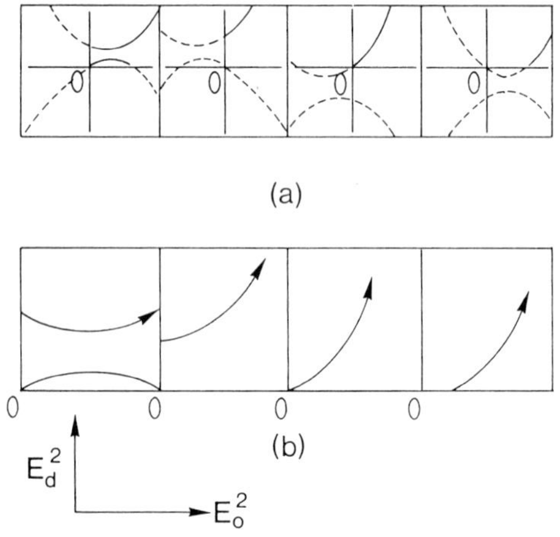

Fig. 3.4. (a) Type II conics. The solid lines show that only real values of E_0 and E_d are acceptable, $|a_c|(\bar{\varepsilon}_f-\bar{\varepsilon}_s)^2<|a_s|(\bar{\varepsilon}_f-\bar{\varepsilon}_c)^2$. (b) schematic drawing of the solid lines in (a). An arrow on a particular curve indicates that it continues to increase without restriction (Boardman and Egan 1985).

the conic section is an ellipse with $\delta_s>0$ and centre $(-\eta_s, -\eta_c)$. Hence when $\eta_s>0$ and $\eta_c>0$ (i.e., $\bar{\varepsilon}_f>\bar{\varepsilon}_s$) there are no real values of E_0 and E_d that will satisfy the conic equation. Hence there are no TE localised waves in this system and the data corresponds to the 'no solutions' part of fig. 3.3.

One of the most striking points to emerge from the theoretical investigation of this type of nonlinear waveguide is the prediction (Akhmediev 1982b) of *asymmetric* TE waves for *symmetric* layered structure consisting of a linear film bounded by two identical nonlinear Kerr media. Their existence is due entirely to the nonlinearity and was first predicted by Akhmediev but with an analysis that is involved, and with little conceptual content. Here, however, a simple, elegant conceptual path (Boardman and Egan 1984a), can be followed. Setting $\bar{\varepsilon}_c = \bar{\varepsilon}_s$, $a_c = a_s$, $a_f = 0$ yields the relationship

$$(E_0^2 - E_d^2)[\bar{\varepsilon}_f - \bar{\varepsilon}_s + (E_0^2 + E_d^2)(a_f - a_s)/2] = 0. \tag{3.31}$$

The solutions $E_0^2 = E_d^2$ correspond to the symmetric and antisymmetric waves expected in a symmetric system. The other physical solution is

$$E_0^2 + E_d^2 = \frac{2(\bar{\varepsilon}_f - \bar{\varepsilon}_s)}{a_s - a_f}, \quad E_0^2 \neq E_d^2, \tag{3.32}$$

and corresponds to *asymmetric* modes. The $E_0^2 = E_d^2$ waves have field distributions that are symmetric with respect to the centre of the film while asymmetric waves do not. Also, asymmetric waves are initiated only above a certain power threshold. This, and other details of the symmetric guiding system, will be discussed in section 5.

It is interesting to see how the (E_0^2, E_d^2) relationship for the symmetric and asymmetric modes of a completely symmetric system develops as an asymmetric system gradually becomes symmetric through parameter changes. This development is shown in fig. 3.5. The two branches of the hyperbola that represent the two sets of asymmetric modes, generally present in a nonlinear asymmetric system, become more and more eccentric as the system approaches the symmetric condition. At the limit $\bar{\varepsilon}_s = \bar{\varepsilon}_c$, $a_s = a_c$ the conic degenerates into the two straight lines given by eq. (3.31).

An interesting question now arises as to whether there are symmetric modes in an asymmetric system. By putting $E_0^2 = E_d^2$ in eq. (3.24) we obtain

$$E_0^2 = \frac{2(\bar{\varepsilon}_s - \bar{\varepsilon}_c)}{a_s - a_s}, \tag{3.33}$$

which is a *constant* of the system. If a line $(E_0^2 = E_d^2)$ is superimposed on the hyperbolic (E_0^2, E_d^2) curves for an asymmetric guide, then any intersection of the line and the curve would represent one of these modes.

Figure 3.5 illustrates the increasing complexity of the (E_0^2, E_d^2) relationship. It has been dealt with at some length because of its power in generating all the major results without recourse to further integration of the field equations. In particular, as will now be appreciated, very concise nonlinear eigenvalue equations can be developed without invoking the field functions (Boardman and Egan 1985), at least when semi-infinite nonlinear media are involved.

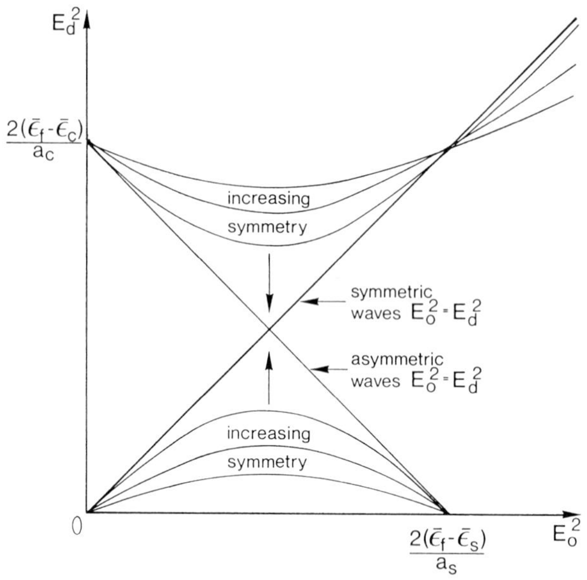

Fig. 3.5. The effect of increasing the symmetry of the waveguide on the type II boundary field amplitude relation. In the limit of a symmetrically loaded film the hyperbolas degenerate to straight lines. Note that each set of hyperbolas has a different vertical scale. The unusual asymmetric waves in a completely symmetric structure are clearly shown here (Boardman and Egan 1985).

3.1.3. *General eigenvalue equation*

For the *m*th nonlinear medium in a waveguide structure the first integral of eq. (3.11) may be expressed as

$$E_m'^2 - k_0^2(q_m^2 - \tfrac{1}{2}a_m E_m^2)E_m^2 = C_m, \tag{3.34}$$

where C_m is a constant of integration that vanishes if the medium is semi-infinite. This equation may be easily factorised and will have Jacobi elliptic function solutions whose precise form depends upon the sign of C_m *and* q_m. Suppose, for example, that $C_m > 0$, $q_m^2 > 0$ and $C_m + k_0^2 q_m^2 > \tfrac{1}{2}k_0^2 a_m E_m^4$ then eq. (3.34) factorises to

$$E_m'^2 = \tfrac{1}{2}k_0^2 a_m \left[\frac{\sqrt{\left(q_m^4 + \dfrac{2a_m}{k_0^2}C_m\right)} + q_m^2}{a_m} - E_m^2 \right] \times \left[\frac{\sqrt{\left(q_m^4 + \dfrac{2a_m}{k_0^2}C_m\right)} - q_m^2}{a_m} + E_m^2 \right], \tag{3.35}$$

which, since each factor is greater than zero, gives

$$\int\left(\frac{\mathrm{d}E_m}{\left[\dfrac{\sqrt{\left(q_m^4+\dfrac{2a_m}{k_0^2}C_m\right)}+q_m^2}{a_m}-E_m^2\right]\left[\dfrac{\sqrt{\left(q_m^4+\dfrac{2a_m}{k_0^2}C_m\right)}-q_m^2}{a_m}+E_m^2\right]}\right)$$

$$=k_0\sqrt{\frac{a_m}{2}}(z+z_0), \tag{3.36}$$

This integral is of the general form

$$I=\int\left(\frac{\mathrm{d}E_m}{(b^2-E_m^2)^{1/2}(a^2+E_m^2)^{1/2}}\right), \tag{3.37}$$

where

$$b^2=\frac{\left(q_m^4+\dfrac{2a_m}{k_0^2}C_m\right)^{1/2}+q_m^2}{a_m}, \tag{3.38}$$

$$a^2=\frac{\left(q_m^4+\dfrac{2a_m}{k_0^2}C_m\right)^{1/2}-q_m^2}{a_m}. \tag{3.39}$$

The standard solution of eq. (3.37), in terms of the cn Jacobi elliptic function is (Akhmediev et al. 1982b, Boardman and Egan 1986b, Chen and Maradudin 1988, Mihalache and Fedyanin 1983)

$$I=\frac{1}{(a^2+b^2)^{1/2}}\operatorname{cn}^{-1}\left(\frac{E_m}{b}\middle|\frac{b^2}{a^2+b^2}\right)=k_0\sqrt{\frac{a_m}{2}}(z+z_0), \tag{3.40}$$

where $b^2/(a^2+b^2)$ is the usual Jacobian modulus.

The field E_m is therefore given by

$$E_m=b\operatorname{cn}\left(\xi(z+z_0)\middle|\frac{\xi^2+q_m^2}{2\xi^2}\right), \tag{3.41}$$

where

$$\xi=\left(q_m^4+\frac{2a_m}{k_0^2}C_m\right)^{1/4}, \tag{3.42}$$

$$b^2=(\xi^2+q_m^2)/a_m, \tag{3.43}$$

and z and z_0 are now measured in units of k_0^{-1}.

Although the cn Jacobi function has emerged in eq. (3.41) it could also be some other elliptic function which is determined by the signs and relative magnitude of the coefficients in the first integral. In semi-infinite media $C_m=0$, and the Jacobi elliptic functions reduce to the simple sech functions of eqs (3.15a) and (3.15b). For the central nonlinear layer of the waveguide structure shown in fig. 3.1 the electric field has the form $J(u|m)$ where u is the argument and m is the modulus, and the Jacobi elliptic function is symbolically represented by J to imply that the number of magnitude and sign options are being kept open.

Hence the field in the central layer may be written, symbolically, as (Boardman and Egan 1984a, 1986b)

$$E_{\rm f} = bJ(\xi z + \xi z_0|m). \tag{3.44}$$

The integration constant z_0 can be eliminated from eq. (3.44) in favour of $\frac{1}{2}a_{\rm s}E_0^2$ by using the inverse Jacobi function as follows. From the application of eq. (3.44) at $z = 0$

$$\xi z_0 = J^{-1}\left(\frac{E_0}{b}\right), \tag{3.45}$$

where, as is often the convention, the function modulus has been suppressed for ease of notation. The central field amplitude then becomes

$$E_{\rm f} = bJ\left[\xi z + J^{-1}\left(\frac{E_0}{b}\right)\right], \tag{3.46}$$

Since $E_{\rm f} = E_d$ at $z = D$, where $D = k_0 d$ is dimensionless, the nonlinear eigenvalue equation is

$$\xi D = J^{-1}\left(\frac{E_d}{b}\right) - J^{-1}\left(\frac{E_0}{b}\right). \tag{3.47}$$

Equation (3.47) together with eq. (3.23) is a formal solution of the problem. This can be usefully exploited even in its most general form. Since the evaluation of inverse Jacobi functions involves incomplete elliptic integrals, eq. (3.47) is not as simple or, indeed, as convenient as it appears to be. Indeed, it would be useful to regain the advantage of using Jacobi functions, since these are much easier to compute through the use of the Landen transformations than elliptic integrals. It is also conceptually more appealing to retain the direct Jacobi function format.

The essential step now is to make ξd the argument of another Jacobi function. Exactly which one to use is, from a formal viewpoint, quite arbitrary so cn is taken here once again. If the modulus of the Jacobi function becomes negative, or greater than unity, then Jacobi's real transformation may be used (Abramowitz and Stegun 1965) to pass to another Jacobi function that will then have a different ξ and m which are now in the correct range. If cn is chosen then the first step is to develop it in the form (Boardman and Egan 1986b)

$$\mathrm{cn}(\xi D) = \mathrm{cn}(\xi z_0 - [\xi D + \xi z_0])$$
$$= \frac{\mathrm{cn}(\xi z_0)\,\mathrm{cn}(\xi D + \xi z_0) + \mathrm{sn}(\xi D)\,\mathrm{dn}(\xi D)\,\mathrm{sn}(\xi D + \xi z_0)\,\mathrm{dn}(\xi D + \xi z_0)}{1 - m\,\mathrm{sn}^2(\xi D)\,\mathrm{sn}^2(\xi D + \xi z_0)}. \tag{3.48}$$

In order to be specific, without significant loss of generality, assume that $C_\mathrm{f} > 0$, $q_\mathrm{f} > 0$ so that

$$E_\mathrm{f} = b\,\mathrm{cn}[\xi(z + z_0)], \tag{3.49}$$

where ξ and b are given by eqs (3.42) and (3.43).

The boundary conditions require that

$$\left(\frac{E_\mathrm{f}'}{E_\mathrm{f}}\right)_{z=0} = -\xi\,\mathrm{sn}(\xi z_0)\frac{\mathrm{dn}(\xi z_0)}{\mathrm{cn}(\xi z_0)} = \pm k_0(q_\mathrm{s}^2 - \tfrac{1}{2}a_\mathrm{s}E_0^2)^{1/2} = \pm k_0 q_{\mathrm{SNL}}, \tag{3.50}$$

$$\left(\frac{E_\mathrm{f}'}{E_\mathrm{f}}\right)_{z=D} = -\xi\,\mathrm{sn}(\xi D + \xi z_0)\frac{\mathrm{dn}(\xi D + \xi z_0)}{\mathrm{cn}(\xi D + \xi z_0)} = \pm k_0(q_\mathrm{c}^2 - \tfrac{1}{2}a_\mathrm{c}E_d^2)^{1/2}$$
$$= \pm k_0 q_{\mathrm{CNL}}. \tag{3.51}$$

These equations define q_{SNL} and q_{CNL} and the $\pm$ signs determine the field profile in the bounding semi-infinite media. For example, if $E_0 > 0$, $+k_0 q_{\mathrm{SNL}}$ implies that the slope of the electric field is positive at the boundary. This can only be the case if the field *decays* exponentially from the lower boundary into the substrate without first developing a self-focussing bulge. If the appropriate sign is $-k_0 q_{\mathrm{SNL}}$ then the field has a negative slope and must develop a bulge to turn the field back and allow it to decay as infinity is approached. It is possible to adopt a slightly more sophisticated approach by writing

$$q_{\mathrm{MNL}} = (-1)^{M_m}(q_m^2 - \tfrac{1}{2}a_m E_{0,d}^2)^{1/2}, \tag{3.52}$$

where $M_m = 0, 1$, $m = \mathrm{s, c}$ are the number of field maxima (bulges) exterior to the guiding film. This approach, which makes q_{SNL} and q_{CNL} carry the sign can be developed further by involving N the number of field pattern nodes within the guiding film. The use of M_m and N, although interesting (Langbein et al. 1985c, d), will not be adopted here so that a direct emphasis may be placed upon the $\pm$ sign allocation.

Equations (3.48), (3.50) and (3.51) yield the general eigenvalue equation in the following compact form

$$\mathrm{cn}(\xi D) = \frac{2E_0E_d(\xi^2 \pm q_{\mathrm{SNL}}q_{\mathrm{CNL}})}{q_{\mathrm{SNL}}^2E_0^2 + q_{\mathrm{CNL}}^2E_d^2 + \xi^2(E_0^2 + E_d^2) + \frac{1}{2}a_\mathrm{f}(E_0^2 - E_d^2)^2}. \tag{3.53}$$

The $\pm$ sign have the following interpretation.

$(+)$: the slopes of the field in the substrate and the cladding have the same sign so that a single maximum or a ‘bulge’ develops in one or other bounding medium;

$(-)$: the slopes of the field in the substrate and the cladding have opposite signs so that either maxima develop in both bounding media (double bulge) or the fields in both bounding media decay in a simple exponential fashion (double attenuation).

If $C_f \neq 0$ makes the guiding becomes more complicated with eq. (3.24) giving the conditions:

(i) $a_f > a_s$ and $\bar{\varepsilon}_f > \bar{\varepsilon}_s$: $C_f > 0$ for all E_0^2 values;
(ii) $a_f > a_s$ and $\bar{\varepsilon}_f < \bar{\varepsilon}_s$: $C_f > 0$ when $E_0^2 > 2(\bar{\varepsilon}_s - \bar{\varepsilon}_f)/(a_f - a_s)$;
(iii) $a_f < a_s$ and $\bar{\varepsilon}_f > \bar{\varepsilon}_s$: $C_f > 0$ when $E_0^2 < 2(\bar{\varepsilon}_s - \bar{\varepsilon}_f)/(a_f - a_s)$;
(iv) $a_f < a_s$ and $\bar{\varepsilon}_f < \bar{\varepsilon}_s$: $C_f < 0$ for all E_0^2 values.

3.1.4. Limiting forms of the eigenvalue equations

3.1.4.1. Nonlinear film bounded by linear semi-infinite dielectrics. For linear bounding media (Akhmediev et al. 1982a, Boardman and Egan 1986b, Chen and Mills 1987, Chen and Maradudin 1988, Holland 1986, Langbein et al. 1983) $a_s \to a_c \to 0$ and $q_{SNL} \to q_s$, $q_{CNL} \to q_c$ so that exponential attenuation of the fields into both semi-infinite media is the only possibility. Provided that $C_f > 0$, i.e.,

$$\tfrac{1}{2}a_f E_0^2 > [\bar{\varepsilon}_c - \bar{\varepsilon}_f] \quad \text{and} \quad q_f^2 > 0,$$

eq. (3.53) then reduces to (Boardman and Egan 1986b)

$$\mathrm{cn}(\xi D) = \frac{2E_0 E_d(\xi^2 - q_s q_c)}{q_s^2 E_0^2 + q_c^2 E_d^2 + \xi^2(E_0^2 + E_d^2) + \frac{1}{2}a_f(E_0^2 - E_d^2)^2}. \tag{3.54}$$

This equation together with the boundary field relationship can be used to obtain direct insight into the variation of $n_x = ck_x/\omega$ with $\frac{1}{2}(a_f E_0^2)$. This information is required before determining the dependence of the total power flow down the guide on n_x. Such a relationship is a strong indicator of possible applications. If, for any other data, the parameter of the Jacobi (elliptic) function becomes negative, or greater than unity, then Jacobi's real transformation must be used to pass to another Jacobi function that will have a different argument ξ and modulus m (Boardman and Egan 1986b, Chen and Maradudin 1988).

Several cases can be identified, some of which are:

(i) $C_f > 0$, $\frac{1}{2}a_f E_0^2 > \bar{\varepsilon}_s - \bar{\varepsilon}_f$, $n_x^2 < \bar{\varepsilon}_f$, $q_f^2 < 0$,

$$\mathrm{cn}(\xi D|m) \to \mathrm{cn}(\xi D|m), \qquad \xi = \left(|q_f^4| + \frac{2a_f}{k_0^2} C_f\right)^{1/4},$$

$$m = \frac{\xi^2 - |q_f^2|}{2\xi^2}. \tag{3.55}$$

No transformation of the Jacobi function is necessary here since ξ is a function of q_f^4 and $m < 1$.

(ii) $C_f < 0$, $\frac{1}{2}a_f E_0^2 < \bar{\varepsilon}_s - \bar{\varepsilon}_f$, $n_x^2 > \bar{\varepsilon}_f$, $q_f^2 > 0$,

$$\mathrm{cn}(\xi D|m) \to \mathrm{dn}(\xi_1 D|m'), \qquad \xi_1 = \left(q_f^4 - \frac{2a_f}{k_0^2}|C_f|\right)^{1/4},$$

$$m' = \frac{2\xi_1^2}{\xi_1^2 + q_f^2}. \tag{3.56}$$

In this case the Jacobi parameter m, in the original cn function is >1 which requires a transformation to m'. This can be seen from the fact that $\xi^4 - q^4 = (\xi^2 - q^2)(\xi^2 + q^2) = -2a_f(|C_f|/k_0^2)$ so that $\xi^2 < q_f^2$.

(iii) $C_f < 0$, $\frac{1}{2}a_f E_0^2 < \bar{\varepsilon}_s - \bar{\varepsilon}_f$, $n_x^2 < \bar{\varepsilon}_f$, $q_f^2 < 0$,

$$\mathrm{cn}(\xi D|m) \to \mathrm{nd}(\xi_2 D|m''), \qquad \xi_2 = \xi_1\sqrt{\frac{|q_f^2| + \xi_1^2}{|q_f^2| - \xi_1^2}},$$

$$m'' = \frac{2\xi_1^2}{\xi_1^2 + |q_f|^2}. \tag{3.57}$$

Here $C_f < 0$, $q_f^2 < 0$ so that $m' = 2/(1 - |q_f|^2/\xi_1^2) < 0$ and $\mathrm{dn}(\xi_1 D|m') \to \mathrm{nd}(\xi_2 D|m'')$ can be used.

The small penalty that must be paid for expressing the eigenvalue equation in the form of eq. (3.54) is that the actual field solution in the field assumes the form, for $C_f > 0$, $n_x^2 > \bar{\varepsilon}_f$,

$$\begin{aligned} E_f &= \frac{b(\mathrm{cn}(\xi z_0)\,\mathrm{cn}(\xi z) - \mathrm{sn}(\xi z)\,\mathrm{dn}(\xi z)\,\mathrm{sn}(\xi z_0)\,\mathrm{dn}(\xi z_0))}{1 - m\,\mathrm{sn}^2(\xi z)\,\mathrm{sn}^2(\xi z_0)} \\ &= \frac{b^2 E_0}{\xi}\left[\frac{\xi\,\mathrm{cn}(\xi z) + q_s\,\mathrm{sn}(\xi z)\,\mathrm{dn}(\xi z)}{b^2 - m(b^2 - E_0^2)\,\mathrm{sn}^2(\xi z)}\right] \\ &= \xi E_0\left[\frac{\xi\,\mathrm{cn}(\xi z) + q_s\,\mathrm{sn}(\xi z)\,\mathrm{dn}(\xi z)}{\xi^2\,\mathrm{dn}^2(\xi z) + \frac{1}{2}a_f E_0^2\,\mathrm{sn}^2(\xi z)}\right]. \end{aligned} \tag{3.58}$$

There are two distinct advantages to having the field amplitude in the film in this form (Boardman and Egan 1986b). The first is that the field in the nonlinear layer is expressed as a function of the optical intensity at the boundary. Another derivation exists in the literature in terms of elliptic integrals (Akhmediev et al. 1982b) so there is a second practical advantage here in that only Jacobi functions are present (Boardman and Egan 1986b, Chen and Maradudin 1988). Furthermore, only a subsequent rather elementary numerical integration is required to yield the total power flow down the guide. Note also that the form of the electric field given above may only be used for $C_f > 0$, $n_x^2 > \bar{\varepsilon}_f$ but for $C_f < 0$ and/or $n_x^2 < \bar{\varepsilon}_f$ the transformations used earlier may be applied. Thus the expression for the field is rather general.

Consider now an air-cladded ($\bar{\varepsilon}_c = 1$, $\bar{\varepsilon}_f = 2.3$) thin nonlinear layer deposited upon a thick (semi-infinite) linear substrate ($\bar{\varepsilon}_s = 2.45$). Since $\frac{1}{2}a_f E_0^2 \geqslant \bar{\varepsilon}_f$, $C_f \geqslant 0$

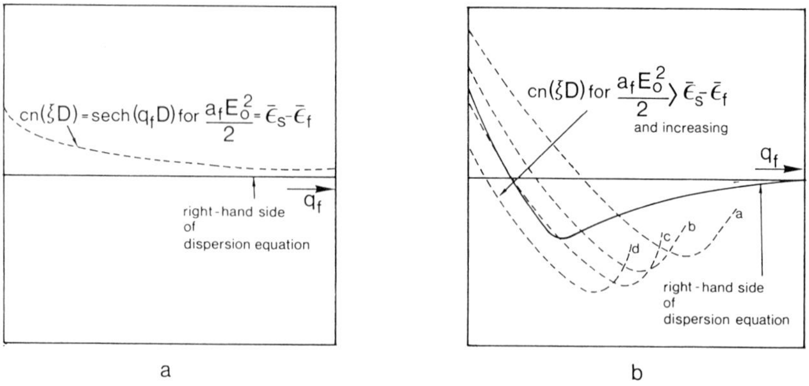

Fig. 3.6. Graphical solution of eq. (3.54). Data: $\bar{\varepsilon}_s = 2.45$; $\bar{\varepsilon}_f = 2.3$; $\bar{\varepsilon}_s = 1$. The labels a, b, s, d denote increasing $a_f E_0^2$ (Boardman and Egan 1986b).

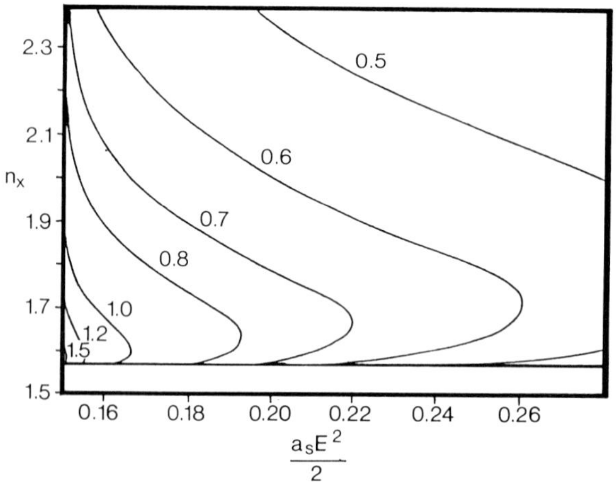

Fig. 3.7. Dependence of the wave index of TE_0 waves in a thin nonlinear film on the optical field strength at the lower boundary. The curves are labelled with values of $\omega d/\pi c$ and they do not exist below the cut-off line $\bar{\varepsilon}_s = n_x^2$ (Boardman and Egan 1986b).

surface wave solutions are given by eq. (3.54). If $\frac{1}{2}a_f E_0^2 = \bar{\varepsilon}_s - \bar{\varepsilon}_f$, then $\frac{1}{2}a_f E_d^2 = 0$ and the right-hand side of eq. (3.54) becomes zero. Also $C_f = 0$ implies that $\xi = q_f$ giving $m = 1$. For such a case the left-hand side of eq. (3.54) becomes $\operatorname{sech}(q_f D)$ but since this is never zero thereto the dispersion equation.

For $\frac{1}{2}a_f E_0^2 > \bar{\varepsilon} - \bar{\varepsilon}_f$ the solutions to eq. (3.54) can be illustrated by plotting the left- and right-hand side against q_f and then searching for any intersections as shown in fig. 3.6. The right-hand side of eq. (3.54) $\rightarrow 0$ as $q_f \rightarrow \infty$, for a fixed value of $\frac{1}{2}a_f E_0^2$, so that there are an indefinite number of intersections as the $\operatorname{cn}(\xi D)$ function continues to cross, and recross, the q_f-axis. Thus, in contrast to the linear film cases, an indefinite number of nonlinear wave solutions exist for any value of $D > 0$. Nonlinear TE_0 waves actually correspond to the first intersections of the left- and right-hand side of eq. (3.54) for $\frac{1}{2}a_f E_0^2$ just greater than $\bar{\varepsilon}_f - \bar{\varepsilon}_s$, for a fixed D. As $\frac{1}{2}a_f E_0^2$ increases, C_f increases and the difference between q_f and ξ increases and thus m becomes smaller. This decrease in the period of the Jacobi function has the effect of moving the $\mathrm{cn}(\xi D)$ curve to the left. Eventually, after a critical value of $\frac{1}{2}a_f E_0^2$ has been reached, no more intersections can occur for the TE_0 wave. This is shown as a bending back in the $(\frac{1}{2}a_f E_0^2, n_x)$ curves in fig. 3.7. A similar result is obtained for the TE_1 wave for which the critical value $\frac{1}{2}a_f E_0^2$, for a particular value of D, is higher than it is for the TE_0 wave.

3.1.4.2. Linear film bounded by nonlinear semi-infinite dielectrics. For this type of waveguide $a_f = 0$ so that $\xi = q_f$ and $m = 1$. Hence the $\mathrm{cn}(\xi D)$ Jacobi function reduces to $\mathrm{sech}(q_f D)$. The $\mathrm{sech}(q_f D)$ arises here because, up to now, it has been convenient to maintain that $q_f^2 = n_x^2 - \bar{\varepsilon}_f > 0$, which characterises a self-focussed surface wave that exponentially decays both into the film and into the nonlinear medium. For guided waves, however, the field inside the film is not hyperbolic and $q_f^2 < 0$. In these cases the definition $p_f^2 = \bar{\varepsilon}_f - n_x^2$ will be used, thus reserving q_f^2 for surface waves.

Guided waves

Since p_f^2 is appropriate here the $\mathrm{sech}(q_f D)$ function transforms to $\cos(p_f D)$ and the guided wave eigenvalue equations are (Boardman and Egan 1985)

$$\cos(p_f D) = \pm \frac{p_f^2 \pm q_{\mathrm{SNL}} q_{\mathrm{CNL}}}{(q_{\mathrm{SNL}}^2 + p_f^2)^{1/2} (q_{\mathrm{CNL}}^2 + p_f^2)^{1/2}}$$

$$= \frac{\pm [p_f^2 \pm (\bar{\varepsilon}_f - \bar{\varepsilon}_s - \frac{1}{2}a_s E_0^2 - p_f^2)^{1/2} (\bar{\varepsilon}_f - \bar{\varepsilon}_c - \frac{1}{2}a_c E_d^2 - p_f^2)^{1/2}]}{[(\bar{\varepsilon}_f - \bar{\varepsilon}_s - \frac{1}{2}a_s E_0^2)^{1/2} (\bar{\varepsilon}_f - \bar{\varepsilon}_c - \frac{1}{2}a_c E_d^2)^{1/2}]}, \tag{3.59}$$

where it should be recalled that q_{SNL}, q_{CNL} are the nonlinear versions of q_s and q_c, defined as $q_{\mathrm{SNL,CNL}}^2 = n_x^2 - \bar{\varepsilon}_{s,c} - \frac{1}{2}a_{s,c}E_{0,d}^2 = \bar{\varepsilon}_f - \bar{\varepsilon}_{s,c} - \frac{1}{2}a_{s,c}E_{0,d}^2 - p_f^2$.

The cosine form of the eigenvalue equation is used here, as opposed to the more familiar tangent formula that appears in the literature on linear waveguides. This is adopted for a number of reasons. Since eq. (3.59) is a nonlinear generalisation of the linear eigenvalue equation it will have physically acceptable roots that would, in a linear system, have to be rejected. In fact, not one but all of our eigenvalue equations are required to complete a description of nonlinear guided/surface wave propagation in place of the single equation permitted in the completely linear case. The complete set of these roots will be discussed here in terms of the intersection of the left- and right-hand sides of eq. (3.59). These roots are much easier to locate with a cosine equation and also to display in terms of odd and even parity solutions, than they would be from the alternative tangent form. Indeed, in the way that we have chosen to represent the nonlinear waves, the tangent form would not distinguish between the inner $\pm$ in eq. (3.59). Also on a purely practical point, the rate at which parts of the tangent curve go to infinity makes it quite difficult to locate, numerically, all the intersection regions.

The four nonlinear eigenvalue equations, implied by eq. (3.59), yield even and odd parity solutions where odd and even means solutions in which E_0 and E_d have opposite or equal signs, respectively. These four equations, with the definition $M = (q_{\mathrm{SNL}}^2 + p_{\mathrm{f}}^2)^{1/2}(q_{\mathrm{CNL}}^2 + p_{\mathrm{f}}^2)^{1/2}$, are.

Even parity:

$(E_0 > 0, E_d > 0 \text{ or } E_0 < 0, E_d < 0)$

$$\text{(i)}\ \cos(p_{\mathrm{f}}D) = (p_{\mathrm{f}}^2 + q_{\mathrm{SNL}}q_{\mathrm{CNL}})/M \tag{3.60}$$

(field maximum (bulge) in only one of the nonlinear media)

$$\text{(ii)}\ \cos(p_{\mathrm{f}}D) = (p_{\mathrm{f}}^2 - q_{\mathrm{SNL}}q_{\mathrm{CNL}})/M \tag{3.61}$$

(simultaneous maxima or pure exponential decay in both nonlinear media)

Odd parity:

$(E_0 > 0, E_d < 0 \text{ or } E_0 < 0, E_d > 0)$

$$\text{(iii)}\ \cos(p_{\mathrm{f}}D) = -(p_{\mathrm{f}}^2 - q_{\mathrm{SNL}}q_{\mathrm{CNL}})/M \tag{3.62}$$

(field maximum in only one of the nonlinear media)

$$\cos(p_{\mathrm{f}}D) = -(p_{\mathrm{f}}^2 - q_{\mathrm{SNL}}q_{\mathrm{CNL}})/M \tag{3.63}$$

(simultaneous maxima or pure exponential decay in both nonlinear media)

All the possibilities concerning the existence of a self-focussing bulge, or otherwise, can be deduced without actually using the quantitative form of the fields in the nonlinear media and are summarised in fig. 3.8.

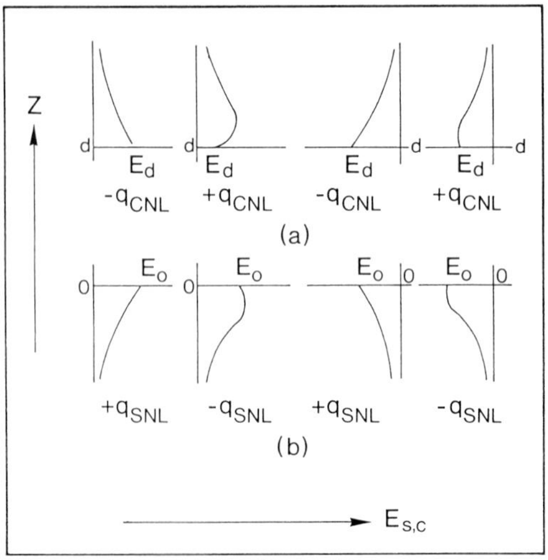

Fig. 3.8. Relationship of the signs of E_0, E_d and the slopes of the fields at the boundaries to the presence of maxima (bulges) or exponential attenuation in the nonlinear media (Boardman and Egan 1985).

The ranges of $p_f D$ and E_d/E_0 over which solutions to eq. (3.59) exist are worth highlighting. These are determined by first applying the boundary conditions at $z=0$ and $z=D$ (in dimensionless coordinates), i.e.,

$$\frac{p_f}{\sin(p_f D)}[E_d - E_0 \cos(p_f D)] = \pm q_{SNL} E_0, \tag{3.64}$$

at $z=0$ and

$$\frac{p_f}{\sin(p_f D)}[E_d \cos(p_f D) - E_0] = \pm q_{CNL} E_d, \tag{3.65}$$

at $z=D$.

Now $\sin(p_f D) > 0$ for $0 < p_f D < \pi$, $2\pi < p_f D < 3\pi$ and, in general, for $4n\frac{1}{2}\pi < p_f D < (4n+2)\frac{1}{2}\pi$ where $n = 0, 1, 2 \ldots$. Also $\cos(p_f D) > 0$ for the general range $4n\frac{1}{2}\pi < p_f D < (4n+1)\frac{1}{2}\pi$. Thus, if the case $\sin(p_f D) > 0$, $E_0 > 0$, $E_d > 0$, $E_d/E_0 > \cos(p_f D) > 0$ is considered, then this corresponds to choosing the + sign in eq. (3.64). For this case, as shown in fig. 3.9, the lower nonlinear medium supports only an exponentially decaying electric field. If, on the other hand, $0 < E_d/E_0 < \cos(p_f D)$ then the choice must be a − sign and fig. 3.9 shows that this can occur only for an electric field that has a bulge in the lower nonlinear medium. Many other possibilities are also shown in fig. 3.9.

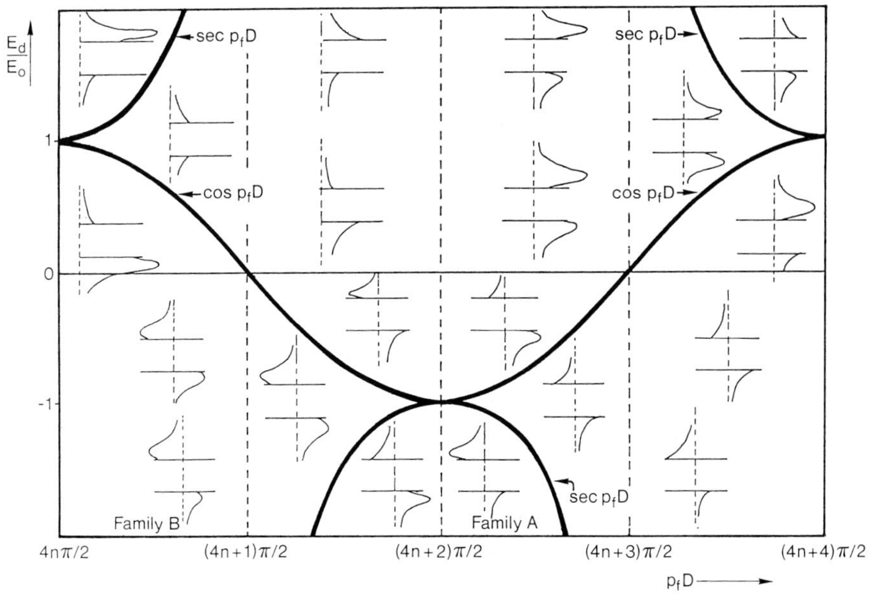

Fig. 3.9. Field pattern distributions across the waveguide. $\bar{\varepsilon}_s = 2.45$; $\bar{\varepsilon}_f = 2.6$; $\bar{\varepsilon}_c = 2.3$; $\beta = a_c/a_s = 1.05$ (Boardman and Egan 1985).

For the even $E_0 > 0$, $4n\frac{1}{2}\pi < p_f D < (4n+1)\frac{1}{2}\pi$ case the conclusions are:

(i) $E_d/E_0 > \sec(p_f D)$: bulge in the upper medium, exponential decay in the lower medium;
(ii) $\cos(p_f D) < E_d/E_0 < \sec(p_f D)$: immediate exponential decay in both lower and upper media;
(iii) $E_d/E_0 < \cos(p_f D)$: exponential decay in the upper medium, bulge in the lower medium.

In this $p_f D$ range the odd solutions for $E_0 > 0$ give a field profile that has a bulge in both nonlinear media.

The analysis can be continued in this manner to cover all the possible guided wave solutions. The resulting field patterns are sketched in fig. 3.9 and their relationships to the four eigenvalue equations in eq. (3.59) are shown in table 3.1 (Langbein et al. 1985c, d).

Surface waves

It is common practice to refer to guided waves as surface waves. This nomenclature is based upon the fact that in a linear guide the fields in the cladding decay in a purely exponential fashion. In the present context this is not a satisfactory terminology because the waveguide can support genuine self-focussing surface waves. This type of wave has a field that is hyperbolic in the film and arises because of the strong optical nonlinearity.

The existence of a bulge in the nonlinear medium is not a sufficient condition, however, for a wave associated with the interface to be classified as a surface wave because a wave only becomes a surface wave when the fields inside the guiding layer become hyperbolic. This occurs when $n_x^2 > \bar{\varepsilon}_f$ so that a switch from p_f^2 to q_f^2 takes place and the surface wave solutions have the following properties.

(i) $E_d/E_0 > \cosh(q_f D)$: exponential decay in the lower medium, bulge in the upper medium;
(ii) $\operatorname{sech}(q_f D) < E_d/E_0 < \cosh(q_f D)$: bulge in both nonlinear media;
(iii) $E_d/E_0 < \operatorname{sech}(p_f D)$: bulge in the lower medium, exponential decay in the upper medium.

For the odd solutions either $E_0 > 0$ and $E_d < 0$ or $E_0 < 0$ and $E_d > 0$. Thus for odd solutions a bulge must exist in both the lower and upper media. In fact, for any surface wave solution there cannot be a field amplitude maximum inside the linear film. This precludes the possibility of matching the field derivatives at the boundaries for an immediate decay into the nonlinear media. Thus a bulge in at least one of the two semi-infinite media is always associated with a surface wave solution. This is a neat way of asserting that TE surface waves cannot exist in the linear limit.

It can be seen that for guided wave solutions $\frac{1}{2}a_s E_0^2$ must be less than $\bar{\varepsilon}_f - \bar{\varepsilon}_s$ so

Table 3.1

	$\frac{1}{2}a_s E_0^2 < \bar{\varepsilon}_f - \bar{\varepsilon}_s$	
	Single bulge	Double bulge/decay
Even parity		
Surface $\cosh(q_f D) =$	$\dfrac{q_{SNL}q_{CNL} - q_f^2}{\sqrt{(q_{SNL}^2 - q_f^2)(q_{CNL}^2 - q_f^2)}}$	No eigenvalue equations
Guided $\cos(p_f D) =$	$\dfrac{p_f^2 + q_{SNL}q_{CNL}}{\sqrt{(q_{SNL}^2 + p_f^2)(q_{CNL}^2 + p_f^2)}}$	$\dfrac{p_f^2 - q_{SNL}q_{CNL}}{\sqrt{(q_{SNL}^2 + p_f^2)(q_{CNL}^2 + p_f^2)}}$

	$\frac{1}{2}a_s E_0^2 > \bar{\varepsilon}_f - \bar{\varepsilon}_s$	
	Single bulge	Double bulge/decay
Even parity		
Surface $\cosh(q_f D) =$	$\dfrac{q_f^2 - q_{SNL}q_{CNL}}{\sqrt{(q_f^2 - q_{SNL}^2)(q_f^2 - q_{CNL}^2)}}$	$\dfrac{q_f^2 + q_{SNL}q_{CNL}}{\sqrt{(q_f^2 - q_{SNL}^2)(q_f^2 - q_{CNL}^2)}}$
Guided	No eigenvalue equations	No eigenvalue equations

	$\frac{1}{2}a_s E_0^2 < \bar{\varepsilon}_f - \bar{\varepsilon}_s$	
	Single bulge	Double bulge/decay
Odd parity		
Surface $\cosh(q_f D) =$	No eigenvalue equations	$\dfrac{q_{SNL}q_{CNL} + q_f^2}{\sqrt{(q_{SNL}^2 - q_f^2)(q_{CNL}^2 - q_f^2)}}$
Guided $\cos(p_f D) =$	$\dfrac{-p_f^2 - q_{SNL}q_{CNL}}{\sqrt{(q_{SNL}^2 + p_f^2)(q_{CNL}^2 + p_f^2)}}$	$\dfrac{-p_f^2 + q_{SNL}q_{CNL}}{\sqrt{(q_{SNL}^2 + p_f^2)(q_{CNL}^2 + p_f^2)}}$

	$\frac{1}{2}a_s E_0^2 > \bar{\varepsilon}_f - \bar{\varepsilon}_s$	
	Single bulge	Double bulge/decay
Odd parity		
Surface $\cosh(q_f D)$	No eigenvalue equations	No eigenvalue equations
Guided	No eigenvalue equations	No eigenvalue equations

that the four eigenvalue equations (3.58) give solutions only in the range $a_s E_0^2 < \bar{\varepsilon}_f - \bar{\varepsilon}_s$. In this range the surface wave solutions are either even, single bulge solutions or odd, double bulge solutions. However, surface wave solutions are also possible for $a_s E_0^2 > \bar{\varepsilon}_f - \bar{\varepsilon}_s$. Analysis of eq. (3.58) with the transformation $p_f D \to q_f D$ shows that for $\frac{1}{2} a_s E_0^2 > \bar{\varepsilon}_f - \bar{\varepsilon}_s$ only even solutions are possible. These may be either single or double bulge solutions. A summary of the eigenvalue equations for surface waves is given in table 3.1.

The numerical examples that are introduced in the next section use typical data such as $\bar{\varepsilon}_s = 2.45$, $\bar{\varepsilon}_f = 2.6$, $\bar{\varepsilon}_c = 2.3$, $\beta = a_c/a_s = 1.05$ with $a_s > 0$ and $a_c > 0$. These values correspond to type II hyperbolic (E_0^2, E_d^2) curves whose centre is in the first quadrant, and are chosen because they appear to lie in the parameter space of current real materials. The type II boundary value curve has two branches that will now be defined as the upper E_d curve and the lower E_d curve, respectively.

It is now possible to develop the infrastructure needed to calculate the dispersion curves of the system. For a linear system a unique dispersion curve exists and this is because the all-linear system gives rise to four unknown amplitudes and there exist exactly four boundary conditions that require the continuity of E_i and E_i' at $z = 0$ and $z = d$, respectively. Also since, in a linear system, the same dispersion takes place at any power level in a linear system no power, or intensity, label is required. In contrast, a nonlinear system exhibits *different* dispersion at each power level so it is characterised by an unbounded, continuous set of dispersion curves. Since in an all-linear waveguide structure the power level is irrelevant and in a nonlinear structure *any* departure from zero power introduces nonlinearity, the zero power level will now be taken as the linear limit. Although the dispersion (ω, k_x) curves in a nonlinear system can be labelled with the optical power density at an interface (e.g., $\frac{1}{2} a_s E_0^2$) the total power flowing down the guide axis is perhaps a more desirable label from an experimental point of view. It is not obvious, however, which label to use or whether the dispersive properties revealed for each type of labelling will be similar.

These points will now be addressed, beginning with the relationship of the guided wave index $n_x = ck_x/\omega$ to $\frac{1}{2} a_s E_0^2$. The relationship of the total power flow to $\frac{1}{2} a_s E_0^2$ and, finally to n_x will be considered in a later section.

The analysis begins with the lower E_d curve, $\frac{1}{2} a_s E_0^2 < \bar{\varepsilon}_f - \bar{\varepsilon}_s$ and the even parity guided wave equations of table 3.1. Only eq. (3.61) has a linear limit which is

$$\cos(p_f D) = \frac{p_f^2 - q_c q_s}{\sqrt{q_s^2 + p_f^2}\sqrt{q_c^2 + p_f^2}} = \frac{p_f^2 (\bar{\varepsilon}_f - \bar{\varepsilon}_s - p_f^2)^{1/2} (\bar{\varepsilon}_f - \bar{\varepsilon}_c - p_f^2)^{1/2}}{(\bar{\varepsilon}_f - \bar{\varepsilon}_s)^{1/2} (\bar{\varepsilon}_f - \bar{\varepsilon}_c)^{1/2}} = L_G. \tag{3.66}$$

Specifically, if $\bar{\varepsilon}_c < \bar{\varepsilon}_s < \bar{\varepsilon}_f$, the linear guided wave is cut-off when $p_f^2 = \bar{\varepsilon}_f - \bar{\varepsilon}_s$ so that L_G, as a function of p_f, sweeps up from -1 up to the point B of fig. 3.10 where $n_x = \sqrt{\bar{\varepsilon}_s}$. All linear guide solutions can then be found from the intersection of $\cos(p_f D)$ curves with the curve L_G for various values of $D = k_0 d$. In the linear system, the general cut-off criterion concerns the rate of change of the fields away from the boundaries so that, when either q_s or q_c becomes imaginary, guiding ceases and radiative waves appear.

If the right-hand side of eq. (3.61) is plotted then, as fig. 3.10a shows, a set of curves for various values of $\frac{1}{2}a_s E_0^2$ exists that, at first sight, appear to have a cut-off when $q_{SNL} = 0$, i.e., $p_f = \bar{\varepsilon}_f - \bar{\varepsilon}_s - \frac{1}{2}a_s E_0^2$. This is a locus that passes from $p_f^2 = \bar{\varepsilon}_f - \bar{\varepsilon}_s$ ($n_x = \sqrt{\bar{\varepsilon}_s}$) at $E_0^2 = 0$ to $p_f^2 = 0$ ($n_x = \sqrt{\bar{\varepsilon}_f}$) at $E_0^2 = 2(\bar{\varepsilon}_f - \bar{\varepsilon}_s)/a_s$. All the nonlinear curves labelled with $\frac{1}{2}a_s E_0^2$ shown in fig. 3.10a end on this apparent cut-off locus and the physically acceptable eigenvalues are found from their intersection points with $\cos(p_f D)$. Figure 3.10a refers to TE_0 waves only, but as D increases the cosine curve crosses the p_f-axis much closer to the origin and higher order solutions appear.

All the solutions, up to the point A, decay exponentially in both nonlinear media. However, the nonlinear cut-off that might be expected is only apparent because, at the point A, a bulge in the field is just on the point of entering the lower nonlinear medium and the following equation holds

$$\cos(p_f D) = \frac{p_f^2}{(\bar{\varepsilon}_f - \bar{\varepsilon}_s - \frac{1}{2}a_s E_0^2)^{1/2}(\bar{\varepsilon}_f - \bar{\varepsilon}_c - \frac{1}{2}a_c E_d^2)^{1/2}} \tag{3.67}$$

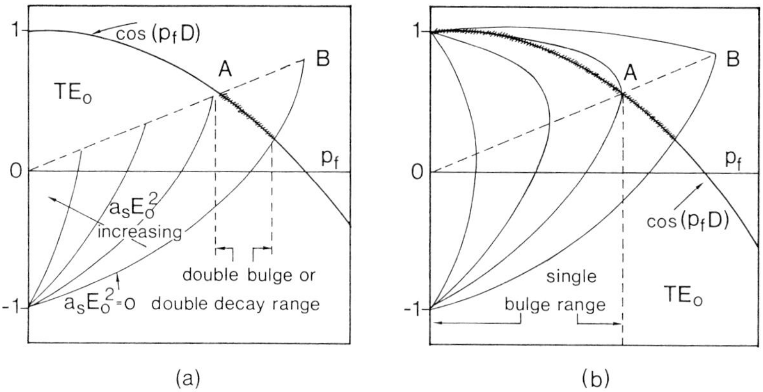

Fig. 3.10. (a) graphical TE_0 guided wave solutions of the second equation in table 3.1 for the lower E_d curve of the boundary field relationship. A typical D is assumed and data of fig. 3.9. The shaded part of the cosine curve denotes the physically acceptable solutions where double exponential decay or double bulges are possible. B is the linear cut-off point. At point A a bulge in the field is on the point of entering the nonlinear medium. (b) graphical TE_0 guided wave solutions for the lower E_d curve of the boundary field relationship of both the second and first equations in table 3.1. The shaded part of the cosine curve shows acceptable bulge guided wave solutions (Boardman and Egan 1985).

Thus there is no cut-off on the locus shown in fig. 3.10a, nor is there an immediate transition to surface waves of the single interface type. In fact the wave index can increase as the field maximum moves smoothly into the nonlinear medium and the propagating wave remains guided. An equally smooth transition to a surface wave will occur as the mode index continues to increase.

All nonlinear fields in the bounding media have a sech shape where the maximum is either real and located in the nonlinear medium (a self-focussing situation for $a_i > 0$), or virtual, in the sense that it is mathematically located outside the nonlinear medium. It may be positioned, for instance, in the linear guide in which case the field in the guide matches smoothly onto the part of the sech function associated with the nonlinear medium that decays away to infinity, without first developing a maximum. As the field maximum in the guide moves into the nonlinear medium it converts a virtual maximum to a real maximum.

Since a point such as A does not lie on a *cut-off locus* the latter will be referred to as a *changeover locus*. These changeover points are exactly where the guided wave eigenvalue equation (3.60) starts to be satisfied. The changeover locus is the locus of all points where q_{SNL} changes sign and creates, initially, exponentially growing fields. This possibility would have to be rejected for a linear structure but now, because of the nonlinearity, exponentially growing fields do not grow indefinitely. They 'turn over' into self-focussing (maxima) bulges, at some distance from the interface. Representative graphical solutions above this changeover locus are shown in fig. 3.10b.

It has also been noted (Langbein et al. 1985c) that the field solutions also fall elegantly into two families A and B. Family A contains all types of nonlinear guided waves that have an invariant virtual field maximum with respect to one interface in the structure. The definition of invariant here is that a field maximum does not change character when the transition point (changeover locus) is reached. In other words a maximum may pass through one interface to become real in the nonlinear medium. The status of the maximum at the other interface remains unchanged. Family B corresponds to all nonlinear guided waves that have an invariant real field maximum.

As the wave index increases, $p_{\mathrm{f}}^2 = \bar{\varepsilon}_{\mathrm{f}} - n_x^2$ decreases and the guided wave equation that is associated with a single bulge (maximum) in the lower nonlinear medium becomes a single bulge surface wave equation. The surface wave solutions then join smoothly onto the guided wave solutions. In the surface wave region n_x increases without bounds. Indeed, for the lower E_d curve, $n_x \to \infty$ as $\frac{1}{2} a_{\mathrm{s}} E_0^2 \to \bar{\varepsilon}_{\mathrm{f}} - \bar{\varepsilon}_{\mathrm{s}}$. This feature is shown in figs 3.11 and 3.12 that also show the changeover locus running from $\sqrt{\bar{\varepsilon}_{\mathrm{f}}}$ when $a_{\mathrm{s}} E_0^2 = 0$ to $\sqrt{\bar{\varepsilon}_{\mathrm{f}}}$ when $p_{\mathrm{f}} = 0$, at which point a transition to surface waves takes place.

Figure 3.12 shows the guided wave region in more detail. For the TE_0 solution shown, it can be seen that, starting at $E_0^2 = 0$, n_x increases with increasing E_0 until the changeover locus is reached. The second guided wave equation in

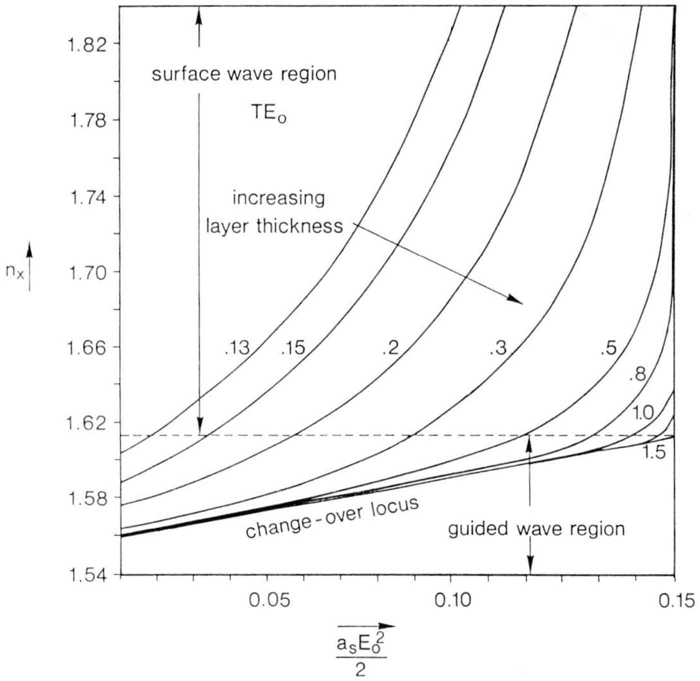

Fig. 3.11. Lower E_d curve, TE_0 solutions. The dotted line is drawn at $n_x = \sqrt{\bar{\varepsilon}_f}$ and is one of the light lines of the linear system. The guided wave region lies in the range $\bar{\varepsilon}_s \leqslant n_x \leqslant \bar{\varepsilon}_f$ and the surface wave region lies in the region $n_x > \bar{\varepsilon}_f$. The data is the same as in fig. 3.9 and is also given in the text. All curves are labelled with $\delta = D/2\pi$ (Boardman and Egan 1985).

table 3.1 gives these solutions. On reaching the changeover locus, the n_x values swap over to those that satisfy the first guided wave eigenvalue equation of table 3.1, and sweep upto $n_x = \sqrt{\bar{\varepsilon}_f}$ where they join smoothly onto the surface wave solutions.

It does not follow, however, as will be shown in the numerical section, that all orders of solution above TE_0 may join onto surface wave solutions. Many of these points will be discussed again in section 5 where detailed numerical results will be presented to support this mathematical section.

Figure 3.12 shows that, as $\frac{1}{2}a_s E_0^2$ increases, for the lower E_d curve no solution exists in the region $\frac{1}{2}a_s E_0^2 > \bar{\varepsilon}_f - \bar{\varepsilon}_s$. Second, it displays the semi-infinite medium behaviour (single interface) in which $a_s E_0^2 = 2(\bar{\varepsilon}_f - \bar{\varepsilon}_s)$ for the lower surface with a consequent lack of n_x dependence. In fact, what is happening is that as $D \rightarrow \infty$ for the lower E_d curve the limit is equivalent to holding the *lower* surface of the guide in place and then taking the *upper* surface to infinity.

The upper E_d curve has quite different properties. For this branch the changeover locus is $p_f^2 = \bar{\varepsilon}_f - \bar{\varepsilon}_s - \frac{1}{2}a_s E_0^2$ or $p_f^2 = \bar{\varepsilon}_f - \bar{\varepsilon}_c - \frac{1}{2}a_c E_d^2$ whichever

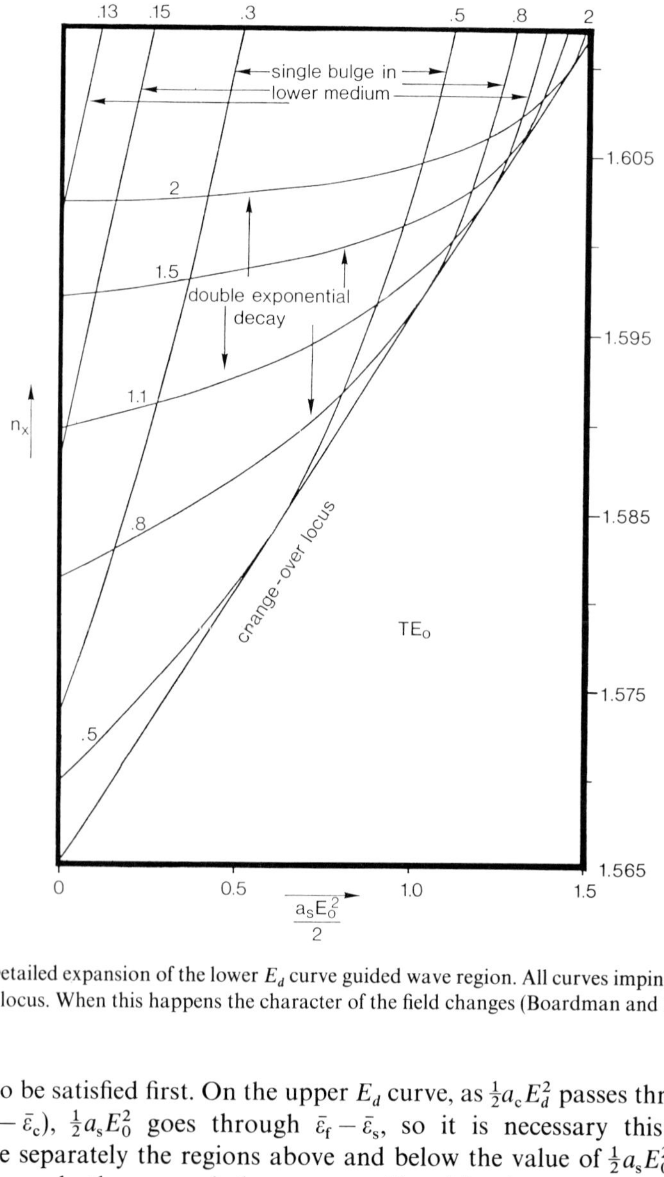

Fig. 3.12. Detailed expansion of the lower E_d curve guided wave region. All curves impinge upon the changeover locus. When this happens the character of the field changes (Boardman and Egan 1985).

happens to be satisfied first. On the upper E_d curve, as $\frac{1}{2}a_c E_d^2$ passes through the value $(\bar{\varepsilon}_f - \bar{\varepsilon}_c)$, $\frac{1}{2}a_s E_0^2$ goes through $\bar{\varepsilon}_f - \bar{\varepsilon}_s$, so it is necessary this time to investigate separately the regions above and below the value of $\frac{1}{2}a_s E_0^2$.

If, for example, the even solutions are considered for the upper E_d curve then, as $\frac{1}{2}a_s E_0^2$ is changed from zero to $(\bar{\varepsilon}_f - \bar{\varepsilon}_s)$, $p_f = 0$ at both $E_0^2 = 0$, where $\frac{1}{2}a_c E_d^2 = \bar{\varepsilon}_f - \bar{\varepsilon}_c$ and at $\frac{1}{2}a_s E_0^2 = \bar{\varepsilon}_f - \bar{\varepsilon}_s$. For the specimen data used to generate figs 3.13 and 3.14, the changeover locus is $p_f^2 = \bar{\varepsilon}_f - \bar{\varepsilon}_c - \frac{1}{2}a_c E_d^2$ and this decreases from $n_x = \sqrt{\bar{\varepsilon}_f}$ to a minimum and then increases back up again to $n_x = \sqrt{\bar{\varepsilon}_f}$. Once again, a single bulge occurs in both the guided and surface wave regions. In the

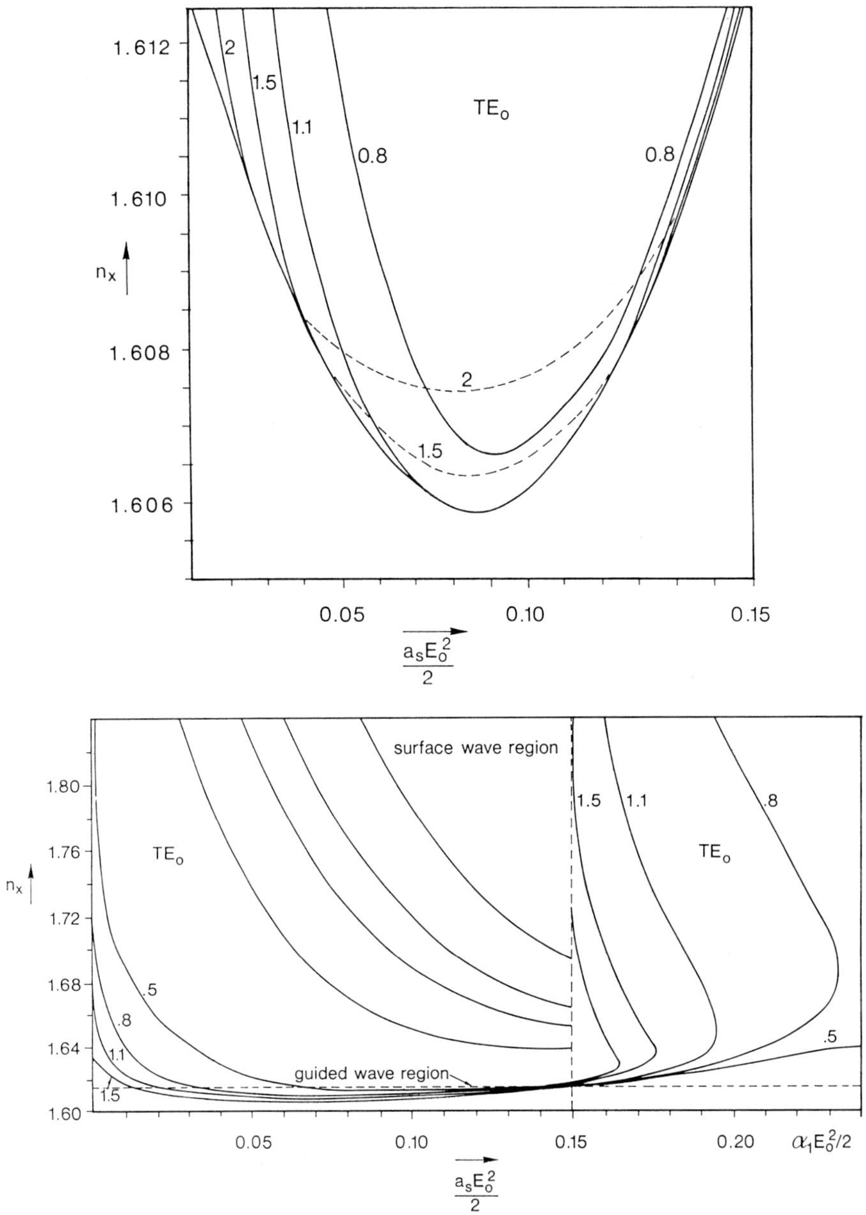

Fig. 3.13. (a) Detailed form of the $(n_x, a_s E_0^2)$ relationship for the upper E_d curve of the boundary field relationship for which the changeover locus runs from $\sqrt{\bar{\varepsilon}_f}$ to $\sqrt{\bar{\varepsilon}_f}$. (b) Full behaviour of the upper E_d curve, TE_0 solutions. Note that guided waves do not exist in the region $a_s E_0^2 > \bar{\varepsilon}_f - \bar{\varepsilon}_s$. For clarity the unlabelled curves, ranging from $\delta = 0.2$ to $\delta = 0.1$ have been terminated at the $\frac{1}{2} a_s E_0^2 = \bar{\varepsilon}_f - \bar{\varepsilon}_s$ line (Boardman and Egan 1985).

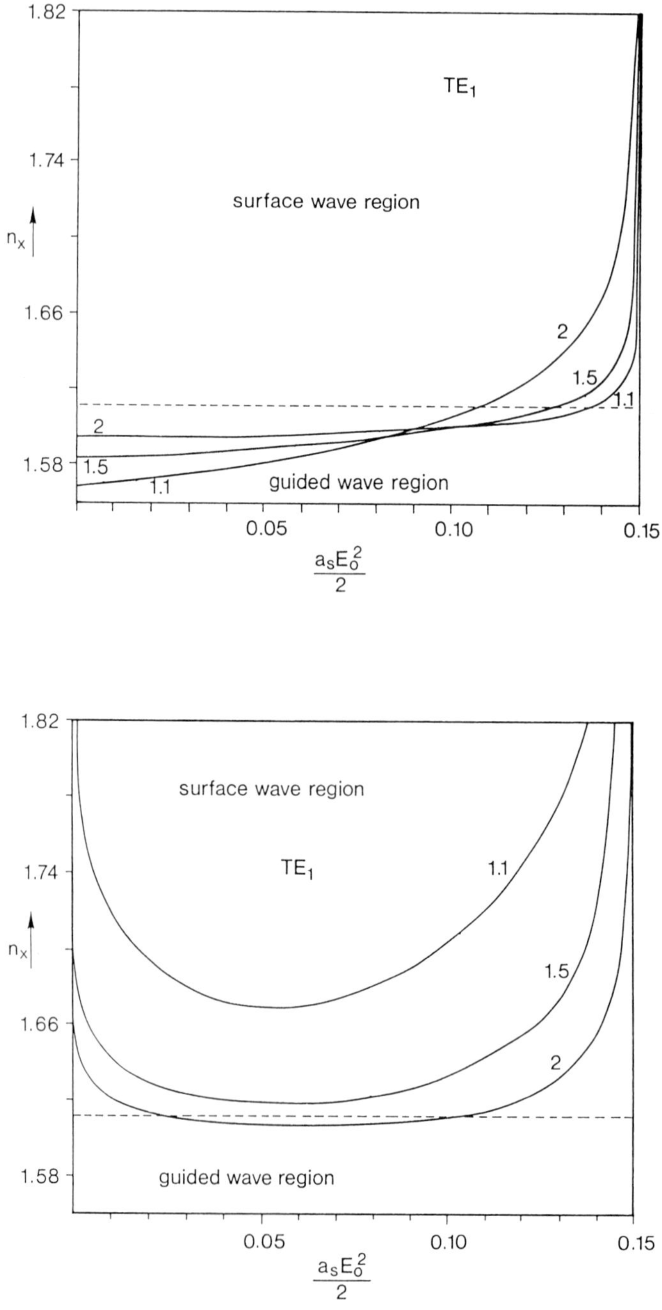

Fig. 3.14. TE$_1$ solutions for lower (a) and upper (b) E_d curves. No solutions exist in the region $\frac{1}{2}a_s E_0^2 > \bar{\varepsilon}_f - \bar{\varepsilon}_s$ (Boardman and Egan 1985).

limit $D \to \infty$, $\frac{1}{2}a_c E_d^2 = \bar{\varepsilon}_f - \bar{\varepsilon}_c$ and the $(n_x, \frac{1}{2}a_s E_0^2)$ relationship degenerates to a single vertical line thus showing, yet again, a lack of dependence on n_x. Since, in this limit, $\frac{1}{2}a_c E_d^2 = \bar{\varepsilon}_f - \bar{\varepsilon}_c$ the interpretation of this is that it corresponds to the *upper* surface of the linear slab remaining fixed while the *lower* surface is removed to infinity.

Figures 3.13a and 3.13b show the $(\frac{1}{2}a_s E_0^2, n_x)$ curves for TE_0 waves. Only surface waves appear in the extended part of fig. 3.13b but to keep some clarity in the figure many of the curves have been terminated at $\frac{1}{2}a_s E_0^2 = \bar{\varepsilon}_f - \bar{\varepsilon}_s$. These curves would carry on into the extended part beyond $\bar{\varepsilon}_f - \bar{\varepsilon}_s$ but would bend back, eventually, to approach infinity at $\bar{\varepsilon}_f - \bar{\varepsilon}_s$. This feature is shown on the curves for larger D.

The odd parity solutions can be analysed with exactly the same techniques. For example, as $n_x \to \sqrt{\bar{\varepsilon}_f}$ for TE_1 waves the field solutions are guided waves that decay in both nonlinear media, or are surface or guided waves with a bulge in both nonlinear media. Hence, as D decreases it is the double decay, or double bulge, solutions only that are encountered. This is in sharp contrast to the even parity solutions. Note that odd parity surface waves must have bulges in both nonlinear media and also, as shown in fig. 3.14, for both the upper and lower E_d curves, $n_x \to \infty$ at both $\frac{1}{2}a_s E_0^2 = \bar{\varepsilon}_f - \bar{\varepsilon}_s$ and $\frac{1}{2}a_s^2 E_0^2 = 0$. This emphasises, quite dramatically, a rather important point, i.e., that odd parity surface wave solutions do not exist beyond $\frac{1}{2}a_s E_0^2 = \bar{\varepsilon}_f - \bar{\varepsilon}_s$.

3.1.4.3. Symmetric systems. It has been shown in eqs (3.31) and (3.32) that nonlinearity introduces the possibility of asymmetric modes existing in a guide that is symmetric with respect to the zero field parameters. Such a symmetric waveguide system is, therefore, a special system worthy of more detailed attention. The nonlinear waves separate out to the expected conventional ones with $E_0^2 = E_d^2$ [classified as symmetric (S) and antisymmetric (AS)] and the unconventional but, with hindsight, rather to be expected ones with $E_0^2 \neq E_d^2$ (classified as asymmetric).

The eigenvalue equation for the symmetric system, obtained from eq. (3.59) by setting $a_s = a_c$, $\bar{\varepsilon}_s = \bar{\varepsilon}_c$, is

$$\cos(p_f D) = \pm \frac{p_f^2 \pm (\bar{\varepsilon}_f - \bar{\varepsilon}_s - \frac{1}{2}a_s E_0^2 - p_f^2)^{1/2}(\bar{\varepsilon}_f - \bar{\varepsilon}_s - \frac{1}{2}a_s E_d^2 - p_f^2)^{1/2}}{(\bar{\varepsilon}_f - \bar{\varepsilon}_s - \frac{1}{2}a_s E_0^2)^{1/2}(\bar{\varepsilon}_f - \bar{\varepsilon}_s - \frac{1}{2}a_s E_0^2)^{1/2}}. \tag{3.68}$$

For symmetric (S) and antisymmetric waves (AS), $E_0^2 = E_d^2$ so that

$$\cos(p_f D) = \pm \frac{p_f^2 \pm (\bar{\varepsilon}_f - \bar{\varepsilon}_s - \frac{1}{2}a_s E_0^2 - p_f^2)}{(\bar{\varepsilon}_f - \bar{\varepsilon}_s - \frac{1}{2}a_s E_0^2)}. \tag{3.69}$$

On considering the inner $\pm$ signs the following relationships are obtained

$$\cos(p_f D) = \pm 1, \tag{3.70}$$

$$\cos(p_f D) = \pm \frac{[2p_f^2 - (\bar{\varepsilon}_f - \bar{\varepsilon}_s - \frac{1}{2}a_s E_0^2)]}{\bar{\varepsilon}_f - \bar{\varepsilon}_s - \frac{1}{2}a_s E_0^2}. \tag{3.71}$$

The solution given by eq. (3.70) shows that no single bulge solution exists which is hardly surprising for a symmetric wave. Equations (3.71) may be usefully rewritten as

$$\tfrac{1}{2}a_s E_0^2 = \bar{\varepsilon}_f - \bar{\varepsilon}_s - p_f^2 \sec^2(\tfrac{1}{2}p_f D) \quad \text{(S)}, \tag{3.72}$$

$$\tfrac{1}{2}a_s E_0^2 = \bar{\varepsilon}_f - \bar{\varepsilon}_s + p_f^2 \operatorname{cosec}^2(\tfrac{1}{2}p_f D) \quad \text{(AS)}. \tag{3.73}$$

The solutions of eqs (3.72) and (3.73) are shown in fig. 3.15.

For asymmetric waves (A) $E_0^2 + E_d^2 = 2(\bar{\varepsilon}_f - \bar{\varepsilon}_s)/a_s$ and the corresponding eigenvalue equation is

$$\cos(p_f D) = \frac{p_f^2 \pm (\frac{1}{2}a_s E_0^2 - p_f^2)^{1/2}(\frac{1}{2}a_s E_d^2 - p_f^2)^{1/2}}{\frac{1}{2}a_s E_0 E_d} \quad \text{(A)}, \tag{3.74}$$

since the outer $\pm$ signs may now be omitted and the signs on E_0 and E_d used explicitly.

It is instructive to examine the connection that the (E_0^2, E_d^2) relationship of the symmetric case has to the more complicated hyperbolic upper and lower E_d curves that are obtained for an asymmetric system. For the symmetric waveguide structure the two straight lines that constitute the (E_0^2, E_d^2) relation-

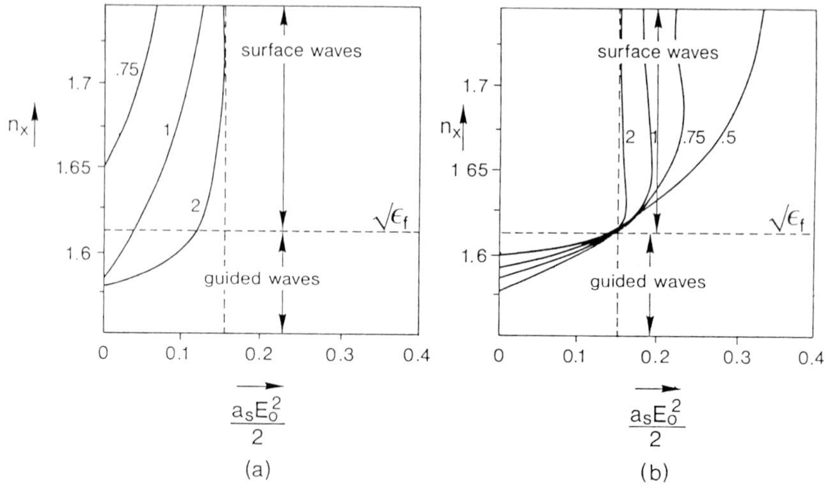

Fig. 3.15. The $(n_x, \frac{1}{2}a_s E_0^2)$ relationship for a symmetric guide. These solutions are obtained from eqs (3.72) and (3.73). (a) TE_1 waves, (b) TE_0 waves. $\bar{\varepsilon}_s = 2.45$; $\bar{\varepsilon}_c = 2.45$; $\bar{\varepsilon}_f = 2.6$; $\beta = 1$.

ship form an upper and lower triangular shape that can be easily identified with the upper and lower E_d curves of an asymmetric structure. This is even more readily appreciated if it is realised that even an infinitesmal asymmetry will lift the degeneracy of the symmetric case and split it into the usual upper and lower E_d curves. Figure 3.16 shows all these features schematically together with sketches of some field distributions.

If the effect of changing E_0^2 on the field shape is considered then, for the upper E_d curve, a field maximum will be pulled into the linear guide. For the lower E_d curve a field maximum will be pushed out into the nonlinear medium. It is therefore possible to see how the symmetric guide will operate. Suppose that a *low* value of E_0^2 exists then this will correspond to a *high* value of E_d^2 on the asymmetric line of the (E_0^2, E_d^2) plot. A field maximum may then exist outside the linear portion of the guide. If E_0^2 is now increased, E_d^2 will start to decrease so that this field maximum, if it exists, is pulled into the linear guide. At the point $E_0^2 = E_d^2$ the field maximum will lie symmetrically within the guide. Any further increase in E_0^2 progresses into the lower E_d segment so that E_d decreases even

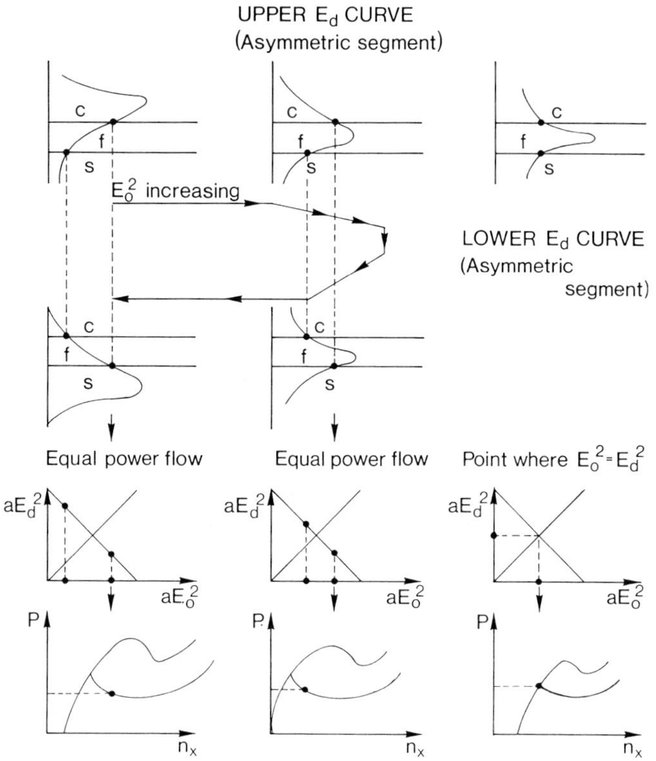

Fig. 3.16. Sketch to illustrate the symmetric nonlinear guide in terms of an approach of the power level to the point of degeneracy.

more. At this stage the field maximum will begin to move out into the lower nonlinear medium.

3.1.4.4. Symmetric modes of an asymmetric system. It is possible, in principle, to have $E_0^2 = E_d^2$ for an asymmetric guiding structure (Boardman et al. 1984b) so that

$$E_d^2 = E_0^2 = \frac{2(\bar{\varepsilon}_s - \bar{\varepsilon}_c)}{a_c - a_s}. \tag{3.75}$$

This exists provided that a real intersection of the $E_0^2 = E_d^2$ line with the conic section of the system exists. Whether such a mode will be a guided or surface mode will be determined by the position of the intersection and that, in turn, will determine which of the general eigenvalue equations of table 3.2 will give a solution for p_f or q_f with the given value of E_0^2. For example, if the intersection is in the region $\frac{1}{2}a_s E_0^2 > \bar{\varepsilon}_f - \bar{\varepsilon}_s$ then the solution will be a surface wave. If $\frac{1}{2}a_s E_0^2 < \bar{\varepsilon}_f - \bar{\varepsilon}_s$ then both surface and guided modes may result.

Putting $E_0^2 = E_d^2$ in an asymmetric system is equivalent to finding a *particular* wave index n_x' such that the magnitudes of the field amplitudes at the boundaries are equal. Varying n_x or E_0^2 will then destroy the symmetry of the field in the linear layer. For this symmetric mode of the asymmetric guide the eigenvalue equations can be written as

$$\beta(\bar{\varepsilon}_f - \bar{\varepsilon}_s) - (\bar{\varepsilon}_f - \bar{\varepsilon}_c) = p_f^2 \times \begin{matrix} (\beta - 1)\sec^2(p_f D/2) & \text{(even)} \\ (1-\beta)\operatorname{cosec}^2(p_f D/2) & \text{(odd)} \end{matrix}, \tag{3.76}$$

where $\beta = a_c/a_s$. The odd and even solutions are mutually exclusive for any given range of system parameters. The type of waves that are possible can in this case be found analytically and the conclusions are summarised in fig. 3.17. Note that in a symmetric guide n_x is a *function* of $\frac{1}{2}a_s E_0^2$, whenever $E_0^2 = E_d^2$. This situation is quite distinct from a *particular* $n_x = n_x'$ being appropriate to $E_0^2 = E_d^2$ in an asymmetric guide.

3.1.4.5. Single interface. This case is the simplest one (Agranovich et al. 1980) that can be projected out of the general formula (3.59) and describes two semi-infinite nonlinear media occupying the substrate region $z < 0$, and the cladding region $z > 0$. If nonlinear waves exist they must now be self-focussing ($a_i > 0$) surface waves that have a field maximum in the upper or lower medium and they have no linear limit.

If, for example, the field amplitude at the boundary is positive (similar arguments will apply to negative fields) then, from the first integral [eq. (3.11)] of the nonlinear differential equations for E_i ($i = $ s or c), a turning point in the magnitude of the field amplitude, for which $E_i' = 0$, can occur if, and only if,

$$E_i = q_i \sqrt{\frac{2}{a_i}}, \quad i = \text{s, c}. \tag{3.77}$$

Table 3.2

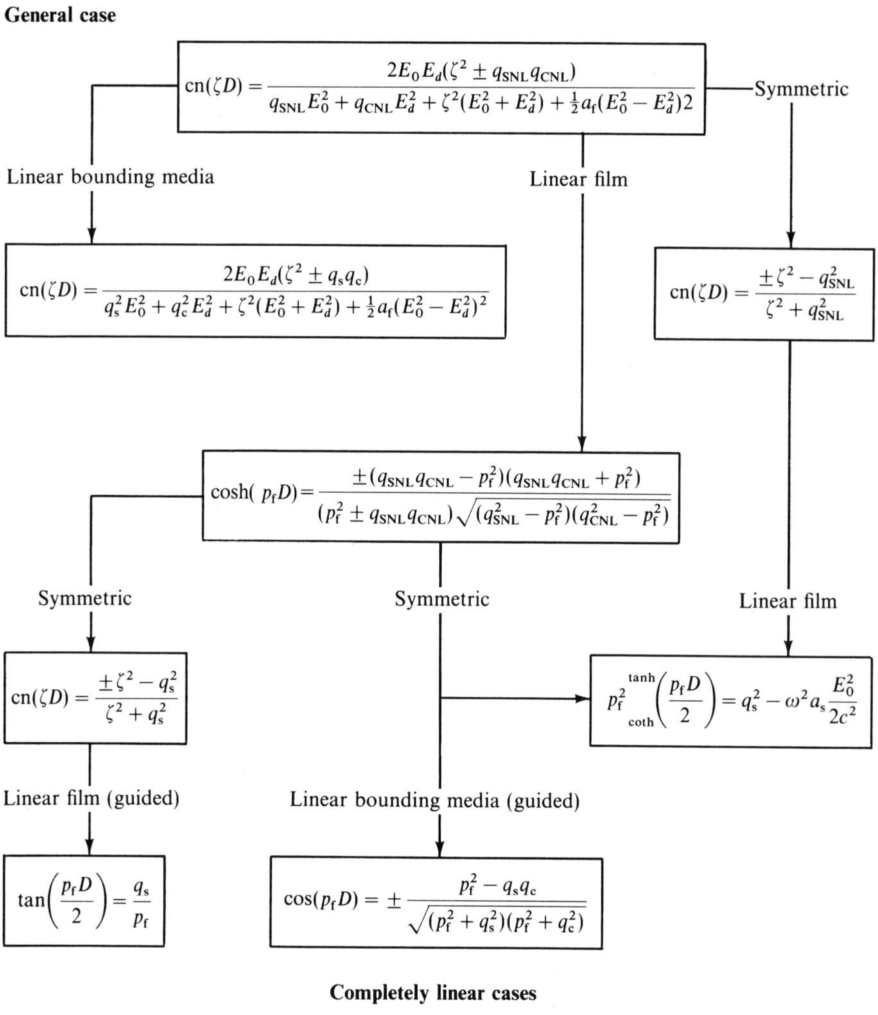

Since E_i is real this requires $a_i > 0$ and from eq. (3.9), with $K = 0$,

$$E_i'' = k_0^2(q_i^2 - 2q_i^2)q_i\sqrt{\frac{2}{a_i}} = -k_0^2\sqrt{\frac{2}{a_i}}q^3 < 0. \tag{3.78}$$

Hence the turning point is a maximum in accord with the earlier analysis.

The boundary conditions at $z = 0$ are

$$E_{\mathrm{s}}(0) = E_{\mathrm{c}}(0) = E_0, \quad E_{\mathrm{s}}'(0) = E_{\mathrm{c}}'(0), \tag{3.79}$$

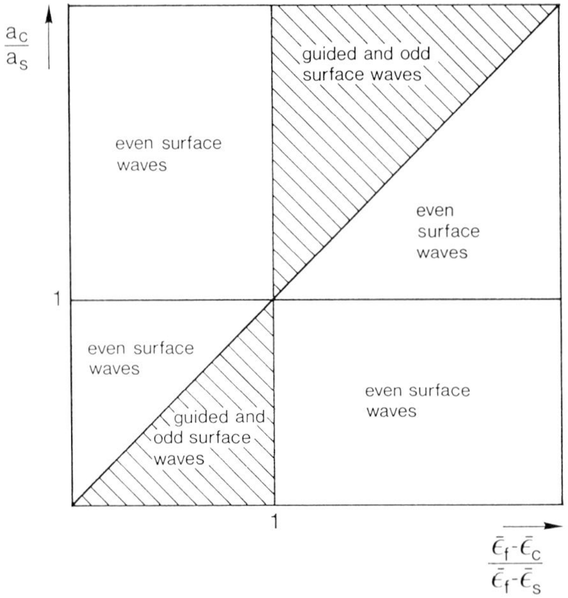

Fig. 3.17. Diagram to illustrate the type of symmetric nonlinear stationary state that can exist in an asymmetric system (Boardman and Egan 1984b).

and the first integral, yields

$$(q_s + q_c)(q_s - q_c) = \tfrac{1}{2}(a_s - a_c)E_0^2. \tag{3.80}$$

The linear limit is $E_0^2 \to 0$ and has no surface-guided wave solution.
This reduces to

$$E_0^2 = \frac{2(\bar{\varepsilon}_s - \bar{\varepsilon}_c)}{a_c - a_s}. \tag{3.81}$$

In all the cases up to now the eigenvalue equation shows that n_x is a function of $\frac{1}{2}a_s E_0^2$. In this single interface it is very interesting that E_0^2 is now fixed and that there appears to be no relationship between n_x and ω. The question now arises as to whether this relationship for E_0^2 will lead in any other way to the conclusion that the surface waves are dispersive.

In order to pursue this point here, without having to invoke the total power flow as a label for dispersion curves, as will be done in the next section, the basic equations can be re-examined at the stage where the boundary conditions are being applied. Also the discussion can be made more intelligible by permitting only the cladding medium to be nonlinear. The TE field (E_y) in each medium is then (Miyagi and Nishida 1972)

$$E_s(z) = E_0 e^{-(k_x^2 - \omega^2\bar{\varepsilon}_s/c^2)_z^{1/2}} = E_0 e^{-q_s z}, \quad z < 0, \tag{3.82}$$

$$E_c(z) = \sqrt{\frac{2}{a_c}}\, q_c \operatorname{sech}[q_c(z - z_0)], \quad z > 0, \tag{3.83}$$

where z is measured in units of k_0^{-1}, $k_0 = \omega/c$, $q_c^2 = n_x^2 - \bar{\varepsilon}_c$ and z_0 locates a real maximum in the field intensity. The application of the boundary conditions then gives

$$E_0 = \sqrt{\frac{2}{a_c}}\, q_c \operatorname{sech}(q_c z_0), \tag{3.84}$$

$$q_s E_0 = q_c^2 \sqrt{\frac{2}{a_c}} \tanh(q_c z_0) \operatorname{sech}(q_c z_0), \tag{3.85}$$

so that

$$\tanh(q_c z_0) = \frac{q_s}{q_c}, \tag{3.86a}$$

$$q_c \tanh(q_c z_0) = \sqrt{q_c^2 + (\bar{\varepsilon}_c - \bar{\varepsilon}_s)}. \tag{3.86b}$$

The bulge (i.e., the maximum of the field) must always be in the nonlinear medium for positive dielectrics and is a self-focussing phenomenon that, for a single interface, has no linear limits. Therefore, $z_0 > 0$ and

$$\tfrac{1}{2} a_c E_0^2 z_0^2 = (\bar{\varepsilon}_s - \bar{\varepsilon}_c) z_0^2 = (q_c z_0)^2 \operatorname{sech}^2(q_c z_0). \tag{3.87}$$

This equation shows that for a given value of $(\bar{\varepsilon}_s - \bar{\varepsilon}_c) z_0^2$ here are *two* values of $(q z_0)$. There are, it seems, two types of surface wave one of which, with the larger $q_c z_0$, has a lower phase velocity and a faster transverse attenuation constant than the other surface wave. For a given system, E_0^2 is *fixed*, however, and *independent* of the power. Thus no matter what power level exists the field solutions *pivot* about the value E_0^2. Also the field maximum gets closer to the interface as the difference $(\bar{\varepsilon}_s - \bar{\varepsilon}_c)$ increases since the range of z_0 that gives the maximum field intensity is

$$0 \leqslant z_0 \leqslant \frac{0.66275}{k_0 \sqrt{\bar{\varepsilon}_s - \bar{\varepsilon}_c}}. \tag{3.88}$$

The decay factor of the field in the linear medium is always less than that in the nonlinear medium but surface waves with a large decay factor have a field distribution that tends towards symmetry about z_0 as the decay factor increases.

3.1.5. Power flow

The linear and nonlinear materials are assumed here to be non-absorptive with the net power flow being parallel to the boundary. This power should, in principle, be calculated by integrating the time-averaged Poynting vector $\langle S \rangle$ over $-\infty \leqslant z \leqslant \infty$ but, as will be seen below, the integration can actually be avoided in a rather striking way.

For TE waves the Poynting vector is

$$\langle \boldsymbol{S} \rangle = \tfrac{1}{2}\,\mathrm{Re}(\boldsymbol{E}\times\boldsymbol{H}) = \frac{c^2\varepsilon_0}{2\omega}\,\mathrm{Re}(k_x E_y^2,\, 0,\, -\mathrm{i}E_y E_y'), \tag{3.89}$$

where Re denotes the real part. The latter point only matters if E_y and k_x are complex which they will not be in this case. The power P transported by the wave, per unit length in the y-direction, is obtained by integrating the x-component of $\langle \boldsymbol{S} \rangle$ to give

$$P = \int_{-\infty}^{\infty} \langle \boldsymbol{S} \rangle \cdot \hat{\boldsymbol{x}} = \frac{c^2\varepsilon_0 k_x}{2\omega}\int_{-\infty}^{\infty} E_y^2\,\mathrm{d}z. \tag{3.90}$$

For guiding structures in which the nonlinear media are semi-infinite the integration constant that appears in the evaluation of the first integral is zero. For these guides P may be evaluated very simply without integration (Boardman and Egan 1984b, 1985). It is done in terms of the boundary fields and their derivatives and is an elegant approach that has been overlooked in a lot of the literature. Naturally, carrying out the actual integrations yields exactly the same results that are derived below.

The first integral for each semi-infinite nonlinear medium, bounding a linear guiding film, is

$$(E_i'/E_i)^2 = k_0^2(q_i^2 - \tfrac{1}{2}a_i E_i^2), \quad i = \mathrm{s,\, c}. \tag{3.91}$$

The differentiation of this expression with respect to z gives

$$2\left(\frac{E_i'}{E_i}\right)\frac{\mathrm{d}}{\mathrm{d}z}\left(\frac{E_i'}{E_i}\right) = -k_0^2 a_i E_i' E_i, \tag{3.92}$$

i.e.,

$$\frac{\mathrm{d}}{\mathrm{d}z}\left(\frac{E_i'}{E_i}\right) = \tfrac{1}{2}k_0^2 a_i E_i^2. \tag{3.93}$$

The power flow integral then becomes

$$\int_{b_1}^{b_2} E_i^2\,\mathrm{d}z = -\frac{2}{k_0^2 a_i}\left[\frac{E_i'}{E_i}\right]_{b_1}^{b_2}, \tag{3.94}$$

where the limits are $b_1 \to -\infty$, $b_2 \to \infty$. This technique transfers attention to (E_i'/E_i) whose limits, as $z \to \pm\infty$, are

$$\lim_{E_i \to 0}\left(\frac{E_i'}{E_i}\right) = \pm k_0 q_i. \tag{3.95}$$

For a single interface, separating a semi-infinite substrate (s) from a semi-infinite cladding (c), the $z \to 0$ limit is

$$\left(\frac{E_i'}{E_i}\right)_{z=0} = \pm k_0 q_{\mathrm{SNL}} = \pm k_0 q_{\mathrm{CNL}}. \tag{3.96}$$

Hence, for such a single interface, the total power flowing in the substrate is

$$P_{\mathrm{s}} = \frac{c^2 \varepsilon_0 n_x}{\omega a_{\mathrm{s}}}(q_{\mathrm{s}} \pm q_{\mathrm{SNL}}), \tag{3.97}$$

where q_{SNL} is evaluated at the boundary field value E_0, the plus sign implies that $(E_{\mathrm{s}}'/E_{\mathrm{s}})_{z=0} < 0$ and the minus sign implies that $(E_{\mathrm{c}}'/E_{\mathrm{c}})_{z=0} > 0$.

The corresponding total power flowing in the cladding is

$$P_{\mathrm{c}} = \frac{c^2 \varepsilon_0 n_x}{\omega a_{\mathrm{c}}}(q_{\mathrm{c}} \pm q_{\mathrm{CNL}}), \tag{3.98}$$

where q_{CNL} is also evaluated at the boundary field value E_0, the plus sign implies that $(E_{\mathrm{c}}'/E_{\mathrm{c}})_{z=0} > 0$ and the minus sign implies that $(E_{\mathrm{c}}'/E_{\mathrm{c}})_{z=0} < 0$. The total power flow is obviously $P = P_{\mathrm{c}} + P_{\mathrm{s}}$ but, because of the sign allocation just discussed, there are two results, one is $P = P_{(+)}^{(\mathrm{N,N})}$ if the gradient at the boundary is positive (with respect to z) and the other is $P_{(-)}^{(\mathrm{N,N})}$ if the gradient $E_{\mathrm{c}}'/E_{\mathrm{c}}$ at the boundary is negative.

Now at the boundary, from eq. (3.33),

$$q_{\mathrm{SNL}} = q_{\mathrm{CNL}} = \left[n_x^2 - \left(\frac{a_{\mathrm{c}}\bar{\varepsilon}_{\mathrm{s}} - a_{\mathrm{s}}\bar{\varepsilon}_{\mathrm{c}}}{a_{\mathrm{c}} - a_{\mathrm{s}}}\right)\right]^{1/2} \equiv q_{\mathrm{NL}}, \tag{3.99}$$

so that $P_{(\pm)}^{(\mathrm{N,N})}$ can be written as

Field maximum in cladding:

$$P_{(+)}^{(\mathrm{N,N})} = \frac{c^2 \varepsilon_0 n_x}{\omega}\left[\frac{q_{\mathrm{s}} - q_{\mathrm{NL}}}{a_{\mathrm{s}}} + \frac{q_{\mathrm{c}} + q_{\mathrm{NL}}}{a_{\mathrm{c}}}\right], \tag{3.100}$$

Field maximum in substrate:

$$P_{(-)}^{(\mathrm{N,N})} = \frac{c^2 \varepsilon_0 n_x}{\omega}\left[\frac{q_{\mathrm{s}} + q_{\mathrm{NL}}}{a_{\mathrm{s}}} + \frac{q_{\mathrm{c}} - q_{\mathrm{NL}}}{a_{\mathrm{c}}}\right]. \tag{3.101}$$

These formulae reflect a precise construction of the field profile, a curious fact first noted for the linear–nonlinear combination. For example, if a field maximum occurs in the substrate then, if the integration range is split up into regions on either side of z_0, where the field maximum occurs,

$$\int_{-\infty}^{z^0} E_{\mathrm{s}}^2 \, \mathrm{d}z = -\frac{2c^2}{\omega^2 a_{\mathrm{s}}}\left[\frac{E_{\mathrm{s}}'}{E_{\mathrm{s}}}\right]_{-\infty}^{z_0} = \frac{2c}{\omega a_{\mathrm{s}}} q_{\mathrm{s}}, \tag{3.102}$$

$$\int_{z_0}^{0} E_s^2 \,\mathrm{d}z = -\frac{2c^2}{\omega^2 a_s}\left[\frac{E_s'}{E_s}\right]_{z_0}^{0} = \frac{2c}{\omega a_s} q_{\mathrm{NL}}. \tag{3.103}$$

A similar result is found if the field amplitude is in the cladding.

For an interface between a linear and a nonlinear medium, assuming that the substrate is nonlinear, the total power flow is $P^{(\mathrm{N,L})}$ where

$$\begin{aligned} P^{(\mathrm{N,L})} &= \frac{c^2 \varepsilon_0 n_x}{\omega}\left[\frac{q_s + q_{\mathrm{NL}}}{a_s} + \lim_{a_c \to 0} \frac{[q_c - q_{\mathrm{NL}}]}{a_c}\right] \\ &= \frac{c^2 \varepsilon_0 n_x}{\omega}\left[\frac{q_s + q_{\mathrm{NL}}}{a_s} + \frac{\bar{\varepsilon}_c - \bar{\varepsilon}_s}{2 a_s q_c}\right]. \end{aligned} \tag{3.104}$$

Now $q_{\mathrm{NL}} = q_c$, since $a_c = 0$, so that $P^{(\mathrm{N,L})}$ simplifies to

$$P^{(\mathrm{N,L})} = \frac{c^2 \varepsilon_0 n_x}{\omega a_s}\left[q_s + q_c + \frac{\bar{\varepsilon}_c - \bar{\varepsilon}_s}{2 q_c}\right]. \tag{3.105}$$

The final term corresponds to power carried in a linear medium.

Equation (3.105) shows that for $q_c \to 0$ (i.e., $n_x \to \sqrt{\bar{\varepsilon}_c}$) large powers are necessary to maintain the surface wave and also that, as $n_x \to \sqrt{\bar{\varepsilon}_c}$, most of the power is carried in the linear medium.

For a three-constituent system consisting of a semi-infinite nonlinear substrate (s), a linear layer (f) and semi-infinite nonlinear cladding (c), of the type discussed extensively in previous sections, the total power flow ($P_s + P_c$) for the semi-infinite constituents is exactly the same as the ($P_c + P_s$) of the foregoing single interface system.

In the linear film guide however, the derivative of the field is, from eq. (3.17),

$$E_f' = \pm (C_f - k_0^2 p_f^2 E_f^2)^{1/2}, \tag{3.106}$$

where C_f, the integration constant, introduced in eq. (3.17), is

$$C_f = k_0^2 (p_f^2 + q_{\mathrm{SNL}}^2) E_0^2 = k_0^2 (p_f^2 + q_{\mathrm{CNL}}^2) E_d^2. \tag{3.107}$$

Equation (3.106) may be squared and multiplied by E_f^2 to give

$$\left[E_f \frac{\mathrm{d}E_f}{\mathrm{d}z}\right]^2 + k_0^2 p_f^2 E_f^4 - C_f E_f^2 = 0. \tag{3.108}$$

This equation, upon differentiation with respect to z, becomes

$$2E_f \frac{\mathrm{d}E_f}{\mathrm{d}z}\frac{\mathrm{d}}{\mathrm{d}z}\left[E_f \frac{\mathrm{d}E_f}{\mathrm{d}z}\right] + 4k_0^2 p_f^2 E_f^3 \frac{\mathrm{d}E_f}{\mathrm{d}z} - 2C_f E_f \frac{\mathrm{d}E_f}{\mathrm{d}z} = 0, \tag{3.109}$$

so that

$$E_f^2 = \frac{1}{2 p_f^2 k_0^2}\left[C_f - \frac{\mathrm{d}}{\mathrm{d}z}\left(E_f \frac{\mathrm{d}E_f}{\mathrm{d}z}\right)\right]. \tag{3.110}$$

The total power flow through the linear layer is, therefore,

$$P_{\mathrm{f}} = \frac{c^2 \varepsilon_0 k_x}{2\omega} \int_0^d E_{\mathrm{f}}^2 \, \mathrm{d}z$$

$$= \frac{c^2 \varepsilon_0 k_x}{4k_0^2 \omega p_{\mathrm{f}}^2} [C_{\mathrm{f}} d - (E_{\mathrm{f}} E_{\mathrm{f}}')_{z=d} + (E_{\mathrm{f}} E_{\mathrm{f}}')_{z=0}]$$

$$= \frac{c^2 \varepsilon_0 n_x}{4\omega p_{\mathrm{f}}^2} [(p_{\mathrm{f}}^2 + q_{\mathrm{SNL}}^2) E_0^2 D \pm q_{\mathrm{SNL}} E_0^2 \pm q_{\mathrm{CNL}} E_d^2], \tag{3.111}$$

where taking the upper signs means that $(E_{\mathrm{f}}'/E_{\mathrm{f}})_{z=0} > 0$ and $(E_{\mathrm{f}}'/E_{\mathrm{f}})_{z=d} > 0$.
The total power flow in the waveguide is (Boardman and Egan 1985)

$$P = \frac{c^2 \varepsilon_0 n_x}{a_{\mathrm{c}} \omega} \Bigg\{ \left(\pm \sqrt{n_x^2 - \bar{\varepsilon}_{\mathrm{s}} - \tfrac{1}{2} a_{\mathrm{s}} E_0^2} + \sqrt{n_x^2 - \bar{\varepsilon}_{\mathrm{s}}} \right) \beta$$
$$+ \sqrt{n_x^2 - \bar{\varepsilon}_{\mathrm{c}}} \pm \sqrt{n_x^2 - \bar{\varepsilon}_{\mathrm{c}} - \tfrac{1}{2} a_{\mathrm{c}} E_d^2}$$
$$+ \frac{1}{4(\bar{\varepsilon}_{\mathrm{f}} - n_x^2)} [(\bar{\varepsilon}_{\mathrm{f}} - \bar{\varepsilon}_{\mathrm{s}} - \tfrac{1}{2} a_{\mathrm{s}} E_0^2) \beta D \tfrac{1}{2} a_{\mathrm{s}} E_0^2$$
$$\mp \tfrac{1}{2} a_{\mathrm{s}} E_0^2 \beta \sqrt{n_x^2 - \bar{\varepsilon}_{\mathrm{s}} - \tfrac{1}{2} a_{\mathrm{s}} E_0^2}$$
$$\mp \tfrac{1}{2} a_{\mathrm{c}} E_d^2 \sqrt{n_x^2 - \bar{\varepsilon}_{\mathrm{c}} - \tfrac{1}{2} a_{\mathrm{c}} E_d^2}] \Bigg\} \tag{3.112}$$

Numerical calculations based upon this result will be presented in section 5.

Since the intensity dependent refractive index of bulk nonlinear material is sometimes expressed in terms of another coefficient n_{2I}, it is useful to establish its relationship to the coefficient a that has been used upto now. If the bulk refractive index is written as $n_0 + n_{2I}$ then I, the intensity of the field, is the time-averaged Poynting vector $(n_0/2\mu_0 c) E^2$. The dielectric function is then, approximately, $n_0^2 + 2n_0 n_{2I} I$ so that the nonlinear coefficient used in this text is $a = n_0^2 n_{2I}/\mu_0 c$ with units of m^2/volt^2.

If the waveguide consists of a nonlinear film, sandwiched between semi-infinite linear media, then the calculation of the power is much more difficult. In this case the use of the dependence of the wave index upon the optical power density at the lower boundary is recommended as being a more physical method than analysing the problem in terms of a dependence of n_x upon the modulus of the Jacobi function. As has already been discussed, the small penalty that is incurred is that E_{f} becomes a combination of Jacobi elliptic functions. This combination, although quite straightforward, is still too complex, however, to permit the kind of analytical solution for the power flow that is enjoyed for the linear guiding layer with nonlinear boundaries. For a nonlinear film, then, the total time-averaged power flow down the guide is

$$P = \frac{(c^2 \varepsilon_0 n_x) a_f}{4\omega} \left[\frac{E_0^2}{q_s} + \frac{E_d^2}{q_c} + 2E_0^2 \int_0^D \xi^2 \left(\frac{\xi \operatorname{cn}(\xi z) + q_s \operatorname{sn}(\xi z) \operatorname{dn}(\xi z)}{\xi^2 \operatorname{dn}^2(\xi z) + \frac{1}{2} a_f E_0^2 \operatorname{sn}^2(\xi z)} \right) \mathrm{d}z \right], \tag{3.113}$$

where z is measured in units of k_0^{-1}.

In general, for a nonlinear layer bounded by dissimilar nonlinear media, the total power flow is

$$\begin{aligned} P = \frac{(c^2 \varepsilon_0 n_x)}{2\omega a_c} \Bigg\{ & \beta [\sqrt{n_x^2 - \bar{\varepsilon}_s} \pm (n_x^2 - \bar{\varepsilon}_s - \tfrac{1}{2} a_s E_0^2)^{1/2}] \\ & + \sqrt{n_x^2 - \bar{\varepsilon}_c} \pm (n_x^2 - \bar{\varepsilon}_c - \tfrac{1}{2} a_c E_d^2)^{1/2} \\ & + \tfrac{1}{2} a_c E_d^2 \int_0^D \xi^2 \frac{\xi \operatorname{cn}(\xi z) + q_s \operatorname{sn}(\xi z) \operatorname{dn}(\xi z)}{\xi^2 \operatorname{dn}^2(\xi z) + \frac{1}{2} a_f E_0^2 \operatorname{sn}^2(\xi z)} \mathrm{d}z \Bigg\}. \end{aligned} \tag{3.114}$$

The power transported down a symmetric guide consisting of a linear layer bounded by identical nonlinear media is obtainable from the simple straight line (E_0^2, E_d^2) relationships and the eigenvalue equations. For the case $E_0^2 = E_d^2$, the total power flow in a symmetric structure is

$$\begin{aligned} P = \frac{(c^2 \varepsilon_0 n_x)}{2\omega a_c} & \{ q_s - p_f \,{}_{(-)}{}^{\tan}_{\cot}(p_f \tfrac{1}{2} D) \} \\ & \times \left\{ 2 + [q_s + p_f \,{}_{(-)}{}^{\tan}_{\cot}(p_f \tfrac{1}{2} D)] \left[\frac{\frac{1}{2} D}{{}^{\cos^2}_{\sin^2}(p_f \frac{1}{2} D)} + {}_{(-)}{}^{\tan}_{\cot}(p_f \tfrac{1}{2} D) \right] \right\}, \end{aligned} \tag{3.115}$$

where $(-)$ denotes the antisymmetric modes. An analytical result may also be developed for the asymmetric modes and the expression (3.115) is identical to one that can be found in the literature, provided that a transformation between surface and guide waves is made.

3.2. *TM waves*

Theoretical formulations of nonlinear TM surface-guided wave propagation that are free of oversimplifying approximations have only recently been produced (Boardman et al. 1987a, b, Boardman and Twardowski 1988a, b). Also the status of the various approximations may only be fully appreciated numerically and a detailed discussion will be deferred until section 5.

Briefly, the nonlinear TM waves, investigated here, propagate in the xz-plane of an isotropic medium and along the x-axis. The wave carries electric field components (E_x, E_z) that are $\frac{1}{2}\pi$ out of phase. The approximations (Agranovich et al. 1980, Agranovich and Chernyak 1982a, b, Ariyasu et al. 1985a, Arutyunan and Barsukov 1985a, Bhat and Tripathi 1975, Boev 1980, Lederer et al. 1983b, Leung 1985b, Lomtev 1981, Mihalache 1986a, Mihalache and Mazilu 1986, Mihalache et al. 1984b, Shivarova and Dimitrov 1985) that have been frequently used in the literature either neglect E_x or E_z, in the nonlinear medium, or give them equal weight in the nonlinear polarisation, even if it is inappropriate to do so. The waveguide structures to which these approximations have been applied consist of linear substrates with positive (inactive) or negative (active) dielectric functions and a nonlinear cladding with a positive linear part to the dielectric function.

For the cladding field components of the kind of surface plasmon-polariton (Seaton et al. 1985d), sustained between a linear negative dielectric substrate and a nonlinear cladding, $E_z \gg E_x$ in the infra-red, whereas $E_z \simeq E_x$, in the visible. It is never the case that $E_z \ll E_x$ for the cladding field. It could be true if the nonlinear material was also an active material such as a plasma. Hence in an isotropic nonlinear medium, a TM wave will, in general, experience a rather complicated third-order nonlinear polarisation that assumes various forms according to the relative strengths of the electric field components and the nonlinear mechanism involved. If E_x and E_z are given equal weight at all frequencies, such an assumption would be consistent with electrostrictive or thermal nonlinearity and has been given quite a lot of previous attention in plasma physics and in some recent solid state literature. For the type of fast response solid state nonlinear materials envisaged for integrated optics, electronic distortion within isotropic materials is a more probable nonlinear mechanism, however, so it is necessary to consider a reasonably full list of nonlinear mechanisms. These include electronic distortion, the molecular Kerr effect and electrostriction (thermal). All of these are accounted for in the theory given below. Early work (Alanakyan 1967) that used a perturbation theory to investigate nonlinear TM waves in materials of lower symmetry has not yet been extended to the strongly nonlinear regime.

On a more general level, a study of nonlinear TM wave propagation is important because it has long been realised that the polarisation of an electromagnetic wave plays a non-trivial role in the self-focussing of light beams (Pohl 1970). For example, stationary nonlinear TM waves in bulk nonlinear media have been analysed in some detail and it has been shown that solitons can be formed with a solitary TM waveguide channel that is wider than the corresponding channel for TE waves. In addition to this kind of polarisation dependence, many systems exist that, in the linear (zero power) limit, only support TM surface-guided waves. These guides have always attracted attention and involve waveguide structures in which there exist one or more components that have negative dielectric functions such as metal films or substrates. These, as hinted at above, also possess a strong frequency dependent dielectric function and behave rather differently in different parts of the electromagnetic spectrum.

3.2.1. The field equations

Suppose that the TM nonlinear electromagnetic waves, propagate, in the xz-plane along the x-axis in an isotropic medium with wave number k_x and angular frequency ω, and have the form

$$\boldsymbol{E} = [E_x(z)]\mathrm{e}^{\mathrm{i}(k_x - \omega t)},$$

$$\boldsymbol{B} = [0, B_y(z), 0]\mathrm{e}^{\mathrm{i}(k_x - \omega t)}, \tag{3.116}$$

where the field components $E_x(z)$ and $E_z(z)$ are related to $B_y(z)$ through the equation

$$\frac{\partial E_x}{\partial z} - \mathrm{i}k_x E_z = \mathrm{i}\omega B_y. \tag{3.117}$$

For notational convenience the dependence of E_x, E_z and B_y upon z is suppressed in eq. (3.117) and in all subsequent expressions. A further improvement in the presentation results from a conversion to dimensionless coordinates. This can be done by replacing z by cz/ω, where c is the velocity of light in vacuo, replacing the electric field components $E_{x,z}$ by $cE_{x,z}$, and then by introducing the dimensionless parameter $n_x^2 = [ck_x/\omega]^2$. After these transformations eq. (3.117) becomes

$$\frac{\partial E_x}{\partial z} - \mathrm{i}n_x E_z = \mathrm{i}B_y. \tag{3.118}$$

This is a nonlinear equation since B_y is related to E_x and E_z through the nonlinear dielectric tensor

$$\varepsilon^{\mathrm{NL}} = \begin{pmatrix} \varepsilon_{xx} & 0 \\ 0 & \varepsilon_{zz} \end{pmatrix}. \tag{3.119}$$

After taking into account the $\frac{1}{2}\pi$ phase difference between E_x and E_z for a guided/surface wave, the components of the nonlinear dielectric tensor are

$$\varepsilon_{xx} = \bar{\varepsilon} + a|E_x|^2 + \gamma a|E_z|^2, \tag{3.120}$$

$$\varepsilon_{zz} = \bar{\varepsilon} + \gamma a|E_x|^2 + a|E_z|^2, \tag{3.121}$$

where $\bar{\varepsilon}$ is the linear dielectric function, a is the usual nonlinear coefficient and, theoretically, γ is $\frac{1}{3}$, $-\frac{1}{2}$ or 1 according to whether the nonlinear mechanism is modelled as electronic distortion, the molecular orientational Kerr effect or electrostriction/thermal. The relationships of B_y to E_x and E_z are

$$E_x = -\frac{\mathrm{i}}{\varepsilon_{xx}}\frac{\partial B_x}{\partial z}\ E_z = -\frac{n_x}{\varepsilon_{zz}}B_y. \tag{3.122}$$

Equations (3.118) and (3.122) lead to

$$\frac{\partial}{\partial z}\left[\frac{1}{\varepsilon_{xx}}\frac{\partial B_y}{\partial z}\right] - \frac{n_x^2 - \varepsilon_{zz}}{\varepsilon_{zz}}B_y = 0. \tag{3.123}$$

3.2.2. The first integral

If eq. (3.123) is multiplied through by $2\partial B_y/\partial z$, the derivative with respect to z is taken in the first term and $(n_x^2/\varepsilon_{zz}^2)(\partial\varepsilon_{zz}/\partial z)B_y^2$ is both added and subtracted then it becomes

$$\frac{n_x^2-\varepsilon_{zz}}{\varepsilon_{zz}}2B_y\left(\frac{\partial B_y}{\partial z}\right)-\frac{2}{\varepsilon_{xx}}\left(\frac{\partial B_y}{\partial z}\right)\frac{\partial^2 B_y}{\partial z^2}+\frac{2}{\varepsilon_{xx}^2}\frac{\partial\varepsilon_{xx}}{\partial z}\left(\frac{\partial B_y}{\partial z}\right)^2$$

$$-\frac{n_x^2}{\varepsilon_{zz}^2}\frac{\partial\varepsilon_{zz}}{\partial z}B_y^2+\frac{n_x^2}{\varepsilon_{zz}^2}\frac{\partial\varepsilon_{zz}}{\partial z}B_y^2=0. \tag{3.124}$$

Equation (3.124) may now be rewritten as

$$\frac{\partial}{\partial z}\left[\frac{n_x^2-\varepsilon_{zz}}{\varepsilon_{zz}}B_y^2\right]+\frac{n_x^2}{\varepsilon_{zz}^2}\frac{\partial\varepsilon_{zz}}{\partial z}B_y^2-\frac{\partial}{\partial z}\left[\frac{1}{\varepsilon_{xx}}\left(\frac{\partial B_y}{\partial z}\right)^2\right]+\left(\frac{1}{\varepsilon_{xx}}\frac{\partial B_y}{\partial z}\right)^2\frac{\partial\varepsilon_{xx}}{\partial z}=0. \tag{3.125}$$

The first integral of eq. (3.125) is

$$\frac{n_x^2-\varepsilon_{zz}}{\varepsilon_{zz}}B_y^2+\int\frac{n_x^2}{\varepsilon_{zz}^2}B_y^2\,\mathrm{d}\varepsilon_{zz}-\frac{1}{\varepsilon_{xx}}\left(\frac{\partial B_y}{\partial z}\right)^2+\int\left(\frac{1}{\varepsilon_{xx}}\frac{\partial B_y}{\partial z}\right)^2\mathrm{d}\varepsilon_{xx}=C, \tag{3.126}$$

where C is an integration constant.

If the $\frac{1}{2}\pi$ phase relationship between E_x and E_z is specifically introduced through the transformations $E_x\to \mathrm{i}|E_x|$, $E_z\to|E_z|$ then

$$\frac{n_x^2-\varepsilon_{zz}}{\varepsilon_{zz}}B_y^2-\varepsilon_{xx}|E_x|2+\int|E_x|^2\,\mathrm{d}\varepsilon_{xx}+\int|E_z|^2\,\mathrm{d}\varepsilon_{zz}=C. \tag{3.127}$$

This equation, after using eqs (3.120) and (3.121), expands to

$$\frac{(n_x^2-\varepsilon_{zz})}{\varepsilon_{zz}}B_y^2-(\bar{\varepsilon}|E_x|^2+a|E_x|^4+\gamma a|E_x|^2|E_z|^2)$$

$$+\int|E_x|^2\{2a|E_x|\,\mathrm{d}|E_x|+2\gamma a|E_z|\,\mathrm{d}|E_z|\}$$

$$+\int|E_z|^2\{2\gamma a|E_x|\,\mathrm{d}|E_x|+2a|E_z|\,\mathrm{d}|E_z|\}=C. \tag{3.128}$$

After adding and taking away $(n_x^2/\varepsilon_{zz})B_y^2$

$$\frac{2n_x^2-\varepsilon_{zz}}{\varepsilon_{zz}}B_y^2-\bar{\varepsilon}E_\mathrm{T}^2-\tfrac{1}{2}a[E_\mathrm{T}^4-2|E_x|^2|E_z|^2+2\gamma|E_x|^2|E_z|^2]=C, \tag{3.129}$$

is obtained, where $E_\mathrm{T}^2=|E_x|^2+|E_z|^2$.

Since B_y is directly related to $\partial E_x/\partial z$, either may be used to define a potential function $V(E_x, E_z)$ with z playing the role of time. This can lead to an alternative representation of nonlinear TM guided waves in terms of 'particle' dynamics within such a potential. A detailed discussion of this approach will be deferred until the next section where it will be seen, at least for thermal nonlinearities, that it can be an interesting way of looking at quite complicated nonlinear problems. It is not a widely used method, however, nor is it necessary to do so to obtain exact numerical solutions. Its limitations will become clearer in the next section.

3.2.3. Nonlinear dispersion equation

3.2.3.1. Single interface. Consider, once again, the $z=0$ interface between a semi-infinite linear cladding ($z>0$) and semi-infinite nonlinear substrate ($z<0$). Although, in general, it will be necessary to consider an interface between two nonlinear media, very little is compromised here by restricting attention to this less involved case. The boundary conditions on the appropriate field components are (Boardman et al. 1987b)

$$[[E_x]]_0=0, \qquad [[B_y]]_0=0, \qquad [[D_z]]_0=0, \tag{3.130}$$

where $[[A]]_0$ denotes the change in quantity A as the interface at $z=0$ is crossed, and D_z is the normal component of the displacement vector. The magnetic flux density in the linear medium is

$$B_y = A\mathrm{e}^{-\sqrt{(n_x^2-\bar{\varepsilon}_c)}z}, \tag{3.131}$$

where A is a constant and $\bar{\varepsilon}_c$ is the linear dielectric function of the $z>0$, cladding medium. The boundary conditions at $z=0$ give

$[[E_x]]_0=0$:

$$|E_x(0)|^2 = \frac{n_x^2-\bar{\varepsilon}_c}{\bar{\varepsilon}_c^2}A^2, \tag{3.132}$$

$[[D_z]]_0=0$:

$$|E_z(0)|^2 = \frac{n_x^2}{\varepsilon_{zz0}^2}A^2, \tag{3.133}$$

where $E_x(0)$ and $E_z(0)$ are the nonlinear field components evaluated at $z=0$ and ε_{zz0} is the z-component of the nonlinear dielectric tensor evaluated at $z=0$. From eqs (3.120) and (3.121),

$$\varepsilon_{zz0} = \bar{\varepsilon}_s + \gamma a_s|E_x(0)|^2 + a_s|E_z(0)|^2. \tag{3.134}$$

Equations (3.132), (3.133) and (3.134) show immediately that ε_{zz0} is the solution of the cubic equation

$$\varepsilon_{zz0}^3 - \left[\bar{\varepsilon}_s + \frac{\gamma a_s(n_x^2-\bar{\varepsilon}_c)}{\bar{\varepsilon}_c^2}A^2\right]\varepsilon_{zz0}^2 - a_s n_x^2 A^2 = 0. \tag{3.135}$$

Since the constant of integration in the first integral is exactly zero for semi-infinite media, the application of the condition $[[B_y]]_0 = 0$ yields the following biquadratic equation for n_x,

$$\left[\tfrac{1}{2}a_s\left(\frac{1}{\bar{\varepsilon}_c^2}+\frac{1}{\varepsilon_{zz0}^2}\right)^2-\frac{a_s(1-\gamma)}{\bar{\varepsilon}_c^2\varepsilon_{zz0}^2}\right]A^2n_x^4$$
$$+\left[\frac{\bar{\varepsilon}_s}{\bar{\varepsilon}_c^2}+\frac{\bar{\varepsilon}_s}{\varepsilon_{zz0}^2}-\frac{a_s}{\bar{\varepsilon}_c}\left(\frac{1}{\bar{\varepsilon}_c^2}+\frac{1}{\varepsilon_{zz0}^2}\right)A^2+\frac{a_s(1-\gamma)}{\bar{\varepsilon}_c\varepsilon_{zz0}^2}A^2-\frac{2}{\varepsilon_{zz0}}\right]n_x^2$$
$$-\frac{\varepsilon_s}{\bar{\varepsilon}_c}+\frac{a_sA^2}{2\bar{\varepsilon}_c^2}+1=0. \tag{3.136}$$

Equations (3.135) and (3.136) thus comprise an algorithm from which the dependence of n_x upon A^2 may be extracted.

3.2.3.2. Linear film bounded by identical semi-infinite nonlinear media. For a layered waveguide structure (Boardman et al. 1987b), such as a symmetric guide consisting of a linear film bounded by identical semi-infinite nonlinear media, the method of first integrals will also lead to a reasonably compact algorithm for the determination of the effective guide index dependence upon the optical power density at a boundary.

An illustration may be derived from a consideration of the B_y associated with the symmetric mode of the film, namely

$$B_y = A\cos(p_f z), \tag{3.137}$$

where $p_f = \sqrt{\bar{\varepsilon}_f - n_x}$ and $\bar{\varepsilon}_f$ is the dielectric constant of the thin linear layer. The boundary conditions at $z = d$ are

$$[[B_y]]_d = 0, \qquad [[D_z]]_d = 0, \tag{3.138}$$

so that, if the nonlinear field component $B_y(d)$, evaluated at d, is defined as B_d then

$$A = \frac{B_d}{\cos(p_f d)}, \tag{3.139}$$

$$E_x(d) = \mathrm{i}\left(\frac{p_f}{\bar{\varepsilon}_f}\right)B_d\tan(p_f d), \tag{3.140}$$

$$\varepsilon_{zzd}E_z(d) = -n_x B_d, \tag{3.141}$$

where ε_{zzd} is now evaluated at $z = d$ and is a solution of the cubic equation

$$\varepsilon_{zzd}^3-\left[\bar{\varepsilon}+p_f^2\frac{\gamma a}{\bar{\varepsilon}_f^2}B_d^2\tan^2(p_f d)\right]\varepsilon_{zzd}^2-an_x^2B_d^2=0, \tag{3.142}$$

where now $\bar{\varepsilon}_s = \bar{\varepsilon}_c = \varepsilon$ and $\bar{a}_s = \bar{a}_c = a$.

The first integral of the wave equation then gives

$$\bar{\varepsilon}\frac{\bar{\varepsilon}_{\mathrm{f}} - n_x^2}{\bar{\varepsilon}_{\mathrm{f}}^2}\tan^2(p_{\mathrm{f}}d) + \frac{n_x^2\bar{\varepsilon}}{\varepsilon_{zzd}^2} + \tfrac{1}{2}a\left[\frac{\bar{\varepsilon}_{\mathrm{f}} - n_x^2}{\bar{\varepsilon}_{\mathrm{f}}^2}\tan^2(p_{\mathrm{f}}d) + \frac{n_x^2}{\varepsilon_{zzd}^2}\right]^2 B_d^2$$

$$-\,a(1-\gamma)\frac{\bar{\varepsilon}_{\mathrm{f}} - n_x^2}{\bar{\varepsilon}_{\mathrm{f}}^2}B_d^2\tan^2(p_{\mathrm{f}}d)\frac{n_x^2}{\varepsilon_{zzd}^2} = \frac{2n_x^2 - \varepsilon_{zzd}}{\varepsilon_{zzd}}. \qquad (3.143)$$

The antisymmetric case may be dealt with in exactly the same way with $-\cot(p_{\mathrm{f}}d)$ replacing $\tan(p_{\mathrm{f}}d)$ everywhere. The asymmetric modes need numerical investigation, and discussion of these waves must be deferred.

As a final comment it should be emphasised that although a certain amount of analytical progress has been made it is necessary to proceed numerically before complete results can be obtained. The presentation of the theory for TM nonlinear waves is therefore not as detailed and elegant in this part of the review, as it is for the TE waves. If the $E_z \gg E_x$ regime is valid, however, then the TM cases can be expressed in an analogous fashion to the TE theory that has gone before.

3.2.4. The power flow

In spite of the difficulty of proceeding to analytical results, some rather surprising progress for the single surface can be made with the power flow integrals, again without compromising the calculations in a significant way. The total power flow P_x parallel to the single interface of case A above is

$$P_x = \frac{c^2 n_x}{2\mu_0\omega}\int_{-\infty}^{\infty}\frac{B_y^2}{\varepsilon_{zz}}\,\mathrm{d}z, \qquad (3.144)$$

where z is measured in units of c/ω. For a completely linear system, this integral evaluates to

$$P_{x,\mathrm{L}} = \frac{c^2 n_x}{4\mu_0\omega}A^2\left[\frac{1}{\bar{\varepsilon}_{\mathrm{s}}\sqrt{n_x^2 - \bar{\varepsilon}_{\mathrm{s}}}} + \frac{1}{\bar{\varepsilon}_{\mathrm{c}}\sqrt{n_x^2 - \bar{\varepsilon}_{\mathrm{c}}}}\right]. \qquad (3.145)$$

Now if both media have positive dielectric functions, and one is optically nonlinear, then self-focussing of the power flux can occur in the nonlinear medium, with a field maximum being established. P_x may, therefore, be estimated by first finding the maximum in B_y^2 in the nonlinear medium and then multiplying by a strip width Δz.

For an active dielectric material with $\bar{\varepsilon}_{\mathrm{c}} < 0$ no power maximum develops in the nonlinear medium, as will be seen in section 5. Here surface plasmons exist in the linear limit and the time-averaged energy flow in the active medium is antiparallel to the wavevector $\boldsymbol{k} = (k, 0, 0)$. For linear surface plasmons the net energy per unit width of the wave is always parallel to $\boldsymbol{k}$ because the contribution from the inactive medium always out-weighs that from the active medium, but with nonlinear surface plasmons this situation can be reversed, resulting in a backward-travelling wave.

In the nonlinear medium the first integral shows that the maximum value of B_y is a solution of

$$X = \frac{\zeta(\bar{\varepsilon}_s + a_s X)}{2n_x^2 - \bar{\varepsilon}_s - a_s X}\left[\frac{\bar{\varepsilon}_s^3 X}{(\bar{\varepsilon}_s + a_s X)^2} + \tfrac{1}{2}a_s X^2\right], \tag{3.146}$$

where $\zeta = (n_x^2/\bar{\varepsilon}_s^2)$ and $X = \zeta B_{ym}^2$ and B_{ym} is the maximum B_y value. The solution of eq. (3.146) to $O(a_s)$ is

$$B_{ym}^2 \simeq \frac{2\bar{\varepsilon}_s^3(\bar{\varepsilon}_s - n_x^2)(2n_x^2 - \bar{\varepsilon}_s)}{a_s n_x^2(3n_x^2 - 4\bar{\varepsilon}_s)}. \tag{3.147}$$

An alternative form, involving a simple binomial expansion of eq. (3.129), is

$$B_{ym}^2 \simeq \frac{2\bar{\varepsilon}_s^3(n_x^2 - \bar{\varepsilon}_s)(2n_x^2 - \bar{\varepsilon}_s)}{n_x^2 a_s(3\bar{\varepsilon}_s - 2n_x^2)}. \tag{3.148}$$

Hence the power flow in the nonlinear halfspace, in the neighbourhood of the maximum in B_{ym}, is

$$P_{x,\mathrm{NL}} \simeq \frac{c^2 n_x}{2\mu_0\omega}\int_{z_1}^{z_2}\frac{B_{ym}^2}{\varepsilon_{zz}}\,\mathrm{d}z. \tag{3.149}$$

If the strip width is $\Delta z = z_2 - z_1$ then

$$P_{x,\mathrm{NL}} \simeq \frac{c^2 n_x}{2\mu_0\omega}\Delta z\, 2\bar{\varepsilon}_s^{-2}\frac{[(2n_x^2 - \bar{\varepsilon}_s)(n_x^2 - \bar{\varepsilon}_s)]}{a_s n_x^2(3\bar{\varepsilon}_s - 2n_x^2)}$$
$$\times \frac{1}{1 + \dfrac{2[(2n_x^2 - \bar{\varepsilon}_s)(n_x^2 - \bar{\varepsilon}_s)]}{n_x^2(3\bar{\varepsilon}_s - 2n_x^2)}}, \tag{3.150}$$

and the total power flux parallel to the interface is

$$P_x = P_{x,\mathrm{L}} + P_{x,\mathrm{NL}}. \tag{3.151}$$

The accuracy of eq. (3.150) will be considered in section 5.

3.3. *Nonlinear interaction between TE and TM waves*

The territory gained in the treatment of the propagation of nonlinear purely TE or TM waves has now been described in detail. This final section deals with the fascinating situation in which coupling of TE and TM waves is permitted, necessitating the simultaneous solution of nonlinearly coupled sets of characteristic differential equations.

The a priori assumption that only one wave polarisation, TE or TM, is present is sustainable for waves travelling in isotropic linear waveguides since the two polarisations decouple completely and can then be treated separately without any loss of generality. In fact, TE–TM coupling, and hence hybrid modes, in linear materials can occur only through anisotropy and off-diagonal elements in the dielectric tensor. In the nonlinear case TE–TM coupling can occur even for isotropic guiding media, however. In general, the waves can be considered as one polarisation being guided by the nonlinear channel created by the other polarisation. Two wavenumbers will exist in this case, one for the TE wave and the other for the TM wave. In other words, they must be considered to be coupled but discrete (Boardman and Twardowski 1988a, b).

The planar guiding structure to be considered here has interfaces that lie in the *xy* Cartesian plane supporting a surface/guided wave whose wavevector is parallel to the *x*-axis, as depicted in fig. 3.1. For CW waves the solutions will be of the general form

$$\mathscr{A}(x,t) = \mathscr{A}(z)\mathrm{e}^{\mathrm{i}(kx-\omega t)} + \mathrm{c.c.} \tag{3.152}$$

where $\mathscr{A}$ is used to denote a component of either the electric or magnetic field. Using this form of the solution, Maxwell's curl equations,

$$\mathrm{curl}\,\mathscr{E} = -\frac{\partial \mathscr{B}}{\partial t}, \tag{3.153}$$

and

$$\mathrm{curl}\,\mathscr{H} = \frac{\partial \mathscr{D}}{\partial t}, \tag{3.154}$$

give, in component form, the following general, and in principle nonlinear, relationships for an isotropic medium

$$\frac{\partial \mathscr{E}_y}{\partial z} = -\mathrm{i}\omega\mu_0\mathscr{H}_x, \tag{3.155a}$$

$$k\mathscr{E}_y = \omega\mu_0\mathscr{H}_z, \tag{3.155b}$$

$$\frac{\partial \mathscr{H}_x}{\partial z} - \mathrm{i}k\mathscr{H}_z = -\mathrm{i}\omega\varepsilon_0\varepsilon_{yy}\mathscr{E}_y, \tag{3.155c}$$

$$\frac{\partial \mathscr{E}_x}{\partial z} - \mathrm{i}k\mathscr{E}_z = \mathrm{i}\omega\mu_0\mathscr{H}_y, \tag{3.155d}$$

$$\frac{\partial \mathscr{H}_y}{\partial z} = \mathrm{i}\omega\varepsilon_0\varepsilon_{xx}\mathscr{E}_x, \tag{3.155e}$$

$$k\mathscr{H}_y = -\omega\varepsilon_0\varepsilon_{zz}\mathscr{E}_z \tag{3.155f}$$

where the substitutions $\mathscr{D}_i = \varepsilon_0\varepsilon_{ij}\mathscr{E}_j$ and $\mathscr{B}_i = \mu_0\mathscr{H}_i$ have been made, ε_0 and μ_0 are the permittivity and permeability of free space and ε_{xx}, ε_{yy} and ε_{zz} are components of the (generally) nonlinear dielectric tensor ε. Equations (3.155) show a division into two natural, but coupled, groups. One contains, explicitly, only the field components $\mathscr{H}_x$, $\mathscr{E}_y$ and $\mathscr{H}_z$ and the other contains explicitly only the components $\mathscr{E}_x$, $\mathscr{H}_y$ and $\mathscr{E}_z$. In a linear isotropic waveguide, the division is complete with the first group being called s- or TE-polarised (Transverse Electric) since its *electric vector* is perpendicular to the direction of propagation, and the second being named p- or TM-polarised (Transverse Magnetic), since it is the *magnetic vector* which is perpendicular to the direction of propagation. If, in a linear guide, off-diagonal elements ε_{xy} and ε_{yz} occur then the coupled TE, TM groups give rise to a single hybrid entity, characterised by a single frequency and wave number. A transfer to a circularly polarised set of basis fields will allow a separation of left and right polarisations, however, for certain propagation directions.

If the TE and TM waves have wave numbers k_{TM} and k_{TE}, respectively, then the total field required to calculate the third-order polarisation is

$$E = E(k_{\mathrm{TE}})\mathrm{e}^{\mathrm{i}(k_{\mathrm{TE}}x - \omega t)} + E(k_{\mathrm{TM}})\mathrm{e}^{\mathrm{i}(k_{\mathrm{TM}}x - \omega t)} + \text{c.c.}, \tag{3.156}$$

where c.c. denotes the complex conjugate.

After considerable manipulation, and the neglect of processes that are not phase matched, the dielectric tensor or a nonlinear material, in the presence of TE and TM waves at two different wave numbers, is

$$\varepsilon = \begin{bmatrix} \bar{\varepsilon} + a(E_x^2 + \eta E_y^2 + \gamma E_z^2) & 0 & 0 \\ 0 & \bar{\varepsilon} + a(E_y^2 + \eta E_z^2 + \eta E_x^2) & 0 \\ 0 & 0 & \bar{\varepsilon} + a(\eta E_y^2 + E_z^2 + \gamma E_x^2) \end{bmatrix}, \tag{3.157}$$

where

$$a = 2\chi_{xxyy} + \chi_{xyyx}, \tag{3.158}$$

$$\gamma = \frac{2\chi_{xxyy} - \chi_{xyyx}}{2\chi_{xxyy} + \chi_{xyyx}}, \tag{3.159}$$

$$\eta = \frac{2\chi_{xxyy}}{2\chi_{xxyy} + \chi_{xyyx}}. \tag{3.160}$$

For the various nonlinear processes we obtain:

- molecular Kerr: $\gamma = -\frac{1}{2}$, $\eta = \frac{1}{4}$;
- electronic: $\gamma = \frac{1}{3}$, $\eta = \frac{2}{3}$;
- thermal: $\gamma = 1$, $\eta = 1$.

Hence, for the commonly encountered thermal nonlinearity, the simple form of ε, derived without introducing the fact that $k_{\mathrm{TE}} \neq k_{\mathrm{TM}}$, is appropriate and has the form

$$\varepsilon = \begin{bmatrix} \bar{\varepsilon} + a(E_x^2 + E_y^2 + E_z^2) & 0 & 0 \\ 0 & \bar{\varepsilon} + a(E_y^2 + E_z^2 + E_x^2) & 0 \\ 0 & 0 & \bar{\varepsilon} + a(E_y^2 + E_z^2 + E_x^2) \end{bmatrix}. \tag{3.161}$$

In the foregoing analysis, the electric and magnetic fields $\boldsymbol{E}$, $\boldsymbol{H}$ were divided into the groups (i) $\mathscr{E}_y$, $\mathscr{E}_z$, $\mathscr{H}_y$ and $\mathscr{H}_z$ and (ii) $\mathscr{E}_x$ and $\mathscr{H}_x$, where the field components within a group lie on the same axis of the complex phase plane and the two groups are orthogonal to each other. The introduction of the substitutions $\mathscr{E}_x = \mathrm{i}E_x$, $\mathscr{E}_y = E_y$, $\mathscr{E}_z = E_z$, $\mathscr{H}_x = \mathrm{i}H_x$, $\mathscr{H}_y = H_y$ and $\mathscr{H}_z = H_z$, where E_i and H_i are real, then permits the analysis to proceed in terms of real quantities alone.

3.3.1. First integral for coupled TE–TM waves

The first integral for coupled TE–TM waves can be obtained in a similar way to that used for deriving the first integral for TM waves. The TM 'set' derived from Maxwell's equations gives

$$\frac{\partial E_x}{\partial z}\frac{\partial^2 E_x}{\partial z^2} = k_{\mathrm{TM}}^2\left(1 - \frac{\varepsilon_{zz}}{n_{\mathrm{TM}}^2}\right)E_z\frac{\partial E_z}{\partial z} - \frac{\omega^2}{c^2}\varepsilon_{xx}E_x\frac{\partial E_x}{\partial z}, \tag{3.162}$$

where $n_{\mathrm{TM}} = k_{\mathrm{TM}}c/\omega$.

Equation (3.162) can be integrated with respect to z to obtain

$$\frac{c^2}{\omega^2}\left(\frac{\partial E_x}{\partial z}\right)^2 + \tfrac{1}{2}a(E_x^4 + E_z^4) + \gamma a E_x^2 E_z^2 + n_{\mathrm{TM}}^2 E_x^2 - (n_{\mathrm{TM}}^2 - \bar{\varepsilon})(E_x^2 + E_z^2)$$
$$= -2a\eta\int(E_y^2 E_z\,\mathrm{d}E_z + \gamma E_y^2 E_x\,\mathrm{d}E_x) + C_{\mathrm{TM}}, \tag{3.163}$$

where the integration constant C_{TM} that has now appeared will only be zero if the field components and their derivatives vanish at infinity.

The TE 'set' that emerges from Maxwell's equations yields the following differential equation for E_y

$$\frac{\partial^2 E_y}{\partial z^2} = \left(k_{\mathrm{TE}}^2 - \frac{\omega^2}{c^2}\varepsilon_{yy}\right)E_y. \tag{3.164}$$

This, when integrated with respect to z, gives

$$\frac{c^2}{\omega^2}\left(\frac{\partial E_y}{\partial z}\right)^2 - (n_{\mathrm{TE}}^2 - \bar{\varepsilon})E_y^2 + \tfrac{1}{2}aE_y^4 = -2a\eta\int(\gamma E_x^2 E_y\,\mathrm{d}E_y + E_z^2 E_y\,\mathrm{d}E_y)$$
$$+ C_{\mathrm{TE}}, \tag{3.165}$$

where another integration constant C_{TE} has been introduced and n_{TE}, the normalised wavenumber is, in general, *not equal* to n_{TM}.

If equations (3.163) and (3.165) are added together, then

$$2\int(E_y^2 E_z\, \mathrm{d}E_z + \gamma E_y^2 E_x\, \mathrm{d}E_x + \gamma E_x^2 E_y\, \mathrm{d}E_y + E_z^2 E_y\, \mathrm{d}E_y)$$

$$= (\gamma E_x^2 E_y^2 + E_y^2 E_z^2). \tag{3.166}$$

Hence, the final result for the first integral for coupled TE–TM waves is

$$a(\eta E_y^2 E_z^2 + \eta E_x^2 E_y^2 + \gamma E_x^2 E_z^2) + \tfrac{1}{2}a(E_x^4 + E_y^4 + E_z^4) + \varepsilon(E_x^2 + E_y^2 + E_z^2)$$

$$= n_{\mathrm{TM}}^2 E_z^2 + n_{\mathrm{TE}}^2 E_y^2 - \frac{c^2}{\omega^2}\left(\left(\frac{\partial E_x}{\partial z}\right)^2 + \left(\frac{\partial E_y}{\partial z}\right)^2\right) + C_{\mathrm{TM}} + C_{\mathrm{TE}}. \tag{3.167}$$

It is, of course, artificial to maintain *two* integration constants but it is interesting to observe that, for semi-infinite media the integration constants C_{TM} and C_{TE} must both *separately* be zero to ensure, absolutely, that all the field components, and their derivatives, are zero at infinity. It is, therefore, clear that the condition that the sum of C_{TE} and C_{TM} is zero is not a sufficient condition to locate a valid solution. This is simply because it does not disallow the possibility that $C_{\mathrm{TE}} = -C_{\mathrm{TM}}$ with both C_{TM} and C_{TE} being finite. The consequence is that the application of single interface boundary conditions to the coupled TE–TM first integral will yield a continuum of eigenvalues that corresponds to the continuum of values that $C_{\mathrm{TE}} = -C_{\mathrm{TM}}$ can take. Only the eigenvalue corresponding to $C_{\mathrm{TE}} = C_{\mathrm{TM}} = 0$ will be physically correct. Since it is not actually possible to treat the two constants separately, the first integral, for this type of problem, is of very limited value. It can, however, serve some purpose as a check on the validity of results obtained from numerical calculations. Indeed, the numerical calculations quickly reveal the existence of the problems outlined above.

3.3.2. TE–TM coupling regions

Although the numerical methods will be presented in section 5, it is worth outlining some of the principal features to be expected. The first step in the calculation of coupled TE–TM surface/guided waves is to determine for what range of n_{TE} and n_{TM} values coupled waves actually exist. It turns out that this can be done without recourse to a coupled wave calculation. There is obviously no coupling of waves at a single interface between a metal and a nonlinear dielectric since this supports only TM-polarised waves. However, a single interface between a nonlinear and linear medium whose dielectric functions are positive will support both TE and TM polarisations, and hence must permit coupling between the two. The important point is that guiding actually takes place due to the nonlinear formation of a channel of relatively high refractive index in the nonlinear medium. It is then reasonable to assume that the channel formed by, for example, a nonlinear TE wave would, if somehow 'frozen' into the nonlinear medium, be able to support a *linear* TM wave with an eigenvalue related (via the shape and size of the channel) to the nonlinear eigenvalue of the TE wave. This example corresponds, in practice, to a nonlinear TE wave with a zero amplitude nonlinear (i.e., effectively linear) TM wave propagating simultaneously with it along a single interface. Such a TM wave is guided by the channel created by the TE wave, but the TE wave is unaffected by the TM wave because of the zero amplitude of the TM field components. It is therefore possible to construct a $(n_{\rm TE}, n_{\rm TM})$ plane and to draw a locus that represents a pure nonlinear TE wave with a zero amplitude TM wave. Obviously, there is also a second locus that corresponds to a pure nonlinear TM wave guiding a zero amplitude TE wave. In principle, solutions to Maxwell's equations should be found within the area between the two loci and these are the desired stationary coupled TE–TM nonlinear surface/guided waves. Points outside this region correspond to nonstationary states and will not be considered here.

The two loci on the $(n_{\rm TE}, n_{\rm TM})$ plane that delineate the nonlinear coupling region can be found by incorporating the expression for the nonlinear guiding channel, created by one polarisation, into the linear dielectric function used to calculate the linear eigenvalue of the other polarisation. For example, the 'nonlinear' dielectric tensor used in calculating the linear eigenvalue of TM waves in a TE induced channel has the form

$$\varepsilon = \begin{pmatrix} \bar{\varepsilon} + \eta a E_y^2(z) & 0 \\ 0 & \bar{\varepsilon} + \eta a E_y^2(z) \end{pmatrix}, \tag{3.168}$$

where, for a nonlinear medium filling the half-space $z < 0$,

$$E_y^2(z) = \frac{q_{\rm TE,s}^2}{\Lambda \cosh^2[q_{\rm TE,s}(z - z_0)]}, \tag{3.169}$$

$$\Lambda = \frac{\omega^2 a}{2c^2}, \tag{3.170}$$

$$\tanh(q_{\rm TE,s} z_0) = -\frac{q_{\rm TE,c}}{q_{\rm TE,s}}, \tag{3.171}$$

and

$$q_{{\rm TE},n}^2 = \frac{\omega^2}{c^2}(n_{\rm TE}^2 - \varepsilon_n), \tag{3.172}$$

where the subscript n refers to medium n. The corresponding dielectric tensor for a TM induced channel cannot be expressed analytically, but, by solving the TM wave equations, can be generated numerically in the same form as the tensor (3.168). The linear eigenvalues for this modified structure can then be calculated using standard ordinary differential equation numerical techniques.

The possibility of nonlinear TE–TM coupling increases if a thin metal film or dielectric layer is inserted between the nonlinear medium and linear dielectric. A metal film structure supports nonlinear long- and short-range plasmons (Burke et al. 1986, Sarid et al. 1982, Stegeman and Burke 1982) as well as TE plasmons and TM pseudoplasmons (to be defined in section 5). It should be possible, therefore, to have nonlinear TE–TM coupling between TE plasmons (Lederer and Mihalache 1986, Stegeman et al. 1984b) and TM pseudoplasmons, TE plasmons and short-range plasmons, and TE plasmons and long-range plasmons. In order to discover if all three types of coupling can occur, the ranges of existence for the various nonlinear waves must be examined. The investigation proceeds in the following way.

Suppose that the substrate is nonlinear and that the cladding is linear, then TE plasmons exist only for metal film thicknesses below a certain critical value that may be determined from a locus of what will be called the upper n_{TE} cut-off, as a function of film thickness. Now the position of the self-focussed peak in the nonlinear field distribution of TE waves is a solution of

$$\tanh(q_{\mathrm{TE,s}}z_0) = \frac{q_{\mathrm{TE,f}}\sigma - 1}{q_{\mathrm{TE,s}}\sigma + 1}, \qquad (3.173\mathrm{a})$$

where

$$\sigma = \frac{q_{\mathrm{TE,f}} - q_{\mathrm{TE,c}}}{q_{\mathrm{TE,f}} + q_{\mathrm{TE,c}}}\mathrm{e}^{-2q_{\mathrm{TE,f}}l}, \qquad (3.173\mathrm{b})$$

and $q_{\mathrm{TE,f}} = n_{\mathrm{TE}}^2 - \bar{\varepsilon}_{\mathrm{f}}$, $q_{\mathrm{TE,s}} = n_{\mathrm{TE}}^2 - \bar{\varepsilon}_{\mathrm{s}}$, $q_{\mathrm{TE,c}} = n_{\mathrm{TE}}^2 - \bar{\varepsilon}_{\mathrm{c}}$. The upper n_{TE} cut-off corresponds to the peak moving an infinite distance away from the metal film so that this locus can be evaluated by solving eqs (3.173) with $z_0 = -\infty$ and $\tanh(q_{\mathrm{TE,s}}z_0) = -1$. It is not possible for the film to have a thickness above a certain critical value, found by locating the intersection of the upper n_{TE} cut-off locus with a lower n_{TE} cut-off, below which the waves become radiative in the linear medium. The lower cut-off, just referred to, occurs for $n_{\mathrm{TE}}^2 = \varepsilon_{\mathrm{c}}$ so that, from eq. (3.173b), the critical film thickness is obtained from

$$\mathrm{e}^{2q_{\mathrm{TE,f}}l} = \frac{q_{\mathrm{TE,f}} + q_{\mathrm{TE,s}}}{q_{\mathrm{TE,f}} - q_{\mathrm{TE,s}}}. \qquad (3.174)$$

Although it seems to have been overlooked in the literature, self-focussed TM waves (pseudoplasmons) in the thin metal film structure also exist and they too have an upper (n_{TM}) cut-off that can be found from the TM first integral at the nonlinear medium/metal film interface, namely

$$n_{\mathrm{TM}}^2\left(1 - \frac{\varepsilon_{zz0}}{n_{\mathrm{TM}}^2}\right)^2 E_{z0}^2 + \tfrac{1}{2}a(E_{x0}^4 + E_{z0}^4) + \gamma a E_{x0}^2 E_{z0}^2 + n_{\mathrm{TM}}^2 E_{x0}^2$$
$$= (n_{\mathrm{TM}}^2 - \bar{\varepsilon}_{\mathrm{s}})(E_{x0}^2 + E_{z0}^2), \qquad (3.175)$$

where E_{x0} and E_{z0} are the amplitudes of E_x and E_z at the interface in the nonlinear medium and ε_{zz0} is defined as

$$\varepsilon_{zz0} = \varepsilon_1 + \gamma a E_{x0}^2 + a E_{z0}^2. \tag{3.176}$$

Since the amplitudes of the field components at the interface go to zero at the cut-off all terms higher than second-order in the field may be neglected and the first integral simplifies to

$$(\bar{\varepsilon}_s - n_{TM}^2) H_{y0}^2 + c^2 \varepsilon_0^2 \bar{\varepsilon}_s^2 E_{x0}^2 = 0, \tag{3.177}$$

where the field amplitudes E_{x0} and H_{y0} at the interface in the nonlinear medium match the field amplitudes in the metal film. Hence, using the linear field profiles, their ratio is given by

$$H_{y0} = \varepsilon_0 \bar{\varepsilon}_f \frac{\omega}{q_{TM,f}} \frac{1+\rho}{1-\rho} E_{x0}, \tag{3.178}$$

where $q_{TM,f,c,s} = n_{TM}^2 - \bar{\varepsilon}_{f,c,s}$ and

$$\rho = \frac{q_{TM,f}\bar{\varepsilon}_c - q_{TM,c}\bar{\varepsilon}_f}{q_{TM,c}\bar{\varepsilon}_f + q_{TM,f}\bar{\varepsilon}_c} e^{-2q_{TM,f}l}. \tag{3.179}$$

Equations (3.177)–(3.179) yield the upper n_{TM} cut-off for self-focussed TM waves (pseudoplasmons) in a metal film structure. The same equations also yield the linear eigenvalues of long- and short-range plasmons since these have identical boundary conditions to those used above for the self-focussed waves at cut-off (i.e., field amplitudes which are zero at infinity and tending to zero at the metal film interfaces). Notice that since none of the equations describing the upper n_{TM} cut-off for self-focussed TE or TM waves contains any nonlinear parameters, the cut-offs are independent of the magnitude and mechanism of the nonlinearity. This is very curious since, near the cut-off point, the waves are very highly nonlinear.

4. *Nonlinear guided waves in non-Kerr media*

4.1. *TE-polarised waves*

In the preceding section a great variety of nonlinear surface and guided waves on thin film structures and single interfaces was discussed. These investigations were limited, however, to Kerr-like media for which the nonlinear wave equation may be solved analytically assuming TE-polarisation. Kerr nonlinearity arises, principally, from electronic distortions, thermal effects or field-induced molecular reorientation and may be expressed mathematically as an expansion of the polarisation field in powers of the local electric field.

Under some experimental conditions the nonlinearity may not be Kerr-like.

In such a case it is necessary at the moment to adopt a simple, yet perhaps not consistent, model. This may take the form of assuming that the nonlinearity varies as $|E|^{\pm p}$ where $1 \leqslant p < 2$ or in some exponential form consistent with a two-level model of the nonlinearity. The essential feature of the non-Kerr model is to include integer and non-integer deviations from the $|E|^2$-Kerr law in the nonlinear polarization and/or some account of the fact that the nonlinearity may saturate at high powers to some maximum value.

Unfortunately using dielectric functions modified to include non-Kerr nonlinearities means that the wave equation is no longer analytically solvable. In this section a generalised approach is given so that eigenvalue equations and the dependence of the propagation constant on power may be determined for arbitrary nonlinearities taking advantage of a recently introduced model (Langbein et al. 1985a). Again this can be done analytically for TE-polarised waves, but the additional existence of an electrical field component in TM waves places it beyond an exact analytical formalism. An approach for TM waves that relies upon a specific form of the dielectric tensor appropriate to thermal nonlinearity (Boardman et al. 1987a–c) will be discussed later on.

4.1.1. The first integral

A more general form for the nonlinear dielectric function of a medium is

$$\varepsilon(|E|) = \bar{\varepsilon} + \varepsilon^{\mathrm{NL}}(|E|). \tag{4.1}$$

$\varepsilon^{\mathrm{NL}}$ is an as yet unspecified nonlinear addition to the linear dielectric background $\bar{\varepsilon}$. Note that if losses are assumed to be zero the fields $E(z)$ are real for guided waves so that $\varepsilon^{\mathrm{NL}}(|E|) \to \varepsilon^{\mathrm{NL}}(E)$.

The modified nonlinear wave equation is therefore

$$\left[\frac{\mathrm{d}^2}{\mathrm{d}z^2} + k_0^2(\bar{\varepsilon} + \varepsilon^{\mathrm{NL}}(E) - n_x^2)\right] E(z) = 0, \tag{4.2}$$

which has the first integral (Langbein et al. 1985a)

$$(\mathrm{d}E/\mathrm{d}z)^2 - k_0^2\left[q^2E^2 - 2\int_0^{E(z)} \varepsilon^{\mathrm{NL}}(\tilde{E})\tilde{E}\,\mathrm{d}\tilde{E} + C\right] = 0, \tag{4.3}$$

where C is a constant of integration, $k_0 = \omega/c$ and $q^2 = n_x^2 - \varepsilon$. This will now be applied to both interface and thin film configurations.

4.1.2. Nonlinear TE waves at single interfaces

Since, in this case, the media are semi-infinite both E and $\mathrm{d}E/\mathrm{d}z$ approach zero as $z \to \pm\infty$. Hence, as shown in section 3, the integration constant C must be zero. This is a convenient point to provide further insight that complements the material in section 3 by presenting the physics of this system in terms of a simple mechanical picture. This interprets E as a coordinate and z as a time. Hence we can identify eq. (4.3) as the law of conservation of energy for a particle of unit mass moving in a potential $V(E, n_x)$ with the velocity dE/dz (Leung 1985a, b), i.e.,

$$\tfrac{1}{2}(dE/dz)^2 + V(E) = 0, \tag{4.4}$$

where

$$V(E) = -\tfrac{1}{2}k_0^2\left[q^2E^2 - 2\int_0^{E(z)} \varepsilon^{NL}(\tilde{E})\tilde{E}\,d\tilde{E}\right].$$

This method of looking at the nonlinear TE waves involves looking for allowed classical particle motions, and hence the $E(z)$ field profiles, influenced by the potential V. This procedure is entirely equivalent to determining the familiar phase portraits $dE/dz(E)$ of the motion (Langbein et al. 1985a).

In this way a qualitative picture of the motion (field profile) can be obtained simply by analyzing V as an alternative to solving eq. (4.2).

First, different potential types V will be discussed in terms of their dependence on ε^{NL}, the nonlinear part of the dielectric function, and the propagation constant n_x. Second, the intersection of the potentials at the interface will be taken into account which will lead to a kind of eigenvalue equation.

Negative nonlinearities give a monotonically decreasing potential (curve $(-)$ in fig. 4.1) whereas positive nonlinearities exhibit a local minimum that will eventually terminate at the $V = 0$ line. The turn back to $V = 0$ can be inhibited

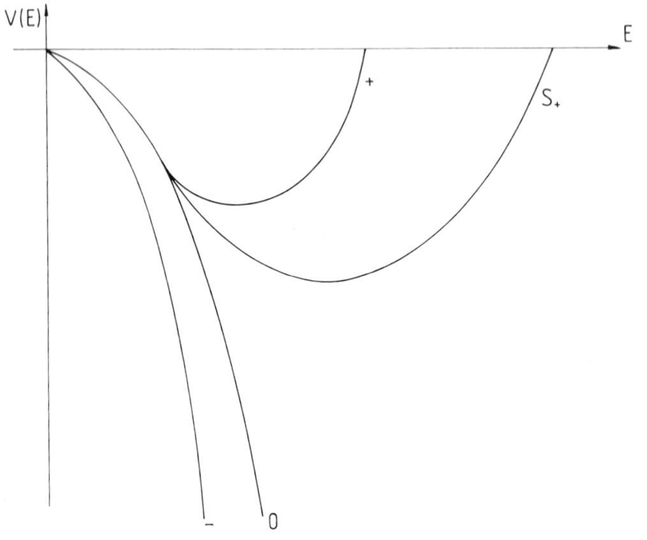

Fig. 4.1. Different potential-type curves $V(E)$. '–' negative nonlinearity; '+' positive nonlinearity; 'S_+' positive nonlinearity that saturates according to eq. (4.5); 'o' vanishing nonlinearity. Parameters: $n = \sqrt{\varepsilon} = 1.55$; $n_x = 1.562$; $\varepsilon_{SAT} = 0.045$.

by saturation effects [curve (S+)] in which the nonlinearity ε^{NL} reaches some maximum value in a sense to be discussed later on. If the nonlinearity vanishes, then the potential is of the type labelled as (0).

Suppose now that two different media are in contact and that they are defined in terms of the potentials V_c and V_s. The boundary conditions are that the field E and its derivative dE/dz are continuous at the interface, placed at $z=0$ for convenience. These conditions force the particle to cross from V_s to V_c at the point where the $V(E)$ curves intersect. This will be indicated by E_0.

As a simple example, consider the system consisting of a linear cladding (c) in contact with a nonlinear substrate (s) that has a nonlinearity which saturates. In order to ensure the existence of an intersection it is necessary to make $\bar{\varepsilon}_c > \bar{\varepsilon}_s$. Also the motion for a small n_x-value ($n_x^2 > \bar{\varepsilon}_c$) will be investigated (fig. 4.2a).

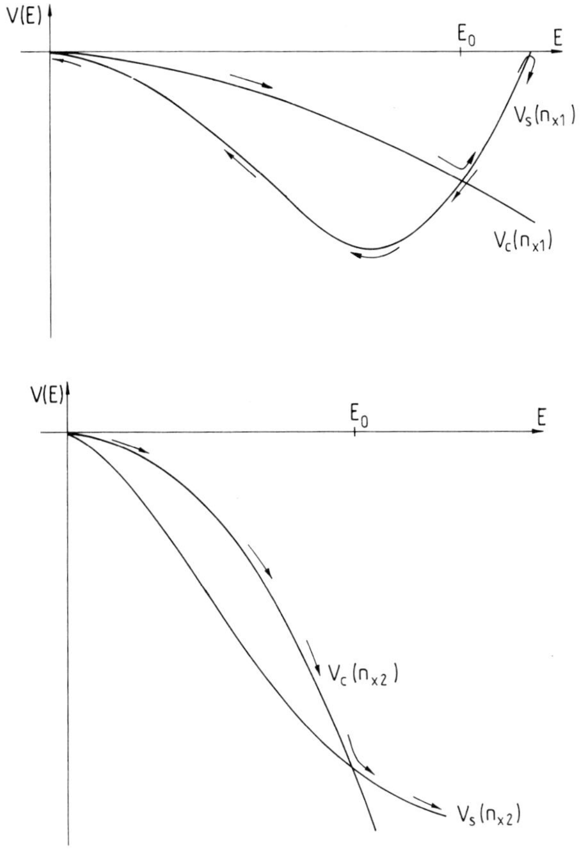

Fig. 4.2. Motion of a 'partical' (indicated by arrows) in a matched potential (single interface configuration: linear cladding, saturable substrate). E_0 is the electric field strength at the interface. Parameters: $n_c = 1.56$; $n_s = 1.55$; $\varepsilon_{SAT} = 0.045$. (a) $n_{x1} = 1.562$, (b) $n_{x2} = 1.565$.

Since $E = \mathrm{d}E/\mathrm{d}z = 0$ at $z = \pm\infty$, the particle may be imagined to start (at $t = -\infty$) from rest at the origin. It then rolls down the potential well V_c to E_0. After crossing to V_s it will roll upwards to reach a maximum value of $E = E_{max}$ before turning back and approaching the origin at $t = +\infty$ ($z = +\infty$) with a velocity $\mathrm{d}E/\mathrm{d}z = 0$.

For larger n_x-values (fig. 4.2b) the situation changes appreciably with the consequence that V_s can no longer reach $V_s = 0$ due to the role played by the saturation of the nonlinearity. The particle will roll down V_c and, after having crossed to V_s, continue down V_s without any opportunity to return to the origin. This implies that the field profile will in this case be unbounded. The transition from a bounded to an unbounded field in its dependences on n_x will occur when $V_s = 0$ cannot be reached for a finite $E_s \neq 0$. An inspection of the dielectric function gives the estimated permitted n_x-range as

$$\bar{\varepsilon}_c \leqslant n_x^2 \leqslant \bar{\varepsilon}_s + \varepsilon_{sat}, \tag{4.5}$$

where

$$\varepsilon_{sat} = \frac{2}{E_s^2}\int_0^{E_s(z)} \varepsilon^{NL}(\tilde{E})\tilde{E}\,\mathrm{d}\tilde{E}, \quad E_s(z) \to \infty,$$

is the saturation limit of the nonlinearity.

Thus, since $\bar{\varepsilon}_c > \bar{\varepsilon}_s$ for an intersection of V_c and V_s to occur, an interface between two linear media as well as between a linear medium and one with a negative nonlinearity will not permit the existence of surface waves (see fig. 4.1). This means, in practice, that a nonlinear surface wave must be always characterized by a bulge in one nonlinear medium whether it is in the cladding or in the substrate in order to fulfil the condition $V_{s,c}(E) = 0$ with $\mathrm{d}E/\mathrm{d}z_E = 0$.

On the other hand the crossing conditions alone do yield via eq. (4.4) a kind of eigenvalue (dispersion) relation, namely

$$-q_c^2 E_0^2 = -q_s^2 E_0^2 + 2\int_0^{E_0} \varepsilon_s^{NL}(\tilde{E})\tilde{E}\,\mathrm{d}\tilde{E},$$

$$(\bar{\varepsilon}_c - \bar{\varepsilon}_s)E_0^2 = 2\int_0^{E_0} \varepsilon_s^{NL}(\tilde{E})\tilde{E}\,\mathrm{d}\tilde{E}. \tag{4.6}$$

If both the cladding and the substrate are nonlinear this relationship generalizes to

$$(\bar{\varepsilon}_c - \bar{\varepsilon}_s)E_0^2 = 2\int_0^{E_0} [\varepsilon_s^{NL}(\tilde{E}) - \varepsilon_c^{NL}(\tilde{E})\tilde{E}]\,\mathrm{d}\tilde{E}. \tag{4.7}$$

This shows that (from fig. 4.2) the medium with the larger dielectric background constant may exhibit a negative nonlinearity or a strong saturation and permit bounded solutions. Obviously some of the results for the Kerr nonlinearity, i.e., $\varepsilon_i^{NL} = a_i E_i^2$, discussed in the previous section follow as a limit of this formalism.

The results obtained above can also be found considering the phase portraits $\mathrm{d}E/\mathrm{d}z(E)$ rather than the potential curves $V(E)$. (See Langbein et al. 1985a). From eq. (4.4)

$$\frac{\mathrm{d}E}{\mathrm{d}z}=\operatorname{sgn}\left(\frac{\mathrm{d}E}{\mathrm{d}z}\right)[-2V(E)]^{1/2}, \tag{4.8}$$

that leads for the example under discussion (i.e., a linear/nonlinear interface) to the phase portraits sketched in fig. 4.3.

The arrows in fig. 4.3 indicate a possible phase path. From a consideration of the motion of a classical particle it is clear that the signs of the velocity, v, and an infinitesimal distance, $\mathrm{d}z$, must be equal to permit a cross-over between two phase portraits. For the fields this condition reads as $(\mathrm{d}E/\mathrm{d}z)\,\mathrm{d}E>0$. The resulting field pattern follows immediately by integrating eq. (4.8)

$$z=\operatorname{sgn}\left(\frac{\mathrm{d}E}{\mathrm{d}z}\right)\int_0^{E(z)}\frac{\mathrm{d}\tilde{E}}{[-2V(\tilde{E})]^{1/2}},$$

where the integration path is fixed by the supposed phase path that will also determine the corresponding sign. The advantage of the phase-portrait technique, as opposed to a method using the motion in a given potential, is that it gives direct access to the velocity, $\mathrm{d}E/\mathrm{d}z$, rather than its square. This can be important in obtaining a qualitative estimate of the actual field profile.

As discussed in the previous section, and shown in detail in section 5, the power flow dependence of the propagation constant provides vital information about the nonlinear wave behaviour. This, as was the case in section 3, can also be calculated here without requiring any information about the explicit form of the field pattern.

The power flow for this single interface case, introduced in the previous section is,

$$P=\tfrac{1}{2}\varepsilon_0 c n_x\int_{-\infty}^{\infty}E^2(z)\,\mathrm{d}z=P_c+P_s, \tag{4.9}$$

where

$$P_s=\tfrac{1}{2}\varepsilon_0 c n_x\int_{-\infty}^{0}E_s^2(z)\,\mathrm{d}z,$$

$$P_c=\tfrac{1}{2}\varepsilon_0 c n_x\int_{0}^{\infty}E_c^2(z)\,\mathrm{d}z. \tag{4.10}$$

In the classical particle picture being discussed here, the integration must be taken along the entire fixed phase path. Since the explicit field pattern $E_{c,s}(z)$ will not be readily accessible we can start with

$$\mathrm{d}z=\operatorname{sgn}\left(\frac{\mathrm{d}E}{\mathrm{d}z}\right)\frac{\mathrm{d}E_{s,c}}{[-2V_{s,c}(E_{s,c})]^{1/2}},$$

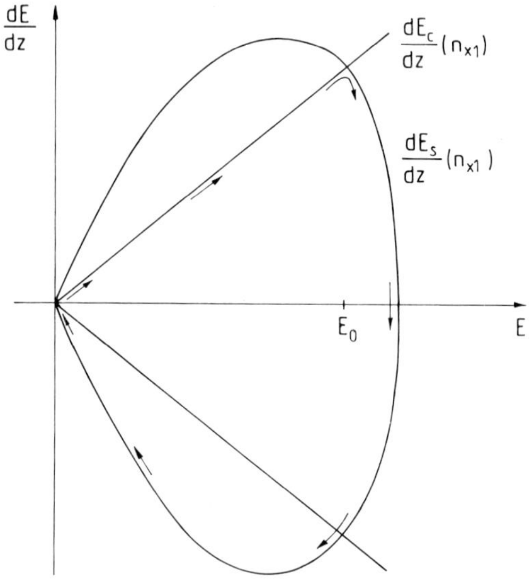

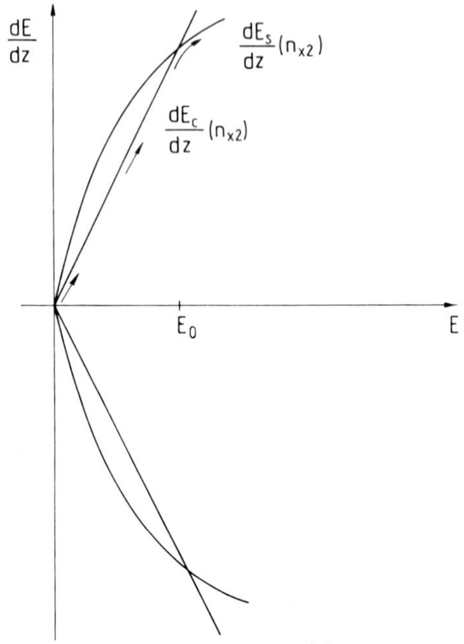

Fig. 4.3. Phase portraits of the motion indicated in fig. 4.2. The arrows indicate the corresponding phase path. The phase path plotted in fig. 4.3a leads to a bounded solution, and that plotted in fig. 4.3b to an unbounded solution.

where the corresponding phase path determines the sign and one field distribution (E_c or E_s) will have a bulge ($dE/dz = 0$). If the sign of the velocity dE/dz changes at $E = E_{max}$ (see fig. 4.3) then, from eq. (4.10),

$$P_s = \tfrac{1}{2}\varepsilon_0 c n_x \left\{ \int_0^{E_s} \frac{E^2\, dE}{[-2V_s(E)]^{1/2}} + (-1)^{M_s} \int_{E_s}^{E_0} \frac{E^2\, dE}{[-2V_s(E)]^{1/2}} \right\}, \tag{4.11}$$

$$P_c = \tfrac{1}{2}\varepsilon_0 c n_x \left\{ \int_0^{E_c} \frac{E^2\, dE}{[-2V_c(E)]^{1/2}} + (-1)^{M_s+1} \int_{E_c}^{E_0} \frac{E^2\, dE}{[-2V_c(E)]^{1/2}} \right\}, \tag{4.12}$$

with $M_s = 1$ for a bulge in the substrate and $M_s = 0$ otherwise. Keeping in mind that only one bulge may exist the exponent in the cladding region M_c must be equal to $M_s + 1$.

4.1.3. *Nonlinear TE-polarised waves in film configurations*

The fundamentals of the method for dealing with non-Kerr media have now been discussed extensively for a single interface configuration. We now turn to a film geometry where the nonlinearities of all media involved may exhibit non-Kerr behaviour and the discussion will concentrate upon some of the peculiarities associated with this geometry.

In the film region, labelled by the subscript f, the integration constant C_f in the first integral of eq. (4.3) is now not zero. As discussed in section 3, the film region is bounded, and the field and its derivative are not obliged to vanish anywhere. Hence the potential V from eq. (4.4) is

$$V_f(E_f) = \tfrac{1}{2}\left[-C_f - q_f^2 E_f^2 + 2 \int_0^{E_f(z)} \varepsilon_f^{NL}(\tilde{E})\tilde{E}\, d\tilde{E} \right]. \tag{4.13}$$

Some forms for the potential V_f within the film region for $C_f > 0$ and different n_x-regions are shown in fig. 4.4. It can be seen here that the cases $n_x^2 < \bar{\varepsilon}_f$, $q_f^2 < 0$, for which guided wave oscillating fields are normally expected, and $n_x^2 > \bar{\varepsilon}_f$, $q_f^2 > 0$, for which surface wave nonoscillating fields are normally expected, can be distinguished.

The curves labelled by + are valid for positive nonlinearities, where, if the nonlinearity is saturable, the potential can be inhibited and induced to turn back to $V_f = 0$, if $q_f^2 > 0$.

The case $C_f < 0$ plotted in fig. 4.5, exhibits some interesting peculiarities. Nonoscillating solutions are permitted for negative nonlinearities even for $n_x^2 < \bar{\varepsilon}_f$ (see fig. 4.5a) and oscillating ones can arise even though the condition $n_x^2 > \bar{\varepsilon}_f$ leads to the expectation of nonoscillating fields (see fig. 4.5b). Note that both types of fields can be suppressed by the existence of a saturated nonlinearity.

The exotic oscillating field has no zeros at all. This, with hindsight, is an expected property of field solutions that are like the Jacobi elliptic functions encountered in the Kerr case and discussed in section 3.

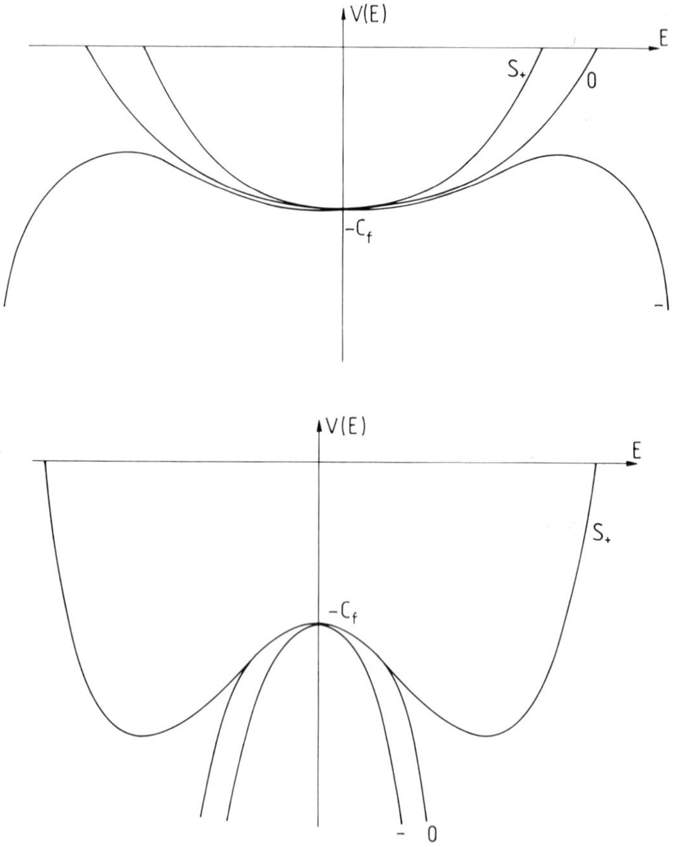

Fig. 4.4. Different potential-type curves for the film region parameters $n_f = 1.57$; $C_f > 0$. (a) $q_f^2 = n_x^2 - \bar{\varepsilon}_f < 0$, $n_x = 1.565$, (b) $q_f^2 > 0$, $n_x = 1.575$.

In a similar manner to the single interface waves, the potential curves V_c, V_f and V_s must be matched. The aim here, however, is not to give a complete discussion of the large number of possible motions but to outline some general rules and to demonstrate the general approach through a simple example. The first rule is that a V_f that is monotonically decreasing must be matched with a V_s and/or V_c that intersects the zero-potential line. Expressed in more physical terms this means that nonlinear waves guided by films with negative nonlinearities, or strongly saturated positive ones, are only permitted if one of the adjacent media can sustain a field that has a bulge (maximum). This rule applies also to a linear film with nonlinear boundaries when n_x^2 exceeds ε_f. In general, it must be emphasized that the matching of the potential curves for a fixed configuration provides useful information concerning the existence of a permitted n_x-range of nonlinear guided waves. This approach constitutes a

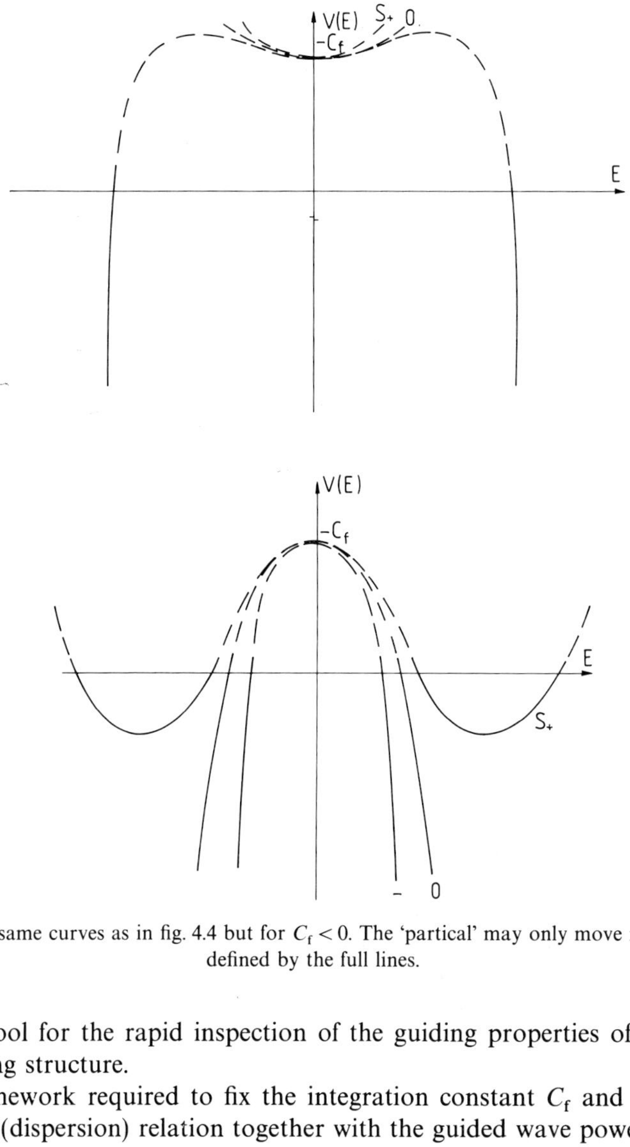

Fig. 4.5. The same curves as in fig. 4.4 but for $C_f < 0$. The 'partical' may only move in the regions defined by the full lines.

powerful tool for the rapid inspection of the guiding properties of any given waveguiding structure.

The framework required to fix the integration constant C_f and obtain the eigenvalue (dispersion) relation together with the guided wave power flow will now be discussed. The analytical formulation will not be restricted to a special example, but, on the contrary, all the sketched results in the figures will apply to a configuration where a linear substrate, a cladding with a positive nonlinearity and a film with a positive, saturable nonlinearity are in contact. In fig. 4.6 two optional motions are sketched, together with the corresponding field patterns.

Up to now, the kind of motions (field patterns) actually permitted have not been determined. In order to do this it is first of all necessary to relate the

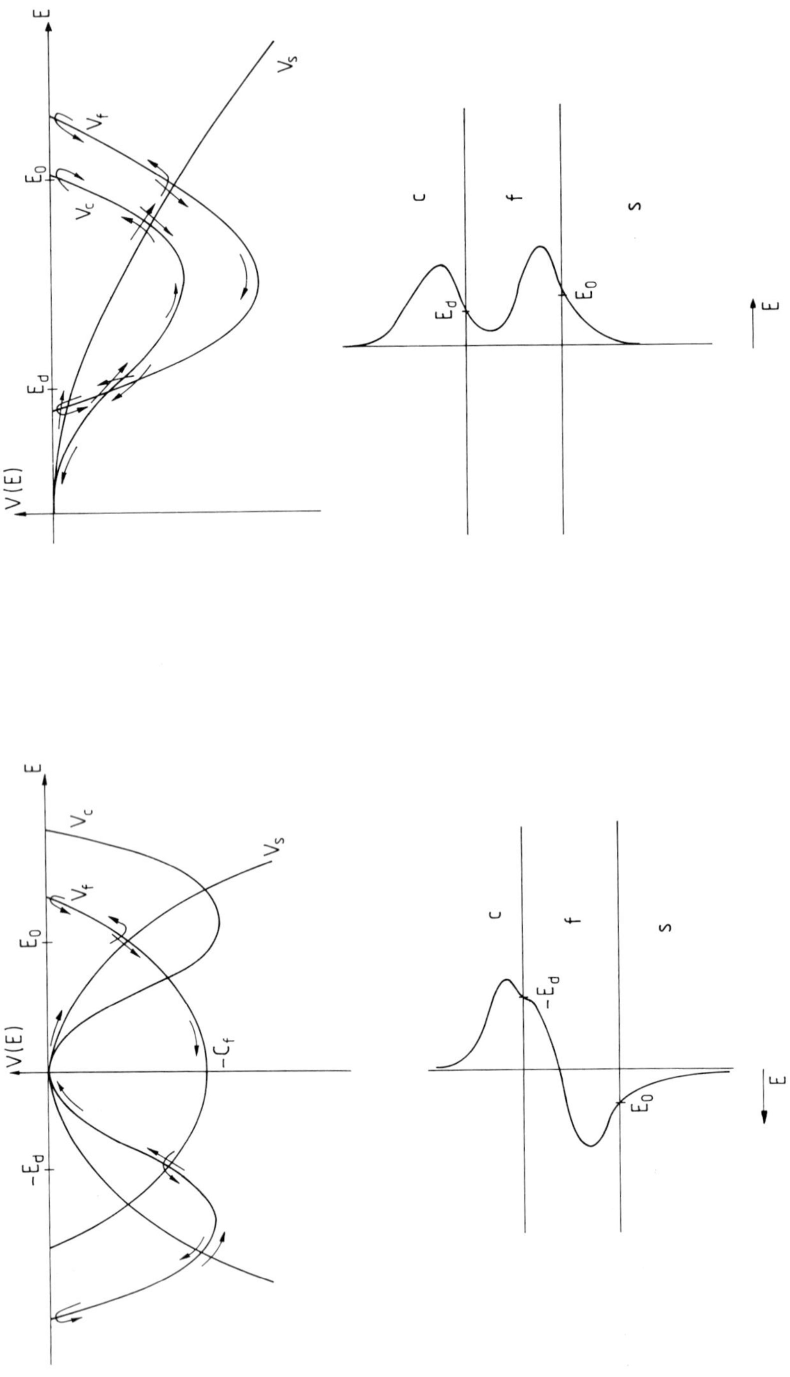

Fig. 4.6. Matched potential curves, optional motions and field patterns for two different film configurations. E and E_d are the electric field strengths at the film/substrate and the film/cladding interfaces respectively. Parameters: (a) $n_c = 1.55$; $n_f = 1.57$; $n_s = 1.56$; $n_x = 1.568$; $C_f > 0$; $q_f^2 < 0$. (b) $n_c = 1.56$; $n_f = 1.55$; $n_s = 1.57$; $n_x = 1.575$; $C_f < 0$; $q_f^2 > 0$.

integration constant C_f to a more physical quantity. From the boundary condition that the E and $\mathrm{d}E/\mathrm{d}z$ are continuous at the interfaces at $z = 0$, d it is clear from fig. 4.6 that if one of the entities E_0, E_d or C_f is fixed then the remaining quantities are entirely determined. If the fixed quantity is selected as E_0 then using

$$\frac{\mathrm{d}}{\mathrm{d}z}E_i = \mathrm{sgn}\left(\frac{\mathrm{d}}{\mathrm{d}z}E_i\right)[-2V_i(E_i)]^{1/2}, \tag{4.14}$$

at the boundaries with V_i given by eqs (4.13) (for V_f) and (4.4) (for both $V_{s,c}$) and a subsequent squaring yields

$$\begin{aligned}\frac{1}{k_0^2}C_f &= (\bar{\varepsilon}_f - \bar{\varepsilon}_s)E_0^2 + 2\int_0^{E_0}[\varepsilon_f^{NL}(\tilde{E}) - \varepsilon_s^{NL}(\tilde{E})]\tilde{E}\,\mathrm{d}\tilde{E} \\ &= (\bar{\varepsilon}_f - \bar{\varepsilon}_c)E_d^2 + 2\int_0^{E_d}[\varepsilon_f^{NL}(\tilde{E}) - \varepsilon_c^{NL}(\tilde{E})]\tilde{E}\,\mathrm{d}\tilde{E}.\end{aligned} \tag{4.15}$$

Note that the relation between E_0 and E_d as well as the integration constant C_f is independent of n_x, being unique for TE-polarised waves. The eigenvalue relation will yield some discrete allowed n_x-values that are dependent upon parameters such as d, E_0 or the power flow P. Such a relationship may be obtained from an inspection of the curves in fig. 4.6. This returns us to the motion picture, discussed above. The distance $(E_d - E_0)$ is fixed by eq. (4.15), but the significant point is, whether the particle is able to cover this distance in a given time $z = d$. This philosophy will lead to an eigenvalue (dispersion) equation because this condition can only be fulfilled for some discrete potential shapes. This means, for $V = V(n_x)$, only for some discrete propagation constants n_x.

In more mathematical terms it is necessary to integrate

$$\frac{\mathrm{d}}{\mathrm{d}z}E_f = \mathrm{sgn}\left(\frac{\mathrm{d}}{\mathrm{d}z}E_f\right)[-2V_f(E_f)]^{1/2}. \tag{4.16}$$

This gives the desired dispersion relation

$$d = \mathrm{sgn}\left(\frac{\mathrm{d}}{\mathrm{d}z}E_f\right)\int_{E_0}^{E_d}\frac{\mathrm{d}\tilde{E}_f}{[-2V_f(\tilde{E}_f)]^{1/2}}, \tag{4.17a}$$

that naturally allows also some periodic motions within the film potential (representing oscillations of the optical field within the film). Taking into account these oscillations and the field pattern shape (bulge or not) in the adjacent media that will fix the sign in eq. (4.17a), the final form of the eigenvalue relation which the permitted n_x must satisfy is

$$\begin{aligned}&(-1)^{M_s}\int_{E_0}^{(-1)^{M_s}E_f}\frac{\mathrm{d}E}{[-2V_f]^{1/2}} + (N - M_s - M_c)\int_{-E_f}^{E_f}\frac{\mathrm{d}E}{[-2V_f]^{1/2}} \\ &+(-1)^{M_c}\int_{E_d}^{(-1)^{M_c}E_f}\frac{\mathrm{d}E}{[-2V_f]^{1/2}} = d.\end{aligned} \tag{4.17b}$$

Here, $M_{c,s}=1$ for a bulge in the cladding and substrate region, respectively, otherwise $M_{c,s}=0$. N is the number of zeros of the inner field and E_f is a solution of $[-V_f(E_f)]^{1/2}=0$, representing the maximum of the inner field. The only exception to this rule arises for the exotic oscillating field that has no zeros (fig. 4.6b). For this case, V_f has two positive zeros and $-E_f$ in eq. (4.17b) is taken as the smaller value. Furthermore, $(N-M_c-M_s)$ has in this case to be replaced by $\bar{N}-1$ where $\bar{N}$ is the number of inner field extrema.

Only the potential of the film region has a marked influence on the eigenvalue equation since the cladding and substrate fields can affect this relation only by their value at the interfaces. These are related by (4.15) which is a part of the entire dispersion relation.

The guided wave power flow is

$$P=P_f+P_c+P_s, \tag{4.18}$$

where

$$P_f=\tfrac{1}{2}\varepsilon_0 c n_x\int_0^d E_f^2(z)\,dz, \tag{4.19}$$

with P_c and P_s being given by eqs (4.11) and (4.12).

The procedure already outlined in the previous calculation of P_c and P_s yields

$$P_f=\tfrac{1}{2}\varepsilon_0 c n_x\left\{(-1)^{M_s}\int_{E_0}^{(-1)^{M_s}E_f}\frac{E^2\,dE}{[-2V_f]^{1/2}}+(N-M_s-M_c)\int_{-E_f}^{E_f}\frac{E^2\,dE}{[-2V_f]^{1/2}}\right.$$
$$\left.+(-1)^{M_c}\int_{E_d}^{(-1)^{M_c}E_f}\frac{E^2\,dE}{[-2V_f]^{1/2}}\right\}. \tag{4.20}$$

Hence, the combination of eqs (4.11), (4.12) and (4.20) gives the total guided wave power flow in a film configuration where the dielectric function of all media involved can be non-Kerr-like. The parameter E_0 can now be removed to evaluate the power dependence of the propagation constant. Equation (4.17) has been solved analytically only in a few outstanding cases, namely for linear and Kerr-like films which results in dispersion relations containing trigonometric and Jacobi functions respectively (see section 3).

The linear film with nonlinear bounding media has attracted the most attention both numerically and experimentally, so this case will now be discussed for non-Kerr bounding media. Equation (4.13), with $\varepsilon_f^{NL}=0$ for the linear film, gives

$$-V_f=\tfrac{1}{2}k_0^2(C_f+q_f^2E_f^2),$$

$$-V_f=\tfrac{1}{2}k_0^2(C_f-p_f^2E_f^2),$$

where

$$p_f^2=\bar{\varepsilon}_f-n_x^2 \quad\text{and}\quad E_f=[C_f]^{1/2}/p_f.$$

The integration in eq. (4.17) can be easily performed to give

$$(1+N-M_s-M_c)\pi-(-1)^{M_s}\arcsin[p_f E_0/(C_f)^{1/2}]$$
$$-(-1)^{M_c}\arcsin(p_f Ed/(C_f)^{1/2}]=k_0 p_f d.$$

After some trigonometric manipulations and taking advantage of eq. (4.15) we arrive at

$$\tan(k_0 p_f d)=\frac{p_f(q_{SNL}+q_{CNL})}{p_f^2-q_{SNL}q_{CNL}}, \tag{4.21a}$$

or the equivalent cosine formula preferred in section 3.

$$\cos(k_0 p_f d)=\frac{p_f^2-q_{SNL}q_{CNL}}{(p_f^2+q_{CNL}^2)^{1/2}(p_f^2+q_{SNL}^2)^{1/2}}, \tag{4.21b}$$

where we have introduced

$$q_{iNL}=(-1)^{M_i}\left[q_i^2-\frac{2}{E_{(i)}^2}\int_0^{E_{(i)}}\varepsilon_i^{NL}(\tilde{E})\tilde{E}\,d\tilde{E}\right]^{1/2}. \tag{4.22}$$

Here, $i=c, s$; $E_{(c)}=E_d$, $E_{(s)}=E_0$.

The guided wave power flow in the film region can be deduced from section 3 to be

$$P_f=\tfrac{1}{4}\varepsilon_0 c n_x E_0^2\left[k_0 d\left(1+\frac{q_{CNL}^2}{p_f^2}\right)+\frac{q_{CNL}}{p_f^2}+\frac{p_f^2+q_{CNL}^2}{p_f^2+q_{SNL}^2}\frac{q_{SNL}^2}{p_f^2}\right]. \tag{4.23}$$

Numerical calculations for non-Kerr media will be discussed in section 5, where will be shown that saturation in the optically induced change of the refractive index can affect the nonlinear guided waves quite strongly. In general, for ε_{sat} larger than $\bar{\varepsilon}_f-\bar{\varepsilon}_c$ or $\bar{\varepsilon}_f-\bar{\varepsilon}_s$ the principal features are retained but for small ε_{sat} they can be destroyed. The conclusion is that the design of nonlinear guided wave devices that rely on power dependent changes in the field distribution may depend upon a non-Kerr form of the nonlinearity. If this is the case this section contains the basic algorithm that can be used to model this possibility.

4.2. *TM-polarised waves*

As shown in the previous chapter for the Kerr case the situation for TM-polarised waves is more involved. This is due to the additional existence of an electric field component (E_x) parallel to the interface. A full numerical solution of this problem has been given very recently (Boardman et al. 1987a) and will be discussed in the next section.

In what follows we present for simplicity an analysis for nonlinearities that arise through electrostriction or thermal effects.

We turn to the principal ideas of an approach for thermal nonlinearities in which $\varepsilon_{xx}(|E_x|^2+|E_z|^2)=\varepsilon_{zz}(|E_x|^2+|E_x|^2)$ (Boardman et al. 1987a, Leung 1985b). We will look for analogues to the procedure introduced in section 4.1 and will include film configurations not considered in Leung (1985b).

4.2.1. The first integral and the eigenvalue equations

The nonlinear material is assumed to be isotropic and the dielectric tensor an arbitrary function of $(|E_x|^2+|E_z|^2)$ so that is is permissible to write

$$\varepsilon_i \equiv \varepsilon_{xxi}+\varepsilon_{zzi}=\bar{\varepsilon}_i+\varepsilon_i^{\mathrm{NL}}(|E_x|^2+|E_z|^2), \tag{4.24}$$

where $i=\mathrm{s, f, c}$.

From this point the medium indexes will be omitted until the interfaces are introduced into the model.

The electric field components have the form

$$A(x,z,t)=\tfrac{1}{2}A(z)\mathrm{e}^{\mathrm{i}(k_0 n_x x-\omega t)}+\mathrm{c.c.}$$

where A stands for E_x, E_z, $B=B_y$ and Maxwell's equations yield

$$n_x B(z)=-\frac{\varepsilon}{c}E_z(z),\quad \frac{\mathrm{d}}{\mathrm{d}z}B(z)=\mathrm{i}\frac{k_0\varepsilon}{c}E_x(z),$$

$$\frac{\mathrm{d}}{\mathrm{d}z}E_x(z)-\mathrm{i}k_0 n_x E_z(z)=\mathrm{i}\omega B(z). \tag{4.25}$$

Equation (4.25) gives the following set of equations for B

$$\frac{\mathrm{d}}{\mathrm{d}z}\left[\frac{1}{\varepsilon}\frac{\mathrm{d}}{\mathrm{d}z}B\right]+\frac{k_0^2}{\varepsilon}[\varepsilon-n_x^2]B=0, \tag{4.26a}$$

$$|E_x|^2+|E_z|^2=\frac{c^2}{\varepsilon^2}\left[n_x^2B^2+\frac{1}{k_0^2}\left(\frac{\mathrm{d}}{\mathrm{d}z}B\right)^2\right], \tag{4.26b}$$

$$\varepsilon=\bar{\varepsilon}+\varepsilon^{\mathrm{NL}}(|E_x|^2+|E_z|^2). \tag{4.26c}$$

The technique now is to introduce $I(\varepsilon-\bar{\varepsilon})$, the inverse function of $\varepsilon^{\mathrm{NL}}$, that is $I(0)=0$ in the linear limit in eq. (4.26c).

This results in the elimination of $|E_x|^2+|E_z|^2$ from eq. (4.26b) through

$$|E_x|^2+|E_z|^2=I(\varepsilon-\varepsilon_0),$$

to yield

$$\left(\frac{\mathrm{d}B}{\mathrm{d}z}\right)^2-k_0^2\left[\frac{\varepsilon^2}{c^2}I(\varepsilon-\varepsilon_0)-n_x^2B^2\right]=0. \tag{4.27}$$

From eq. (4.26a) the first integral

$$B^2=\frac{\varepsilon}{c^2}\frac{1}{2n_x^2-\varepsilon}[\varepsilon I(\varepsilon-\bar{\varepsilon})-G(\varepsilon-\bar{\varepsilon})], \tag{4.28}$$

can be obtained, where

$$G'(\varepsilon-\bar{\varepsilon}) = \int_0^{\varepsilon-\bar{\varepsilon}} I(\bar{\varepsilon})\,\mathrm{d}\bar{\varepsilon}.$$

The final insertion of eq. (4.28) into eq. (4.27) permits a return to the classical motion picture discussed earlier, viz.

$$\left(\frac{\mathrm{d}B}{\mathrm{d}z}\right)^2 - \frac{k_0^2}{c^2}\frac{\varepsilon}{2n_x^2-\varepsilon}[eI(\varepsilon-\bar{\varepsilon})(n_x^2-\varepsilon)+n_x^2G(\varepsilon-\bar{\varepsilon})], \tag{4.29a}$$

or

$$\left(\frac{\mathrm{d}B}{\mathrm{d}z}\right)^2 + V(\varepsilon) = 0. \tag{4.29b}$$

This again can be interpreted as the motion of a point mass identifying ε as a coordinate and z as a time in a potential

$$V(\varepsilon) = -\tfrac{1}{2}\frac{k_0^2}{c^2}\frac{\varepsilon}{2n_x^2-\varepsilon}\left(\frac{\mathrm{d}B}{\mathrm{d}z}\right)^2[\varepsilon I(\varepsilon-\bar{\varepsilon})(n_x^2-\varepsilon)+n_x^2G(\varepsilon-\bar{\varepsilon})]. \tag{4.29c}$$

Equation (4.29) can serve as a starting point for a qualitative discussion of permitted motions as did eq. (4.4) for the TE case. It is not necessary to go into further details since the method was presented in section 4.1.

Equation (4.29b) gives $\varepsilon(z)$ through

$$z = \mathrm{sgn}\left(\frac{\mathrm{d}\varepsilon}{\mathrm{d}z}\right)\int_{\varepsilon(0)}^{\varepsilon(z)} \frac{\mathrm{d}\bar{\varepsilon}}{[-2V(\bar{\varepsilon})]^{1/2}}, \tag{4.30}$$

after using the first integral [eq. (4.28)] to obtain $B^2(z)$.

Attention will be focussed on the derivation of the eigenvalue (dispersion) relations. For this purpose we use eqs (4.28) and (4.29a) to give

$$\frac{\mathrm{d}B/\mathrm{d}z}{B} = \mathrm{sgn}\left(\frac{\mathrm{d}B}{\mathrm{d}z}\right)k_0\left[\frac{\varepsilon I(\varepsilon-\bar{\varepsilon})(n_x^2-\varepsilon)+n_x^2G(\varepsilon-\bar{\varepsilon})}{\varepsilon I(\varepsilon-\bar{\varepsilon})-G(\varepsilon-\bar{\varepsilon})}\right]^{1/2}. \tag{4.31}$$

As a first example the single interface problem will be investigated, where, in the TM case $(1/\varepsilon)\mathrm{d}B/\mathrm{d}z$ and B must be continuous across the interface, situated for convenience at $z=0$. Also from this point onwards, the entities ε, I, G will have the medium subscripts c and s respectively. The entity $q_{i\mathrm{NL}}$ can be introduced through eq. (4.31) and is

$$q_{i\mathrm{NL}} = (-1)^{M_i}\left[\frac{e_iI_i(\varepsilon-\bar{\varepsilon})(n_x^2-\varepsilon_i)+n_x^2G(\varepsilon-\bar{\varepsilon})}{\varepsilon_iI_i(\varepsilon-\bar{\varepsilon})-G(\varepsilon-\bar{\varepsilon})}\right]^{1/2}, \tag{4.32}$$

where all ε, I, G must be taken at the interface, e.g.,

$$\varepsilon_i|_0 = \bar{\varepsilon}_i + \varepsilon_i^{\mathrm{NL}}(|E_{0x}|^2+|E_{0z}|^2).$$

Equations (4.31) and (4.32) can be exploited together with the boundary conditions to give immediately the eigenvalue equation

$$q_{\mathrm{SNL}}/(\varepsilon_{\mathrm{s}}|_0) + q_{\mathrm{CNL}}/(\varepsilon_{\mathrm{c}}|_0) = 0. \tag{4.33a}$$

Note that in eq. (4.33a) the actual field-dependent dielectric functions at the interface occur and not just the background constants $\bar{\varepsilon}_i$. Also from the dependence on $\varepsilon_{\mathrm{s}}|_0$ and $\varepsilon_{\mathrm{c}}|_0$ the permitted n-values can be calculated, at least in principle. The point is that both quantities may be chosen independently and the calculation, say $\varepsilon_{\mathrm{c}}|_0 = F(\varepsilon_{\mathrm{s}}|_0)$, via the crossing conditions for E_x and E_z is rather involved for general functions $\varepsilon^{\mathrm{NL}}$. Hence, it seems that the model must be restricted in practice to a configuration where a linear medium is in contact with a nonlinear one, that gives, instead of eq. (4.33a)

$$q_{\mathrm{s}}/\bar{\varepsilon}_{\mathrm{s}} + q_{\mathrm{CNL}}/(\varepsilon_{\mathrm{c}}|_0) = 0, \tag{4.33b}$$

with $q_{\mathrm{s}}^2 = n_x^2 - \bar{\varepsilon}_{\mathrm{s}}$ as explicitely evaluated by Leung (1985b).

The approach will now be extended to configurations where a nonlinear film is sandwiched between a nonlinear cladding and nonlinear substrate. As before the nonlinearity of all media involved is assumed to be an arbitrary function of $|E_x|^2 + |E_z|^2$. The eigenvalue (dispersion) relation resembles that given in eq. (4.17a) for the TE case and comes from eq. (4.30) by setting $z = d$ and integrating between $\varepsilon|_0$ and $\varepsilon|_d$, i.e.,

$$d = \operatorname{sgn}\left(\frac{\mathrm{d}\varepsilon}{\mathrm{d}z}\right) \int_{\varepsilon(0)}^{\varepsilon(d)} \frac{\mathrm{d}\bar{\varepsilon}}{[-2V(\bar{\varepsilon})]^{1/2}}. \tag{4.34}$$

In order to fix the actual integration path similar investigations must be done to those leading to eq. (4.17b). It should be noted here that the restrictions of the formalism pointed out in evaluating eq. (4.33) lead to the condition that in practice two nonlinear media are not allowed to be in contact at all.

As a further example the dispersion relation for the linear film geometry will be derived. Using eqs (4.29), (4.31), (4.32) and (4.34), and the well-known field of a linear film at the interfaces $z = 0, d$ to ensure that the boundary conditions are satisfied, gives the eigenvalue (dispersion) relation

$$\tan(k_0 p_{\mathrm{f}} d) = \frac{\bar{\varepsilon}_{\mathrm{f}} p_{\mathrm{f}} (\varepsilon_{\mathrm{s}}|_0 q_{\mathrm{CNL}} + \varepsilon_{\mathrm{c}}|_d q_{\mathrm{CNL}})}{\varepsilon_{\mathrm{c}}|_d \varepsilon_{\mathrm{s}}|_0 p_{\mathrm{f}}^2 - \bar{\varepsilon}_{\mathrm{f}}^2 q_{\mathrm{CNL}} q_{\mathrm{SNL}}} \tag{4.35a}$$

Obviously, $\varepsilon_{\mathrm{s}}|_0$ and $\varepsilon_{\mathrm{c}}|_d$ are mutually dependent since for a linear film the fields B_0 and B_d are related through

$$B_d^2 = \frac{p_{\mathrm{f}}^2 + \bar{\varepsilon}_{\mathrm{f}}^2/\varepsilon_{\mathrm{s}}^2|_0\, q_{\mathrm{SNL}}^2}{p_{\mathrm{f}}^2 + \bar{\varepsilon}_{\mathrm{f}}^2/\varepsilon_{\mathrm{c}}^2|_0\, q_{\mathrm{CNL}}^2} B_0^2. \tag{4.35b}$$

Inserting the first integral (4.28) for both boundary fields we can find $\varepsilon_{\mathrm{c}}|_d$ in dependence on $\varepsilon_s|_0$. In summary, then, eqs (4.46a, b) together with eq. (4.39) constitute the dispersion relation that gives the dependence of n_x on the parameter $\varepsilon_s|_0$.

The relationship of the propagation constant to the guided power flow is the final goal here and will be obtained without explicit information about the field structure. This can be achieved by rewriting the power flow as

$$P_i = \frac{cn_x}{2\mu_0\varepsilon_i}\int B_i^2(z)\,\mathrm{d}z,$$

and employing eq. (4.29b)

$$\mathrm{d}z = \mathrm{sgn}\left(\frac{\mathrm{d}\varepsilon}{\mathrm{d}z}\right)\frac{\mathrm{d}\varepsilon}{[-2V(\varepsilon)]^{1/2}},$$

and the first integral (4.28) to get

$$P_i = \mathrm{sgn}\left(\frac{\mathrm{d}\varepsilon}{\mathrm{d}z}\right)\frac{n_x}{2\mu_0 c}\int_{\bar{\varepsilon}_i}^{\varepsilon_i|_0 \mathrm{or} d}\frac{\tilde{\varepsilon}_i I_i(\tilde{\varepsilon}-\bar{\varepsilon}) - G_i(\tilde{\varepsilon}-\bar{\varepsilon})}{[-2V_i(\tilde{\varepsilon})]^{1/2}(2n_x^2-\tilde{\varepsilon})}\,\mathrm{d}\tilde{\varepsilon}_i. \tag{4.36}$$

The power flow supported by the linear film derived elsewhere, is given by

$$P_\mathrm{f} = \frac{n_x c}{4\mu_0\bar{\varepsilon}_\mathrm{f}}B_0^2\left\{\left(1+\frac{q_\mathrm{CNL}^2}{p_\mathrm{f}^2}\frac{\bar{\varepsilon}_\mathrm{f}}{\varepsilon_\mathrm{c}A_d''}\right)d + \frac{q_\mathrm{SNL}^2}{k_0 p_\mathrm{f}^2}\frac{\bar{\varepsilon}_\mathrm{f}}{\varepsilon_\mathrm{s}A''} + \frac{q_\mathrm{CNL}^2}{k_0 p_\mathrm{f}^2}\frac{\bar{\varepsilon}_\mathrm{f}}{\varepsilon_\mathrm{c}A_d''}\frac{B_d^2}{B_0^2}\right\}, \tag{4.37}$$

where for an explicit calculation eqs (4.32) and (4.35b) must be used.

Hence, a final combination of eq. (4.33) (for the single interface) or of eq. (4.35) with eqs (4.36) and (4.37) yields, at least in principle, the power flow dependence of the propagation constant.

In particular, for Kerr media

$$\varepsilon = \bar{\varepsilon} + a(|E_x|^2 + |E_z|^2),$$

or

$$I(\varepsilon-\bar{\varepsilon}) = \frac{\varepsilon-\bar{\varepsilon}}{a} \quad \text{and} \quad G(\varepsilon-\bar{\varepsilon}) = \frac{(\varepsilon-\bar{\varepsilon})^2}{2a},$$

where advantage is taken of the definition of I and G, the dispersion relation can easily be written down (Leung 1985b).

Numerical results are accessible through a procedure that derives $n_x(P)$ in the following algorithmic manner:

- fix the nonlinear dielectric function $\varepsilon_c^{\mathrm{NL}}$ and $\varepsilon_s^{\mathrm{NL}}$;
- assume $\varepsilon_s|_d$ and n_x;
- calculate $\varepsilon_c|_d$ via eqs (4.35b) and (4.28);
- determine a suitable d by eq. (4.35) to get $n_x(\varepsilon_s|_0, d)$;
- calculate $P(n_x, \varepsilon_s|_0, d)$;
- match both results to gain $n_x(P)$.

In conclusion, an approach has been presented that permits, given a special form of the dielectric tensor, a derivation of the power flow dependence of the propagation constant for a quite general dependence of the dielectric function on the field intensity. A final decision concerning the performance efficiency of this approach will depend on numerical results. Exact numerical solutions now exist for TM waves (Boardman et al. 1987a) and these will be presented in the next section. The results there do not use the method just described but nevertheless the present section contains an illuminating approach that is of value for this type of special dielectric tensor. It will be interesting to see it used in the future for comparison with the more direct approach used by Boardman et al. (1987a).

5. *Numerical studies of nonlinear TE and TM stationary waves at planar interfaces*

This section complements the theory presented in sections 3 and 4 by providing a comprehensive set of numerical examples of stationary wave propagation. As we have already seen, in the earlier sections, two basic structures have commanded a lot of attention in the literature. The first consists of a linear film bounded by semi-infinite, self-focussing, nonlinear media that can be analysed in terms of elementary functions. In consequence, there has been a considerable growth of activity in this area. The second consists of a self-focussing nonlinear film, bounded by linear media, solutions to which involve Jacobi elliptic functions. Since the latter are much more complicated to handle, even numerically, relatively little work has been completed on nonlinear films, producing a dependence here upon the work of Akhmediev et al. (1982a), Boardman and Egan (1986b) and Chen and Maradudin (1988).

It would seem, at first sight, that there is a good case for presenting symmetric guides independently of the asymmetric guides, as indeed would be done for completely linear structures. If nonlinear media are involved, however, regarding a symmetric guiding structure as a degenerate asymmetric system can lead to a deeper insight into the effect of the nonlinearity, especially since a nonlinear symmetric guiding system can sustain asymmetric field distributions. Also it is convenient to regard the single interface between a nonlinear and a linear medium or between two different nonlinear media as limiting cases of a three layer system with a central linear film. The former is approached as the film thickness becomes infinite, while the latter is approached as the film thickness tends to zero.

Finally, in the majority of this section, all the nonlinear media are undamped, unsaturated and Kerr-like, with positive third-order nonlinear coefficients. Self-defocussing, attenuation, and the possibility of non-Kerr behaviour will be dealt with separately.

5.1. Linear film with nonlinear bounding Kerr media: TE waves

5.1.1. Total power flow

For a nonlinear structure the total power flow parallel to the interfaces varies with the guide index $n_x = ck_x/\omega$. Typical variations of the important TE_0 and TE_1 waves in the asymmetric guiding structure of fig. 3.1 are shown in fig. 5.1. The results given there are for typical data that permit distortions of the modal fields due to the power level. This data ranges over the values $0.2 \leqslant \delta \leqslant 2$ where $\delta = D/2\pi = \omega d/2\pi c$ is to be regarded as the width of the film in wavelengths, for a given frequency, or as the normalised frequency for a given film thickness. This range of δ classifies the film as thin and, as has been used by Stegeman (1982a, b) to emphasise nonlinear waveguiding features such as threshold power values, switching and mode isolation. These, in turn, have important physical and technological applications. For example, within certain ranges of wave index, the nonlinearity may allow a variety of field profile shapes for one particular power level, thus opening up the possibility of multistable states that can be used in switching devices. In such a case the power curves exhibit local maxima and/or multiple branches. The other characteristics of the power curves are just as important and will be discussed as the interpretation of fig. 5.1 is developed.

All the curves of fig. 5.1 eventually show a monotonic increase in power with n_x so that an absolute maximum power level does not exist. This is to be

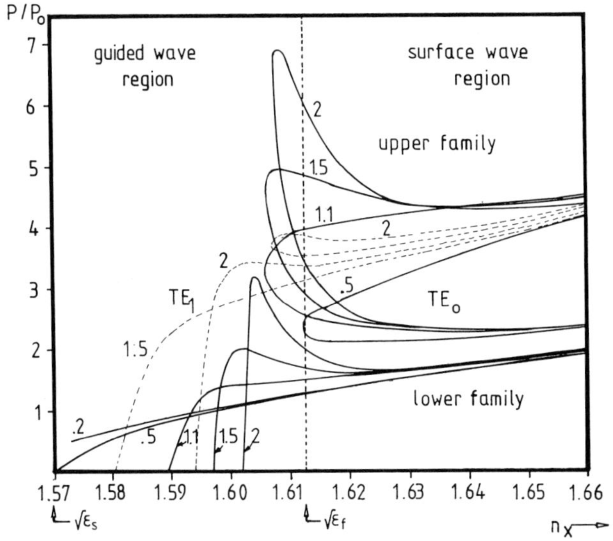

Fig. 5.1. Variation of power (energy) flow down an asymmetric nonlinear guide with wave index n_x, for both lower and upper (E_0, E_d) curves. The power flow is written as the normalised flow P/P_0. The surface wave region begins at $n_x = \sqrt{\bar{\varepsilon}_f}$. TE_0 and TE_1 solutions are shown for the data: $\bar{\varepsilon}_c = 2.45$; $\bar{\varepsilon}_f = 2.6$; $\bar{\varepsilon}_f = 2.3$; $\beta = a_c/a_s = 1.05$; $a_c > 0$; $a_s > 0$. (Boardman and Egan 1985.)

expected since a strictly Kerr medium does not place any limit on the attainable nonlinearity, i.e., it is not saturable so the power may increase indefinitely with n_x. Even in saturable non-Kerr media, however, this is still a possibility and will be discussed in a later subsection.

The TE_0 and TE_1 waves are special cases for films with a nonlinear cladding and nonlinear substrate in that they are the only waves that can, in principle, satisfy both the surface and guided wave criteria elaborated in section 3. Indeed, for very thin films on which only surface waves are possible, only these solutions will exist. For films with only a nonlinear cladding or substrate the TE_1 waves have an extinction point at a certain value of n_x and will not satisfy the surface wave criterion. The discussion of the physics of this point will be deferred, for the moment. Much of the presentation that follows will be concerned with the two principal regimes that involve guided and surface waves and access to them through TE_0 and TE_1 waves. Higher order waves will be discussed towards the end of this section.

It may not be inappropriate here to note that there is no indication of any sudden change in physical properties at $n_x = \sqrt{\varepsilon_f}$ where the guided wave changes to a surface wave. The power curves indicate a smooth transition as the curvature of the guided wave profile changes sign, through the value zero, to the surface wave profile. This can be seen in fig. 5.1, in which there are two families of curves with each value of D giving both a lower and an upper member. This is a particular characteristic of the asymmetric system with the families arising from the two branches of the conic section given by eq. (3.25) that, with the parameters used here, is a type II hyperbola with separated branches and a centre in the first quadrant. The continuity of the (E_0^2, E_d^2) curve on each branch of the hyperbola has its counterpart in the continuity of the power–n_x curves. For data used here, there is no point at which the upper and lower curves of the type II hyperbola meet and this results in the total separation of the upper and lower families of power curves in fig. 5.1. An increase in the asymmetry of the system (i.e., an increase in the difference between the nonlinear refractive indices of the substrate and cladding) increases the separation of the upper and lower families of power curves. Decreasing the asymmetry decreases the separation until in the limit, when the system becomes symmetric, the upper family coalesces with the lower one. In this case, for any particular value of D, the characteristic form of the power curve, as shown by Akhmediev (1982b), and others, is given in fig. 5.2. In principle, this is a very interesting power–guide index dependence that possesses a branch point at which different types of nonlinear waves become degenerate. This degeneracy is easily lifted, however, and will merit a more detailed discussion later on.

In the TE_0 waves of the lower family of curves of fig. 5.1 the effect of the nonlinearity in the cladding may be neglected in the general discussion, since the value of E_d, in this case, is relatively small. Since $\bar{\varepsilon}_s > \bar{\varepsilon}_c$, for the data assumed, then $n_x = \sqrt{\bar{\varepsilon}_s}$ is the linear cut-off eigenvalue of the waveguide. Also, since in the

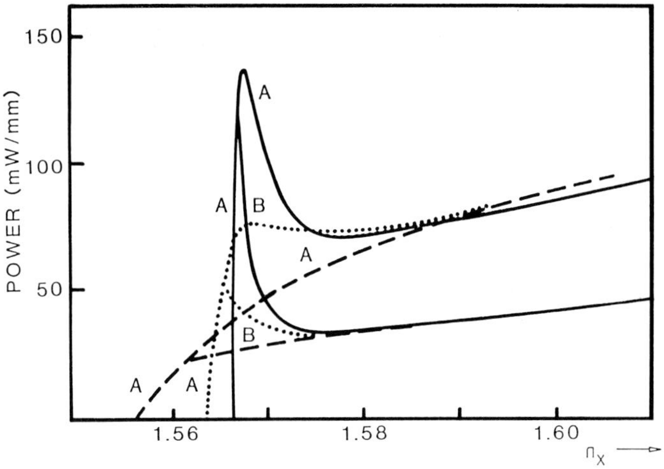

Fig. 5.2. Power flow, calculated in units mW/mm, for a symmetric nonlinear guide for three film thicknesses. Dashed line: $d = 0.5$ μm; dotted line: $d = 1.25$ μm; solid line: $d = 2.0$ μm. A is the symmetric branch. B is the nonlinearly induced asymmetric branch. Data: $\bar{\varepsilon}_s = \bar{\varepsilon}_c = 2.4025$; $\bar{\varepsilon}_f = 2.4649$; $a_s = a_c = c\varepsilon_0\bar{\varepsilon}_s n_{2c}$. $n_{2c} = 10^{-9}$ m^2 V^{-2}. (Seaton et al. 1985c.)

absence of nonlinearity, the system is linear at *all* power levels, the zero-power level in fig. 5.1 will be taken as the linear limit reference level. All finite power variations of n_x express the presence of nonlinearity. This means simply that intersections of the power curves with the n_x-axis are the linear eigenvalues of the system.

These eigenvalues characterise linear guided waves that lie on the range $\sqrt{\bar{\varepsilon}_s} \leqslant n_x \leqslant \sqrt{\bar{\varepsilon}_f}$. This range is accessible provided the film thickness $D > D_L^{(1)}$, where $D_L^{(1)}$ is to be calculated later on and is, obviously, called the cut-off film thickness. The notation here uses a subscript L to label results that refer to the lower family of (E_d^2, E_0^2) curves and a label U for those that refer to the upper family. Even though there are no linear eigenvalues, and hence no zero power limits, below the cut-off value $\sqrt{\bar{\varepsilon}_s}(D < D_L^{(1)})$, or above $\sqrt{\bar{\varepsilon}_f}$ it is clear from fig. 5.1 that the nonlinearity still permits the existence of power–n_x curves. The range $n_x > \sqrt{\bar{\varepsilon}_f}$ is easier to explain since it is a surface wave region characterised by self-focussing that arises strictly because of the nonlinearity.

Figure 5.3a shows qualitatively what will happen to a linear TE_0 wave as the power is increased. Since there is only one *fixed* n_x solution, for the linear eigenvalue equation, increasing the power will increase E_0^2 and field amplitude but it will not change its field shape, i.e., in the linear case the field maximum remains in a fixed position. In contrast fig. 5.3b shows that for a typical nonlinear system the field shape changes as E_0^2 (power) changes and the field maximum in this case may eventually cross a guide boundary. In the example

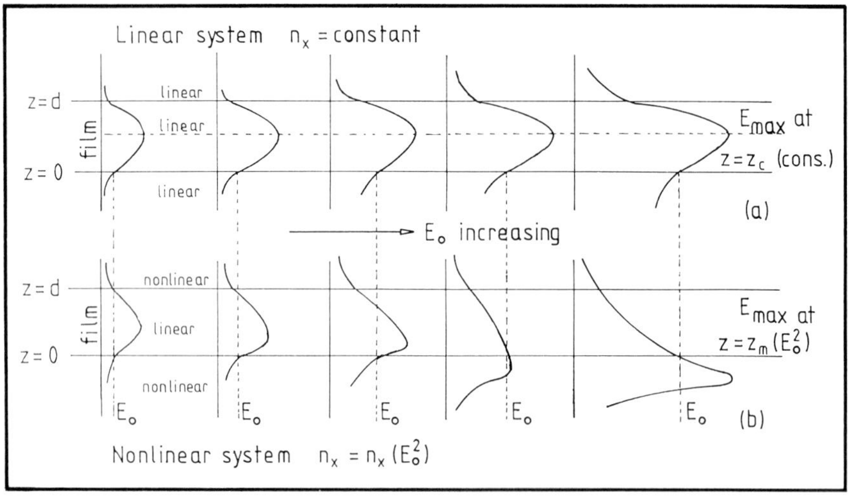

Fig. 5.3. (a) variation of a typical linear TE_0-mode as the power flow down the guide is increased. (b) variation in field shape of a typical nonlinear TE_0 wave as the power level is raised.

shown this happens to occur at $z = 0$. The constancy of n_x, as the power increases in a linear guiding system, is further emphasised by fig. 5.4. This figure shows that the projection onto the power–n_x plane is a set of vertical lines, demonstrating that the system is locked onto its eigenvalues for all possible power levels. Thus if, for example, the eigenvalue is 1.57 it will not vary in a linear guide as the power increases. This explains, in more detail, why in fig. 5.1 we can take *zero power* as the linear reference level with any departure from the n_x-axis as the power increases taking us into the nonlinear regime. Note that the field shape may change perceptibly with optical power density only if the nonlinear change in the refractive index of the nonlinear medium matches the change in the linear refractive indices across the linear/nonlinear boundary.

In a nonlinear guiding system them, increasing the power results in the field maximum (the bulge) being drawn closer to and eventually being pushed through the film/substrate boundary. A further property of the nonlinear system now emerges, however. Once the bulge has passed into the substrate, increasing the power to higher levels has little effect on the amplitude profile of the guided wave inside the film. The intensity at the boundary approaches a limiting value, equal to $2(\bar{\varepsilon}_f - \bar{\varepsilon}_s)/a_s$, that leads to a decrease in the power transported by the linear film as the curvature of the amplitude profile decreases. Eventually the curvature becomes zero and changes sign. The field profile then exhibits a surface wave pattern of behaviour with the bulge in the substrate increasing in size with n_x. Thus, at high power levels, considering fig. 5.1 once again, most of the energy for TE_0 (and TE_1) waves is carried outside the linear film. This will

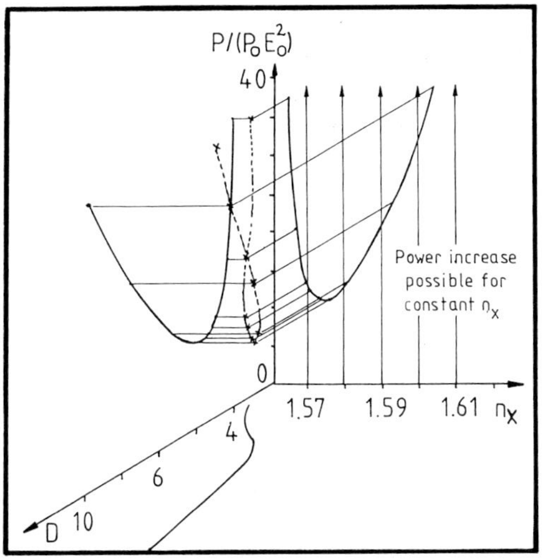

Fig. 5.4. Normalised power flow $P/P_0E_0^2$ of a TE_0-mode of an asymmetric guide plotted as both a function of n_x and normalised guide thickness D. The projection onto the (n_xD) plane gives the unique linear dispersion curve. The fixed E_0^2–n_x dependence is quite different from the result of varying E_0^2 from zero up to any desired value. It is interesting to observe that as $D+0$ there is no allowed n_x value because of the familiar cutoff due to asymmetry. In the nonlinear case guided waves could occur at smaller values of D. (Boardman and Egan 1985.)

be very significant if the nonlinear substrate happens to be lossy, since it gives a possible mechanism for the dumping of power from a linear guide into a nonlinear environment.

In the lower TE_0 family of fig. 5.1, the monotonic increase of power over the whole range of allowed n_x values, for the smaller values of film thickness, is easily identified with the increase in area under the amplitude profile curves of fig. 5.3b. This change in area is not so straightforward for larger values of film thickness. In order to illustrate this consider the $\delta = 2$ curve extracted from fig. 5.1 and reproduced, on an extended scale, together with the corresponding amplitude profiles in fig. 5.5. Here it can be seen that, after an initial increase, the decrease in the area under the part of the field profile residing in the film is much more rapid than in the small δ cases. This is due to the rapidly changing shape across the wider film. This rapid decrease dominates the low rate of area increase under the part of the field profile residing in the substrate. Therefore an overall decrease in power with increasing n_x occurs, before, on further increasing n_x, the rate of increase in the area under the profile in the substrate begins to dominate, and the total power again begins its monotonic increase with n_x. It is clear from the numerical work, that another critical film thickness $D_L^{(2)}$ exists for the lower family of curves, above which the power–n_x curve displays a local maximum in the guided wave region. This critical thickness cannot be found analytically but it is important since it indicates the possibility of multistable states associated with only one particular power curve. If $D > D_L^{(2)}$ then above a minimum power level and below the local power maximum there are three possible values of n_x corresponding to a single power level. These n_x values are associated with three different electric field amplitude profiles, as shown in fig. 5.5. One of these is always in the guided wave region and one is always in the surface wave region. The third may be in either region, its position depending on the power level chosen.

It is informative now to reconsider the $\delta = 2$ member of the lower TE_0 family of fig. 5.1 in terms of the effect that increasing the value of the boundary intensity has upon the nonlinear refractive index distribution of the guide. If the nonlinear refractive index of the substrate and the cladding is written as $\sqrt{\varepsilon_{NLS}}$ or $\sqrt{\varepsilon_{NLC}}$, respectively, then it is the changing value of $\sqrt{\varepsilon_{NLS}}$ that determines the shape of the field profile since, for the lower family of power curves, there is very little change in $\sqrt{\varepsilon_{NLC}}$. The field profiles considered in fig. 5.5 together with the corresponding nonlinear refractive index distribution are shown in fig. 5.6.

The curvature of the field profile cannot change sign within the linear film but can do so on crossing into the nonlinear media. In fact, the curvature of the field profile in the substrate will remain positive provided that $n_x > \varepsilon_{NLS}$. At the value $n_x = \varepsilon_{NLS}$ the curvature in the substrate will then change sign. However, the field in the substrate is, by definition, $E_s = \sqrt{(\varepsilon_{NLS} - \bar{\varepsilon}_s) a_s^{-1}}$, so that the field profile cannot extend beyond the boundary formed by the plot of ε_{NLS}. Since the extension of this boundary beyond the film substrate boundary is relatively small, as shown in figs 5.6a, d, there is a squeezing effect on the field profile just

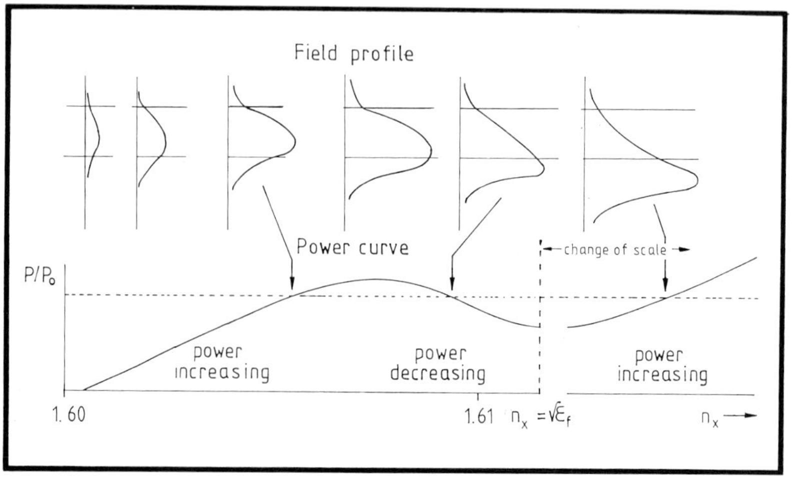

Fig. 5.5. The TE_0 powers–n_x $\delta = D/2\pi$ curve of fig. 5.1 plotted on an extended scale to illustrate the behaviour of the field profile at various power levels.

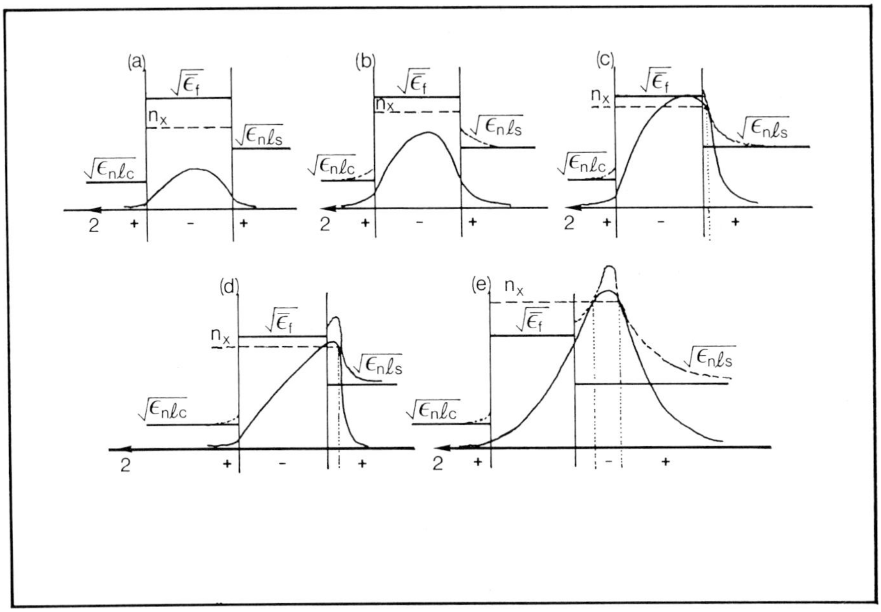

Fig. 5.6. The field profiles considered in fig. 5.5 together with the corresponding refractive index distributions.

as the bulge (maximum in the field) is leaving the film. This causes a reduction in the total power flow along the guide. As n_x is increased further the intensity self-focussed in the substrate continues to increase but the intensity at the film/substrate boundary then approaches the limit $2(\bar{\varepsilon}_f - \bar{\varepsilon}_s)a_s^{-1}$. This relieves the squeezing effect and a surface wave profile develops as n_x increases beyond $\bar{\varepsilon}_f$ as sketched in fig. 5.6e. After this, the power carried by the wave begins to increase monotonically with n_x.

Since the discussion, so far, has concentrated upon the lower family of fig. 5.1, for which the nonlinearity in the cladding is negligible, it might be expected that the power–n_x curves for a linear film with a nonlinear substrate and a linear cladding would behave in the same way. This is indeed the case for TE_0 waves but for TE_1 waves there is an interesting difference as shown in fig. 5.7.

The power–n_x curve for a TE_1 wave in a linear cladding-linear film-nonlinear substrate waveguide structure has an extinction point at a value of n_x above the zero power (linear) eigenvalue. The TE_1 power curves for the structure used to generate fig. 5.1 do not have such extinction points. In that type of waveguide it is the TE_2 waves that have extinction points. Such features arise in the following way.

The field profile of a TE_1 wave crosses the zero field axis once inside the linear film. For the guiding structure shown in fig. 5.7, the field extrema may only move

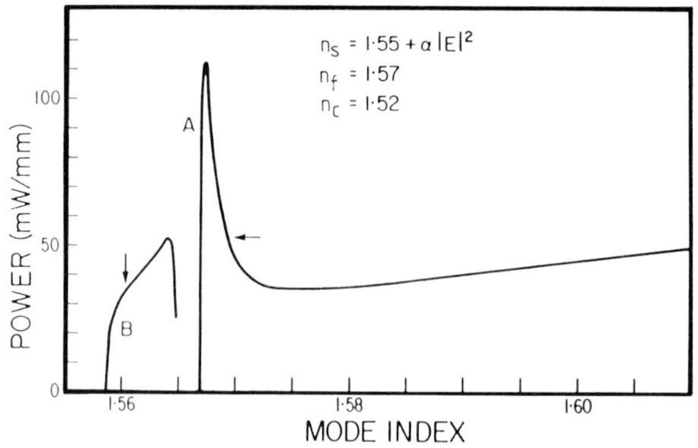

Fig. 5.7. Power–n_x curves for an asymmetric nonlinear guide consisting of a linear substrate, a linear film and a nonlinear cladding. Data: $\bar{\varepsilon}_s = 2.4025$; $\bar{\varepsilon}_f = 2.4649$; $\bar{\varepsilon}_c = 2.4025$; $a_c = c\varepsilon_0\varepsilon_c n_{2s}$ and $n_{2s} = 10^{-9}$ m^2 W; $d = 2$ μm. (Stegeman et al. 1984a.)

into the nonlinear substrate so the crossing point of the field inside the film moves towards the film substrate interface as the power increases. At a certain value of n_x the zero of the field will reach the film-cladding interface. The TE_1 field can then exist at higher n_x values only if it crosses the z-axis outside the film. This is actually forbidden for a planar structure because once the field has crossed the axis externally it must repeatedly cross the guide axis all the way out to infinity. In other words it is simply not a physical solution.

If the waveguide is a linear film with a cladding and substrate that are both nonlinear, as it is for the curves in fig. 5.1 then TE_1 waves do not have such extinction points. This is because there are two maxima associated with the TE_1 wave in the film and these may move in opposite directions as the power increases. Hence the field extrema can move both into the cladding and the substrate with the zero field crossing point still located (i.e., trapped) at the centre of the film. This prevents the appearance of an extinction point.

For TE_2 waves there are two crossing points inside the linear film. In the nonlinear-linear-nonlinear waveguide structure of fig. 5.1 although extinction points do not exist for TE_1 waves they do exist for TE_2 waves. This is because the crossing points will move in opposite directions towards the film boundaries. As a crossing point reaches a boundary the nonlinear wave is extinguished. At a higher value of n_x the nonlinear wave will be re-established but will again be extinguished as the second crossing point reaches an interface. These features are shown in fig. 5.8.

It is convenient, at this stage, to determine what would happen to the power–n_x curves of fig. 5.1 in the limiting case when the film thickness is large enough to result in decoupled, single interface behaviour. It is already apparent in fig. 5.1 that as δ increases, both the maximum in the power–n_x curves, and the

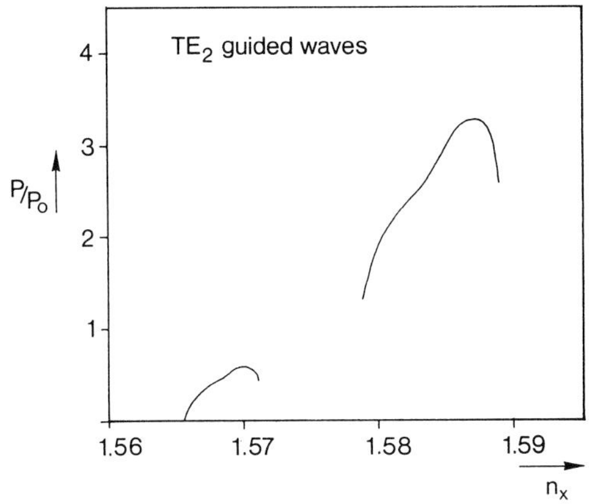

Fig. 5.8. Illustration of extinction points for TE_2 nonlinear guided waves for the asymmetric guide data of fig. 5.1.

linear eigenvalue are approaching $n_x = \sqrt{\bar{\varepsilon}_s}$. In the limit $\delta \to \infty$ only single surface linear/nonlinear interface solutions remain and no guided TE waves can exist. Furthermore, only the TE_0 and TE_1 can progress from guided waves to surface waves. Even this is true only for the nonlinear-linear-nonlinear waveguide structures. In the linear-linear-nonlinear guide the TE_1 wave, as was discussed above, is extinguished before the surface wave region can be reached while for fig. 5.1 it is TE_n, $n \geqslant 2$ that are extinguished before this region is encountered.

Single interfaces require a critical power level to be reached before they can support surface waves. The surface wave region, $n_x > \sqrt{\bar{\varepsilon}_f}$, shows that, as $\delta \to \infty$, the TE_0 power curves develop three minima, while the portion of the curves to the left of $\sqrt{\bar{\varepsilon}_f}$ becomes asymptotic to the vertical line through $n_x = \sqrt{\bar{\varepsilon}_f}$. The upper minimum is associated with separating the interfaces at an equal rate and is the sum of the powers carried by surface waves at the upper and lower single surfaces that begin to decouple from each other as $\delta \to \infty$. The next lowest minimum is associated with power carried by a surface wave along the single interface between the upper nonlinear medium and the linear medium. The lowest minimum corresponds to power being carried along a single interface between the lower nonlinear medium and the linear medium. The TE_1 waves are all associated with the upper minimum. This is because, in the TE_1 waves, bulges develop in both media so the $\delta \to \infty$ limit must be associated with the sum of surface waves power at both interfaces. It is clear then that single surface behaviour as shown in fig. 5.9, is characterised by a power–n_x curve that is asymptotic to the ordinate at $n_x = \sqrt{\bar{\varepsilon}_f}$. It possesses a shallow minimum close to the right of this value and increases and then increases slowly and montonically with n_x.

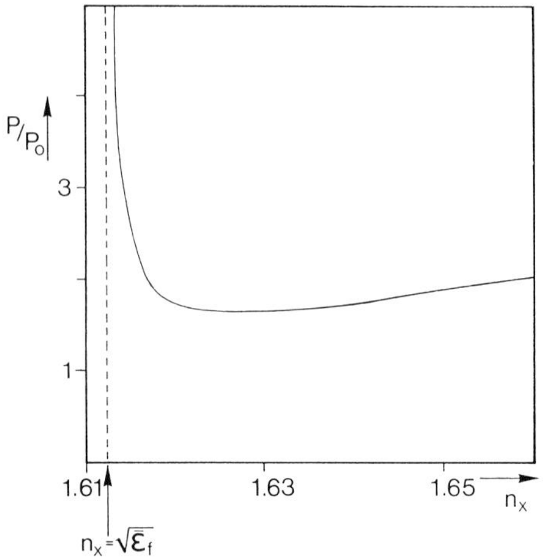

Fig. 5.9. Typical single surface power–n_x behaviour. Notice the existence of a threshold power and the asymptotic approach to $n_x = \sqrt{\bar{\varepsilon}_f}$.

The final point to be made here about the lower family of power curves shown in fig. 5.1 concerns the appearance of extinction points. For example, a TE_0 curve with an extinction point appears for linear film thicknesses $D < D_L^{(1)}$ where $D_L^{(1)}$ is the linear cut-off thickness of the film. If the dispersion equation given by eq. (3.61) is considered, then its linear limit $(a_s E_0^2 = 0,\ a_c E_d^2 = 0)$ is eq. (3.66) and a linear guided wave cut-off occurs when q_s or q_c becomes imaginary. This linear cut-off is at $p_f^2 = \bar{\varepsilon}_f - \bar{\varepsilon}_s$ or $p_f^2 = \bar{\varepsilon}_f - \bar{\varepsilon}_c$, whichever is the small cut-off thickness from the equation

$$\cos(p_f D) = \frac{\bar{\varepsilon}_f - \bar{\varepsilon}_s}{\sqrt{(\bar{\varepsilon}_f - \bar{\varepsilon}_s)(\bar{\varepsilon}_f - \bar{\varepsilon}_c)}} \quad \text{or} \quad \frac{\bar{\varepsilon}_f - \bar{\varepsilon}_c}{\sqrt{(\bar{\varepsilon}_f - \bar{\varepsilon}_s)(\bar{\varepsilon}_f - \bar{\varepsilon}_c)}}. \tag{5.1}$$

The value of $D_L^{(1)}$ is therefore given by

$$D_L^{(1)} = \frac{1}{(\bar{\varepsilon}_f - \bar{\varepsilon}_s)^{1/2}} \cos^{-1}\left(\frac{\bar{\varepsilon}_f - \bar{\varepsilon}_s}{\bar{\varepsilon}_f - \bar{\varepsilon}_c}\right)^{1/2}, \tag{5.2}$$

or

$$D_L^{(1)} = \frac{1}{(\bar{\varepsilon}_f - \bar{\varepsilon}_c)^{1/2}} \cos^{-1}\left(\frac{\bar{\varepsilon}_f - \bar{\varepsilon}_c}{\bar{\varepsilon}_f - \bar{\varepsilon}_s}\right)^{1/2}. \tag{5.3}$$

For film thicknesses $D < D_L^{(1)}$ there are no linear eigenvalues but valid nonlinear solutions are now given by eq. (3.60), whose solutions show the existence of a field maximum in the substrate. Figure 5.10, which is a further development of fig. 3.10, is a sketch designed to show the linear cut-off as D becomes small (fig. 5.10a) and the fact that even for $D < D_L^{(1)}$ a nonlinear wave can still exist. As D decreases, the cosine curve lifts above the intersection point

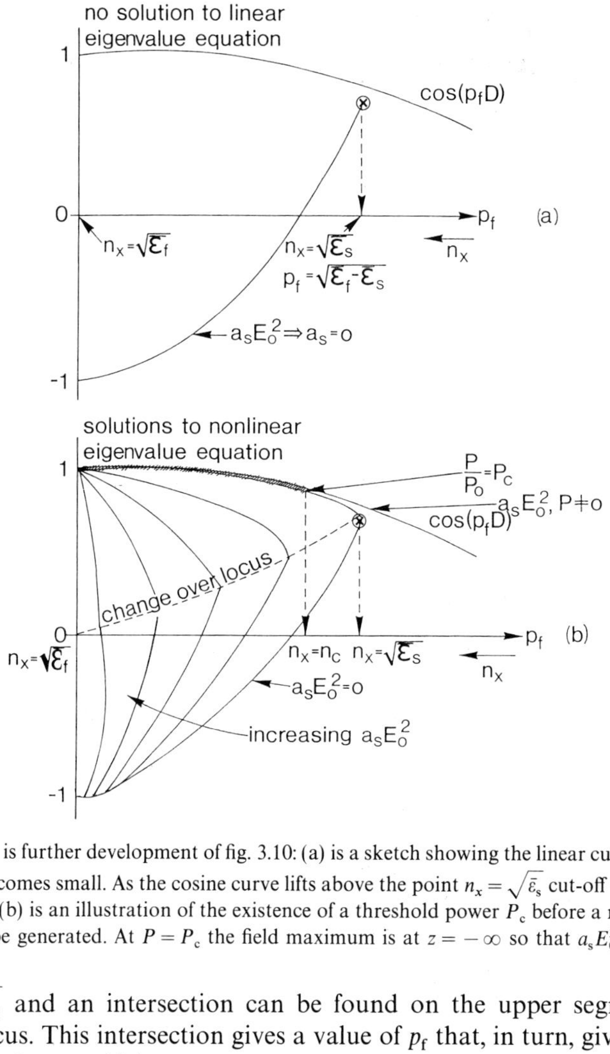

Fig. 5.10. This is further development of fig. 3.10: (a) is a sketch showing the linear cut-off as the film thickness D becomes small. As the cosine curve lifts above the point $n_x = \sqrt{\bar{\varepsilon}_s}$ cut-off would occur in a linear guide; (b) is an illustration of the existence of a threshold power P_c before a nonlinear wave can be generated. At $P = P_c$ the field maximum is at $z = -\infty$ so that $a_s E_0^2 = 0$.

at $n_x = \sqrt{\bar{\varepsilon}_s}$ and an intersection can be found on the upper segment of the $a_s E_0^2 = 0$ locus. This intersection gives a value of p_f that, in turn, gives the minimum value of n_x at which a wave can propagate at that value of D. This is not a cut-off value but an extinction point, as is shown in fig. 5.1.

These extinction points lie on a locus in the power–n_x plane that can be found from the general power equation (3.112). If, in this equation, we set $a_s E_0^2 = 0$, $a_c E_d^2 = 0$ then the threshold power, P_{th}, for guided waves, for a film thickness $D_L^{(3)} < D < D_L^{(1)}$, where it will be shown soon that $D_L^{(3)}$ is the absolute minimum film thickness that will support guided waves, is given by

$$P_{th} = P_0 n_x \left\{ \beta[\sqrt{(n_x^2 - \bar{\varepsilon}_s)} \pm \sqrt{(n_x^2 - \bar{\varepsilon}_s)}] + \sqrt{(n_x^2 - \bar{\varepsilon}_f)} \pm \sqrt{(n_x^2 - \bar{\varepsilon}_f)} \right\}, \quad (5.4)$$

where $P_0 = 2c^2\varepsilon_0/\omega^2 a_c$, $\beta = a_c/a_s$ and n_x is the value at the intersection point. The + sign is taken when the bulge happens to be in the medium to which the subscript refers. Since the lower of the (E_0^2, E_d^2) curves is being considered here the bulge is in the film or the substrate and, in fact, in fig. 5.10b, is in the substrate. Equation (5.4) therefore reduces to

$$\frac{P_{\text{th}}}{P_0} = 2\beta n_x\sqrt{(n_x^2 - \bar{\varepsilon}_s)}, \tag{5.5}$$

and the value of n_x associated with it is

$$n_x = \tfrac{1}{2}\bar{\varepsilon}_s + \left[(\tfrac{1}{2}\bar{\varepsilon}_s)^2 + \left(\frac{P_{\text{th}}}{2\beta P_0}\right)^2\right]^{1/2} \tag{5.6}$$

An example of a critical power is shown in fig. 5.1 for $\delta = 0.2$. For smaller values of δ, the critical power levels are higher. After the threshold power is reached a single self-focussed bulge is established in the substrate. At the extinction point the stationary state has a field profile that goes to zero at interface between the nonlinear substrate and the film and it sustains this zero value right across the film and throughout the cladding. This shows that the field maximum (bulge) has at this point moved to $z_0 = -\infty$, i.e., it is an infinite distance away from the substrate/film boundary and this another way to express the extinction condition.

As D decreases further the intersection point shown in fig. 5.10b moves to the left as the cosine curve lifts upwards. This will continue until $p_f = 0(n_x = \sqrt{\bar{\varepsilon}_d})$, at some value $D = D_L^{(3)}$. Below this film thickness, guided waves are no longer possible and the film will be relatively much too thin to support a wave whose field profile has a bulge in the substrate and a positive curvature inside the film. Only a profile with a negative curvature in the film (i.e., a surface wave) can be accommodated. This kind of effect is only possible within a nonlinear system.

An analytical expression for $D_L^{(3)}$ can easily be found, since this is the minimum value of D for which an intersection between $\cos(p_f D)$ and the upper $a_s E_0^2 = 0$ of fig. 5.10b can occur. The small D limit of eq. (3.60) gives, as $p_f^2 \to 0$,

$$1 - p_f^2(D_L^{(3)})^2 \simeq 1 - \tfrac{1}{2}p_f^2\left[\frac{(\sqrt{\bar{\varepsilon}_f - \bar{\varepsilon}_s} - \sqrt{\bar{\varepsilon}_f - \bar{\varepsilon}_c})}{\sqrt{\bar{\varepsilon}_f - \bar{\varepsilon}_s}\sqrt{\bar{\varepsilon}_f - \bar{\varepsilon}_c}}\right]. \tag{5.7}$$

The expression for $D = D_L^{(3)}$ is therefore

$$D_L^{(3)} = \left|\frac{[\sqrt{\bar{\varepsilon}_f - \bar{\varepsilon}_s} - \sqrt{\bar{\varepsilon}_f - \bar{\varepsilon}_c}]}{[\sqrt{\bar{\varepsilon}_f - \bar{\varepsilon}_s}\sqrt{\bar{\varepsilon}_f - \bar{\varepsilon}_c}]}\right|^{1/2}. \tag{5.8}$$

Since this expression is for the absolute minimum thickness that will sustain an extinction point with $n_x < \sqrt{\bar{\varepsilon}_f}$, it does not contain any nonlinear parameters. In a purely linear system, however, $D_L^{(3)} < D_L^{(1)}$, where $D_L^{(1)}$ is the linear cut-off thickness.

The TE_1 waves, described by eqs (3.62) and (3.63), also have interesting properties that are, in fact, characteristic of all waves of higher order than TE_0. Figure 5.11a provided that, for that particular waveguide,

$$(\bar{\varepsilon}_f - \bar{\varepsilon}_s)^{1/2} < \frac{2\pi}{D} - \frac{1}{D}\cos^{-1}\left[\left(\frac{\bar{\varepsilon}_f - \bar{\varepsilon}_s}{\bar{\varepsilon}_f - \bar{\varepsilon}_c}\right)^{1/2}\right]_{min}, \tag{5.9}$$

the power–n_x curves for the TE_1 waves of fig. 5.1 are similar to those for the TE_0 waves. However, as shown in fig. 5.11b, for

$$(\bar{\varepsilon}_f - \bar{\varepsilon}_s)^{1/2} > \frac{2\pi}{D} - \frac{1}{D}\cos^{-1}\left(\frac{\bar{\varepsilon}_f - \bar{\varepsilon}_s}{\bar{\varepsilon}_f - \bar{\varepsilon}_c}\right)^{1/2}, \tag{5.10}$$

a second set of eigenvalues exists. This second set is obtained by 'pulling in' the $\cos(p_f D)$ curve to produce intersections with the $a_s E_0^2 = 0$ locus, as shown in

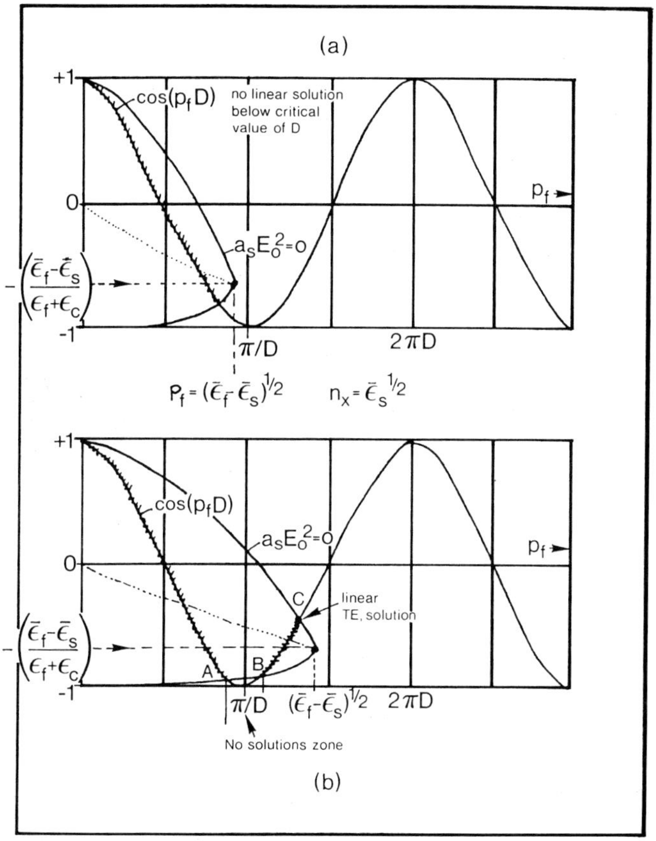

Fig. 5.11. Illustration of the solution of the eigenvalue equation for TE_1 nonlinear waves in an asymmetric guide. Two sets of solutions are evident depending upon the value of D.

fig. 5.11b. This can be achieved by making D larger, as indicated by the inequalities defined above. For asymmetric systems, $[(\bar{\varepsilon}_f - \bar{\varepsilon}_s)/(\bar{\varepsilon}_f - \bar{\varepsilon}_c)]^{1/2} \neq 1$, and the second set is separated from the first set by a range of values of n_x that are not solutions of the eigenvalue equation. This is shown graphically in fig. 5.11b as a region where no intersections of the curves are possible. This means that, for values of D that are large enough, a second TE_1 branch to the power–n_x curve will appear. This will form a loop that starts and finishes on the n_x-axis. The lowest value of n_x on this loop is the linear eigenvalue. Since the appearance of this structure in the TE_1 power curve requires a thicker film than $\delta = 2$, it does not, for simplicity, appear in fig. 5.1.

The solutions shown in fig. 5.11b can be demonstrated by the TE_1 field profile behaviour, given in fig. 5.12, in which the sequence (c), (b), (a), (d), (e), (f) is in order of increasing n_x. Figure 5.12c represents the linear limit, point C in fig. 5.11b. As a progression from the linear limit (c) is made, the power increases from zero, rises to a maximum, and then falls again to zero at (a) [point B in fig. 5.11b]. After a jump in n_x occurs, profile (d) is established [point A in fig. 5.11b] and the power then increases gradually forcing the bulge out of the film. Increasing the power from profile (d), however, can force a bulge into the nonlinear cladding and the nonlinear substrate. After these events occur, further increases in power will eventually create typical self-focussing surface wave behaviour. The power will then be carried, to a large extent, in the nonlinear bounding media.

The outstanding difference between the upper and lower families of power curves is the double-valued nature of the upper family. On the lower curves, to

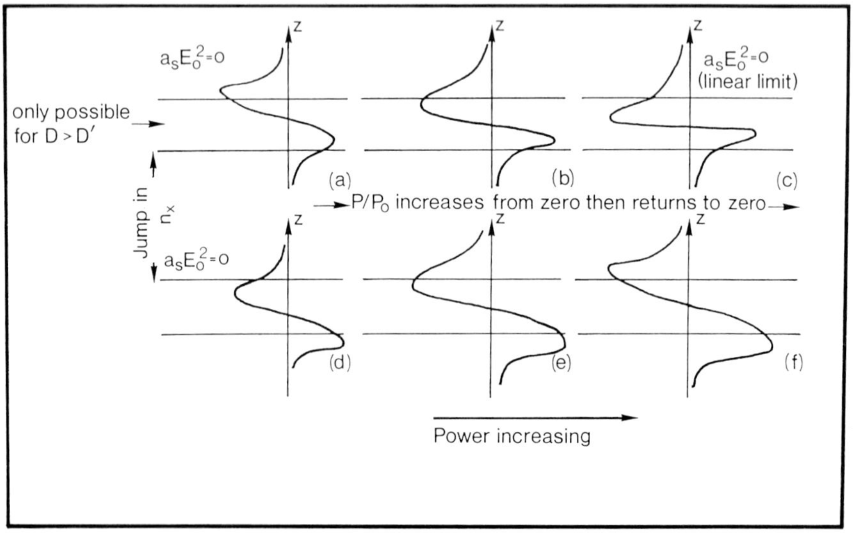

Fig. 5.12. Sketch of the field profile evolution for the TE_1 wave of fig. 5.11.

each allowed n_x, there is a unique value of power. On the upper curves, with the exception of the turning point of the curves, there are two distinct power values, for each allowed n_x. This property is really exposed by the relationship between the optical intensity at the film/substrate boundary and the wave index shown in fig. 3.13, where, except at the minimum point on the curves, there are two values of $a_s E_0^2$ for each value of n_x. These lead to two different power values corresponding to different amplitude profiles for the same value of wave index. For TE_0 waves, starting with values of optical intensity at the film/substrate boundary that approach zero, a check with fig. 3.5 shows that the optical intensity at the other boundary approaches a finite limit $2(\bar{\varepsilon}_f - \bar{\varepsilon}_c)/a_c$. Hence, on the upper family of TE_0 curves in fig. 5.1, $a_s E_0^2 \rightarrow 0$ corresponds to the development of a surface wave in the cladding with a growing field amplitude. For this wave P/P_0 becomes larger and larger, eventually approaching infinity as $a_s E_0^2 \rightarrow 0$ and $a_c E_d^2 \rightarrow \bar{\varepsilon}_f - \bar{\varepsilon}_c$. As shown in fig. 5.13, for the upper loop in fig. 5.1, increasing the intensity at the film/substrate boundary from zero, first causes the intensity at the film/cladding boundary to decrease. It then increases again, this time without limit so that $a_s E_0^2 \rightarrow \infty$, $a_c E_d^2 \rightarrow \infty$ and $P/P_0 \rightarrow \infty$. This accounts for the two segments of each curve in the upper family of fig. 5.1 being able to show a monotonic increase of power with wave index. The interesting detail in the region $n_x \simeq \sqrt{\bar{\varepsilon}_f}$, where the two segments join to form a continuous curve is determined by the particular physical parameters of the system. The field amplitude profiles of fig. 5.13a are produced for $\delta = D/2\pi = 2$. As indicated above it shows that, starting with the optical intensity, at the film/substrate boundary, in the region of zero, the bulge is in the cladding. Increasing the intensity at the film/substrate boundary causes an initial decrease in the intensity at the film/cladding boundary and the bulge is drawn into the film. This stage describes the lower part of the upper power loop of fig. 5.1 for which the power transported by the wave is falling, and is carried mainly in the film. A further increase in the intensity at the film/substrate boundary results in an increasing intensity at the film/cladding boundary, and the bulge moves back out into the cladding. The intensity at both boundaries may then increase together without limit and is a property of the upper part of the upper power loop of fig. 5.1.

For the parameters used up to now, however, $a_c E_d^2 > a_s E_0^2$. If the refractive index in the cladding is less than that in the substrate, an increase in the intensity at the lower boundary will reach a value where, for the upper family of power–n_x curves, $E_d^2 < E_0^2$. In this situation a bulge that starts off in the cladding passes first into the film and, as the intensity at the lower boundary increases further, passes from the film into the substrate. Figure 5.13b shows the developing amplitude profile of such a bulge as n_x, and hence the power, increases.

Figure 5.13c contains a useful summary of the behaviour of both guided waves and surface waves. It is an extension of the description initiated in fig. 3.9 and is a plot of E_d/E_0 against $p_f D$ and $q_f D$. From eqs (3.64) and (3.65), it is apparent that the loci $E_d/E_0 = \sec(p_f D)$ or $\cos(p_f D)$, for guided waves, and

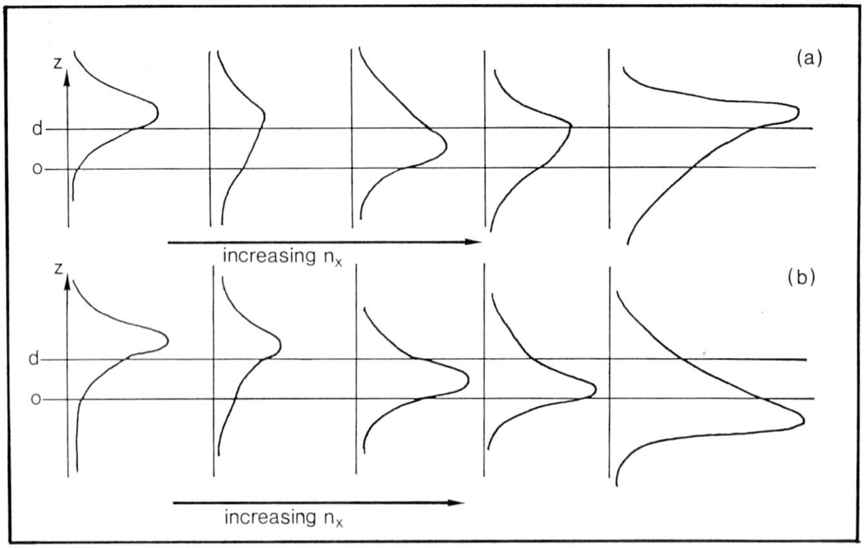

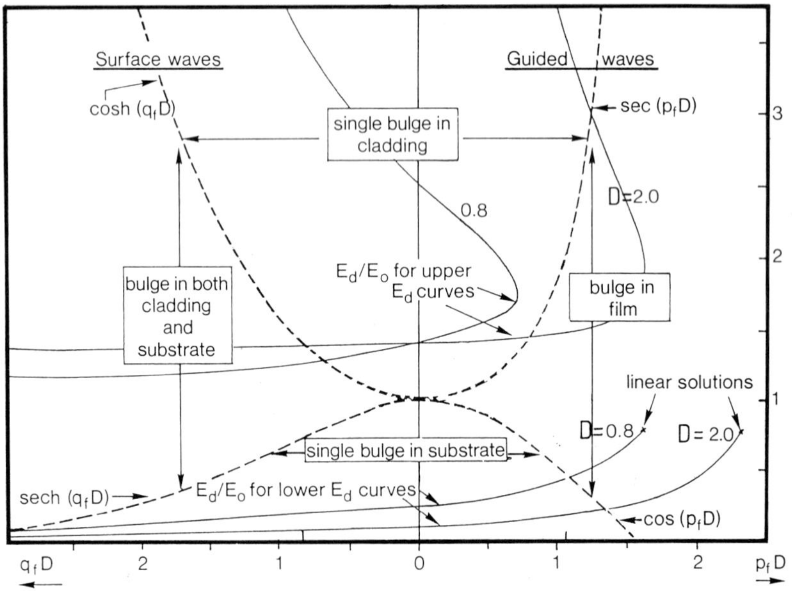

Fig. 5.13. (a) Field profile development for the upper part of the $\delta = 2$ member of the upper TE_0 family of loops in fig. 5.1. Initially the optical intensity at the substrate boundary is near to zero and the bulge moves back into the cladding. (b) Field profile development for the lower part of the same family loop member showing that a bulge that starts off in the cladding passes through to the substrate. (c) Summary of the behaviour of guided and surface waves.

$\operatorname{sech}(p_f D)$ and $\cosh(q_f D)$ for surface waves, determine the regions of the parameter space that are associated with bulges (maxima) in the film or the cladding/substrate media.

For the upper family in fig. 5.1 the changeover locus is

$$p_f^2 = \bar{\varepsilon}_f - \bar{\varepsilon}_s - \tfrac{1}{2} a_s E_0^2, \tag{5.11}$$

or

$$p_f^2 = \bar{\varepsilon}_f - \bar{\varepsilon}_c - \tfrac{1}{2} a_s E_d^2, \tag{5.12}$$

depending upon which one is satisfied first. The upper E_d curve sets out with $\frac{1}{2} a_c E_d^2 = \bar{\varepsilon}_f - \bar{\varepsilon}_c$ as opposed to the zero value it would have on the lower branch. $\frac{1}{2} a_c E_d^2$ on this upper branch falls to a minimum, rises again to the value $\bar{\varepsilon}_f - \bar{\varepsilon}_c$ and then continues to increase indefinitely. Since $\frac{1}{2} a_c E_d^2$ passes through the value $\bar{\varepsilon}_f - \bar{\varepsilon}_s$, it is necessary to discuss the regions above and below $\frac{1}{2} a_s E_0^2 = \bar{\varepsilon}_f - \bar{\varepsilon}_s$.

There is a critical cut-off thickness $D_U^{(1)}$, below which no purely exponential decay attenuation occurs in the upper medium. This is different from the one derived for the lower E_d curves (family). For example, if $\beta < 2$ which it is for the data being used for fig. 5.1, then eq. (5.12) is satisfied before eq. (5.11) and $D_U^{(1)}$ is obtained from

$$\cos(p_f D_U^{(1)}) = \frac{p_f}{(\bar{\varepsilon}_f - \bar{\varepsilon}_s - \frac{1}{2} a_s E_0^2)^{1/2}}. \tag{5.13}$$

For the upper family, p_f is a maximum when $\frac{1}{2} a_c E_d^2$ falls to a minimum at $a_s E_0^2 = \bar{\varepsilon}_f - \bar{\varepsilon}_s$, in which case

$$\cos(p_f D_U^{(1)}) = \frac{\sqrt{2}(\bar{\varepsilon}_f - \bar{\varepsilon}_c - a_c E_d^2)^{1/2}}{(\bar{\varepsilon}_f - \bar{\varepsilon}_s)^{1/2}}, \tag{5.14}$$

where

$$E_d^2 = \frac{\bar{\varepsilon}_f - \bar{\varepsilon}_c}{a_c} + \left[\frac{a_s \eta_c^2 - a_c \eta_s^2}{a_s a_c} \right]^{1/2}. \tag{5.15}$$

Hence $D_U^{(1)}$ is

$$D_U^{(1)} = \frac{2}{\bar{\varepsilon}_f - \bar{\varepsilon}_c - [(\bar{\varepsilon}_f - \bar{\varepsilon}_c)^2 - \beta(\bar{\varepsilon}_f - \bar{\varepsilon}_s)^2]^{1/2}} \times \cos^{-1}\left\{ \frac{\bar{\varepsilon}_f - \bar{\varepsilon}_c}{\bar{\varepsilon}_f - \bar{\varepsilon}_s} - \left[\frac{(\bar{\varepsilon}_f - \bar{\varepsilon}_c)^2}{(\bar{\varepsilon}_f - \bar{\varepsilon}_s)^2} - \beta \right]^{1/2} \right\}^{1/2}. \tag{5.16}$$

The value of D, below which no guided wave solution can occur, is

$$D_U^{(3)} = \left| \frac{x - y}{xy} \right|, \tag{5.17}$$

where

$$x = [\tfrac{1}{2}(\bar{\varepsilon}_{\mathrm{f}} - \bar{\varepsilon}_{\mathrm{s}})]^{1/2}, \tag{5.18}$$

$$y = \sqrt{\tfrac{1}{2}\{\bar{\varepsilon}_{\mathrm{f}} - \bar{\varepsilon}_{\mathrm{c}} - [(\bar{\varepsilon}_{\mathrm{f}} - \bar{\varepsilon}_{\mathrm{c}})^2 - \beta(\bar{\varepsilon}_{\mathrm{f}} - \bar{\varepsilon}_{\mathrm{c}})^2]^{1/2}\}}. \tag{5.19}$$

Note that there are no extinction points on the upper TE_0 power curves. Decreasing the value of D merely pushes the curves into the surface wave region. In the limit as $D \to 0$, the power curve for the single interface between two nonlinear media emerges. Also the power curves for nonlinear waves with symmetry higher than TE_1 are restricted to values of n_x less than the linear eigenvalue for the principal wave, where TE_0 is the principal wave for TE_{2n} waves and TE_1 is the principal wave for TE_{2n+1} waves with $n = 0, 1, 2, \ldots$. In all cases, however, $n_x > \sqrt{\bar{\varepsilon}_{\mathrm{s}}}$. For the thin film considered so far the permitted region of n_x for these higher-order waves is extremely narrow. Even within this narrow region, as was discussed earlier on, there can be a forbidden band of n_x that is associated with much more complicated field profile changes than has been discussed upto now. The general principles exposed here can be readily applied to the higher orders so it is not considered to be necessary to supply any further numerical evidence here.

5.1.2. Dispersion relations

For planar systems in which at least one of the media is nonlinear, two distinct sets of dispersion curves can be identified. In the theoretical development of section 3, for a three-layer system in which the refractive indices and the film thickness are given, there are four fundamental parameters: the normalised power written as P/P_0, the boundary field value expressed as $a_{\mathrm{s}}E_0^2$, the dimensionless 'frequency' of film 'thickness' written as $D = (\omega d)/c$ and wave index n_x. Two-dimensional sections through the $(P/P_0, D, n_x)$ curves, parallel to the D–n_x-plane will produce curves labelled with power but similar sections through the $(a_{\mathrm{s}}E_0^2, D, n_x)$ spaces will produce curves labelled with the optical nonlinearity at the film/substrate boundary. For each type of section a continuous family of curves is obtained as opposed to the unique, and often, single branched dispersion curve of the linear system. Both types of nonlinear dispersion curve have appeared in the literature, sometimes with the implicit assumption that the particular label used (i.e., P/P_0 or $a_{\mathrm{s}}E_0^2$ or their equivalents) gives a unique frequency/wave index relation. The dispersion curves are not unique, however, for nonlinear waves as be seen from a comparison of the power-labelled and the $a_{\mathrm{s}}E_0^2$ intensity-labelled dispersion curves. If the typical numerical values that lead to fig. 5.1 are used then the results shown in fig. 5.14 and fig. 5.15 for the lower E_d curves and in fig. 5.16 and fig. 5.17 for the upper E_d curves are obtained. The linear light lines $n_x = \sqrt{\bar{\varepsilon}_{\mathrm{s}}}$ and $n_x = \sqrt{\bar{\varepsilon}_{\mathrm{f}}}$ are also shown in the figures. For a completely linear system curves close to those labelled quasi-linear (a notation meant to indicate that it is numerically linear for all

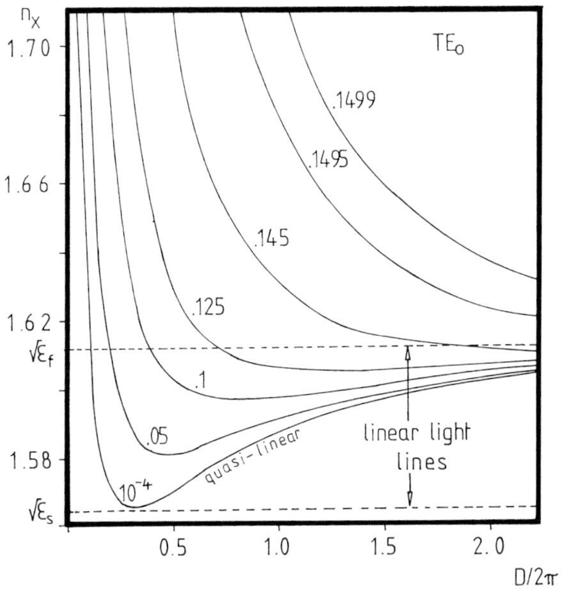

Fig. 5.14. Dispersion curves, for the lower branch of the (E_d, E_0) curves, labelled with $a_s E_0^2$. Data as in fig. 5.1. (Boardman and Egan 1985.)

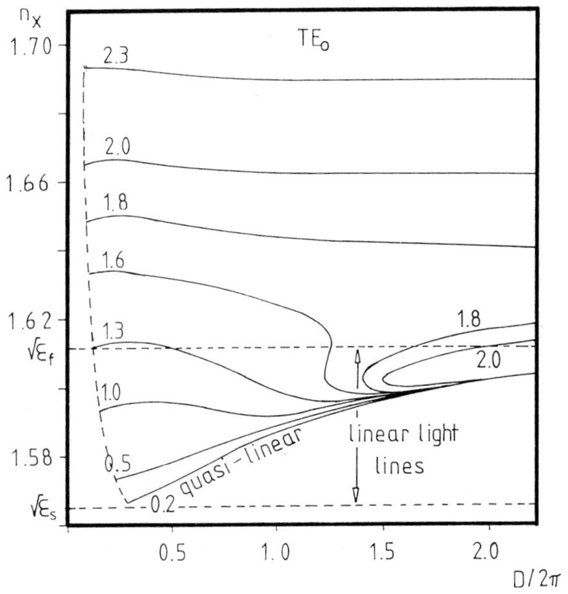

Fig. 5.15. Dispersion curves, for the lower E_d branch, labelled with power flow down the guide. Data as in fig. 5.1. (Boardman and Egan 1985.)

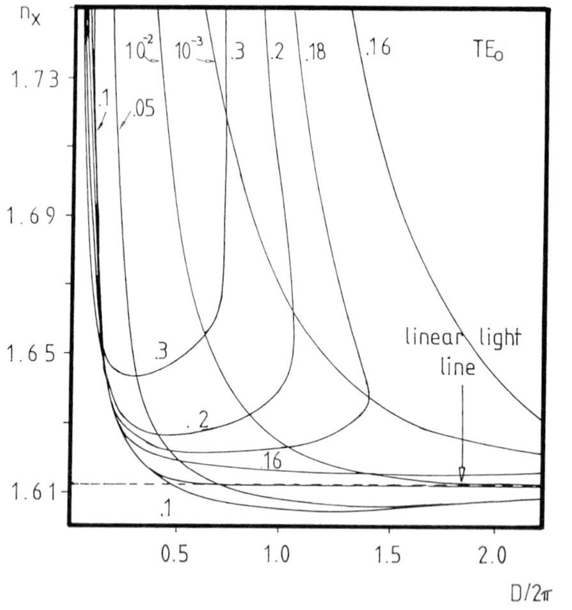

Fig. 5.16. Dispersion curves, for the upper branch of the E_d curve, labelled with $a_s E_0^2$. Data as in fig. 5.1. (Boardman and Egan 1985.)

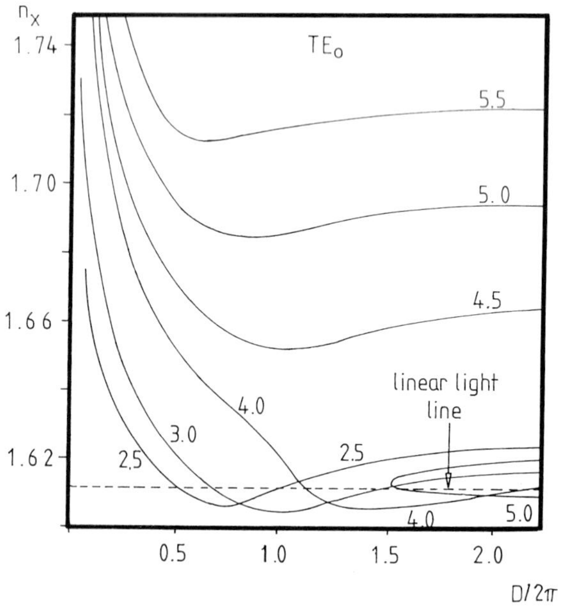

Fig. 5.17. Dispersion curves, for the upper E_d branch, labelled with the power flow down the guide. Data as in fig. 5.1. (Boardman and Egan 1985.)

practical purposes) in fig. 5.14 and fig. 5.15 are obtained. In this limit the power-labelled and boundary field-labelled curves become identical and lie completely between the two light lines and to the *right* of the minimum in fig. 5.14. The *left* of the minimum in fig. 5.14, for $a_s E_0^2 = 0$, also exists for a nonlinear medium and corresponds to solutions in which the field maximum has moved into the substrate. In fact, this has been discussed earlier and is a purely nonlinear phenomenon in which $a_s \neq 0$ and the $a_s E_0^2 = 0$ curve gives values of D below the linear cut-off value. They are, of course, the extinction thicknesses that were derived above. In the limit of very weak nonlinearity the dispersion curve is almost vertical in the region beyond the minimum showing that the group velocity of the waves is approaching zero. Hence the power necessary to excite waves in the quasi-linear, weakly nonlinear limit approaches infinity and the nonlinear branch becomes inaccessible.

The choice of the power flow as a label for dispersion curves leads to an entirely new family. This is shown in fig. 5.15 which also features the linear light lines. For each power level, this figure shows the minimum frequency/film thickness that is required for a wave to be excited. Figure 5.15 also shows that, as δ becomes large, with, say, the film thickness becoming large for a given frequency, the single surface behaviour can be seen. For a single surface it should be remembered that the power–n_x curve will lie to the right of $n_x = \sqrt{\bar{\varepsilon}_f}$ which is the upper light line of fig. 5.15. From the contour labels it can be seen that, for $n_x > \sqrt{\bar{\varepsilon}_f}$ the power does rise steeply near to $\sqrt{\bar{\varepsilon}_f}$. After falling to a minimum the power rises again slowly, as n_x increases. At a large enough value of D the sharp rise in power will be asymptotic to $\sqrt{\bar{\varepsilon}_f}$, from above as predicted in fig. 5.9. The upper E_d curve (upper family in fig. 5.1) also leads to two distinct sets of dispersion curves that can be labelled with optical surface power density, or nonlinearity, the power flow down the guide. These are shown in figs 5.16 and 5.17 and exhibit far more complexity than those for the lower E_d curve. One feature that can be identified, however, is that in the limit of very small nonlinearity there are regions that approach the linear limit.

5.1.3. Limiting cases

5.1.3.1. Symmetric systems. A symmetric system is defined here as one for which the linear film has a substrate identical to the cladding (Akhmediev 1982a). As shown in fig. 3.5 the (E_0^2, E_d^2) plot for this system consists of two straight lines at right angles to each other. This feature shows, immediately, that there are symmetric and antisymmetric modes with $E_0^2 = E_d^2$ and a nonlinearly induced asymmetric mode with $E_0^2 \neq E_d^2$. The TE_0 power–n_x curves for a typical symmetric system is shown in fig. 5.2. Obviously, for TE_0 waves for instance, the branches of the power curve on which $E_0^2 = E_d^2$ have field distributions that are symmetric with respect to the centre of the film while the asymmetric waves do not. Also the asymmetric waves are initiated only above a certain power threshold.

In order to understand the power–n_x curves of the symmetric system it is instructive to consider it as the limiting case of an asymmetric system. An asymmetric system would have an upper and a lower (E_0^2, E_d^2) curve that, in the limit, degenerates to the two sections with the common point A. These are sketched in fig. 5.18a. Figure 5.19b shows how the distinct loops of the power curves of fig. 5.1 degenerate as the limit of a symmetric system is approached. It also shows clearly that the origin of the power threshold is related to the self-focussing surface wave behaviour and is therefore a nonlinear effect that has no linear limit.

The changes in the TE_0 field profile geometry in relation to the straight line (E_0^2, E_d^2) diagram are sketched in fig. 5.19. The TE_0 wave profile develops, with increasing $a_s E_0^2 (a_c = a_s)$ along each line and a reference to fig. 5.19a shows that the top two segments are analogous to the upper E_d curve of an asymmetric system.

It should be emphasised here that the point A in fig. 5.18 is pivotal in the symmetric case since this is where the upper and lower line segments join. Even for infinitesimal differences in the nonlinear refractive indexes of the substrate and cladding, however, the degeneracy in the (E_0^2, E_d^2) relation will be lifted and the intersecting straight lines will cease to exist. This is a most important observation and any stability analysis, based upon expansions around this point are suspect.

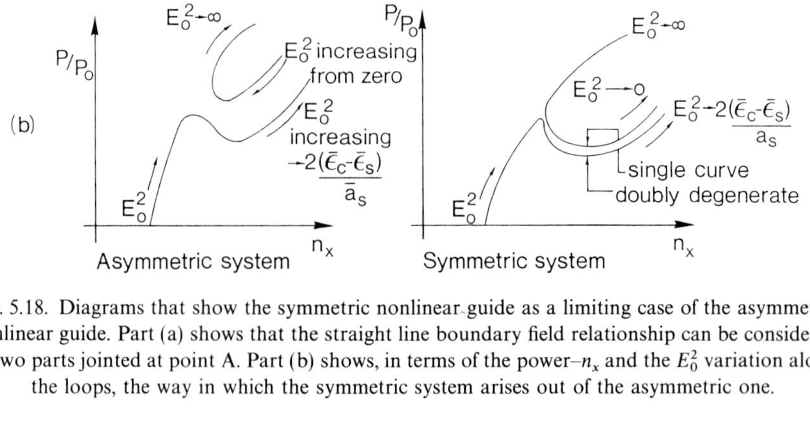

Fig. 5.18. Diagrams that show the symmetric nonlinear guide as a limiting case of the asymmetric nonlinear guide. Part (a) shows that the straight line boundary field relationship can be considered as two parts jointed at point A. Part (b) shows, in terms of the power–n_x and the E_0^2 variation along the loops, the way in which the symmetric system arises out of the asymmetric one.

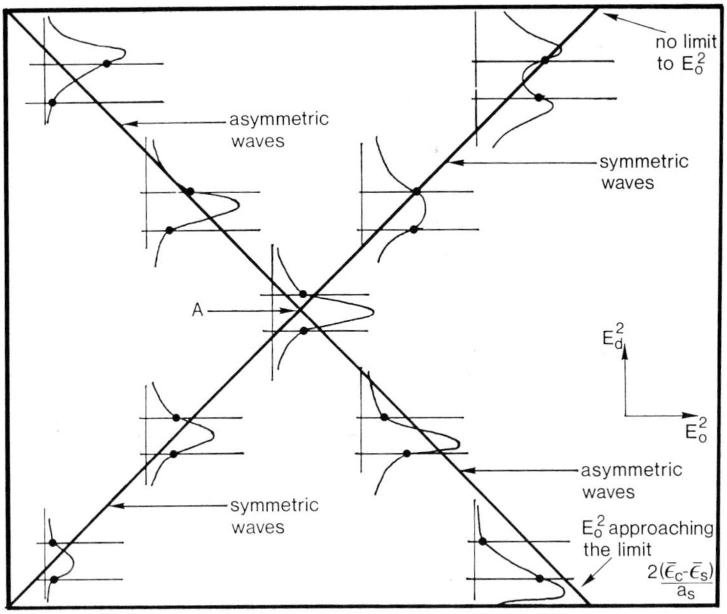

Fig. 5.19. Changes in the field profiles along the straight line (E_0^2, E_d^2) relationship for a symmetric guide.

The power–n_x curves and the field profiles for TE_1 waves in a symmetric structure are shown, respectively, in figs 5.20 and 5.21. The power–n_x curves in this case are quite complex and have many sections. Nevertheless, they have been given a good qualitative interpretation by Seaton et al. (1985c). Branch C is the most straightforward since it originates from the linear eigenvalue for which, as shown in fig. 5.21a, the field maxima are inside the film, as the power increases with n_x, these maxima gradually move out into the nonlinear bounding media. Branch C, for $n_x > \sqrt{\bar{\varepsilon}_f}$, evolves, in the way that was discussed earlier, into single interface surface waves. Figure 5.21b shows the field profiles associated with branch D which is degenerate with C at high power values. Branch D is associated with asymmetric field profiles that have no linear limit. Figure 5.20 also shows an extinction point for a film thickness below the linear cut-off value. This extinction point moves, as shown earlier, to higher powers as the thickness of the film is progressively reduced. In the light of the earlier discussion, concerning another type of extinction for TE_1 waves in linear–linear–nonlinear structures, with a film thickness above the linear cut-off value, it is obvious for this symmetric structure that there is no extinction of this kind. The other branch E consists of two solutions that are degenerate with respect to power but not with respect to field profile. This branch corresponds to a non-dispersive, nonlinear wave since it does not vary with n_x, and is the unusual solution $p_f D = \pi$.

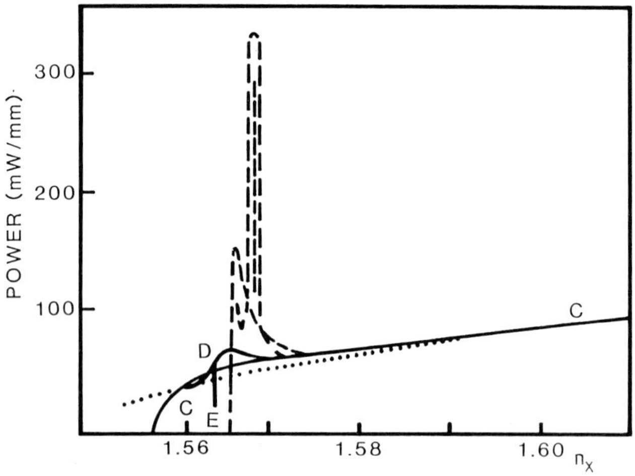

Fig. 5.20. TE_1 guided wave power–n_x curves for a symmetric guide. Dotted curve: $d = 0.8$ μm; solid curve: $d = 2$ μm; dashed curve: $d = 4$ μm. Cut-off thickness > 0.8 μm. $\bar{\varepsilon}_s = \bar{\varepsilon}_c = 2.4025$; $\bar{\varepsilon}_f = 2.2649$; $a_c = a_s = c\varepsilon_0\bar{\varepsilon}_c n_{2s}$ and $n_{2s} = 10^{-9}$ m^2 W. (Seaton et al. 1985a.)

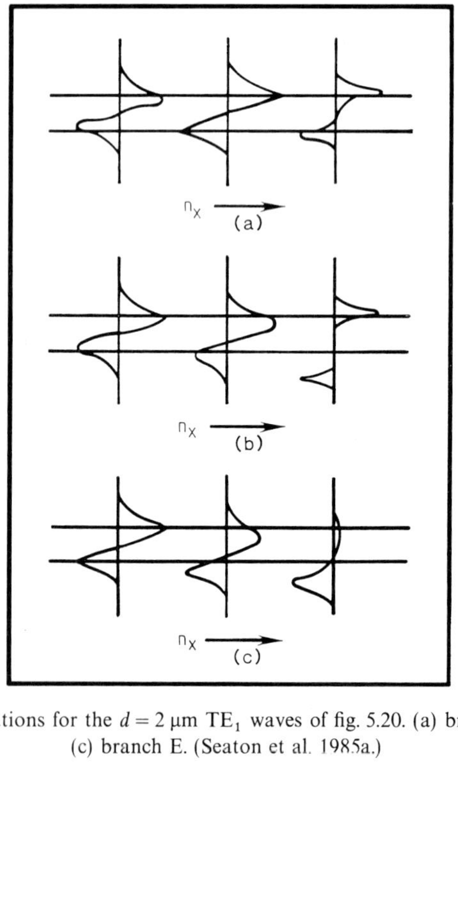

Fig. 5.21. Field distributions for the $d = 2$ μm TE_1 waves of fig. 5.20. (a) branch C; (b) branch D; (c) branch E. (Seaton et al. 1985a.)

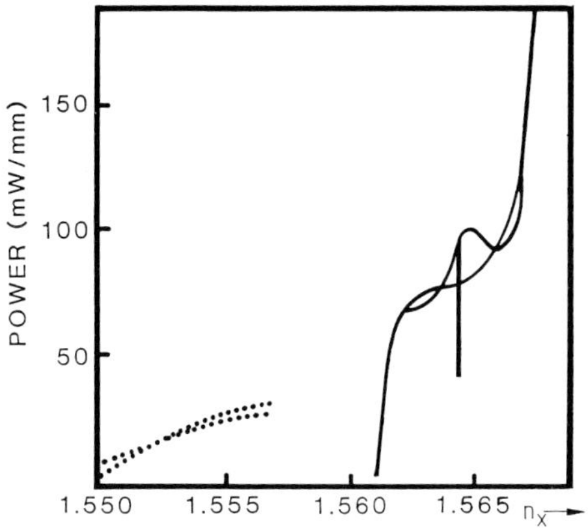

Fig. 5.22. Power–n_x curves for nonlinear TE_2 waves in a symmetric guide with $\bar{\varepsilon}_c = \bar{\varepsilon}_s = 2.4025$ and $\bar{\varepsilon}_f = 2.2649$ and $n_{2s} = n_{2c} = 10^{-9}\ m^2\ W$. Solid line: $d = 4$ μm. Dotted lines: $d = 1.96$ mm and $d = 2.06$ μm. Cut-off thickness > 2.06 μm. (Seaton et al. 1985a.)

Figure 5.22 shows the power–n_x curves for the TE_2 waves in a symmetric system. There are three branches and two thresholds. There is also a dispersionless branch ($p_f D = 2\pi$) of the type that appeared in fig. 5.20. In this case there is no surface wave branch with $n_x > \sqrt{\bar{\varepsilon}_f}$. It is fairly easy to see from a simple sketch, that this is because the TE_2 guided wave field profile cannot be matched to self-focussed nonlinear surface waves.

As with other structures and other nonlinear guided wave orders, the symmetric system supports waves for films that are thinner than the linear cut-off thickness. It is evident from fig. 5.22 that for TE_2 waves only a limited power range is permissible and this range decreases as the film gets thinner.

5.1.3.2. Single interface. The single interface has a power–n_x variation of the form shown in fig. 5.9. Its principal characteristic, which has already been discussed, is that it has a minimum. Hence, for positive dielectrics, a threshold power exists and a single interface may only support a self-focussed surface wave (Egan 1984). There is no linear limit for this type of wave.

At various points during the development of the preceeding sections it was emphasised that for thick films the power curves develop the characteristics of fig. 5.9. Thus various single interface limits may be easily identified by extrapolating to large D both for the asymmetric and the symmetric structures.

5.2. Nonlinear slab: Kerr media: TE waves

It has already been shown that the properties of guided and surface waves in the three-layer system are determined to a large extent by the geometry of the electric field wave profiles. Now a change in the sign of the curvature of the profile from positive to negative, or vice-versa, cannot occur in a linear medium, so that for a system in which the film is linear, any changes in the sign of curvature of the profile necessary to satisfy the boundary conditions, and at the same time ensure that the field decays exponentially to zero at infinite distances from the film, must occur in the nonlinear bounding layers. There is no such restriction when the film itself is nonlinear. Changes in the sign of the curvature of the field profile are possible as the waves are described by Jacobi functions (Akhmediev et al. 1982a, Boardman and Egan 1986b, Chen and Maradudin 1988). This allows for a much greater variety of field profile geometries in the film, particularly if the substrate and cladding are also nonlinear. Also the difference between surface and guided waves becomes merely a mathematical formality, since oscillatory solutions are possible for both wave types. Some simplification does occur if the substrate and cladding are both linear media since, in this case, no field maxima can occur outside the film. Numerical analysis for this case has been reported both by Akhmediev et al. (1982a), Boardman and Egan (1986b) and Chen and Maradudin (1988). In the Boardman and Egan paper, similar values of refractive indices for the linear substrate and nonlinear film were used (i.e., $\bar{\varepsilon}_s = 2.45$, $\bar{\varepsilon}_f = 2.3$, $\bar{\varepsilon}_c = 1$ and $a_{s,c} > 0$). They define a nonlinear film with an air cladding. Since the zero-power refractive index of the film is less than that of the substrate, the system has no linear limit and therefore, as expected, will not act as a waveguide until a minimum value of optical intensity at the film/substrate boundary has been reached. This value must be sufficient to ensure that the effective nonlinear refractive index in the film is greater than that in the substrate. This is neatly illustrated by constructing the (E_0^2, E_d^2) relation for the parameters listed above. Here $\bar{\varepsilon}_s > \bar{\varepsilon}_f > \bar{\varepsilon}_c$ so that the relevant (E_0^2, E_d^2) curve is a type II hyperbola with the centre in the fourth quadrant (cf., fig. 3.4). Solutions are only possible, therefore, for a $a_f E_0^2 \geqslant 2(\bar{\varepsilon}_s - \bar{\varepsilon}_f)$. The eigenvalue equation for the nonlinear film is eq. (3.54) and for TE_0 waves yields the intensity-wave index curves of fig. 3.7. For higher-order waves, the shape of these curves is still preserved, but, for any particular film thickness, the values of both n_x and $a_f E_0^2$ are very much higher, and increase with increasing order. After the intensity-wave index relationship is calculated, the power-wave index curves can be found by a direct integration over the total field. For TE_0 waves this procedure leads to fig. 5.23. This figure can be given a clear interpretation by comparing the power carried in the film to that carried in the substrate, as the wave index changes. This separation is performed in fig. 5.24, for two different values of film thickness. The appropriate field amplitude profiles in the film are given in fig. 5.25. For any particular value of film thickness, as $n_x \to \sqrt{\bar{\varepsilon}_s}$, the gradient of the profile at the film/substrate boundary approaches zero, so that the field in the substrate approaches a constant value and hence requires an infinite power level. As n_x increases from the value $\sqrt{\bar{\varepsilon}_s}$, the field in the substrate has an exponentially decaying profile and the power transported in this medium falls rapidly, as does the total power.

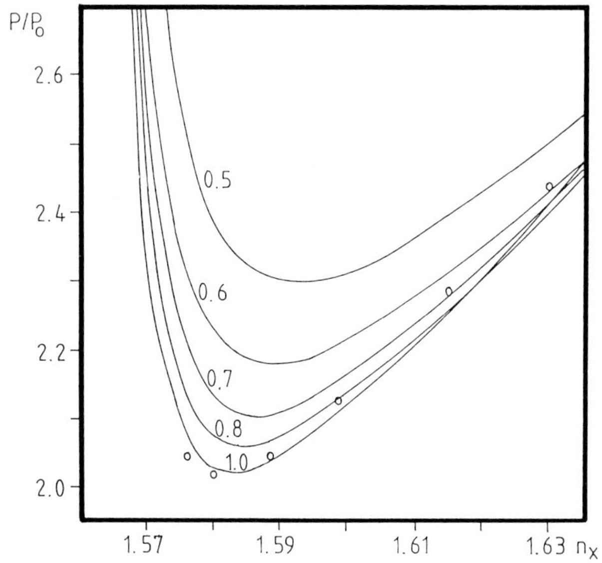

Fig. 5.23. Normalised power flow for the TE_0 wave in a nonlinear film as a function of n_x. Each curve is labelled with $D = \omega d/c$. The circles show the, relatively, unchanged position of the power curve for the much larger D value ($D = 2$). (Boardman and Egan 1986b.)

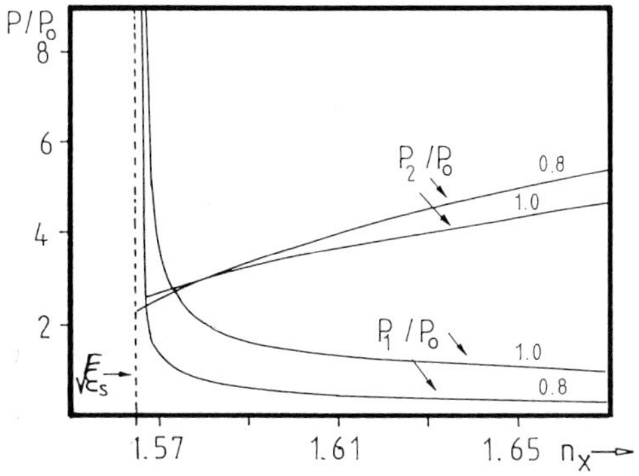

Fig. 5.24. Separation of the total power flow of a TE wave in a nonlinear film into constituent power flows in the substrate (P_1) and the power flow in the film (P_2). P_0 is the usual normalising power defined in the text. (Boardman and Egan 1986b.)

At the same time the field amplitude in the film is developing an increasingly larger maximum and so becomes responsible for transporting an increasingly larger portion of the total power. At some critical value of n_x, the rate of increase in the power transported by the film becomes larger than the rate of decrease in the power transported by the substrate and the total power transported begins to increase again.

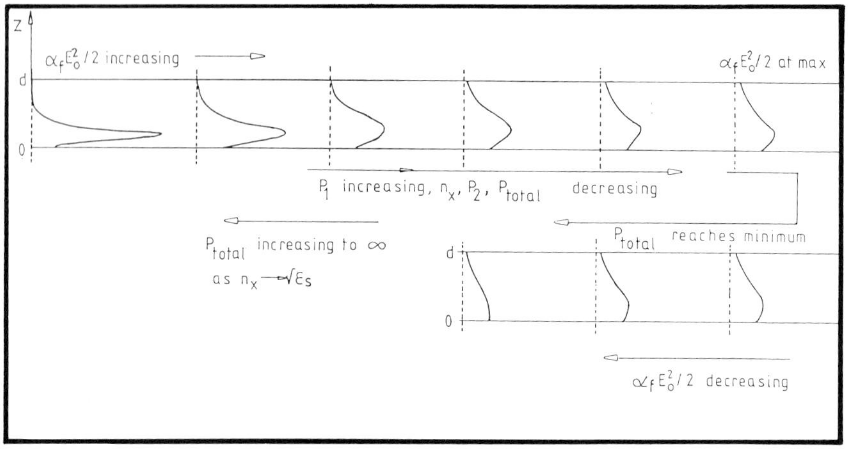

Fig. 5.25. Field profiles (Jacobian elliptic functions) of the TE_0 wave inside the nonlinear film, as the power changes with wave index. Note that as $n_x \to \sqrt{\bar{\varepsilon}_f}$ the gradient of the field amplitude at the boundary approaches zero. (Boardman and Egan 1986b.)

Figure 5.26 for TE_0 waves, also shows that, although the minimum value of P/P_0 decreases with increasing values of D, there appears to be a limiting value of D above which the change in the minimum value of power (P_{min}) is insignificant. This is also the case for higher-order waves, as is shown in fig. 5.26. Here $P^{(2)}_{min} > P^{(1)}_{min} > P^{(0)}_{min}$ by significant margins and leads to the possibility of separating the lower-order waves by control of the available power.

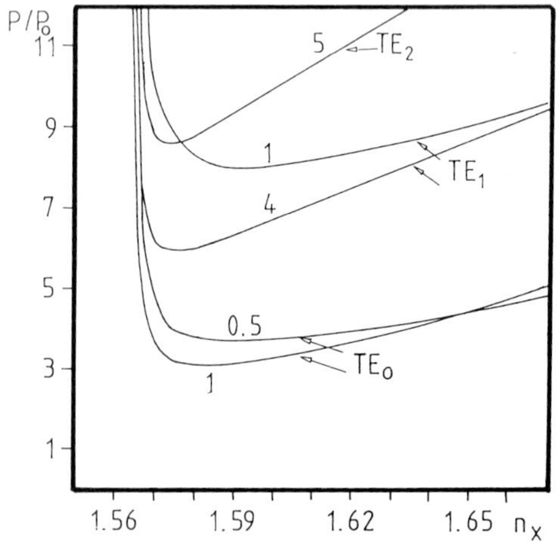

Fig. 5.26. Illustration of the minimum power levels for TE_0, TE_1 and TE_2 waves in a nonlinear thin film. Further increases in D will not significantly lower the minima.

Other detailed calculations have also been performed by Akhmediev et al. (1982) in which a nonlinear film bounded on one or both sides by air has been analysed. For the former the data adopted ($\bar{\varepsilon}_s = 2.56$, $\bar{\varepsilon}_f = 4.84$, $\bar{\varepsilon}_c = 1$) describes a system that has linear eigenvalues, for which a critical power threshold does not exist. There are, however, higher-order waves that do not have a linear limit and, as shown many times above, these can exist only after the nonlinearity reaches a threshold value. Such peculiarly nonlinear waves have power–n_x curves like those in fig. 5.26 or fig. 5.23. For a given power level above the threshold, as shown by Akhmediev, the larger n_x value is associated with power density that is largely concentrated within the film. In contrast, the lower n_x value has most of its field energy residing outside the film. The choice of an air-bounded nonlinear film, with a dielectric function 4.84×10^{-11} $(\mathrm{cm}^3/\mathrm{erg})|E|^2$, by Akhmediev et al., results in very high power levels being required to reach the strongly nonlinear regime. This is true even if the substrate refractive index is 1.6. As we have seen previously, however, by selecting data that is realistically appropriate to nonlinear integrated optics the power levels predicted by Akhmediev fall by several orders of magnitude. The theory then predicts nonlinear refractive index changes that can be sustained by a growing range of modern optical materials.

5.3. Self-defocussing, damping and non-Kerr media

In the previous sections it has been assumed that the nonlinear media are not absorptive, are self-focussing and have Kerr-like non-saturable nonlinearity. This section will now address the question of what will happen when these assumptions are relaxed. It is necessary to do this since, in principle, the foregoing work may need to be applied to real materials in regimes where it may not be possible, or even desirable, to avoid damping or saturation. The numerical work in this area of nonlinear guided waves is not as extensive as it is in some others, nor is it necessary to be so. The three topics, therefore, have been collected here under one heading.

The bulk of the work reported so far concerns self-focussing in Kerr media for which the nonlinear coefficient in the intensity-dependent refractive index is positive. If this coefficient is negative, then self-defocussing occurs. As expected, this will lead to rather different power–n_x curves and these will be discussed in section 5.3.1.

The role of damping has been neglected up to now. Its inclusion in any dispersion equation, linear or nonlinear, can involve grave mathematical difficulties. This is particularly true in trying to solve the nonlinear CW equations. The main hope is that the lossless cases will yield results that give a correct insight into the behaviour of real systems. If that is not possible then, provided that the propagation distance is not too large, it may be the case that a weak absorption theory will suffice. One of the approaches, therefore, is to include absorption within an eigenvalue equation formulation and then to determine the real and imaginary parts of the wave number n_x, for a given real frequency. One of the problems that can be anticipated, however, is that as the wave propagates the absorption reduces the power of the wave, and therefore reduces the nonlinearity. This leads to what can be called propagation effects but these can only be treated within a slowly varying envelope formalism. The latter will be analysed numerically in detail in section 6, where nonstationary states are investigated and the stability of nonlinear waves is discussed.

The final topic of this section concerns non-Kerr media. Most of the literature deals with media for which not only does the nonlinear part of the dielectric function vary as $|E|^2$ but no upper limit, or saturation level, is imposed. The consequences of relaxing these assumptions must also be assessed numerically since it is expected that if the nonlinearity saturates very quickly the system will, for practical purposes, appear to be linear.

5.3.1. Self-defocussing

Since only the effect of self-defocussing on the power–n_x behaviour needs to be highlighted, the simplest structure will serve as an illustration. This consists of a linear film bounded above by a self-defocussing nonlinear cladding and below by a linear substrate. The main conclusion is that the effect of self-defocussing is to drive the effective guide index n_x towards linear cut-off value, as the power increases (Akhmediev et al. 1982b, Seaton et al. 1984, 1985c). The physical behaviour of the guided waves in the cut-off region is, however, dramatically different from the linear wave case. In fact, in a linear system, at cut-off, the field in the bounding medium with the highest refractive index will spread over the whole medium, assuming the form of plane waves. Thus, at the cut-off point, in a completely linear system the power required to drive the system becomes infinite and such a situation is physically impossible.

In the nonlinear case under consideration here, two cases can be distinguished. If $\sqrt{\bar{\varepsilon}_s} \geqslant \sqrt{\bar{\varepsilon}_c}$ then $n_x \to \sqrt{\bar{\varepsilon}_s}$ as cut-off is approached. For these conditions, it is easy to show that although the power carried by the substrate becomes infinite, the contributions from the cladding and the film remain finite. Nevertheless the expected conventional behaviour occurs and the power becomes infinite, as shown in fig. 5.27. On the other hand, if $\sqrt{\bar{\varepsilon}_s} < \sqrt{\bar{\varepsilon}_c}$ then $n_x \to \sqrt{\bar{\varepsilon}_c}$ as cut-off is approached and all the contributions to the total power flow remain finite. This is also shown in fig. 5.27 where the field for this second type of cut-off can also be seen.

The linear film, bounded on both sides by self-defocussing nonlinear media, behaves in much the same way, with n_x being driven to either $\sqrt{\bar{\varepsilon}_s}$ or $\sqrt{\bar{\varepsilon}_c}$ according to whichever is the larger. If the cladding is self-focussing and the substrate self-defocussing, or vice-versa, then naturally one nonlinear action competes with the other. For TE_0 waves, for example, there is an initial tendency

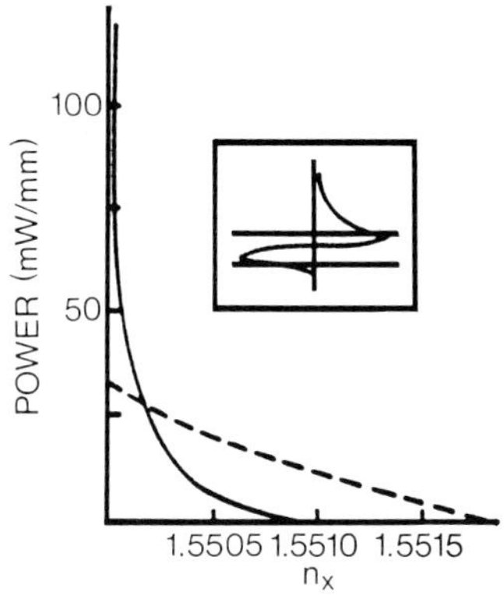

Fig. 5.27. Power–n_x variation for TE_1 waves in a nonlinear guide with a linear substrate, a linear guiding film and a nonlinear self-defocussing cladding. Dashed line: $d = 1.6$ μm. Solid line: $d = 1.2$ μm. Cut-off thickness $\leqslant 1.2$ μm. The data is the same as fig. 5.20 except that the nonlinear coefficient is now negative. (Seaton et al. 1984.)

for n_x to decrease, as the power level increases from the zero level, but this is overwhelmed by the self-focussing action as the power gets high enough. This permits the TE_0 waves to evolve to self-focussing nonlinear surface polaritons. The higher-order nonlinear waves do not evolve in this way but have a behaviour pattern like that of a linear substrate-linear-film-nonlinear cladding system. Even for this system there is still a competition between self-focussing and self-defocussing. In the self-focussing dominated region, however, the behaviour leads to the kind of extinction points that have been discussed before.

A self-defocussing nonlinear layer (Akhmediev et al. 1982a) with an air cladding and a dielectric substrate has also been considered. For this case as well, as $n_x \rightarrow \sqrt{\bar{\varepsilon}_s}$, the field outside the nonlinear film assumes the form of a plane wave and the field is ejected from the film.

In the case of a single surface if the interface is between a linear and nonlinear semi-infinite half-spaces then no surface waves may exist. This is also true for the interface between two semi-infinite self-defocussing nonlinear media. The situation changes for an interface separating a semi-infinite self-defocussing medium from a semi-infinite self-focussing medium. Here, a self-focussed surface wave exists and this narrows, in the normal way, as n_x increases. More interesting, perhaps, is the fact that as $n_x \rightarrow \sqrt{\bar{\varepsilon}_s}$ the power remains finite. Hence there is cut-off at a finite power, of the type shown in fig. 5.27.

5.3.2. Effects of absorption

In linear surface (and guided) waves the power carried by the wave is not related to the wave eigenvalue so that the loss of power from the wave does not affect the wave propagation characteristics. Thus the effect of loss on a linear wave can be described simply in terms of an exponential decrease in the wave amplitude with the distance travelled. This is no longer true of nonlinear waves because the power dispersion uniquely fixes the field amplitudes of the nonlinear stationary states. This means that the inclusion of loss in the formalism for CW nonlinear surface/guided waves strictly speaking requires a propagation study of the waves. It is still possible to include loss within the current framework, however, if it is assumed that the stationary field distributions obtained in the absorption-free case remain valid in the presence of weak loss. This is a reasonable assumption for propagation distances that are small compared to the attenuation coefficient, or in practice, for small attenuation per unit guided wavelength.

The expression for the attenuation coefficient, given by Ariyasu et al. (1985a, b) lacks detail, so a complete derivation will be given here (Twardowski 1988). It can be derived in the following way. The divergence of the Poynting vector, $\boldsymbol{E} \times \boldsymbol{H}$, can be expanded using a vector identity to give

$$\begin{aligned} \nabla \cdot (\boldsymbol{E} \times \boldsymbol{H}) &= \boldsymbol{H} \cdot \operatorname{curl} \boldsymbol{E} - \boldsymbol{E} \cdot \operatorname{curl} \boldsymbol{H} \\ &= -\boldsymbol{H} \cdot \frac{\partial \boldsymbol{B}}{\partial t} - \boldsymbol{E} \cdot \frac{\partial \boldsymbol{D}}{\partial t}. \end{aligned} \tag{5.20}$$

This is simply a differential form of Poynting's theorem, where the left-hand side represents the rate at which energy flows out from a point in a medium and the right-hand side represents the rate of change of energy density at that point. Hence it is an expression of the conservation of energy. Substituting for $\boldsymbol{B}$ and $\boldsymbol{D}$ using the constitutive relations for an isotropic material, $\boldsymbol{B} = \mu_0 \boldsymbol{H}$ and $\boldsymbol{D} = \varepsilon\varepsilon_0 \boldsymbol{E}$ together with a time variation of the form $e^{-i\omega t}$ gives

$$\nabla \cdot (\boldsymbol{E} \times \boldsymbol{H}) = i\omega(\mu_0 |H|^2 + \varepsilon\varepsilon_0 |E|^2). \tag{5.21}$$

Note here that it is being assumed that the dielectric tensor has approximately equal elements ε. This simplifies the treatment and allows the loss to be included by using

$$\varepsilon = \varepsilon_R + i\varepsilon_I, \tag{5.22}$$

where ε_R is now the real part of the type of nonlinear dielectric functions used up to now, and ε_I is an imaginary part giving rise to the absorption. Equating the real parts of eq. (5.21) gives a useful result, provided the calculation is performed to an approximation that the fields used in the evaluations of the *complex* wave number are the zero absorption values. This result is

$$-\nabla \cdot (\boldsymbol{E} \times \boldsymbol{H}) = \omega\varepsilon_0\varepsilon_I |E|^2, \tag{5.23}$$

where ε_I will come from the linear part of the refractive index. If this equation is integrated over a small volume V bounded by a surface $\boldsymbol{S}$, then

$$-\int \nabla \cdot (\boldsymbol{E} \times \boldsymbol{H})\, \mathrm{d}V = \omega\varepsilon_0 \int \varepsilon_{\mathrm{I}} |E|^2\, \mathrm{d}V, \tag{5.24}$$

where $\mathrm{d}V = \mathrm{d}x\, \mathrm{d}y\, \mathrm{d}z$.

Hence, the power loss per unit length along the x- and y-axes can be written as

$$\frac{\partial P_x}{\partial x} = -\tfrac{1}{2}\omega\varepsilon_0 \int \varepsilon_{\mathrm{I}} |E|^2\, \mathrm{d}z, \tag{5.25}$$

where P_x is defined as the total time-averaged power carried by the wave per unit length along the y-axis. This power loss can also be expressed in terms of an attenuation coefficient, if a complex wavenumber k is defined as

$$k_x = k_{x\mathrm{R}} + \mathrm{i}k_{x\mathrm{I}}, \tag{5.26}$$

where the presence of k_{I} produces the attenuation. The average power flow per unit length along the y-axis can then be written as

$$P_x = -\tfrac{1}{2}\mathrm{e}^{-2k_{x\mathrm{I}}x} \int_{-\infty}^{\infty} E_z H_y\, \mathrm{d}z. \tag{5.27}$$

Equations (5.20) and (5.18) therefore give

$$k_{x\mathrm{I}} = \frac{\omega\varepsilon_0}{4P_x} \int_{\infty}^{\infty} \varepsilon_{\mathrm{I}} |E|^2\, \mathrm{d}z, \tag{5.28}$$

and, for a system with j layers, each with a different ε_{I},

$$k_{x\mathrm{I}} = \frac{\omega\varepsilon_0}{4P_x} \sum_j \varepsilon_{\mathrm{I}j} \int |E|^2\, \mathrm{d}z_j, \tag{5.29}$$

where $\mathrm{d}z_j$ signifies an integration over the contributing layer j, and P_x is the total lossless power flow. For TE waves, where E_y is the only electric field component, this simplifies to

$$k_{x\mathrm{I}} = \frac{\omega}{2\beta c P_x} \sum_j \varepsilon_{\mathrm{I}j} P_j, \tag{5.30}$$

where P_j is the time-averaged power flow in layer j.

The imaginary part of the normalised wavenumber is

$$n_{x\mathrm{I}} = \frac{k_{x\mathrm{I}}c}{\omega}, \tag{5.31}$$

which permits the number of guided wavelengths, in which the field amplitudes fall to half their initial value, to be written as

$$n_\lambda = \frac{n_{x\mathrm{R}} \ln 2}{2\pi n_{x\mathrm{I}}}, \tag{5.32}$$

where n_{xR} is the real part of the complex normalised wavenumber, and is equal to its undamped value.

It can be seen in fig. 5.28 that, for a linear substrate supporting a linear film with a nonlinear cladding, the form of the n_{xR} variation with power is the same as it would be expected to be in the absence of damping (Ariyasu et al. 1985a) but a substantial variation of n_{xI} with power exists, so much so that in TE_0 wave propagation, for example, the guided wave attenuation should be expected to vary by about a factor of 30 over the complete power range. For TE_1 nonlinear waves n_{xI} differs by a factor of 3 between the positive and negatively sloped branches. As the guided wave moves away from the linear cut-off region the dependence of n_{xI} upon power becomes stronger.

Self-defocussing guides and structures with both nonlinear cladding and substrate have also been considered. Similar conclusions emerge namely that attenuation can be strongly power-dependent. Hence great care must be exercised in any attempt to transfer lossless theory into the domain of device design.

In an experimental situation propagation effects, in the sense of following the evolution of, initially, stationary states in an absorptive medium cannot be completely ignored as was done above. For instance, the evolved absorptive profile may have focussing properties. This could occur, for example, in a structure containing a linear film that is only weakly absorbing compared to the nonlinear cladding. If a self-focussed peak occurs in the cladding then it may be attracted towards the less absorbing film, as propagation proceeds. Since the previous work given here precludes such a feature, it is a useful prelude to the

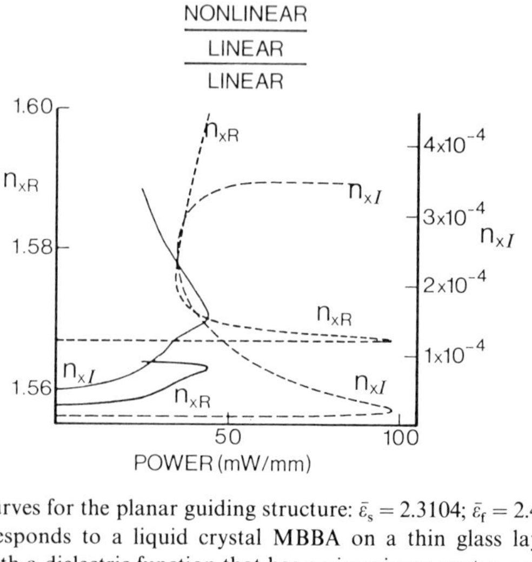

Fig. 5.28. Power–n_x curves for the planar guiding structure: $\bar{\varepsilon}_s = 2.3104$; $\bar{\varepsilon}_f = 2.4649$; $\bar{\varepsilon}_s = 2.4025$ and $n_{2c} = 10^{-6}$. This corresponds to a liquid crystal MBBA on a thin glass layer. The cladding is therefore absorbing with a dielectric function that has an imaginary part $\varepsilon_{Ic} = 0.001$. (Ariyasu et al. 1985a, b.)

next section to make some slightly more detailed assessment of the influence of propagation.

As we have seen above the nonlinear wave travels down the x-axis then the x-dependent flux is proportional to

$$P_x = \int_{-\infty}^{\infty} |E(x, z)|^2 \, dz. \tag{5.33}$$

Also the equation for the slowly varying envelope $E(x, z)$ is

$$2i\frac{\omega}{c} n_x \frac{\partial E}{\partial x} + \frac{\partial^2 E}{\partial z^2} - \frac{\omega^2}{c^2}(n_x^2 - \varepsilon_R)E + i\frac{\omega^2}{c^2}\Gamma(z)E = 0, \tag{5.34}$$

where the introduction of $\Gamma(z)$ allows an effective, or local, absorption to be defined as

$$\Gamma_e(x) = \frac{1}{P_x}\int_{-\infty}^{\infty} \Gamma(z)|E(x, z)|^2 \, dz, \tag{5.35}$$

Examples of P_x and $\Gamma_e(x)$ are shown in fig. 5.29. Branch I is associated with a field that is concentrated mainly in the low absorption linear film. In this case the absorption, essentially, introduces the normal exponential decay and the

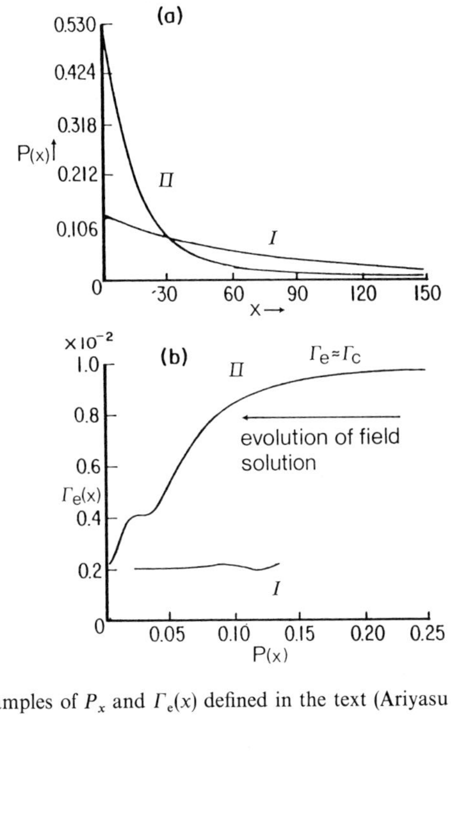

Fig. 5.29. Examples of P_x and $\Gamma_e(x)$ defined in the text (Ariyasu et al. 1985a, b).

propagation can be discussed in terms of the CW stationary state theory. In this case, therefore, there are no 'peculiar' propagation distortions of the field profiles. Branch II, on the other hand, corresponds to nonlinear fields that are concentrated, initially, in the cladding where, in this case, the absorption is high. On branch I the field retains its shape so that $\Gamma_e(x)$ does not depend very much on P_x but on branch II the field evolves towards the film and eventually penetrates it so that the overall absorption decreases. Indeed, after a single oscillation it settles nicely into the film.

From these brief results it can be concluded that for weak absorption nonlinear fields remain stable and have properties given by the lossless theory. More precisely, for a guide length of 300 wavelengths and with $\Gamma_c = 10^{-5}$ or 10^{-4}, basic lossless nonlinear guided wave phenomena do survive, in spite of absorption. For $\Gamma_c = 10^{-3}$, however, this is not the case. Although it is obvious that device length is limited by absorption, for nonlinear waves it appears that absorption can also introduce instability and behaviour that is outside the stationary state regime. The choice of optimum guide length must also take this feature into account.

5.3.3. Non-Kerr media

In a real material the optical nonlinearity cannot grow without limit and many materials will only permit nonlinear changes in refractive index that saturate quite rapidly with power. Saturation often limits refractive index changes to an upper limit of 0.1 but apparently 0.001 is fairly common with a lower limit being typically 0.0001. Clearly the saturated refractive index change varies significantly from material to material and even the typical 0.1 upper limit may be breached as new intrinsic or artificial dielectric materials become available. It should be emphasised, however, that saturation at modest values of nonlinearity will allow strongly nonlinear changes in the wave fields, if the saturation level is the order of the change in the linear refractive indices across the interface in the guiding system.

As is often the case, other complications will also occur in real nonlinear materials and one of these is the fact that some dielectrics, even without saturation, exhibit a different type of non-Kerr behaviour (Langbein et al. 1985a, Mihalache and Mazilu 1987c, Stegeman et al. 1986a, c). This means that their refractive index does not have the familiar Kerr quadratic dependence upon electric field component. In a material such as a semiconductor, for example, it is not uncommon to find a refractive index that varies as E^p where $p < 2$. Some interesting recent work has assessed numerically both saturated and non-Kerr nonlinearity and a brief account will now be given of the principal results.

Basically the role of saturation is quite easy to understand. Since it limits the achievable nonlinear change in the refractive index saturation tends to keep the system close to the linear regime. Suppose that the nonlinear dielectric function of a saturable medium is $\bar{\varepsilon} + \varepsilon_{sat}[1 - \exp(-aE^2/\varepsilon_{sat})]$, where a is the nonlinear coefficient, then $\varepsilon_{\text{sat}} = \infty$ corresponds to a Kerr medium with a dielectric function $\bar{\varepsilon} + aE^2$ that has been considered extensively throughout this article. On the other hand $\varepsilon_{\text{sat}} = 0$ implies that the dielectric constant is simply $\bar{\varepsilon}$ and the medium is therefore completely linear. A knowledge of these simple facts is almost enough to predict what will happen to the power–n_x curves of the pure Kerr cases if saturation is present.

For a self-focussing cladding on a linear film attached to a linear substrate fig. 5.30 shows the influence of saturation on both the TE_0 and TE_1 members of a family of power–n_x curves. The saturation is expressed through the parameter $n_{\text{sat}} = \sqrt{\bar{\varepsilon}_c + \varepsilon_{\text{sat}}} - \sqrt{\bar{\varepsilon}_c}$ and, as expected, for small values of n_{sat} the system behaves more or less linearly. For a perfectly linear system the eigenvalues are, of course, locked onto single values for all power levels. As the permitted nonlinearity is allowed to rise and n_{sat} increases the curves begin to exhibit their more familiar features including a local maximum. The extinction points on the TE_1 curves occur at higher and higher values as n_{sat} decreases and the familiar surface mode limit of the TE_0 waves is only achievable at $n_{\text{sat}} = \infty$. In fact, for the TE_0 waves, $n_x \to n_c + n_{\text{sat}}$ on the branch with increasing power in fig. 5.30. Also self-focussed peaks are established at greater distances from the boundary than they would be in the non-saturated case.

Solutions have also been given for a linear film bounded by two dissimilar self-focussing media. In this case separate loops that may overlap in the unsaturated case can separate as the nonlinear media become saturated. This considerably changes the expectations for the switching behaviour in any device that will saturate.

Some power-law results for $p \neq 2$, and nonlinearity proportional to E^p are shown in fig. 5.31 for the linear–linear–nonlinear layer system. It is clear that

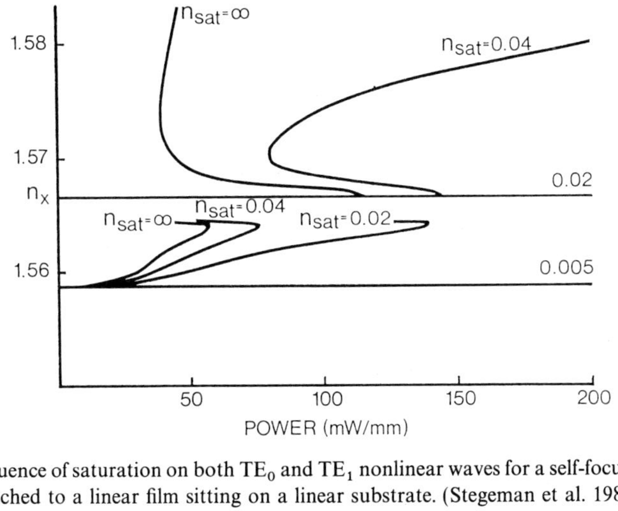

Fig. 5.30. Influence of saturation on both TE_0 and TE_1 nonlinear waves for a self-focussing cladding attached to a linear film sitting on a linear substrate. (Stegeman et al. 1986c.)

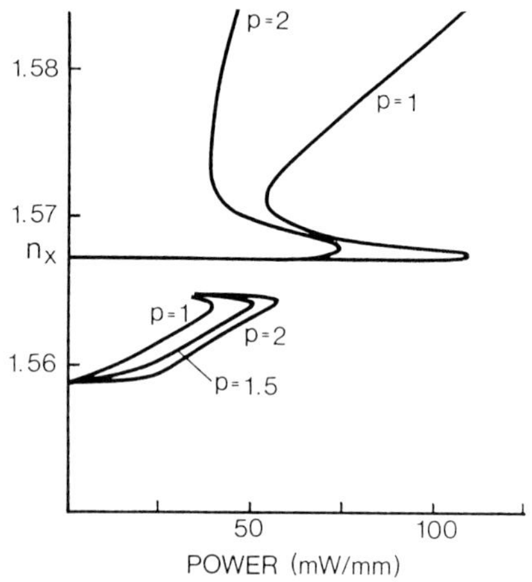

Fig. 5.31. Influence of deviation away from Kerr to non-Kerr power law (E^p) nonlinearity on nonlinear waves in the linear-linear-nonlinear layer system. (Stegeman et al. 1986c.)

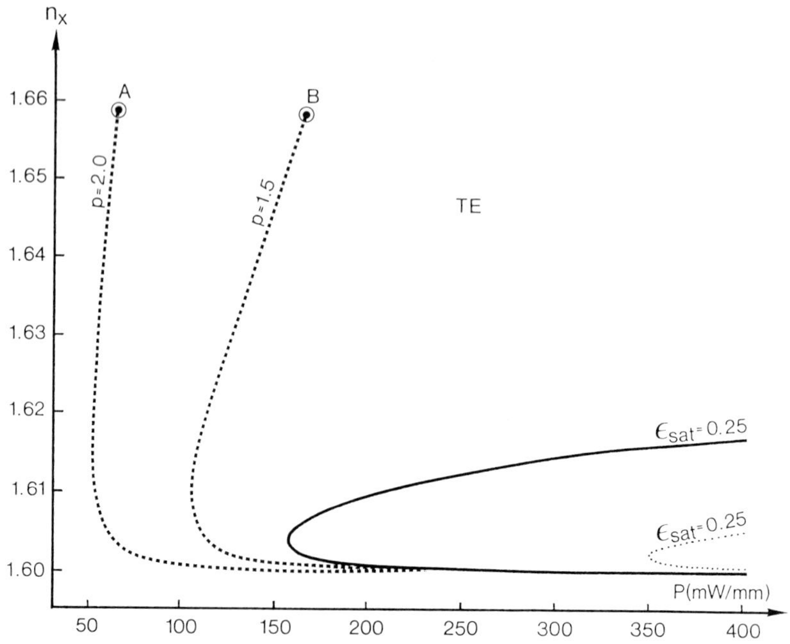

Fig. 5.32. Power–n_x curves for an E^p non-Kerr nonlinear medium cladding a linear metal film sitting on a linear substrate. (Lederer and Mihalache 1986.)

many features of the Kerr, $p = 2$, case are retained. For example, the extinction point of the TE_1 branch remains at roughly the same power level for $1 < p < 2$ but, for $p = 1$, the power beyond the TE_0 local maximum rises sharply in a way that does not allow the usual surface wave limit to be established. Here also a deeper penetration of the nonlinear medium is achieved for $p < 2$. Self-defocussing media, such as multiple-quantum-well (MQW) materials, have also been studied for both saturation and $p \neq 2$ effects. Similar features to the Kerr case are found when $n_{sat} \to 0$ drives the system into a linear state.

It is an intriguing fact that TE nonlinear waves can also propagate on a metal film (negative dielectric) bounded by semi-infinite nonlinear media. Such waves obviously have no linear limit and must be characterised by a power threshold. They can, of course, exist on a film bounded by identical nonlinear media and for a metal film that is only bounded by a nonlinear medium on one side. In the latter case, as we have seen earlier on, the cladding and substrate are interchangeable. Suppose then that only the cladding is nonlinear then fig. 5.32 shows that the power–n_x curves, for an E^p non-Kerr medium exhibits not only a distinct minimum but also an upper n_x limit or extinction point. This minimum occurs at higher power values both as the n_s increases and for $p < 2$.

In fact the effect of either saturation or an E^p law is what can now be expected. The minimum moves to higher power values for $p < 2$ and drops to lower ones for $p > 2$. As $n_{sat} \to 0$, in the saturation case, the minimum again rises with the limit $n_x \to \sqrt{\varepsilon_c} + \varepsilon_{sat}$ being approached as the power increases.

5.4. *TM waves in Kerr and non-Kerr media*

The general theory of nonlinear TM wave propagation in isotropic media was discussed in section 3, where it was indicated, briefly, that the relative status of certain approximations (Akhmediev 1982a, Akhmediev et al. 1982a, b, Boardman and Egan 1986a, b, Chen and Maradudin 1988, Litvak and Mironov 1968, Seaton et al. 1984, 1985c) that exist in the literature has only recently been investigated (Akhmediev et al. 1982b, Ariyasu et al. 1985a). The difficulty with nonlinear TM waves is that they possess two field components, as opposed to TE waves that have only one. It is therefore quite difficult to incorporate the nonlinearity due to TM waves propagating in an isotropic medium into Maxwell's equations. In this discussion it will be assumed that the TM waves propagate along the x-axis, carrying electric field components E_x and E_z that are $\frac{1}{2}\pi$ out of phase with each other. More limited models exist and have attracted a lot of recent attention, but they either neglect E_x or E_z (Akhmediev 1982a, Seaton et al. 1984), or give them equal weight in the nonlinear polarisation, *even if this is inappropriate* (Langbein et al. 1985a, Mihalache and Mazilu 1987c, Stegeman et al. 1986a, c). The type of waveguide of interest here will include both positive and negative dielectric functions. The nonlinear waves will be discussed almost entirely in numerical terms with carefully structured examples being used to expose the nonlinear TM wave properties.

5.4.1. Single interface

The generation of numerical results, based upon the first integral given in section 3, seems, at first sight, to be a satisfactory way of obtaining from first principles exact solutions to Maxwell's equations for p-polarised nonlinear waves that do not presuppose a dominant field component or that only thermal nonlinearity is possible. However, structures for which such a semi-analytical approach is not worthwhile are quickly encountered, so an alternative direct numerical approach that does not invoke the first integral has actually been used to generate the principal results here. It is a spline collocation method that uses the transformation $Z = 1 - \exp(-z/L)$ to convert the range $0 < z < \infty$ to $0 < Z < 1$ where L is a scaling factor (Akhmediev et al. 1982b, Ariyasu et al. 1985a). The scaling is an important step in the numerical work since it permits numerical determination over a finite range and exploits the linearity of the solutions as $z \to \infty$.

For a single interface between a nonlinear substrate and a linear cladding the shape of the TM nonlinear field distributions can be obtained from a 'phase plot' of E_x *versus* H_y using values obtained numerically from the first integral (Akhmediev et al. 1982a, Mihalache and Mazilu 1987c, Twardowski 1988). If the linear solution of Maxwell's equations is also plotted on the same axes, then the point of intersection of the two phase curves gives the field amplitudes of the surface or guided wave at the interface. By proceeding along one of these 'phase paths' a qualitative idea of the shape of the field profiles can be obtained. This is demonstrated in fig. 5.33 for two cases of a single interface between semi-infinite linear and nonlinear media.

Although the nonlinear medium is a dielectric with a positive nonlinearity (i.e., $\bar{\varepsilon}_s > 0$ and $a_s > 0$), the linear medium can be either a positive or negative dielectric cladding, i.e., $\bar{\varepsilon}_c < 0$ or > 0. The origin of the E_x/H_y plot occurs where both these fields are zero. It therefore corresponds to either $z = -\infty$ or $z = +\infty$, according to the path direction in the phase plane. Since the gradient of the magnetic field is

$$\frac{\partial H_y}{\partial z} = -\omega\varepsilon_0\varepsilon_{xx}E_x, \tag{5.36}$$

where ε_{xx} an element of the sub-tensor (3.119), the following conclusions can be drawn:

(1) nonlinear-metal interface:

$$\text{substrate } (z \to -\infty)\text{:} \frac{\partial H_y}{\partial z} > 0\text{:} E_x < 0\text{:} \frac{\partial H_y}{\partial z} = -\omega\varepsilon_0\varepsilon_{xx}E_x, \tag{5.37}$$

$$\text{cladding: } \frac{\partial H_y}{\partial z} < 0\text{:} E_x < 0\text{:} \frac{\partial H_y}{\partial z} = \omega\varepsilon_0|\bar{\varepsilon}_c|E_x, \tag{5.38}$$

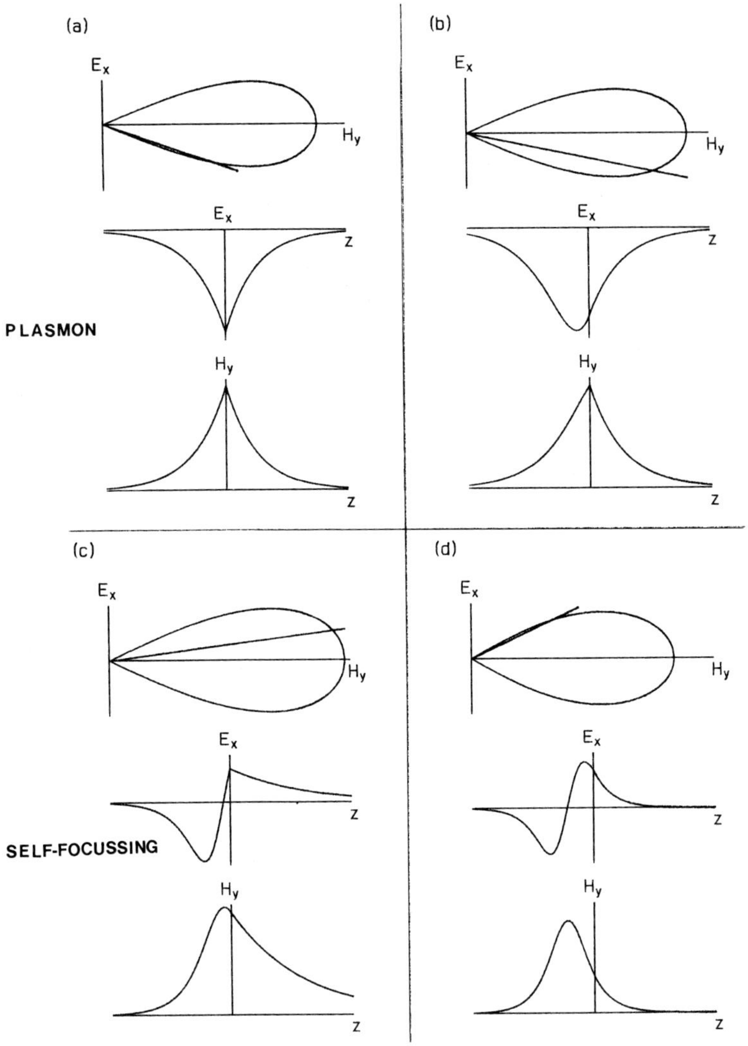

Fig. 5.33. Phase trajectories for nonlinear TM surface waves. (Twardowski 1988.)

(2) nonlinear-positive dielectric interface:

$$\text{substrate } (z \to -\infty)\text{: } \frac{\partial H_y}{\partial z} > 0 : E_x < 0 : \frac{\partial H_y}{\partial z} = -\omega\varepsilon_0\varepsilon_{xx}E_x, \tag{5.39}$$

$$\text{cladding: } \frac{\partial H_y}{\partial z} < 0 : E_x > 0 : \frac{\partial H_y}{\partial z} = -\omega\varepsilon_0\bar{\varepsilon}_c E_x. \tag{5.40}$$

The phase plot will be traced from $z = -\infty$ to $z = +\infty$, so if ε_{xx} is positive, a positive gradient of H_y implies that E_x will begin to decrease as the phase plot evolves from the origin at $z = -\infty$. Figure 5.33 shows four examples, with the first two being for surface plasmons ($\varepsilon_c < 0$) and the second pair being for positive dielectrics. In all the plots in fig. 5.33 the phase trajectories follow the negative E_x loop until the intersection with the straight line linear solution. This intersection corresponds to the interface position at $z = 0$ and the trajectory joins the line at this point and stays on it until it reaches the origin again at $z = +\infty$. In fig. 5.33 it will be seen that the nonlinear phase trajectory has been kept fixed but different data has been used for the linear claddings. The straight line linear trajectories have been selected to intersect the nonlinear phase-path either before its minimum in the $E_x < 0$ space or after it. From all these plots the shape of the field components can be readily deduced. For example, in (a) E_x goes more negative as we move away from the origin, reaches a negative peak and then goes more positive. H_y just keeps rising, however, until the interface is reached. As the interface is crossed E_x, since it is continuous there continues to go more positive until $z = \infty$ is reached, where it returns to zero. All the figures can be read in exactly the same way and the solutions divide into two main groups:

(i) If the linear medium has $\varepsilon_c < 0$, i.e., it is a metal, the H_y component peaks at the interface and can be classed as a surface wave. This type of wave has a linear limit that is the familiar surface plasmon-polariton.

(ii) If the linear medium has $\varepsilon_c > 0$, a bulge occurs in the H_y field component simular to that described in the section on TE waves, and a nonlinear wave is guided by the interface. In common with single interface guided TE waves of section 3, this type of wave has no linear limit and is a purely nonlinear phenomenon. Additionally, as shown in the figure, each main group can be further subdivided into two groups according to the number of maxima that occur in E_x in the nonlinear medium.

The approximate analytical result, eq. (3.150), obtained earlier for positive dielectrics, gives the total power flow along a single interface separating a nonlinear from a linear medium. Figure 5.34 shows that simply taking the exact numerical maximum value of H_{ym} and some sensible strip width Δz gives a fair representation (Ariyasu et al. 1985a) of the exact numerical solution of eq. (3.144). If the analytical result given by eq. (3.150) is used then a good representation is also obtained. Figure 5.35 shows that below the minimum in the total power flow the bulk of the power is carried in the linear medium. This explains why agreement between the exact result and the approximate analytical results is so strong in this region. Nevertheless, the analytical result, since it displays the parameter dependence, is quite useful in the study of self-focussing of TM guided waves in a nonlinear medium that has a positive linear part to its dielectric function and is interfaced to a positive dielectric.

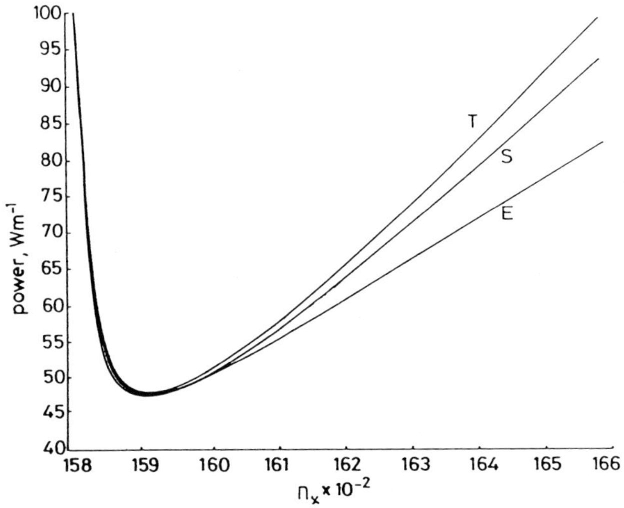

Fig. 5.34. Power flow for a single interface as function of effective guide index for nonlinear TM waves flowing in materials with positive dielectric functions. E: exact numerical solution; S: solution based upon finding the exact value of the maximum magnetic field and then using a sensible strip width Δz; T: analytical solution based upon the formula given by eq. (3.150). Data: $\bar{\varepsilon}_s = 2.4025$; $\bar{\varepsilon}_c = 2.5$; $a_s = 6.379 \times 10^{-12}$ m^2 V^{-2}; $= 3.66 \times 10^{-15}$ rad s^{-1}; $\Delta z = 0.38$ μm. (Boardman et al. 1987b.)

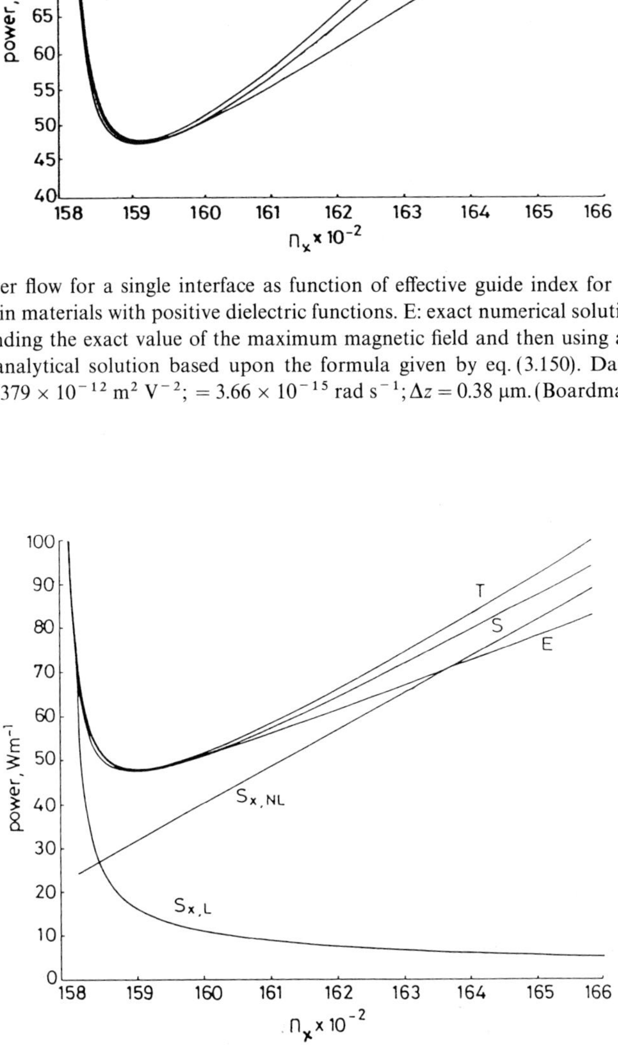

Fig. 5.35. Power flow curves of fig. 5.34 on a scale that enables a breakdown of the analytical result into nonlinear ($P_{x,\mathrm{NL}}$) and linear ($P_{x,\mathrm{L}}$) contributions. (Boardman et al. 1987b.)

For the single interface, the value of ε_{zz0} satisfies a cubic equation that together with eq. (3.136) enables $H_y(0)$ to be determined as a function of the effective index n_x. For TE waves, the equivalent relationship is that $E_y(0)$ is a constant, for constant materials parameters, but for TM waves $H_y(0)$ is a function of n_x. It is questionable whether $H_y(0)$ should be used to label (ω, n_x) curves, since the total power flow appears to be a more suitable label. Although the variation of $H_y(0)$ with n_x is small it is significant and will lead to quantitatively different self-focussing from that experienced by TE waves. The qualitative differences between the TM and TE power flow–n_x curves are not very great, however, for positive dielectrics.

In the past the major developments of the theory of p-polarised nonlinear waves have either rested upon two quite different uniaxial assumptions (Akhmediev 1982a, Seaton et al. 1984) or have assumed that the form of nonlinearity that is appropriate to electrostriction or heating is generally applicable. The latter is often acceptable (Mihalache and Mazilu 1987c, Stegeman et al. 1986a) but the first kind of uniaxial assumption (Akhmediev 1982a), which will be labelled as U_x, neglects the nonlinearity in the ε_{zz} component of the dielectric sub-tensor (3.119) and retains only the E_x contribution to the nonlinearity in the ε_{xx} component, i.e., the dielectric tensor in the U_x approximation has the form, for a nonlinear substrate,

$$\varepsilon = \begin{pmatrix} \bar{\varepsilon}_s + aE_x^2 & 0 \\ 0 & \bar{\varepsilon}_s \end{pmatrix}, \tag{5.41}$$

that allows Maxwell's equations to be written, entirely in terms of E_x, as

$$\frac{\partial^2 E_x}{\partial z^2} = q_s^2 E_x - 2\Lambda E_x^3, \tag{5.42}$$

where

$$q_s^2 = \left(k_x^2 - \frac{\omega^2}{c^2} \bar{\varepsilon}_s \right), \tag{5.43}$$

and

$$\Lambda = -\frac{q_s^2 a}{2\bar{\varepsilon}_s}. \tag{5.44}$$

Equation (5.42), for the TM waves, is now identical in form to the TE wave equation derived earlier. Physically, however, there is an important difference between the solutions for E_x and the one for E_y that is at the centre of the TE theory. This is because of the difference in sign in the definition of the nonlinear parameter given by eq. (5.44), as opposed to the corresponding term in eq. (3.7). The TE solution had a sech variation for E_y with $a_s > 0$ but TM solutions of eq. (5.42) have a sech dependence for E_x when $a_s/\varepsilon < 0$ and a cosech dependence when $a_s/\bar{\varepsilon}_s > 0$. The solutions in full are:

$$E_x(z) = \frac{1}{\sqrt{|\Lambda|}} \frac{q_s}{\sinh[q_s(z - z_0)]} \quad \text{for} \quad \Lambda < 0, \tag{5.45}$$

$$E_x(z) = \frac{1}{\sqrt{\Lambda}} \frac{q_s}{\cosh[q_s(z - z_0)]} \quad \text{for} \quad \Lambda > 0, \tag{5.46}$$

and

$$H_y(z) = \frac{\omega\varepsilon_0\bar{\varepsilon}_s}{q_s} \coth[q_s(z - z_0)] E_x(z) \quad \text{for} \quad \Lambda < 0, \tag{5.47}$$

$$H_y(z) = \frac{\omega\varepsilon_0\bar{\varepsilon}_s}{q_s} \tanh[q_s(z - z_0)] E_x(z) \quad \text{for} \quad \Lambda > 0. \tag{5.48}$$

For $\Lambda < 0$, a negative value of z_0 causes the E_x field solution to diverge and is unphysical. In fact, from eq. (5.44), it is clear that if the nonlinear coefficient is $s_s > 0$, as it will be for self-focussing, then it is necessary to have $\bar{\varepsilon}_s < 0$ to achieve this. Since the metal media are hardly likely to be nonlinear in a guiding format it is difficult to find an application for the U_x approximation, except perhaps to gaseous plasmas (Shivarova and Yu 1986).

The second uniaxial approximation, U_z, assumes that the dominant electric field component is E_z (Seaton et al. 1984) and it will be seen that this approximation applies rather well to surface plasmon-polaritons on metals that are bounded by nonlinear media for which $\bar{\varepsilon}_s > 0$. The nonlinear dielectric sub-tensor tensor in this case has the form

$$\varepsilon = \begin{pmatrix} \bar{\varepsilon}_s & 0 \\ 0 & \bar{\varepsilon}_s + aE_z^2 \end{pmatrix}. \tag{5.49}$$

In common with the dielectric tensor used for TE waves, the nonlinearity arises here from a transverse (to the direction of propagation) field component, whereas in the U_x approximation it was due to a longitudinal field component. This suggests that a close similarity between the solutions from the U_z scheme and those of the TE case will be established.

This approximation leads, very simply, to

$$\frac{\partial^2 H_y}{\partial z^2} = q_s^2 H_y - 2\Lambda H_y^3, \tag{5.50}$$

where q_s^2 has the same definition as for the U_x approximation but this time

$$\Lambda = \frac{k^4 a_s}{2\omega^2 \varepsilon_0^2 \varepsilon_{zz}^3}. \tag{5.51}$$

Equation (5.50) has exactly the same form as the nonlinear TE wave equation but a possible complication is the dependence of Λ on the nonlinear dielectric tensor element ε_{zz}, which also depends on E_z. A sweeping approximation here would be to set ε_{zz}^3 equal to $\bar{\varepsilon}_s$, which is thought to be justified on the basis that $a_s E_z^2$ is small compared to $\bar{\varepsilon}_s$ and that A, being the nonlinear term, is small anyway. The assumption $E_z \gg E_x$ will now be discussed.

It is straightforward to show for the cladding that in the linear limit (Akhmediev et al. 1982b)

$$|E_z/E_x| = \sqrt{|\bar{\varepsilon}_s/\bar{\varepsilon}_c|}, \tag{5.52}$$

where $\bar{\varepsilon}_c$ and $\bar{\varepsilon}_s$ are, respectively, the cladding and substrate dielectric constants. In the substrate this ratio is inverted. For a metal substrate, such as copper at $\lambda \simeq 5.5$ μm, that has a large negative dielectric constant ($\varepsilon_s = -1000$), and with InSb as a cladding ($\varepsilon_c = 16$), it is clear that E_z is indeed dominant. At frequencies in the visible, however, this is not the case and $|E_z/E_x|$ can approach unity. Some minor corrections to the predictions of the U_z model may be expected, then, in the infra-red and major ones are anticipated in the visible (Akhmediev et al. 1982b, Ariyasu et al. 1985a).

The other uniaxial approximation, labelled as U_x, attempts to control the nonlinearity through a dominant E_x component so that ε_{zz} has only its linear value. For solid state materials and frequencies in the infra-red this virtually eliminates the nonlinearity since it is E_z that is really important. The U_x approximation must therefore be expected to give eigenvalues that are nearly locked onto their linear values, for all reasonable infra-red power levels. In the visible the predictions of the U_x and the U_z models are expected to be almost the same since E_x and E_z have equal weight. They will both be significantly different from the exact results, however.

Figure 5.36 shows the variation, with effective guide index, of the total power flow along the interface between a metal that has a large negative dielectric constant and a nonlinear cladding with a positive dielectric constant. Several approximations and the influence of three different types of nonlinearity are contrasted. The wavelength used is in the infra-red and there is a large mismatch between the absolute values of the linear dielectric constants of the two media in contact. The linear eigenvalue, at the zero power level, is a solution of the familiar surface plasmon-polariton dispersion equation. For a practical range of n_x near to the linear eigenvalue it can be seen that there is little to choose between the various approximations with the exception of U_x. In this case, as intimated earlier on, the neglect of the strongest field component that contributes to the nonlinearity has made the system essentially linear. The power in the U_x approximation is therefore 'locked' onto the linear solution and cannot vary in any detectable way with guide index. The other uniaxial approximation, U_z, is quite close to the exact case. The variation of total power flow with effective guide index is quite significant compared with the U_x case, even though the important power range is still fairly close to the coupled-mode regime. Figure 5.37 shows the profiles (Twardowski 1988) of the field components E_x, H_y

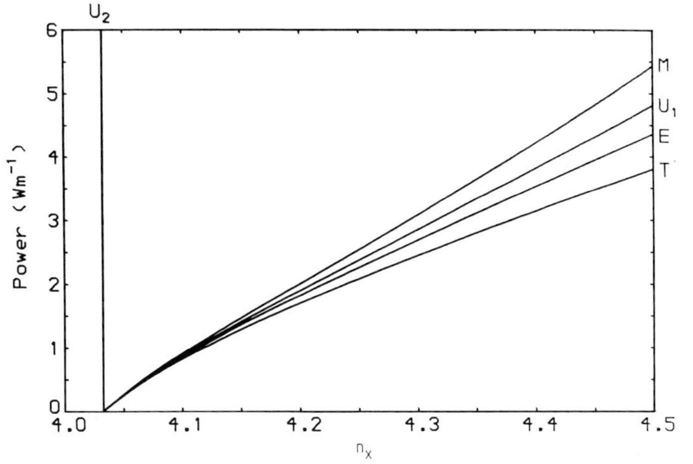

Fig. 5.36. Total power TM power flow along the interface between a semi-infinite metal and a semi-infinite nonlinear medium with a positive dielectric constant. $\omega = 3.425 \times 10^{14}$ rad s^{-1}. Dielectric constant of the metal is $\bar{\varepsilon}_s = -1000$. Dielectric constant of the nonlinear cladding is $\bar{\varepsilon}_c = 16$. $a_c = 4.24 \times 10^{-9}$ m^2 V^{-2}. $U_{x,z}$ are the uniaxial approximations. M, T and E label molecular orientation Kerr effect, electrostriction (thermal), and nonlinear electronic distortion nonlinearities, respectively. (Boardman et al. 1987a.)

and E_z as a function of the perpendicular distance from the interface for three nonlinear eigenvalues n_x. In contrast to the linear case, each value of n_x has associated with it a unique field profile. As the field amplitudes at the boundary increase, the nonlinear eigenvalue increases and the fields become more localised at the interface.

In principle, in fig. 5.36 the curves could be extended to very large n_x and eventually even the U_x approximation will exhibit a maximum, as would all the other cases. Such a regime can only be reached, as will be seen below, through the generation of enormous unphysical nonlinearities, however. The visible range of wavelengths is in sharp contrast to this because this maximum can be reached for reasonably modest nonlinear changes in the refractive index. This fact could be exploited experimentally, provided some reasonable matching of linear index change across the boundary to this nonlinear change can be achieved. In the power curve shown in fig. 5.38, the uniaxial approximations are almost indistinguishable from each other and from the curve corresponding to nonlinearity arising from molecular orientation. The curves corresponding to nonlinear electronic distortion and electrostriction have much lower maxima and are fairly close together. In the uniaxial cases the extent of the nonlinear change in the refractive index is underestimated, compared to the exact case, to almost the same degree. This explains both why they are close together and why they stay close to the linear eigenvalue for longer, as the power changes. The

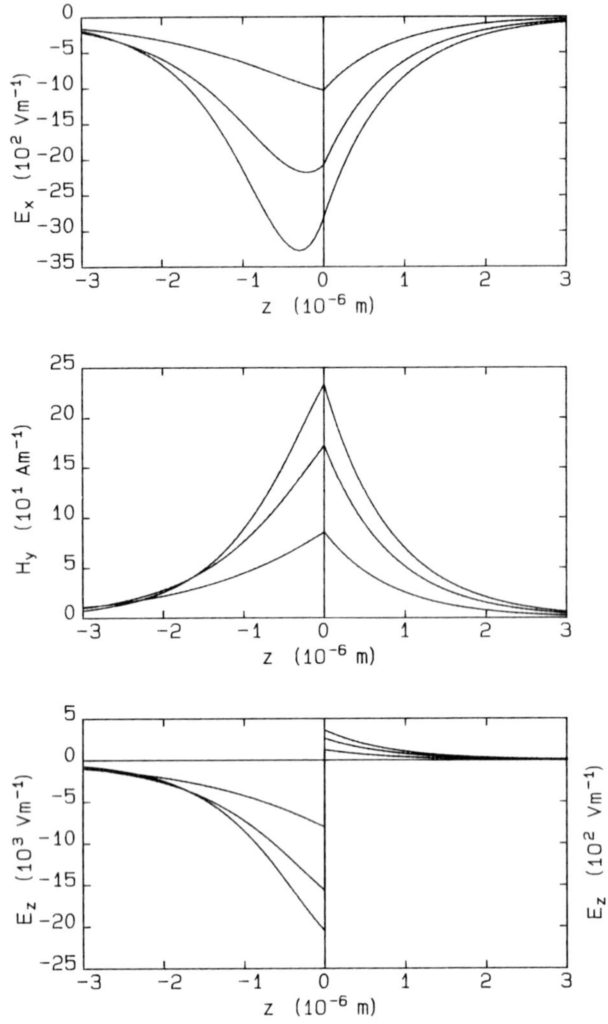

Fig. 5.37. Profiles of the field components of nonlinear TM waves at a single interface as a function of the perpendicular distance from the interface, for three values of the nonlinear eigenvalues n_x. (Twardowski 1988.)

molecular orientation Kerr-nonlinearity case, although it is exact, has a negative γ in eqs (3.120) and (3.121). It, therefore, involves a smaller effective nonlinearity and thus lies close to the U curves.

At large n_x, in fig. 5.38 shows another curious feature. This is that the power flow becomes negative. Even for linear surface polaritons, the power flow in the medium on either side of the interface is in the opposite direction. At low powers

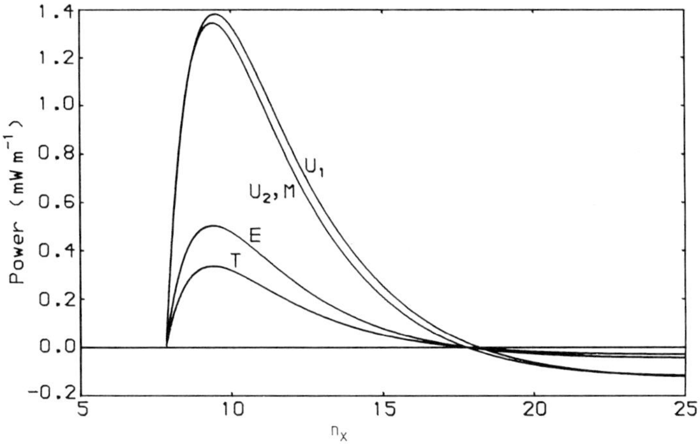

Fig. 5.38. Total power flow variation with effective guide index n_x for $\omega = 3.66 \times 10^{15}$ rad s^{-1}; $\bar{\varepsilon}_s = -2.5$; $\bar{\varepsilon}_c = 2.405$; $a_c = 6.4 \times 10^{-12}$ m^2 V^{-2}. The labels are as for fig. 5.37. (Boardman et al. 1987a.)

(the linear regime) this does not provoke any comment since the net flow is positive. In the problem studied here one of the media is now nonlinear. As the value of n_x increases a strongly nonlinear regime is entered creating strong powers that are contributed in opposition from each side of the interface. The nonlinearity then causes the net direction to be eventually reversed.

Figure 5.39 shows the Z-variation of the ratio of the electric field components in the nonlinear medium for case of electronic distortion. It can be seen that for data used for fig. 5.36 a different asymptotic value is reached for each value of n_x. Again for data used for fig. 5.36 the asymptotic limit is virtually independent of n_x. The formula for the limit is

$$|E_z/E_x| = \frac{n_x}{n_x^2 - \bar{\varepsilon}_c}, \tag{5.53}$$

where n_x is matched through to the nonlinear region of the interface. n_x is therefore not the linear value although it has the same form in the linear approximation. The curves (a), (b) and (c) in fig. 5.39 achieve the limits 7.61, 6.45 and 4.56, respectively. For curve (d) the limit is $\simeq 1$ and virtually independent of n_x.

The nonlinear contribution to the refractive index in the range $0 < Z < 1$ is presented, for the exact electronic polarisability case E, in fig. 5.40. For the data used in fig. 5.36 it can be seen that as n_x moves away from the linear eigenvalue an enormous change in the nonlinearity occurs. These changes are unrealisable in real materials and are therefore of academic interest only. The nonlinearity changes from the data of fig. 5.38 are within what could be expected for some optically nonlinear substances even within and beyond the maximum.

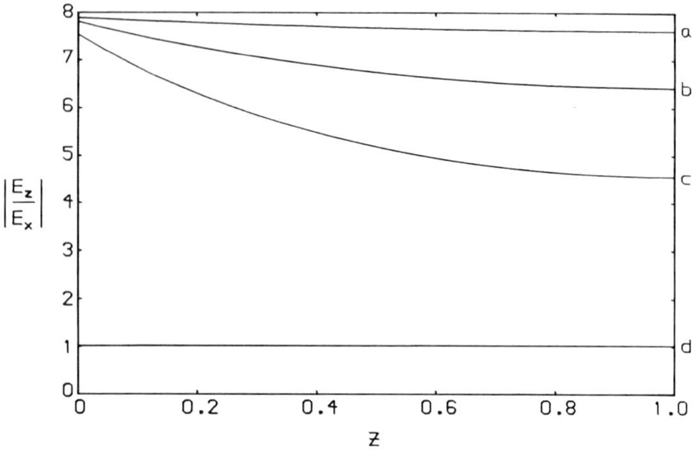

Fig. 5.39. Variation of the nonlinear cladding-field component ratio with Z for case E with fig. 5.36 data: (a) $n_x = 4.035$, (b) $n_x = 4.05$, and (c) $n_x = 4.10$. Figure 5.38 data: (d) almost independent of n_x. (Boardman et al. 1987a.)

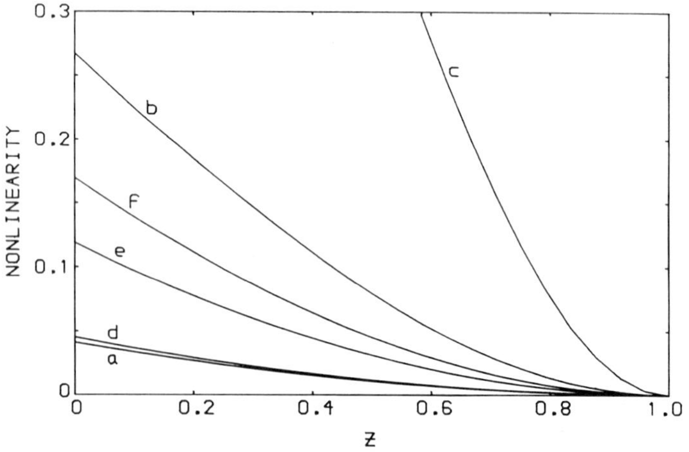

Fig. 5.40. Nonlinearity as a function of Z for case E with fig. 5.36 data: (a) $n_x = 4.035$, (b) $n_x = 4.05$, and (c) $n_x = 4.10$. Figure 5.38 data: (d) $n_x = 9$, (e) $n_x = 13$, and (f) $n_x = 25$. (Boardman et al. 1987a.)

5.4.2. Layered structures

The basic physics and the detailed form of the power–n_x curves, for layered structures, with only positive dielectrics is much the same for TM waves as it is for TE waves (Ariyasu et al. 1985a).

Waveguide structures that contain components with negative dielectric functions, such as a metal film bounded by identical nonlinear dielectrics, are more interesting since they support TM surface plasmon-polaritons in the linear limit. The power flow for long range plasmons on such a metal film is shown in figs 5.41 and 5.42, as the film thickness is varied. Figure 5.41 shows the power dispersion curves for the smaller film thicknesses (50–55 nm). These have a pronounced maximum in the power flow, beyond which the total power flow actually reverses direction before passing through a minimum. After rising again to another, smaller, maximum the power flow has another reversal and eventually becomes very small. The magnitudes of the initial maximum and minimum decline as the film thickness is increased. This is because the interfaces are becoming uncoupled and the single interface behaviour is being approached. The power flow reversal has been discussed earlier and arises because of the interplay between the power flow in the film and the bounding dielectrics. In the nonlinear system the power balance between positive and negative contributions is disturbed by the nonlinearity leading to a net reversal or negative flow. The question of whether or not such a power flow will turn out to be heavily damped remains a topic for further investigations. Figure 5.42 shows power dispersion curves for the larger film thicknesses (55–100 nm). These represent decreasing coupling between the two interfaces and display the corresponding approach to single interface behaviour.

Figures 5.43 and 5.44 show the field profiles for film thicknesses of 0.05 μm and 0.1 μm for selected values of n_x that span the power $-n_x$ curve ranges of

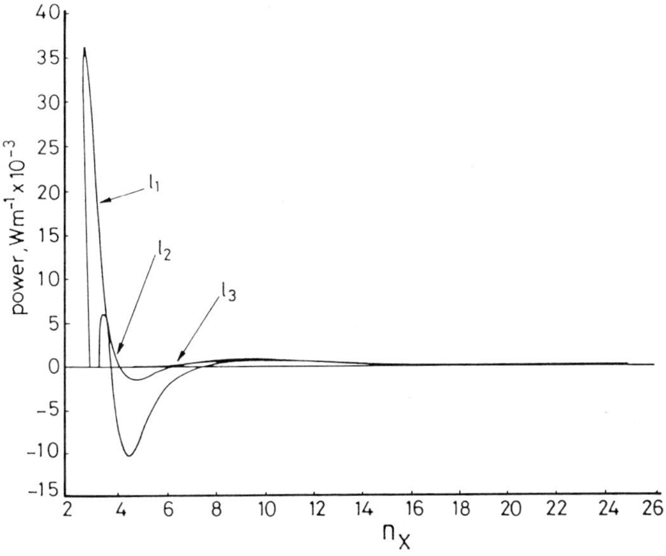

Fig. 5.41. Long-range surface plasmon power-flow $-n_x$ curves for a metal film bounded by identical nonlinear media. Film thicknesses $l_1 = 50$, $l_2 = 52.5$ and $l_3 = 55$ nm are illustrated for the data $\bar{\varepsilon}_c = \bar{\varepsilon}_s = 2.4025$; $a_c = a_s = 6.379 \times 10^{-12}$ m^2 V^{-2}; $\bar{\varepsilon}_f = -2.5$ and $\omega = 3.658 \times 10^{15}$ rad s^{-1}. (Boardman et al. 1987b.)

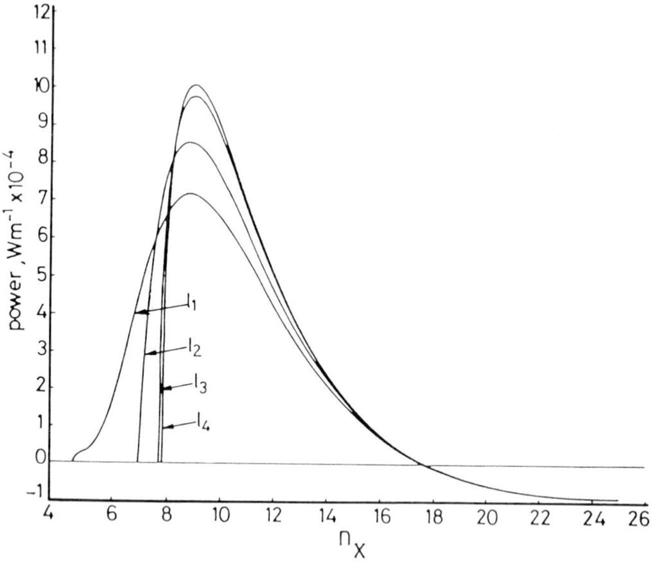

Fig. 5.42. Expanded portion of fig. 5.41 showing maxima in power curves at larger values of n_x. Film thicknesses $l_1 = 55$, $l_2 = 60$, $l_3 = 75$ and $l_4 = 100$ nm. (Boardman et al. 1987b.)

figs 5.41 and 5.42. No self-focussing occurs in this negative linear dielectric function case and it can be seen that as n_x increases $H_y(0)$ grows initially and then begins to fall off. This is responsible for the drop in the magnitude of the power flow at higher n_x. At high values of n_x, for all film thicknesses, the fields are strongly peaked at the two interfaces and correspond to two decoupled surface plasmons. For small film thicknesses at lower values of n_x the field magnitudes within the metal film are relatively large giving rise to the large initial power flows (Akhmediev et al. 1982b, Ariyasu et al. 1985a).

5.4.3. Non-Kerr behaviour: saturation

Since it is the effect of saturation that is of interest here, only single interface behaviour will be considered. A study of more complicated saturated guides will reveal nothing that is essentially different. Basically, as was shown in the earlier TE work, the presence of strong saturation drives the nonlinear guide towards linearity by simply holding down the nonlinearity. The role of saturation can be modelled by writing the dielectric tensor elements as

$$\varepsilon_{xx} = \bar{\varepsilon}_s + a_{xx}A_{xx}^2\left[1 - \exp\left(-\frac{E_x^2}{A_{xx}^2}\right)\right] + a_{xz}A_{xz}^2\left[1 - \exp\left(-\frac{E_z^2}{A_{xz}^2}\right)\right], \quad (5.54)$$

$$\varepsilon_{zz} = \bar{\varepsilon}_s + a_{zz}A_{zz}^2\left[1 - \exp\left(-\frac{E_z^2}{A_{zz}^2}\right)\right] + a_{zx}A_{zx}^2\left[1 - \exp\left(-\frac{E_x^2}{A_{zx}^2}\right)\right], \quad (5.55)$$

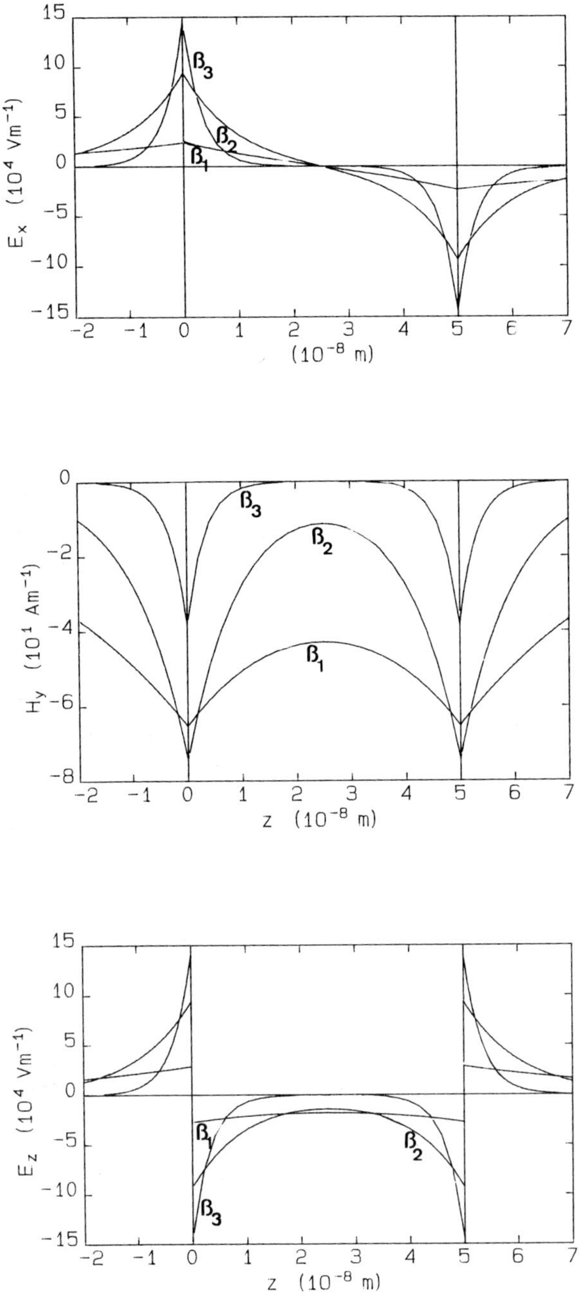

Fig. 5.43. Long-range plasmon field profiles for a symmetrically loaded metal film of thickness $l = 50$ nm. Data as for fig. 5.41. $n_{x1} = 2.8$; $n_{x2} = 8.3$; $n_{x3} = 25$. (Boardman et al. 1987b.)

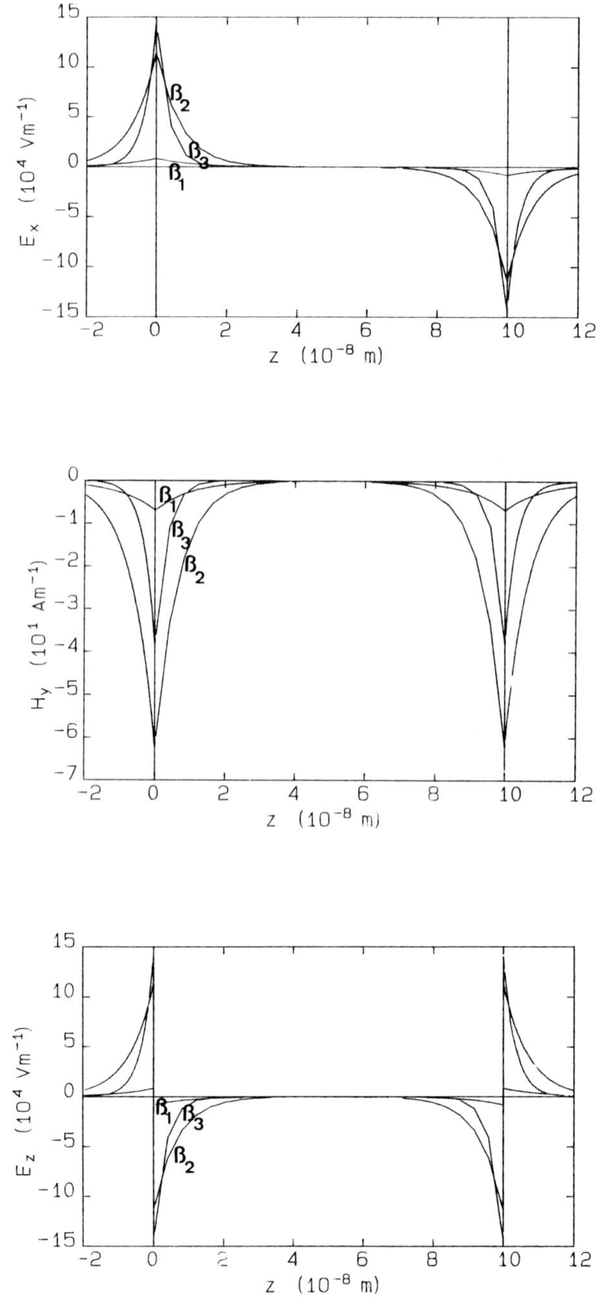

Fig. 5.44. Long-range plasmon field profiles for a symmetrically loaded metal film of thickness $l = 100$ nm. Data as for fig. 5.41. $n_{x1} = 7.85$; $n_{x2} = 12.1$; $n_{x3} = 25$. (Boardman et al. 1987b.)

where the nonlinear mechanisms are assigned as follows

electronic distortion (E): $a_{xx} = a_{zz} = a$, $a_{xz} = a_{zx} = \frac{1}{3}a$;

molecular orientation (M): $a_{xx} = a_{zz} = a$, $a_{xz} = a_{zx} = -\frac{1}{2}a$;

electrostriction or thermal (T): $a_{zz} = a$, $a_{xx} = a_{zz} = a_{xz} = a_{zx} = 0$

and A_{xx}, A_{zz}, A_{xz} and A_{zx} are constants that can be adjusted to given experimental conditions. These equations show that the maximum nonlinearities are

$$\varepsilon^{\mathrm{NL}}_{xx,\max} = a_{xx}A^2_{xx} + a_{xz}A^2_{xz}, \tag{5.56}$$

$$\varepsilon^{\mathrm{NL}}_{zz,\max} = a_{zz}A^2_{zz} + a_{zx}A^2_{zx}. \tag{5.57}$$

This, of course, is not the only way to model saturation effects but it will suffice for the present purpose (Akhmediev et al. 1982b). Figure 5.45 shows the role of saturation for the case E with fig. 5.38 data. Each curve corresponds to setting all the aA^2 terms in eqs (5.56) and (5.57) equal to the same value. It is interesting here to see that, apart from the fact that a fairly low value for the maximum achievable nonlinearity will drive the curves towards the linear values, it is possible for media with maximum nonlinearities in the region of 0.1 still to show the maximum in the power curves. This is interesting from the point of view of possible experiment with proof-of-concept materials such as artificial nonlinear dielectrics.

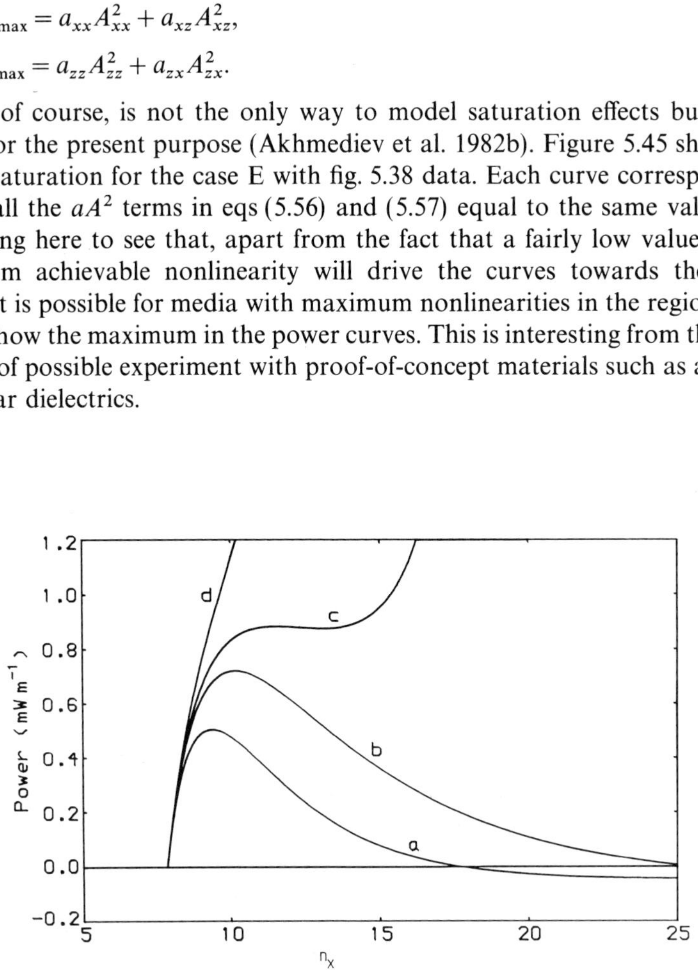

Fig. 5.45. The effect of saturation (case E) on the power–n_x curve for TM nonlinear waves flowing along a single interface. Data as for fig. 5.38. Maximum nonlinearity: (a) unlimited, (b) 0.10, (c) 0.08, and (d) 0.06. (Boardman et al. 1987a.)

5.5. TE–TM coupling

The numerical results given here will illustrate the TE–TM coupling discussed in section 3. Since the discussion will use the terminology of nonlinear channel creation, however, it does make contact with some recent work of Shen et al. (1988) that considered the propagation of a linear TM wave in a self-focussed nonlinear channel. Shen et al. showed that the threshold power needed to support weak TM waves is larger than the threshold power for strong TE wave propagation. This is an interesting conclusion since it implies the possibility of modulating the threshold for nonlinear TE waves to switch a weak TM wave from nonpropagation to propagation. The whole area of nonlinear channel creation in condensed matter is, therefore, a topic of potential importance. The examples to be given below represent the first steps towards a more general understanding of TE–TM coupling.

Figure 5.46 shows the region on the $(n_{\mathrm{TE}}, n_{\mathrm{TM}})$ plane in which nonlinear TE–TM coupling occurs for a single interface between nonlinear substrate like a liquid crystal that exhibits a thermal nonlinearity, with $\gamma = 0$, $\bar{\varepsilon}_c = 2.4025$, $a_s = 6.379 \times 10^{12}\ \mathrm{m^2\ V^{-2}}$, and a linear dielectric cladding with $\bar{\varepsilon}_c = 2.5$. The loci, discussed in section 3, converge to a point at $n_{\mathrm{TE}}^2 = n_{\mathrm{TM}}^2 = \bar{\varepsilon}_c$, below which the nonlinear waves become oscillatory in the linear dielectric and the power flow becomes infinite. This feature of the figure makes it apparent that no stationary coupled TE–TM wave solutions can exist for $n_{\mathrm{TE}} = n_{\mathrm{TM}}$. This is rather an important distinction since it shows that hybrid modes are not created by this type of nonlinear coupling. For this case the coupled TE–TM waves maintain

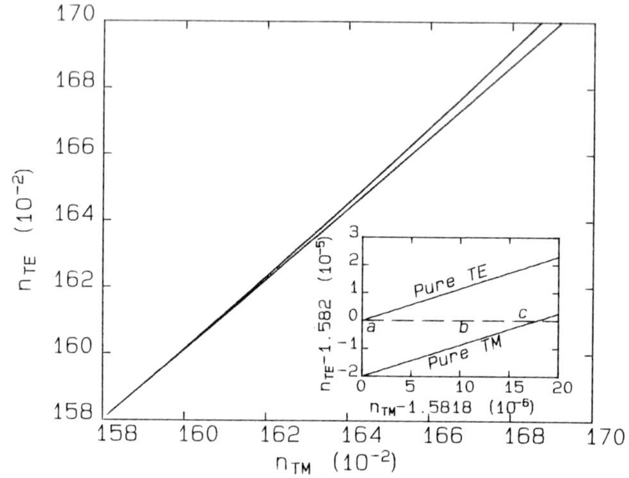

Fig. 5.46. n_{TE}–n_{TM} plane for nonlinearwaves guided by a single interface between a nonlinear liquid crystal (thermal) type medium (data given earlier) and a positive dielectric $\bar{\varepsilon}_c = 2.5$. The inset is an expansion about $n_{\mathrm{TE}} = 1.582$. (Boardman and Twardowski 1988a.)

their separate identities by having different wave numbers (Boardman and Twardowski 1988a, b). This kind of coupling is unique to nonlinear guided waves and arises through one polarised wave creating a channel for the other. The inset in fig. 5.46 shows the expanded region around $\beta_{TE} = 1.582$. The upper locus gives the linear eigenvalues of linear TM waves in the nonlinear TE wave channel and the lower locus gives the eigenvalues of zero amplitude (linear) TE waves in the nonlinear TM wave channel. The dashed line is the locus of eigenvalues for a fixed $n_{TE} = 1.582$.

Figure 5.47 shows, for a metal film ($\bar{\varepsilon}_f = -10$) with the same nonlinear substrate and the same linear cladding, the variation of the upper β cut-off with the metal film thickness for both TE plasmons and TM pseudoplasmons. The same figure also shows the linear eigenvalues of long- and short-range plasmon polaritons. As the film thickness is increased, the upper n_x cut-off for TE plasmons decreases. Eventually it meets the lower cut-off at a critical film thickness. The short-range plasmon eigenvalue will increase monotonically, as the film thickness is reduced, until it reaches values that overlap the nonlinear TE plasmon range. It is therefore expected that short-range plasmons will be able to couple to nonlinear TE plasmons. However, since the long-range plasmon eigenvalue decreases with decreasing film thickness it reaches the low n_x cut-off before the maximum thickness for nonlinear TE plasmons to exist can be reached. The means that coupling between long-range plasmons and TE plasmons *cannot occur* and, in thin metal film structures, nonlinear coupling can only occur between TE plasmons and TM pseudoplasmons, and between TE plasmons and short-range plasmons. It should be pointed out that the linear eigenvalues for the long- and short-range plasmons, shown in fig. 5.46, neglect the effect of the TE wave on the dielectric function of the nonlinear medium. The

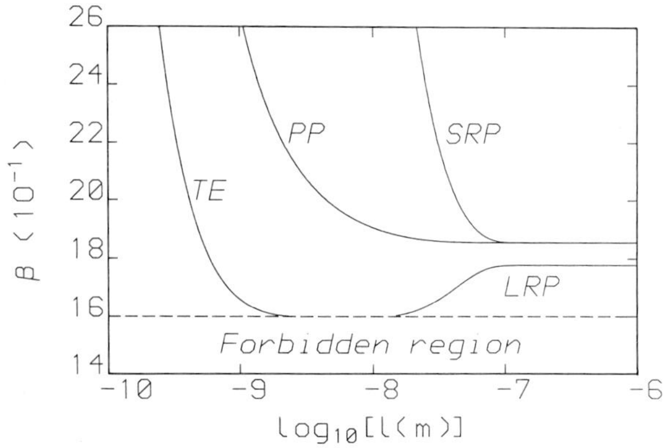

Fig. 5.47. Variation of the linear long- and short-range eigenvalues with metal film thickness together with the loci of the upper n_x cut-offs of the TM pseudoplasmons, the TE plasmons and the lower $n_x = 1.6$ cut-off. (Boardman and Twardowski 1988a.)

presence of a TE wave will shift these linear eigenvalues slightly but this fact does not alter the above conclusions about the regimes of TE–TM coupling.

Figure 5.48 shows the loci on the n_{TE}/n_{TM} plane between which there exist stationary coupled TE-plasmon/TM-pseudoplasmon states corresponding to the TE–TM coupled states at a single interface. In contrast with the single interface results, however, the loci, and hence the coupling region, are finite. It can be seen that the pure TE locus that corresponds to a linear TM wave flowing in a nonlinear TE channel has its extinction point within the scale of the figure but the one for the TM channel is off the scale. The origin of the figure is to the right of the low n_{TM} cut-off point so the loci above the low n_{TM} cut-off. This implies that, for low values of n_{TM}, the guiding channel created by the TM pseudoplasmon is not large enough to support a zero amplitude TE wave. Hence the stationary states along the line joining the two loci at the low n_{TE} cut-off will consist of mixed TE–TM states. Similarly, there is an upper boundary to the coupling region, also consisting of mixed states.

The locus of linear short-range plasmon eigenvalues in the presence of a TE plasmon is shown in fig. 5.49, where it can be seen that nonlinear stationary coupled TE-plasmon/short-range plasmon states exist in a region to the right of the pure TE locus. Any increase in the value of n_{TE} produces a decrease in the TE–TM coupling due to the movement of the self-focussed peak of the TE wave away from the interface. At the cut-off point the self-focussed peak has moved out to infinity. For this case there is no locus corresponding to nonlinear short-range plasmons carrying zero amplitude TE pseudoplasmons. Since the latter are entirely nonlinear entities the short-range plasmons cannot form a guiding channel for zero amplitude TE waves.

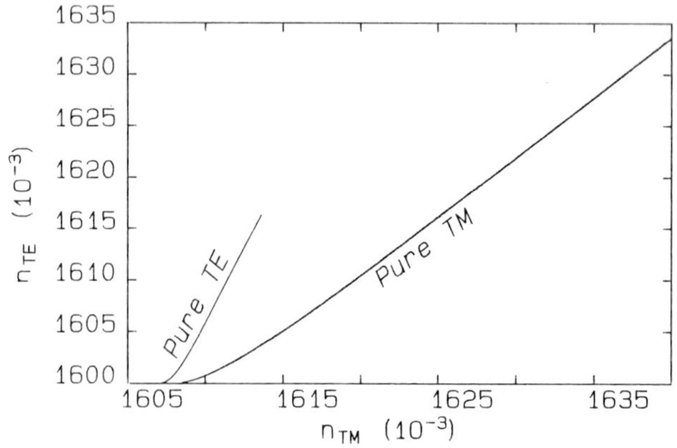

Fig. 5.48. n_{TE}–n_{TM} plane for nonlinear waves in a liquid crystal/metal/linear dielectric structure. The data is for MBBA liquid crystal, aluminium of thickness 1.5 nm and $\bar{\varepsilon}_c = 2.5$. The curves show loci corresponding to linear TE plasmons in a nonlinear TM pseudoplasmon channel and to linear pseudoplasmons in a nonlinear TE channel. (Boardman and Twardowski 1988a.)

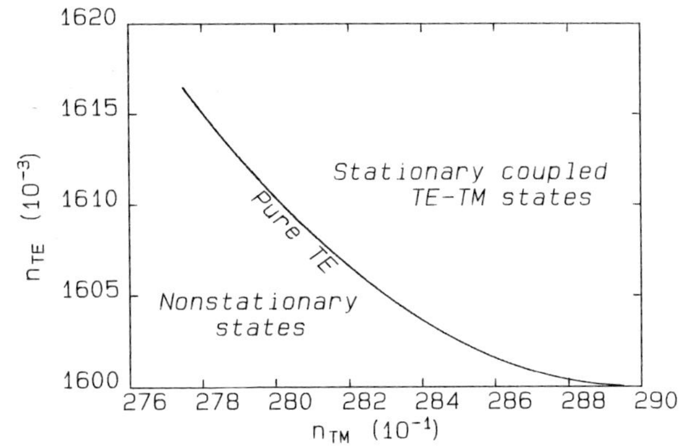

Fig. 5.49. Locus of linear short-range plasmon eigenvalues in a nonlinear TE plasmon channel for the MBBA/Al/linear dielectric structure. Film thickness 1.5 nm. (Boardman and Twardowski 1988a.)

Figure 5.50 shows the relative TE–TM content of a coupled wave as a function of n_{TM}, for a fixed n_{TE}, in terms of the fraction of the total power carried by the TM wave. The total power flow, as a function of n_{TM} is also shown. This plot corresponds to crossing from the pure TE locus to the pure TM locus along the dashed line shown in the inset of fig. 5.46.

Figure 5.51 shows how the varying contributions to the total power flow from the TE and TM waves is reflected in the changing field profiles of the two polarisations. These are given for a fixed n_{TE} value and three values of n_{TM}: one near to the pure TE locus, one approximately in the centre of the coupling

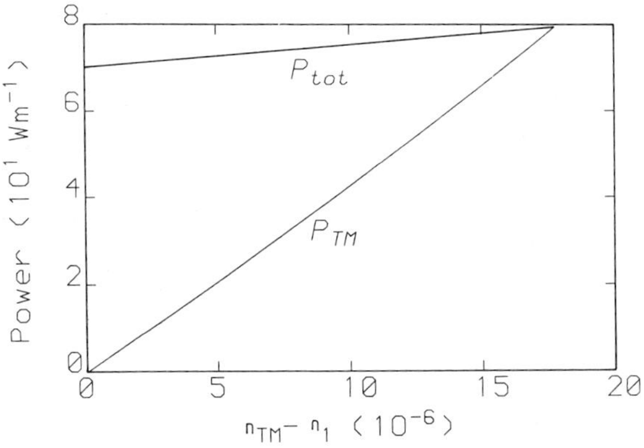

Fig. 5.50. Power flow division, for coupled TE–TM surface/guided waves for a single MBBA/linear dielectric structure. T labels the total power flow curve and R labels the power carried by the TM wave. (Boardman and Twardowski 1988a.)

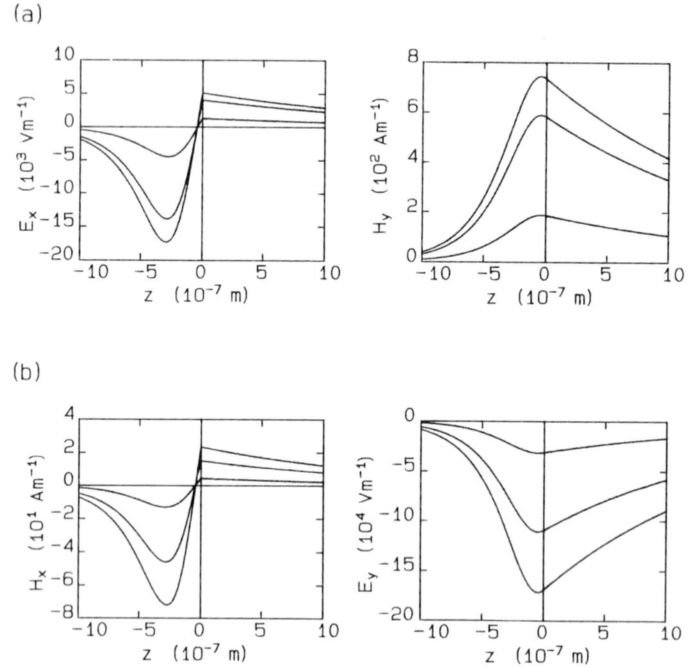

Fig. 5.51. Field profiles for (a) TM components (b) TE components of single interface coupled TE–TM waves. n_{TE} is fixed at 1.582, n_{TM} takes on the values 1.581800, 1.581811 and 1.581817 respectively and corresponds to the labels a, b and c of fig. 5.46. (Boardman and Twardowski 1988a.)

region and the third close to the pure TM locus. As anticipated, the amplitudes of the TM field components increase from zero to their maximum values as n_{TM} goes from the pure TE locus to the pure TM locus, and those of the TE field components decrease from their maximum values to zero.

Figure 5.52 shows the power carried by the short-range plasmon as a function of n_{TM}, for a fixed value of n_{TE}, and compares it to the power carried by an uncoupled short-range plasmon in exactly the same guiding structure. Also shown is the power flow carried by an uncoupled short-range plasmon if the linear part of the nonlinear dielectric function is increased to $\bar{\varepsilon}_s = 2.58$. Since the latter matches quite closely the power flow for the coupled wave, it is evident that, for short-range plasmons, the only effect that the TE wave to which it is coupled exerts is a small shift in the effective value of the linear part of the nonlinear dielectric function.

The effect of the short-range plasmon on the TE plasmon is similarly quite small as can be seen from fig. 5.53, which shows the power carried by the TE plasmon in the presence of the short-range plasmon. The power carried by the TE wave at the linear limit of the short-range plasmon is exactly equal to its

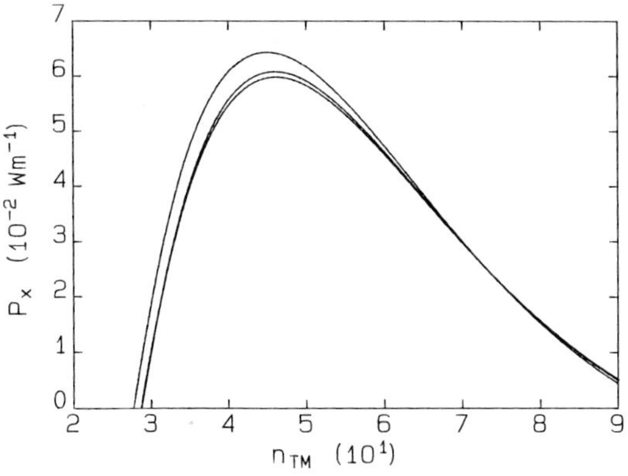

Fig. 5.52. Power–n_{TM} curves for short-range plasmons using the same data as before. Curve 1 shows the power flow of short-range plasmons coupled to TE plasmons at $n_{TE} = 1.601$. Curve 2 shows the uncoupled power flow and curve 3 shows the uncoupled power flow after $\bar{\varepsilon}_s$ is adjusted to the value 2.58. (Boardman and Twardowski 1988a.)

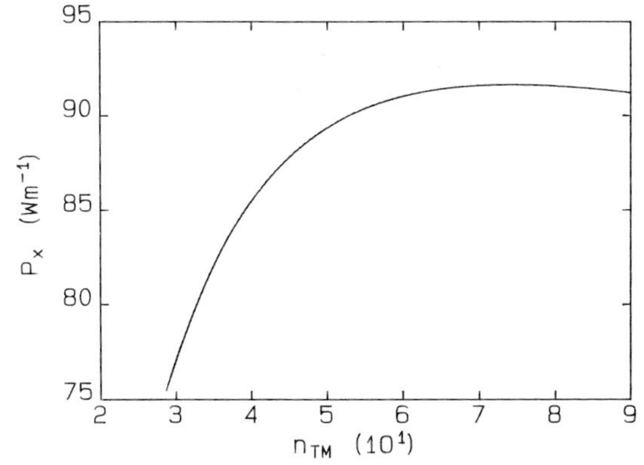

Fig. 5.53. Power flow for nonlinear TE plasmons coupled to short-range plasmons for $n_{TE} = 1.601$ in the MBBA/Al/linear dielectric structure given above. Film thickness = 1.5 nm. (Boardman and Twardowski 1988a.)

uncoupled value. As n_{TM} is increased and the TM field amplitudes grow, the TE power increases at first but then begins to fall off again. This is because, although the TM field components increase, as n_{TM} increases, they become more localised at the surfaces of the metal film. The influence on the dielectric function, due to the initial increase in the TM field components, is therefore reversed by their increased confinement.

The difference between the effects observed in TE-plasmon/TM-pseudoplasmon coupling and TE-plasmon/short-range plasmon coupling is actually a result of the different length scales involved in the two cases. For

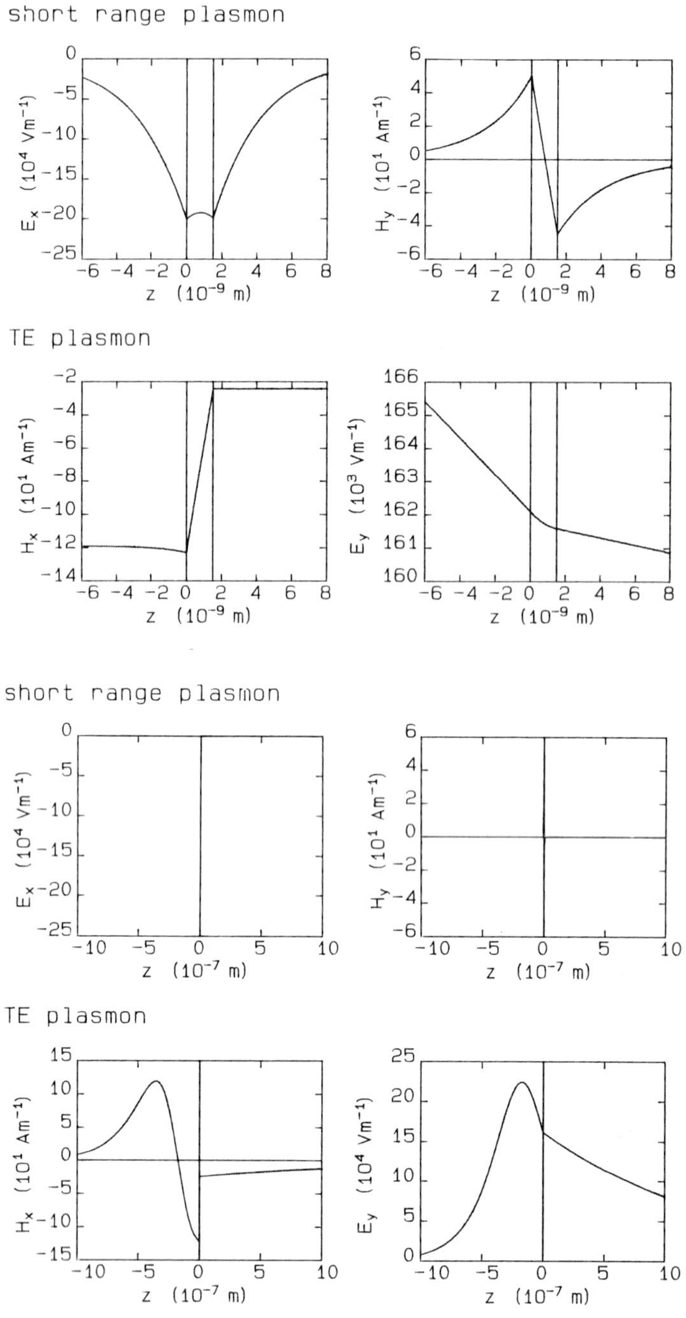

Fig. 5.54. (a) Field profiles of coupled short-range plasmons and TE plasmons at $n_{TM} = 30.02$, $n_{TE} = 1.601$, in the thin metal film (1.5 nm) structure given above, on the scale of the short-range plasmons. (b) The field profiles of (a) on the scale of the TE plasmons. (Boardman and Twardowski 1988a.)

TE-plasmon/TM-pseudoplasmon coupling, the fields of the two polarisations extend to more or less the same distance from the metal film. On the other hand, for TE-plasmon/short-range plasmon coupling the field profiles of the TE and TM waves extend over vastly differing length scales. This is demonstrated in figs 5.54a and 5.54b, where it is clear that, as far as the short-range plasmon is concerned, the TE plasmon is simply a uniform field. On the other hand the TE plasmon sees the short-range plasmon as only a 'blip' at the interface between the nonlinear medium and metal film.

6. *Numerical studies of nonstationary NGW-phenomena*

6.1. Motivation

Obviously, the NGWs* dealt with in the preceding sections represent a specific class of solutions to the three-dimensional wave equation. These solutions describe steady-state plane guided waves revealing characteristic but fixed amplitude profiles perpendicular to the interface(s) forming the planar waveguide geometry. Owing to their fixed amplitude profiles, both with respect to the direction of wave propagation (x-axis) and in the transverse direction along the plane phase fronts (y-axis), these wavefields will be labelled 'stationary' or 'equilibrium' NGWs.

In physical terms, stationary NGWs manifest themselves through a global, overall equilibrium between three propagation mechanisms, namely, diffraction, the guiding effect of the film interfaces, and the trapping effect by self-focussing. Owing to the nonlinear origin of the latter effect, this global equilibrium need not be stable. The presence of field distortions of deterministic and/or stochastic origin may break this equilibrium, at least locally, leading to 'nonstationary' NGW-field distributions, the transverse field pattern of which is no longer fixed down the waveguiding interface or film, but may reveal dramatic changes. Also, the assumed steady-state temporal behaviour may be altered by self-switching effects.

In the remainder of this section all potentially non-steady-state questions will be excluded from our considerations for they are still too far away from any serious assessment. Moreover, we will focus our attention to TE-polarised fields being subject of the following *basic problem.*

Consider a fixed cross-section taken perpendicular to the interfaces of a given configuration that can support NGWs. If one selects a certain field distribution at this cross-section, how will this field distribution evolve when proceeding away from the initial cross-section?

It is evident that any initial field distribution differing from a 'stationary' NGW may generate a 'nonstationary' field evolution down the waveguide.

*Unless otherwise stated in this section, the term 'NGW' shall cover both film-guided waves and interface-guided waves (NSWs).

According to the intentions of this chapter, the context of stationary NGWs with respect to nonstationary field distributions is of particular interest, i.e., the effect of small field distortions on an initially stationary NGW ('stability problem'), the evolution of unstable NGW-profiles over large propagation distances, the excitation of different stationary NGWs by appropriately tailored initial field profiles, and the effect of both material losses and saturation on the performance of the resulting phenomena.

The mathematical description of our basic problem calls for sufficiently general solutions to the appropriate three-dimensional nonlinear wave equation in order to match the boundary-value problem posed by fixing the initial field pattern. Unfortunately, such nonstationary NGW-solutions are not known analytically. Until now analytical attempts have been successful only in two limiting cases.

The first one is the low power limit, when the nonlinear polarisation effects the mode amplitudes only. This so-called 'weak nonlinear regime' has been the subject of section 2 of this chapter.

The second limiting case applies to initial field profiles which deviate from a stationary NGW-solution by a small perturbation only. Under this condition the nonlinear wave equation can be linearized leading to a complicated, but linear eigenvalue problem. For initial field profiles, such as TE-polarised NSWs as well as TE_0–NGWs, this eigenvalue problem has been solved analytically (Akhmediev et al. 1984, 1985, Akhmediev (ch. 3, this volume), Jones and Moloney 1986a, b) leading to a set of 'perturbational modes'. The knowledge of these modes together with the corresponding eigenvalues permits clarification of whether a given stationary NGW remains stationary down the waveguide when being affected by small distortions ('stability problem'). Second, these modes may provide insight into the onset of various types of nonstationary propagation phenomena. A detailed presentation of the perturbational analysis can be found in the chapter by Akhmediev et al. (ch. 3).

The present section gives a survey on a large number of numerical studies on nonstationary NGW-evolution. These studies rely on numerical solutions of the mixed linear/nonlinear Schrödinger equation which is essentially the nonlinear wave equation in a slowly varying envelope approximation. The development of some powerful and flexible numerical techniques permits the simulation of the evolution of a large class of prescribed initial field profiles over distances up to a few thousand wavelengths.

In this way various nonstationary propagation phenomena in the so-called 'strong nonlinear regime', i.e., for highly nonlinear materials and/or high input powers, could be predicted. These phenomena lie well beyond the validity range of the analytically treated limiting cases mentioned above. Also the question of stationary NGW excitation by appropriately tailored initial field profiles could be addressed in this way.

Additionally, the numerical simulation procedure performs automatically, stability tests of the field profiles under consideration by the permanent introduction of numerical noise. These stability tests complement the existing analytical conclusions but provide also results where the analytical apparatus fails (e.g., higher-order NGWs and NGWs supported by nonlinear films). The flexibility of the numerical technique also permits the inclusion of important effects such as material losses and saturable nonlinearities in a straightforward manner.

6.2. Theoretical background

6.2.1. The basic evolution equation

Our subsequent discussions will refer to planar geometries which can be characterized by a complex dielectric function of the type

$$\varepsilon(x, |E|^2) = \varepsilon'(x, |E|^2) + \mathrm{i}\varepsilon''(x, |E|^2). \tag{6.1}$$

The x-coordinate in ε shall indicate the relevant spatial profile and E is the relevant component of a steady-state TE-polarised electric field

$$\boldsymbol{E}(x, z, t) = \tfrac{1}{2}\boldsymbol{e}_y E(x, z)\mathrm{e}^{-\mathrm{i}\omega t} + \text{c.c.}, \tag{6.2}$$

$$E(x, z) = E(x, z)\mathrm{e}^{\mathrm{i}\beta z} = |E(x, z)|\mathrm{e}^{\mathrm{i}[\phi(x,z) + \beta z]}. \tag{6.3}$$

For the sake of an enhanced transparency of the subsequent formulae and the displayed numerical results, all spatial coordinates are scaled in units of $k_0^{-1} = c/\omega$.

The basic dielectric function according to eq. (6.1) is rather general, allowing both for non-cubic Kerr nonlinearities and for absorption mechanisms of linear and/or nonlinear origin. Note, on the other hand, that any potential intrinsic dynamics of the carriers responsible for the nonlinear response of the waveguide materials has been excluded from our discussion. This applies in particular for memory effects or nonlocal effects due to carrier diffusion.

The field ansatz according to eq. (6.2) displays the restricted complexity of the evolution phenomena to be studied in this section. In detail:

(a) All wavefields are strictly monochromatic (frequency: ω), i.e., all potential temporal effects such as temporal instabilities, self-switching etc., are not subject of this survey.

(b) All field distributions have been assumed homogeneous with respect to one coordinate, say y. This assumption assures the separate treatment of TE- and TM-polarised field distributions, but it shifts the consideration of modulational instabilities as well as any potential nonstationary effects with respect to this coordinate outside the scope of this section.

(c) Only TE-polarised fields will be considered, which can be described by a complex scalar function $E(x, z)$.

(d) Every field distribution $E(x, z)$ shall reveal a rapidly oscillating phase βz

(β = effective index) fixing the principal propagation direction of the wavefield. Consequently, $E(x, z)$ shall be a slowly varying envelope with respect to z.

The potential complexity of the phenomena covered by the ansatz according to eqs (6.2) and (6.3) reflects the present state-of-the-art in this field. On the other hand, all mentioned restrictions indicate areas of future research interest in order to gain a deeper insight into the physical relevance of the basic phenomena which have been identified in the existing studies. Some work extending beyond the scope of this section, e.g., the consideration of a potential modulational instability with respect to the y-coordinate (Moloney 1987) will be briefly mentioned.

The equation for the envelope $E(x, z)$ which is central to our consideration, is

$$2i\beta\partial E/\partial z + \partial^2 E/\partial x^2 - (\beta^2 - \varepsilon'(x, |E|^2))E + i\varepsilon''(x, |E|^2)E = 0, \tag{6.4}$$

a modified nonlinear Schrödinger equation (NLS).

In contrast to the usual NLS (see, e.g., Boullough 1977), the 'potential function' $\varepsilon'(x, |E|^2)$ depends on the x-coordinate explicitly. For that reason the powerful inverse scattering methods can not be applied. In many cases the dielectric function $\varepsilon(x, |E|^2)$ will differ in adjoining layers in a step-like fashion, and crossing conditions have to ensure the continuity of $E(x, z)$ and its derivative $\partial E/\partial x$ at each interface where the dielectric function goes through a step jump.

Equation (6.4) is a direct consequence of Maxwell's equations under conditions fixed by eqs (6.1) through (6.3). The neglecting of the second derivative with respect to z implies

$$|\partial^2 E/\partial z^2| \ll |2\beta\partial E/\partial z|, \qquad |\partial^2 E/\partial z^2| \ll |(\beta^2 - \varepsilon(x, |E|^2))E|, \tag{6.5}$$

which is the slowly varying envelope approximation (SVA). Roughly speaking, conditions (6.5) state that any characteristic field displacement along the direction of wave propagation will evolve slowly in a $c/\omega \simeq \lambda_0$ scale. This is consistent with the requirement of paraxial wave propagation and applies in many cases of interest. Nevertheless, the application of this approximation calls for both a proper choice of the boundary conditions to eq. (6.4) (initial field distributions) and the avoidance of abrupt changes of the material parameters down the waveguide. In particular, the consideration of reflected field tails that may arise, e.g., where two different waveguide sections have been stuck together via their endfaces, has been ruled out.

Equation (4) gives rise to the rate equation

$$\operatorname{div} S = -\varepsilon''(x, |E|^2)|E|^2, \tag{6.6}$$

where

$$S = |E|^2(\partial\phi/\partial x, O, \beta), \tag{6.7}$$

multiplied by $\frac{1}{2}(\varepsilon_0/\mu_0)^{1/2}$, can be interpreted as the time-averaged Poynting vector in the SVA. S differs from the exact Poynting vector in its z-component by the amount $|E|^2\partial\phi/\partial z$ which is a consequence of the SVA.

In what follows we will focus our attention to spatially localized wavefields in the sense

$$\lim_{x\to\pm\infty} E=0, \qquad \lim_{x\to\pm\infty} \partial E/\partial x=0, \tag{6.8}$$

i.e., we will disregard any radiative field tails, coming from or going to infinity. Under this condition, according to eq. (6.6), the integral power flux in the z-direction

$$P(z)\int_{-\infty}^{\infty} S_z\,\mathrm{d}x=\beta J(z)=\beta\int_{-\infty}^{\infty}|E(x,z)|^2\,\mathrm{d}x, \tag{6.9}$$

through an infinite strip of unit width in y-direction, obeys the relation

$$\mathrm{d}J/\mathrm{d}z=-1/\beta\int_{-\infty}^{\infty}\varepsilon''(x,|E|^2)\,\mathrm{d}x. \tag{6.10}$$

In the absorption-free case the 'energy flux', $J(z)$, represents an important first integral to eq. (6.4).

In the more familiar problem of optical pulse propagation in fibres, the x-variable is replaced by time and the second derivative term arises from an expansion of the dispersion relation about some carrier frequency. In contrast to these latter time domain pulses, the nonlinear waves in the present context will refer to spatially-localized $c\omega$ field distributions propagating along an optical interface or waveguide.

Two particular types of solutions to eq. (6.4) have to be pointed out, both of them representing distinct equilibria of the different propagation mechanisms covered by our NLS. One type comprises the stationary NGWs, which have been the subject of the preceding sections of this chapter. They manifest themselves as z-independent solutions of the NLS, eq. (6.4), applied to the lossless case. Indeed, for $E(x,z)=E(z)$, and $\varepsilon''(x,|E|^2)=0$ the familiar nonlinear eigenvalue equation

$$\mathrm{d}^2E(x)/\mathrm{d}x^2+\varepsilon'(x,|E|^2)E(x)=\beta^2E(x), \tag{6.11}$$

is recovered.

The second type applies to propagating nonstationary NGWs which may become well separated from any interface or graded index inhomogeneity, i.e., where $\varepsilon(x,|E|^2)\to\varepsilon(|E|^2)$ loses its explicit x-dependence. In this case the self-focussing nonlinearity may balance diffraction alone, giving rise to spatial solitons or soliton-like waves of the appropriate regular NLS with a dielectric function $\varepsilon(|E|^2)$ appropriate to the medium in which the wave is localized. Such spatial solitons represent travelling self-focussed channels feeling no longer the influence of the interface, thus seeing an homogeneous nonlinear medium. Spatial soliton formation at some distance from any interface is consistent with the fact that any arbitrary initial data will, under propagation, decompose into its soliton and radiation components.

For lossless Kerr-media, i.e., $\varepsilon(x, |E|^2) = \varepsilon'(|E|^2) = \varepsilon + a|E|^2$, the self-focussed channel can be described by the familiar one-soliton solution of the appropriate regular NLS, see, e.g., Boullough (1977)

$$E^s(x, z) = \sqrt{(2/a)}\sigma[\cosh(\sigma(x - x_0 + \tau z))]^{-1} \times \exp[i\sqrt{\varepsilon(\tau x + (\sigma^2 - \beta 2\tau 2)z/(2\varepsilon))}], \quad (6.12)$$

where β has been replaced by $\sqrt{\varepsilon}$ and both ε and a apply to the medium where the channel is formed. The real parameters σ, x_0 and τ determine respectively, the channel width, the situation of the (virtual) channel maximum at $z = 0$, and the branching angle θ between the channel axis and the z-axis: $\theta = \arctan(\tau)$.

6.2.2. *The numerical implementation*

The three optical interface geometries shown in fig. 6.1 will be of central interest in the subsequent discussion.

In the single interface case (fig. 6.1a), only the substrate layer is assumed nonlinear. By now, this case is well understood in many respects although we will briefly refer to the generalized version where both adjoining media are assumed nonlinear. In the thin film geometry of fig. 6.1b the nonlinear cladding

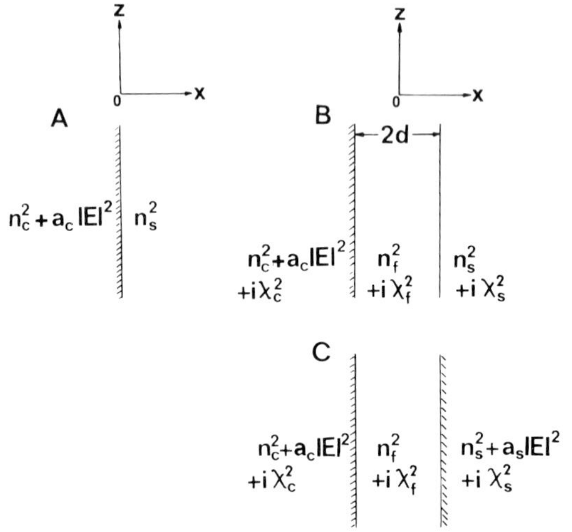

Fig. 6.1. Nonlinear single interface (a) and thin film geometries of interest (b), (c); (b) may also be regarded as a thin film loaded nonlinear interface. Cubic Kerr-nonlinearities are assumed, in partial combination with linear losses in the film geometries. The thin-film guiding layers of width '$2d$' are assumed linear. The subscripts c, f, s refer to, respectively, cladding, film and substrate region.

is separated from the substrate layer by a thin guiding film both assumed linear. In the thin film geometry of fig. 6.1c we assume that the bounding cladding and substrate layers are nonlinear, while the guiding film is linear.

Both symmetric ($n_c = n_s$, $a_c = a_s$) and asymmetric ($n_c = n_s$, $a_c \neq a_s$) configurations will be considered. All three media may reveal linear losses. The film width, $2d$, will be typically of the order of some wavelengths. All of the subsequent analysis will assume positive Kerr-coefficients a_i leading to a much wider phenomenology than for the negative self-defocussing case. The latter has been studied in the literature with specific device applications in mind (Seaton et al. 1985a). Some extended versions of thin film geometries where, for example, material saturation is taken into account or the guiding film is nonlinear, will be briefly mentioned.

Any solution $E(x, z)$ of the parabolic partial differential equation (6.4) is determined uniquely by fixing the solution of an open boundary, say, at a transverse cross-section $z = 0$, $E(x, 0)$, and at $|x| = D \gg d$ in the sense:

$$\lim_{|x| \to D} E(x, z) = 0 \quad \text{for all } z.$$

While $E(x, 0)$ plays the role of an initial field distribution, the latter condition states the desired localization of the field distribution to the vicinity of the interfaces. On the other hand, this means for any numerical procedure, that the choice of the basic transverse interval D depends on the performance of the actual field evolution.

There are different numerical techniques that can be used to simulate the evolution of initial field data $E(x, 0)$ according to the NLS, eq. (6.4). A powerful procedure that has been successfully used to study transverse effects in various fields on nonlinear optics is due to Fleck et al. (1978). This procedure, which is called 'propagating beam method' (PBM), can be well adapted to planar geometries. With respect to eq. (6.4) the basic formula of the standard PBM is given by

$$\begin{aligned} E(x, z + \Delta z) \simeq{}& \exp[\mathrm{i}\Delta z/(4\beta)\partial^2/\partial x^2] \\ & \times \exp[-\mathrm{i}\Delta z(\beta^2 - \varepsilon(x, |E(x, z + \tfrac{1}{2}\Delta z)|^2)/(2\beta)] \\ & \times \exp[\mathrm{i}\Delta z/(4\beta)\partial^2/\partial x^2] E(x, z), \end{aligned} \tag{6.13}$$

where the transverse field pattern $E(x, z + \Delta z)$ is approximately expressed in terms of $E(x, z)$ within an accuracy of $O(\Delta z)^3$ (Hermansson et al. 1978). The $\partial^2/\partial x^2$ operation is to be performed in the Fourier-space with respect to x which calls for successive application of a Fast Fourier Transform procedure. The processing speed of these Fourier transforms determines effectively the overall speed of the complete method.

In detail, the appropriate selection of the basic transverse interval $2D$, in combination with the number of transverse sampling points, as well as the determination of an optimum step width Δz, is a delicate task when the computing time is to be minimised. This task also includes the handling of field tails which reach the boundaries of the transverse interval at $|x| = D$, i.e., the formation of field tails reflected back into the interval must be suppressed.

In the lossless case, the first integral $J(z) = \text{const.}$, see eq. (6.10), and other first integrals of the NLS, can be used for accuracy tests.

Another numerical technique, applied successfully to planar geometries, takes advantage of the Crank–Nicolson Scheme (Ames 1965) in order to get a difference approximation of eq. (6.4). The resulting system of nonlinear equations has been solved by the Newton–Picard Algorithm (Ortega and Rheinboldt 1970).

Both numerical methods can be used to study the *stability* of stationary (equilibrium) initial NGW-profiles against propagation.

The onset of instabilities is caused by fluctuations in the numerics. The introduction of such random field distortions may correspond to actual experimental situations where the required fluctuations will be caused by material inhomogeneities, or by the excitation geometry (Moloney et al. 1986a).

An unsatisfactory feature of any numerical stability analysis is the possibility of weakly unstable solutions requiring impractically long propagation distances to identify. For all practical purposes, however, such a wave can be considered stable.

Our subsequent discussion is organised as follows.

First, we will give a survey on the existing stability predictions mainly with respect to the geometries of fig. 6.1. These data will be complemented by illustrating the different types of nonstationary NGW-propagation associated with the decay of unstable NGWs. First attempts towards a three-dimensional stability analysis will be briefly mentioned.

Second, the potential end-fire excitation of NGWs will be discussed, when both linear guided modes or Gaussian beams are launched through the entrance face of a film guide. The NSW-excitation by narrow light beams, totally internally reflected at a nonlinear interface, is related to the nonlinear surface reflectivity problems considered elsewhere in this volume.

Third, the effect of both material losses and material saturation on the phenomena mentioned above will be addressed.

Eventually, one particular nonstationary phenomenon, namely the formation of soliton-like self-focussed channels will be discussed with respect to the different excitation and propagation conditions that have been studied so far.

The numerically based stability analysis contrasts to the analytical approach, presented in a forthcoming chapter of this volume. There, the dominant perturbational modes for TE–NSWs and TE_0–NGWs have been determined from a linearized version of the NLS. In this way stable and unstable sections of the NGW-dispersion curves could be clearly separated. The selective addition of the various perturbational but deterministic modes to unstable NGW-profiles permits also the identification of different nonstationary propagation phenomena, at least in their initial state. It is of particular interest that a given unstable NGW-profile can decay in different ways, depending on the dominant perturbational mode that forms the actual perturbation. This question has also been highlighted by Moloney 1987), using an equivalent particle description to relate the angular deflection of a laser beam incident as self-focussed channels on a nonlinear interface, with the motion of a particle in an effective potential; see also the chapter 'Nonlinear Surface Reflectometry' of this volume.

The first analytical stability predictions on TE_0–NGWs confirmed by numerics, have been performed by Jones and Moloney using topological arguments with respect to the composite phase portraits of the NGW under discussion (Jones and Moloney 1986a, b).

6.3. Stability predictions and decay features of unstable NGWs

6.3.1. The single nonlinear interface

Undoubtedly, the stability properties of the single nonlinear interface are best understood with respect to all other planar geometries. A full account of the corresponding analytical investigations can be found elsewhere in this volume. For the sake of completeness, we will briefly summarize the relevant stability predictions, referring to Vysotina et al. (1984a, b), Akhmediev et al. (1984, 1985) and Moloney (1987).

Figure 6.2 shows a graph of β versus the energy flux characteristic [nonlinear dispersion curve (NDC)], which is typical for all NSWs supported by single nonlinear interfaces. In this section the 'energy flux', J, instead of the integral power flow, P, see eq. (6.9), will be preferably used as the control parameter. This is because the analytical stability studies have identified $J(\beta)$ as the relevant parameter. In particular, the stability criterion for all single interface geometries according to fig. 6.1a is

$$\mathrm{d}J(\beta)/\mathrm{d}\beta > 0, \tag{6.14}$$

see, e.g., Akhmediev et al. (1984, 1985).

According to the analytical stability studies of Akhmediev et al. (1984, 1985) and Moloney (1987), an unstable NSW can decay essentially in two different ways, depending on the structure of the perturbation that is added to the unstable equilibrium NWS. A small initial drift towards the nonlinear half-space leads to the formation of a stable low flux NSW off-radiating the excess energy flux (see arrow and inset to fig. 6.2). An initial drift in the opposite direction leads to a nearly complete emission of the NSW-flux via a beam-shaped field distribution (see right-hand inset to fig. 6.2). Note that the decay features displayed in the insets to fig. 6.2 have been provoked by the addition of small deterministic disturbances rather than stochastic ones in the subsequent numerical studies. All potential consequences of these instability predictions, in

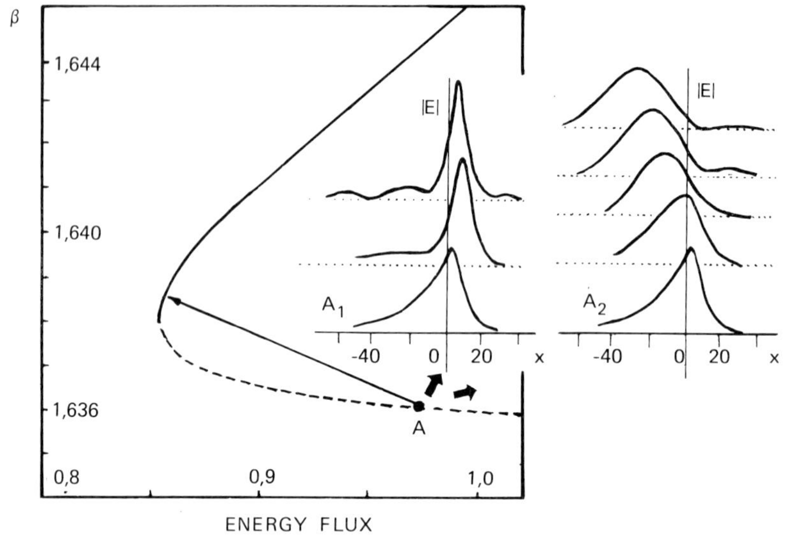

Fig. 6.2. Graph of the β vs. J characteristic (nonlinear dispersion curve) for equilibrium NSWs at a linear/nonlinear interface according to fig. 6.1a. Solid section indicates stable NSWs; dotted branches unstable NSWs. Insets illustrate the two basic decay channels of the unstable NSW excited at point 'A' of the nonlinear dispersion curve. The arrow indicates the low flux NSW-formation shown in inset A_1. (After Akhmediev et al. 1985.)

particular with respect to NSW-excitation and the switching behaviour of nonlinear interfaces can be found in the appropriate chapters of this volume.

A very recent stability analysis of equilibrium NSWs at the interface between *two* self-focussing media indicates that the condition (6.14) is not sufficient for stability (Moloney 1987). In this geometry the unstable NDC-branch is bent upwards, acquiring a positive slope but remaining unstable.

Vysotina et al. (1984a) also addressed the NSW-stability against small three-dimensional perturbations, i.e., NSW-stability under propagation with respect to the y-coordinate. They found that all NSWs will decay into separate beams for propagation lengths exceeding

$$L_y \simeq \lambda n_c^2[(n_s^2 - n_c^2)\pi], \tag{6.15}$$

where λ is the vacuum-wavelength, L_y is typically of the order of a millimeter. This filamentation phenomenon should be avoidable by using strip loaded interfaces rather than flat ones.

6.3.2. The thin film loaded nonlinear interface

In fig. 6.3 a typical NDC for a lossless geometry according to fig. 6.1b is depicted. Both TE_0– and TE_1–NGWs will be considered.

TE_0–NGWs and NSWs are of similar origin, and the analytical stability

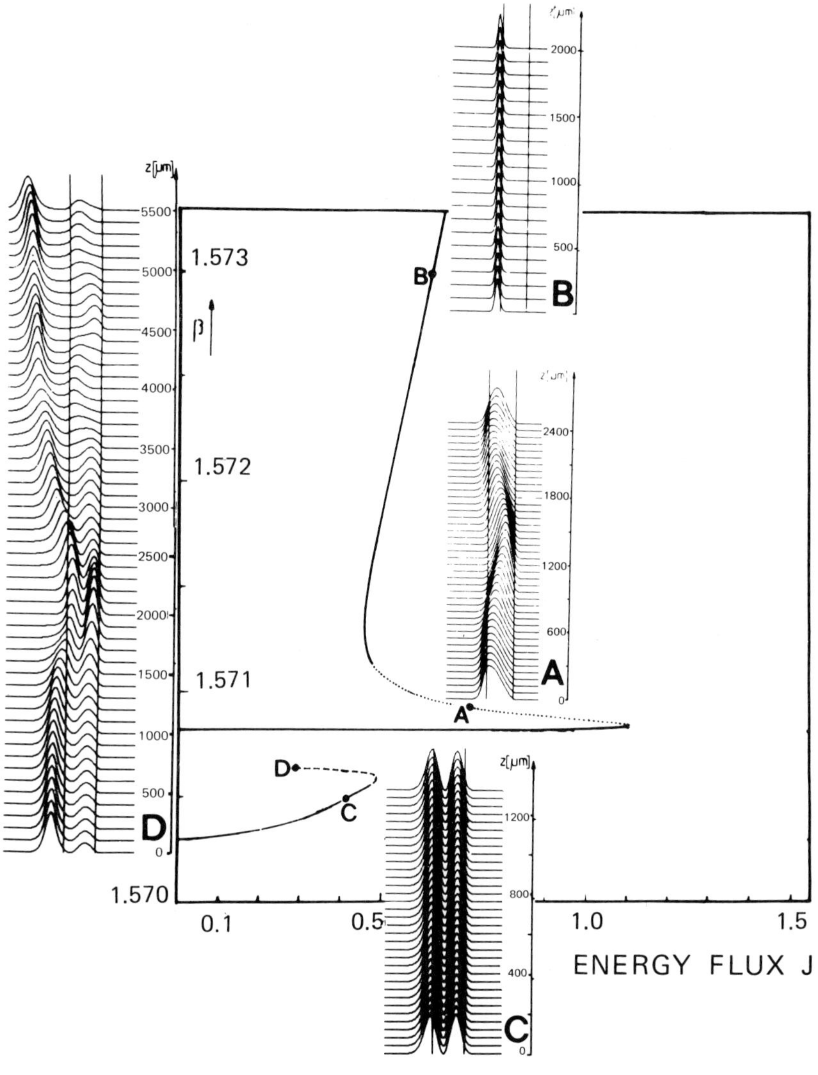

Fig. 6.3. Characteristic graph of a nonlinear dispersion curve for equilibrium NGWs at a (lossless) film loaded nonlinear interface depicted in fig. 6.1b. (After Leine et al. 1987a.) Solid lines denote stable sections, dashed lines indicate unstable sections. Insets illustrate the evolution of some typical NGWs (squared amplitude profiles) initially excited at the indicated points. Inset D illustrates a long distance instability.

predictions for NSWs apply also for TE_0–NGWs. The numerical studies have complemented this picture by giving insight into the decay features of unstable TE_0–NGWs.

The low flux TE_0–NGWs show nearly linear features and are stable. The numerical stability tests of the unstable, i.e., backbent TE_0-branch, led to the identification of a weak oscillational instability, displaying a periodic bouncing of the field maximum between both film interfaces [see inset (a) to fig. 6.3]. At least in its initial state this nonstationary phenomenon resembles the NSW decay via an off-branching beam formation [inset (b) in fig. 6.2]. When approaching the second interface, the beam gets back-reflected until it acquires nearly its initial shape, thus completing one oscillation period (Leine et al. 1987a).

A survey on the stability characteristics of other geometries reveals that the oscillational instability is of a universal nature. Apparently, it is always connected with the backbent section of the TE_0–NDC that reveals a linear limit, i.e., starting at $J = 0$ (Jones and Moloney 1986a, b, Moloney et al. 1986a, b, Leine et al. 1987a, b).

There are numerically based indications that under certain conditions, TE_0–NGWs of the unstable NDC-section under discussion may also decay by emitting a spatial soliton instead of oscillating down the film (Mihalache and Mazilu 1987b). A clarifying analytical analysis, at least in a perturbational framework, is still in order.

The polariton-like upper TE_0-branch has the same stability properties as in the single interface geometry. The equilibrium waves are stable against two-dimensional perturbations (as assumed throughout this subsection), but suffer from a modulational instability (with respect to the y-coordinate) when three-dimensional perturbations apply. This has been numerically tested by Moloney (1987). Note that this modulational instability does not occur when the second film interface becomes significant, i.e., for β values below the polariton branch.

For equilibrium TE_1–NGWs no analytical stability predictions are available. Numerical stability studies indicate that the TE_1-branch depicted in fig. 6.3 is nearly stable within a basic propagation length of 1500 μm. On the negatively sloped branch, a slight drift of the self-focussed cladding field maximum towards the film cladding interface can be observed. For much larger propagation lengths this drift leads to an impressive run-away effect. In some intermediate regions the nonlinear wave is nearly completely captured by the film. Eventually the wave decays into an off-branching self-focussed channel and some low power field tails oscillating between both film boundaries. Inset D to fig. 6.3 displays this behaviour provoked by the TE_1–NGW excited at the NDC stop point D.

Note that the initial equilibrium NGW under discussion feels two refractive index channels, i.e., the film profile and a self-focussed channel, well separated from the film profile. The local intensity at the film cladding interface does not increase the local refractive index beyond that of the linear film.

The preliminary TE_1-stability results do not permit an identification of a stability criterion that is independent of a concrete geometry. Obviously the criterion $dJ/d\beta$ is only of orientational value (see also next subsection). In particular, TE_1–NDCs revealing a backbent section below J_{max} (see section 3) should display subtle stability features.

6.3.3. The thin film bound by two nonlinear media

This geometry, oriented at the lossless geometry depicted in fig. 6.1c, reveals a rich set of equilibrium NGWs associated with distinct NDCs.

Our survey on the stability features of TE_0–NGWs starts with the completely symmetrical configuration $n_c = n_s$, $a_c = a_s$. There are clear analytical stability predictions for this case, based both on topological arguments (Jones and Moloney 1986a, b), and on the explicit determination of the perturbation modes of the linearized problem (Akhmediev, ch. 3, this volume).

The outcome is: symmetric TE_0–NGWs are stable on their low power NDC-branch until the bifurcation point is reached where the asymmetric NDC-branch branches off.

For this asymmetric branch the stability condition (6.14), $dJ/d\beta > 0$, applies again. The validity of this criterion reflects the physically reasonable fact that the present asymmetric TE_0–NGWs, the single interface NSWs, and the thin film loaded interface NGWs are of similar origin: exceeding some characteristic flux level, self-focussing occurs only in one of the nonlinear bounding media.

Numerical stability investigations confirmed the analytical predictions but disclosed also two principal decay features of unstable TE_0–NGWs (Jones and Moloney 1986a, b, Leine et al. 1986, Akhmediev, ch. 3, this volume).

(i) Asymmetric NGWs on the negatively sloped NDC-branch show an oscillational behaviour similar to that observed for the thin film loaded interface.

(ii) Unstable symmetric NGWs decay into stable asymmetric NGWs. Depending on the onset of the numerical fluctuations, the field maximum is formed either in the (identical) substrate or cladding region. In fig. 6.4 both decay features are illustrated.

Both decay patterns reported so far will be complemented by additional decay phenomena when the underlying NDC possesses a richer structure than that one in fig. 6.4. For instance, the increase of both the film thickness and/or the refractive index mismatch across the boundaries may lead to a pronounced backbending even of the symmetric NDC-branch (see section 3 of this chapter). Equilibrium NGWs of the resulting negatively sloped symmetric NDC-section may also decay either by emitting spatial solitons symmetrically into both bounding media, or by evolving into stable equilibrium NGWs below the bifurcation point. The concrete decay performance will depend on the definite perturbational modes occurring in the initial field disturbance. These findings have been pointed out by Akhmediev (ch. 3, this volume) and will be detailed elsewhere in this volume.

The introduction of *dissimilar* bounding media, e.g., by selecting different nonlinear parameters a_c, a_s, leads to a splitting of the former doubly degenerated asymmetric NDC-branch. The bifurcation point disappears and two completely separate NDC-branches arise. Figure 6.5 shows a typical NDC-example in the β-region where the backbending is significant.

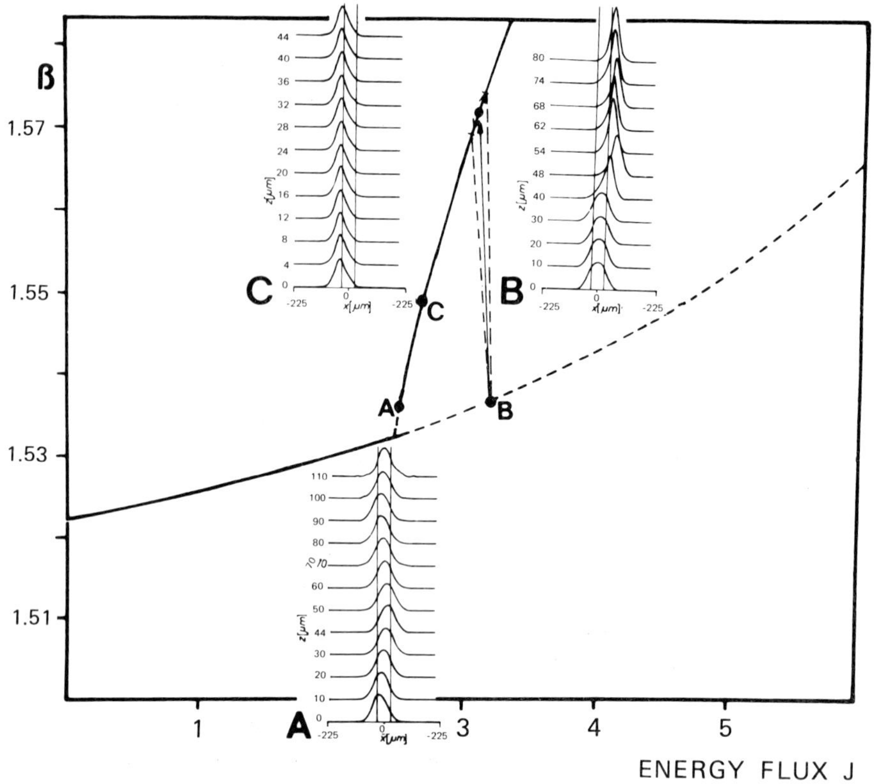

Fig. 6.4. Graph of a nonlinear dispersion curve for equilibrium TE_0–NGWs at a (lossless) symmetric nonlinear film configuration according to fig. 6.1c. (After Leine et al. 1986.) Same notation as in fig. 6.3. The arrow marks the evolution of the unstable symmetric NGW into an asymmetric NGW, inset B. In spite of its positive slope, the asymmetric branch just above the bifurcation point is slightly unstable. This behaviour reflects the extreme sensitivity of this (nearly vertical section) against numerical fluctuations.

At large β values, the 'J'-branch exhibits self-focussing in the medium with the largest nonlinearity. The stability properties of this branch are completely similar to those of the film loaded nonlinear interface, i.e., stability for $\mathrm{d}J/\mathrm{d}\beta > 0$; oscillational instability otherwise (see fig. 6.5b). This type of instability tends to occur at low energy fluxes.

The twin branch IIa to NDC-branch I corresponds to equilibrium NGWs that are localized near the boundary between the film and the bounding medium with the smallest nonlinearity. Here again the positively sloped part is stable (not indicated in fig. 6.5a), however, the decay features observed on branch IIa are different from those of branch I. The field maxima of unstable NGWs drift gradually toward the opposite interface at which point it breaks into a wave

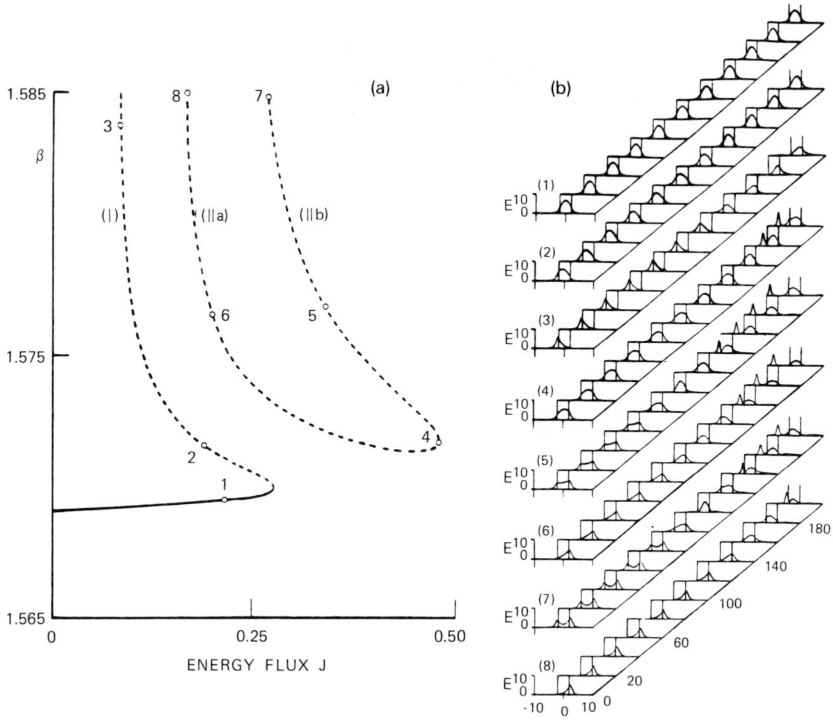

Fig. 6.5. Graph of a nonlinear dispersion curve for equilibrium TE_0–NGWs at a (lossless) asymmetric film configuration according to fig. 6.1c. (After Moloney et al. 1986b.) Solid part corresponds to stable, dashed parts to unstable NGWs. The numbers 1–8 identify positions for which the evolution of the guided waves with propagation distance is shown in fig. 6.5b. (b) The evolution of the NGWs for positions 1–8 on the nonlinear dispersion curve depicted in fig. 6.5a. The propagation coordinate is in units of wavelength. The vertical lines are the waveguide boundaries. (After Moloney et al. 1986b.)

localized in the film and a second soliton-like wave which propagates into the cladding, i.e., the medium with the largest nonlinearity.

The high flux part of branch II and IIb, is completely unstable. The corresponding equilibrium field distributions are localized initially near both film boundaries. Both field maxima collapse gradually to a common peak which bounces against the film–cladding interface, where it separates again into a wave guided primarily inside the film and a soliton-like component branching off into the cladding (see fig. 6.5b). Note that both wave fractions remain phase locked and exchange some energy, at least in the initial state, as displayed in fig. 6.5b.

The latter kind of instability is reminiscent of what occurs when a Gaussian beam, close to the critical angle for total internal reflection, is reflected off a nonlinear interface. In this latter situation, when the local intensity at the interface is large enough to compensate for the refractive index difference between the two media, part of the beam is refracted into the nonlinear medium and propagates away as a self-focussed channel. The numerics suggest that the energy flux confined by the linear film is bound so that the system behaves as an optical limiter (Moloney et al. 1986b).

The explicit analytical stability analysis of the asymmetric geometry is still missing. The knowledge of the relevant perturbational modes will complete the survey on the potential decay channels of unstable equilibrium NGWs.

Due to the lack of analytical results, the stability properties of equilibrium TE_1–NGWs are not understood completely. Figures 6.6a, b give a survey on the numerical findings concerning a highly structured TE_1–NDC (fig. 6.6a). The underlying asymmetric film geometry is the same as in figs 6.5a, b.

Here the low β-branch is completely stable, which accords approximately with the thin film loaded interface configuration and moderate propagation

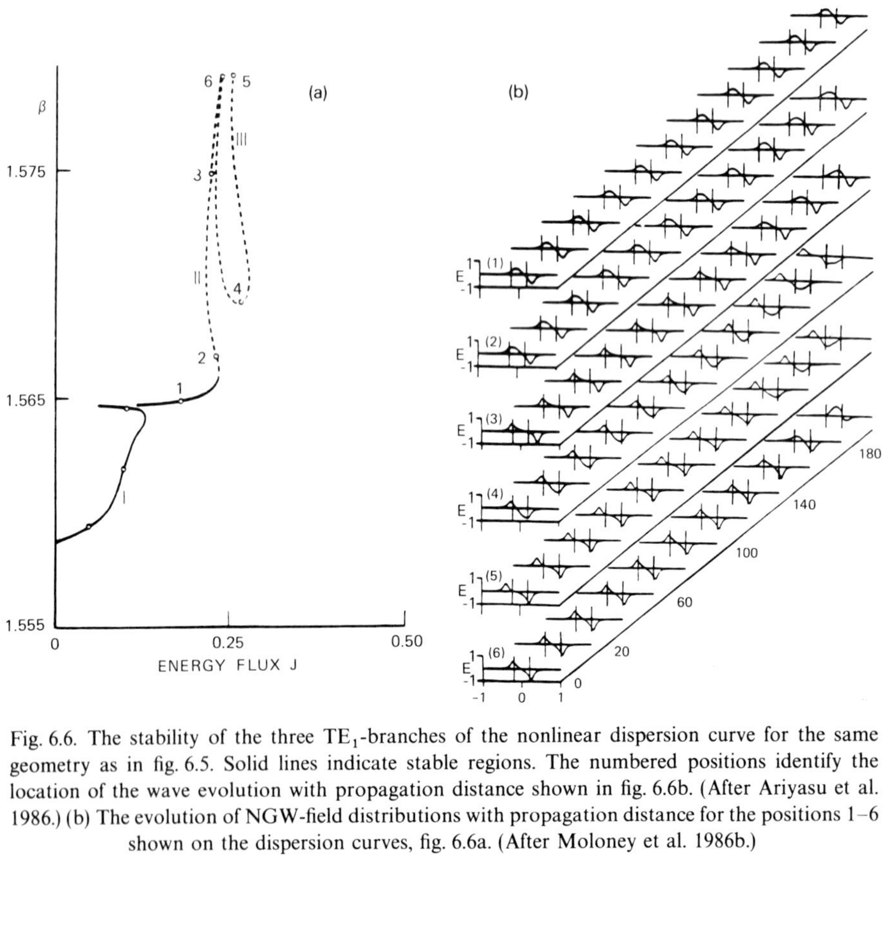

Fig. 6.6. The stability of the three TE_1-branches of the nonlinear dispersion curve for the same geometry as in fig. 6.5. Solid lines indicate stable regions. The numbered positions identify the location of the wave evolution with propagation distance shown in fig. 6.6b. (After Ariyasu et al. 1986.) (b) The evolution of NGW-field distributions with propagation distance for the positions 1–6 shown on the dispersion curves, fig. 6.6a. (After Moloney et al. 1986b.)

lengths (see fig. 6.3). In both cases, the local intensity at the nonlinear interfaces did not increase the local refractive index beyond that of the linear film.

The waves on the curve labelled II in fig. 6.6 are stable on the positively sloped region and unstable on the negatively sloped part [see fig. 6.6b (Ariyasu et al. 1986)]. The latter authors argue that the instability of the NGW at higher β values which corresponds to a region of extreme positive slope may be due to the very high sensitivity to numerical fluctuations.

The NDC-branch labelled III is completely unstable. At least for the highest power NGWs there is a suggestion of a soliton-like component leaving the thin film region, leaving behind a TE_0-like wave guided by the film. This corresponds to the findings at the thin film loaded nonlinear interface (see fig. 6.3).

In summary, the stability problem for nonlinear thin film geometries is far from being completely solved. The lack of analytical predictions with respect to higher-order NGWs prevent a proper interpretation of the numerous numerical data. There are only some very preliminary stability predictions for TE_2–NGWs (Ariyasu et al. 1986). Current numerical results for the nonlinear film bound by two linear or nonlinear media indicate also a rich set of nonstationary phenomena (Mann 1987).

The stability of equilibrium NGWs against three-dimensional perturbations has been addressed only for the single nonlinear interface polariton case. Eventually, the analytical description of nonstationary wave evolution along long distances needs to be addressed. Unfortunately, the powerful 'Inverse Scattering Method' (Zakharov and Shabat 1972), does not apply when interfaces come into the play.

6.4. *Preliminary studies on NGW-excitation*

The central question of this subsection is: how can one excite a definite point on the NDC, i.e., an appropriate equilibrium NGW, by means of a suitably tailored external wavefield?

This question is closely related to the more general problem: which field solutions will be excited when an arbitrary external field is 'launched' into the nonlinear waveguide?

Due to the lack of a suitable analytical calculus this question is almost unsolvable. However, some application oriented numerical studies could provide some interesting conclusions.

6.4.1. *NGW-excitation by distributed coupling*

In principal, the lateral excitation of NGWs by distributed coupling, i.e., by launching the incident wave through a properly prepared interface into the waveguide, is much more involved than in linear waveguide optics. The distributed coupling between the incident wave and the arising NGW manifests itself in a complicated interplay of both transversal and longitudinal nonstationary effects driven by the nonlinear properties of the whole configuration. This applies also for the resonant transient NGW-excitation in an attenuated total internal reflection (ATR) geometry.

A very promising approach takes advantage of the decay features of equilibrium NGWs. The potential emission of spatially localized field components both in interface and film geometries (see figs 6.2, 6.3, 6.5, 6.6) suggests thinking of the reversed situation, where the off-branching decay channel serves as an incident beam. This point of view leads directly to the nonlinear surface reflectivity problem addressed elsewhere in this volume. In particular the NSW-excitation at nonlinear single interfaces by totally internally reflected Gaussian beams (Rozanov 1979, Marcuse 1980, Tomlinson et al. 1982, Akhmediev et al. 1985), or soliton-shaped light beams (Moloney 1987) has been extensively studied.

Until now, the same approach has not been systematically applied to the more involved thin film geometries. Essentially, nonlinear prism- or grating-couplers have been analysed properly only in two limiting cases:

(i) In the weakly nonlinear regime where efficient coupling can be reduced to a proper phase matching of both fields. This model, which has been detailed in the initial part of this chapter, is appropriate only when the low energy flux NDC-branches near their linear limits are excited.

(ii) For ATR-geometries a second special case arises, when plane incident waves are assumed. In this limit all propagation effects are ruled out and the response characteristic is determined by the resonant excitation of stationary NGWs which may lead to optical bistability (OB) (Martinot et al. 1984, Vincent et al. 1985, Reinisch et al. 1985, Montemayor and Deck 1985, 1986, Agarwal and Gupta 1986a, b, Arlot et al. 1986, Arlot and Vitrant 1987, Gupta and Agarwal 1986a, b, Hickernell and Sarid 1986).

The predicted bistable optical response with respect to variations of the incident intensity and/or the angle of incidence is, in these models, due to the interplay of optical feedback, provided by the film/prism boundaries and the (instantaneous) Kerr-like nonlinearity of some waveguide material. There are actual discussions of whether this OB-mechanism applies to recent experimental situations, see, e.g., Fortenberry et al. (1986), Lukosz et al. (1986) and Dannberg and Boese (1987).

In principle, the absence of any propagation effect places the latter models outside the scope of this section. Additional informations can be found in the initial subsections of this chapter as well as elsewhere in this volume.

6.4.2. NGW-excitation by endfire coupling

'Endfire' coupling means the launching of appropriately tailored localized wavefields through one of the transverse endfaces of the waveguide. This geometry covers both the excitation by external light beams and by auxillary, in general linear, waveguides mounted together with the nonlinear waveguide of interest, via their endfaces.

The current theoretical knowledge in this field is determined by a numerical study of Wright et al. (1986a). The authors solved essentially the NLS, eq. (6.4), with initial data corresponding to a collimated Gaussian input beam

$$E(x) = \tilde{E} \exp[-(x - x_0)^2/w_0^2], \tag{6.16}$$

where $\tilde{E}$ is used to adjust the input beam flux, x_0 is the displacement of the beam centre from the guide centre and w_0 is the Gaussian spot size. This boundary condition corresponds to endfiring a Gaussian beam onto the waveguide when any nonlinear reflection effects can be ignored. This will be a reasonable approximation because the input beam is incident normally onto the air–waveguide interface whereas many nonlinear reflection effects occur only for angles close to the critical angle for total internal reflection, see, e.g., the surface reflectivity chapter of this volume.

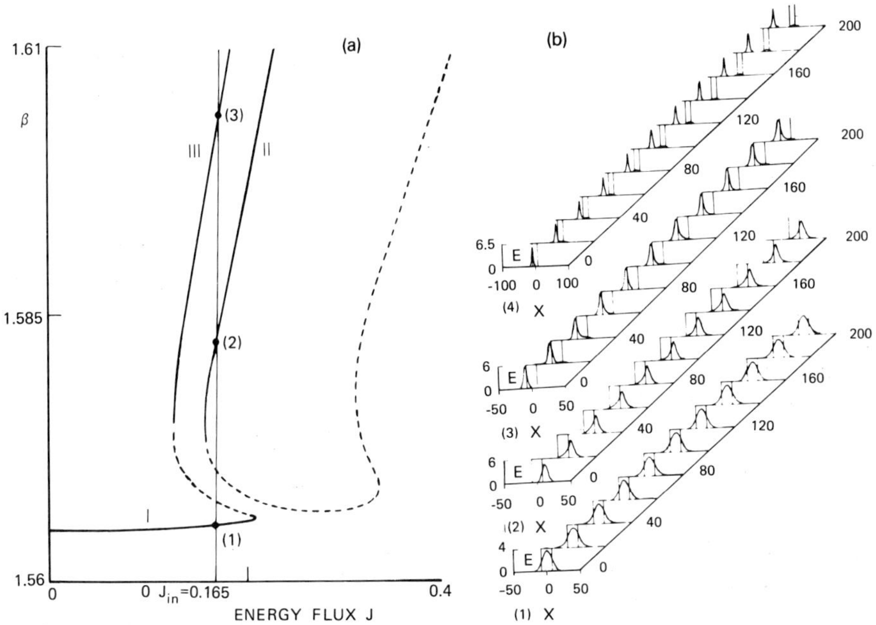

Fig. 6.7. (a) Graph of a nonlinear dispersion curve for two dissimilar self-focusing bounding media. The solid and dashed lines indicate stable and unstable regions respectively. Vertical line marks the fixed input flux of the Gaussian input beam. Parameter values are: $n_c = n_s = 1.55$; $n_f = 1.57$; $a_c = 10^{-2}$, $a_s = 8.10^{-3}$, $a_f = 0$; $d = 8$. (Wright et al. 1986a.) (b) Evolution of Gaussian input beams for $J_{in} = 0.165$ and: (1) $x_0 = 0$, $w_0 = 10.0$; (2) $x_0 = 10.3$, $w_0 = 6.0$; (3) $x_0 = 10.6$, $w_0 = 4.0$; (4) $x_0 = -9.0$, $w_0 = 4.0$. All coordinates are in units of free space wavelength; the vertical lines are the waveguide boundaries. (After Wright et al. 1986a.)

The numerical results indicate that, for a fixed input flux, selective excitation of the corresponding TE_0–NGWs should be possible (see fig. 6.7).

Quite obviously, the input flux alone is not sufficient to characterise the input. On the contrary, the input beam profile is a significant control parameter. The desired excitation takes place only when the values for x_0 and w_0 have been chosen such that the input Gaussian profile would resemble the corresponding NGW (see fig. 6.7).

Though no exceptional sensitivity to initial conditions was found, the excited wave will decay mainly via emission of radiation and/or solitons if the input beam profile differs too much from an equilibrium NGW available at the given input flux (see fig. 6.7). On the other hand, this process produces threshold behaviour and optical limiting which can be used for device design concepts (see next subsection).

The results of Wright et al. (1986a), also give a clear estimate of what will happen when the initial input beam profiles will hold fixed at some TE-mode of the linear guide. This situation is approximately met experimentally when, e.g., only a certain waveguide section is covered by a nonlinear cladding.

Evidently in this way only the low flux NDC-branches, having still nearly linear characteristics, can be excited when the input power will be increased. This has been numerically verified both for linear TE_0-input profiles (Wright et al. 1986b, Leine et al. 1987a) and for TE_0-input profiles (Leine et al. 1987a).

Generally, the stable low power TE_0–NGW-branch can be excited in this way; this does not apply for the TE_0-branch. After exceeding some critical deviation between the input profile and the NGW-profile with the same energy flux, soliton emission sets in. This phenomenon has interesting implications which will be addressed in subsection 6.6.

6.5. NGW-propagation and excitation in lossy waveguides

Owing to the intimate connection of dispersive and absorptive properties in highly optically nonlinear materials, the consideration of material losses is a further essential step towards a realistic description of experimental situations.

Essentially, the basic NGW-phenomenon is of dispersive origin. The presence of absorption will deteriorate NGW-features. Consequently, an important task is to estimate the extent of absorption effects which can be tolerated, having a definite phenomenon in mind.

In contrast to linear waveguide optics, absorption causes a complicated interplay of field energy dissipation and redistribution effects down the non-linear waveguide, which is not completely understood theoretically. In principle, all endfired field distributions will become intrinsically nonstationary and the total energy flux decays according to eq. (6.10). In particular, equilibrium NGWs according to eq. (6.11) no longer exist.

The existing studies are of preliminary character and are restricted to the consideration of linear absorption. Both the thin film loaded nonlinear interface

(Gubbels et al. 1987, Leine et al. 1987a) and the symmetric film configuration (Leine et al. 1987b), have been analysed numerically. Here, we will illustrate the essential predictions at the film loaded interface configuration according to fig. 6.1b.

The comparison of the propagation pattern displayed in figs 6.8 and 6.3

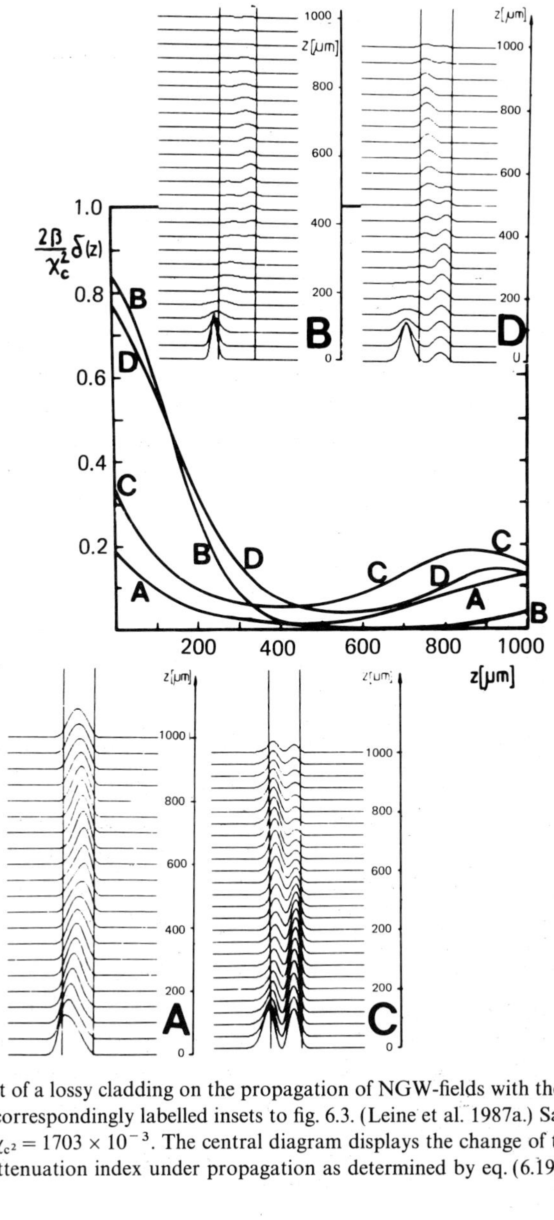

Fig. 6.8. The effect of a lossy cladding on the propagation of NGW-fields with the same initial field profiles as in the correspondingly labelled insets to fig. 6.3. (Leine et al. 1987a.) Same configuration as in fig. 6.3, but $\chi_{c^2} = 1703 \times 10^{-3}$. The central diagram displays the change of the corresponding attenuation index under propagation as determined by eq. (6.19).

reveals that after some initial propagation distance, the field tails in the lossy cladding tend to move into the film region, thereby reducing their attenuation considerably. This 'focusing property' was first pointed out by Gubbels et al. (1987), who also tested the reversed situation where the film losses exceed the losses of the bounding media. In this case the film 'repels' the field. The initial TE_0-field breaks into two side waves, evolving away from the film.

In general, the absorption effect cannot be characterised by a global attenuation constant, implying an overall exponential field decay under propagation. The rate equation (6.10), can be used to introduce an attenuation index $\delta(z)$ in the sense

$$J(z) = J(0)\exp\left[-2\int_0^z \delta(z')\,\mathrm{d}z'\right], \tag{6.17}$$

where $\delta(z)$ is determined by

$$\delta(z) = \int_{-\infty}^{\infty} \varepsilon''(x)|E(x, z)|^2\,\mathrm{d}x \Big/ \left(2\beta\int_{-\infty}^{\infty} |E(x, z)|^2\,\mathrm{d}x\right). \tag{6.18}$$

For the configuration of fig. 6.8, $\delta(z)$ reads as

$$\delta(z) = (\chi_c^2/2\beta)\int_{-\infty}^{0} |E(x, z)|^2\,\mathrm{d}x \Big/ \int_{-\infty}^{\infty} |E(x, z)|^2\,\mathrm{d}x, \tag{6.19}$$

i.e., $\delta(z)$ is determined by the product of the bulk attenuation constant and the relative actual energy flux fraction localized in the lossy cladding.

According to the actual field evolution, $\delta(z)$ may change considerably down the waveguide. This is illustrated in fig. 6.8 for the four decay pattern depicted in insets A through D. The loss reduction under propagation is clearly visible, in particular for the cases B and D, where initially most of the flux was concentrated in the cladding. The more or less pronounced single $\delta(z)$-oscillations are due to a secondary field enhancement at the cladding–film interface.

Apparently the $\delta(z)$-evolution can be interpreted also in terms of the actual energy flux $J(z)$, leading to the formulation of a flux dependent attenuation index $\delta = \delta(J)$. This concept has been detailed by Gubbels et al. (1987). An early approach to describe NGE-attenuation by a flux dependent attenuation index stems from Ariyasu et al. (1985). However, this approach neglects any propagation effect.

The impact of attenuation on specific decay features of unstable equilibrium NGWs has been investigated numerically by Gubbels et al. (1987) and Leine et al. (1987a), for the thin film loaded nonlinear interface and by Leine et al. (1987b) for the symmetric film configuration. The question of whether the onset of a specific instability feature (e.g., oscillational instability, symmetric and asymmetric NGW-transformation, soliton formation), will survive for some propagation distance depends on the characteristic evolution length the phenomenon of interest takes to evolve. If this length becomes comparable with some effective absorption length, roughly $\min[\delta(z)^{-1}]$, the specific effect will be seriously distorted.

Initially stable NGW-field profiles appear to move adiabatically down their associated NDC-branches because of absorption, as long as the corresponding NDC-branch decreases monotonically with decreasing energy flux. This applies, e.g., for the surface polariton decay displayed in fig. 6.8a (Leine et al. 1987a, b).

In principle, the transient excitation of NGW-like field pattern by endfiring appropriately tailored Gaussian beams is also possible in lossy configurations. This has been numerically tested by Gubbels et al. (1987), who showed that the basic NGW-phenomenon can survive the absorption along some finite propagation lengths. However, as indicated above, the impact of absorption will strongly depend on the actual field profile versus the given absorption profile of the geometry. The realization of some desired effective propagation lengths will call for a careful adjustment of all relevant parameters.

6.6. *Soliton emission from nonlinear waveguides*

Soliton-like solutions of the NLS manifest themselves as self-focussing channels of invariant shape where nonlinear refraction balances diffraction.

In planar single or multiple interface geometries the transmission of solitons through such interfaces represents an important nonstationary phenomenon. Since the pioneering work of Rozanov (1979), the role of spatial solitons in nonlinear interface reflectivity configurations has been extensively studied.

In thin waveguiding film geometries, the emission of spatial solitons has been identified as an important decay route both of certain unstable equilibrium NGWs (Akhmediev et al. 1987a, Ariyasu et al. 1986, Jones and Moloney 1986a, b, Leine et al. 1987a, Mihalache et al. 1987, Moloney 1987), and of endface excited field profiles which deviate from stable equilibrium NGWs (Wright et al. 1986a, b, Leine et al. 1987a, Gubbels et al. 1987). Potential applications of this effect include optical limiting, coupling of adjacent waveguides in the absence of evanescent field overlap, and a light-driven angular scanning element (Wright et al. 1986b).

For that reason, the soliton emission is of interest in its own right and will be briefly analysed in this subsection, thereby following essentially the basic papers of Wright et al. (1986b) and Gubbels et al. (1987).

An efficient way of provoking soliton emission proved to be to launch an input beam of fixed profile but variable energy flux onto the waveguide. Such a situation could be realised experimentally by, say, prism coupling into the TE_0-mode of a linear waveguide section and interfacing this onto the nonlinear system.

The NDC of a film loaded nonlinear interface is depicted in fig. 6.9a. For $J < J_c$ the input beam very closely matches the NGW on branch I where the system retains essentially its linear characteristic. This leads to a very effective

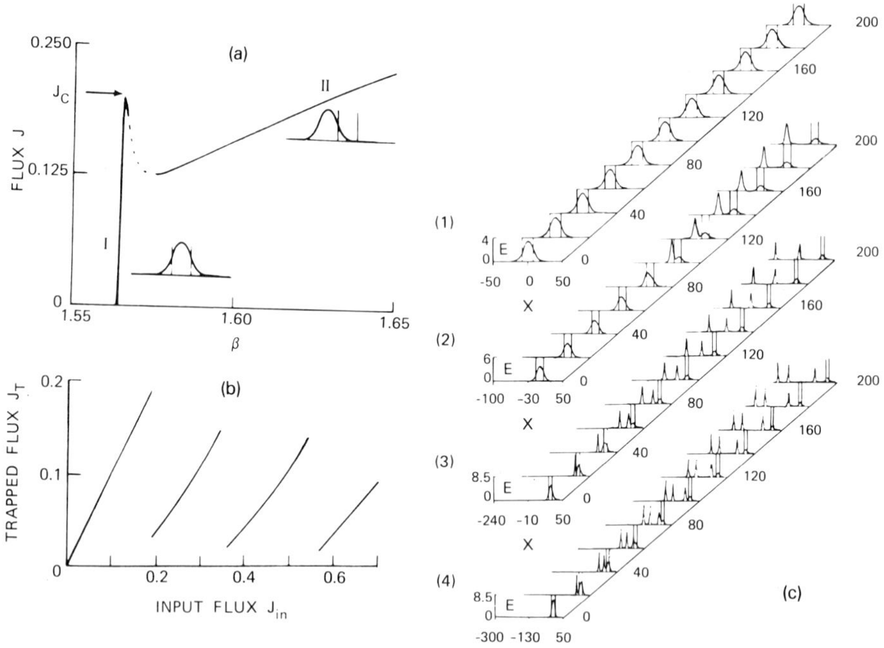

Fig. 6.9. (a) Nonlinear dispersion curve for a configuration: $n_c = n_s = 1.55$, $n_f = 1.57$, $a_c = 0.02$, $a_s = 0$, $d = 8$. The dashed region is unstable and the insets indicate the general nature of the NGWs on branches I and II. (b) Flux trapped in the guide J_T vs. the input flux J_{IN}. The input beam profile is held fixed as the linear TE_0-guided mode. (c) Evolution of the input fields for flux levels: (1) 0.15; (2) 0.2; (3) 0.4; (4) 0.6. The propagation coordinate is in units of free-space wavelength. (After Wright et al. 1986b.)

coupling of the input into these NGWs as already pointed out in subsection 6.4.2. For $J > J_c$ only branch II-NGWs are available to the system and the input beam matches these NGWs very poorly. The excess energy is emitted by a spatial soliton leaving a branch I-NGW captured by the film (see fig. 6.9c).

As the input flux is further increased, the flux trapped in the branch I-NGW increases accordingly, leading eventually to the emission of a second soliton when the trapped flux, J_T, will exceed J_c again (Wright et al. 1986b, Leine et al. 1987a). In this way several solitons can be emitted until material saturation sets in. The corresponding threshold behaviour of the energy flux, J, trapped by the film, versus the input flux of the fixed TE_0-mode is shown in fig. 6.9b. The reduced higher thresholds are due to a still existing interaction between the solitons and the low power NGWs trapped by the film.

Similar effects have been found when the initial profile is a linear TE_0-mode of variable energy flux. Here the field maximum next to the nonlinear cladding interface evolves continuously into the soliton channel leaving some captured low power field tails which oscillate between both film boundaries (Leine et al. 1987a).

The potential exploitation of the first (optimised) J-threshold for optical limiting applications has been proposed by Wright et al. (1986b). This threshold characteristic may potentially be combined with a directional coupler action without the need for evanescent field overlap of the guides. A soliton emitted from one nonlinear waveguide can be at least partially trapped by a second adjacent guide (Wright et al. 1986b). The angle of soliton emission has been found to depend on the input flux (Tomlinson et al. 1982, Wright et al. 1986b, Leine et al. 1987a).

This effect should have potential use as a light driven angular scanning element (Tomlinson et al. 1982).

In a very recent paper Gubbels et al. (1987), studied numerically the effects of linear absorption, input beam misalignments and nonlinear saturation on the basic phenomenon of soliton emission described so far.

The basic evolution equation was essentially eq. (6.4) with

$$\varepsilon'(x, |E|^2) = n_\gamma^2 + (a_\gamma|E|^2)/(1 + \rho_\gamma|E|^2), \quad \gamma = \mathrm{c, f, s}, \quad a_\mathrm{f} = a_\mathrm{s} = 0, \tag{6.20}$$

where the 'ideal case', i.e., $\rho_\gamma = 0$, $\varepsilon'' = 0$, coincides with the configuration of fig. 6.9. Of course, in the presence of linear absorption and nonlinear saturation, only soliton-like waves reminiscent of the ideal case can be expected.

The basic phenomenon was found to survive in the presence of these complications within a suitable range of configuration parameters. Input beam alignment is generally accompanied by significant amounts of radiation. Cladding losses cause a broadening of the localised wavepacket accompanied by a reduced off-branching from the interface. The number of emitted waves will be reduced. The waves emitted in the presence of absorption are solitary waves which was verified by a numerical collision experiment (Gubbels et al. 1987).

The potential effect of modulational instabilities, which are highly likely, needs to be clarified by including the third space dimension (y) into the analysis. Such a more involved analysis should cover also the very important channel waveguide geometry including the implications on the phenomena considered above.

References

Abramowitz, A., and I.A. Stegun, 1965, in: Handbook of Mathematical Functions (Dover, New York).

Aceves, A., J.V. Moloney and A.C. Newell, 1988, J. Opt. Soc. Am. B **5**.

Agarwal, G.S., and S.D. Gupta, 1986a, Phys. Rev. B **34**, 5239–5243.

Agarwal, G.S., and S.D. Gupta, 1986b, Optical Bistability with Surface Plasmons Beyond Plane Waves in a NonLinear Dielectric, in: Optical Bistability III, eds H.M. Gibbs, P. Mandel, N. Peyghambarian and S.D. Smith (Springer-Verlag, Berlin) pp. 117–119.

Agranovich, V.M., 1980, Some Linear and Nonlinear Properties of Surface Polaritons in the Region of Resonance with Vibrations in the Transition Layer, in: Proc. 7th Int. Conf. on Raman Spectroscopy, Linear and Nonlinear Processes, ed. W.F. Murphy (North-Holland) p. 372–375.

Agranovich, V.M., and V.Ya. Chernyak, 1982a, Phys. Lett. A **88**, 423–426.

Agranovich, V.M., and V.Ya. Chernyak, 1982b, Solid State Commun. **44**, 1309–1311.

Agranovich, V.M., and D.L. Mills, eds, 1982, Surface Polaritons: Electromagnetic Waves at Surfaces and Interfaces (North-Holland, Amsterdam).

Agranovich, V.M., V.S. Babichenko and V.Ya. Chernyak, 1980, JETP Lett. **32**, 512–515.

Aitchison, J.S., J.D. Valera, A.C. Walker, S. Ritchie, P.M. Rogers and P. McIlroy, 1987, Appl. Phys. Lett. **51**, 561–563.

Akhmediev, N.N., 1982a, Sov. Tech. Phys. Lett. **198**, 249–250.

Akhmediev, N.N., 1982b, Sov. Phys.-JETP **56**, 299–303.

Akhmediev, N.N., 1983, Sov. Phys.-JETP **57**, 1111–1116.

Akhmediev, N.N., K.O. Boltar and V.M. Eleonskii, 1982a, Opt. & Spectrosc. **53**, 654–658.

Akhmediev, N.N., K.O. Boltar and V.M. Eleonskii, 1982b, Opt. & Spectrosc. **53**, 540–542.

Akhmediev, N.N., V.I. Korneev, Yu.V. Kuzmenko, 1984, Sov. Tech. Phys. Lett. **10**, 327–329.

Akhmediev, N.N., V.I. Korneev, Yu.V. Kuzmenko, 1985, Sov. Phys.-JETP **61**, 62–67.

Al-Bader, S., and H.A. Jamid, 1987, IEEE J. Quantum Electron. **QE-23**, 1947–1955.

Alanakyan, Yu.R., 1967, Sov. Phys.-Tech. Phys. **12**, 587–589.

Aliev, Yu.M., and S.V. Kuznetsov, 1984, Sov. Phys.-Lebedev Inst. Rep. **3**, 5–8.

Ames, W.F., 1965, Nonlinear Partial Differential Equations in Engineering (Academic Press, New York).

Arakelian, S.M., L.S. Aslanyan, G.L. Grigorian, S.Ts. Nersisian and Yu.S. Chilingaryan, 1984, Optical Bistability on Excitation of Surface Plasmons at Metal–Nematic Liquid Crystal Interface, in: Proc. 6th All-Union Conf. on the Nonresonant Interaction of Optical Radiation with Matter, Palanga, USSR.

Arakelian, S.M., G.L. Grigorian, L.M. Kocharian, S.Ts. Nersisian and Yu.S. Chilingaryan, 1986, Phys. Lett. A **118**, 254-259.

Ariyasu, J., C.T. Seaton and G.I. Stegeman, 1985a, Appl. Phys. Lett. **47**, 355–357.

Ariyasu, J., C.T. Seaton, G.I. Stegeman, A.A. Maradudin and R.F. Wallis, 1985b, J. Appl. Phys. **58**, 2460–2466.

Ariyasu, J., C.T. Seaton, G.I. Stegeman and J.V. Moloney, 1986, IEEE J Quantum Electron. **QE-22**, 984–987.

Arlot, P., and G. Vitrant, 1987, Appl. Phys. Lett. **50**, 650–652.

Arlot, P., E. Pic, R. Reinisch and G. Vitrant, 1986, Angle and frequency optical bistability: Resonant excitation of a nonlinear guided mode, in: Optical Bistability III, eds H.M. Gibbs, P. Mandel, N. Peyghambarian and S.D. Smith (Springer, Berlin) pp. 113–116.

Arutyunan, Kh.S., and K.A. Barsukov, 1985a, Sov. J. Contemp. Phys. **20**, 198–215.

Arutyunan, Kh.S., and K.A. Barsukov, 1985b, Opt. & Spectrosc. **58**, 650–652.

Askaryan, G.A., 1962, Sov. Phys.-JETP **15**, 1088–1090.

Assanto, G., B. Svensson, D. Kuchibhatla, U.J. Gibson, C.T. Seaton and G.I. Stegeman, 1986, Opt. Lett. **11**, 644–666.

Atanassov, V., E. Mateev and I. Zhelyazkov, 1981, Phys. Lett. A **1986**, 414–416.

Banyai, L., and S.W. Koch, 1986, Z. Phys. B **63**, 283–291.

Bennion, I., M.J. Goodwin and W.J. Stewart, 1985, Electron. Lett. **21**, 41–42.

Bergou, J., and M.O. Scully, 1985, Acta Phys. Austriaca **57**, 139–144.

Berkhoer, A.L., and V.E. Zakharov, 1970, Sov. Phys.-JETP **31**, 486–490.

Bertolotti, M., C. Sibilia and I. Anselmi, 1985, Bistability in nonlinear waveguides, in: Guided Optical Structures in the Military Environment.

Bhat, B., and V.K. Tripathi, 1975, J. Appl. Phys. **46**, 1141–1145.

Boardman, A.D., and P. Egan, 1984a, J. Phys. Colloq. (France) **45**, C5-291–303.

Boardman, A.D., and P. Egan, 1984b, Philos. Trans. R. Soc. London A **313**, 363–369.
Boardman, A.D., and P. Egan, 1985, IEEE J. Quantum Electron. **QE-21**, 1701–1713.
Boardman, A.D., and P. Egan, 1986a, Nonlinear electromagnetic surface and guided waves: Theory, in: Surface Waves in Plasmas and Solids, ed. S. Vulkovic (World Scientific, Singapore) pp. 3–77.
Boardman, A.D., and P. Egan, 1986b, IEEE J. Quantum Electron. **QE-22**, 319–324.
Boardman, A.D., and T. Twardowski, 1988a, J. Opt. Soc. Am. B **5**.
Boardman, A.D., and T. Twardowski, 1989, Phys. Rev. **39**, 2481–2492.
Boardman, A.D., G.S. Cooper and P. Egan, 1984a, J. Phys. Colloq. (France) **45**, C5-197–205.
Boardman, A.D., P. Egan and A. Shivarova, 1984b, Appl. Sci. Res. **41**, 345–353.
Boardman, A.D., G.S. Cooper, P. Egan and T. Twardowski, 1986a, Properties of nonlinear surface waveguides, in: Optical Bistability III, eds H.M. Gibbs, P. Mandel, N. Peyghambarian and S.D. Smith (Springer, Berlin) p. 1974–1978.
Boardman, A.D., G.S. Cooper, A.A. Maradudin and T.P. Shen, 1986b, Phys. Rev. B **34**, 198273–198278.
Boardman, A.D., A.A. Maradudin and T.P. Shen, 1986c, Surface polariton solitons, in: Surface Waves in Plasmas and Solids, ed. S. Vulkovic (World Scientific, Singapore) pp. 522–525.
Boardman, A.D., G.S. Cooper and D.J. Robbins, 1986d, Opt. Lett. **11**, 112–114.
Boardman, A.D., A.A. Maradudin, G.I. Stegeman, T. Twardowski and E.M. Wright, 1987a, Phys. Rev. A **35**, 1159–1164.
Boardman, A.D., T. Twardowski, A. Shivarova and G.I. Stegeman, 1987b, IEE Proc. J. **134**, 152–160.
Boardman, A.D., G.S. Cooper and M. Wilkins, 1987c, SPIE 19800, 106–116.
Boev, A.G., 1980, Sov. Phys.-JETP **50**, 47–51.
Boev, A.G., 1982, Dopov. Akad. Nauk Ukrsr. Ser A **6**, 55–57.
Boev, A.G., and A.V. Propkov, 1976, Sov. Phys.-JETP **42**, 617–621.
Bolshinskii, L.G., and A.I. Lomtev, 1985, Sov. Tech. Phys. Lett. **11**, 148–149.
Bolshinskii, L.G., and A.I. Lomtev, 1986, Sov. Phys.-Tech. Phys. **31**, 499–502.
Born, M., and E. Wolf, 1968, in: Principles of Optics (Pergamon Press, Oxford).
Boullough, R.K., 1977, Solitons, in: Interaction of Radiation with Matter, Vol. 1 (IAEA, Vienna) pp.446–449.
Briguet, V., P. Pirani and W. Lukosz, 1987, Opt. Lett. **12**, 525–527.
Burke, J.J., G.I. Stegeman and T. Tamir, 1986, Phys. Rev. B **33**, 5186–5201.
Butcher, P.N., 1965, Bulletin 200, Eng. Expl. Station (Ohio State University, Columbus, OH).
Cada, M., R.C. Gauthier, B.E. Paton and J. Chrostkowski, 1986, Appl. Phys. Lett. **49**, 19755–19757.
Caglioti, E., S. Trillo, S. Wabnitz, B. Diano and G.I. Stegeman, 1987, Appl. Phys. Lett. **51**, 293–295.
Carter, G.M., and V.J. Chen, 1983a, Appl. Phys. Lett. **42**, 643–645.
Carter, G.M., V.J. Chen and S.K. Tripathy, 1983b, Appl. Phys. Lett. **43**, 891–892.
Chen, W., and A.A. Maradudin, 1988, J. Opt. Soc. Am. B **5**.
Chen, W., and D.L. Mills, 1987, Phys. Rev. B **35**, 524–532.
Chen, Y.J., and G.M. Carter, 1982, Appl. Phys. Lett. **41**, 307–309.
Chen, Y.J., and G.M. Carter, 1983, Solid State Commun. **45**, 277–280.
Chen, Y.J., and G.M. Carter, 1984, J. Phys. Colloq. (France) **45**, C5-261–267.
Chen, Y.J., G.M. Carter, G.J. Sonek and J.M. Ballantyne, 1986, Appl. Phys. Lett. **48**, 272–274.
Chiao, R.Y., E. Garmire, C.H. Townes, 1964, Rev. Lett. **13**, 479–482.
Chrostkowski, J., and S. Chelkowski, 1987, Opt. Lett. **12**, 528–530.
Crosignani, B., A. Cutolo and P. DiPorto, 1982, J. Opt. Soc. Am. **72**, 1136–1141.
Cullen, T.J., C.N. Ironside, C.T. Seaton and G.I. Stegeman, 1986, Appl. Phys. Lett. **49**, 1403–1405.
Dannberg, P., and E. Broese, 1987, Observation of optical bistability by prism excitation of a nonlinear film guided wave, submitted to Appl. Opt.

Diano, B.G.G., and S. Wabnitz, 1985, J. Appl. Phys. **58**, 4512–4514.
Dohi, H., Y. Kuwamura, M. Fukui and O. Tada, 1984, J. Phys. Soc. Jpn **53**, 2828–2832.
Egan, P., 1984, Ph.D. Thesis (University of Salford, UK).
Eleonskii, V.M., L.G. Oganesyants and V.P. Silin, 1972, Sov. Phys.-JETP **35**, 44–47.
Fedyanin, V.K., and D. Mihalache, 1982, Z. Phys. B **47**, 167–173.
Finlayson, N., E.M. Wright, C.T. Seaton, G.I. Stegeman and Y. Silberg, 1987, Appl. Phys. Lett. **50**, 1562–1564.
Fleck Jr, J.R., J.R. Morros and E.S. Bliss, 1978, IEEE J. Quantum Electron. **QE-14**, 353–363.
Flytzanis, C., 1984, Bistability, instability and chaos in passive nonlinear optical systems, in: Proc. 3rd Inst. School on Condensed Matter Physics, Varna, Bulgaria, pp. 268–314.
Fortenberry, R.M., R. Moshrefzadeh, G. Assanto, Xu Mai, E.M. Wright, C.T. Seaton and G.I. Stegeman, 1986, Appl. Phys. Lett. **49**, 687–689.
Fukui, M., V.C.-Y. So and G.I. Stegeman, 1979, Solid State Commun. **20**, 683–687.
Fuzi, Y., C. Zhuangqi and P. Tunxin, 1985, J. Shanghai Chiao Tung University **19**, 1–8.
Gibbons, W.M., and D. Sarid, 1987, Appl. Phys. Lett. **51**, 403–405.
Goldburt, E.S., and P.St.J. Russell, 1985, Appl. Phys. Lett. **46**, 338–340.
Grozev, D., and A. Shivarova, 1984, J. Plasma Phys. **31**, 177–191.
Grozev, D., A. Shivarova and A.D. Boardman, 1987, J. Plasma Phys. **38**, 427–437.
Gubbels, M., E.M. Wright, G.I. Stegeman, C.T. Seaton and J.V. Moloney, 1987, Opt. Commun. **61**, 357–362.
Gubbels, M.A., E.M. Wright, G.I. Stegeman, C.T. Seaton and J.V. Moloney, 1987a, Opt. Commun. **61**, 357–361.
Gupta, S.D., and G.S. Agarwal, 1986a, Optical bistability with surface plasmons beyond plane waves in a nonlinear dielectric, in: Optical Bistability III, eds H.M. Gibbs, P. Mandel, N. Peyghambarian and S.D. Smith (Springer, Berlin) pp. 117–119.
Gupta, S.D., and G.S. Agarwal, 1986b, J. Opt. Soc. Am. B **3**, 236–238.
Gusovskii, D.D., E.M. Dianov, A.A. Maier, V.B. Neustruev, E.I. Shklovskii and I.A. Shcherbakov, 1985, Sov. J. Quantum Electron. **QE-15**, 1523–1526.
Gusovskii, D.D., E.M. Dianov, A.A. Maier, V.B. Neustruev, W.W. Osiko, A.M. Prochorov, K.Y. Sitarski and I.A. Shcherbakov, 1987, vant Elektr. **14**, 1144–1147.
Hagata, K., M. Nagai and M. Koshiba, 1987, Electron. Lett. **23**, 1307.
Haus, H.A., 1966, Appl. Phys. Lett. **8**, 128.
Haus, H.A., and N.A. Whitaker Jr, 1984, Phil. Trans. R. Soc. (London) A **313**, 311–319.
Haus, H.A., N.A. Whitaker Jr and M.C. Gabrial, 1985, Proc. SPIE **578**, 122–129.
Hayata, K., M. Koshiba and M. Suzuki, 1987, Electron. Lett. **23**, 429–431.
Hermansson, B., D. Yevick and L. Thylen, 1978, Opt. Quantum Electron. **16**, 525–529.
Hey, J.D., and F.A. Hopf, 1982, Non-linear optics, in: Quantum Optics: Proc. South African Summer School in Theoretical Physics, ed. C.A. Englebrecht (Springer, Berlin) pp. 211–329.
Hickernell, R.K., and D. Sarid, 1986, J. Opt. Soc. Am. B **3**, 1059–1069.
Hoffe, R., and J. Chrostowski, 1986a, Opt. Commun. **57**, 34–38.
Hoffe, R., and J. Chrostowski, 1986b, Proc. SPIE **667**, 192–196.
Holland, W.R., 1986, J. Opt. Soc. Am. B **3**, 1529–1534.
Hwang, W.L., 1984, J. Appl. Phys. **55**, 3956–3960.
Ironside, C.N., T.J. Cullen, R.H. Hutchins, W.C. Banyia, C.T. Seaton and G.I. Stegeman, 1985, Proc. SPIE **578**, 162–165.
Ivanov, S.T., O.V. Dolgenko and A.A. Rukhadze, 1975, J. Phys. A: Math. Gen. **198**, 585–593.
Jackson, J.D., 1975, Classical Electrodynamics, 2nd Ed. (Wiley, New York).
Jakubczyk, Z., H. Jerominek, S. Patela, R. Tremblay and C. Delisle, 1987, IEEE J. Quantum Electron. **QE-23**, 1921–1927.
Jensen, S.M., 1982, IEEE J. Quantum Electron. **QE-18**, 1580–1583.
Jerominek, H., S. Patela, C. Delisle and R. Tremblay, 1986, Proc. SPIE **667**, 115–121.

Jones, C.K.R.T., and J.V. Moloney, 1986a, Phys. Lett. A **117**, 175–180.
Jones, C.K.R.T., and J.V. Moloney, 1986b, Stability and instability of nonlinear standing waves in planar optical waveguides, in: Optical Bistability III, eds H.M. Gibbs, P. Mandel, N. Peyghambarian and S.D. Smith (Springer, Berlin) pp. 98–101.
Karpman, V.I., 1973, in: Nonlinear Waves in Dispersive Media (Pergamon Press).
Kelly, P.L., 1965, Phys. Rev. Lett. **15**, 1005–1009.
Kim, B.S., E. Garmire, N. Shibata and S. Zembutsu, 1987, Appl. Phys. Lett. **51**, 475–477.
Kinzer, D.J., and G.M. Morris, 1986, Appl. Opt. **25**, 1335–1338.
Kitayama, K., and S. Wang, 1983, Appl. Phys. Lett. **43**, 17–19.
Kocharyan, L.M., 1984, Law of Dispersion of p-polarised surface electromagnetic waves in low nonlinearity three layer system, in: Proc. 6th All-Union Conf. on the Nonresonant Interaction of Optical Radiation with Matter, Palanga, USSR.
Kogelnik, H., 1975, Integrated Optics, Vol. 7, Topics in Applied Physics, ed. T. Tamir (Springer, Berlin).
Korneev, V.I., N.N. Akhmediev and Yu.V. Kuzmenko, 1986, Radiophys. & Quantum Electron. **29**, 344–349.
Kumar, A., and M.S. Sodha, 1987, Opt. Lett. **12**, 352–354.
Kushwaha, M.S., 1987, Jpn. J. Appl. Phys. **26**, L1035–1037.
Lambsdorf, M., C. Dörnfeld and C. Klingshirn, 1986, Z. Phys. B **64**, 409–416.
Langbein, U., and F. Lederer, 1986, Ph.D. Thesis, University of Jena.
Langbein, U., F. Lederer and H.E. Ponath, 1983, Opt. Commun. **46**, 167–169.
Langbein, U., F. Lederer and H.E. Ponath, 1984a, Opt. Quantum Electron. **16**, 251–259.
Langbein, U., F. Lederer and H.E. Ponath, 1984b, Opt. Quantum Electron. **16**, 261–266.
Langbein, U., F. Lederer and H.E. Ponath, 1984c, Opt. Acta **31**, 1141–1149.
Langbein, U., F. Lederer, H.E. Ponath and U. Trutschel, 1984d, J. Mol. Structure **115**, 493–496.
Langbein, U., F. Lederer and H.E. Ponath, 1985a, Opt. Commun. **53**, 417–420.
Langbein, U., F. Lederer, T. Peschel and H.E. Ponath, 1985b, Opt. Lett. **10**, 571–573.
Langbein, U., F. Lederer, H.E. Ponath and U. Trutschel, 1985c, Appl. Phys. B **36**, 187–193.
Langbein, U., F. Lederer, H.E. Ponath and U. Trutschel, 1985d, Appl. Phys. B **38**, 263–268.
Langbein, U., F. Lederer, H.E. Ponath and U. Trutschel, 1985e, IEEE J. Quantum Electron. **QE-21**, 1639–1643.
Lardner, R.W., and G.E. Tupholme, 1986, J. Elasticity **16**, 251–265.
Lattes, A., H.A. Haus, F.J. Leonberger and E.P. Ippen, 1983, IEEE J. Quantum Electron. **QE-19**, 1718–1723.
Lederer, F., and D. Mihalache, 1986, Solid State Commun. **59**, 151–153.
Lederer, F., U. Langbein and H.E. Ponath, 1983a, Appl. Phys. B **31**, 69–73.
Lederer, F., U. Langbein and H.E. Ponath, 1983b, Appl. Phys. B **31**, 187–190.
Leine, L., C. Wachter, U. Langbein and F. Lederer, 1986, Opt. Lett. **11**, 590–592.
Leine, L., C. Wachter, V. Langbein and F. Lederer, 1987a, The evolution of nonlinear guided optical fields down a dielectric film with a nonlinear cladding, to be published in J. Opt. Soc. Am. B.
Leine, L., C. Wachter, V. Langbein and F. Lederer, 1987b, Opt. Lett. **12**.
Leine, L., C. Wachter, U. Langbein and F. Lederer, 1988, J. Opt. Soc. Am. B **5**.
Leung, K.M., 1985a, Phys. Rev. A **31**, 1189.
Leung, K.M., 1985b, Phys. Rev. B **32**, 5093–5101.
Leung, K.M., 1985c, Phys. Rev. A **31**, 1189–1192.
Li Kam Wa, P., J.M. Marsh, P.N. Robson, J.S. Roberts and N.J. Mason, 1985a, Electron. Lett. **21**, 26–27.
Li Kam Wa, P., J.M. Marsh, P.N. Robson, J.S. Roberts and N.J. Mason, 1985b, Proc. SPIE **578**, 110–116.
Liao, C., and G.I. Stegeman, 1984, Appl. Phys. Lett. **44**, 164–166.

Liao, C., G.I. Stegeman, C.T. Seaton, R.L. Shoemaker, J.D. Valera and H.G. Winful, 1984, Opt. Soc. Am. A **2**, 590–594.
Likhanskii, V.V., and B.P. Rysev, 1984, Sov. J. Quantum Electron. **14**, 19718–19720.
Litvak, A.G., and V.A. Mironov, 1968, Izv. Vuz. Radiofiz. **11**, 1911–1912.
Lomtev, A.I., 1981, JETP Lett. **34**, 60–63.
Lomtev, A.I., 1984, Opt. & Spectrosc. **55**, 656–657.
Lomtev, A.I., and L.G. Bolshinskii, 1985, Soc. Tech. Phys. Lett. **11**, 338–339.
Lomtev, A.I., and L.G. Bolshinskii, 1986, Ukr. Fiz. Zh. **31**, 34–36 (in Russian).
Love, O.J., G.H.C. New and M.A. Lauder, 1986, Opt. Commun. **59**, 177–182.
Lukosz, W., and V. Briguet, 1986, Light-induced desorption: A new mechanism for bistability in integrated optical devices, in: Optical Bistability III, eds H.M. Gibbs, P. Mandel, N. Peyghambarian and S.D. Smith (Springer, Berlin) pp. 87–90.
Lukosz, W., P. Pirani and V. Briguet, 1986, Bistability and self-pulsing using input couplers on nonlinear planar waveguides, in: Optical Bistability III, eds H.M. Gibbs, P. Mandel, N. Peyghambarian and S.D. Smith (Springer, Berlin) pp. 109–112.
Lukosz, W., P. Pirani and V. Briguet, 1987, Opt. Lett. **12**, 263–265.
Maier, A.A., 1982, Sov. J. Quantum Electron. **12**, 1490–1496.
Maier, A.A., 1984, Sov. J. Quantum Electron. **14**, 101–106.
Maier, A.A., N.Y. Serdynchenko, K.Y. Sitarskii and M.Y. Shcherbakov, 1987, Kwant. Elektr. **14**, 1157–1159.
Maker, P.D., and R.W. Terhune, 1965, Phys. Rev. A **137**, 801–818.
Mann, M., 1987, Diploma work (in preparation).
Maradudin, A.A., 1981, s-Polarised nonlinear surface polaritons, Z. Phys. B **41**, 341–344.
Maradudin, A.A., 1983, Nonlinear surface electromagnetic waves, in: Optical and Acoustic Waves in Solids – Modern Topics. Proc. 2nd Int. School on Condensed Matter Physics, ed. M. Borissov (World Scientific, Singapore) pp. 1972–2142.
Marcuse, D., 1974, in: Theory of Dielectric Optical Waveguides (Academic Press, New York).
Marcuse, D., 1980, Appl. Opt. **10**, 3130–3139.
Martinot, P., 1983, Optical bistability by resonant excitation of a surface plasmon, Thèse 3eme cycle (Université de Paris Sud, Orsay).
Martinot, P., S. Laval and A. Koster, 1984, J. Phys. **45**, 597–600.
Martinot, P., A. Koster and S. Laval, 1985, IEEE J. Quantum Electron. **QE-21**, 1140–1143.
Mihalache, D., 1985, Rev. Roum. Phys. **30**, 433–443.
Mihalache, D., 1986a, Solid State Commun. **58**, 125–127.
Mihalache, D., 1986b, Rev. Roum. Phys. **31**, 987–991.
Mihalache, D., and V.K. Fedyanin, 1983, Theor. & Math. Phys. **54**, 289–297.
Mihalache, D., and D. Mazilu, 1985, TM polarised nonlinear waves guided by asymmetric dielectric layered structures, Appl. Phys. B **37**, 107–113.
Mihalache, D., and D. Mazilu, 1986, Appl. Phys. B **41**, 119–123.
Mihalache, D., and D. Mazilu, 1986, Solid State Commun. **60**, 397–399.
Mihalache, D., and D. Mazilu, 1987a, Rev. Roum. Phys. **32**, 343–351.
Mihalache, D., and D. Mazilu, 1987b, Phys. Lett. A **122**, 381–384.
Mihalache, D., and D. Mazilu, 1987c, Solid State Commun. **63**, 215–217.
Mihalache, D., and H. Totia, 1984, Rev. Roum. Phys. **29**, 365–374.
Mihalache, D., and H. Totia, 1985, Solid State Commun. **54**, 175–177.
Mihalache, D., D. Mazilu and H. Totia, 1984a, Phys. Scr. **30**, 335–340.
Mihalache, D., R.G. Nazmitdinov and V.K. Fedyanin, 1984b, Phys. Scr. **29**, 269–275.
Mihalache, D., D. Mazilu and F. Lederer, 1986, Opt. Commun. **59**, 391–394.
Mihalache, D., G.I. Stegeman, C.T. Seaton, E.M. Wright, R. Zanoni, A.D. Boardman and T. Twardowski, 1987, Opt. Lett. **12**, 187–189.
Miyagi, M., and S. Nishida, 1972, Rep. Res. Inst. Electron. Commun. (Tohoku University) **24**, 53–67.

Miyagi, M., and S. Nishida, 1973, Rep. Res. Inst. Electron. Commun. (Tohoku University) **25**, 53–66.
Miyagi, M., and S. Nishida, 1974, Proc. IEEE **62**, 1284–1285.
Miyagi, M., and S. Nishida, 1975a, Electron. & Commun. Jpn. (USA) **58**, 104–110.
Miyagi, M., and S. Nishida, 1975b, Radio Sci. **10**, 19833–19838.
Moloney, J.V., 1985, IEEE J. Quantum Electron. **QE-21**, 1393–1398.
Moloney, J.V., 1987, Transverse effects in nonlinear optics, in: Proc. NATO ASI School, Lucca, Italy.
Moloney, J.V., J. Ariyasu, C.T. Seaton and G.I. Stegeman, 1986a, Opt. Lett. **11**, 315–317.
Moloney, J.V., J. Ariyasu, C.T. Seaton and G.I. Stegeman, 1986b, Appl. Phys. Lett. **48**, 19826–19828.
Montemayor, V.J., and R.T. Deck, 1985, J. Opt. Soc. Am. B **2**, 1010–1013.
Montemayor, V.J., and R.T. Deck, 1986, J. Opt. Soc. Am. B **3**, 1211–1220.
Nickerson, S.B., and T.W. Johnston, 1979, Phys. Lett. A **1973**, 20–22.
Nozaka, M., and S. Nishida, 1971, Rep. Res. Inst. Electron. Commun. (Tohoku University) **23**, 31–38.
Ogusu, K., 1987a, Opt. Commun. **63**, 380–384.
Ogusu, K., 1987b, Opt. & Quantum Electron. **19**, 65–72.
Ogusu, K., 1987c, Opt. Commun. **64**, 425–430.
Ortega, J.M., and W.C. Rheinboldt, 1970, Iterative Solution of Nonlinear Equations in Several Variables (Academic Press, New York).
Pardo, F., A. Koster, H. Chelli, N. Paraire and S. Laval, 1986, Bistable optical device using guided mode excitation in silicon on sapphire, in: Optical Bistability III, eds H.M. Gibbs, P. Mandel, N. Peyghambarian and S.D. Smith (Springer, Berlin) pp. 83–86.
Patela, S., H. Jerominek, C. Delisle and R. Tremblay, 1986, J. Appl. Phys. **60**, 1591–1594.
Peneva, Y.K., 1985, Bulg. J. Phys. **12**, 235.
Permyakov, V.A., 1985, Radiophys. & Quantum Electron. **28**, 101–107.
Pohl, D., 1970, Phys. Rev. A **5**, 1906–1909.
Ponath, H.E., 1979, Ann. Phys. **36**, 253–265.
Quail, J.C., and H.J. Simon, 1984, J. Opt. Soc. Am. B **1**, 317–319.
Reinisch, R., P. Arlot and G. Vitrant, 1985, Appl. Phys. Lett. **47**, 1248–1250.
Reinisch, R.P., and M. Neviere, 1986, Nonlinear Surface Polariton Interactions: Surface Enhanced Nonlinear Optical Effects, in: Electromagnetic Surface Excitations, eds R.F. Wallis and G.I. Stegeman, Springer Series on Wave Phenomena, Vol. 3 (Springer, Berlin) pp. 232–260.
Robbins, D.J., 1983, Opt. Commun. **47**, 309–312.
Romagnoli, M., and G.I. Stegeman, 1987, Opt. Commun. **64**, 343–346.
Rozanov, N.N., 1979, Opt. & Spectrosc. **47**, 335–339.
Sakakibara, T., and N. Okamoto, 1987, IEEE J. Quantum Electron. **QE-23**, 2084–2088.
Sarid, D., 1981, Appl. Phys. Lett. **39**, 19889–19891.
Sarid, D., R.T. Deck and J.J. Fasano, 1982, J. Opt. Soc. Am. **1972**, 1345–1347.
Satomura, Y., 1986, Tech. Res. Rep. Inst. Electron. Commun. Eng. Jpn. (in Japanese) 1974–1979.
Scheik, R., 1984, NTZZ Arch. (in German) **6**, 309–315.
Seaton, C.T., J.D. Valera, R.L. Shoemaker, G.I. Stegeman, J.T. Chilwell and S.D. Smith, 1984, Appl. Phys. Lett. **45**, 1162–1163.
Seaton, C.T., Xu Mai, G.I. Stegeman and H.G. Winful, 1985a, Opt. Eng. **24**, 593–599.
Seaton, C.T., G.I. Stegeman and H.G. Winful, 1985b, SPIE **578**, 143–149.
Seaton, C.T., G.I. Stegeman, W.M. Hetherington III and H.G. Winful, 1985c, Nonlinear third-order integrated optics, preprint.
Seaton, C.T., J.D. Valera, R.L. Shoemaker, G.I. Stegeman, J.T. Chilwell and S.D. Smith, 1985d, IEEE J. Quantum Electron. **QE-21**, 19774–19783.
Seaton, C.T., J.D. Valera, B. Svenson and G.I. Stegeman, 1985e, Opt. Lett. **10**, 149–150.
Shen, T.P., G.I. Stegeman and A.A. Maradudin, 1988, Appl. Phys. Lett. **52**, 1–3.

Shen, Y.R., 1982, in: The Principles of Nonlinear Optics (Wiley, New York).
Shivarova, A., and N. Dimitrov, 1985, Plasma Phys. & Controlled Fusion **27**, 219–224.
Shivarova, A., and M.Y. Yu, 1986, p-polarised nonlinear electromagnetic waves in cylindrical layered structures, in: Surface Waves in Plasmas and Solids, ed. S. Vukovic (World Scientific) pp. 634–637.
Sibilia, C., and M. Bertolotti, 1986, Nonlinear waveguide couplers, in: Optical Bistability III, eds H.M. Gibbs, P. Mandel, N. Peyghambarian and S.D. Smith (Springer, Berlin) pp. 105–108.
Sibilia, C., M. Bertolotti and D. Sette, 1984, Philos. Trans. R. Soc. (London) **313**, 361–362.
Silberberg, Y., and G.I. Stegeman, 1987, Appl. Phys. Lett. **50**, 19801–19803.
Sipe, J.E., and G.I. Stegeman, 1979, J. Opt. Soc. Am. **69**, 1676–1693.
Sitenko, A.G., V.N. Pavlenko and S.M. Revenchuk, 1981, Sov. J. Plasma Phys. **197**, 907–915.
Smith, P.W., 1974, Philos. Trans. R. Soc. London A **313**, 349–355.
Smith, P.W., and W.J. Tomlinson, 1984, IEEE J. Quantum Electron. **QE-20**, 30–36.
Stegeman, G.I., 1982a, IEEE J. Quantum Electron. **QE-18**, 1610–1618.
Stegeman, G.I., 1982b, IEEE Trans. Microwave Theory & Tech. **MTT-30**, 1598–1607.
Stegeman, G.I., and J.J. Burke, 1982, Appl. Phys. Lett. **41**, 906–908.
Stegeman, G.I., and C.T. Seaton, 1984, Opt. Lett. **9**, 235–237.
Stegeman, G.I., and C.T. Seaton, 1985a, Nonlinear surface polaritons, in: Phenomena at Surfaces, Interfaces and Superlattices, eds F. Nizzoli, Rieder and Willis (Springer, Berlin) pp. 266–275.
Stegeman, G.I., and C.T. Seaton, 1985b, J. Appl. Phys. **58**, R57–R79.
Stegeman, G.I., C.T. Seaton and H.G. Winful, 1984a, Philos. Trans. R. Soc. (London) **313**, 321–326.
Stegeman, G.I., J.D. Valera, C.T. Seaton, J.E. Sipe and A.A. Maradudin, 1984b, Solid State Commun. **52**, 293–297.
Stegeman, G.I., C.T. Seaton, J.T. Chilwell and S.D. Smith, 1984c, Appl. Phys. Lett. **44**, 19830–19832.
Stegeman, G.I., J. Ariyasu, C.T. Seaton, T.P. Shen and J.V. Moloney, 1985a, Appl. Phys. Lett. **47**, 1254–1256.
Stegeman, G.I., C.T. Seaton, J. Ariyasu, R.F. Wallis and A.A. Maradudin, 1985b, J. Appl. Phys. **58**, 2453–2459.
Stegeman, G.I., C.T. Seaton, J. Ariyasu, T.P. Shen and J.V. Moloney, 1986a, Opt. Commun. **56**, 365–368.
Stegeman, G.I., C.T. Seaton, W.M. Hetherington III, A.D. Boardman and P. Egan, 1986b, Nonlinear guided waves, in: Electromagnetic Surface Excitations, eds R.F. Wallis and G.I. Stegeman (Springer, Berlin) pp. 261–300.
Stegeman, G.I., E.M. Wright, C.T. Seaton, J.V. Moloney, T.P. Shen, A.A. Maradudin and R.F. Wallis, 1986c, IEEE J. Quantum Electron. **QE-22**, 977–983.
Stegeman, G.I., C.T. Seaton, C.N. Ironside, T. Cullen and A.C. Walker, 1987a, Appl. Phys. Lett. **50**, 1035–1037.
Stegeman, G.I., C.T. Seaton and C.N. Ironside, 1987b, Opt. Commun. **61**, 277–281.
Stegeman, G.I., E. Caglioti, S.Trillo and S. Wabnitz, 1987c, Proc. ECOC '87, 25–29.
Stegeman, G.I., E. Caglioti, S.Trillo and S. Wabnitz, 1987d, Opt. Commun. **63**, 281–284.
Stenflo, L., M.Y. Yu and I. Zhelyazkov, 1983, Beitr. Plasmaphyzik **23**, 621–623.
Stenflo, L., I. Zhelyazkov and M.Y. Yu, 1987, Beitr. Plasmaphyzik **27**, 1985–1986.
Svelto, O., 1974, Progress in Optics, Vol. 12, ed. E. Wolf (North-Holland, Amsterdam) pp. 3–51.
Thylen, L., E.M. Wright, G.I. Stegeman, C.T. Seaton and J.V. Moloney, 1986, Opt. Lett. **11**, 19739-19741.
Tomlinson, W.J., 1980, Opt. Lett. **5**, 323–325.
Tomlinson, W.J., J.P. Gordon, P.W. Smith and A.E. Kaplan, 1982, Appl. Opt. **21**, 2041–2051.
Trillo, S., and S. Wabnitz, 1986, Appl. Phys. Lett. **49**, 752–754.
Twardowski, T., 1988, Ph.D. Thesis (University of Salford, UK).

Ulrich, R., 1970, J. Opt. Soc. Am. **60**, 1337–1351.
Vach, H., C.T. Seaton, G.I. Stegeman and I.C. Khoo, 1984, Opt. Lett. **9**, 238–240.
Valera, J.D., C.T. Seaton, G.I. Stegeman, R.L. Shoemaker, Xu Mai and C. Liao, 1984, Appl. Phys. Lett. **45**, 1013–1015.
Valera, J.D., B. Svenson, C.T. Seaton and G.I. Stegeman, 1986, Appl. Phys. Lett. **48**, 573–574.
Varshney, R.K., M.A. Nehme, R. Srivastava and R.V. Ramaswamy, 1986, Appl. Opt. **25**, 3899–3902.
Vincent, P., N. Paraire, M. Neviere, A. Koster and R. Reinisch, 1985, J. Opt. Soc. Am. B **2**, 1106–1116.
Vitrant, G., and P. Ariot, 1987, J. Appl. Phys. **61**, 4744–4748.
Vladimirov, V.S., 1986, Sov. J. Plasma Phys. **12**, 552–556.
Vysotina, N.V., N.N. Rozanov and V.A. Smirnov, 1984a, Pis'ma Zh. Tekhn. Fiz. **10**, 1206–1210.
Vysotina, N.V., N.N. Rozanov and V.A. Smirnov, 1984b, Sov. Tech. Phys. Lett. **10**, 510–511.
Vysotina, N.V., N.N. Rozanov and V.A. Smirnov, 1986, Zh. Tekh. Fiz. **56**.
Wabnitz, S., E.M. Wright, C.T. Seaton and G.I. Stegeman, 1986, Appl. Phys. Lett. **49**, 838–840.
Wachter, C., U. Langbein and F. Lederer, 1987, Appl. Phys. B **42**, 161–164.
Walker, A.C., J.S. Aitchison, J.T. Chilwode, S.T.D. Ritchie and M. Rogers, 1986, Intrinsic optical bistability in a passive GaAlAs waveguide, in: Optical Bistability III, eds H.M. Gibbs, P. Mandel, N. Peyghambarian and S.D. Smith (Springer, Berlin) pp. 94–97.
Welford, K.R., J.R. Sambles and M.G. Clark, 1987, Liquid Crystals **2**, 91–105.
Wendler, L., 1983, Phys. Status Solidi B **118**, 205–210.
Wendler, L., 1986, Phys. Status Solidi B **135**, 19759–19774.
Wendler, L., and R.J. Haupt, 1986, J. Appl. Phys. **59**, 3289–3291.
Wright, E.M., G.I. Stegeman, C.T. Seaton and J.V. Moloney, 1986a, Appl. Phys. Lett. **49**, 435–436.
Wright, E.M., G.I. Stegeman, C.T. Seaton, J.V. Moloney and A.D. Boardman, 1986b, Phys. Rev. A **34**, 4442–4444.
Wysin, G.M., H.J. Simon and R.T. Deck, 1981, Opt. Lett. **6**, 50–52.
Yu, M.Y., 1983, Phys. Rev. A **28**, 1855–1856.
Zakharov, V.E., and A.B. Shabat, 1972, Sov. Phys.-JETP **34**, 62–66.
Zhao, Y., 1986, IEEE J. Quantum Electron. **QE-22**, 1012.
Zhelyazkov, I., T. Vodenicharova and M.Y. Yu, 1986, J. Plasma Phys. **36**, 143–154.
Zyss, J., 1985, J. Mol. Electron. **1**, 25–45.

CHAPTER 3

The Problem of Stability and Excitation of Nonlinear Surface Waves

N.N. AKHMEDIEV

F.V. Lukin Research and Development Institute of Physical Problems (Zelenograd), 103460 Moscow, USSR

Nonlinear Surface Electromagnetic Phenomena
Edited by
H.-E. Ponath and G.I. Stegeman

© *Elsevier Science Publishers B.V., 1991*

Contents

Introduction

Nonlinear surface waves propagating along the interface of linear and nonlinear media have a number of novel extraordinary properties which attracted attention of many investigators [see, e.g., Agranovich et al. 1980, Tomlinson 1980, Litvak and Mironov 1968, Akhmediev 1982, 1983, Seaton et al. 1985, Boardman and Egan 1985, Lomtev 1981, Lederer et al. 1983, and Mihalache et al. 1984; more complete lists of references can be found in the other chapters of this book and in the excellent review articles of Seaton et al. (1985) and Boardman and Egan (1985)]. In recent years, in general, the dispersion relations for different types of structures have been considered and the possible modes of these structures have been analyzed.

For any type of nonlinear waves the problem of their stability is of a special interest. In the case of nonlinear waves in an unbounded medium this problem have been solved (e.g., Zakharov and Rubenchik 1973, Kolokolov 1973, Vakhitov and Kolokolov 1973). Unsufficient attention has been given to this question in the case of nonlinear surface waves and the problem of their stability to date cannot be considered as fully solved. Nevertheless, there are a number of approaches to the problem both using numerical simulations (Akhmediev et al. 1984, 1985, Moloney et al. 1986a, Leine et al. 1986), and analytical methods (Ariyasu et al. 1986, Moloney et al. 1986b), which we shall try to illuminate to some extent. We shall consider both the waves guided by a single interface and by a symmetric waveguide. However, we shall restrict ourselves to the consideration of the simplest case of s-polarized waves, for which the vector problem is reduced to the scalar one.

Complementary to the problem of stability is the problem of excitation of nonlinear surface waves by external beams. Here we should go beyond the frames of the linear perturbation theory and should seek solutions of the evolutional equation at large lengths. A certain particular solution of this problem by means of the numerical analysis will be also considered in this chapter.

1. Formulation of the problem of stability

In this chapter we shall consider monochromatic fields $\exp(-i\omega t)$ only and the stability within the limits of a stationary problem. Let us consider a layered structure characterized by the dielectric constant quadratically dependent on a wave field

$$\varepsilon(z, |E|^2) = \varepsilon(z) + \alpha(z)|E|^2, \qquad (1)$$

the z-axis of the Cartesian system is assumed to be normal to the plane of layers and $\varepsilon(z)$ and $\alpha(z)$ are constant inside each layer. In particular, $\alpha(z) = 0$ when the medium is linear.

Let us assume that the direction of the wave propagation coincides with the x-axis. Then the Maxwell equations for s-polarized waves are reduced to the scalar equation for the field component $E = E_y$,

$$\Delta E + \varepsilon(z, |E|^2) E = 0, \tag{2}$$

where the coordinates are normalized by the factor ω/c. It is also convenient to normalize the fields by the factor $\alpha_0^{1/2}$, where α_0 is the nonlinearity constant of one of the layers. Further, we can write the field in the form

$$E_y(x, y, z) = \alpha_0^{1/2} A(x, y, z) \exp(inx), \tag{3}$$

where $A(x, y, z)$ is a slowly varying field envelope, n is the effective refractive index.

Substituting eq. (3) into eq. (2) we obtain the following equation for the function $A(x, y, z)$,

$$2inA_x + A_{xx} + A_{yy} + A_{zz} - \gamma^2(z) A + \beta(z)|A|^2 A = 0, \tag{4}$$

where $\gamma^2(z) = n^2 - \varepsilon(z)$, $\beta(z) = \alpha(z)/\alpha_0$. In a number of cases we can restrict ourselves to the solution of a parabolic equation while neglecting the second derivative A_{xx} because the function $A(x, y, z)$ can be regarded as slowly varying with x. Moreover, we can restrict ourselves to a two-dimensional problem and consider the fields independent of y ($A_{yy} = 0$) at the initial step of the stability investigation. Then eq. (4) is simplified,

$$2inA_x + A_{zz} - \gamma^2(z) A + \beta(z)|A|^2 A = 0. \tag{5}$$

The analysis of the field evolution can be carried out numerically on the basis of eq. (5). This equation has been solved numerically for specific structures (Akhmediev et al. 1984, 1985, Moloney et al. 1986a,b, Leine et al. 1986, Ariyasu et al. 1986).

The first integral of eq. (5) is of decisive importance in the investigation of stability

$$I = \int_{-\infty}^{\infty} |A(x, z)|^2 \, dz. \tag{6}$$

This integral coincides with the energy flow along the x-axis within a constant factor.

Until recently, the primary interest of investigators has been focused on looking for the steady-state solutions $A(x, y, z) = A_0(z)$ of eq. (5) for particular layered structures. If the steady-state solution $A_0(z)$ is known, then the analytical investigation of its stability is possible on the basis of eq. (4). An investigation of stability of waves propagating along the interface between a linear and nonlinear medium has been carried out by Rosanov (Vysotina et al. 1984, 1987).

Let us seek the solution of eq. (4) in the form

$$A(x, y, z) = [A_0(z) + \mu f(x, y, z)] \exp(\mathrm{i}nx), \tag{7}$$

where $f(x, y, z)$ is a perturbation function of the steady-state solution $A_0(z)$, μ is a small parameter. When we substitute eq. (7) into eq. (4) and retain only the terms of first order, we obtain the following equation for the perturbation function,

$$2\mathrm{i}nf_x + f_{xx} + f_{yy} + f_{zz} - \gamma^2(z) f + \beta(z) A_0(z)^2 (2f + f^*) = 0, \tag{8}$$

where the asterisk denotes complex conjugate. We shall consider the y-dependence of the perturbation function of a modulational type, so that the function $f(x, y, z)$ can be written in the form

$$f(x, y, z) = \tfrac{1}{2}[(u + v) \exp(\delta x + \mathrm{i}\kappa y) + (u^* + v^*) \exp(\delta^* x - \mathrm{i}\kappa y)], \tag{9}$$

where u and v are functions of z only. Substituting eq. (9) into eq. (8), we obtain the set of coupled differential equations for these functions

$$\begin{aligned} u_{zz} - a^2(z) u + 3\beta(z) A_0^2(z) u &= -\mathrm{i}\xi' v, \\ v_{zz} - a^2(z) v + \beta(z) A_0^2(z) v &= -\mathrm{i}\xi' u, \end{aligned} \tag{10}$$

where

$$a^2(z) = \gamma^2(z) + \kappa^2 - \delta^2, \qquad \xi' = 2n\delta.$$

Let us note some of the properties of the functions u and v and the eigenvalue δ, which will be useful in what follows. The set of equations (10) is equivalent to the equation of the fourth order with the real eigenvalue δ^2. So the eigenvalue $\pm\delta$ of eqs. (10) is either real or imaginary. In the case when $\pm\delta$ is real we can make the function u real ($u = u_0$) and the function v imaginary ($v = \mathrm{i}v_0$) by choosing the proper phase factor. Then for $x = 0$ and $\kappa = 0$ the perturbation function takes the form

$$f(x, y, z) = f_0(z) = u_0(z) + \mathrm{i}v_0(z). \tag{11}$$

The function $u_0(z)$ is then orthogonal to $A_0(z)$. Indeed, if we substitute eq. (7) into eq. (6) and take into account that the integral I is independent of x then we get, with the use of eq. (9),

$$\int_{-\infty}^{\infty} A_0(z) u_0(z) \, \mathrm{d}z = 0.$$

It is important that two complex conjugated eigenfunctions f_0 and f_0^* correspond to two eigenvalues $+\delta$ and $-\delta$, respectively, so if there is a growing perturbation $f(x, y, z)$, there is, obviously, the solution $f^*(x, y, z)$ which converges towards $A_0(z)$.

If we take $\delta^2 = 0$ and $\kappa = 0$ then the solution of the second equation of the set (10) will be $v_0 = A_0$ for any value of n, while the solution of the first equation of the set (10) will be the function $u_0 = (dA_0/dz)$ but for discrete values of n only, as the second derivative (d^2A_0/dz^2) is a discontinuous function.

If we take a parabolic equation and the two-dimensional perturbations we shall have $a^2(z) = \gamma^2(z)$, as $\kappa = 0$ and δ^2 can be neglected being small. We shall obtain the same result simply by taking $\delta^2 = \kappa^2$. This means that the parabolic equation yields the exact solution in the case of three-dimensional perturbations of a particular form when $\delta^2 = \kappa^2$ (Vysotina et al. 1987). Generally, when working out the solution of the parabolic equation the applicability conditions should be tested, i.e., $A_{xx} \ll 2inA_x$ and $A_{xx} \ll \gamma^2(z)A$ (or $\delta \ll 2n$ and $\delta^2 \ll \gamma^2(z)$). Fortunately, when $a^2 = \gamma^2$, the exact solution of eqs. (10) was obtained by Vysotina et al. (1987), and this solution makes analytical calculations possible, at least when using the parabolic equation and two-dimensional perturbations. Generally, the set of differential equations (10) can be solved numerically. In what follows, we shall consider two specific structures: the single interface of linear and nonlinear media and the symmetric waveguide with a nonlinear substrate and cladding.

2. Nonlinear surface waves on a single interface

Let the plane $z = 0$ be an interface of linear and nonlinear media. Then for $\varepsilon(z)$ and $\beta(z)$ in eq. (4) we have

$$\varepsilon(z) = \begin{cases} \varepsilon_L, & z < 0, \\ \varepsilon_0, & z > 0, \end{cases} \qquad \beta(z) = \begin{cases} 0, & z < 0, \\ 1, & z > 0. \end{cases} \tag{4a}$$

Equation (5) has a steady-state solution of the form of a nonlinear surface wave (Tomlinson 1980, Litvak and Mironov 1968); when $\Delta\varepsilon = \varepsilon_L - \varepsilon_0 > 0$

$$A(z) = \begin{cases} (2\Delta\varepsilon)^{1/2} \exp(\gamma_L z), & z < 0, \\ 2^{1/2}\gamma_{NL} \cosh^{-1}(\gamma_{NL}(z - z_0)), & z > 0, \end{cases} \tag{12}$$

where $\gamma_L = (n^2 - \varepsilon_L)^{1/2}$, $\gamma_{NL} = (n^2 - \varepsilon_0)^{1/2}$ and the position of the field maximum is determined by

$$z_0 = \frac{1}{2\gamma_{NL}} \ln \frac{\gamma_{NL} + \gamma_L}{\gamma_{NL} - \gamma_L}$$

(note that this function is a single-valued function of n). In an analysis of the stability, the convenient parameter of a nonlinear surface wave is $\eta = \tanh z_0 = \gamma_L/\gamma_{NL}$, which varies from 0 to 1 when $\varepsilon_L < n < \infty$.

For the steady-state solution (12) the integral (6) becomes

$$I(n) = \frac{\Delta\varepsilon}{\gamma_L} + 2(\gamma_L + \gamma_{NL}) = \frac{(\Delta\varepsilon)^{1/2}(1 + \eta)^2}{\eta(1 - \eta^2)^{1/2}}. \tag{13}$$

Figure 1 shows this integral as a function of n for $\varepsilon_0 = 2.647$ and $\varepsilon_L = 2.674$. The integral I has a minimum at a certain n_{cr} which can be easily calculated from eq. (13),

$$n_{cr} = \left(\frac{4\varepsilon_L - \varepsilon_0}{3}\right)^{1/2}.$$

For this value of n we have $I(n_{cr}) = 3[3(\varepsilon_L - \varepsilon_0)]^{1/2}$, and the parameter $\eta = 0.5$. The numerical value of n_{cr} for the case under consideration is $n = 1.638$.

Kolokolov (1973) has shown that the stability of a self-focused solution in an unbounded nonlinear medium is determined by the sign of the derivative $\mathrm{d}I/\mathrm{d}n$. It turns out that the same criterion of stability takes place for the interface of linear and nonlinear media. This result we shall discuss in detail below.

First let us consider the case when $\kappa^2 = \delta^2$. The set of equations (10) in this case can be rewritten as

$$u'' - u + \frac{6}{\cosh^2 \tilde{z}} u = -\mathrm{i}\xi v,$$

$$v'' - v + \frac{2}{\cosh^2 \tilde{z}} v = -\mathrm{i}\xi u, \qquad (14)$$

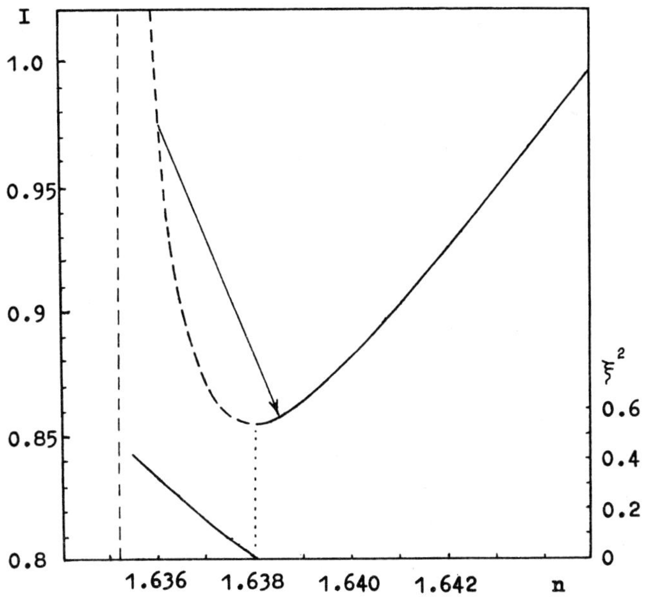

Fig. 1. The integral I (the upper curve) and the square of the normalized instability growth rate ξ versus n. The branch of the unstable nonlinear surface wave is denoted by the dashed line.

where $\tilde{z}=\gamma_{NL}(z-z_0)$, $\xi=\xi'/\gamma_{NL}^2$, and primes denote the derivatives with respect to $\tilde{z}$. The solutions of eqs. (14) which decay as $\tilde{z}\to\infty$ are given by (Vysotina et al. 1987)

$$u=C_1\exp(-p\tilde{z})(-\mathrm{i}\xi+2p\tanh\tilde{z}+2\tanh^2\tilde{z})$$
$$+C_2\exp(-p\tilde{z})(\mathrm{i}\xi+2p\tanh\tilde{z}+2\tanh^2\tilde{z}),$$
$$v=C_1\exp(-p\tilde{z})(2-\mathrm{i}\xi+2p\tanh\tilde{z})$$
$$-C_2\exp(-p\tilde{z})(2+\mathrm{i}\xi+2p\tanh\tilde{z}), \qquad (15)$$

where $p=(1-\mathrm{i}\xi)^{1/2}$, C_1 and C_2 are constants to be determined from the boundary conditions.

In a linear medium the set of equations (10) takes the form

$$u''-u\gamma_L^2/\gamma_{NL}^2=-\mathrm{i}\xi v,$$
$$v''-v\gamma_L^2/\gamma_{NL}^2=-\mathrm{i}\xi u, \qquad (16)$$

where for the convenience of further calculations we have chosen the variable $\bar{z}=\gamma_{NL}z$ as an argument of the functions. The solutions of eqs. (16), decaying as $\bar{z}=-\infty$, are

$$u=A_1\exp(s\bar{z})+A_2\exp(s^*\bar{z}),$$
$$v=A_1\exp(s\bar{z})-A_2\exp(s^*\bar{z}), \qquad (17)$$

where $s=(\gamma_L^2/\gamma_{NL}^2-\mathrm{i}\xi)^{1/2}$ and the constants A_1 and A_2 are determined from the boundary conditions.

Matching the functions u and v and their first derivatives at $z=0$ we obtain a set of four equations for the coefficients A_i and C_i. The dispersion relation for ξ^2 is obtained by equating to zero the determinant of the coefficients in these equations:

$$|F_3-F_1^*-2s^*F_2^*|^2-|F_3+F_1^*+2s(1-\eta^2)|^2=0, \qquad (18a)$$

or, after some rearrangements,

$$s^2((1-\eta^2)^2-F_2^2)+\mathrm{Re}[s(F_2(F_3-F_1^*)+(1-\eta^2)(F_1+F_3^*))+F_1F_3]=0, \qquad (18b)$$

where

$$F_1=-\mathrm{i}\xi p-2p^2\eta^2+2p\eta^2,$$
$$F_2=(p-\eta)^2,$$
$$F_3=(2-\mathrm{i}\xi)p^*-2p^2\eta-4p^*\eta^2+4\eta^3. \qquad (19)$$

It is convenient to take $C_2=C_1^*$ and $A_2=A_1^*$ in calculating eigenfunctions. In this case $u=u_0$ and $v=\mathrm{i}v_0$.

Equation (18) has a root $\xi_0 = 0$ for any value of η. The mode $u_0 = 0$ and $v_0 = A_0(z)$ which corresponds to this eigenvalue is indifferently stable. It can be shown by means of numerical methods, or by a power series expansion up to second order in ξ^2, that, in addition to $\xi_0 = 0$, eq. (18) has a root $\xi_1(\eta)$, which is real for $0 < \eta < 0.5$ and imaginary for $0.5 > \eta > 1$. The value $\eta = 0.5$ corresponds to the critical value of the effective refractive index n_{cr} for which $dI/dn = 0$. Thus, in the two-dimensional problem for nonlinear surface waves on a single interface Kolokolov's criterion is valid, viz. the nonlinear surface wave is stable for $dI/dn > 0$ and unstable for $dI/dn < 0$. The same result has been obtained by Akhmediev et al. (1984, 1985) by a direct numerical integration of the set of equations (10) when $a^2 = 1$.

To illustrate the above-mentioned points, fig. 1 plots the function $\xi_1(n)$ together with the function $I(n)$. The true growth rate δ is related with ξ_1 by $\delta = \xi_1 \gamma_{NL}^2/2n$, so the condition of applicability of the parabolic equation $\delta \ll 2n$ in this case takes the form $\xi_1 \ll 4n^2/\gamma_{NL}^2$. In the most unfavourable situation, when $n^2 \approx \varepsilon_L$, we have $\xi_1 \ll 4\varepsilon_L/\Delta\varepsilon$, so even if $\Delta\varepsilon \approx \varepsilon_L$ this condition is fulfilled, to say nothing of the case when $\Delta\varepsilon \ll \varepsilon_L$ or $n \to n_{cr}$. The second condition $\delta^2 \ll n^2 - \varepsilon_L$ imposes certain limitations on the range of n: $n^2 - \varepsilon_L \gg \frac{1}{4}(\Delta\varepsilon/\varepsilon)^2 \xi_1^2$.

Vysotina et al. (1987) have investigated also the stability of nonlinear surface waves in the three-dimensional problem beyond the limits of the parabolic equation. They have analyzed numerically the set of equations (10) for arbitrary values of κ and δ. The results are shown in fig. 2. As it is seen from this figure

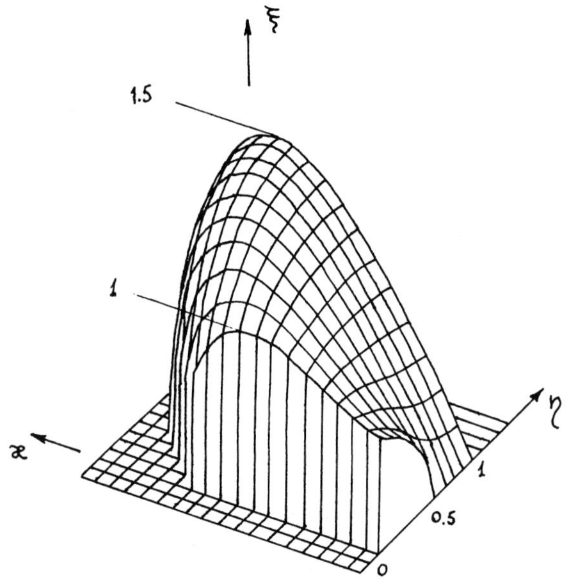

Fig. 2. The normalized instability growth rate ξ versus κ and η (Vysotina et al. 1987).

there are real nonzero values of the growth rate ξ. Therefore, the plane nonlinear surface waves are always unstable with respect to three-dimensional perturbations. If there are any arbitrarily small perturbations, the nonlinear surface waves will decay into separate beams propagating away along the surface and the distance between them will be $\sim\kappa$.

The region of unstability of the nonlinear surface waves in the (η, κ)-plane has been calculated by Vysotina et al. (1984) and is shown in fig. 3. The upper bound of the unstability with respect to κ satisfies to the relation

$$3[2\eta-(1+4\kappa^2/B)^{1/2}](1-\eta^2)+[(1+4\kappa^2/B)^{1/2}+(\eta^2+4\kappa^2/B)^{1/2}] \times [4\kappa^2/B+3\eta^2-3\eta(1+4\kappa^2/B)^{1/2}]=0,$$

where $B=2\gamma_{\mathrm{NL}}^2/\varepsilon_0$. For any η there exists a limiting value of κ, above which the nonlinear surface wave is stable. For any η there exists also the value of κ for which the growth rate is a maximum. The maximum growth rate increases monotonously with increasing η.

The results for ξ obtained in the approximation $\alpha^2=\delta^2$ correspond to the intersection of the surface $\xi(\kappa, \eta)$ in fig. 2 with the surface $\kappa^2=\gamma_{\mathrm{NL}}^2\xi^2/4n^2$. For small $\Delta\varepsilon/\varepsilon$ this surface is close to the plane $\kappa=0$, so the curve $\xi(n)$ in fig. 1 (when reconstructed as a function of η) almost coincides with the curve $\xi(\kappa=0, \eta)$ in fig. 2.

The length of the decay of the nonlinear surface waves into the separate beams can be estimated as $L_c \approx \lambda/(2\pi\delta) \approx n\lambda/(\pi\xi\gamma_{\mathrm{NL}}^2) \approx \varepsilon\lambda/(\pi\xi\Delta\varepsilon)$. For $\xi \simeq 1$ and $\Delta\varepsilon/\varepsilon=10^{-4}$ we have $L_c=0.3$ cm. The decay length decreases with increasing $\Delta\varepsilon/\varepsilon$.

The circumstance that the plane nonlinear surface waves of eq. (12) are always unstable does not mean, however, that the stable stationary field distribution

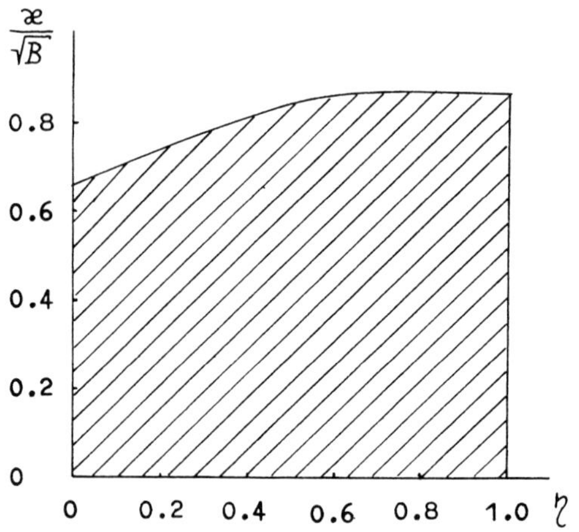

Fig. 3. The region of unstability of nonlinear surface waves in the (κ, η)-plane (Vysotina et al. 1984).

localized at the interface does not exist at all in the three-dimensional problem. The stationary solution

$$E_y(x, y, z) = \sqrt{\alpha} A(y, z) \exp(inx),$$

of the three-dimensional wave equation

$$A_{yy} + A_{zz} - \gamma^2(z) A + \beta(z) \frac{A}{1 + \mu A^2} = 0, \tag{20}$$

with a saturation of the nonlinear dielectric constant was found by Akhmediev et al. (1988a). The quantity μ is a saturation parameter. The analytical solution of eq. (20) does not, apparently, exist and numerical simulations of solutions of eq. (20) of the form of nonlinear surface waves were carried out in this work. Such three-dimensional solutions were called stripe nonlinear surface waves.

In numerical calculations the parameters $\varepsilon_L = 2.647$ and $\varepsilon_0 = 2.674$ were used. The values of the saturation parameter used were $\mu = 0$ (the cubic nonlinearity), 0.3, 1 and 3.

The example of the field distribution in the (y, z)-plane is given in fig. 4a. The lines of the equal values of the function $A(y, z)$ normalized by its maximum value, A_{max}, are shown in fig. 4b for $n = 1.6355$, $\mu = 0$. We can see from this figure that the maximum value of the field of the self-focused solution is located inside the nonlinear medium, and the field has no cylindrical symmetry due to the presence of the interface. The calculations show that with increasing n at fixed μ the field distribution is sharpened and approaches the self-focused solution with

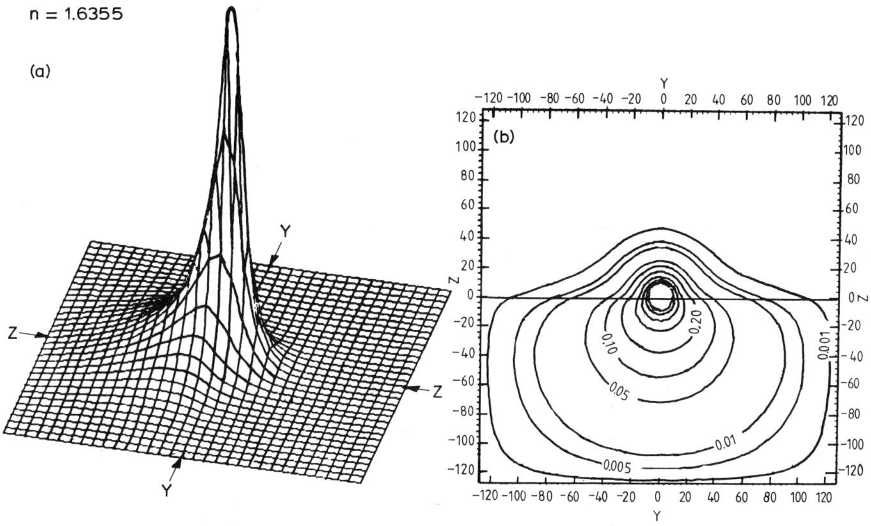

Fig. 4. The field distribution of the nonlinear surface wave (a) and the lines of the equal values of the function $A(y, z)$ normalized by the maximum value of the function (b) for $n = 1.6355$ and $\mu = 0$.

the cylindrical symmetry. Moreover, the energy stored in the field in the nonlinear medium increases with increasing n. This process is seen from fig. 5, where the relative energy fraction in the nonlinear medium versus n is plotted. Qualitatively the same dependence takes place in the case of the two-dimensional nonlinear surface waves.

The behaviour of the function $A(y, z)$ as a function of n in the case of saturated nonlinearity is qualitatively the same as when $\mu = 0$. However, when n is a constant, the transverse dimensions of the beams and the maximum field amplitude increase with increasing μ. As an example, the field distribution and the lines of the equal values of the function $A(y, z)$ for $n = 1.67$ at $\mu = 0$ and $\mu = 3$ are shown in figs. 6 and 7, respectively. The fraction of the energy in the nonlinear medium increases also with increasing μ (see fig. 5).

The dependences of the energy integral

$$I = \int_{-\infty}^{\infty} \int_{-\infty}^{\infty} |A|^2 \, dy \, dz$$

on n for different values of μ are shown in fig. 8 by the solid curves. The energy integral increases as n approaches $\sqrt{\varepsilon_L}$, as the energy in the linear medium grows rapidly $\gamma^2(z < 0) \to 0$. All the curves, therefore, have a common vertical asymptote, $n = \sqrt{\varepsilon_L}$.

For the case of a cubic nonlinearity the energy integral decreases monotonously with increasing n up to a certain value of I, $I = I_0$, which corresponds to the energy integral for the self-focused solution in an infinite nonlinear medium. Indeed, as was shown above, with increasing n the field is concentrated almost

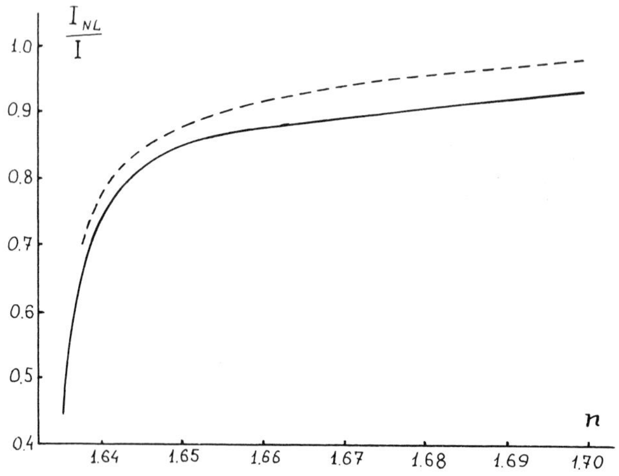

Fig. 5. The dependence of the relative fraction of the energy in the nonlinear medium on the effective refractive index n for $\mu = 0$ (solid line) and $\mu = 3$ (dashed line).

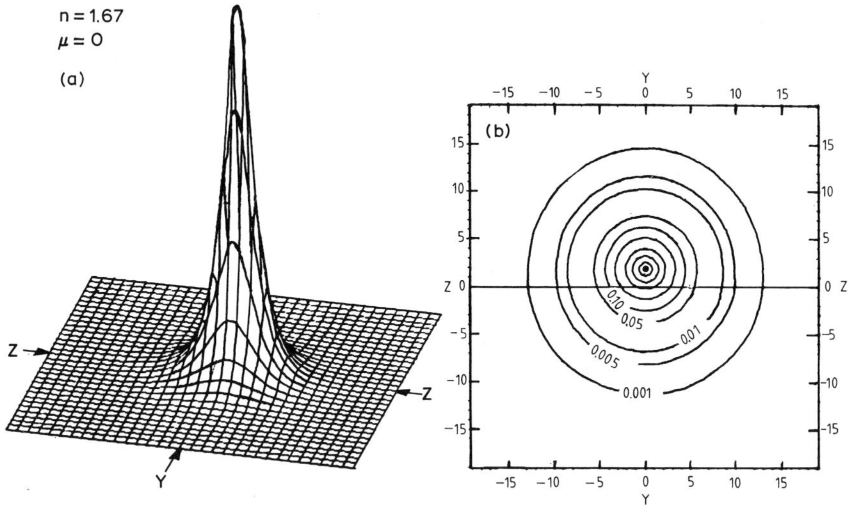

Fig. 6. The same as in fig. 4 for $n = 1.67$ and $\mu = 0$.

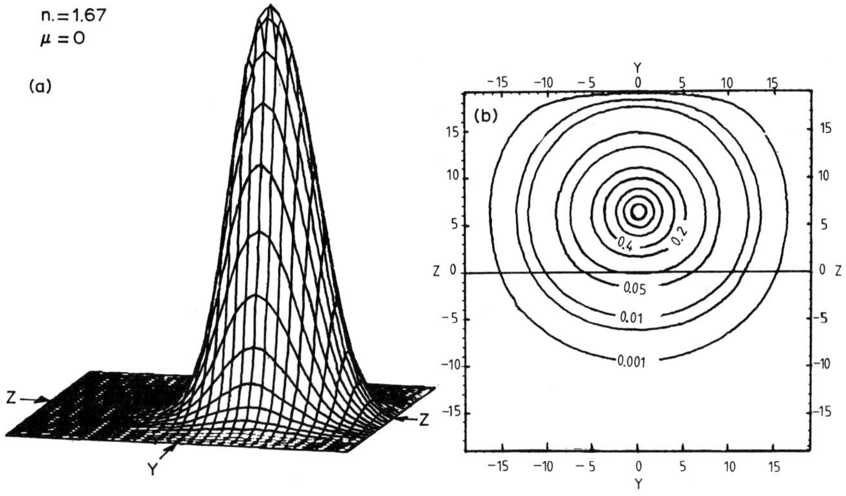

Fig. 7. The same as in fig. 4 for $n = 1.67$ and $\mu = 3$.

entirely in the nonlinear medium. The energy integral I_0 is independent of n for the cubic nonlinearity and its value is an asymptote for the curve $I(n)$.

The energy integral I for the self-focused solution in the infinite nonlinear medium increases monotonously for $\mu \neq 0$ with increasing n. These dependences are shown in fig. 8 by the dashed lines for $\mu = 0.3$, 1 and 3. These lines are asymptotes for the curves $I(n)$ in the case of the stripe nonlinear surface waves.

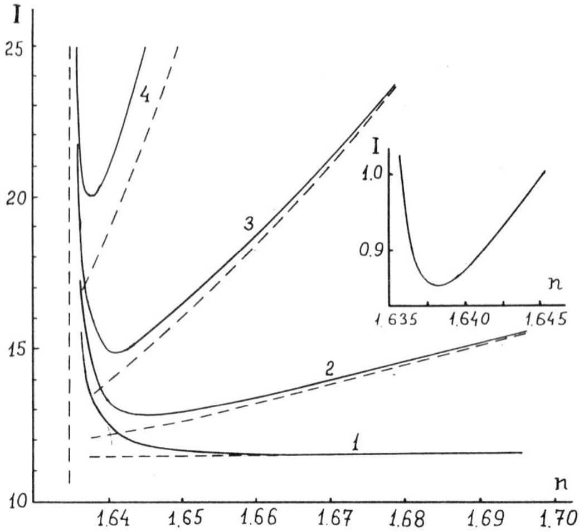

Fig. 8. The dependence of the energy integral of the stripe nonlinear surface waves (solid lines) on the effective refractive index n for different values of the saturation parameter: $\mu = 0$ (curve 1); 0.3 (curve 2); 1 (curve 3); 3 (curve 4). Inset: the plot of I versus n for the two-dimensional nonlinear surface waves ($\mu = 0$).

On the whole, the curves $I(n)$ for the stripe nonlinear surface waves at saturated nonlinearity have the same form as in the case of the two-dimensional nonlinear surface waves. For a comparison the curve presented in fig. 2 is shown again in the insert of fig. 8.

Any special investigations into the stability of the stripe nonlinear surface waves have not been performed because of a great amount of calculations required for such a problem. It has been shown previously that the two-dimensional nonlinear surface waves are stable when $\mathrm{d}I/\mathrm{d}n > 0$, and are unstable when $\mathrm{d}I/\mathrm{d}n < 0$. Such a stability criterion can be broken when there are several branches of the dispersion curves of nonlinear waves. This situation occurs for plane waveguides and will be considered later. Only one branch of the dispersion curve exists in the case of a single interface under consideration here. So we can suppose that the abovementioned stability criterion is valid for the stripe nonlinear surface waves. From fig. 8 we can see that the stripe nonlinear surface waves are always unstable for the cubic nonlinearity. The curves $I(n)$ at $\mu \neq 0$ have descending as well as ascending branches, exactly as in the case of the two-dimensional nonlinear surface waves. So the branches of the stable as well as unstable stripe nonlinear surface waves exist in this case.

In connection with the above speculations it should be emphasized that although the nonlinear surface waves in the two-dimensional problem are, in some sense, the convenient idealization, we can obtain for them plenty of analytical results, which is of great importance by itself. As for the possibility to extend these results to the three-dimensional (stripe) nonlinear surface waves this is to be the subject of further investigations (see, e.g., Akhmediev et al. 1988a).

3. Evolution of nonlinear surface waves beyond the framework of the linear approximation

Using the linear approximation of the theory of stability, we can calculate the instability growth rate δ which is very important for initial estimations. However, the problem of a further evolution, when we cannot restrict ourselves to the terms linear in μ in eq. (7), is of no less importance. So far this problem has been solved only on the basis of the parabolic equation and for two-dimensional perturbations (Akhmediev et al. 1985).

In principle, some predictions can be made on the basis of the linear approximation if the form of the functions $u_0(z)$ and $v_0(z)$ is known. Indeed, the addition μu_0 to $A_0(z)$ shows us how the nonlinear surface wave field distribution varies at initial moments of evolution, while the quantity $\mu A_0(\mathrm{d}v_0/\mathrm{d}z)$, which is proportional to the z-component of the Poynting vector, shows us the direction of the subsequent redistribution of the energy in the cross-section of the nonlinear surface wave. For a particular sign of $f(z)$ in eq. (7) we would be equally justified in assigning the opposite sign to the parameter μ. By virtue of the asymmetry of the problem with respect to the sign of z, these two choices lead to different final results for a given initial perturbation mode $f(z)$.

As an example, we show in fig. 9 the perturbation mode u_0, v_0 in the case when $n = 1.636$ for the above-used values of the dielectric constants. The growth rate

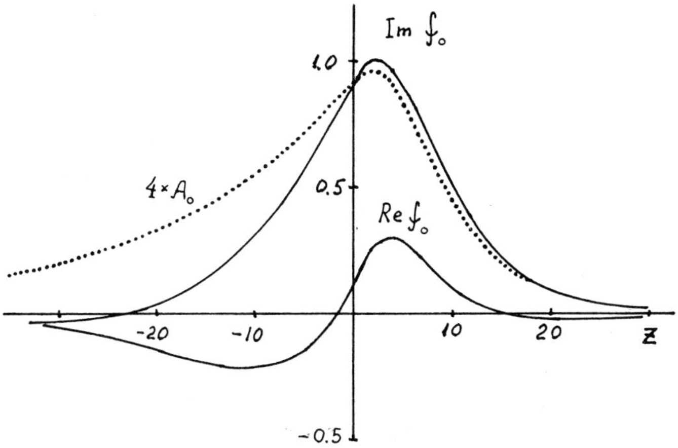

Fig. 9. The real and imaginary parts of the perturbation mode $f_0 = u_0 + iv_0$ of the nonlinear surface waves for $n = 1.636$, normalized by the maximum value of the imaginary part. The stationary solution $A_0(z)$ is shown (in the scale 1:4) by points.

here is $\delta = 0.00412\ldots$. The field distribution of the nonlinear surface wave is shown in the same figure by the dotted line. We see from this figure that for negative μ the field of the nonlinear surface wave will move from the nonlinear medium and will diffuse, while for positive μ the field will be concentrated near its maximum. This conclusion is confirmed by the numerical solution of eq. (5) with the initial condition $A = A_0 + \mu f_0(z)$.

If we choose the negative sign for μ, then we obtain that after the initial stage of exponential growth of the perturbation mode, the nonlinear surface wave is ejected into the linear medium as a result of the further evolution. It propagates away from the interface in the form of a beam at an angle greater than the critical angle for the total reflection. This process is illustrated in fig. 10a. We note, however, that the larger fraction of the energy of the nonlinear surface wave is converted only into the energy of the beam. As the beam emerges from the nonlinear medium a small fraction of its energy is reflected backward as can be clearly seen from curve 5 in fig. 10a. The energy of the beam reflected into the nonlinear medium is about 2–3% of the total energy.

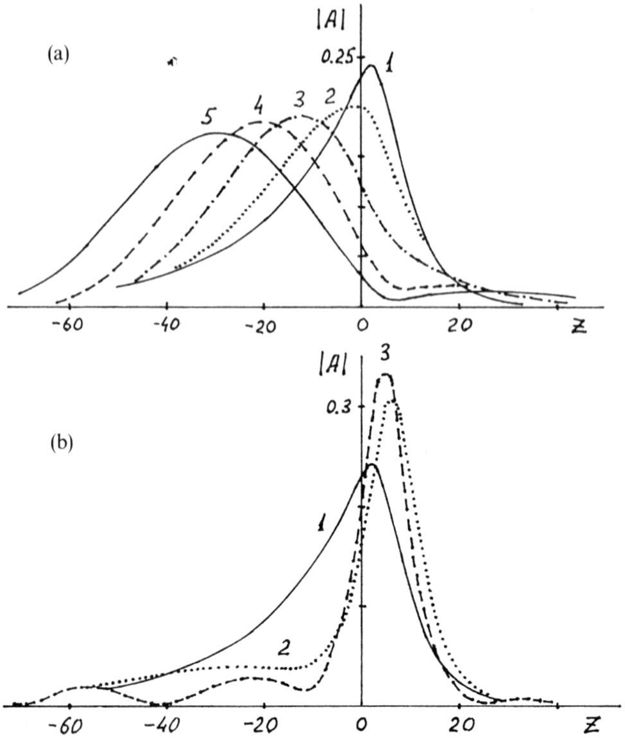

Fig. 10. Decay of the nonlinear surface wave of the unstable branch ($n = 1.636$) for different signs of the initial perturbation. (a) $\mu < 0$: (1) $x = 0$, (2) 900, (3) 1200, (4) 1400, (5) 1600. (b) $\mu > 0$: (1) $x = 0$, (2) 1600, (3) 3300.

When we choose the positive sign for μ, the growth of the perturbations leads to a shift of the field energy into the nonlinear medium, with the subsequent excitation of the nonlinear surface waves of the stable branch; simultaneously, a small fraction (3–5%) of the energy splits off into the energy of the beam propagating away from the interface into the nonlinear medium. This process is illustrated in fig. 10b. Curve 2 shows the part of the beam which initially splits off into the linear medium. The stable nonlinear surface wave is excited with a perturbation which oscillates along the x-axis. As the nonlinear surface wave propagates along the interface, during each oscillation a small fraction of its energy splits off into the beam propagating away from the interface. In curve 3 of fig. 10b we can see a sequence of two such beams. After splitting off the amplitude of the oscillations of the perturbation decreases and the energy of the next beam split off turns out to be substantially smaller. Thus, as a result of the evolution of the perturbation with the positive μ the nonlinear surface wave of the unstable branch is transformed into the nonlinear wave of the stable branch. The process is shown schematically by the arrow in fig. 1.

It is clear that for random initial perturbations after some transients the mode with a nonzero growth rate is formed and the decay of the nonlinear surface wave of the unstable branch with probability 0.5 proceeds in one of the two manners described above. So the considerations carried out above practically cover all the possibilities.

It is of interest to solve the three-dimensional problem of nonlinear surface wave evolution. Unfortunately, such calculations require large expenditures of computer time and, as far as we know, have not been performed. It should be noted that in the three-dimensional case in the numerical calculations nonlinearity with quadratic dependence on the field can lead to an infinite growth of the field at the self-focusing point (Konno and Suzuki 1979).

4. *On the stability of modes in a symmetric waveguide*

Let us now turn to the analysis of stability of nonlinear waves in the symmetric layered structure shown in fig. 11. There are asymmetric modes in such a structure (Akhmediev 1982) which complicate considerably the analysis of stability, and the criterion $\mathrm{d}I/\mathrm{d}n > 0$ is broken in some cases. We shall show this breakdown in the example of the TE_0-mode only.

The stationary solution of eq. (5) for the structure under consideration can be written in the form:

$$A_0(z) = \begin{cases} \sqrt{2}\gamma_{NL} \cosh^{-1}(\gamma_{NL}(z - z_1)), & z < -d, \\ C \cosh(\gamma_L(z - d_0)), & |z| < d, \\ \sqrt{2}\gamma_{NL} \cosh^{-1}(\gamma_{NL}(z - z_2)), & z > d, \end{cases} \tag{21}$$

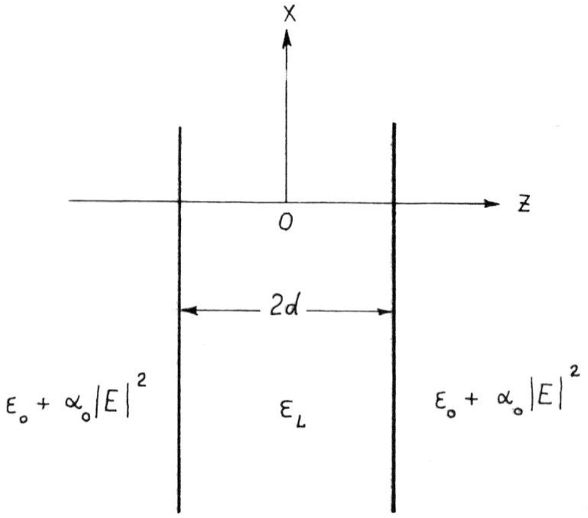

Fig. 11. The symmetric layered structure geometry.

where $2d$ is the layer thickness, z_1, z_2 and d_0 are the parameters whose relation was given by Akhmediev (1982). For the TE_0-mode we have $d_0 = 0$, $z_2 = -z_1 = z_0$, $C = \sqrt{2}\gamma_{\mathrm{NL}} \cosh^{-1}(\gamma_{\mathrm{L}} d) \cosh^{-1}(\gamma_{\mathrm{NL}}(d - z_0))$. In the case when $\varepsilon_0 < n^2 < \varepsilon_{\mathrm{L}}$, $\gamma_{\mathrm{L}} = (\gamma_{\mathrm{L}} - n^2)^{1/2} = \mathrm{i}\Gamma_{\mathrm{L}}$, and the function $\cosh(\gamma_{\mathrm{L}} z)$ becomes $\cos(\Gamma_{\mathrm{L}} z)$, so that, if γ_{L} is regarded as a complex function, the dispersion curve on the whole can be determined by a single relation. The dispersion relation for solution (21) is

$$\tanh(\gamma_{\mathrm{NL}}(z_0 - d)) = \frac{\gamma_{\mathrm{L}}}{\gamma_{\mathrm{NL}}} \tanh(\gamma_{\mathrm{L}} d). \tag{22}$$

We shall consider the problem of stability of the TE_0-mode within the limits of the parabolic equation and for two-dimensional perturbations only. It is convenient to introduce parallel with n the parameter $\rho = \tanh(\gamma_{\mathrm{NL}}(z_0 - d))$, varying from -1 to $+1$ while n varies over the region of the TE_0-mode existence. The parameter ρ is equal to minus unity at the cutoff point of the mode, and $\rho = (\gamma_{\mathrm{L}}^2 + \gamma_{\mathrm{NL}}^2)^{1/2}/(\sqrt{2}\gamma_{\mathrm{NL}})$ at the point $(n = n_{\mathrm{A}})$ of splitting off the asymmetric mode A [in view of the definition of Akhmediev (1982)], and $\rho = 1$ as $n + \infty$. The dispersion curves in variables (n, I) for the symmetric mode and the mode A are shown in figs. 12a, b. The values of the parameters used were $\varepsilon_0 = 2.25$, $\varepsilon_{\mathrm{L}} = 4.0$, $2d/\lambda = 0.4$ and 0.6.

We can somewhat simplify the problem and seek even and odd perturbation functions separately since the function $A_0(z)$ is an even function. Owing to this fact we can restrict ourselves to the analysis of zeros of the determinant of the fourth order. Generally, the dispersion relation is obtained by equating to zero the determinant of the eighth order.

The solutions of the set of equations (14) in the space region $z > d$ have the

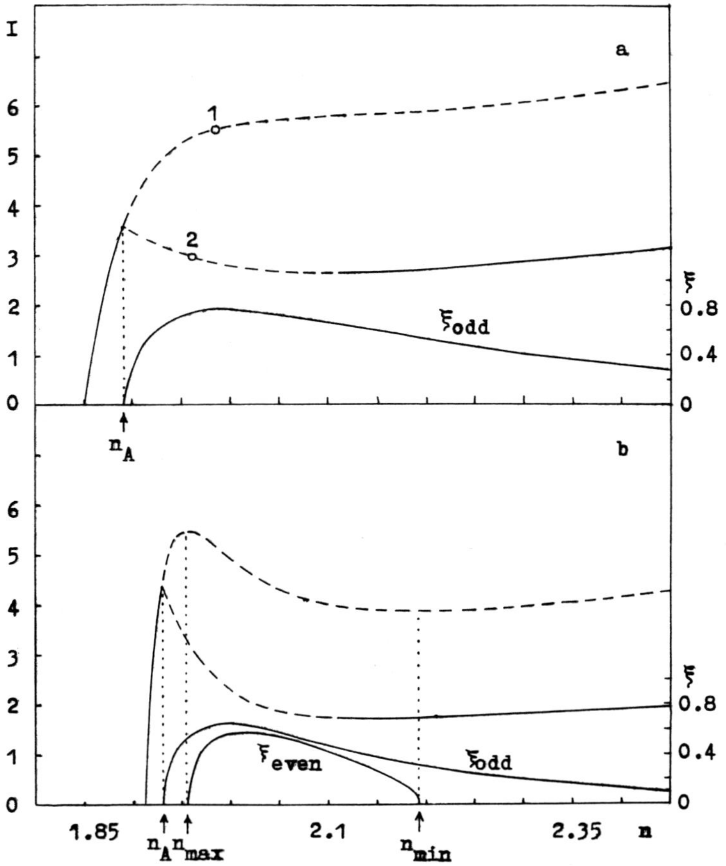

Fig. 12. The integral I and the normalized instability growth rate versus n for even and odd perturbation modes. The values of $2d/\lambda$ are 0.4 (a) and 0.6 (b). The branches of unstable waves are denoted by dashed lines.

form of eq. (15) just like for the nonlinear surface wave. But in the region $|z| < d$ we have both the even solutions of eqs. (16),

$$u = A_1 \cosh(s\bar{z}) + A_2 \cosh(s^*\bar{z}),$$

$$v = A_1 \cosh(s\bar{z}) - A_2 \cosh(s^*\bar{z}), \tag{23}$$

and the odd solutions,

$$u = B_1 \sinh(s\bar{z}) + B_2 \sinh(s^*\bar{z}),$$

$$v = B_1 \sinh(s\bar{z}) - B_2 \sinh(s^*\bar{z}). \tag{24}$$

It is clear that in this case it is sufficient to match the functions across the boundary $z=d$, then the boundary conditions at the surface $z=-d$ are satisfied automatically. Thus, equating the functions and their derivatives at the boundary $z=d$ we obtain the dispersion relation for the even modes:

$$|2sF_2-(F_3^*-F_1)\coth(\gamma_{\mathrm{NL}}sd)|^2$$
$$-|(F_3+F_1^*)\coth(\gamma_{\mathrm{NL}}sd)-2s(\rho^2-1)|^2=0, \tag{25}$$

where the functions F_1, F_2 and F_3 are given by eqs. (19), in which the parameter η must be replaced by the parameter ρ. The dispersion relation for the odd modes can be written in terms of the same functions

$$|2sF_2-(F_3^*-F_1)\tanh(\gamma_{\mathrm{NL}}sd)|^2$$
$$-|(F_3+F_1^*)\tanh(\gamma_{\mathrm{NL}}sd)-2s(\rho^2-1)|^2=0. \tag{26}$$

In finding the eigenfunctions it is convenient also to set $A_2=A_1^*$, $B_2=B_1^*$ and $C_2=C_1^*$. In this case $u=u_0$, $v=\mathrm{i}v_0$, where u_0 and v_0 are real functions.

First let us analyze eq. (25). It has the solution $\xi_0=0$ for any $|\rho|>1$, which corresponds to the indifferently stable perturbation mode $u_0=0$, $v_0=A_0(z)$. Equation (25) has no other roots for small thicknesses d, when the integral I is a monotone function of n (for example, $2d/\lambda=0.4$ in fig. 12a). With increasing thickness $d\,(2d/\lambda=0.6$ in fig. 12b), the function $I(n)$ becomes the N-shaped function and eq. (25) has the real root $\xi_{\mathrm{even}}(n)$ in the region of the negative slope. This root vanishes at the points $n_{\max}$ and $n_{\min}$, where $\mathrm{d}I/\mathrm{d}n=0$. The curve $\xi_{\mathrm{even}}(n)$, obtained by means of numerical methods on the basis of eq. (25), is shown in fig. 12b together with the curve $I(n)$. As an example, the even perturbation function is shown in fig. 13a; for $n=2.07$ and $2d/\lambda=0.6$. The instability growth rate ξ_{even} in this case equals 0.504 ....

The initial solution $A_0(z)$ remains even with the further evolution of the nonlinear wave if the perturbation function is even. Therefore, the criterion of stability $\mathrm{d}I/\mathrm{d}n>0$ for the TE_0-mode of the symmetric waveguide takes place in the class of even functions only. Indeed, by means of the numerical simulation of the solutions of eq. (5) with the initial condition $A=A_0+\mu f_{\mathrm{e}}'$, where f_{e}' is an even function, we have found that the perturbations grow only when n is within the interval $(n_{\max}, n_{\min})$: $n_{\max}<n<n_{\min}$. The stationary solution $A_0(z)$ remains stable beyond this interval.

However, an odd perturbation function exists also within the considered range of n. Indeed, eq. (26) for any d has a real root $\rho_{\mathrm{odd}}(n)$, which equals zero at the points $\rho_1=(\gamma_{\mathrm{L}}^2+\gamma_{\mathrm{NL}}^2)^{1/2}/(\sqrt{2}\gamma_{\mathrm{NL}})$ and $\rho_2=\gamma_{\mathrm{L}}/\gamma_{\mathrm{NL}}$, as can be easily seen from eq. (26). The value $\rho=\rho_1$, corresponds to the point of splitting off the asymmetric mode (Akhmediev 1982, Korneev et al. 1986) (the eigenfunctions in this case are $u_0=\mathrm{d}A_0/\mathrm{d}z$ and $v_0=0$), whereas the value $\rho=\rho_2$ is approached as $n\to\infty$. The solution ξ_{odd} as a function of n, obtained by means of the numerical solution of eq. (26) for the two values of $2d/\lambda$, is shown in fig. 12a, b. The eigenfunction corresponding to $\xi_{\mathrm{odd}}(n)$ is odd and gets the initial field distri-

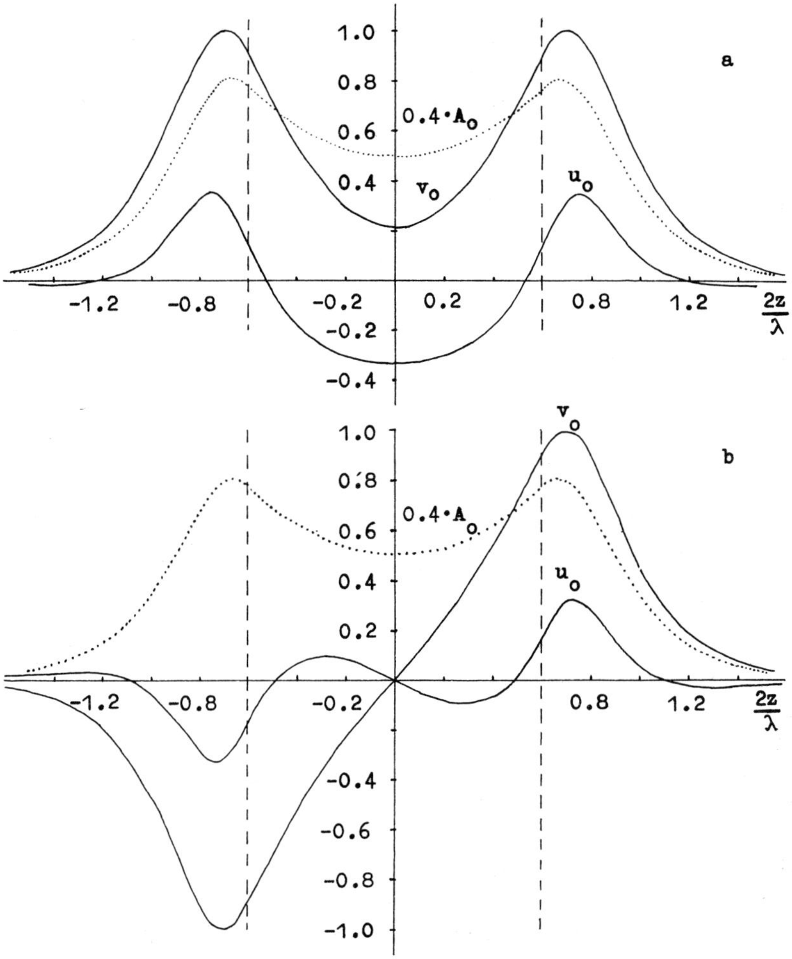

Fig. 13. Even (a) and odd (b) perturbation functions $f_0 = u_0 + iv_0$ of the TE_0-mode for $2d/\lambda = 0.6$, $n = 2.07$, normalized by the value of the maximum of the imaginary part v_0. The stationary solution $A_0(z)$ is shown (in the scale 5:2) by points.

bution $A_0(z)$ out of the class of even functions. The odd perturbation function calculated for $n = 2.07$ and $2d/\lambda = 0.6$ is shown in fig. 13b. In this case $\xi_{odd} = 0.536\ldots$.

Thus, if we do not restrict ourselves to the class of even functions, the symmetric TE_0-mode turns out to be stable only up to the point of splitting off of the asymmetric mode A. The stable branch of the TE_0-mode is plotted in fig. 12 by the solid line. We can see from the foregoing consideration that if the several different modes of the nonlinear waveguide correspond to one and the same value of n, the criterion of stability $\mathrm{d}I/\mathrm{d}n > 0$ is not valid and the stability of every branch of the dispersion curve of the nonlinear waves has to be analyzed in the manner described above. In particular, for the asymmetric mode A, existing when $n_A < n < \infty$ the stability criterion is the inequality $\mathrm{d}I/\mathrm{d}n > 0$, as in the case of nonlinear surface waves.

The result obtained for the stability of the TE_0-mode is in agreement with the data of Moloney et al. (1986a), where the stability has been studied by means of the numerical simulation of solutions of the parabolic equation, eq. (5), with the initial condition $A(z) = A_0(z) +$ random perturbations. The stability of the mode A and the waves in an asymmetric waveguide with different nonlinearity coefficients α in the substrate and the cladding has been investigated also (Moloney et al. 1986a). Computer simulation studies for other modes of a plane waveguide on the basis of eq. (5) have been carried out in a number of works (Leine et al. 1986, Ariyasu et al. 1986, Moloney et al. 1986b). We shall discuss their results in detail in the next section. Here we note only, that the criterion of stability $\mathrm{d}I/\mathrm{d}n > 0$ contains the first integral (eq. (6)) of eq. (5), but not the energy flux which differs from I by the factor n. This difference might be overlooked in the numerical simulation of solutions of eq. (5), whereas the analytical results enable us to clear up this circumstance.

Up to now, the three-dimensional problem for the case of a plane waveguide has not been practically investigated. Akhmediev and Ostrovskaja (1988b) have shown with the use of the methods developed by Vysotina et al. (1987) that the solution, eq. (21), of the form of the TE_0-mode in the three-dimensional case is unstable with respect to perturbations periodical along y and it breaks into separate beams. The numerical analysis of the set of equations (10) for arbitrary κ and n for a symmetric waveguide considered in this section has been carried out.

The plot of the instability growth rate δ versus κ and n for even and odd (with respect to z) perturbation functions is shown in fig. 14. The parameters $\varepsilon_L = 4.0$, $\varepsilon_0 = 2.25$ and $2d/\lambda = 0.6$ were used. It is seen from this figure that for $\kappa = 0$, $\delta(n)$ dependences for even and odd perturbation functions are the same as the curves in fig. 12. This fact indicates that the parabolic approximation used above is valid. Indeed, the conditions of applicability of the parabolic approximation $A_{xx} \ll 2inA_x$ and $A_{xx} \ll \gamma^2 A$ (or $\delta \ll 2n$ and $\delta^2 \ll \gamma^2$) are fulfilled for the parameters chosen.

The growth rate δ for nonzero values of κ turns out to be real and nonzero for even and odd perturbation functions. The regions in the (n, κ)-plane where the growth rate of even and odd perturbation functions are real are shown in fig. 15. The lines of the maximum growth rate in this figure are shown by the dashed lines. These values of κ for every n allow to determine the most probable dimensions of the beams into which the solution (21) breaks upon the further evolution. The stationary solution of eq. (4) limited in the y direction for the symmetric waveguide must be, clearly, of the same order in their magnitude. Such stationary solutions of the stripe nonlinear surface wave type have been found recently for waveguides by Akhmediev et al. (1989).

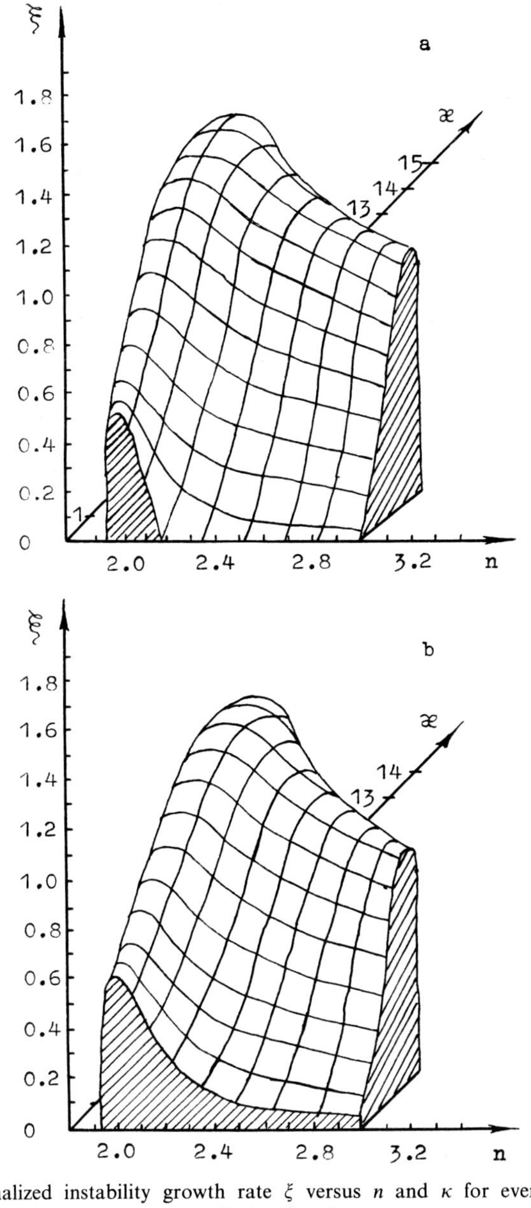

Fig. 14. The normalized instability growth rate ξ versus n and κ for even (a) and odd (b) perturbation functions.

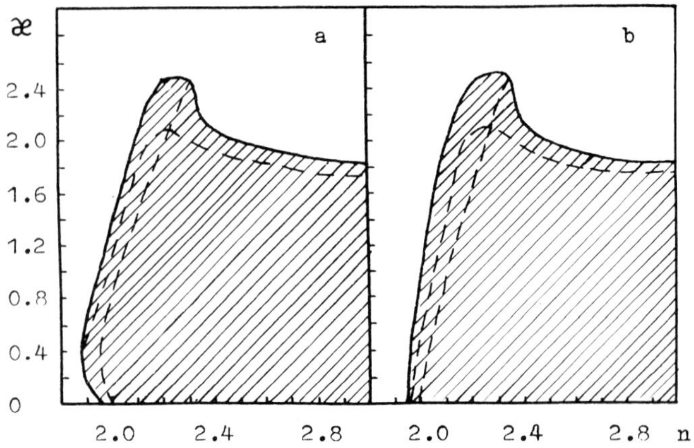

Fig. 15. The region of unstability of the TE_0-mode in the (n, κ)-plane relative to even (a) and odd (b) perturbations. The curves of the maximal value of the growth rate in two orthogonal sections are shown by dashed lines.

5. *On the evolution of unstable modes of the symmetric waveguide*

Let us now turn to the question of a field evolution beyond the linear approximation limits. We have seen that in the case of nonlinear surface waves there is the only perturbation mode which leads to two different ways of going out from the stationary state $A_0(z)$. In the case of the symmetric TE_0-mode within the intervals $n_A < n < n_{max}$ and $n_{min} < n$ (or $n_A < n < \infty$ at small d) there exists also the only perturbation mode (the odd mode) with the nonzero instability growth rate and two ways of evolution of the stationary solution $A_0(z)$, distinguished by the sign of μ. Both ways of evolution lead the initial function $A(z)$ out of the class of even functions and the further evolution along these two ways are identical within the transformation $z \to -z$ due to the structure symmetry.

The knowledge of the linear perturbation function $f_0(z)$ enables us to predict the basic features of the TE_0-mode evolution. Actually, the odd perturbation function $f_0(z)$ has the same form as in fig. 13b for all values of n within the interval $n_A < n < \infty$. We can see from this figure that upon the evolution the TE_0-mode will be transformed into the asymmetric mode A for $\mu > 0$ with the field maximum near the right boundary and the beam propagating away into the nonlinear medium at the opposite side of the layer. These speculations are confirmed by the numerical results of Moloney et al. (1986a). The graphs of the energy density as a function of z for different values of x given in this work (fig. 16a) show that at the instability points of the dispersion curve the TE_0-mode becomes asymmetrical with increasing x.

For the thick layers, when the curve $I(n)$ becomes N-shaped, the TE_0-mode evolution are enriched considerably. In this case at the initial step of the perturbation evolution the function $A(x, z)$ can be written (for $n_{max} < n < n_{min}$) in the form

$$A(x, z) = [A_0(z) + \mu_e f_{even} \exp(\xi_{even} x) + \mu_o f_{odd} \exp(\xi_{odd} x)] \exp(inx), \quad (27)$$

where μ_e and μ_o are small initial amplitudes of these modes of perturbation. It is convenient to carry out the numerical analysis of eq. (5) using the initial conditions of eq. (27) at $x = 0$, since the functions f_{even} and f_{odd} can be obtained from the set of equations (10). The instability growth rate of the even and odd perturbation modes are, as can be seen in fig. 12, approximately the same in a wide range of n, so these modes will grow with the same rates. In this case the evolution of the symmetric mode depends strongly on the initial value of the ratio of the amplitudes μ_e and μ_o of the even and odd modes.

In particular if we set $\mu_o = 0$ and restrict ourselves to the consideration of the class of even functions only then we can see from fig. 13a that the subsequent evolution of the TE_0-mode will be determined by the sign of μ_e. For a positive μ_e, the growth of the field maxima in the nonlinear media and the symmetrical splitting off of two solitons leaving the layer region will be observed. A small fraction of the energy of the initial wave remains in the layer in the form of the TE_0-mode with a small amplitude. An example of this process (for the other parameters of the structure) is shown in fig. 16. In the case when $\mu_e < 0$, we shall observe the concentration of the energy in the layer which corresponds to the transformation of the initial field distribution into the nonlinear wave of the

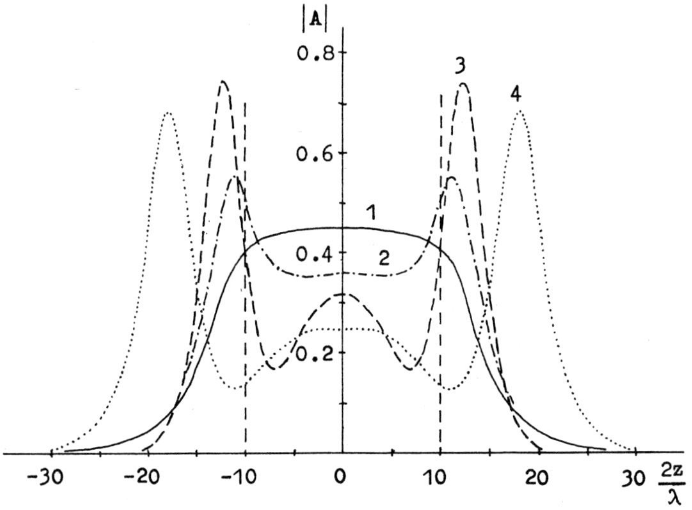

Fig. 16. Evolution of the TE_0-mode with an even perturbation in the case of the positive μ: $x = 0$ (curve 1), 220 (2), 260 (3), 340 (4). The parameters used are $\varepsilon_0 = 2.71$, $\varepsilon_L = 2.8$, $n = 1.6735$, $2d/\lambda = 10$.

stable branch with simultaneous symmetric splitting off two beams of low energy into the nonlinear media.

In a general case for the nonzero μ_e and μ_o to clear up all the possible ways of the TE_0-mode evolution, it is sufficient to investigate the field evolution for a fixed amplitude of the odd mode μ_o and different (including the change of the sign) values of an initial amplitude of the even mode μ_e. The choice of the initial conditions in the form of eq. (27) with perturbation modes permits us to get rid of ambiguities and arbitrariness of the results obtained under the initial conditions with a random perturbation function. And what is more, we shall know the probability of either one or another final result if the probability of each value of the ratio μ_e/μ_o in the initial conditions with random perturbations is known.

In the case of the asymmetric mode A the perturbation function $f_0(z)$ either looks like the perturbation function of the nonlinear surface waves shown in fig. 9 (but displaced along the z-axis over a distance d) or is symmetric to the latter, depending on the position of the field maximum of the mode A. There are also two ways of the field evolution for the unstable branch of the mode A to be determined by the sign of μ. For a positive μ we shall observe the energy concentration in the region of the field maximum of the mode A and the transformation into the mode A of the stable branch. At a negative μ the energy of the mode A will cross the film and excite the mode A on the second interface. The process may repeat many times. Just this situation has occurred in numerical simulations of the evolution of the mode A in the work of Moloney et al. (1986a) (see fig. 17b). The analogous oscillations (for the other parameters of the waveguide) have been observed also by Leine et al. (1986).

Recently articles have been published in which eigenmodes of a cylindrical waveguide with a cladding characterized by the nonlinear dielectric constant of the form of eq. (1) have been obtained (Korneev et al. 1984, Boardman et al. 1986). It has been shown that the N-shaped dependence of the energy flux on the propagation constant takes place for the stationary solutions of the structure, just as for the plane waveguide considered above. An analog of the asymmetric mode A in the case of a fiber is a mode depending on the angular variable φ. Asymmetric modes have not been sought in these works because of the difficulties in solving the three-dimensional wave equation in cylindrical coordinates, although their existence in a cylindrical fiber is doubtless. It can be seen, for example, from the solutions of the three-dimensional self-focusing problem in an unbounded nonlinear medium in the cylindrical coordinate system (Konno and Suzuki 1979). In particular, it follows from the results of the work of Konno and Suzuki (1979) that even if the initial field distribution has cylindrical symmetry, the presence of the asymmetric perturbation modes leads it out of the class of symmetric functions as a result of the evolution. Therefore, the problem of stability in fibers with a nonlinear cladding have to be considered from the start with regard to the angular dependence rather than within the class of cylindrically symmetric functions.

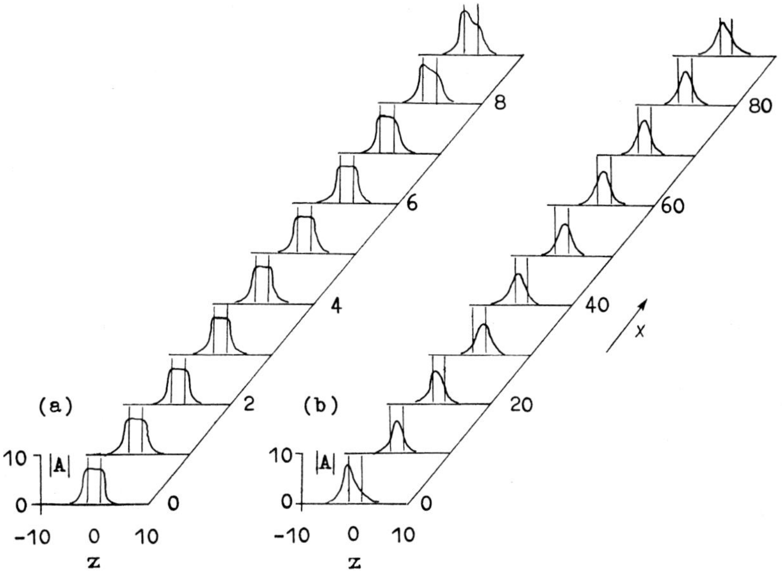

Fig. 17. Evolution of the unstable TE_0-mode (a) and the symmetric mode A (b) in a symmetrical layered structure (Moloney et al. 1986a).

6. *Excitation of nonlinear surface waves*

Many different ways of linear surface wave excitation are known (Agranovich and Mills 1982). There are, for example, the attenuated total reflection technique, the methods of grating-coupling, the edge-coupling technique, etc. All these methods are equally useful in generating nonlinear surface waves. Some experiments of this kind have been carried out (Valera et al. 1986, Chen et al. 1986, Vach et al. 1984, Bennion et al. 1985, Valera et al. 1984). However, the excitation of nonlinear surface waves can give rise to some peculiarities which require theoretical analysis. It is known, for example, that linear surface waves cannot be excited by radiation incident onto a flat interface since the tangential components of the wave vectors of the surface wave and the incident radiation cannot be matched. However, as was shown by Akhmediev et al. (1985), nonlinear surface waves can be excited by light beams incident onto a flat surface at grazing angles, close to the critical angle of total reflection.

The problem of excitation of nonlinear surface waves by an optical field is closely related with the problem of their stability. Actually, let a stationary solution $A_0(z)$ of eq. (4) be unstable and the perturbation mode f of this solution be growing as $\exp(+\delta x)$. It should be kept in mind, however, that parallel with the divergent solution $\exp(+\delta x)$ there is the convergent to $A_0(z)$ solution $\exp(-\delta x)$. So if the external beam with optimal parameters is formed, then at least on the distances $l_x \sim \delta^{-1}$ the field distribution close to that of the nonlinear surface wave of the unstable branch will exist. The stable nonlinear surface waves can be produced, as we have seen, as a result of decay of the unstable nonlinear surface waves.

This idea can be clearly illustrated by the example of the two-dimensional problem. Let us return to the question of the evolution of the unstable nonlinear surface wave (section 3). If the wavefront is inverted, $A \to A^*$ at $x = 1600$ in fig. 10a (curve 5), then the beam propagates in the opposite direction, and this is equivalent to the change of sign of the growth rate δ. This means that the surface wave of the unstable branch can, in principle, be excited by a light beam of an optimal shape and given density which is incident onto the boundary of the nonlinear medium at a certain angle greater than the critical angle for total internal reflection. As a result of evolution, however, the nonlinear surface wave decays into two beams with different energy. To show that the excitation can take place with one beam we need, upon the wavefront inversion, to discard at the same time the part of the beam which splits off into the nonlinear medium. Our numerical simulations, in fact, show that if the field in the nonlinear medium is set equal to zero at the wavefront inversion, and if the beam amplitude is increased by a factor of 1.05 for all $z < 0$ in order to conserve the integral I, then the beam produced in this manner excites the nonlinear surface waves of the unstable branch as it strikes the interface. After propagating some distance through the nonlinear medium, it is ejected back into the linear medium due to the instability of the nonlinear surface waves excited.

During the wavefront inversion, however, the beam thus formed has a certain shape which, although approximately Gaussian, cannot be described analytically. To study the possibility of exciting nonlinear surface waves by real beams we use the Gaussian model for the incident beam,

$$A = A_{\mathrm{m}} w \exp\{w^{-2}(-(z-z_0)^2 + \mathrm{i} w_0^2 n_1 (z-z_0)\sin\theta - (x-x_0)^2\tan^2\theta)\}, \tag{28}$$

where $w^2 = w_0^2 + 2\mathrm{i}(x-x_0)/n_1\cos\theta$, w_0 is the width of the beam at its narrowest point, θ is the angle of incidence, measured in this case from the interface, $n_1 = \sqrt{\varepsilon_{\mathrm{L}}}$, (x_0, z_0) are the coordinates of the focus, and the beam amplitude A_{m} is related to the integral I by

$$I = \frac{A_{\mathrm{m}}}{w_0}\left(\frac{\pi}{2}\right)^{1/2}. \tag{29}$$

The parameters of the Gaussian beam are chosen so that the distribution of the field is, for all x, as close as possible to the field distribution of the beam leaving the nonlinear medium, shown in fig. 10a, at the corresponding values of x. The parameters found for the beam by this fitting procedure are listed in the caption of fig. 18. Figure 18 shows the evolution of the beam with these parameters as it interacts with the interface. We can see that the Gaussian beam does, in fact, excite the nonlinear surface waves of the unstable branch with the field distribution exponentially approaching the steady-state solution. However, the unavoidable error in the simulation of the beam in fig. 13a by the Gaussian beam has the consequence that near the steady-state solution $A_0(z)$ the perturbation does not vanish but is a linear combination of positive and negative exponential functions with the two parameters μ_1 and μ_2,

$$A(x, z) = A_0(z) + \mu_1 f_0 \exp(\delta x) + \mu_2 f_0^* \exp(-\delta x), \tag{30}$$

where $f_0(z)$ is the perturbation mode of the nonlinear surface wave.

It can be seen from eq. (30) that in excitation of the nonlinear surface waves of the unstable branch by Gaussian beams with an approximately optimum shape there are two possible ways in which the steady-state solution can be approached. The situation is governed by the relation between the two parameters. The solution in fig. 18 corresponds to the case when $\mu_1 < 0$ and $\mu_2 < 0$. In this case the sign of the real part of the perturbation in eq. (30) does not change at any time during the process. After an asymptotically exponential approach to the steady-state solution the perturbation grows again with the same sign of the real part, and the field is ultimately ejected into the linear medium in the form of a beam. This beam can be regarded as a reflected beam since it contains most of the beam energy. The dashed line in fig. 18 shows the path traced in the (x, z)-plane by the maximum value of the beam field during this process. We see from this figure that the light beam emerges from the nonlinear medium at a distance

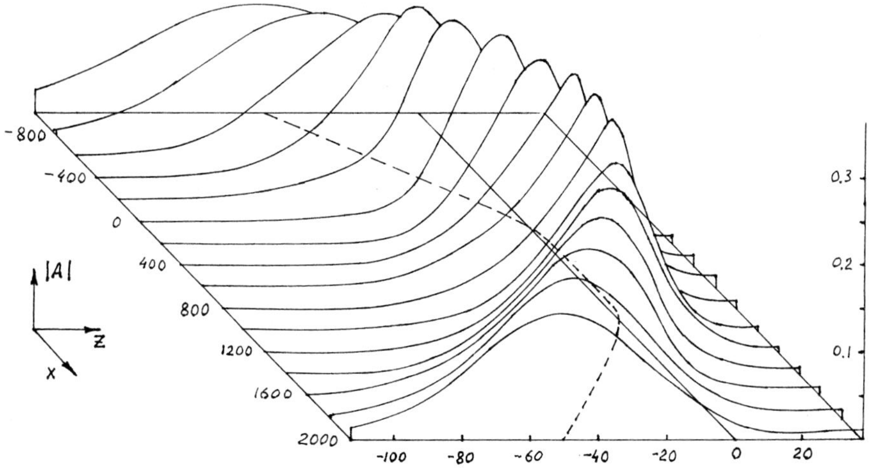

Fig. 18. Evolution of the transverse field distribution during the incidence of a Gaussian beam with the following parameters on the interface: $I = 1.03$, $a = 25$, $z_0 = 0$, $x_0 = -2000$, $\theta = 2.57°$. The dashed line in the (x, z)-plane shows the path traced out by the maximum value of the field.

from the entry point several times greater than the width of the beam. This displacement of the reflected beam is an analog of the Goos–Hänchen effect for the case of an interface between linear and nonlinear media. We see that this displacement can be many orders of magnitude greater than a linear displacement, i.e., we are dealing with a giant effect in this case. The effect is observed even when the parameters of the Gaussian beam differ slightly from those given above. The value of the integral I, of the angle of incidence, and of the beam width, on the other hand, can differ by no more than 3–5% from the optimum values listed in the caption of fig. 18. The value of x_0 must be chosen so that most of the field energy (up to 98%) is initially in the linear medium. In this case the result is, of course, independent of x_0. The magnitudes given here for the parameters are the optimum for the case when $\varepsilon_L = 2.647$ and $\varepsilon_0 = 2.674$. For the other values of the dielectric constants the optimum parameters will be different. A more accurate fit of the parameters would seem to be possible; a more accurate fit would lead to smaller values of μ_1 and μ_2 and, thus, to the stronger Goos–Hänchen effect.

If $\mu_1 < 0$ and $\mu_2 > 0$, the sign of the real part of the perturbation in eq. (30) will change as the beam evolves after the steady-state unstable solution $A(z) = A_0(z)$ is reached, so that the subsequent process will lead to the excitation of a nonlinear surface wave of the stable branch, as in the case of a positive μ in the study of stability in section 3. The nonlinear surface wave of the stable branch is ultimately excited by the Gaussian beam. At relatively large values of μ_1 and μ_2 for which the parameters of the Gaussian beam differ significantly from the optimum, the excitation stage for a nonlinear surface wave of the unstable branch can be extremely brief or missing altogether. In the latter case as a result of the interaction of the Gaussian beam with the interface the beam decays directly into a nonlinear surface wave of the stable branch and the beam which propagates away from the interface. Figure 19 shows an example of the excitation of a nonlinear surface wave of the stable branch by the Gaussian beam. The parameters of the initial beam are listed in the figure caption. A necessary condition for the excitation of a nonlinear surface wave of the stable branch is the larger value of the integral than in the preceding case. The energy of the initial beam must be sufficient for both the nonlinear surface wave and the new beam. The angle of incidence must still be near the optimum value found above. The dashed line in fig. 19 shows the path traced out by the field maximum. We see that beyond $x = 2000$, the value of z_{max} corresponding to the field maximum stabilizes. We pursued the calculations up to $x = 6000$, successively equating to zero the field of the reflected beam which reaches the boundary $z = -L$. Up to these values of x the excited wave remains stable; the values of z_{max} and of the part of the integral I, which remains in the nonlinear surface wave, correspond to the stationary solution with $n \approx 1.6382\ldots$, i.e., to the stable branch of the nonlinear surface waves.

Since nonlinear surface waves of the stable branch are excited initially with an

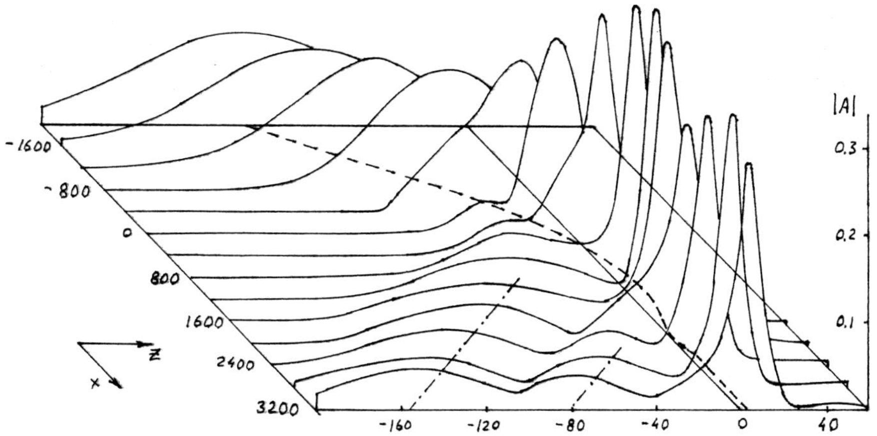

Fig. 19. The same as in fig. 18 for the beam parameters $I = 1.1$, $a = 35$, $z_0 = 0$, $x_0 = -2000$ and $\theta = 3°$.

oscillatory perturbation, as a rule, small beams may split off successively into the linear medium upon each oscillation, so that the wave reflected from the nonlinear medium will have, in addition to the main beam, a tail stretched out along the interface. The beam tail, in its turn, breaks up into separate beams (or streams). The splitting off of the second reflected beam can be clearly seen in fig. 19. The dot-dashed lines in the (x, z)-plane show the paths traced out by the maximum value of the field of the reflected beams. The onset of the tail of the reflected wave here is somewhat analogous to the lateral-wave effect in the linear optics of bounded media (Brekhovskikh 1960), where the reflected beam is followed by a tail of decreasing intensity which propagates at the critical angle for total internal reflection. In the linear case we know that there is no surface wave of s-polarization, and the reason for the appearance of a tail in the reflected beam is a backward diffraction (into the first medium) of the transmitted beam grazing along the surface. In the situation under discussion here the role of this transmitted beam is played by the nonlinear surface wave, and the secondary reflected beams arise due to oscillatory relaxation to the stationary solution. The 'nonlinear lateral wave' is of a pulsating nature, stretches out over substantially greater distances along the surface and has a higher intensity than that of the linear lateral wave.

We carried out also the numerical experiments with the self-focused beams,

$$A(x, z) = \frac{\sqrt{2}\,\gamma_{\mathrm{NL}}}{\cosh \gamma_{\mathrm{NL}}(z - x \tan \theta)} \exp(inx \tan \theta(1 - \tfrac{1}{2}\tan^2 \theta)), \tag{31}$$

incident onto the interface at small angles from the nonlinear medium. The nonlinear waves of the stable branch are also excited in this case if the angle of incidence $\theta \approx 2^\circ$. The ordinary reflection and transmission of the beam occur at large angles of incidence.

We must remember, however, that we restricted ourselves to the two-dimensional case in which the stable nonlinear surface waves are excited. The plane nonlinear surface wave [eq. (12)] is unstable in the three-dimensional case and the Gaussian beam with a wide front along the y-axis will decay into isolated beams for distances of the order of L_c. Nevertheless, in the three-dimensional problem the solution which is limited in the y-direction (stripe nonlinear surface waves) can exist. This kind of nonlinear surface wave can be excited by the beam limited in the transverse direction incident onto the interface from either a linear or a nonlinear medium. To verify this assumption we should solve the three-dimensional problem of interaction with the interface of the beam limited in the transverse direction.

7. Conclusion

The investigations of the stability of nonlinear waves in layered structures are, at present, in the very early stages. The stability of plane nonlinear surface waves has been investigated with a sufficient fullness. In the two-dimensional case, the inequality $\mathrm{d}I/\mathrm{d}n > 0$ serves as a criterion of stability for nonlinear surface waves. The possibility of obtaining analytical results in the two-dimensional case is most interesting and this is of special importance for nonlinear guided waves in plane layered structures where several different modes can correspond to one and the same value of the propagation constant n, so the simple criterion of stability $\mathrm{d}I/\mathrm{d}n > 0$ is not valid in this case.

The idea of excitation of nonlinear waves by external beams incident onto the interface at grazing angles, to our opinion, deserves further investigations. An important point is that it is possible to excite in this manner surface waves of both the stable and unstable branches.

The investigation of the three-dimensional problem of stability of striped nonlinear surface waves is in progress since plane nonlinear surface waves turned out to be unstable with respect to modulations in the transverse direction. The evolutional problems for nonlinear surface waves in the three-dimensional case have not been solved at all, although at present there are opportunities to solve these problems.

Thus, there are at present the solved problems of stability and excitation of nonlinear waves, but many unanswered questions also remain. The author shall consider his mission completed if this paper promotes to some extent the statement of future investigations.

Acknowledgment

The author wishes to thank Prof. H. Ponath and Prof. G.I. Stegeman for many helpful discussions.

References

Agranovich, V.M., and D.L. Mills, eds, 1982, Surface Polaritons (North-Holland, Amsterdam).
Agranovich, V.M., V.S. Babichenko and V.Ya. Chernyak, 1980, Pis'ma Zh. Eksp. & Teor. Fiz. **32**, 535.
Akhmediev, N.N., 1982, Sov. Phys.-JETP **56**, 299.
Akhmediev, N.N., 1983, Sov. Phys.-JETP **57**, 1111.
Akhmediev, N.N., and N.V. Ostrovskaya, 1988, Sov. Phys.-Tech. Phys. **33**, 1333.
Akhmediev, N.N., V.I. Korneev and Yu.V. Kuz'menko, 1984, Sov. Tech. Phys. Lett. **10**, 327.
Akhmediev, N.N., V.I. Korneev and Yu.V. Kuz'menko, 1985, Sov. Phys.-JETP **61**, 62.
Akhmediev, N.N., R.F. Nabiev and Yu.M. Popov, 1988, Solid State Commun. **66**, 981.
Akhmediev, N.N., R.F. Nabiev and Yu.M. Popov, 1989, Opt. Commun. **69**, 247.
Ariyasu, J., C.T. Seaton, G.I. Stegeman and J.V. Moloney, 1986, IEEE J. Quantum Electron. **QE-22**, 984.
Bennion, I., M.J. Goodwin and W.J. Stewart, 1985, Electron. Lett. **21**, 41.
Boardman, A.D., and P. Egan, 1985, IEEE J. Quantum Electron. **QE-21**, 1701.
Boardman, A.D., G.S. Cooper and D.J. Robbins, 1986, Opt. Lett. **11**, 114.
Brekhovskikh, L.M., 1960, Waves in Layered Media (Academic Press, New York).
Chen, Y.J., G.M. Carter, G.J. Sonek and J.M. Ballantyne, 1986, Appl. Phys. Lett. **48**, 272.
Kolokolov, A.A., 1973, Zh. Prikl. Mekh. Tekh. Fiz. **3**, 153.
Konno, K., and H. Suzuki, 1979, Phys. Scr. **3–4**, 382.
Korneev, V.I., N.N. Akhmediev and T.D. Shermergor, 1984, Izv. Vyssh. Uchebn. Zaved. Radiofiz. **27**, 1525.
Korneev, V.I., N.N. Akhmediev and Yu.V. Kuz'menko, 1986, Izv. Vyssh. Uchebn. Zaved. Radiofiz. **29**, 456.
Lederer, F., U. Langbein and H.E. Ponath, 1983, Appl. Phys. B **31**, 69.
Leine, L., Ch. Wachter, U. Langbein and F. Lederer, 1986, Opt. Lett. **11**, 590.
Litvak, A.G., and V.A. Mironov, 1968, Izv. Vyssh. Uchebn. Zaved. Radiofiz. **11**, 1911.
Lomtev, A.I., 1981, Pis'ma Zh. Eksp. & Teor. Fiz. **34**, 645.
Mihalache, D., D. Mazilu and H. Totia, 1984, Physica Scripta **30**, 335.
Moloney, J.V., J. Ariyasu, C.T. Seaton and G.I. Stegeman, 1986a, Appl. Phys. Lett. **48**, 826.
Moloney, J.V., J. Ariyasu, C.T. Seaton and G.I. Stegeman, 1986b, Appl. Phys. Lett. **48**, 573.
Seaton, C.T., J.D. Valera, R.L. Shoemaker, G.I. Stegeman, J.T. Chiwell and S.D. Smith, 1985, IEEE J. Quantum Electron. **QE-21**, 774.
Tomlinson, W.J., 1980, Opt. Lett. **5**, 323.
Vach, H., C.T. Seaton, G.I. Stegeman and I.C. Khoo, 1984, Opt. Lett. **9**, 238.
Vakhitov, N.G., and A.A. Kolokolov, 1973, Izv. Vyssh. Uchebn. Zaved. Radiofiz. **16**, 1020.
Valera, J.D., C.T. Seaton, G.I. Stegeman, R.L. Shoemaker, Xu Mai and C. Liao, 1984, Appl. Phys. Lett. **45**, 1013.
Valera, J.D., B. Svensson, C.T. Seaton and G.I. Stegeman, 1986, Appl. Phys. Lett. **48**, 573.
Vysotina, N.V., N.N. Rosanov and V.A. Smirnov, 1984, Pis'ma Zh. Tekhn. Fiz. **10**, 1206.
Vysotina, N.V., N.N. Rosanov and V.A. Smirnov, 1987, Zh. Tekhn. Fiz. **57**, 173.
Zakharov, V.E., and A.M. Rubenchik, 1973, Zh. Eksp. & Teor. Fiz. **65**, 997.

CHAPTER 4

Nonlinear Waves and Switching Effects at Nonlinear Interfaces

A.E. KAPLAN

Department of Electrical and Computer Engineering
The Johns Hopkins University
Baltimore, MD 21218, USA

and

P.W. SMITH and W.J. TOMLINSON

Bell Communications Research
Red Bank, NJ 07701-7020, USA

Nonlinear Surface Electromagnetic Phenomena
Edited by
H.-E. Ponath and G.I. Stegeman

© *Elsevier Science Publishers B.V., 1991*

Contents

1. *Introduction*

Over the past several years, a new direction in nonlinear optics has arisen, which can be described as self-action at nonlinear interfaces (i.e., an interface between linear and nonlinear transparent media). In the most studied configuration, light is incident from a linear medium (with susceptibility ε_0) onto a plane surface of nonlinear medium with a nonlinear refractive index (fig. 1). The studies of such an interface have followed an almost classical progression of alternating theoretical and experimental advances, which have resulted in an increasingly detailed understanding of the behavior, features, and limitations of nonlinear interfaces, and have brought into focus the remaining unsolved problems.

Studies of the behavior of nonlinear interfaces originated in theoretical analyses (Kaplan 1976, 1977, 1978, 1979, 1981a,b) which assumed that the incident and reflected fields (i.e., the fields in the linear medium) can be represented as infinite plane waves, and that the field in the nonlinear medium is either an infinite plane wave or a nonlinear evanescent wave, or some mix of both. These studies resulted in closed-form analytical expressions. We refer to them as the 'plane-wave theory'. A major result of the plane-wave theory was that under appropriate conditions the reflectivity of a nonlinear interface would exhibit a threshold behavior as a function of the intensity of the incident field: for intensities below the threshold value the interface would be totally reflecting, and for intensities above the threshold the interface would have a significant transmission. The plane-wave theory also predicted that for a range of input intensities just below the threshold value the reflectivity of the interface can be bistable.

The behavior predicted by the plane-wave theory suggested several applications (e.g., ultra-fast optical switches, scanners, and logic elements), and experiments were initiated to study these predicted phenomena. The initial experimental results were obtained using CS_2 as the nonlinear medium (with the so called positive or self-focusing nonlinearity, see below), and a high-power

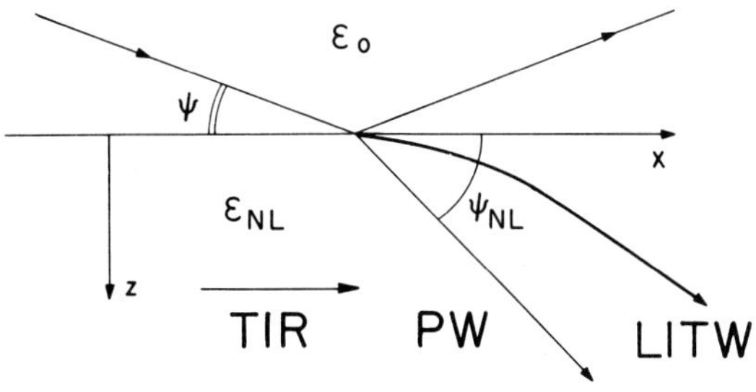

Fig. 1. Wave diagram of the nonlinear interface. Ray trajectories in nonlinear medium: (1) traveling plane wave (PW), (2) longitudinally inhomogeneous traveling wave (LITW), (3) total internal reflection (TIR).

mode-locked ruby laser as the source (Smith et al. 1979, 1981, Smith and Tomlinson 1981a). The experimental results clearly displayed the predicted threshold behavior. They also seemed to show certain evidence of the predicted bistability (Kaplan 1976, 1977). However, because of the limited time resolution of the detection system, and the fact that the experiments were done with only a single input pulsewidth, this evidence was inconclusive.

The interpretation of the experimental results was also complicated by the fact that the experiments were done with a moderately tightly focused Gaussian beam (not an infinite plane wave). Various numerical simulations have been reported for a one-dimensional Gaussian input beam (i.e., a beam with a Gaussian distribution in the plane perpendicular to the interface, but which extends uniformly to infinity in the direction perpendicular to the plane of incidence) (Smith et al. 1981, Smith and Tomlinson 1981a, Rosanov 1979, Marcuse 1980, Kolokolov and Sukov 1978, Tomlinson et al. 1982), with contradictory results. Some of them (Rosanov 1979, and to some extent Marcuse 1980) endorsed the existence of bistability for the Gaussian-beam case, while others (Kolokolov and Sukov 1978) disputed it. Although some numerical simulations (Smith et al. 1981, Marcuse 1980, Tomlinson 1980) predicted the appearance of nonlinear surface waves, later results (Tomlinson et al. 1982, Kaplan 1981c), showed that these surface waves are prohibited under conditions implied in the works of Smith et al. (1981), Rosanov (1979), Marcuse (1980), and Kolokolov and Sukov (1978). The most recent modeling study confirmed the threshold behavior, but showed that the detailed behavior of the interface is quite complex (Tomlinson et al. 1982). The reflectivity exhibits not a single threshold, but a series of thresholds at increasingly higher input intensities. The transmitted light was shown to form self-trapped channels, which can be easily identified with the two-dimensional self-trapping (Askar'yan 1962, Talanov 1964, Chiao et al. 1964, Kelley 1965) well-known in the theory of the light propagation in a semi-infinite medium with positively nonlinear refractive index. In the case of the grazing incidence, though, these channels propagate along complicated curved paths, with each step in the reflectivity associated with the formation of an additional channel. The simulations used a steady-state model, which was not capable of proving, or disproving, the existence of bistability, but it was postulated that the nonlinear interface could not exhibit bistability, because of the lack of any memory mechanism to store the information on the previous field configuration (Tomlinson et al. 1982).

The next set of experiments made use of an 'artificial' nonlinear medium, with a very large, but also very slowly responding, effective nonlinearity (Smith and Tomlinson 1984). This made it possible to do experiments with a CW laser, and relatively detailed data were obtained on the reflectivity of a nonlinear interface as a function of the input-beam intensity. The results clearly displayed the multiple thresholds, and were in good agreement with the (one-dimensional) Gaussian-beam simulations. Bistability was *not* observed. However, it was found that when the input intensity was lowered through a threshold value, the high transmission state would persist for the order of 300 times the response time of the nonlinearity. This quasi-bistability has not yet been accounted for theoretically. Hysteresis and, in fact, quasi-stability has also been recently reported (Altshuler et al. 1985) in an experiment on nonlinear reflection of picosecond pulses from a glass–liquid interface; Altshuler et al. (1985) attribute this effect to probable thermal changes in the near-surface layers in both glass and liquid.

The case of a negative (self-defocusing) nonlinearity has been much less thoroughly explored in experiment, but for this case, the plane-incident-wave theory predicts (Kaplan 1978, 1979, 1981a) an excitation of nonlinear waves of a new kind (described by Kaplan 1978, 1979, 1981a) as longitudinally inhomogeneous traveling waves which possess interesting physical features.

All of the nonlinear interface effects are caused by the competition between the mismatch in linear susceptibilities $\Delta\varepsilon_L$ and the nonlinear component $\Delta\varepsilon_{NL}$. If the linear difference is sufficiently small, this competition can result in a strong change of reflection even if the nonlinear component is small as well (as it usually is in optical range). As the intensity of the incident field increases, the penetrating (or evanescent) field in the nonlinear medium changes the index difference at the interface. At a critical intensity, an abrupt switch from total internal reflection (TIR) to partial transmission occurs if the nonlinearity is positive, or vice versa, if the nonlinearity is negative. In order for these effects to occur, it is necessary to have $|\Delta\varepsilon_L/\varepsilon_0| \ll 1$, and the angle of incidence, $\psi \ll 1$ (Kaplan 1976, 1977, 1978, 1979) (see below, section 2). The lack of these conditions leads to absence of the phenomenon. This may be the reason why such effects were not discovered earlier, either in theoretical or experimental research, e.g., in the work of Bloembergen and Pershan (1962), Bloembergen and Ducuing (1963), Bloembergen and Lee (1967) devoted to harmonic generation due to reflection of light from nonlinear dielectrics.

The optical hysteresis and bistable behavior suggested by the plane-wave theory (Kaplan 1976, 1977) can be viewed as phenomena quite typical for many nonlinear systems. The concept of bistability (multistability, in general) implies that the system can assume one of two (or more) stable stationary states under the same external conditions. Usually it correlates with hysteresis which occurs as some of the external parameters of the system (e.g., the intensity of incident radiation) vary. This phenomenon is well-known in physics, e.g., in the field of phase transitions, in the theory of nonlinear mechanical or electronic oscillators (Stoker 1950, Kaplan et al. 1966), etc.

One of the reasons that the predictions of bistable and hysteretic reflectivity at a nonlinear interface attracted attention is that in recent years optical bistability has become a very promising field in nonlinear optics (Gibbs 1985). Elements based on optical bistability can be utilized for optically controlled memory and logic operations in high-speed optical signal processing systems (Smith and Tomlinson 1981b). Most known bistable optical devices comprise a Fabry–Pérot resonator filled with a nonlinear medium, as first proposed in Seidel (1969) and Szöke et al. (1969); bistability in these systems was first observed in Gibbs et al. (1976). In these systems, bistability is due to the presence of a resonator which provides a feedback. The media used might have a resonant saturated absorption (Seidel 1969, Szöke et al. 1969, Gibbs et al. 1976) or a nonresonant Kerr nonlinearity (Felber and Marburger 1976). The use of Fabry–Pérot resonators causes these devices to be strongly selective to the frequency of the incident light. In contrast to these devices, a nonlinear interface does not employ any resonators. Therefore, no resonant tuning of the laser frequency is required; moreover, a broad spectrum of input light can be used.

A number of new effects, which do not exist in resonators, have also been predicted for nonlinear interfaces, and have stimulated various proposals for applications. New optical switching devices related to the nonlinear interfaces have been proposed by Kaplan (1981b) and Smith and Tomlinson (1981a). It was suggested by Kaplan (1976, 1977) and studied in detail by Dlodlo (1981, 1983), to use nonlinear interfaces as bistable mirrors for self-pulsing lasers. Most recently, reversible optical computing circuits based on the plane-wave switching effect at nonlinear interfaces were proposed (Cuykendall and Andersen 1987). Interesting modifications of the nonlinear interface idea have appeared, including optical bistability in a frustrated-total-reflection optical cavity (Bosacchi and Narducci 1983), and in a dielectric cladded thin film near the TIR state (Khoo and Hou 1985). An enhancement of nonlinear interface effects by exploiting surface plasmons at metallic surface was proposed by Vysin et al. (1981) and experimentally explored by Martinot et al. (1985). Optical bistability using long-range surface plasmons was proposed (Hickernell and Sarid 1986), and surface resonance modes (attributed to dielectric spheres placed on a metal substrate) (Inoue 1987). An interesting analogy of bistable TIR at nonlinear interfaces was found (Slepyan 1985) for bistability of the reflection of light from metallic diffraction gratings (positioned at the nonlinear interface) in the vicinity of the Wood anomaly. One trend in the theory and applications of nonlinear interfaces is the study of nonlinear waveguides, e.g., dielectric waveguide formed by a linear dielectric layer sandwiched between two nonlinear layers or vice versa (see, e.g., Smith and Tomlinson 1981a, Stegeman 1982, Seaton et al. 1985, Vach et al. 1984 and references therein) which have promising characteristics.

Despite its deficiencies, the plane-wave theory of nonlinear interfaces has proven to be a useful tool, and a powerful stimulant to research in this field. The theory has suggested a number of interesting nonlinear effects:

(a) multistability and hysteresis jumps in the reflection coefficient (Kaplan 1976, 1977, 1978, 1979, 1981a,b, Rosanov 1977, 1978, Permyakov and Bagdasaryan 1978);

(b) change and scanning of the refraction angle and reflection coefficient by varying the intensity of incidence light (Kaplan 1976, 1977, 1979);

(c) optically induced transparency of the interface by incident light with specific intensity (Kaplan 1976, 1977, 1979);
(d) change of penetration depth of the field into a reflecting medium in the regime of TIR (Kaplan 1977);
(e) excitation of nonlinear waves of a new kind (Kaplan 1977, 1978, 1979, 1981a,b) (longitudinally inhomogeneous traveling waves (LITW)) and effects related to LITW; namely
(f) strong nonlinear self-parallax of refracted rays along the interface (Kaplan 1979); and
(g) self-limitation of the energy flux of the refracted light (Kaplan 1979, Permyakov and Bagdasaryan 1978).

The theory has also been used to describe other situations, for example, bistability at an electro-optically driven interface (Kaplan 1981) which has recently been observed experimentally (Smith et al. 1982) and bistable reflection of light at the nonlinear optical waveguide junction (Kaplan 1981b). The main results of the plane-wave theory are described in section 2. The effects predicted by this theory in the particular cases of positive and negative nonlinearities are described in sections 3 and 4, respectively. The experimental and computer simulation results are discussed in section 5.

2. *Plane-wave theoretical formalism*

Let a plane wave with amplitude E_{in} be incident from a linear medium with susceptibility ε_0 at the glancing angle ψ (fig. 1) onto the boundary of a nonlinear medium whose susceptibility, ε_{NL}, depends on the field amplitude E in the medium according to the equation:

$$\varepsilon_{\text{NL}}(E) = \varepsilon_0 + \Delta\varepsilon_{\text{L}} + \Delta\varepsilon_{\text{NL}}(|E|^2), \tag{1}$$

where

$$\Delta\varepsilon_{\text{NL}}(|E|^2) = \varepsilon_2|E|^2, \tag{2}$$

ε_2 is the nonlinear coefficient (this is usually referred to as a Kerr nonlinearity), and $\Delta\varepsilon_{\text{L}}$ is the field-independent mismatch between susceptibilities of the two media. The most interesting self-action effects are expected when the nonlinear component $\Delta\varepsilon_{\text{NL}}$ (even if small) can compete with linear mismatch $\Delta\varepsilon_{\text{L}}$. Only then can the nonlinear component (caused by the field penetrating into the nonlinear medium) dramatically change the total difference $\Delta\varepsilon = \Delta\varepsilon_{\text{L}} + \Delta\varepsilon_{\text{NL}}$ between the susceptibilities of the two media. Such a change, in turn, can result in an abrupt jump of the reflectivity of the interface at some critical incident intensity, e.g., the switch from the TIR state into the transmission state. For the same reason, the incidence of the light has to be almost grazing, because the required magnitude of the glancing angle ψ should be of the order of the critical angle of TIR, which results in the above-mentioned conditions (Kaplan 1976, 1977),

$$\left|\frac{\Delta\varepsilon_{\mathrm{L}}}{\varepsilon_0}\right| \sim \left|\frac{\Delta\varepsilon_{\mathrm{NL}}}{\varepsilon_0}\right| \ll 1, \qquad \psi \sim \left(\frac{\Delta\varepsilon}{\varepsilon_0}\right)^{1/2} \ll 1. \tag{3}$$

These conditions give $\psi \sim 1^\circ$ for $|\Delta\varepsilon_{\mathrm{NL}}/\varepsilon_0| \sim 10^{-4}$. For most of the rest of this chapter we will assume that conditions (3) are valid. This will simplify the theory and make the phenomena independent of the polarization of the incident light (Kaplan 1976, 1977, 1979).

The field in the linear medium is represented in the form of two homogeneous plane waves, one of which is the incident wave and the other is the reflected wave, determined by an unknown complex reflection coefficient r. The wave equation for the complex amplitude of the field, E, in a nonlinear medium in the two-dimensional case can be written as

$$\frac{\mathrm{d}^2 E}{\mathrm{d}z^2} + k_0^2 E\left[\frac{\varepsilon_{\mathrm{NL}}(|E|^2)}{\varepsilon_0} - \cos^2\psi\right] = 0, \tag{4}$$

where $k_0 = \omega\sqrt{\varepsilon_0}/c$, and the z-axis is taken to be perpendicular to the boundary. The total field is $\frac{1}{2}E\exp(-\mathrm{i}\omega t) + \mathrm{c.c.}$ By equalizing tangential components of the fields on both sides of the boundary, the generalized boundary condition for the refracted wave E can be obtained (Kaplan 1977, 1978, 1979),

$$\mathrm{i}\frac{\mathrm{d}E(0)}{\mathrm{d}z} + k_0 \sin\psi[2E_{\mathrm{in}} - E(0)] = 0. \tag{5}$$

The expression for the reflection coefficient r is

$$r = \frac{E(0)}{E_{\mathrm{in}}} - 1, \tag{6}$$

which also follows from the continuity of tangential components of the fields at the interface. The boundary conditions are a very important part of any problem related to the nonlinear interface. The failure to define these conditions adequately could result in missing important effects. For example, inadequate conditions for the refracted plane wave in an earlier work (Boiko et al. 1975) made it impossible to determine the wave amplitude and refraction angle separately in the case of $\varepsilon_2 > 0$. This led Boiko et al. (1975) to the controversial conclusions regarding the existence of optical bistability even within plane-wave approach.

The field E in the nonlinear medium is written in the form (Kaplan 1977, 1978, 1979, 1981a)

$$E = u(z)\exp\left[\mathrm{i}k_0\int_0^z \xi(z)\,\mathrm{d}z + \mathrm{i}\phi + \mathrm{i}k_0 x\cos\psi\right], \tag{7}$$

where $u(z)$ and $\xi(z)$ are real quantities, u being the real amplitude of the field, and ξ being the angle formed by the rays and the x-axis at a given point z; ϕ is a constant phase term. For plane waves (PW), u and ξ are constant, and for surface (or evanescent) waves (SW), when TIR occurs, $\xi = 0$. In general, u and ξ are not constant.

In the theory of reflection from a semi-infinite medium, the conditions at infinity must be defined. Since there are no sources inside the nonlinear medium, only traveling waves should propagate *away* from the interface in the nonlinear medium (or SW, which do not carry any energy along the z-axis at all). Therefore, in eq. (7) for $z \to \infty$, the following conditions have to be satisfied (Kaplan 1976, 1977, 1978):

$$u \to \text{const} \equiv u_\infty \geqslant 0, \qquad \xi \to \text{const} \equiv \xi_\infty \geqslant 0, \tag{8}$$

which is, essentially, the Sommerfeld radiation condition (the absence of the backward-traveling wave). Substituting the field in the form of eq. (7) into the wave equation [eq. (4)], we obtain its first integral,

$$I = \xi u^2 = \text{const} \equiv \xi_\infty u_\infty^2 \geqslant 0, \tag{9}$$

(which expresses the conservation of the energy flux) as well as the equation for the real amplitude u,

$$\frac{\mathrm{d}^2 u}{\mathrm{d}z^2} + k_0^2 u\left(\frac{\varepsilon_{\mathrm{NL}}(u^2)}{\varepsilon_0} - \cos^2\psi - \frac{I^2}{u^4}\right) = 0. \tag{10}$$

The first integral of this equation, which satisfies the radiation condition [eq. (8)], can be written in the form (Kaplan 1978):

$$\left(\frac{\mathrm{d}u}{\mathrm{d}z}\right)^2 = k_0^2 \int_{u_\infty}^{u} [F(u) - F(u_\infty)]\, \mathrm{d}\left(\frac{1}{u^2}\right), \tag{11}$$

where the nonlinear 'characteristic' function $F(u)$ is introduced,

$$F(u) = u^4\left[\frac{\varepsilon_{\mathrm{NL}}(u^2)}{\varepsilon_0} - \cos^2\psi\right]. \tag{12}$$

Integrating eq. (11) and taking the boundary condition given by eq. (5) into account, one can obtain all possible wave solutions for the problem.

3. *Positive nonlinearities*

In the simplest case of cubic nonlinearity [eq. (2)] with $\varepsilon_2 > 0$ (i.e., Kerr nonlinearity), only two kinds of wave modes, with their intensities depending on z only, can exist:

(a) the homogeneous plane wave (PW), $u(z) = \text{const.}$, which corresponds to the transmission regime ($|r| < 1$); and

(b) the surface wave (SW), or evanescent wave, $\xi = 0$, which corresponds to TIR ($|r| = 1$).

Both of these waves are nonlinear analogs of corresponding linear waves, but differ now from those in their dependence on the incidence light intensity. (We should note that in two, or more, dimensions, a homogeneous plane wave is not a stable solution in a medium with a positive Kerr nonlinearity (Talanov 1964), and thus we would only expect the plane-wave theory to apply to the transmitted field in the region sufficiently close to the interface.) Using eqs. (4)–(7) with u and ξ constant, one readily obtains 'Snell's nonlinear formula' for the transmission angle ξ (Kaplan 1976, 1977):

$$\left(1 + \frac{\xi}{\psi}\right)^2 \left(\xi^2 - \psi^2 - \frac{\Delta\varepsilon_L}{\varepsilon_0}\right) = 4\frac{\varepsilon_2}{\varepsilon_0}|E_{in}|^2, \tag{13}$$

or 'Fresnel's nonlinear formula' for the reflection coefficient r,

$$4r\psi^2 + \frac{\Delta\varepsilon_L}{\varepsilon_0}(1+r)^2 + \frac{\varepsilon_2}{\varepsilon_0}|E_{in}|^2(1+r)^4 = 0 \tag{14}$$

(see fig. 2). It was shown by Kaplan (1981a) that in the general case of arbitrary magnitude of nonlinearity $\varepsilon_2|E_{in}|^2$ and the incident angle ψ [i.e., in the case when eq. (3) is not satisfied any more], the 'Snell formula' for the angle ψ_{NL} of a refracted plane wave is

$$\left(\frac{\tan^2 \psi_{NL}}{\tan^2 \psi} - 1 - \frac{\Delta\varepsilon_L}{\varepsilon_0 \sin^2 \psi}\right) = \frac{4\varepsilon_2|E_{in}|^2}{\varepsilon_0 \sin^2 \psi}, \tag{15}$$

(where now $\tan \psi_{NL} = \xi/\cos \xi$), while the 'Fresnel formula' for the amplitude u of reflected plane wave is

$$4 \sin^2 \psi \left(1 - \frac{u}{|E_{in}|}\right) = \frac{u^2}{\varepsilon_0|E_{in}|^2}(\Delta\varepsilon_L + \varepsilon_2 u^2). \tag{16}$$

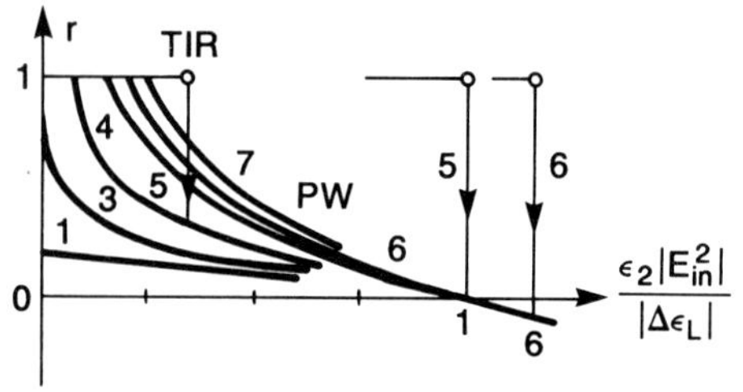

Fig. 2. Reflection coefficient r versus the incident light intensity $|E_{in}|^2$ at different glancing angles ψ in the case of negative linear mismatch of susceptibilities ($\Delta\varepsilon_L < 0$) and positive nonlinearity ($\varepsilon_2 > 0$). Curves: (1) $\psi > \psi_{cr}$; (2) $\psi = \psi_{cr}$; (3, 4) $\psi/\psi_{cr} > 1/2\sqrt{2}$; (5) $\psi = \psi_{cr}/2\sqrt{2}$ (nonlinear bleaching); (6, 7) $\psi < \psi_{cr}/2\sqrt{2}$. (After Kaplan 1976.)

The formula for the reflection coefficient r follows from eq. (16) via eq. (6) which again yields eq. (14) where one has to replace ψ^2 by $\sin^2 \psi$.

The nonlinear TIR state can be excited if linear mismatch is negative ($\Delta\varepsilon_L < 0$) and $\psi < \psi_{cr}$, where $\psi_{cr} = (|\Delta\varepsilon_L|/\varepsilon_0)^{1/2}$ is a critical angle of linear TIR. The profile of wave intensity now is not exponential; it is obtained from eq. (11) with $u_\infty = 0$. In the case of Kerr nonlinearity [eq. (2)], the general equation [eq. (11)] under the condition of eq. (3), can then be written in the form

$$\left(\frac{du}{dz}\right)^2 = k_0^2 u^2(\psi_{cr}^2 - \psi^2 - \tfrac{1}{2}u^2). \tag{17}$$

Integration of this equation yields (Kaplan 1976, 1977) the amplitude profile $u(z)$ of a TIR evanescent field (surface wave) which has a 'self-channel' shape well-known in the soliton theory of two-dimensional self-trapping,

$$u = \left(\frac{2\varepsilon_0}{\varepsilon_2}\right)^{1/2} \frac{\gamma}{\cosh(k_0 \gamma z + C)}, \tag{18}$$

where $\gamma = (\psi_{cr}^2 - \psi^2)^{1/2}$, and constant C is determined from the boundary conditions [eq. (5)], and can have from one to four values (for different ψ and E_{in}) which causes multistability and hysteretic jumps (see fig. 2). Indeed, from the complex boundary conditions [eq. (5)], one obtains a condition for the real amplitude $u(0)$ of an evanescent field at the boundary:

$$\left[\frac{du(0)}{dz}\right]^2 + k_0^2\psi^2[u^2(0) - 4|E_{in}|^2] = 0, \tag{19}$$

as well as a formula for the phases of the evanescent field ϕ and reflected wave ϕ_r [where ϕ_r is determined by the relationship $r = \exp(i\phi_r)$],

$$\phi = \tfrac{1}{2}\phi_r = \mp \arccos[2u(0)/E_{in}], \tag{20}$$

where the upper sign corresponds to $u(0) < 0$ and the lower to $u'(0) > 0$. Now, eliminating $du(0)/dz$ from eq. (19), and from eq. (17) (taken at $z = 0$), one obtains an amplitude $u(0)$ (Kaplan 1977),

$$|u(0)|^2 = \frac{\varepsilon_0}{\varepsilon_2}\left[\psi_{cr}^2 \pm \left(\psi_{cr}^4 - 8\psi^2 \frac{\varepsilon_2|E_{in}|^2}{\varepsilon_0}\right)^{1/2}\right]. \tag{21}$$

Stipulating that the amplitude $u(0)$, eq. (21), and the phase ϕ, eq. (20), be real quantities, one obtains the conditions for the nonlinear TIR to occur,

$$0 \leqslant u^2(0) \leqslant 4|E_{in}|^2, \qquad 8\psi^2\varepsilon_2|E_{in}|^2 \geqslant \psi_{cr}^4\varepsilon_0. \tag{22}$$

Based on these conditions, one can show that the threshold grazing angle ψ_{th}, at which the switch from nonlinear TIR to partial transmission occurs, is given by (Kaplan 1976, 1977)

$$\begin{aligned}(\psi_{th}/\psi_{cr})^2 &= 1 - 2\varepsilon_2|E_{in}|^2/|\Delta\varepsilon_L|, \quad 4\varepsilon_2|E_{in}|^2 < |\Delta\varepsilon_L| \\ &= \tfrac{1}{8}|\Delta\varepsilon_L|/\varepsilon_2|E_{in}|^2, \qquad 4\varepsilon_2|E_{in}|^2 > |\Delta\varepsilon_L|.\end{aligned} \tag{23}$$

Therefore, for a fixed grazing angle ψ, the threshold input intensity $|E_{\text{th}1}|^2$, at which such a switch occurs, is given by

$$\frac{\varepsilon_2|E_{\text{th}1}|^2}{\Delta\varepsilon_L} = \begin{cases} \frac{1}{2}(1-\psi/\psi_{\text{cr}})^2, & \frac{1}{2}\sqrt{2} \leqslant \psi/\psi_{\text{cr}} < 1, \\ \frac{1}{8}(\psi_{\text{cr}}/\psi)^2, & 0 < \psi/\psi_{\text{cr}} \leqslant \frac{1}{2}\sqrt{2}. \end{cases} \tag{24}$$

Thus, if starting from TIR at the low-input intensity and gradually increasing the intensity, the interface switches from the TIR state ($|r| = 1$) to the partial-transmission state ($|r| \leqslant = 1$, see fig. 2) at some intensity given by eq. (24). The reflection coefficient r at this point is determined by eq. (14), where the value of $\varepsilon_2|E_{\text{th}1}|^2$ is to be substituted from eq. (24). In the backward process (i.e., when the input intensity is decreased), the switch from the partial-transmission state to the TIR state occurs at the lower intensity, $|E_{\text{th}2}|^2$, which is determined by eq. (13) [or eq. (14)] by assuming $\xi = 0$ (or $r = 1$),

$$\frac{\varepsilon_2|E_{\text{th}2}|^2}{\Delta\varepsilon_L} = \tfrac{1}{4}[1-(\psi/\psi_{\text{cr}})^2], \tag{25}$$

i.e., $|E_{\text{th}2}|^2 < |E_{\text{th}1}|^2$ (see fig. 2). The jump of the reflection coefficient only occurs during the switch from TIR to transmission; when moving in the backward direction, there is only a jump in the phase of reflection.

One of the interesting features of nonlinear reflection is what can be called nonlinear bleaching (or optically induced transparency) (Kaplan 1976, 1977). It is seen from eq. (14) that at $\Delta\varepsilon_L < 0$, a 'bleaching' input intensity, $|E_{\text{bl}}|^2$ exists:

$$|E_{\text{bl}}|^2 \equiv |\Delta\varepsilon_L/\varepsilon_2|, \quad \text{i.e.,} \quad \Delta\varepsilon_L = -\Delta\varepsilon_{\text{NL}}, \tag{26}$$

at which there is no reflection at all ($r = 0$) for any glancing angle ψ (see fig. 2). This phenomenon cannot be regarded as a nonlinear analog of the Brewster bleaching (since it does not depend on either the angle or polarization of incident light). The nonlinear bleaching is attributed to the fact that field equalizes the susceptibilities of the two media and, therefore, makes the boundary completely transparent. In this case, the jump from total reflection to total transmission occurs at $\psi = \psi_{\text{cr}}/2\sqrt{2}$ (see curve 5 in fig. 2). For CS_2, where $n_2 \sim 10^{-11}$ cgs esu, we have $E_{\text{bl}} \sim 1.8 \times 10^6$ V/cm at $\Delta\varepsilon_L \sim -10^{-3}$, in which case, $\psi = 0.54°$.

4. *Negative nonlinearities; longitudinally inhomogeneous traveling waves (LITW)*

In a number of physical situations, the nonlinearity can be negative (i.e., $\varepsilon_2 < 0$), for instance, due to nonlinear resonant interactions (see, e.g., Butylkin et al. 1990), or due to giant nonlinearities, recently observed in some semiconductors (e.g., in InSb (Miller et al. 1980)). In such a case, the nonlinear analogs of linear waves (i.e., PW and TIR) can exist as well. For plane waves (PW), the Snell [eqs (13), (15)], as well as the Fresnel [eqs (14), (16)], formulae remain valid (one has to take into consideration that now the sign of ε_2 is changed). In this regime it is also possible to vary the value of the transmission angle ξ and reflection r by changing the incident light intensity (fig. 3). Similarly to the 'positive' nonlinearity, a nonlinear total bleaching of the interface also exists; the 'bleaching' incident light intensity is given by the same expression, eq. (26), but now the value of the linear mismatch should be positive (fig. 3, curves 3 and 4).

In contrast to the 'positive' nonlinear case, the surface (evanescent) waves at 'negatively' nonlinear interface in the TIR regime exist in exactly the same range of linear mismatches and glancing angles [i.e., $\Delta\varepsilon_L < 0$ and $\psi < \psi_{cr} = |\Delta\varepsilon_L/\varepsilon_0|^{1/2}$] as in the linear case. However, now the amplitude profile of the SW, given by integration of eq. (11), is (Kaplan 1977)

$$u = \left[2\frac{\varepsilon_0}{|\varepsilon_2|}\right]^{1/2} \frac{\gamma}{\sinh(k_0\gamma z + C)}, \tag{27}$$

where $\gamma = (\psi_{cr}^2 - \psi^2)^{1/2}$, and C is a constant determined by boundary conditions.

The amplitude of the field $u(0)$ at the boundary is given by eq. (21), where one has to take the negative sign in front of the radical. As $z \to \infty$, the evanescent field, eq. (27), decreases as $\sim\exp(-K_0\gamma z)$, just as in the linear case. But the depth of field penetration into the medium (at the half-intensity level), L_{NL}, which is determined by the shape, eq. (27), and the value of $u(0) \equiv u_0$, decreases (Kaplan 1977)

$$L_{NL} = L_L\left[1 + \frac{2}{\ln 2}\ln\frac{1+(1+u_0^2/4\gamma^2)^{1/2}}{1+(1+u_0^2/2\gamma^2)^{1/2}}\right], \tag{28}$$

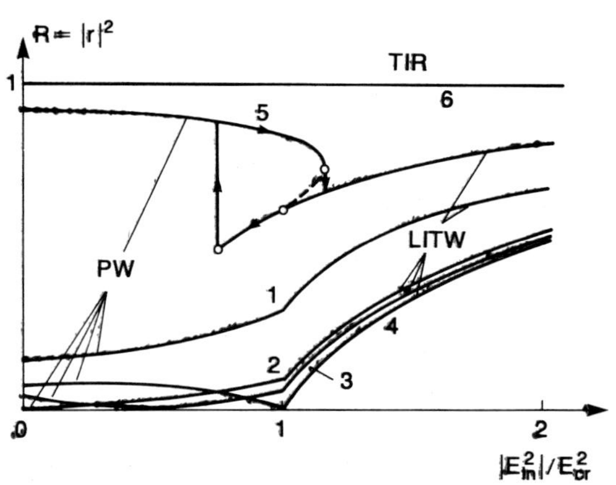

Fig. 3. Reflectivity $R = |r|^2$ as a function of the incident light intensity $|E_{in}|^2$, for a negative nonlinearity ($\varepsilon_2 < 0$), and $D \equiv \Delta\varepsilon_L/\varepsilon_0$. Curves: (1) $D < 0$, $\psi^2 > |D|$; (2) $D = 0$; (3) $\psi^2 > \frac{1}{2}D > 0$; (4) $\psi^2 = \frac{1}{2}D$; (5) $\psi^2 < \frac{1}{2}D$; (6) $D < 0$, $\psi^2 < |D|$. (After Kaplan 1979.)

where $L_{\rm L} = (\ln 2)/2k_0\gamma$ is the depth of the linear penetration. In a strong field, as $u_0^2 \gg \gamma^2$, one has $L_{\rm NL} \approx 0.6\gamma L_{\rm L}/u_0 \ll L_{\rm L}$.

The most interesting feature in the case of negative nonlinearity is the feasibility of excitation (under some special conditions) of waves of a new kind, the so-called longitudinally inhomogeneous traveling waves (LITW), obtained analytically by Kaplan (1978, 1979, 1981a). Being inhomogeneous and non-planar near the boundary, these waves reduce to plane waves sufficiently far from the boundary. The concept of longitudinal inhomogeneity was introduced by Kaplan (1978) to distinguish this phenomenon from the self-actions due to transversely inhomogeneous waves, such as self-focusing (Askar'yan 1962, Talanov 1964, 1965, Chiao et al. 1964, Kelley 1965) and self-bending (Kaplan 1969) of the beams with limited cross-section.

Examination of eq. (11) shows that under the conditions

$$\tfrac{2}{3}q < \frac{|\varepsilon_2|}{\varepsilon_0} u_\infty^2 < q, \quad q \equiv \frac{\Delta\varepsilon_{\rm L}}{\varepsilon_0} + \sin^2\psi, \tag{29}$$

excitation of LITW is possible. Integration of eq. (11) yields two possible intensity profiles of LITW (Kaplan 1978, 1979):

$$u^2 = u_\infty^2 \pm 2B^2 \begin{Bmatrix} \sinh^{-2} \\ \cosh^{-2} \end{Bmatrix} (Bk_0 z + C), \quad B \equiv \left[\frac{3}{2}\frac{|\varepsilon_2|}{\varepsilon_0} u_\infty^2 - q\right]^{1/2}. \tag{30}$$

For the minimum possible energy density at infinity,

$$u_\infty^2 = u_{\rm m}^2 = \frac{2}{3}\frac{\varepsilon_0}{|\varepsilon_2|} q, \tag{31}$$

only one (limiting) type of LITW remains:

$$u_{\rm lim}^2 = 2\frac{\varepsilon_0}{|\varepsilon_{\rm s}|}\left[\tfrac{1}{3}q + \frac{1}{(k_0 z + C)^2}\right]. \tag{32}$$

The unique 'continuum' problem arises pertinent to the new solutions: we have only two conditions [which are equivalent to a single complex boundary condition eq. (5)] to determine three unknown constants [u_∞ and C in eq. (32), and ϕ in eq. (7)]. The radiation conditions, eq. (8), have already been used for the construction of LITW [eqs. (11) and (30)]. Therefore, there is a continuum of solutions even when boundary and radiation conditions are completely specified.

This situation differs drastically from the situation in a linear medium and a 'positive' nonlinear medium. An energy criterion for selecting a unique kind of physically realized LITW was proposed (Kaplan 1979) which is based on the minimization of wave energy density (simultaneously, it results in a maximum energy flux). The only 'surviving' kind of LITW occurs to be a limiting type, eq. (32), in which case u_∞^2 reduces to the minimum possible value for LITW equal to u_m^2 [eq. (31)]. This choice was verified analytically (Kaplan 1979) by an asymptotic method whereby a weak absorption is introduced and the asymptotic results are obtained when the absorption tends to zero. A small parameter of this limiting transition occurred to be $(k_0 L_{ab})^{-2/5}$, where L_{ab} is a characteristic length of the wave absorption.

Now, when the unique 'surviving' *type* of LITW (but not unique *solution* yet), is found, one can obtain all the characteristics of these waves, in particular, their amplitude at the boundary, u_0 (Kaplan 1979, 1981a). Based on this result and using the relationship for u_0 for PW, eq. (16), it is easy to find the domains of existence of all of the possible waves (i.e., PW, LITW and TIR) in the system (Kaplan 1981a). These domains are shown in fig. 4 in the space of generalized parameters P and Q,

$$P = \frac{\varepsilon_2 |E_{in}|^2}{\varepsilon_0 \sin^2 \psi}, \qquad Q = \frac{\Delta\varepsilon_L}{\varepsilon_0 \sin^2 \psi} + 1. \tag{33}$$

The LITW excitation gives rise to new effects. First of all, as in the 'positive' nonlinear case, it leads to the hysteretic behavior of a system under definite conditions, but now the hysteresis jump occurs between states of two different traveling waves, PW and LITW (in contrast to 'positive' nonlinearity, where they occur between PW and TIR). Hysteresis occurs only under the conditions (Kaplan 1979, 1981a)

$$|\varepsilon_2|\,|E_{in}|^2 > \Delta\varepsilon_L > 2\varepsilon_0 \sin^2 \psi.$$

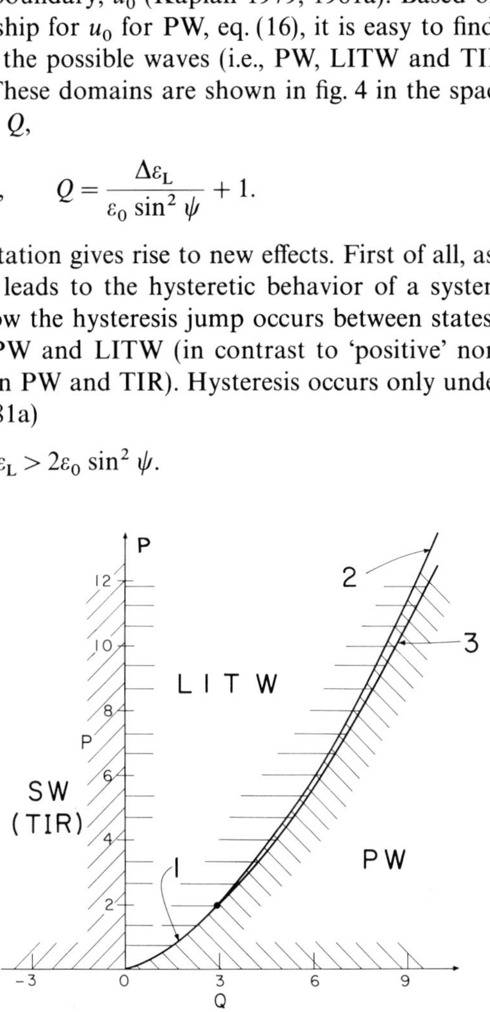

Fig. 4. Diagram of wave states in the space of parameters P and Q. Curve (1) is the boundary between PW and LITW states; curve (2) is the boundary of transition jump PW → LITW; curve (3) is the boundary of reverse transition jump LITW → PW. The space between curves (2) and (3) is the hysteresis region. (After Kaplan 1981a.)

The behavior of reflectivity for different parameters of a system is shown in fig. 3 (in particular, one can see hysteresis, curve 5, and self-bleaching, curves 3 and 4). In this figure, E_{cr}^2 is the critical intensity of incident light required for excitation of LITW for each particular case, and $D \equiv \Delta\varepsilon_L/\varepsilon_0$.

Another effect which is of interest for applications, consists of the self-limitation of the energy flux of LITW penetrating into a non-linear medium (Kaplan 1979, Permyakov and Bagdasaryan 1978). This is a direct consequence of the selection of a unique value $u_\infty = u_m$ [eq. (31)]. Behavior of the energy flux I is shown in fig. 5 for hysteretic and non-hysteretic situations.

One of the most interesting effects attributed to the excitation of LITW is self-parallax (Kaplan 1979), i.e., displacement of the refracted rays along the interface. This effect is not a nonlinear analog of Goos–Hänchen displacement of Gaussian beams under TIR. This is because, in contrast to Goos–Hänchen, self-parallax is now valid for plane waves and for the transmission regime, rather than for TIR. It corresponds to the initial 'sticking' of the rays to the interface (fig. 1, curve 2) which occurs because of conservation of the energy flux [eq. (9): the angle ξ should increase as the intensity u^2 decreases], and is not related to the existence of hysteresis. The ray trajectory in space is determined by (Kaplan 1979)

$$x = z(3/Q)^{1/2} + l(1 + z_l/z)^{-1}, \tag{34}$$

where l is the parallax (ray displacement) for $z \to \infty$ and z_l is the characteristic depth for half of the parallax; l and z_l are determined by E_{in}, ψ and $\Delta\varepsilon_L$. Let $k_0 = 10^5\ \mathrm{cm}^{-1}$, $\Delta\varepsilon_L/\varepsilon_0 = -10^{-4}$, $\psi \sim 0.7°$ and $|\varepsilon_2/\varepsilon_0| = 10^{-10}$ cgs esu; then the critical value of E_{in} for excitation of LITW is $E_{cr} = 1.15 \times 10^5$ V/cm, and if $E_{in} = 2E_{cr}$, one obtains $l = 1.2$ cm and $z_l = 0.01$ mm; the parallax is very strong and occurs at a very small depth.

The existence of LITW in a system with 'negative' nonlinearity and its absence in a 'positive' nonlinear system poses a number of questions related to the kind of nonlinearity of the medium, which may cause LITW, such as: is it possible to predict what kinds of nonlinearities and system characteristics give rise to

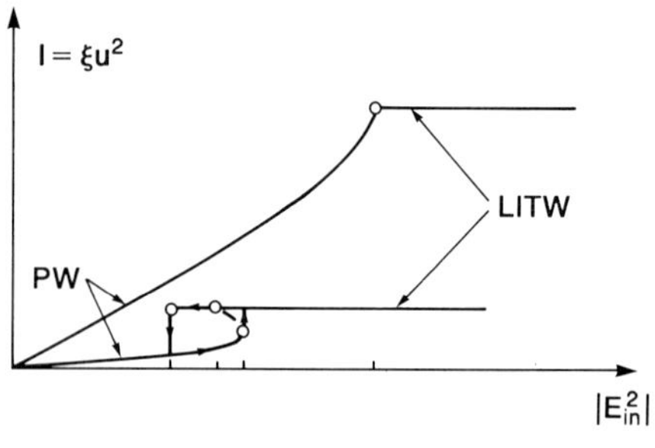

Fig. 5. Energy flux I of transmitted waves as a function of the incident light intensity $|E_{in}|^2$ for hysteresis and nonhysteresis situations. (After Kaplan 1979.)

LITW and what kinds of nonlinearities do not allow LITW? What are the parameters of possible LITW if allowed?

These questions have been addressed by a theorem formulated and proved by Kaplan (1978, 1979). This theorem relates the existence of different kinds of traveling waves to the behavior of a 'characteristic' nonlinear function $F(u)$, eq. (12) above. The main conclusion of this theorem is that the existence of LITW in a transparent medium is possible if, and only if:

(a) there is a range V, in which $F(u) > 0$, $u \in V$;

(b) $F(u)$ falls at least somewhere in V (i.e., there is at least one interval $W \subset V$ where $F(u)$ falls monotonically);

(c) if these conditions are satisfied, the LITW, if excited, can have a value of u_∞ which must belong only to this falling interval, i.e., always $u_\infty \in W$.

The proof of this theorem is based on an investigation of the behavior of the integral in the right-hand side of eq. (11).

One of the main implications of this theorem is that LITW can be excited only in a nonlinear medium whose nonlinearity $\varepsilon_{NL}(u^2)$ has at least one interval of fall. On the other hand, LITW are prohibited in the medium with increasing function $\varepsilon_{NL}(u^2)$. This is the reason why excitation of LITW is prohibited for 'positive' nonlinearity.

The 'continuum problem' arises for all kinds of LITW. Therefore, the special issue in the theory of two-dimensional nonlinear waves is related to 'survival' of nonlinear transmitted waves in the case of arbitrary nonlinearity. Several theorems related to this problem were formulated by Kaplan (1979). In particular, a complete class of functions $F(u)$ were found which allow for the 'surviving' LITW, and it was proved that the principle of minimization of wave energy remains valid throughout this class.

5. *Experiments*

For the purpose of describing the experiments, it is convenient to use the refractive index, n, rather than the susceptibility, ε. We write the index of the nonlinear medium as

$$n(I) = n_0 + \Delta + n_2 I \tag{35}$$

where n_0 is the index of the linear medium, Δ is the mismatch between the refractive indices of the two media, which does not depend on the light intensity, I, and n_2 is the nonlinear index. In terms of the parameters used earlier in this paper, we can write (for $\Delta \ll 1$; $n_2 I \ll 1$)

$$\Delta = \Delta\varepsilon_L / 2n_0, \qquad n_2 I = \varepsilon_2 |E|^2 / 2n_0. \tag{36}$$

The first experiment to observe the effects suggested by theory (Kaplan 1976,

1977) was reported by Smith et al. (1979), and Smith and Tomlinson (1981a). In this experiment, the liquid CS_2 was utilized as a nonlinear medium with positive nonlinearity ($\varepsilon_2 > 0$), and glass was used as a linear medium. The temperature regulation of the cell served as a fine control of the difference $\Delta\varepsilon_L$ between both susceptibilities required by relationship, eq. (3). The experiment is briefly described below following (Smith et al. 1979, 1981, Smith and Tomlinson 1981a).

Experiments were performed with the apparatus shown in fig. 6. The input pulse was generated by a mode-locked traveling-wave ruby laser. Internal Fabry–Pérot mode selection was employed so that the laser output consisted of a train of ≈ 1 ns pulses. The absence of internal structure in these pulses was verified with a streak camera. A single pulse was selected from this train and directed onto the setup shown in fig. 6. A portion of the beam was sampled and directed with an optical delay of ≈ 6 ns onto a fast photodiode. The remainder of the beam was focused into the glass cell containing CS_2. The glass was chosen to have an index of refraction close to that of CS_2 at room temperature in order to satisfy the conditions of eq. (3). The entire cell was placed in a temperature-controlled holder and the temperature was adjusted so that for the 694.3 nm wavelength of the ruby laser, the index difference $\Delta \approx 10^{-3}$. This corresponded to $\psi_{cr} \approx 2.0°$ in the glass cell.

The reflected beam was monitored by the same photodiode used to monitor the incident pulse, and the diode output was displayed on a fast oscilloscope. The detector–oscilloscope combination had a measured response time of 320 ps. Figure 7a shows the measured incident and reflected pulse shapes for an incident intensity slightly above the threshold for hysteresis. The dashed curve is from calculations, based on the plane-wave theory (Kaplan 1976, 1977) [see eq. (14)], and was fitted to the data by adjusting the ratio of the peak intensity to the threshold intensity $I_0 = |E_{th1}|^2$, which corresponds to the switch from TIR to partial transmission. The fit becomes less good as the peak input intensity is increased (see fig. 8), however, the plane-wave theory (Kaplan 1977) still

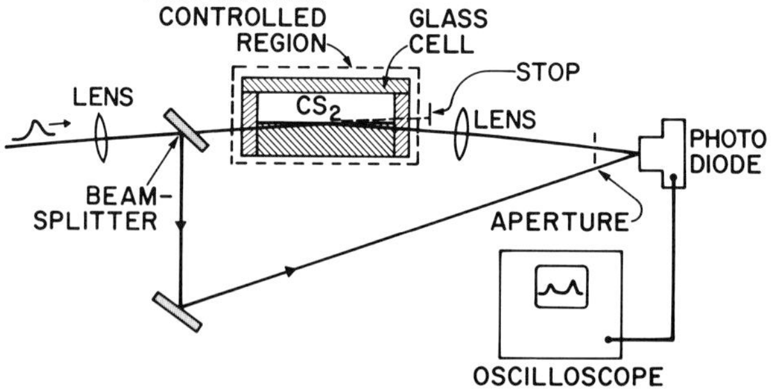

Fig. 6. Experimental set-up. (After Smith et al. 1979.)

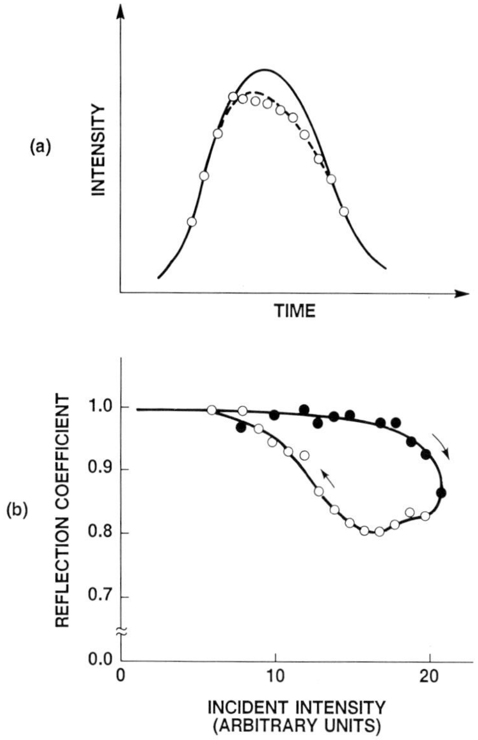

Fig. 7. Experimental measurements of incident and reflected pulse shapes for $\psi/\psi_c = \frac{1}{2}\sqrt{2}$. (a) The solid curve is the incident pulse, and the dots are the experimental measurements of the reflected pulse. The dashed line is the reflected pulse calculated from an incoherent plane-wave analysis for $I/I_0 = 1.125$ (see text). (b) The experimental data in (a) plotted in a way that demonstrates the optical hysteresis observed. (After Smith et al. 1979.)

accurately predicts the threshold intensity $I_0 = |E_{\text{th}1}|^2$ [eq. (24) above] for the onset of the switch from TIR to transmission. Moreover, one can see from figs. 7 and 8, that always (Kaplan et al. 1982)

$$I_2 \approx \tfrac{1}{2} I_0, \tag{37}$$

which is also consistent with the plane-wave theory prediction [see eq. (25) as compared with eq. (24)].

From the published value $n_2 = 3 \times 10^{-8}\ (\text{MW/cm}^2)^{-1}$ for CS_2 (Moran et al. 1975), the theoretical value of $I_0 = 8.1 \times 10^9\ \text{W/cm}^2$ was computed based on eq. (24) above. The experimentally measured value was $I_0 = 7.5 \times 10^9\ \text{W/cm}^2$, in very good agreement with the theoretical value. In fig. 7b, the same experimental data as in fig. 7a are shown but replotted in the form of reflectivity versus input intensity. This form emphasizes the optical hysteresis observed. However, because of the limited time resolution of the detection system, and the

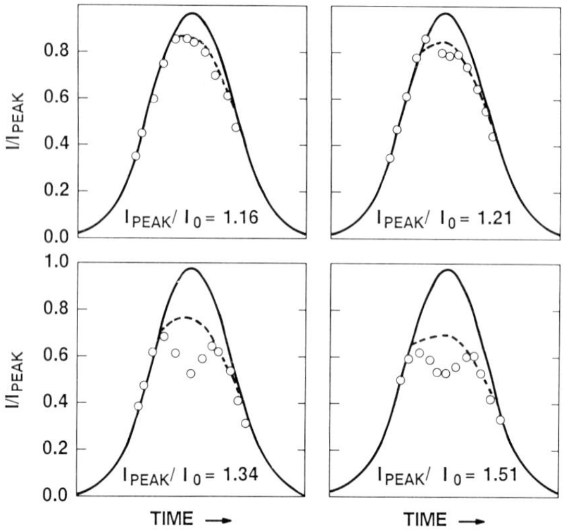

Fig. 8. Experimental measurements of incident and reflected pulse shapes for $\psi/\psi_{cr} = \frac{1}{2}\sqrt{2}$. The solid curves are the incident pulse and the dots are the experimental measurements of the reflected pulses. The dashed lines are from the incoherent plane-wave analysis fitted to all four curves with a single value of the adjustable parameter, I_0. (After Smith et al. 1981.)

fact that the experiments were only done for a single input pulsewidth, the observed hysteresis does not prove that the reflectivity is bistable.

Finally, the measured values of threshold intensity I_0 as a function of angle of incidence ψ are shown in fig. 9. We see that the results are again in good agreement with the predictions of the plane-wave theory (Kaplan 1977) [see eq. (24) above].

Some of the experimental results show substantial disagreement with plane-wave calculations. For instance, for sufficiently large peak input intensities, the measured reflected pulse shapes deviate significantly from those obtained from calculations. It was believed and seemingly backed by computer simulations by Smith et al. (1981), Smith and Tomlinson (1981a), Marcuse (1980) that at least some of these disagreements could be caused by the excitation of a nonlinear surface wave propagating along the interface with no decay, beginning from the point of reflection of the Gaussian beam. Indeed, it was shown by Tomlinson (1980) that a surface nonlinear wave in the form of a stationary channel can propagate along the interface, traveling from $x = -\infty$ to $+\infty$, in the absence of any wave incident from either linear or nonlinear media. [Similar waves were discussed previously (Litvak and Mironov 1968, Alanakyan 1967) in application to plasmas]. However, using a general theory of the stability of solitons of a cubic-nonlinear Schrödinger equation (Zakharov and Shabat 1971), it was pointed out by Kaplan (1981c) that if the surface nonlinear wave (in fact, a soliton solution) propagates in the direction $x \to \infty$ beyond the point where the

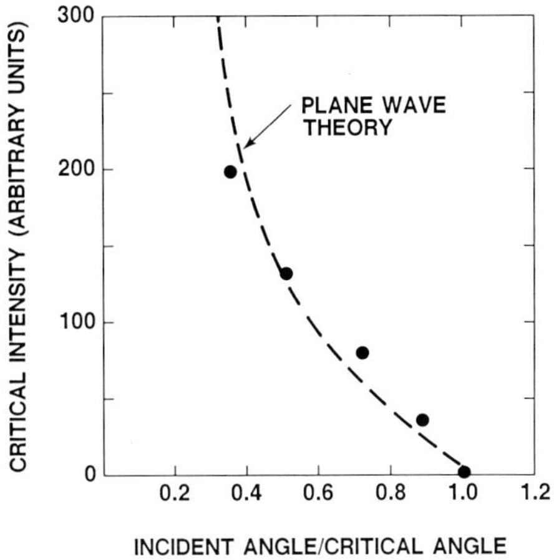

Fig. 9. Experimentally measured critical intensity, I_0, as a function of ψ/ψ_c. The dashed curve shows the critical intensity predicted by the plane-wave theory. (After Smith et al. 1981, Smith and Tomlinson 1981a.)

incident Gaussian beam hits the interface, then the same soliton-like surface wave (albeit with a spatially shifted maximum) must exist even before this point, i.e., this wave can be observed even when $x \to -\infty$. This would be inconsistent with the formulation of the problem (indeed, it is presumed that at $x \to -\infty$, the entire energy of the field is concentrated in the linear semi-space); therefore the 'semi-infinite' nonlinear surface wave (Smith and Tomlinson 1981a, Marcuse 1980) is prohibited.

The computer simulation (Tomlinson et al. 1982) confirmed this result. Furthermore, it showed that when the intensity of a laser Gaussian beam exceeds a critical level, the TIR state is drastically disturbed. The laser beam instead of being totally reflected, breaks through the interface and propagates in the nonlinear material away from the interface. The intensity of the beam at this moment is sufficiently high for the beam to form a self-trapped channel, as shown in fig. 10. If the incident intensity increases further, a second channel is formed (Tomlinson et al. 1982). Similar multisoliton emission from a nonlinear waveguide has recently been demonstrated in a computer simulation by Gubbels et al. (1987). It is natural to expect formation of more and more channels as the intensity increases, which would correspond to the multisoliton solution (Zakharov and Shabat 1971). The formation of each self-trapped channel is accompanied by a downward jump in the reflection coefficient (Tomlinson et al. 1982). Since the self-trapped channels form right at the interface, resulting in a strong spatial modulation of the intensity at the interface, it is clear that this phenomenon cannot be described by the plane-wave theory, in which it is assumed that the transmitted field is a homogeneous plane wave. The simula-

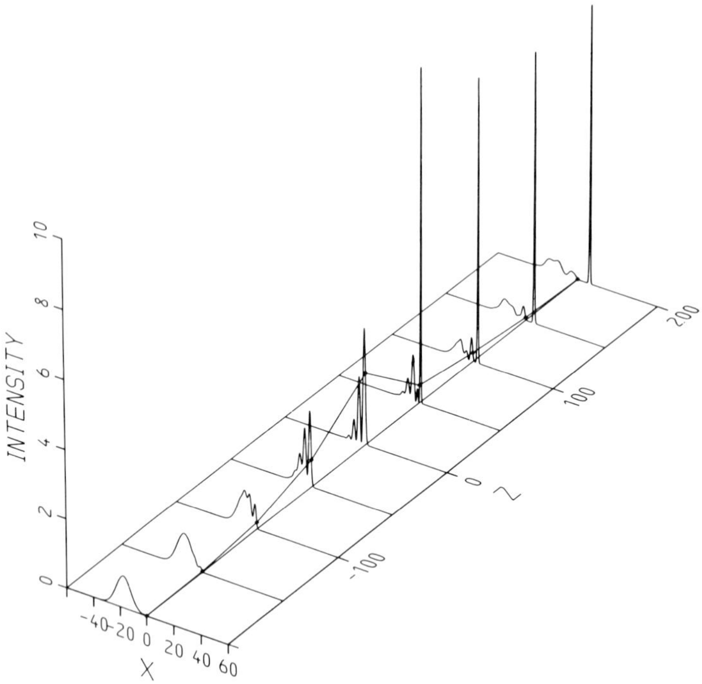

Fig. 10. Perspective plot showing the results of numerical simulations of the behavior of a nonlinear interface for an incident one-dimensional Gaussian beam. This plot is for an input intensity such that a self-focused channel forms in the nonlinear medium. For this figure (from Tomlinson et al. 1982), the definitions of the x- and z-coordinates are interchanged from those used in the rest of this paper.

tions used a steady-state model, which is not capable of proving, or disproving, the existence of bistability, but it was postulated that a nonlinear interface cannot exhibit bistability, because of the lack of any memory mechanism to store the information on the previous field configuration. (This argument does not exclude bistable behavior for the case of other nonlinearities, such as either thermal or diffusion nonlinearities, that are either non-instantaneous and/or non-local.)

In later experimental research (Smith and Tomlinson 1984), an 'artificial' nonlinear medium (dielectric spheres suspended in a liquid) was used in order to observe switching at a nonlinear interface in the cw regime. The results of these observations were very close to the computer results (Tomlinson et al. 1982). Figure 11 shows the results of the computer calculations for a one-dimensional Gaussian beam. The experimental reflectivity data shown in fig. 12 are in good qualitative agreement with this model. Perhaps more striking is the excellent agreement between the experimental data for the critical intensity for reflectivity 'jumps' and the predictions of the Gaussian-beam model (fig. 13).

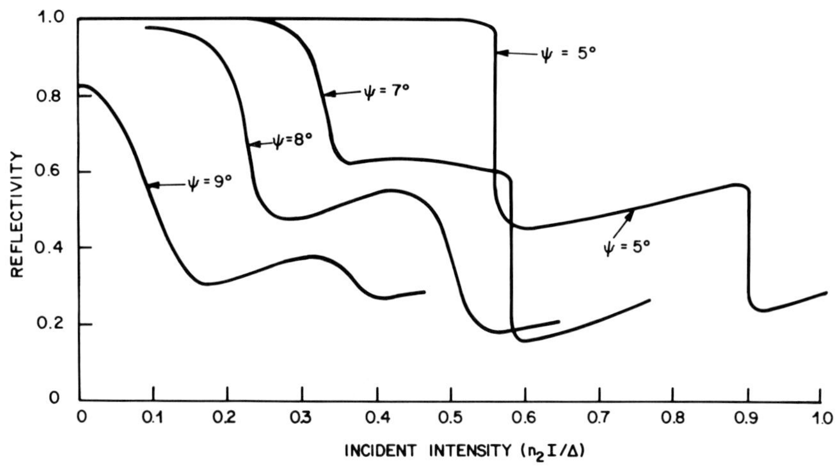

Fig. 11. Reflectivity versus intensity for a nonlinear interface: one-dimensional Gaussian-beam theory. (After Smith and Tomlinson 1984.)

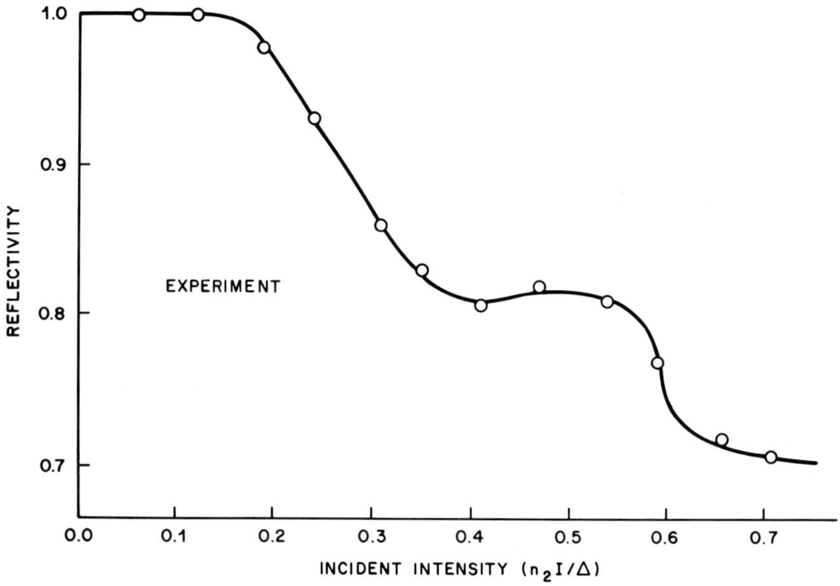

Fig. 12. Experimental plot of reflectivity versus incident intensity for $\psi = 6°$. (After Smith and Tomlinson 1984.)

For short observation times, hysteresis was observed. Only the upper branch of the reflection hysteresis characteristic was stable, however. The lower branch would decay to the upper branch with a surprisingly long time-constant of $\sim 10^3 \times$ (the response time of the nonlinear medium). The nature of this effect

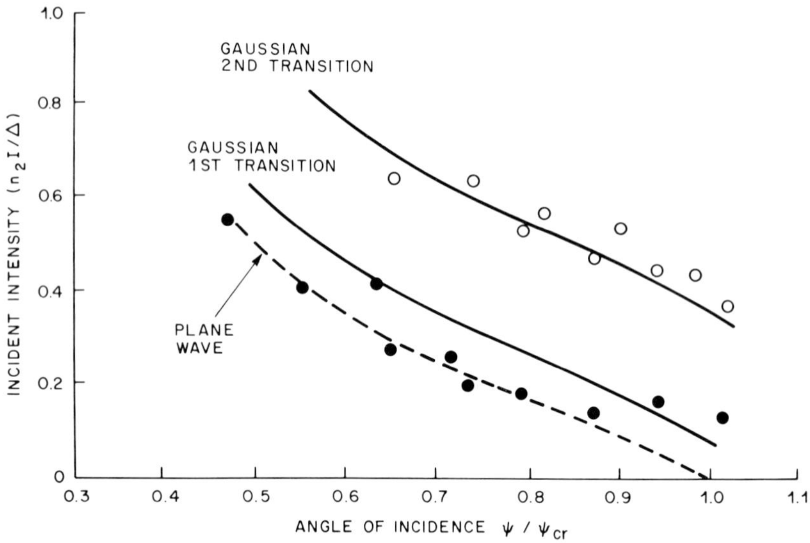

Fig. 13. Critical intensity for reflectivity 'jumps' as a function of angle of incidence. (After Smith and Tomlinson 1984.)

remains unclear, although it suggests, at least, that for any application with a pulse mode of operation, the nonlinear interface will demonstrate hysteresis. Indeed, the more recent experimental work (Delfino and Mormile 1985) done with essentially the same kind of media as that of Smith and Tomlinson (1984) using the long-pulse mode of operation, revealed very distinct hysteretic jumps. Furthermore, another more recent experiment (De Nicola et al. 1989) showed that under certain conditions the lower branch of the hysteretic curves can also be stable (see note added in proof). Delfino and Mormile (1985) have also experimentally observed the large nonlinear Goos–Hänchen effect suggested by Kaplan (1977) and obtained in a computer stimulation (Tomlinson et al. 1982). The Goos–Hänchen effect consists in the displacement of a reflected beam along the interface with regard to the point of incidence (see, e.g., Lotsch 1970, 1971, 1972). It was shown by Tomlinson et al. (1982) that for the intensities of the incident beam slightly lower than the critical intensity of formation of the first self-trapped channel, the point of reflection at the nonlinear interface is moving very far away from the point of incidence as the incident intensity increases.

6. *Conclusions*

In this chapter, we have attempted to review the present status of the theoretical analysis of nonlinear interface phenomena, and to present some of the available experimental data. In general, the experimental results are in good agreement with theoretical predictions. The major area where uncertainty remains is the question of optical bistability for a fast Kerr-nonlinear material and a single nonlinear interface, with the local nonlinearity. No clear evidence from either theory or experiment exists for the presence of two completely stable states when finite-size beams are involved.

There are many phenomena predicted of the theoretical analyses that have not yet been observed experimentally. This is primarily due to the difficulty of finding suitable nonlinear media. Recent studies of nonlinear coefficients in highly transparent optical glasses (Friberg and Smith 1990) may lead to all-glass interface structures that will allow detailed experimental verification of many of the novel phenomena predicted by the analysis presented in this paper.

Acknowledgment

The work by A.E. Kaplan is supported by AFOSR.

Note added in proof

Since the original draft of this chapter was submitted, quite a few publications on the subject have appeared. Space does not allow us to review them in detail or even make a complete listing of them; we will briefly survey only those recent publications that are closely related to the issue discussed here.

Since the very existence of CW hysteresis and optical bistability, as well as a nonlinear surface wave in the situation when the incident beam has limited size, remained in question, numerous studies have been undertaken to clarify this situation. Computer simulation (Andersen 1988) showed that if the EM wave interaction with a Kerr-nonlinear material at a nonlinear interface is diffusive (or nonlocal), a nonlinear surface wave can couple to an incident Gaussian beam, a situation prohibited for a regular Kerr-nonlinear interface (Kaplan 1981c, Tomlinson et al. 1982). In an experiment (De Nicola et al. 1989) on the reflection of laser beam from the interface between a linear (glass) and an "artificial" nonlinear material (i.e. dielectric spheres suspended in liquid) similar to that of Smith and Tomlinson (1984), the reflection versus incident intensity curve exhibited hysteresis, and the lower branch of the hysteresis loop could be very stable and robust for optimized conditions. This indicated that a real CW hysteresis (and, therefore, optical bistability) can exist in the system, and it was attributed by the authors to the nonlocal property of interaction. In another experiment (Strobl and Cuykendall 1989, 1990), switching at a nonlinear interface (also with an "artificial" nonlinear material) was observed with enhanced contrast, and this was attributed by the authors to the saturation of the nonlinearity, which results in suppression of the multiple self-trapped channels and associated multiple thresholds observed earlier (Smith and Tomlinson 1984), see figs. 11 and 12 here. An apparently related effect of stable (i.e. not broken into self-trapped channels) propagation of the beam refracted from a nonlinear interface between glass and a Kerr-nonlinear liquid has been reported by Li et al. (1990). Self-induced reflection (Mohebi et al. 1989) as well as energy limiting (Mohebi et al. 1990) have also been experimentally observed at a nonlinear interface between a dye solution and fused silica, using the large resonant nonlinearity of the saturably-absorbing dye solution.

Nonlinear interfaces with liquid-crystal layers have attracted much interest recently, apparently because of relative experimental simplicity: CW lasers with moderate power and almost any wavelength usually suffice as source of radiation. In most cases the originally expected nonlinearity due to reorientation of molecules in the optical field did not produce any significant contribution; the observed optical bistability was basically due to a thermal effect. Analytical as well as experimental investigations (Peschel et al. 1988, Boiko et al. 1988, Khoo et al. 1989, Haelterman and Waelbroek 1990) of optical transmission (and reflection) through a nonlinear layer, in particular a liquid-crystal layer sandwiched by glass prisms in a configuration similar to that of Khoo and Hou (1985), confirmed the existence of large hysteresises and optical bistability (with the nonlinearity apparently attributed to thermal effects); computer simulation of its dynamics was done by Khoo and Zhou (1989). Bistability due to surface plasmon-polaritons at a liquid-crystal/glass interface has been observed and interpreted by Martinot et al. (1985) and Innes and Sambles (1989). Arakelian et al. (1986) experimentally observed bistable excitation of a nonlinear surface waves in a prism-metal liquid-crystal film using a coupling technique originally described by Chen and Carter (1982, 1983). Although CW bistable effects in liquid crystals are unusually well pronounced, they seem to have marginal application potential, aside from demonstration purposes, because of the very slow switching time of the thermal nonlinearity.

Some new configurations related to nonlinear interfaces have been proposed and investigated. Theoretical work by Voronko et al. (1986) discussed a multistable excitation of a channeled surface polariton in a nonlinear resonator with mirrors reflecting the surface wave back on itself. Bistability and redirecting of optical beams has been observed near Brewster's angle by Schülzgen et al. (1988) using a nonlinear ZnSe prism. The possibility of bistable effects using normal incidence at a nonlinear dispersive interface has been discussed by Cao and Banerjee (1989) and Banerjee et al. (1990). The theoretical and computer simulation aspects of the physics of nonlinear interfaces are of significant interest, especially for the realistic configurations. Extensive computer simulations of a three-dimensional Gaussian beam incidence on a nonlinear interface (Andersen and Regan 1989) revealed behavior patterns some of which differ qualitatively from those for the two-dimensional simulation (Tomlinson et al. 1982). However, for Kerr-nonlinear material these simulations verified the absence of a nonlinear surface wave coupled to the incident beam. Some interesting results were obtained in computer simulation (Aceves et al. 1989) of reflection of soliton-like light beams at nonlinear–nonlinear interfaces, although these results cannot be extended to the linear–nonlinear interfaces. An analytical theory of radiative losses of solitons in these systems has been developed by Kivshar et al. (1990).

References

Aceves, A.B., J.V. Moloney and A.C. Newell, 1989, Phys. Rev. A **39**, 1809.

Alanakyan, Yu.R., 1967, Sov. Phys.-Tech. Phys. **12**, 587.

Altshuller, G.B., V.S. Ermolaev, K.I. Krylov, M.A. Makarov and L.I. Pavlov, 1985, Opt. Commun. **56**, 131.

Andersen, D.A., 1988, Phys. Rev. A **37**, 189.

Andersen, D.A., and J.J. Regan, 1989, J. Opt. Soc. Am. B **6**, 1484.

Arakelian, S.M., G.L. Grigorian, L.M. Kocharian, S.Ts. Nersissaian and Yu.S. Chilingarian, 1986, Phys. Lett. A **118**, 254.

Askar'yan, G.A., 1962, Sov. Phys.-JETP **15**, 1088.

Banerjee, P.P., M.R. Chatterjee and M. Maghraoui, 1990, J. Opt. Soc. Am. **7**, 21.

Bloembergen, N., and J. Ducuing, 1963, Phys. Lett. **6**, 5.

Bloembergen, N., and C.H. Lee, 1967, Phys. Rev. Lett. **19**, 835.

Bloembergen, N., and P.S. Pershan, 1962, Phys. Rev. **128**, 606.

Boiko, B.B., I.Z. Dzhilavdari and N.S. Petrov, 1975, Appl. Spectrosc. **23**, 1511.

Boiko, B.B., N.S. Petrov, V.A. Shakin and A.B. Zimin, 1988, Phys. Status Solidi b **150**, 527.

Bosacchi, B., and L.N. Narducci, 1983, Opt. Lett. **8**, 324.

Butylkin, V.S., A.E. Kaplan, Yu.G. Khronopulo and E.I. Yakubovich, 1990, in Resonant Nonlinear Interactions of Light With Matter, to be published (Springer, Berlin).

Cao, G., and P.P. Banerjee, 1989, J. Opt. Soc. Am. B **6**, 191.

Chen, Y.J., and G.M. Carter, 1982, Appl. Phys. Lett. **41**, 307.

Chen, Y.J., and G.M. Carter, 1983, Solid State Commun. **45**, 643.

Chiao, R.Y., E. Garmire and C.H. Townes, 1964, Phys. Rev. Lett. **13**, 479.

Cuykendall, R., and D.R. Andersen, 1987, Opt. Lett. **12**, 542.

Cuykendall, R., and K. Strobl, 1989, J. Opt. Soc. Am. **6**, 877.

De Nicola, S., A.E. Kaplan, S. Martelluci, P. Mormile, G. Pierattini and J. Quartieri, 1989, Appl. Phys. B **49**, 441.

Delfino, G., and P. Mormile, 1985, Opt. Lett. **10**, 618.

Dlodlo, T.S., 1981, Phys. Lett. A **84**, 107.

Dlodlo, T.S., 1983, Phys. Rev. A **27**(2).

Felber, F.S., and J.H. Marburger, 1976, Appl. Phys. Lett. **28**, 731.

Friberg, S., and P.W. Smith, 1987, IEEE J. Quantum Electron. **QE-23**, 2089.

Gibbs, H.M., 1985, Optical Bistability (Academic Press, New York).

Gibbs, H.M., S.L. McCall and T.N.C. Venkatesan, 1976, Phys. Rev. Lett. **36**, 1135.

Gubbels, M.A., E.M. Wright, G.I. Stegeman, C.T. Seaton and J.V. Moloney, 1987, Opt. Lett. **4**, 1837.

Haelterman, M., and C. Waelbroeck, 1990, Appl. Phys. Lett. **56**, 512.

Hickernell, R.K., and D. Sarid, 1986, J. Opt. Soc. Am. B **3**, 1059.

Innes, R.A., and J.R. Sambles, 1989, J. Phys. Condens. Matter **1**, 6231.

Inoue, M., 1987, Phys. Rev. Lett. **58**, 871.
Kaplan, A.E., 1969, JETP Lett. **9**, 33.
Kaplan, A.E., 1976, JETP Lett. **24**, 114.
Kaplan, A.E., 1977, Sov. Phys.-JETP **45**, 896.
Kaplan, A.E., 1978, Sov. J. Quantum Electron. **8**, 95.
Kaplan, A.E., 1979, Radiophys. & Quantum Electron. **22**, 229.
Kaplan, A.E., 1981a, IEEE J. Quantum Electron. **QE-17**, 336.
Kaplan, A.E., 1981b, in: Optical Bistability, eds C.W. Bowden, M. Ciftan and M.R. Robl (Plenum Press, New York) p. 447.
Kaplan, A.E., 1981c, J. Opt. Soc. Am. **71**. 1640.
Kaplan, A.E., 1981d, Appl. Phys. Lett. **38**, 67.
Kaplan, A.E., Yu. A. Kravtsov and V.A. Rylov, 1966, Parametric Oscillators and Frequency Dividers (Sov. Radio, Moscow) in Russian.
Kaplan, A.E., P.W. Smith and W.J. Tomlinson, 1982, Proc. SPIE **317**, 305.
Kelley, P.L., 1965, Phys. Rev. Lett. **15**, 1005.
Khoo, I.C., and J.Y. Hou, 1985, J. Opt. Soc. Am. B **2**, 761.
Khoo, I.C., and P. Zhou, 1989, J. Opt. Soc. Am. **6**, 884.
Khoo, I.C., P. Zhou, R.R. Michael, R.G. Lundquist and R. Mansfield, 1989, IEEE J. Quantum Electron. **QE-25**, 1755.
Kivshar, Yu.S., A.M. Kosevich and O.A. Chubykalo, 1990, Phys. Rev. A **41**, 1677.
Kolokolov, A.A., and A.I. Sukov, 1978, Radiophys. & Quantum Electron. **21**, 1013.
Li, H.W., G. Martinelli, Cl. Froehly and A. Lacourt, 1990, Opt. Commun. **75**, 321.
Litvak, A.G., and V.A. Mironov, 1968, Izv. Vyssh. Uchebn. Zaved. Radiofiz. **11**, 1911.
Lotsch, H., 1970 and 1972, Optik **32**, 776, 189, 299, 553.
Marcuse, D., 1980, Appl. Opt. **19**, 3130.
Martinot, P., A. Koster and S. Laval, 1985, IEEE J. Quantum Electron. **QE-21**, 1140.
Martinot, P.A., A. Koster and L. Laval, 1985, IEEE J. Quantum Electron. **QE-21**, 1140.
Miller, D.A.B., S.D. Smith and B.S. Wherett, 1980, Opt. Commun. **35**, 221.
Mohebi, M., B. Jean-Jean and J.-C. Diels, 1989, Opt. Lett. **14**, 1353.
Mohebi, M., M. Lai, B. Jean-Jean and J.-C. Diels, 1990, Opt. Commun. **75**, 149.
Moran, M.J., C.-Y. She and R.L. Carmen, 1975, IEEE J. Quantum Electron. **QE-11**, 259.
Permyakov, V.A., and O.V. Bagdasaryan, 1978, Radiophys. & Quantum Electron. **21**, 92.
Peschel, Th., P. Danberg, U. Langbein and F. Lederer, 1988, J. Opt. Soc. Am. B **5**, 29.
Rosanov, N.N., 1977, Sov. Tech. Phys. Lett. **3**, 583.
Rosanov, N.N., 1978, Sov. Tech. Phys. Lett. **4**, 30.
Rosanov, N.N., 1979, Opt. Spectrosc. **47**, 335.
Schülzgen, A., F. Henneberger and H. Rossmann, 1988, Phys. Status Solidi b **150**, 495.
Seaton, C.T., J.D. Valera, R.L. Shoemaker, G.I. Stegeman, J.T. Chilwell and S.D. Smith, 1985, IEEE J. Quantum Electron. **QE-21**, 774.
Seidel, H., 1969, U.S. Patent No. 3, 610, 731.
Slepyan, G.Ya., 1985, Radio Eng. & Electr. Phys. **8**.
Smith, P.W., and W.J. Tomlinson, 1981a, in: Optical Bistability, eds C.W. Bowden, M. Ciftan and M.R. Robl (Plenum Press, New York) p. 463.
Smith, P.W., and W.J. Tomlinson, 1981b, IEEE Spectrum **18**(6), 26.
Smith, P.W., and W.J. Tomlinson, 1984, IEEE J. Quantum Electron. **QE-20**, 30.
Smith, P.W., J.-P. Hermann, W.J. Tomlinson and P.J. Maloney, 1979, Appl. Phys. Lett. **35**, 846.
Smith, P.W., W.J. Tomlinson, P.J. Maloney and J.-P. Hermann, 1981, IEEE J. Quantum Electron. **QE-17**, 340.
Smith, P.W., W.J. Tomlinson, P.J. Maloney and A.E. Kaplan, 1982, Opt. Lett. **7**, 57.
Stegeman, G.I., 1982, IEEE Trans. Microwave Theory & Tech. **MTT-30**, 1598.
Stoker, J.J., 1950, Nonlinear Vibrations in Mechanical and Electrical Systems (Interscience).

Strobl, K.H., and R. Cuykendall, 1989, Phys. Rev. A **40**, 5143.
Strobl, K.H., and R. Cuykendall, 1990, Opt. Lett. **15**, 96.
Szöke, A., V. Daneu, T. Goldhar and N.A. Kurnit, 1969, Appl. Phys. Lett. **15**, 376.
Talanov, V.I., 1964, Izv. Vuzov. Radiofizika **7**, 564 [1965, JETP Lett. **2**, 138].
Tomlinson, W.J., 1980, Opt. Lett. **5**, 323.
Tomlinson, W.J., J.P. Gordon, P.W. Smith and A.E. Kaplan, 1982, Appl. Opt. **21**, 2041.
Vach, H., C.T. Seaton, G.I. Stegeman and I.C. Khoo, 1984, Opt. Lett. **9**, 238.
Voronko, A.I., L.G. Klimova and G.N. Shkerdin, 1986, Sov. Phys.-Solid State **28**, 602.
Wysin, G.M., H.J. Simon and R.T. Deck, 1981, Opt. Lett. **6**, 30.
Zakharov, V.E., and A.B. Shabat, 1971, Sov. Phys.-JETP **38**, 62.

CHAPTER 5

Second-Order Nonlinear Optical Effects at Surfaces and Interfaces

T.F. HEINZ

IBM Research Division
T.J. Watson Research Center
Yorktown Heights, NY 10598, USA

Nonlinear Surface Electromagnetic Phenomena
Edited by
H.-E. Ponath and G.I. Stegeman

© *Elsevier Science Publishers B.V., 1991*

Contents

1. Introduction

Second-order nonlinear optical effects in centrosymmetric media offer a challenging and intriguing area of study. On the one hand, problems falling within this rubric include such fundamental issues as the lowest-order nonlinear response of free electrons. On the other hand, the process is symmetry forbidden in the electric-dipole approximation and is, consequently, governed by the subtle higher-order response in the bulk of a material and by the delicate influence of surfaces and interfaces of the medium. These considerations alone make the second-order nonlinear optical effects of second-harmonic generation (SHG) and sum-frequency generation (SFG) or difference-frequency generation well worthy of study.

An additional impetus also exists for obtaining a complete understanding of second-order nonlinear optical effects in centrosymmetric media: the processes of SHG and SFG have been developed into powerful and versatile probes of surfaces and interfaces. From fundamental symmetry considerations, the nonlinear signals exhibit a strong sensitivity to surface and interface properties on an atomic scale. Yet as a purely optical probe, the technique offers several advantages not found in conventional methods of surface analysis, such as applicability to a wide range of surfaces and interfaces and the possibility for measurements with superlative resolution in time and energy. Of course, to exploit the technique effectively, an appropriate theoretical foundation is necessary.

1.1. Scope of chapter

It is the aim of this chapter to outline key aspects of the theory of second-order nonlinear optical effects in centrosymmetric media. No attempt is made to review systematically the applications of SHG or SFG in studies of the chemical and physical properties of surfaces or interfaces. Even within this more limited scope the macroscopic and microscopic aspects of second-order nonlinear optical effects are not discussed in full. From the macroscopic perspective of wave propagation effects, essentially no treatment of surface or guided wave effects has been given. Excellent reviews of work in this area up until the early 1980s by Sipe and Stegeman (1982), Shen and de Martini (1982) and Fukui and Stegeman (1982) are available. An overview of surface nonlinear optical effects, including SHG by surface plasmons, can also be found in the text by Shen (1984). The more recent work on surface enhancement effects for second-order nonlinear optical processes is not examined in this chapter. On the other hand, a rather detailed phenomenological description is presented of the simple problem of the second-order nonlinear response at an interface between two flat centrosymmetric media under excitation by waves propagating in the bulk. The effect of the bulk and surface terms is discussed both for isotropic and anisotropic (cubic) media. Explicit formulas are derived and the relation between this description of the nonlinear process and experimental measurements is developed; experimental technique is examined in this context. In regard to the microscopic theory, the principal approaches applied to analyze the second-order nonlinear processes in centrosymmetric media are outlined. The prime emphasis has been placed on relating these microscopic descriptions to one another, discussing their general range of application, and integrating them into the common phenomenological description. Given the broad range of microscopic schemes, from aligned molecular adsorbates to density-functional calculations on a surface of jellium, the discussion of each microscopic model is, of necessity, rather limited.

Several recent review articles have dealt with second-order nonlinear optical effects as probes of surface or interface properties. Applications are surveyed in current reviews by Shen (1989a), McGilp (1989), and by Heinz and Reider (1989), and more complete discussions of the technique by Shen (1986a,b, 1989b). A very comprehensive review of SHG at surfaces and interfaces has been published by Richmond et al. (1988), which provides an especially extensive discussion of its applications in electrochemistry. Two recent papers also describe progress in the area of SHG by the surfaces of semiconductors, with particular emphasis on anisotropic effects (Akhmanov et al. 1985, Heinz and Loy 1988).

1.2. Historical perspective

Optical second-harmonic generation was initially observed by Franken et al. (1961) in a noncentrosymmetric quartz crystal. Second-harmonic generation in the centrosymmetric calcite crystal was reported shortly thereafter by Terhune et al. (1962). Over the following few years, SHG was observed in a wide range of centrosymmetric materials, including metals (Brown et al. 1965, Brown and Parks 1966, Bloembergen et al. 1966, Sonnenberg and Heffner 1968, Krivoshchekov and Stroganov 1968, 1969), semiconductors (Bloembergen et al. 1966, Bloembergen and Chang 1966), and insulating solids (Wang and Duminski 1968) and liquids (Wang 1969). Much of the early experimental and theoretical work is summarized in a major paper by Bloembergen et al. (1968). The theoretical framework developed by Bloembergen et al. involved a weak (generalized) polarization in the bulk arising from the (leading-order nonlocal) magnetic-dipole and quadrupole terms and a thin strongly polarized layer in the perturbed surface region. This still remains the scheme for a phenomenological description of SHG in centrosymmetric media. The radiation arising from such a source of polarization could be determined from the appropriate limits of radiation from a polarized slab using the analysis of Bloembergen and Pershan (1962). The nonlinear response of the surface region was deduced from the discontinuity of the pump fields in conjunction with the quadrupole terms. The surface contribution was estimated with coefficients for the quadrupole response based on bulk properties. This theoretical approach, based on bulk properties, was consistent with the apparent lack of sensitivity of the SH response to surface conditions.

The high sensitivity of the SH process to surface conditions was first noted by Brown and Matsuoka (1969). In contrast to earlier work under ambient conditions, they observed a much stronger SHG from a freshly evaporated Ag film held under vacuum than from a film exposed to air. An even more dramatic dependence of the SH on the atomic nature of the surface was seen by Chen and co-workers for Na deposition on clean Ge surfaces (Chen et al. 1973, 1974). Although these investigations clearly demonstrated the potential of SHG for surface studies of centrosymmetric media, this avenue was not pursued vigorously until the 1980s.

In the intervening period, interest was focussed primarily on metal surfaces. Important theoretical developments included the recognition by Rudnick and Stern (1971, 1974) that the broken inversion symmetry of the surface (as distinct from the rapid variation of the pump electric field) should constitute an important source of surface SHG. Subsequently, a hydrodynamic model was developed by Sipe and Stegeman (1982) to describe the surface response with a more realistic (semi) classical model than had been used previously. A further significant development was the excitation of SH radiation by surface plasmons, as first demonstrated by Simon et al. (1974). Much of the subsequent work along these lines has been reviewed by Sipe and Stegeman (1982), Shen and de Martini (1982), Fukui and Stegeman (1982).

The recent interest in second-order nonlinear optical effects at surfaces was triggered by the investigations of surface enhancement effects. Just as in Raman scattering from Ag surfaces (Chang and Furtak 1982), it was found that the surface SHG process could be increased in strength by many orders of magnitude (Chen et al. 1981a). [Subsequent studies of surface-enhanced SHG are reviewed by Richmond et al. (1988) and will not be discussed in this chapter.] It was also recognized (Chen et al. 1981b, Heinz et al. 1981) that SHG may serve as a probe of surface reactions and of adsorption/desorption phenomena with sensitivity on the level of a fractional monolayer even in an electrochemical environment. The development of second-order nonlinear optical effects as probes of surfaces and interfaces has followed rapidly since that time, yielding both new analytic capabilities and an improved understanding of the nonlinear process itself.

The possibility of examining relative adsorbate density by SHG was exploited immediately in measurements of equilibrium and dynamic aspects of adsorption from solution (Chen et al. 1981c). The technique was demonstrated to have the potential for surface-specific spectroscopy through the resonant SHG process (Heinz et al. 1982). When the SH response could be attributed to the nonlinear polarizability of a molecular monolayer, it was realized that surface nonlinear susceptibility depended sensitively not only on the nonlinear polarizability of the molecules, but also on their alignment. This led to a scheme for inferring molecular orientation of adsorbed molecules from measurements of the SH polarization dependence, i.e., through a determination of the independent elements of the surface nonlinear susceptibility (Heinz et al. 1983). Of course, the second-order nonlinear response of the surface should reflect not only the structure of isotropic media, but that of materials with long-range order as well. That such is indeed the case was proved by the marked dependence of the SH response on crystallographic orientation observed for oxidized Si surfaces (Guidotti et al. 1983, Driscoll and Guidotti 1983, Tom et al. 1983). As we discuss below, this effect can arise from anisotropic terms in either the surface or the bulk. The importance of the surface terms was demonstrated in investigations of SHG from clean Si(111) faces with different surface reconstructions (Heinz et al. 1985a). This development opened the way to studies of surface symmetry and order, as exemplified by investigations of the phase transformation between the two different reconstructions of the Si(111) surface (Heinz et al. 1985a). The remarkable sensitivity of the SH process to the detailed nature of the surface has also been exemplified in a variety of studies of clean and adsorbate-covered surfaces under ultrahigh vacuum conditions (Tom et al. 1984, 1986a, b, Zhu et al. 1985, 1988, 1989, Mate et al. 1986, Song et al. 1987, 1988, Heskett et al. 1986, 1988, Heuer et al. 1987, Grubb et al. 1988a, Anderson and Hamilton 1988, Hamilton and Anderson 1988, Rosenzweig and Asscher 1988, Li et al. 1988, 1989, Holdering et al. 1990a, b).

In the past few years, SHG has served for investigations of the surface properties on the atomic and molecular scale in a wide range of physical systems, including the electrochemical environment (Richmond et al. 1988, Murphy et al. 1983, Chen et al. 1984, Corn et al. 1984a, b, Campbell and Corn 1987, 1988, Voss et al. 1986, Furtak et al. 1987, Miragliotta and Furtak 1988, Marshall and Korenowski 1986, 1987, Biwer et al. 1986, 1988), surfaces and interfaces of liquids (Rasing et al. 1985a, b, 1988, 1989, Berkovic et al. 1986, Guyot-Sionnest et al. 1986a, Grubb et al. 1988b, Hicks et al. 1986, Kemnitz et al. 1986, Goh et al. 1988, Goh and Eisenthal 1989, Bhattacharyya et al. 1987, 1988, Castro et al. 1989), molecular adsorbates on solid surfaces (Aktsipetrov et al. 1983, 1985, 1987a, b, Reider et al. 1983, Marowsky et al. 1985, 1987, 1988a, b, Dick et al. 1985, Felderhof and Marowsky 1987a, Epperlein et al. 1987, Marowsky and Steinhoff 1988, Chen et al. 1985, van Wyck et al. 1985, di Lazzaro et al. 1985, Girling et al. 1985, 1987) and solid/solid interfaces (Heinz et al. 1989). One of the distinctive advantages of a purely optical technique is the possibility for extremely high time resolution. This characteristic has already been exploited in a few studies of interesting physical processes occurring on rapid time scales (Shank et al. 1983, Tom et al. 1988, Arjavalingam et al. 1986, Sitzman and Eisenthal 1988, 1989, Hicks et al. 1988, Robinson and Richmond 1989). A very important recent development has been the successful demonstration of infrared-visible SFG. This permits the surface-specific technique to be extended to perform vibrational spectroscopy by resonant infrared excitation (Zhu et al. 1987, Hunt et al. 1987, Guyot-Sionnest et al. 1987a, 1988, Superfine et al. 1988, Harris et al. 1987, Harris and Levinos 1989).

In this chapter, we focus on the unifying theoretical developments that have accompanied this surge of interest and activity in second-order nonlinear optical effects at surfaces. As the reader will see, significant advances have been made in a variety of areas. For example, the macroscopic description of the nonlinear process in a centrosymmetric medium has been refined to include anisotropic effects, and the manner in which the bulk longitudinal polarization enters into SHG measurements has been clarified. Significant progress has also been made in the development of the microscopic theory, as we outline in section 4.

2. *Macroscopic description*

2.1. *Phenomenological model*

To describe a second-order nonlinear optical response at an interface between two centrosymmetric media, it is useful to divide the matter into three regions: the interfacial zone and the two bulk media. From the standpoint of developing a phenomenological model, we wish to include in the interfacial region all of the material contributing strongly to the nonlinear response. One suitable definition of the interface is the zone where either the material properties (electronic structure, molecular orientation, etc.) or the electromagnetic fields differ appreciably from those found in bulk media. For most systems, this transitional behavior will occur within a distance of a few Ångstroms. Thus, as far as nonlinear radiation is concerned, we can treat the nonlinearity of the interface as a sheet of current or, equivalently, as a sheet of (generalized) nonlinear source polarization $\boldsymbol{P}_s^{\text{nls}}$, with $\partial \boldsymbol{P}^{\text{nls}}/\partial t = \boldsymbol{J}_s^{\text{nls}}$.

This sheet of polarization is determined assuming a second-order nonlinear response to the electric fields on either side of the interface. Thus a surface nonlinear susceptibility can be defined by*

$$\boldsymbol{P}_s(\Omega) = \chi_s^{(2)}(\Omega = \omega_1 + \omega_2) : \boldsymbol{E}(\omega_1)\boldsymbol{E}(\omega_2). \tag{1}$$

The actual numerical values for $\chi_s^{(2)}$ will reflect the convention adopted for the location where the electric fields in eq. (1) are measured. Differing approaches have appeared in various works, depending on the physical system being treated. Although this situation can be somewhat confusing, it does not present any fundamental difficulty, since the pump electric fields on either side of the interface are identical for the tangential components and differ only by well-defined ratios of the dielectric constants for the normal components. To

*Fourier components in this chapter are related to the physical quantities by $\boldsymbol{E}(t) = \boldsymbol{E}(\omega)e^{-i\omega t} + \text{c.c.}$

eliminate this ambiguity, however, it has been suggested by Guyot-Sionnest et al. (1986b) that the surface nonlinear susceptibility be defined with respect to tangential components of $\boldsymbol{E}$ and normal components of the displacement vector $\boldsymbol{D}$. Another common convention (Mizrahi and Sipe 1988), originally introduced in modeling the free-electron response of a metal surface, assumes the pump fields to be measured inside the medium, but treats the polarized sheet as if radiating in vacuum.

A useful, widely adopted phenomenological model (Heinz 1982, Heinz et al. 1983, Aktsipetrov et al. 1985, Dick et al. 1985, Felderhof and Marowsky 1987, Girling et al. 1985, 1987, Guyot-Sionnest et al. 1987b) consists of relating the surface nonlinear polarization to the fields present in an interfacial region characterized by a dielectric constant ε'. In the model, it is further assumed that the nonlinear source polarization of the interfacial region sees a linear dielectric response characterized by ε'*. The complete phenomenological model is summarized in fig. 1. The bulk linear dielectric response is taken to be isotropic and described by ε_1 and ε_2, which will generally exhibit dispersion. For centrosymmetric media, the bulk nonlinearity is absent in the electric-dipole approximation. The leading-order nonlocal response, characterized by the fourth-rank tensor $\chi_q^{(2)}$, must, however, be considered. It is introduced in section 2.5.

As we shall discuss in section 2.6, we expect this phenomenological description of the properties of the interface interacting with radiation to be valid provided that the interfacial layer is much less than a wavelength in thickness. This does not imply that the actual spatial variation of the linear optical fields is abrupt. By the same token, the nonlinear response will generally have a complex nonlocal and spatially varying nature. The quantities ε' and $\chi_s^{(2)}$ simply parameterize the measurable response in optical experiments. Microscopic schemes for determining ε' and $\chi_s^{(2)}$ will be outlined in section 4.

2.2. Boundary conditions of a polarized sheet

The solution for the transmitted and reflected nonlinear radiation from the interface of fig. 1 can be determined by solving Maxwell's equations with a nonlinear source polarization occupying a finite volume and then taking the limit as the thickness goes to zero. The complete problem for a slab of finite thickness was solved early in the history of nonlinear optics by Bloembergen and Pershan (1962). An alternative approach is to derive boundary conditions to match electric and magnetic fields across the polarized sheet. With these

*The convention of Guyot-Sionnest et al. (1986b) is equivalent to taking $\varepsilon' = 1$ at all frequencies. The scheme of measuring the fields in the medium and placing the polarized sheet in vacuum corresponds to setting $\varepsilon' = \varepsilon_2$ at the pump frequencies and $\varepsilon' = 1$ at the second-harmonic or sum frequency.

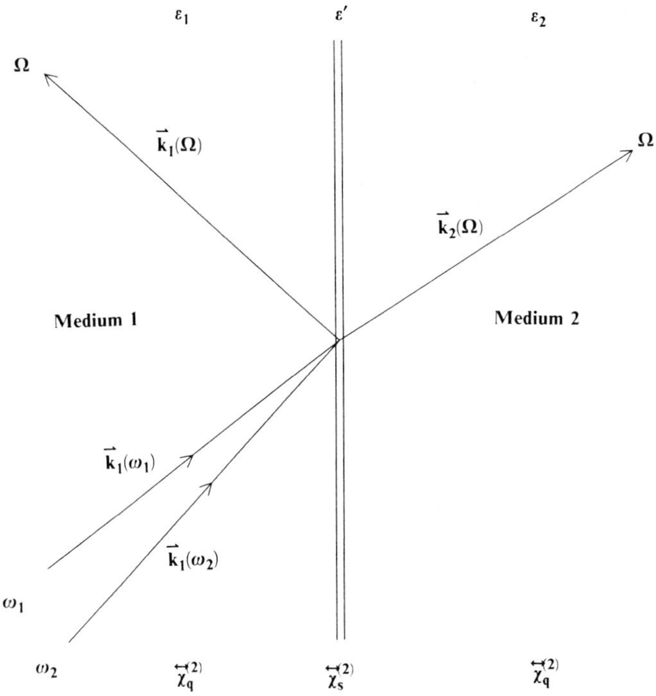

Fig. 1. Schematic representation of the phenomenological model for second-order nonlinear optical effects at the interface between two centrosymmetric media. Pump waves at frequencies ω_1 and ω_2 are assumed to approach the interface from medium 1 with wave vectors $\boldsymbol{k}_1(\omega_1)$ and $\boldsymbol{k}_2(\omega_2)$. Radiated sum-frequency radiation at frequency Ω is emitted in directions described by wave vectors nonlocal response is designated by $\chi_q^{(2)}$. The figure shows the xz-plane with z increasing from left to right and $z = 0$ defining the interface.

boundary conditions in hand, the radiated fields can be determined immediately.

The starting point of the derivation of the boundary conditions is the set of Maxwell's equations with a nonlinear source polarization $\boldsymbol{P}^{\text{nls}}$,

$$\nabla \cdot \boldsymbol{D} = -4\pi \nabla \cdot \boldsymbol{P}^{\text{nls}}, \tag{2a}$$

$$c\nabla \times \boldsymbol{E} + \frac{\partial \boldsymbol{B}}{\partial t} = 0, \tag{2b}$$

$$\nabla \cdot \boldsymbol{B} = 0, \tag{2c}$$

$$c\nabla \times \boldsymbol{H} - \frac{\partial \boldsymbol{D}}{\partial t} = 4\pi \frac{\partial \boldsymbol{P}^{\text{nls}}}{\partial t}. \tag{2d}$$

Here the fields $\boldsymbol{D}$ and $\boldsymbol{H}$ are related to the $\boldsymbol{E}$ and $\boldsymbol{B}$ fields by the usual linear response, $\boldsymbol{D} = \varepsilon \boldsymbol{E}$ and $\boldsymbol{H} = \mu^{-1} \boldsymbol{B}$, and do not include any contribution from the nonlinear source terms. We consider the case of a polarized sheet represented in fig. 1, but with a polarized region of a finite thickness. In order to derive boundary conditions to match fields across the polarized region, we need to establish what fields are present inside the thin layer. For the case of a polarized sheet of finite thickness, the usual boundary conditions derived from eq. (2) require continuity of $(D_z + 4\pi P_z^{\rm nls})$, $\boldsymbol{E}_{\rm t}$, B_z and $\boldsymbol{H}_{\rm t}$, where the subscripts z and t refer, respectively, to components perpendicular and parallel to the surface in question. Since the fields outside the polarized slab are always finite, the continuity relations above imply that in the limit of a vanishing thickness for the polarized sheet

$$\int_{0^-}^{0^+} (D_z + 4\pi P_z^{\rm nls})\,{\rm d}z = \int_{0^-}^{0^+} \boldsymbol{E}_{\rm t}\,{\rm d}z = \int_{0^-}^{0^+} B_z\,{\rm d}z = \int_{0^-}^{0^+} \boldsymbol{H}_{\rm t}\,{\rm d}z = 0. \tag{3}$$

Thus, we see that $\boldsymbol{E}_{\rm t}$ and $\boldsymbol{B}$ remain finite inside the polarized sheet, but $D_z = \varepsilon' E_z$ increases as the slab is compressed, so that

$$\int_{0^-}^{0^+} D_z\,{\rm d}z = -4\pi \int_{0^-}^{0^+} P_z^{\rm nls}\,{\rm d}z \equiv -4\pi P_{{\rm s},z}^{\rm nls}. \tag{4}$$

This equation, which agrees with what would be deduced from electrostatics, states that when a constant surface polarization is maintained in a slab of decreasing thickness, the normal component of the electric field becomes singular. With the aid of eqs (2) and (3), we can derive the matching conditions across the slab by the traditional approach of integrating Maxwell's equations over suitable surfaces and volumes.

The matching conditions for $\boldsymbol{B}$ and $\boldsymbol{H}$ can be handled easily, since $\boldsymbol{B}$ and $\boldsymbol{H}$ remain finite in the polarized sheet. From eq. (2c), we obtain immediately

$$\Delta B_z = B_z(z=0^+) - B_z(z=0^-) = 0, \tag{5}$$

by applying the divergence theorem to the volume shown in fig. 2a. For the matching condition on $\boldsymbol{H}_{\rm t}$, let us consider the line integral of fig. 2b. Taking

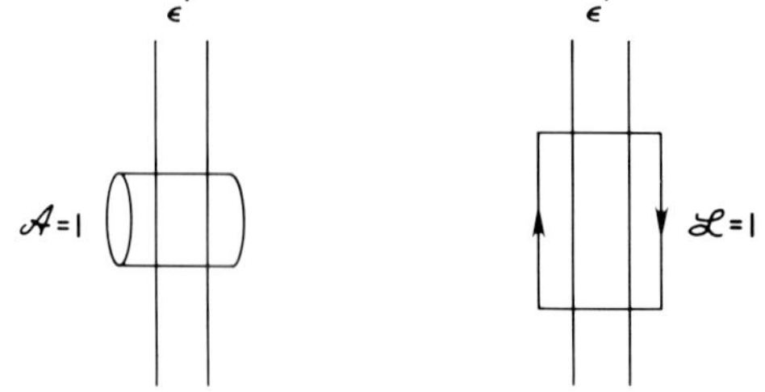

Fig. 2. Cross-section of polarized sheet with pillbox and circuit used in the derivation of the boundary conditions.

length $l = 1$, we have

$$\Delta H_t = \oint_{\mathscr{C}} \boldsymbol{H} \cdot d\boldsymbol{l} = \int_{\mathscr{A}} (\nabla \times \boldsymbol{H}) \cdot (\hat{z} \times \hat{t})\, d\mathscr{A}. \tag{6}$$

From eq. (2d) and eq. (3), we find

$$\Delta \boldsymbol{H}_t = (4\pi/c)\partial \boldsymbol{P}_s^{nls}/\partial t \times \hat{z}. \tag{7}$$

To determine the boundary condition on D_z, we integrate Maxwell's eq. (2a) over a volume including the polarized sheet, as in fig. 2(a):

$$\int_{\mathscr{V}} \nabla \cdot \boldsymbol{D}\, d^3x = -4\pi \int_{\mathscr{V}} \nabla \cdot [\boldsymbol{P}_s^{nls}(x, y)\, \delta(z)]\, d^3x, \tag{8}$$

where the explicit form for the polarized sheet has been used. Realizing that $\boldsymbol{E}_t \propto \boldsymbol{D}_t$ is finite in the polarized sheet, we find

$$\Delta D_z = -4\pi \nabla_t \cdot \boldsymbol{P}_s^{nls}(x, y), \tag{9}$$

with $\nabla_t = \hat{x}(\partial/\partial x) + \hat{y}(\partial/\partial y)$.

By integrating Maxwell's eq. (2b) over the area shown in fig. 2b, we see that

$$\oint_{\mathscr{C}} \boldsymbol{E} \cdot d\boldsymbol{l} = 0. \tag{10}$$

Stokes' theorem has been applied, taking account of the finiteness of $\boldsymbol{B}_t$ [eq. (3)]. Unlike the familiar boundary problem, eq. (10) does not imply the continuity of $\boldsymbol{E}_t$. Allowance must be made for the finite integral of E_z obtained in passing through the slab. This contribution is determined by applying eq. (4) with $D_z = \varepsilon' E_z$ inside the slab. The resulting matching condition is

$$\Delta \boldsymbol{E}_t = -(4\pi/\varepsilon')\nabla_t P_{s,z}^{nls}. \tag{11}$$

In summary, the boundary conditions for electric and magnetic fields on either side of a polarized sheet $\boldsymbol{P}_s^{nls}(x, y)$ [with a volume polarization $\boldsymbol{P}_s^{nls}(x, y)\delta(z)$] are

$$\Delta D_z = -4\pi \nabla_t \cdot \boldsymbol{P}_s^{nls}, \tag{12a}$$

$$\Delta \boldsymbol{E}_t = -(4\pi/\varepsilon')\nabla_t P_{s,z}^{nls}, \tag{12b}$$

$$\Delta B_z = 0, \tag{12c}$$

$$\Delta \boldsymbol{H}_t = (4\pi/c)\partial \boldsymbol{P}_s^{nls}/\partial t \times \hat{z}. \tag{12d}$$

Here ε' represents the linear dielectric constant associated with the polarized sheet. The same derivation applies, it should be noted, when ε' is not isotropic, but is diagonal in the (x, y, z) coordinate system. In this case, we set $\varepsilon' = \varepsilon'_{zz}$ in the relations above. These boundary conditions have recently been derived in more formal treatments by Sipe (1987) and Felderhof and Marowsky (1987b).

They are of course, consistent with the result for radiation from a finite polarized slab with a prescribed (oscillatory) spatial variation when the thickness is allowed to approach zero (Bloembergen and Pershan 1962). Finally, it should be stressed that our derivation of the boundary conditions relied on the complete, time-dependent Maxwell equations and, hence, leads to equations valid for arbitrary frequencies.

2.3. Radiation from a polarized sheet

To represent SHG by an interface under plane-wave excitation, we need to consider a polarized sheet

$$\boldsymbol{P}^{\mathrm{nls}}(x, y, z, t) = \boldsymbol{P}_{\mathrm{s}}^{\mathrm{nls}}(x, y, t)\, \delta(z), \tag{13a}$$

with

$$\boldsymbol{P}_{\mathrm{s}}^{\mathrm{nls}}(x, y, t) = \boldsymbol{P}_{\mathrm{s}} \mathrm{e}^{\mathrm{i}px - \mathrm{i}\Omega t} + \mathrm{c.c.} \tag{13b}$$

Later we shall identify this source term as arising from a surface nonlinear susceptibility $\chi_{\mathrm{s}}^{(2)}$ under the influence of plane-wave pump radiation. For now we need only consider a spatially varying sheet of polarization described by wave vector $p\hat{x}$ and frequency Ω. Such a polarized sheet will give rise to transmitted and reflected beams propagating through media 1 and 2 described by (isotropic) dielectric constants ε_1 and ε_2, respectively. We assume here and henceforth that the magnetic properties of media can be neglected, i.e., $\mu = 1$. The direction of the beam in medium 1 (the reflected beam for the usual measurement) is given by the wave vector

$$\boldsymbol{k}_1 = p\hat{x} - q_1\hat{z}, \tag{14a}$$

with

$$q_1 = [\varepsilon_1 K^2 - p^2]^{1/2} \tag{14b}$$

to satisfy the dispersion relation in medium 1. Here $K = \Omega/c$ is the wave vector in free space and $\varepsilon_1 = \varepsilon_1(\Omega)$ is the dielectric constant evaluated at the appropriate frequency. For ε_1 real and $p^2 < \varepsilon_1 K^2$, the positive value for q_1 should be chosen, corresponding to a wave propagating away from the interface. More generally, the wave may decay as it leaves the surface, either because $p^2 > \varepsilon_1 K^2$ or because ε_1 is complex. In eq. (14b), q_1 should then have a non-negative imaginary part. Analogously, for the (transmitted) wave in medium 2,

$$\boldsymbol{k}_2 = p\hat{x} + q_2\hat{z}, \tag{15a}$$

with

$$q_2 = [\varepsilon_2 K^2 - p^2]^{1/2}. \tag{15b}$$

and q_2 positive or, more generally, $\mathrm{Im}(q_2) \geqslant 0$.

The expression for the radiated fields from the interface follows directly from application of the boundary conditions of eq. (12) with plane wave characterized by wave vectors $\boldsymbol{k}_1$ and $\boldsymbol{k}_2$. If we assume initially $\varepsilon_1 = \varepsilon_2 = \varepsilon' = 1$, we obtain an electric field just beyond the polarized sheet in either medium $i = 1$ or $i = 2$ of

$$\boldsymbol{E}_i(x, y, z, t) = \boldsymbol{E}_i(\Omega)\mathrm{e}^{\mathrm{i}\boldsymbol{k}_i\cdot\boldsymbol{x}-\mathrm{i}\Omega t} + \text{c.c.}, \tag{16}$$

with $\boldsymbol{E}_i = \boldsymbol{E}_i(\Omega)$ given by

$$\boldsymbol{E}_i = \frac{2\pi\mathrm{i}K^2}{q_i}[\boldsymbol{P}_\mathrm{s} - \hat{\boldsymbol{k}}_i(\hat{\boldsymbol{k}}_i\cdot\boldsymbol{P}_\mathrm{s})]. \tag{17}$$

In this relation, $\hat{\boldsymbol{k}}_i = \boldsymbol{k}_i/(\boldsymbol{k}_i\cdot\boldsymbol{k}_i)^{1/2}$ a unit vector along the direction of propagation. The equation shows that the radiated field is proportional to the transverse component of $\boldsymbol{P}_\mathrm{s}$. A convenient alternative form of eq. (17) can be written in terms of a unit polarization vector $\hat{\boldsymbol{e}}_i$ satisfying $\hat{\boldsymbol{k}}_i\cdot\hat{\boldsymbol{e}}_i = 0$:

$$\hat{\boldsymbol{e}}_i\cdot\boldsymbol{E}_i = 2\pi\mathrm{i}q_i^{-1}K^2(\hat{\boldsymbol{e}}_i\cdot\boldsymbol{P}_\mathrm{s}). \tag{18}$$

Once again, the same form applies for radiation on either side of the sheet; allowable vectors $\hat{\boldsymbol{e}}$ will differ in general for the radiated fields on the two sides of the polarized sheet.

The presence of a linear response in the bulk and interfacial region ($\varepsilon_1, \varepsilon_2, \varepsilon' \neq 1$) can be incorporated into the analysis by means of the relevant Fresnel factors. In order to treat the excitation of the surface and radiation by the surface symmetrically, it is convenient to introduce the Fresnel corrections for *incoming* waves, i.e., for waves propagating towards the thin layer described by dielectric constant ε' from the two bulk media*. We continue to denote a polarization vector of an electric field in media $i = 1, 2$ by $\hat{\boldsymbol{e}}_i$ (with $\hat{\boldsymbol{k}}_i\cdot\hat{\boldsymbol{e}}_i = 0$). Then a wave with electric field $E_0\hat{\boldsymbol{e}}_i$ in medium i propagating towards medium j will lead to an electric field $E_0\boldsymbol{e}$ in the interfacial region with

$$\boldsymbol{e} = \mathsf{F}_{i\to j}\hat{\boldsymbol{e}}_i. \tag{19}$$

The Fresnel transformation $\mathsf{F}_{i\to j}$ can be written in terms of transmission coefficients for s- and p-polarized light. If $\varepsilon' = \varepsilon_1$ or $\varepsilon' = \varepsilon_2$, this involves only the usual Fresnel coefficients appropriate for wave propagation between two dielectric media. With an arbitrary value for ε', it is convenient to express $\mathsf{F}_{i\to j}$ in terms of the cartesian coordinates of fig. 1. For isotropic bulk media, $\mathsf{F}_{i\to j}$ has only diagonal elements:

$$\mathsf{F}_{i\to j} = F^{xx}_{i\to j}\hat{\boldsymbol{x}}\hat{\boldsymbol{x}} + F^{yy}_{i\to j}\hat{\boldsymbol{y}}\hat{\boldsymbol{y}} + F^{zz}_{i\to j}\hat{\boldsymbol{z}}\hat{\boldsymbol{z}}, \tag{20a}$$

*In the analysis of Mizrahi and Sipe (1988), Fresnel factors for the outgoing field are considered, which differ slightly from the factors introduced here.

with

$$F^{xx}_{i\to j} = 2\varepsilon_i q_j/(\varepsilon_j q_i + \varepsilon_i q_j), \tag{20b}$$

$$F^{yy}_{i\to j} = 2q_i/(q_i + q_j), \tag{20c}$$

$$F^{zz}_{i\to j} = 2(\varepsilon_1\varepsilon_2/\varepsilon')q_i/(\varepsilon_j q_i + \varepsilon_i q_j). \tag{20d}$$

Here q_i ($i = 1, 2$) denote the z-component of the wave vectors, as specified by eqs (14b) and (15b).

Making use of the transformed polarization vector $\boldsymbol{e}$, we can express the radiated field from the polarized sheet in a concise form even when the linear dielectric properties are taken into account:

$$\hat{\boldsymbol{e}}_i \cdot \boldsymbol{E}_i = 2\pi \mathrm{i} q_i^{-1} K^2(\boldsymbol{e} \cdot \boldsymbol{P}_\mathrm{s}). \tag{21}$$

In this generalized form of eq. (18), the component of the radiated electric field in medium i along polarization vector $\hat{\boldsymbol{e}}_i$ is specified in terms of $K = (\Omega/c)$, wave vector component q_i [eqs (14) or (15)], and the transformed polarization vector $\boldsymbol{e}$ defined by eqs (19) and (20).

A useful alternative form of eq. (21),

$$\hat{\boldsymbol{e}}_i \cdot \boldsymbol{E}_i = 2\pi \mathrm{i} \varepsilon_i^{-1/2} K \sec\theta_i(\boldsymbol{e} \cdot \boldsymbol{P}_\mathrm{s}), \tag{22a}$$

can be written by defining

$$\sec\theta_i = (\varepsilon_i^{1/2} K)/q_i. \tag{22b}$$

In a lossless medium, θ_i represents the angle between the direction of propagation of the radiation and the surface normal.

2.4. Surface nonlinear response

To complete the formulation of second-order surface nonlinear optical effects within our phenomenological model, we treat the nonlinear polarized sheet as arising from the surface nonlinear susceptibility tensor $\chi_\mathrm{s}^{(2)}$ introduced above. For the general case of non-frequency generation, we have

$$\boldsymbol{P}_\mathrm{s}(\Omega) = \chi_\mathrm{s}^{(2)}(\Omega = \omega_1 + \omega_2) : \boldsymbol{E}(\omega_1)\boldsymbol{E}(\omega_2). \tag{23}$$

Here ω_1 and ω_2 are frequencies of the two pump fields and $\Omega = \omega_1 + \omega_2$ is the sum-frequency; the $\boldsymbol{E}$ fields are evaluated in the interfacial medium described by dielectric constant ε'. This expression can be inserted into the equations for the radiation from a polarized sheet given above. Before doing so, we relate the electric fields in the dielectric medium ε' to the applied fields. We assume plane wave radiation for each pump beam. For a pump beam incident from medium i, the electric field is expressed as

$$\begin{aligned} \boldsymbol{E}_i(x, y, z, t) &= \boldsymbol{E}_i(\omega)\mathrm{e}^{\mathrm{i}\boldsymbol{k}_i \cdot \boldsymbol{x} - \mathrm{i}\omega t} + \mathrm{c.c.} \\ &= E_i(\omega)\hat{\boldsymbol{e}}_i(\omega)\mathrm{e}^{\mathrm{i}\boldsymbol{k}_i \cdot \boldsymbol{x} - \mathrm{i}\omega t} + \mathrm{c.c.}, \end{aligned} \tag{24}$$

where ω represents either ω_1 or ω_2 for sum-frequency generation and $\hat{e}_i(\omega)$ is a polarization vector describing the pump field with amplitude $E_i(\omega)$. The wave vector $\boldsymbol{k}_i$ for pump radiation propagating towards the interface can be expressed as

$$\boldsymbol{k}_1 = p\hat{x} + q_1\hat{z}, \tag{25a}$$

$$\boldsymbol{k}_2 = p\hat{x} - q_2\hat{z}. \tag{25b}$$

Here q_i is the (positive) z-component of the wave vector in medium i,

$$q_i = [\varepsilon_i(\omega/c)^2 - p^2]^{1/2}, \tag{26}$$

and it has been assumed for simplicity that both pump waves propagate in the xz-plane.

With these definitions, the surface polarization $\boldsymbol{P}_s(\Omega)$ is related to the incident pump fields $\boldsymbol{E}_{i_1}(\omega_1)$ and $\boldsymbol{E}_{i_2}(\omega_2)$ by

$$\boldsymbol{P}_s(\Omega) = \chi_s^{(2)}(\Omega = \omega_1 + \omega_2) : \boldsymbol{e}_{i_1}(\omega_1)\boldsymbol{e}_{i_2}(\omega_2)E_{i_1}(\omega_1)E_{i_2}(\omega_2). \tag{27}$$

Here $\boldsymbol{e}_{i_1}(\omega_1)$ and $\boldsymbol{e}_{i_2}(\omega_2)$ represent polarization vectors of the pump fields corrected for the Fresnel factor describing propagation into the interfacial region. In terms of our previous notation, we have

$$\boldsymbol{e}(\omega) = \boldsymbol{F}_{i\to j}\hat{e}_i(\omega), \tag{28}$$

with $i = i_1$ or i_2 being the medium through which the wave propagates toward the interface and beyond into medium j. The explicit expression for $\boldsymbol{F}$ of eq. (20) can be applied with the q_i now specified by eq. (26) and all dielectric constants evaluated at the appropriate frequency.

Assembling our results [eqs (22) and (27)], we obtain the following expression for sum-frequency generation at an interface in terms of pump and radiated fields in the bulk media:

$$\hat{e}_i(\Omega) \cdot \boldsymbol{E}_i(\Omega) = \frac{2\pi \mathrm{i}\Omega \sec\theta_i(\Omega)}{c\varepsilon_i^{1/2}(\Omega)}[\boldsymbol{e}(\Omega) \cdot \chi_s^{(2)} : \boldsymbol{e}(\omega_1)\boldsymbol{e}(\omega_2)]E_{i_1}(\omega_1)E_{i_2}(\omega_2). \tag{29}$$

This relation is valid for the radiated field $\boldsymbol{E}_i(\Omega)$ with polarization vector $\boldsymbol{e}_i(\Omega)$ in either medium $i = 1$ or 2 produced by excitation from fields $\hat{e}_{i_1}(\omega_1)E_{i_1}(\omega_1)$ and $\hat{e}_{i_2}(\omega_2)E_{i_2}(\omega_2)$ propagating towards the interface in media i_1 and i_2. The transformed polarization vectors $\boldsymbol{e}(\Omega)$, $\boldsymbol{e}(\omega_1)$, and $\boldsymbol{e}(\omega_2)$ are obtained by propagating $\hat{e}_i(\Omega)$, $\hat{e}_{i_1}(\omega_1)$, and $\hat{e}_{i_2}(\omega_2)$ to the interfacial region. Explicit forms are given by eqs (19) and (20) with dielectric constants evaluated at the appropriate frequencies and normal components of the wave vectors determined by the dispersion relations [eqs (14), (15), and (26)] for components of the wave vectors parallel to the surface of $p(\omega_1)$, $p(\omega_2)$ and $p(\Omega) = p(\omega_1) + p(\omega_2)$. The factor $\sec\theta_i(\Omega) = [\varepsilon_i^{1/2}(\Omega)\Omega]/[cq_i(\Omega)]$ which, for a lossless medium, implies that $\theta_i(\Omega)$

represents the angle between the direction of the wave at frequency Ω and the (unsigned) surface normal. For the case of collinear excitation beams in a dispersionless medium, $\theta_i(\Omega)$ for the reflected nonlinear wave will match the angle of incidence of the pump beam.

Equation (29) is suitable for describing a variety of geometries of excitation and detection. It is valid for any complex isotropic dielectric constants ε_1, ε_2, and ε'. (The formulas are also valid with $\varepsilon' = \varepsilon'_{zz}$ for a diagonal, but non-isotropic dielectric response of the interface, $\boldsymbol{\varepsilon}' = \varepsilon'_{xx}\hat{x}\hat{x} + \varepsilon'_{yy}\hat{y}\hat{y} + \varepsilon'_{zz}\hat{z}\hat{z}$.) When absorption is present in the bulk media, the electric field amplitudes in eq. (29) correspond to incoming or outgoing fields near the interface.

While it has been presumed that the two pump beams both lie in the same plane of incidence, eq. (29) retains the same form for more general excitation geometries. The modification required is in the explicit form of the Fresnel transformation $\boldsymbol{F}_{i\to j}$, which is defined in eq. (20) for light incident in the xz-plane. The direction of the radiated beam is determined for an arbitrary arrangement of pump beams by the in-plane wave vector components

$$\boldsymbol{p}(\Omega) = \boldsymbol{p}(\omega_1) + \boldsymbol{p}(\omega_2), \tag{30}$$

in conjunction with the linear dispersion relation in medium i [eqs (14) and (15) with $p^2 = \boldsymbol{p}(\Omega)\cdot\boldsymbol{p}(\Omega)$]. This form of the nonlinear Snell's law (Bloembergen and Pershan 1962) has been recently verified by Muenchausen et al. (1987) for surface second-harmonic and sum-frequency generation in a reflection.

To obtain a relation for the irradiance at Ω in terms of irradiances at ω_1 and ω_2, we use the relation $I = c\varepsilon^{1/2}|E|^2/2\pi$ for a plane wave in a medium with dielectric constant ε. Equation (29) then yields

$$I_i(\Omega) = \frac{8\pi^3\Omega^2 \sec^2\theta_i(\Omega)}{c^3[\varepsilon_i(\Omega)\varepsilon_{i_1}(\omega_1)\varepsilon_{i_2}(\omega_2)]^{1/2}} \times |\boldsymbol{e}(\Omega)\cdot\boldsymbol{\chi}_s^{(2)}{:}\boldsymbol{e}(\omega_1)\boldsymbol{e}(\omega_2)|^2 I_{i_1}(\omega_1) I_{i_2}(\omega_2). \tag{31}$$

The quantities $I_{i_1}(\omega_1)$ and $I_{i_2}(\omega_2)$ denote the irradiances of the pump beams at frequencies ω_1 and ω_2 incident from media i_1 and i_2, respectively; $I_i(\Omega)$ is the irradiance of the nonlinear radiation at frequency Ω in medium i with a polarization $\hat{\boldsymbol{e}}_i(\Omega)$ (corresponding to $\boldsymbol{e}(\Omega)$ after the appropriate Fresnel correction). All other symbols are as defined above.

For comparison with experiment, it is helpful to have an expression for the total radiated power at the nonlinear frequency for pump beams of finite spacial extent. This can be obtained immediately from eq. (31) in the limit of beams sufficiently large to be measured by geometrical optics. Consider first the case of uniform, perfectly overlapping pump beams giving rise to a radiated beam of area A (measured normal to the direction of propagation). Then the total power radiated into medium i with polarization $\hat{e}_i(\Omega)$ is given by

$$\mathscr{P}_i(\Omega) = \frac{8\pi^3\Omega^2 \sec\theta_{i_1}(\omega_1)\sec\theta_{i_2}(\omega_2)}{c^3[\varepsilon_i(\Omega)\varepsilon_{i_1}(\omega_1)\varepsilon_{i_2}(\omega_2)]^{1/2}} \times |\boldsymbol{e}(\Omega)\cdot\chi_s^{(2)}:\boldsymbol{e}(\omega_1)\boldsymbol{e}(\omega_2)|^2 \frac{\mathscr{P}_{i_1}(\omega_1)\mathscr{P}_{i_2}(\omega_2)}{A}, \tag{32}$$

where $\mathscr{P}_{i_1}(\omega_1)$ and $\mathscr{P}_{i_2}(\omega_2)$ denote the total power in each pump beam and $\theta_{i_1}(\omega_1)$ and $\theta_{i_2}(\omega_2)$ represent angles of the pump beams incident from media i_1 and i_2. (For simplicity, the dielectric media ε_1 and ε_2 have been assumed to be lossless.) If the pump beams have an arbitrary spatial profile, we have the generalized relation

$$\mathscr{P}_i(\Omega) = \frac{8\pi^3\Omega^2 \sec^2\theta_i(\Omega)}{c^3[\varepsilon_i(\Omega)\varepsilon_{i_1}(\omega_1)\varepsilon_{i_2}(\omega_2)]^{1/2}} \times |\boldsymbol{e}(\Omega)\cdot\chi_s^{(2)}:\boldsymbol{e}(\omega_1)\boldsymbol{e}(\omega_2)|^2 \frac{f\mathscr{P}_{i_1}(\omega_1)\mathscr{P}_{i_2}(\omega_2)}{A^{1/2}(\omega_1)A^{1/2}(\omega_2)}. \tag{33a}$$

In this relation each effective area $A(\omega)$ is defined in terms of the total power of the beam $\mathscr{P}(\omega)$ and of a representative intensity $I^0(\omega)$ (e.g., the central or peak intensity):

$$A(\omega) = \mathscr{P}(\omega)/I^0(\omega). \tag{33b}$$

The dimensionless factor f describes the overlap of the two pump beams:

$$f = \cos\theta_i(\Omega)\frac{[A(\omega_1)A(\omega_2)]^{1/2}}{\mathscr{P}_{i_1}(\omega_1)\mathscr{P}_{i_2}(\omega_2)}\int I_{i_1}(\omega_1)I_{i_2}(\omega_2)\,\mathrm{d}x\,\mathrm{d}y, \tag{33c}$$

where the integral of the incident intensity profiles is carried out over the plane of the nonlinear layer. A particular example of some interest is that of collinear Gaussian pump beams with the waists of the pump beams focussed on the nonlinear layer. Equation (21) can be applied (within its approximation) with $A(\omega_1) = \pi r_1^2$ and $A(\omega_2) = \pi r_2^2$ for the effective beam areas (computed with respect to an intensity measured at the center of the beam) corresponding to intensity profiles of the form $\sim \mathrm{e}^{-r^2/r_0^2}$. The overlap factor is then given by $f = [(r_1/r_2) + (r_2/r_1)]^{-1}[\cos\theta_i(\Omega)/\cos\theta(\omega)]$, with $\theta(\omega)$ denoting the common angle of incidence of the two pump beams.

2.5. Bulk nonlinear response

In the bulk of the material surrounding the interface, we can express the (generalized) nonlinear source polarization in a multipole expression in successive degrees of nonlocality:

$$\begin{aligned}\boldsymbol{P}^{\mathrm{nls}}(\Omega) &= \chi^{(2)}(\Omega=\omega_1+\omega_2):\boldsymbol{E}(\omega_1)\boldsymbol{E}(\omega_2)\\ &\quad+\tfrac{1}{2}\chi_{\mathrm{q}}^{(2)}(\Omega=\omega_1+\omega_2):\boldsymbol{E}(\omega_1)\nabla\boldsymbol{E}(\omega_2)\\ &\quad+\tfrac{1}{2}\chi_{\mathrm{q}}^{(2)}(\Omega=\omega_1+\omega_2):\boldsymbol{E}(\omega_2)\nabla\boldsymbol{E}(\omega_1)+\ \dots. \end{aligned} \tag{34}$$

The relative contribution of the successive terms in such an expression is expected to vary like (a/l), where a is a typical atomic dimension and l denotes the length scale over which the electric field changes. In the homogeneous bulk (unlike in the surface or interface region), this expression will converge rapidly, since $a/l \sim ka \ll 1$ for a typical atomic dimension $a \sim$ Å and a wave vector in the visible $k = 10^{-3}$ Å. Thus, an accurate description of the bulk response will be given by the leading-order nonvanishing term in the expansion. For studies of surfaces and interfaces, the interesting case occurs when the first (electric-dipole) term vanishes. This condition is satisfied for all centrosymmetric media, i.e., for all materials possessing a point with respect to which the material remains unchanged under the operation of spatial inversion $(\boldsymbol{x} \rightarrow -\boldsymbol{x})$*. For the case of SHG (but not SFG), the electric-dipole contribution will actually vanish for all isotropic media, whether or not they are centrosymmetric, and for all crystals with 432 point-symmetry.

It should be noted that in some instances the bulk electric-dipole response may be eliminated by an appropriate choice of polarizations. This approach for studying the surfaces and interfaces of non-centrosymmetric materials is particularly attractive if the bulk material possesses high symmetry and, hence, few nonzero tensor elements of $\chi_{\mathrm{s}}^{(2)}$. The $\bar{4}$3m-symmetry class, which includes materials such as GaAs, is one such example. For this symmetry, $\chi^{(2)}(\Omega = \omega_1 + \omega_2)$ has just a single independent nonvanishing tensor element. As Stehlin et al. (1988) have demonstrated, the bulk electric-dipole nonlinearity does not occur for many combinations of the pump and probe polarization, permitting surface-sensitive measurements to be performed without a strong bulk nonlinear response.

Assuming no bulk electric-dipole terms are present, the leading-order response consists of the electric-quadrupole and magnetic-dipole terms comprised of products of $\boldsymbol{E}$ and spatial derivatives of $\boldsymbol{E}$. In keeping with the available discussion in the literature, we shall consider the influence of bulk terms only for

*For materials with long-range order, this criterion applies as stated. For disordered materials, such as liquids or gases, we compare the medium and its image under inversion averaged over a macroscopic volume. Thus, an ensemble of randomly oriented non-centrosymmetric molecules will exhibit macroscopic inversion symmetry if and only if the molecules are not chiral, i.e., if and only if an inversion operation about some point followed by an appropriate rotation constitutes a symmetry operation for the molecule. It should be noted that a different analysis applies if one considers incoherent nonlinear radiation. This is expected whenever *locally* non-centrosymmetric structures are present, as in liquids (Maker 1970) and amorphous solids (DiVincenzo 1987).

the case of SHG; analogous considerations will apply for SFG. The bulk contributions to the (generalized) nonlinear source polarization for SHG can be written as

$$\boldsymbol{P}(\Omega) = \chi_{\mathrm{q}}^{(2)}(\Omega = \omega + \omega) : \boldsymbol{E} \nabla \boldsymbol{E}. \tag{35}$$

For an arbitrary material, the fourth-rank tensor $\chi_{\mathrm{q}}^{(2)}$ will have 81 independent elements. In this chapter, we restrict our attention to materials with cubic symmetry. While most of the discussion in the literature relates to cubic materials, it should be observed that studies have also been performed on trigonal centrosymmetric crystals of $\bar{3}$m-symmetry, such as calcite (Terhune et al. 1962, Bjorkholm and Siegman 1967, Dinev et al. 1978). In these birefringent materials, it has been shown that phase-matching of the bulk SHG process may be achieved, thus permitting the weak quadrupole terms to produce relatively strong SH signals.

For cubic media of m3m- or 432-symmetry*, we have the following independent, nonvanishing tensor elements: $\chi_{\mathrm{q},iiii}^{(2)}$, $\chi_{\mathrm{q},iijj}^{(2)}$, $\chi_{\mathrm{q},ijij}^{(2)}$, and $\chi_{\mathrm{q},ijji}^{(2)}$, where the tensor is expressed in terms of the crystallographic axes, and indices i and j are distinct. The i-th component of the effective SH polarization can then be written explicitly in terms of the pump field as

$$P_i(\Omega) = \gamma \nabla_i(\boldsymbol{E} \cdot \boldsymbol{E}) + (\delta - \beta - 2\gamma)(\boldsymbol{E} \cdot \nabla) E_i + \beta(\nabla \cdot \boldsymbol{E}) E_i + \zeta E_i \nabla_i E_i. \tag{36}$$

Here we have introduced the phenomenological parameters β, γ, δ, and ζ defined by

$$\beta = \chi_{\mathrm{q},iijj}^{(2)}, \tag{37a}$$

$$\gamma = \tfrac{1}{2} \chi_{\mathrm{q},ijij}^{(2)}, \tag{37b}$$

$$\delta = \chi_{\mathrm{q},ijji}^{(2)} - \chi_{\mathrm{q},iijj}^{(2)} - \chi_{\mathrm{q},ijij}^{(2)}, \tag{37c}$$

$$\zeta = \chi_{\mathrm{q},iiii}^{(2)} - (\chi_{\mathrm{q},ijji}^{(2)} + \chi_{\mathrm{q},iijj}^{(2)} + \chi_{\mathrm{q},ijij}^{(2)}). \tag{37d}$$

The three parameters β, γ, and δ describe an isotropic response, which is independent of sample orientation. These coefficients, already introduced by Bloembergen et al. (1968), are expected to appear in all media. For purposes of characterizing the bulk response of a homogeneous medium, however, we can always neglect the third term in eq. (36), since it includes the factor $\nabla \cdot \boldsymbol{E} = \nabla \cdot \varepsilon^{-1} \boldsymbol{D}$, which vanishes according to Maxwell's equations. Frequently, the

*The two cubic point groups corresponding to *centrosymmetric* structures are m3 and m3m. Studies of surfaces and interfaces of crystals with either of these point groups by second-order nonlinear processes are attractive, since the leading-order bulk terms are absent. The case of m3-symmetry, for which the tensor $\chi_{\mathrm{q}}^{(2)}$ has 7 independent elements, has not yet been treated in the literature and will not be discussed here. Materials exhibiting 432-symmetry class, on the other hand, lack inversion symmetry and may support dipole-allowed second-order nonlinear effects; however, the dipole contribution is absent for the SHG process (as opposed to SFG), and the leading-order bulk response is as given in the text.

second term of eq. (36) may also be omitted from consideration. It cannot contribute to the bulk SH polarization when only a single plane wave is present in the medium, as follows immediately from the transverse nature of the pump electric field. For the simple case of a SH measurement in reflection with a plane-wave pump this contribution vanishes, provided the medium into which the reflection enters has no nonlinear response. If this is not the case, then the incident and reflected pump waves may excite a polarization associated with $(\boldsymbol{E}\cdot\nabla)\boldsymbol{E}$. The first term of eq. (36) will be present under most circumstances, although it will disappear if the pump field is circularly polarized. This contribution to the nonlinear source polarization has the interesting attribute of being a longitudinal field for any pump field $\boldsymbol{E}$. For excitation by a single plane wave, this implies that only p-polarized radiation will be produced. More importantly, as will be discussed below, the longitudinal character of the γ term implies that it will give rise to radiation fields indistinguishable from those of an equivalent surface nonlinear response. In an isotropic material, we have the relation $\zeta = \chi^{(2)}_{\mathrm{q},iiii} - (\chi^{(2)}_{\mathrm{q},ijji} + \chi^{(2)}_{\mathrm{q},iijj} + \chi^{(2)}_{\mathrm{q},ijij}) = 0$. In a cubic material, however, ζ can be finite and the fourth term in eq. (36) will give rise to a nonlinear polarization sensitive to the orientation of the crystal. The influence of the bulk anisotropic response in SHG measurements is considered in detail in section 3.2.

Although we shall not present explicit formulas for SH radiation from all of the different bulk terms in eq. (36), it is instructive to compare the nature of radiation from a bulk source polarization with that from a surface polarization sheet. In general, the angular and polarization dependence of the bulk terms will be distinct from that exhibited by the surface terms, as one would expect given their different physical origins. (As we make clear below, this statement does *not* apply to the longitudinal bulk polarization.) A convenient comparison can be made if we consider a bulk nonlinear source polarization described by a single wave vector $\boldsymbol{k}_\mathrm{b}$:

$$\boldsymbol{P}(x, y, z, t) = \boldsymbol{P}_\mathrm{b}(\Omega, \boldsymbol{k}_\mathrm{b})\mathrm{e}^{\mathrm{i}\boldsymbol{k}_\mathrm{b}\cdot\boldsymbol{x} - \mathrm{i}\Omega t}\theta(z) + \text{c.c.}, \tag{38}$$

where we have assumed for simplicity that there is no bulk nonlinear response in medium 1, as indicated by the step function $\theta(z)$. A bulk polarization of this type would arise for any $\chi^{(2)}_\mathrm{q}(\Omega = \omega + \omega)$ under excitation by a single plane wave with wave vector $\frac{1}{2}\boldsymbol{k}_\mathrm{b} = \boldsymbol{k}_2(\omega)$, or it could correspond to a wave vector produced by mixing two distinct pump beams with a total wave vector $\boldsymbol{k}_\mathrm{b}$. The direction of the SH wave propagating in medium 1 will be determined by matching of the component of wave vector $\boldsymbol{k}_\mathrm{b}$ parallel to the interface. Thus we expect to be able to describe the radiation from the bulk nonlinear source polarization by an *equivalent* nonlinear polarization sheet located at the interface between medium 1 and medium 2:

$$\boldsymbol{P}^\mathrm{eq}_\mathrm{s}(x, y, z, t) = \boldsymbol{P}^\mathrm{eq}_\mathrm{s}(\Omega, \boldsymbol{k}_\mathrm{b})\mathrm{e}^{\mathrm{i}\boldsymbol{k}_\mathrm{b}\cdot\boldsymbol{x} - \mathrm{i}\Omega t}\delta(z) + \text{c.c.}, \tag{39}$$

Considering the bulk polarization as a succession of polarized sheets and taking into account the phase shifts arising from the source polarization and from propagation of the nonlinear wave in medium 2, we find that the equivalent polarization sheet is given by

$$\boldsymbol{P}_{\mathrm{s}}^{\mathrm{eq}}(\Omega, \boldsymbol{k}_{\mathrm{b}}) = \mathrm{i}(q_{\mathrm{b}} + q_2)^{-1}[P_{\mathrm{b},x}(\Omega, \boldsymbol{k}_{\mathrm{b}})\hat{\boldsymbol{x}} + P_{\mathrm{b},y}(\Omega, \boldsymbol{k}_{\mathrm{b}})\hat{\boldsymbol{y}} + (\varepsilon'(\Omega)/\varepsilon_2(\Omega))P_{\mathrm{b},z}(\Omega, \boldsymbol{k}_{\mathrm{b}})\hat{\boldsymbol{z}}]. \tag{40}$$

In this relation, q_{b} denotes the z-component of $\boldsymbol{k}_{\mathrm{b}}$ and q_2 is the (positive) z-component of the freely propagating wave of frequency Ω in medium 2, as defined in eq. (15); the factor involving the ratio of dielectric constants arises from the assumption that the polarized sheet is embedded in a dielectric medium described by ε'. The complete expression for radiation from the bulk terms is given by inserting the expression for $\boldsymbol{P}_{\mathrm{s}}^{\mathrm{eq}}$ into the equations, such as eq. (21), for radiation from a polarized sheet*.

The form of eq. (40) can be understood as an effective nonlinear surface polarization given by the bulk nonlinear polarization contributing over a coherence length $l_{\mathrm{c}} = |q_{\mathrm{s}} + q_2|^{-1}$. For the reflection geometry, q_{s} and q_2 will both be positive, leading to the expected short coherence length. As an estimate of this coherence length, we take $q_{\mathrm{s}} = q_2$, as would occur in a dispersionless medium. We can then write the coherence length as $l_{\mathrm{c}} = |\Lambda/[2\cos\theta_2(\Omega)]|$, where $\Lambda = [2\pi c]/[\Omega\varepsilon_2^{1/2}(\Omega)]$ denotes the wavelength of a freely propagating wave in medium 2 and $\cos\theta_2(\Omega) = q_2/[\varepsilon_2^{1/2}(\Omega)(\Omega/c)]$ represents the angle of this wave with respect to the surface normal. If the pump beam propagates towards the interface from medium 2, the coherence length can become significantly longer (infinite in principle), as reflected in the formula for the coherence length with $\mathrm{Re}(q_{\mathrm{s}}) < 0$. If we consider the reflection geometry, however, the coherence length will remain $\sim\Lambda$. In this case, the magnitude of the equivalent surface polarization of eq. (40) will be given by $P_{\mathrm{s}}^{\mathrm{eq}} \sim \Lambda P_{\mathrm{b}}$. On the other hand, we expect the strength of the quadrupole contribution to be comparable to that from a dipole-allowed polarization times the factor (a/Λ), where a is a length on the order of an atomic dimension. It follows that the equivalent surface polarization for the bulk terms in the reflection geometry will be $P_{\mathrm{s}}^{\mathrm{eq}} \sim aP_{\mathrm{d}}$, with P_{d} representing a dipole-allowed nonlinear source polarization. Now, the quantity (aP_{d}) should be comparable to the contribution from the non-centrosymmetric surface region. Thus, we conclude that for the reflected SH signal, the surface and bulk contributions may be of comparable importance. This rough discussion does

*An equivalent surface nonlinear susceptibility tensor $\chi_{\mathrm{s,eq}}^{(2)}$ may be introduced to describe the SHG radiation associated with the bulk quadrupole nonlinear susceptibility tensor $\chi_{\mathrm{q}}^{(2)}$. This tensor may be inserted formally into the equations of section 2.4 for the radiated SH fields. For excitation by a single plane wave with wave vector $\boldsymbol{k}_2(\omega)$, we have $\chi_{\mathrm{s,eq}}^{(2)} = -[\varepsilon'(\omega)/\varepsilon_2(\omega)]^{(\delta_{j,z}+\delta_{k,z})}[\varepsilon'(\Omega)/\varepsilon_2(\Omega)]^{\delta_{i,z}}[k_{2,l}(\omega)]/[q_2(\omega)+(1/2)q_2(\Omega)]\chi_{\mathrm{q},ijkl}^{(2)}$. In this expression, the components of $\chi_{\mathrm{q}}^{(2)}$ refer to the laboratory (x, y, z) coordinates, into which the tensor elements of eq. (36) must be transformed before the relation can be applied.

not take the linear dielectric properties into account. For materials with high dielectric constants, one may expect that the bulk contributions will be decreased with respect to the surface contributions because of the screening of the fields in the bulk medium (Guyot-Sionnest et al. 1986b).

The presence of the factor for the coherence length in eq. (40) leads us to expect that measurements of the dependence of the SH radiation on the angle of incidence of the pump should allow us to separate the bulk and surface contributions. Practically speaking, this may be difficult if the coherence length does not vary strongly over the accessible angles of incidence, as will occur for reflection measurements from materials with high refractive indices. Interestingly, however, for a longitudinal polarization [as is produced by the $\gamma\nabla(\boldsymbol{E}\cdot\boldsymbol{E})$ term of eq. (36)], this separation is impossible in *principle*.

To see how this works within the framework of the present discussion, let us recall that the *p*-polarized radiation produced by the longitudinal polarization will have an amplitude proportional to $\boldsymbol{e}_{\mathrm{p}}\cdot\boldsymbol{P}_{\mathrm{s}}^{\mathrm{eq}}$, where $\boldsymbol{e}_{\mathrm{p}}$ is the polarization vector for p-polarized light in the interfacial region specified by eqs (19) and (20), and $\boldsymbol{P}_{\mathrm{s}}^{\mathrm{eq}}$ is the effective surface polarization for a bulk nonlinear polarization $\boldsymbol{P}_{\mathrm{b}}(\Omega)=\mathrm{i}\gamma\boldsymbol{k}_{\mathrm{s}}[\boldsymbol{E}(\omega)\cdot\boldsymbol{E}(\omega)]$. More explicitly, we find

$$\begin{aligned}\boldsymbol{e}_{\mathrm{p}}\cdot\boldsymbol{P}_{\mathrm{s}}^{\mathrm{eq}}(\Omega) &= C[q_2\hat{\boldsymbol{x}}+(\varepsilon_2/\varepsilon')p\hat{\boldsymbol{z}}]\\ &\quad\cdot\{[-\gamma(\boldsymbol{E}\cdot\boldsymbol{E})(q_{\mathrm{s}}+q_2)^{-1}][p\hat{\boldsymbol{x}}+(\varepsilon'/\varepsilon_2)q_{\mathrm{s}}\hat{\boldsymbol{z}}]\}\\ &= -C\gamma p(\boldsymbol{E}\cdot\boldsymbol{E})\\ &= \boldsymbol{e}_{\mathrm{p}}\cdot[-\gamma(\varepsilon'/\varepsilon_2)(\boldsymbol{E}\cdot\boldsymbol{E})\hat{\boldsymbol{z}}]=\boldsymbol{e}_{\mathrm{p}}\cdot\boldsymbol{P}_{\mathrm{s}}^{\gamma}(\Omega),\end{aligned}\tag{41}$$

where we have defined a surface polarization

$$\boldsymbol{P}_{\mathrm{s}}^{\gamma}(\Omega)=-\gamma[\varepsilon'(\Omega)/\varepsilon_2(\Omega)][\boldsymbol{E}(\omega)\cdot\boldsymbol{E}(\omega)]\hat{\boldsymbol{z}}.\tag{42}$$

Thus, we see that by specifying the radiated field in terms of a surface polarization $\boldsymbol{P}_{\mathrm{s}}^{\gamma}\parallel\hat{\boldsymbol{z}}$ we can completely reproduce the angular and polarization dependence of the bulk longitudinal polarization without introducing the vector $\boldsymbol{k}_{\mathrm{s}}$. It follows that it is impossible to separate the bulk longitudinal polarization from the *z*-component of the surface polarization. This result has been demonstrated only for excitation of the longitudinal polarization by a single plane wave. We now treat this issue in somewhat more detail and show this finding holds regardless of the nature of the pump excitation.

Let us write a general expression for the contribution of the first term in eq. (36) including a possible nonlinear response in both media 1 and 2:

$$P^{\gamma}(\Omega)=\gamma_1\nabla[\boldsymbol{E}(\omega)\cdot\boldsymbol{E}(\omega)]\theta(-z)+\gamma_2\nabla[\boldsymbol{E}(\omega)\cdot\boldsymbol{E}(\omega)]\theta(z),\tag{43}$$

where the pump field at the fundamental frequency, $\boldsymbol{E}(\omega)$, has an arbitrary spatial variation, and γ_i represents the coefficient in medium *i*. Even with an arbitrary pump field, the induced SH polarization has the important property that it is longitudinal, i.e., $\nabla \times \boldsymbol{P}^{\gamma}(\Omega) = 0$. In a homogeneous dielectric medium, we can find the solution to Maxwell's equations by adding waves satisfying the homogeneous equation (with $\boldsymbol{P}^{\mathrm{nls}} = 0$) to a particular solution of the full equations including the source term. For a longitudinal source polarization, the latter can be written immediately as polarization as

$$\boldsymbol{E}_{\mathrm{d}}(\Omega) = -(4\pi/\varepsilon)\boldsymbol{P}^{\gamma}(\Omega), \tag{44a}$$

$$\boldsymbol{B}_{\mathrm{d}}(\Omega) = 0. \tag{44b}$$

Thus, for an infinite medium the longitudinal source polarization will not give rise to radiation fields. In the presence of an interface, however, the driven longitudinal field will couple to transverse fields, as required by the boundary conditions. In this problem no singular sheet of polarization is present at the interface, so the usual boundary conditions requiring the continuity of $\boldsymbol{E}_{\mathrm{t}}$, D_z, B_z, and $\boldsymbol{H}_{\mathrm{t}}$ must be obeyed. Taking into account the known values of the driven fields in the two bulk media, we deduce the following relations for the freely propagating waves across the interface at $z = 0$:

$$\Delta D_z = \Delta B_z = \Delta \boldsymbol{H}_{\mathrm{t}} = 0, \tag{45a}$$

$$\begin{aligned}\Delta \boldsymbol{E}_{\mathrm{t}} &= [4\pi/\varepsilon_2(\Omega)]\boldsymbol{P}_{\mathrm{t}}^{\gamma}(\Omega, z = 0^+) - [4\pi/\varepsilon_1(\Omega)]\boldsymbol{P}_{\mathrm{t}}^{\gamma}(\Omega, z = 0^-)\\ &= [4\pi/\varepsilon_2(\Omega)]\gamma_2 \nabla_{\mathrm{t}}[\boldsymbol{E}(\omega, z = 0^+) \cdot \boldsymbol{E}(\omega, z = 0^+)]\\ &\quad - [4\pi/\varepsilon_1(\Omega)]\gamma_1 \nabla_{\mathrm{t}}[\boldsymbol{E}(\omega, z = 0^-) \cdot \boldsymbol{E}(\omega, z = 0^-)].\end{aligned} \tag{45b}$$

Returning to the boundary conditions derived for a polarized sheet, [eq. (12)], we note that the stipulation of eq. (45) is equivalent to that for a polarized sheet

$$\begin{aligned}\boldsymbol{P}_{\mathrm{s}}^{\gamma}(\Omega) &= \gamma_1\left[\frac{\varepsilon'(\Omega)}{\varepsilon_1(\Omega)}\right][\boldsymbol{E}(\omega, z = 0^-) \cdot \boldsymbol{E}(\omega, z = 0^-)]\hat{z}\\ &\quad - \gamma_2\left[\frac{\varepsilon'(\Omega)}{\varepsilon_2(\Omega)}\right][\boldsymbol{E}_2(\omega, z = 0^+) \cdot \boldsymbol{E}(\omega, z = 0^+)]\hat{z},\end{aligned} \tag{46}$$

where ε' denotes the linear dielectric constant of the polarized sheet. This equation constitutes a complete generalization of eq. (42), since here nothing is assumed about the form of the pump electric field $E(\omega)$ and the longitudinal nonlinear response is included for both medium 1 and medium 2. As far as radiated fields at the SH frequency are concerned, we conclude that bulk terms of the form $\gamma\nabla(\boldsymbol{E} \cdot \boldsymbol{E})$ are strictly equivalent to a polarized sheet with $\boldsymbol{P}_{\mathrm{s}} = P_{\mathrm{s},z}^{\gamma}\hat{z}$ at the interface.

The influence of the bulk longitudinal nonlinear source polarization can be conveniently represented as an equivalent surface nonlinear susceptibility with nonvanishing elements:

$$(\chi_{s,\gamma}^{(2)})_{\perp\parallel\parallel} = \gamma_1 \frac{\varepsilon'(\Omega)}{\varepsilon_1(\Omega)} - \gamma_2 \frac{\varepsilon'(\Omega)}{\varepsilon_2(\Omega)}, \tag{47a}$$

$$(\chi_{s,\gamma}^{(2)})_{\perp\perp\perp} = \gamma_1 \frac{\varepsilon'(\Omega)}{\varepsilon_1(\Omega)} \left[\frac{\varepsilon'(\omega)}{\varepsilon_1(\omega)}\right]^2 - \gamma_2 \frac{\varepsilon'(\Omega)}{\varepsilon_2(\Omega)} \left[\frac{\varepsilon'(\omega)}{\varepsilon_2(\omega)}\right]^2. \tag{47b}$$

Here $\perp$ corresponds to the z-direction and $\parallel$ denotes the equivalent x- and y-directions in the plane of the interface. The additional factors involving ratios of the dielectric constants at the fundamental frequency arise when the z-components of the pump fields in media 1 and 2 are expressed in terms of the field in the interfacial layer.

The equivalence of the radiation from the γ term in the bulk and an effective surface nonlinear susceptibility was already noted by Wang (1969) for the case of plane-wave excitation of a single nonlinear medium. The result of eq. (47) was found to obtain for plane-wave excitation of two adjoining nonlinear media (including the influence of the reflected fundamental wave) by subsequent workers (Heinz 1982, Shen 1984). The general result was first demonstrated by Sipe et al. (1987a). The derivation presented here is analogous to their construction, in which a Green function formalism was applied.

For the problem of the analysis of the surface or interface nonlinear response, the bulk contribution of eq. (47) cannot be separated from the surface nonlinear susceptibility by any conventional measurement. Only by directly measuring the longitudinal polarization inside the bulk media can γ_1 and γ_2 be determined independently of surface terms. From a practical point of view, however, it may be possible to separate the surface and bulk contributions by modification of the surface of the sample.

To summarize, we have seen that the bulk nonlinear source polarization associated with leading-order nonlocal terms consists of two parts in a homogeneous isotropic medium: $\boldsymbol{P}(\Omega) = \gamma\nabla(\boldsymbol{E}\cdot\boldsymbol{E}) + (\delta - \beta - 2\gamma)(\boldsymbol{E}\cdot\nabla)\boldsymbol{E}$. The second term is absent for excitation of the medium by a single plane wave, but can in principle be identified in a suitable multi-beam measurement. (Explicit expressions for this term for mixing by an incident and reflected wave are given by Guyot-Sionnest et al. (1986b).) The first term, associated with a longitudinal polarization, always radiates in a manner indistinguishable from that of the surface or interface term. Thus, whenever the $(\boldsymbol{E}\cdot\nabla)\boldsymbol{E}$ can be neglected, the SH response from an interface between two bulk centrosymmetric media can be described by the effective surface nonlinear susceptibility tensor

$$\boldsymbol{\chi}_{s,\text{eff}}^{(2)} = \boldsymbol{\chi}_s^{(2)} + \boldsymbol{\chi}_{s,\gamma}^{(2)}, \tag{48}$$

consisting of the surface nonlinear susceptibility $\boldsymbol{\chi}_s^{(2)}$ describing the currents localized near the surface of interface and the contribution $\boldsymbol{\chi}_{s,\gamma}^{(2)}$ from the bulk longitudinal polarization, as specified by eq. (47). The radiated SH fields are still represented by the equations of section 2.4 with $\boldsymbol{\chi}_s^{(2)}$ replaced by $\boldsymbol{\chi}_{s,\text{eff}}^{(2)}$.

2.6. Discussion

Before continuing with our treatment of second-order surface nonlinear optical effects, it is perhaps worthwhile to offer a few comments concerning the phenomenological framework developed above. First, it should be observed that the scheme of a surface nonlinear polarization (or current) and a bulk quadrupole and magnetic-dipole response is common to essentially all of the literature on surface second-harmonic or sum-frequency generation*. On the other hand, few experiments have been performed to test explicitly the accuracy of this formulation.

From the theoretical standpoint, the approximations implicit in the model appear to be relatively accurate. The simplifications consist in failing to treat the detailed current distribution in the surface or interface region and in neglecting higher-order terms in the multipole expression in the bulk. The use of a leading-order description of the bulk response omits corrections only of the order $(ka)^2$, where k represents a wave vector for an optical field and a refers to a typical atomic dimension. This factor enters to the second power because alternate terms in a multipole expression vanish in a centrosymmetric medium. In principle, higher-order terms could be included in a straightforward fashion. As for the surface region, we expect the phenomenological model to require corrections of a magnitude (kl), where in this case, l characterizes the depth of the surface region. This estimate is based on the importance of wave propagation effects over a length l. For typical abrupt interfaces with the bulk material structure and electromagnetic fields present except in a region of a few Ångstroms, we expect the phenomenological model to describe radiated fields to better than 1% accuracy. It should be stressed, however, that conditions may arise in which a more diffuse interface is present, with material properties changing over a longer length scale. In this event, the model of a sheet of polarization will naturally become less appropriate.

In a recent study of SHG from thin metallic films on metal substrates, Song et al. (1987, 1988) found strong oscillations in the SH response as a function of the film thickness over a range of 10's of Ångstroms. This fascinating result was explained in terms of the oscillations in the longitudinal component of the pump electric field inside the film. At finite frequencies (especially at frequencies approaching the plasma frequency) the spatially varying longitudinal fields may be present at appreciable distances from a surface or interface. This constitutes a case in which the electric fields do not change abruptly to their bulk values. To treat this example within the framework of the phenomenological model, we must include the entire region in which the fields vary as part of the

*In the work of Corvi and Schaich (1986) an explicit solution of Maxwell's equations with appropriate boundary conditions has been given for a hydrodynamic model involving several steps at the surface. It is found that the angular and polarization dependence can be parameterized accurately in the manner prescribed by the phenomenological model.

'interface'. For an extended region, the accuracy of the macroscopic model of a polarized sheet will be diminished.

3. Experiment

3.1. Isotropic media

A wide range of studies has been conducted on isotropic surfaces and interfaces with SHG and, more recently, with SFG. These developments have been sketched in section 1.2 and treated extensively in other review articles. In this section, we restrict ourselves to some of the general considerations relevant in such work.

The form of the surface nonlinear susceptibility tensor is indicated in table 1 for surfaces with crystallographic and continuous symmetries. Isotropic surfaces may be divided into two classes, depending on whether or not they exhibit a mirror plane. In the literature, it appears that mirror symmetry has always been assumed. For this case, it is convenient to denote the independent nonvanishing tensor elements as $\chi^{(2)}_{s,\perp\perp\perp}$, $\chi^{(2)}_{s,\perp\|\|}$, $\chi^{(2)}_{s,\|\perp\|}$, and $\chi^{(2)}_{s,\|\|\perp}$, where $\perp$ corresponds to the normal (z) direction and $\|$ to the equivalent (x and y) directions in the plane. For SHG, there are just three independent elements, since $\chi^{(2)}_{s,\|\perp\|} = \chi^{(2)}_{s,\|\|\perp}$.

Table 1

Independent nonvanishing elements of $\boldsymbol{\chi}^{(2)}_s$ for crystallographic and continuous point groups for a surface in the xy-plane (Giordmaine 1965). When mirror planes are present, one of them is always perpendicular to $\hat{y}$. For SHG, $\chi^{(2)}_{s,ijk} = \chi^{(2)}_{s,ikj}$ and elements related by permutation of the last two indices have been omitted. For SFG, these elements are generally distinct; any symmetry constraints are indicated in parentheses. The terms enclosed entirely in parentheses are antisymmetric elements only present for SFG.

Symmetry class	Independent non-vanishing elements of $\boldsymbol{\chi}^{(2)}_s$:
1	$xxx, xxy, xyy, yxx, yxy, yyy$
	$xxz, xyz, yxz, yyz, zxx, zxy, zyy, xzz, yzz, zxz, zyz, zzz$
m	$xxx, xyy, xzz, xzx, yzy, yxy, zxx, zyy, zxz, zzz$
2	$xzx, xyz, yxz, yzy, zxx, zyy, zxy, zzz$
mm2	xzx, yzy, zxx, zyy, zzz
3	$xxx = -xyy = -yyx\ (= -yxy)$
	$yyy = -yxx = -xyx\ (= -xxy),\ yzy = xzx,\ zxx = zyy,\ xyz = -yxz$
	$zzz,\ (zxy = -zyx)$
3m	$xxx = -xyy = -yxy\ (= -yyx)$
	$xzx = yzy$
	$zxx = zyy,\ zzz$
4, 6, ∞	$xxz = yyz,\ zxx = zyy,\ xyz = -yxz,\ zzz,\ (zxy = -zyx)$
4mm, 6mm, ∞m	$xxz = yyz,\ zxx = zyy,\ zzz$

Before treating the important case of ∞m-symmetry in more detail, let us consider the isotropic surface lacking mirror symmetry. From table 1, we find the following additional independent tensor elements for SFG: $\chi^{(2)}_{s,xyz} = -\chi^{(2)}_{s,yzx}$, $\chi^{(2)}_{s,xzy} = -\chi^{(2)}_{s,yxz}$, and $\chi^{(2)}_{s,zxy} = -\chi^{(2)}_{s,zvx}$. For SHG, only one independent tensor element remains, $\chi^{(2)}_{s,xyz} = \chi^{(2)}_{s,xzy} = -\chi^{(2)}_{s,yzx} = -\chi^{(2)}_{s,yxz}$. These elements yield SH and SF signals for polarization combinations not allowed for an isotropic surface with mirror planes. For example, p-polarized pump excitation would produce s-polarized SH radiation. It should be observed that a surface with isotropic symmetry but no mirror plane perpendicular to the surface may occur in real physical systems. An obvious situation where this behavior is expected would be the surface of a liquid of chiral molecules, since chiral molecules cannot be restored to their original configuration by means of any rotation after a reflection operation.

When a mirror plane is present, SHG from an isotropic surface or interface is fully characterized by the three independent elements $\chi^{(2)}_{s,\perp\perp\perp}$, $\chi^{(2)}_{s,\perp\parallel\parallel}$, and $\chi^{(2)}_{s,\parallel\perp\parallel}$, with all other elements vanishing. It follows that all angular and polarization dependence should be fully determined by these three (complex) quantities and the linear dielectric properties of the bulk media. In the treatment here and below, we can include the longitudinal bulk polarization in elements $\chi^{(2)}_{s,\perp\perp\perp}$ and $\chi^{(2)}_{s,\perp\parallel\parallel}$ to obtain the complete response for plane-wave excitation. Among the consequences of these general considerations is a well known selection rule: s-polarized SH radiation is allowed only under pump excitation of mixed p- and s-polarization, not for pure s- or p-polarized light. This behavior has been noted repeatedly in experimental investigations. As has also been consistently reported, the SH signal is predicted to vanish for both ∞m- and ∞-symmetry for excitation at normal incidence. While relatively little effort has been devoted to assessing the accuracy of more complete polarization dependencies explicitly, it has been possible to match measured data to the expected form. One simple test of the predicted variation of the SH intensity can be performed with a purely s-polarized pump beam. In this case, p-polarized radiation is expected and its strength will be governed only by $\chi^{(2)}_{s,\perp\parallel\parallel}$. Figure 3 illustrates SHG under these conditions for a rhodamine dye layer on a fused silica substrate in air (Guyot-Sionnest et al. 1987b). In this measurement, a total reflection geometry was employed and the angle of incidence refers to that inside the glass. The shape of the theoretical prediction, determined entirely by linear optical properties*, agrees well with the experimental results, even in the rapidly changing region near the critical angle for total internal reflection. Related studies for external reflection have been performed by Hollering and Teeselink (1990).

In experimental measurements of SHG from clean glass surfaces with s-polarized pump excitation, Sipe et al. (1987a) demonstrated in a convincing

*The bulk nonlinear polarization of the form $(\boldsymbol{E}\cdot\nabla)\boldsymbol{E}$ does not occur for s-polarized pump radiation even when a reflected beam is present.

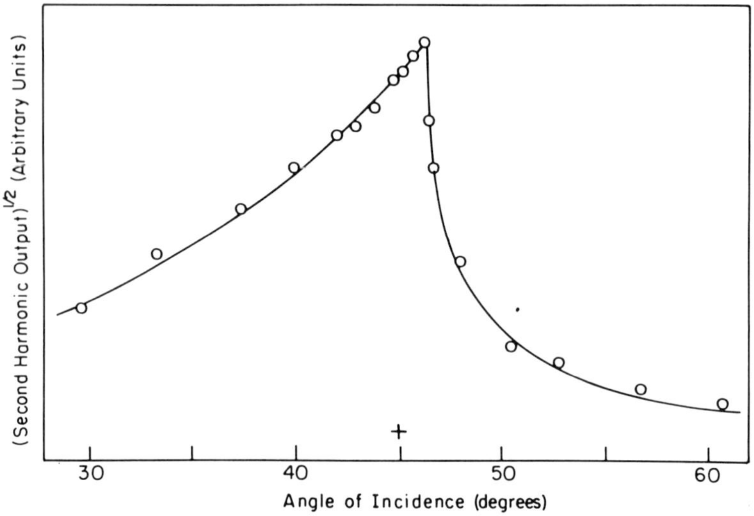

Fig. 3. SH field strength versus angle of incidence from the surface of a fused silica prism covered with a monolayer of rhodamine 6G dye molecules. The pump excitation is directed from the silica side of the interface. The circles are experimental data and the solid line is a fit to theory based on the phenomenological model. (After Guyot-Sionnest et al. 1987b.)

fashion that the bulk longitudinal polarization does, indeed, behave like a surface term. It was found that similar SH signals arose for transmission and reflection measurements. Again the relative intensities expected for different geometries could be predicted, solely on the basis of the linear optical properties of the materials. The results provided direct experimental evidence that the $\nabla(\boldsymbol{E}\cdot\boldsymbol{E})$ bulk polarization does not lead to SH radiation with an intensity continuing to grow over a coherence length. If this had been the case, the bulk contribution in transmission would have exceeded that in the reflection geometry (for which the coherence length is much shorter) by $>10^3$, in contradiction to the experimental findings.

We now address the question of the determination of $\chi^{(2)}_{s,\perp\perp\perp}$, $\chi^{(2)}_{s,\perp\parallel\parallel}$, and $\chi^{(2)}_{s,\parallel\perp\parallel}$ from experimental measurements of surface SHG. The elements $\chi^{(2)}_{s,\perp\parallel\parallel}$ and $\chi^{(2)}_{s,\parallel\perp\parallel}$ can be isolated in a direct manner by the proper choice of pump and probe polarizations. For $\chi^{(2)}_{s,\perp\parallel\parallel}$, this requires s-polarized pump radiation, which will lead to p-polarized SH radiation; $\chi^{(2)}_{s,\parallel\perp\parallel}$ can be separated by detecting the s-polarized radiation, which will be present only for a pump beam of mixed polarization. The $\perp\perp\perp$-component could in principle be isolated if the pump field has only significant components normal to the surface. This generally requires light traveling at grazing incidence in the medium of higher refractive index and is usually difficult to achieve in practice.

A more tractable approach to this problem lies in performing additional measurements of the polarization dependence of the SHG process with a fixed angle of incidence. (Similar information can also be obtained by varying the angle of incidence; for reflection measurements, this approach, although frequently applied in the past, may be somewhat more cumbersome.) Let us first assume that all of the tensor elements are real (or all differ from real elements by an identical phase factor). In this case, the following simple scheme may be applied to establish the magnitudes and relative phases of the three independent tensor elements. First separate intensity measurements with the polarization configurations detailed above yield values for $|\chi^{(2)}_{s,\perp\parallel\parallel}|$ and $|\chi^{(2)}_{s,\parallel\perp\parallel}|$. The remaining independent tensor element $\chi^{(2)}_{s,\perp\perp\perp}$, and the unknown relative sign of $\chi^{(2)}_{s,\perp\parallel\parallel}$ and $\chi^{(2)}_{s,\parallel\perp\parallel}$ can then be established by measuring the output polarization of the SH radiation for pump radiation with two different mixed polarizations. To explain this method, let us recall that we can express the s- and p-polarized SH electric fields in the form

$$E_p = A_p(\hat{e}_\omega)\chi^{(2)}_{s,\parallel\perp\parallel} + B_p(\hat{e}_\omega)\chi^{(2)}_{s,\perp\parallel\parallel} + C_p(\hat{e}_\omega)\chi^{(2)}_{s,\perp\perp\perp}, \tag{49a}$$

$$E_s = A_s(\hat{e}_\omega)\chi^{(2)}_{s,\parallel\perp\parallel}. \tag{49b}$$

Explicit expressions for the coefficients A_p, B_p, C_p, and A_s can be derived from eq. (29). These coefficients depend on the (known) linear optical properties of the bulk media and the (known or assumed) linear optical properties of the interfacial layer characterized by ε'. The factors also clearly depend on the characteristics of the pump laser beam, including its frequency, intensity, and angle of incidence (considered fixed). The dependence on the polarization of the pump radiation, $\hat{e}_\omega$, has been explicitly noted. The polarization of the radiated SH beam at frequency Ω can be described by the vector $\hat{e}^{\text{null}}_\Omega$ for which no radiation passes through the analyzer. For this polarization we can write

$$\hat{e}^{\text{null}}_\Omega \cdot (E_p\hat{p}_\Omega + E_s\hat{s}_\Omega) = 0, \tag{50}$$

where $\hat{p}_\Omega$ and $\hat{s}_\Omega$ refer to unit polarization vectors for p- and s-polarized SH radiation. Refering to eq. (49), we can cast the null condition of eq. (50) in the form

$$A(\hat{e}^{\text{null}}_\Omega, \hat{e}_\omega)\chi^{(2)}_{s,\parallel\perp\parallel} + B(\hat{e}^{\text{null}}_\Omega, \hat{e}_\omega)\chi^{(2)}_{s,\perp\parallel\parallel} + C(\hat{e}^{\text{null}}_\Omega, \hat{e}_\omega)\chi^{(2)}_{s,\perp\perp\perp} = 0, \tag{51a}$$

with coefficients

$$A(\hat{e}^{\text{null}}_\Omega, \hat{e}_\omega) = (\hat{p}_\Omega \cdot \hat{e}^{\text{null}}_\Omega)A_p(\hat{e}_\omega) + (\hat{s}_\Omega \cdot \hat{e}^{\text{null}}_\Omega)A_s(\hat{e}_\omega), \tag{51b}$$

$$B(\hat{e}^{\text{null}}_\Omega, \hat{e}_\omega) = (\hat{p}_\Omega \cdot \hat{e}^{\text{null}}_\Omega)B_p(\hat{e}_\omega), \tag{51c}$$

$$C(\hat{e}^{\text{null}}_\Omega, \hat{e}_\omega) = (\hat{p}_\Omega \cdot \hat{e}^{\text{null}}_\Omega)C_p(\hat{e}_\omega), \tag{51d}$$

specified by the known values of A_p, B_p, C_p, A_s, and the experimentally determined null polarization vector, $\hat{e}^{\text{null}}_\Omega$. Equation (51) can be solved for $\chi^{(2)}_{s,\perp\perp\perp}$, provided $\chi^{(2)}_{s,\parallel\perp\parallel}$ and $\chi^{(2)}_{s,\perp\parallel\parallel}$ are available. For the case under discussion, we have yet to determine the relative sign of $\chi^{(2)}_{s,\parallel\perp\parallel}$ and $\chi^{(2)}_{s,\perp\parallel\parallel}$. This can be accomplished by measuring the SH polarization for two distinct pump polarizations, leading to two independent equations of the form of eq. (51), which then yield the magnitude of $\chi^{(2)}_{s,\perp\perp\perp}$ and all of the relative signs of the tensor elements.

In the simplest implementation, the procedure outlined above will permit the determination of the relative magnitudes and signs of the three independent tensor elements of $\chi^{(2)}_s$. The absolute magnitudes and phases of the tensor elements are most readily obtained by comparison with a non-centrosymmetric material with a well-characterized $\chi^{(2)}$. A schematic representation of an apparatus suitable for such absolute magnitude and phase measurements is displayed in fig. 4. The calibration of the magnitude of the elements of $\chi^{(2)}_s$ can be accomplished by replacing the sample by a bulk material with known $\chi^{(2)}$ and performing a comparison of SH intensities. This method circumvents the difficult practical problem of fully characterizing the pump laser beam and the detection sensitivity, since all these factors enter in the same fashion in both measurements. A convenient alternative approach suitable in many circumstances consists of inserting a reference nonlinear material in the beam path of either the incident or the reflected beam.

The absolute phase of the tensor elements of $\chi^{(2)}_s$ relies on interference techniques analogous to those developed by Bloembergen and co-workers (Chang et al. 1965, Wynne and Bloembergen 1969) for bulk materials. The scheme consists of producing an interference between the SH wave from the sample and a second nonlinear material excited by the same pump beam. If

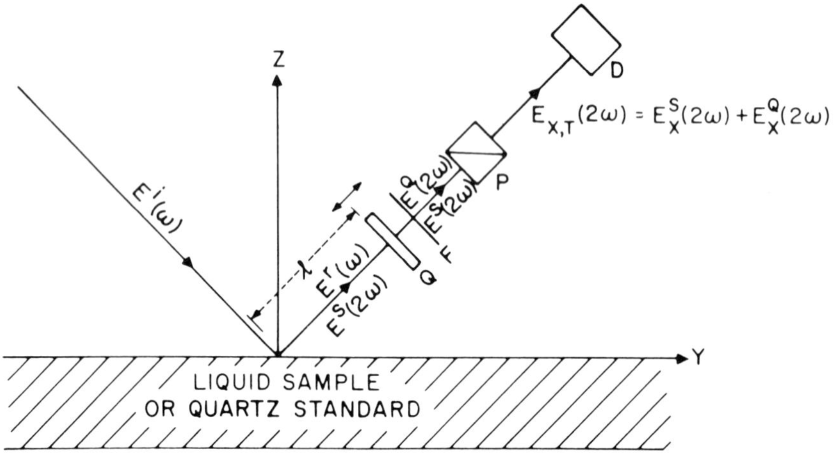

Fig. 4. Experimental scheme suitable for measurements of the magnitude and phase of elements of $\chi^{(2)}_s$. The pump wave incident on the sample leads to reflected linear and nonlinear waves. The secondary nonlinear material Q generates an additional wave at the SH frequency. After passing through filter F and analyzer P, the two SH waves interfere at detector D. The relative phase of the two SH waves can be adjusted by translating Q, as indicated. The sample of interest can be replaced by a material with a known bulk nonlinearity to set a standard for measurements of the magnitude and the phase of the signals. (After Kemnitz et al. 1986.)

the magnitudes of the two second-harmonic waves (generated by the sample and the auxiliary nonlinear material) are comparable, phase changes between the two waves will lead to a strong modulation of the observed SH intensity. In the implementation of fig. 4, the auxiliary second-harmonic wave is produced by a nonlinear material excited by the reflected pump radiation. The phase of this second-harmonic wave arriving at the detector can be altered simply by translating the nonlinear material through air. This effect, which can be reproduced by changing the pressure of a gas for a fixed geometry, arises from the difference in the refractive indices at frequencies ω and 2ω. Figure 5 shows SH interference patterns obtained with this apparatus. One curve corresponds to a reflection from the sample of interest (phenol molecules at a water surface) and the other data, obtained with a quartz plate in place of the sample, provide a calibration of the nonlinear interferometer. Similar measurements for resonant SH response have been performed by Berkovic et al. (1989).

The data of fig. 6, from work by Goh et al. (1988), illustrate the application the three independent elements of $\chi_s^{(2)}$ has been determined separately for the

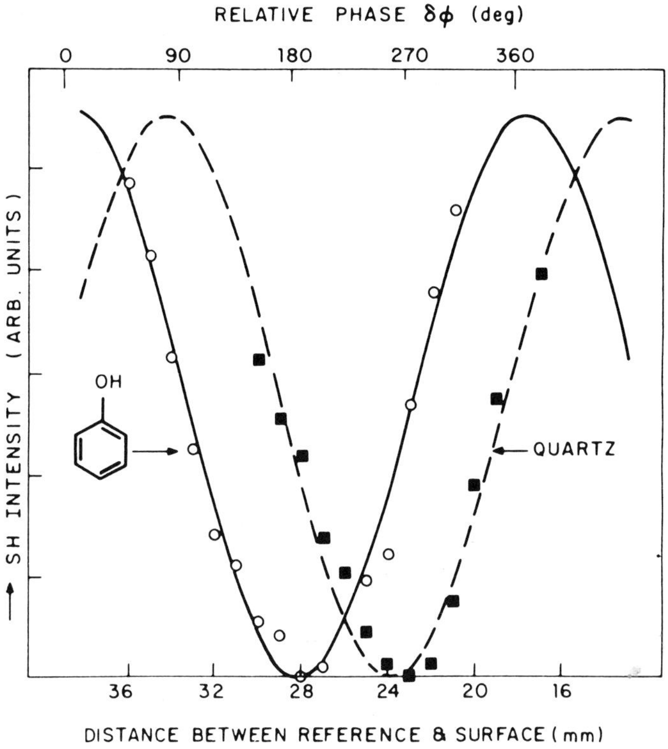

Fig. 5. Interference pattern of the SH fields using the apparatus of fig. 4 for the sample (phenol segregated on a water surface) and a quartz standard. (After Kemnitz et al. 1986.)

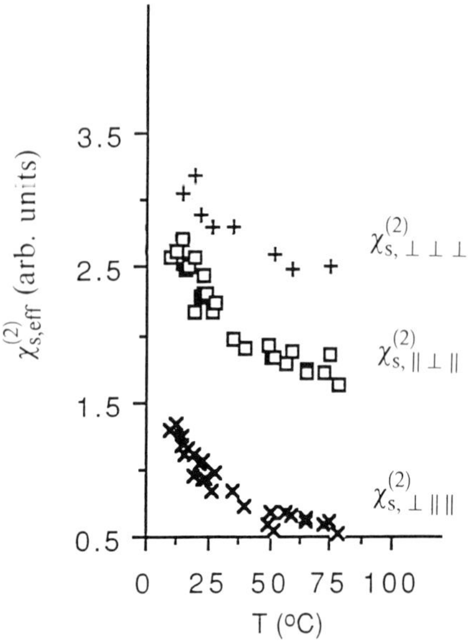

Fig. 6. Example of the complete determination of the (effective) surface nonlinear susceptibility tensor $\chi_s^{(2)}$ for SHG by an isotropic surface. The three independent tensor elements are shown as a function of the temperature of the water/vapor interface. For an outward-going surface normal, all tensor elements were found to have a phase of 180°. (After Goh et al. 1988.)

interface for various temperatures. Although the absolute magnitude of the elements of $\chi^{(2)}$ was not established, the phase of the elements was determined (Goh et al. 1988).

We now consider briefly the more general problem of a complete determination of $\chi_s^{(2)}$ for an isotropic surface (with mirror symmetry) when the elements $\chi^{(2)}_{s,\parallel\perp\parallel}$, $\chi_{s,\perp\parallel\parallel}$, and $\chi^{(2)}_{s,\perp\perp\perp}$ are complex. It should be noted that the method sketched above can be applied with either real or complex *linear* dielectric response. The only modification of the prescription is the need to deal with SH radiation having elliptical polarization; this implies the use of a compensator and analyzer and a complex $\hat{e}_{\text{null}}$ vector. When the elements of $\chi_s^{(2)}$ also have arbitrary phases, the problem becomes more difficult. If accurate phase measurements can be performed, the following procedure may be appropriate. One isolates $\chi^{(2)}_{s,\parallel\perp\parallel}$ and $\chi^{(2)}_{s,\perp\parallel\parallel}$ with the appropriate choice of pump or detected polarization. Then, in addition to determining the magnitude of these elements, one establishes the phase of the two terms. In this event, two (or more) measurements of the SH polarization for different mixed-polarization pump beams will provide independent equations of the form of eq. (51) to determine both the magnitude and phase of $\chi^{(2)}_{s,\perp\perp\perp}$. Although this procedure is systematic, it may be difficult to utilize in practice. It may be desirable to determine $|\chi^{(2)}_{s,\parallel\perp\parallel}|$

and $|\chi^{(2)}_{s,\perp\parallel\parallel}|$ as indicated, but to infer the remaining parameters by fitting continuous measurements of the SH intensity as a function of pump polarization (and angle of incidence).

3.2. Anisotropic media

Although the possibility of observing effects associated with the orientation of cubic centrosymmetric materials in SHG was discussed in the early work of Bloembergen et al. (1968), it was not until relatively recently that experimental evidence for this effect was obtained. In studies of high-symmetry faces of (oxidized) Si crystals, it was found by Guidotti and co-workers (Guidotti et al. 1983, Driscoll and Guidotti 1983) and Tom et al. (1983) that the SH intensity was strongly modulated as the sample was rotated about its surface normal. These results were subsequently confirmed and extended by Litwin et al. (1983), Aktsipetrov et al. (1986) and Hollering (1990) for oxidized semiconductor surfaces. It has also been shown that strong anisotropies in SHG are not limited to semiconductors, but may be present in single-crystal metals, as was first demonstrated for Cu surfaces by Tom and Aumiller (1986). Subsequent studies have shown an observable anisotropy to be present for a variety of other metals (Richmond et al. 1988, Anderson and Hamilton 1988, Hamilton and Anderson 1988, Miragliotta and Furtak 1988, Shannon et al. 1987a–c, 1989, Akhmanov et al. 1989, Friedrich et al. 1989), insulators (Reif et al. 1988), and even in partially-ordered organic (Aktsipetrov et al. 1985, Mizrahi et al. 1989, Chen et al. 1989) and inorganic (Mizrahi et al. 1987) materials.

While the nature of an anisotropic response in the bulk of centrosymmetric materials is undoubtedly an important fundamental problem, the possibility of applying nonlinear techniques to examine the symmetry and order of the top atomic layer of a surface or an interface is particularly intriguing. The investigations of Heinz and co-workers (Heinz et al. 1985a, b, 1987, Iyer et al. 1987) on Si(111) surfaces in ultrahigh vacuum demonstrate very clearly the power of the technique for surface studies. For metals, the anisotropic response is also strongly influenced by the detailed surface structure, as was first shown in investigations of clean and oxidized Cu surfaces by Tom and Aumiller (1986). This sensitivity to surface structure on an atomic level has already been exploited in a range of studies of physically interesting phenomena, such as surface phase transformations (Heinz et al. 1985a), deposition and growth of semiconductor (Iyer et al. 1987, Heinz et al. 1987) and metal (McGilp and Yeh 1986, McGilp 1987) films on semiconductor surfaces; and molecular adsorption (Anderson and Hamilton 1988, Hamilton and Anderson 1988) and growth of metal films on metal surfaces (Richmond et al. 1988, Miragliotta and Furtak 1988, Koos 1989). Measurements of bulk effects can also be performed, as was illustrated in subpicosecond time-resolved measurements of melting of Si crystals (Shank et al. 1983, Tom et al. 1988).

In our present discussion, we shall again emphasize the basic principles and not review the diverse applications of the method. From this point of view, a common description can be given within the framework of the phenomenological model developed earlier in this chapter, which has been found adequate to explain existing results on SHG. (No analogous SFG measurements have been performed to date.) The discussion follows along the treatment of Tom et al. (1983), as subsequently codified by Tom (1984), Aktsipetrov et al. (1986, 1987c), Guyot-Sionnest et al. (1986b), and Sipe et al. (1987b). We first treat the bulk response and then turn to the influence of the surface.

The nonlocal bulk response has only been discussed extensively for cubic centrosymmetric materials (m3m- or 432-symmetry), to which we confine our discussion. As introduced in section 2.5, the bulk nonlinear polarization for SHG arising from leading-order nonlocal terms in a homogeneous medium can be cast in the form

$$P_i(\Omega) = \gamma \nabla_i(\boldsymbol{E}\cdot\boldsymbol{E}) + (\delta - \beta - 2\gamma)(\boldsymbol{E}\cdot\nabla)E_i + \zeta E_i \nabla_i E_i, \tag{52}$$

where it is assumed, as in eq. (36), that the cartesian coordinate axes coincide with the principal crystallographic axes and $\Omega = 2\omega$ denotes the SH frequency. The first and second terms, present in isotropic media, clearly do not depend on crystallographic orientation; the third term, however, reflects the cubic symmetry of the material.

The analytic form for SH radiation associated with the isotropic and anisotropic bulk terms in cubic materials can be determined for any crystal orientation from eq. (52) in conjunction with eqs (40) and (18). To apply eq. (52), it is necessary to transform quantities in the laboratory or beam coordinates into coordinates associated with the principal axes of the crystal, or equivalently, to transform $\chi_q^{(2)}$ into the lab coordinates. This procedure has been carried out systematically by Tom (1984), Aktsipetrov et al. (1986) and Sipe et al. (1987b) for high-symmetry crystal faces rotated by an arbitrary angle φ about the surface normal. For either p- or s-polarized pump radiation, the variation of the SH fields for (001) and (111) crystal faces can be summarized succinctly as

$$E_p(\Omega) = a + b^{(m)}\cos(m\varphi), \tag{53a}$$

$$E_s(\Omega) = c^{(m)}\sin(m\varphi), \tag{53b}$$

where $m = 3, 4$ pertain to the (111) and (001) crystal faces, respectively. The angle of rotation φ is measured between the plane of incidence and the $[2\bar{1}\bar{1}]$ direction in the (111) surface and the [100] direction in the (001) surface. The SH electric field is found to exhibit the expected 2-, 3-, and 4-fold symmetry as a function of rotation about the surface normal for (110), (111), and (001) faces, respectively. Convenient explicit expressions for the coefficients in these formulas are given for both s- and p-polarized pump radiation by Sipe et al. (1987b). The corresponding result for rotation about the surface normal of the (110) face of a cubic centrosymmetric medium is

$$E_p(\Omega) = a + b^{(2)}\cos(2\varphi) + b^{(4)}\cos(4\varphi), \tag{54a}$$

$$E_s(\Omega) = c^{(2)}\sin(2\varphi) + c^{(4)}\sin(4\varphi). \tag{54b}$$

For the (110) face, the angle φ is zero when the [100] crystal axis lies in the plane of incidence. The dependence of the SH radiation in eqs (53) and (54) can be understood by recalling that the fourth-rank tensor $\chi_q^{(2)}$ can be decomposed into spherical tensors of rank $m \leqslant 4$, with each spherical tensor component of rank m transforming like $e^{im\varphi}$ under rotation (Edmonds 1960). Since cubic materials have 3- and 4-fold rotational axes about the [111] and [001] directions, respectively, only the corresponding $m = 3$ and 4 components in the decomposition into spherical tensors can survive for the (111) and (001) surfaces. For the (110) surface, the crystal has only 2-fold rotational symmetry, which is compatible with both $m = 2$ and $m = 4$ components.

The anisotropic response of the bulk can be isolated by performing measurements on crystal faces for which the surface contribution from $\chi_s^{(2)}$ is isotropic. Referring to table 1, we find this to be the case for crystal faces with 4mm-symmetry. As we shall discuss below, (001) faces of cubic crystals will generally exhibit this (macroscopic) symmetry. Figure 7 displays the SH intensity from the face of an oxidized Si(001) surface in reflection as a function of the angle of rotation φ about the surface normal (Tom et al. 1983, Tom 1984). Since the surface is expected to exhibit 4mm-symmetry, the variation in the SH intensity can be attributed to the anisotropic bulk term. In fig. 7, both sets of data correspond to p-polarized excitation at the fundamental frequency, with the s-polarized SH intensity shown in (a) and p-polarized output in (b). The dashed curves are fit to the experimental data using the form from the symmetry analysis of eq. (53) (with $m = 3$). For the s-polarized SH radiation, only the anisotropic response is present, while the p-polarized signal includes a substantial contribution from the isotropic (surface and bulk) terms. In both cases, good agreement is found between the experimental results and the predictions of eq. (53). It should be noted that the bulk anisotropic term is sufficiently strong to be detected despite the low coupling efficiency expected for a (001) surface of any material with a large dielectric constant. The poor coupling arises from the fact that the pump beam after being refracted in the medium will be directed nearly along the $\hat{z}$ direction independent of the angle of incidence from any low-index material; but for pump radiation traveling strictly along a crystal axis, the anisotropic term vanishes, since $P_i(\Omega) = \zeta E_i \nabla_i E_i = \zeta E_i \nabla \cdot \boldsymbol{E} = 0$.

We now turn to the more distinctive aspect of the anisotropic nonlinear response, namely that associated with surfaces and interfaces. Table 1 enumerates the independent nonvanishing tensor elements of $\chi_s^{(2)}$ for all point groups compatible with long-range order of a surface. It should be observed that all symmetry classes except 4, 6, and 4mm, 6mm exhibit tensor elements not

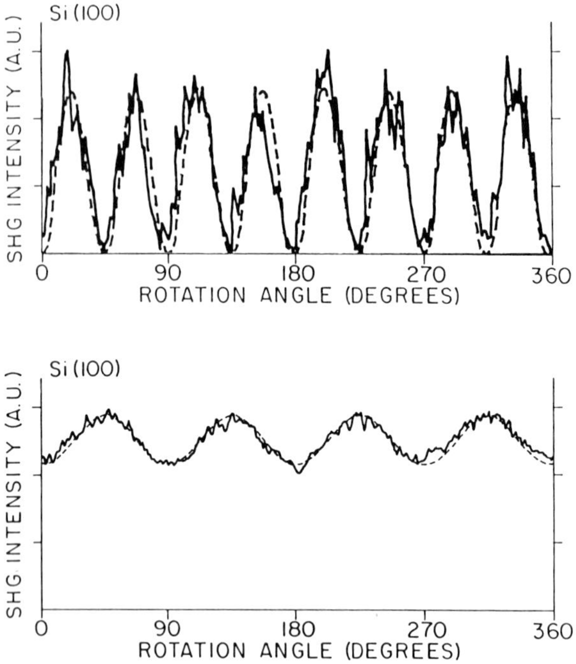

Fig. 7. Second-harmonic intensity from oxidized Si(100) surfaces as a function of rotation about the surface normal. The dashed lines are fits to the symmetry analysis considering only the bulk nonlinear response for a cubic crystal. The pump radiation is p-polarized; the s-polarized SH signal is shown in (a), the p-polarized component in (b). (After Tom et al. 1983, 1984.)

found for an isotropic surface. For 4- and 6-fold rotational axes, the response of the surface is isotropic, but it is still distinguishable from the case of highest symmetry in which mirror planes are present perpendicular to the surface. Thus, an ordered crystal surface of any symmetry other than 4mm or 6mm will exhibit a surface nonlinear susceptibility tensor $\chi_s^{(2)}$ of a form differing from that of a disordered surface.

SHG measurements on Si(111) surfaces under ultrahigh vacuum conditions (Heinz et al. 1985a, b, 1987, Iyer et al. 1987, McGlip and Yeh 1986, McGilp 1987) illustrate how the surface structure is reflected in SH response. Clean Si(111) can be prepared with two different ordered surface structures: the metastable 2×1 reconstruction and the equilibrium 7×7 reconstruction (Haneman 1987). (The designation $n \times m$ refers to the size of the surface unit cell after reconstruction with respect to the unit cell of the ideal unreconstructed surface.) Figure 8 displays the SH intensity for normal incidence excitation of Si(111)-2×1 and 7×7 surfaces as a function of the orientation of the linearly polarized pump beam (Heinz et al. 1985a). The upper panels correspond to the

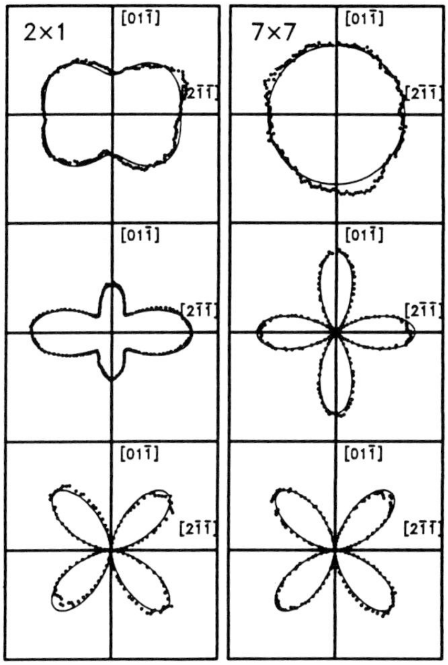

Fig. 8. Second-harmonic intensity from clean Si(111)-2 × 1 and 7 × 7 surfaces as a function of the polarization of a normally incident pump beam. The top panels display the total SH intensity; the SH signals resolved into components polarized along the [2$\bar{1}\bar{1}$] and [01$\bar{1}$] directions are shown in the middle and lower panels, respectively. The solid curves are based on the symmetry analysis discussed in the text (Heinz et al. 1985a).

total SH intensity, while the components polarized along the [2$\bar{1}\bar{1}$] and [01$\bar{1}$] directions are shown in the middle and lower panels, respectively. Although the bulk structure of the crystals is, of course, identical for the Si(111)-2 × 1 and 7 × 7 structures, the SH response is decidedly different. In fact, by modifying the surface conditions by oxidation (Heinz et al. 1985b), for example, it can be shown that the anisotropic surface contributions are $\gtrsim 10^2$ times stronger than the bulk terms for the 1.06 μm pump wavelength utilized in these measurements.

The expected form for the polarization dependence for a SH signal associated only with the surface nonlinear susceptibility has a particularly simple form for excitation at normal incidence. From eqs (29) and (20), we obtain for the SH field with polarization $\hat{e}(\Omega)$ in medium 1:

$$\hat{e}(\Omega) \cdot \boldsymbol{E}(\Omega) = 2\pi i \varepsilon_1^{-1/2}(\Omega)(\Omega/c)F(\Omega)F^2(\omega)[\hat{e}(\Omega) \cdot \chi_s^{(2)} : \hat{e}(\omega)\hat{e}(\omega)]E^2(\omega), \tag{55a}$$

for pump radiation with polarization $\hat{e}(\omega)$ and electric field amplitude $E(\omega)$

incident on the interface from medium 1. The Fresnel factors are

$$F = 2\varepsilon_1^{1/2}/(\varepsilon_1^{1/2} + \varepsilon_2^{1/2}), \tag{55b}$$

with the dielectric constants evaluated at the appropriate frequencies. With the aid of this expression, we can write immediately the SH intensity expected for surfaces with different symmetry as a function of the fundamental and SH polarizations. It should be noted that SHG (or SFG) is permitted in the normal incidence geometry only for surfaces with structure, since none of the isotropic terms in $\boldsymbol{\chi}_s^{(2)}$ are accessed.

We illustrate this elementary analysis for a surface with 3m-symmetry. From table 1, we find that $\boldsymbol{\chi}_s^{(2)}$ will have only one independent element, $\chi_{s,xxx}^{(2)} = -\chi_{s,xyy}^{(2)} = -\chi_{s,yxy}^{(2)}$, where we have assumed the mirror plane to lie perpendicular to $\hat{y}$*. The irradiance of the reflected SH for light polarized in the x and y directions can then be written immediately from eq. (55) as

$$I_x(\Omega) = A|\chi_{s,xxx}^{(2)}|^2 \cos^2 2\varphi, \tag{56a}$$

$$I_y(\Omega) = A|\chi_{s,xxx}^{(2)}|^2 \sin^2 2\varphi, \tag{56b}$$

where φ is the angle of the linearly polarized pump beam with respect to the x-axis, and the (polarization-independent) factor

$$A = |8\pi^3 \varepsilon_1^{-1/2}(\Omega)\varepsilon_1^{-1}(\omega)c^{-3}\Omega^2 F^2(\Omega) F^4(\omega) I^2(\omega)|, \tag{56c}$$

is proportional to the square of the pump laser irradiance, $I^2(\omega)$. The form prescribed by eq. (56) can be applied to describe the experimental data on the Si(111)-7 × 7 surface in fig. 8. The x- and y-axes in eq. (56) are defined in terms of the crystallographic axes as $\hat{x} = [2\bar{1}\bar{1}]$ and $\hat{y} = [01\bar{1}]$. The very good agreement between the experimental data and the predictions of eq. (56) (solid lines in the figure) implies that the Si(111)-7 × 7 exhibits 3m-symmetry, as would be expected based on the symmetry of the underlying Si crystal.

The predictions for a surface with 3m-symmetry clearly do not match the experimental results for the Si(111)-2 × 1 surface shown in fig. 8. In this case, the highest symmetry compatible with the results is that of a single mirror plane m oriented normal to $\hat{y} = [01\bar{1}]$. The expression for the two independently polarized SH irradiances corresponding to eq. (56) involve three independent tensor elements, $\chi_{s,xxx}^{(2)}$, $\chi_{s,xyy}^{(2)}$ and $\chi_{s,yxy}^{(2)}$:

$$I_x(\Omega) = A|\chi_{s,xxx}^{(2)} \cos^2 \varphi + \chi_{s,xyy}^{(2)} \sin^2 \varphi|^2, \tag{57a}$$

$$I_y(\Omega) = A|\chi_{s,xxx}^{(2)}|^2 \sin^2 2\varphi, \tag{57b}$$

with A and φ as defined in eq. (56). These equations are capable of accurately describing the experimental observations for the Si(111)-2 × 1 surface, as shown

*The independent element $\chi_{s,xxx}^{(2)}$ is often denoted in the literature as $\chi_{s,\xi\xi\xi}^{(2)}$ to emphasize the distinction between the surface coordinates and the crystallographic axes.

in fig. 8. This fitting procedure also yields values for all of the magnitudes of the in-plane tensor elements of $\chi_s^{(2)}$ and some of the relative phases (Heinz et al. 1985a). The physical origin of the lowered symmetry of the Si(111)-2 × 1 surface arises from the asymmetric nature of the reconstructed unit cell. For this to be manifested in the SH measurement, however, it is necessary that the reconstructed surface form in a single domain (rather than in a combination of regions with different domain orientation, leading to macroscopic 3m-symmetry). Experimentally, a reconstructed surface consisting of a single domain could indeed be obtained by a proper cleaving procedure (Heinz et al. 1985a).

As a final example of the sensitivity of the SHG process to the symmetry and order of the surface, let us consider the process of the deposition of atomic Si on a clean Si(111)-7 × 7 surface. If the deposition process is carried out at sufficiently high temperature, epitaxial growth will occur and the Si(111)-7 × 7 surface will be preserved during deposition. On the other hand, at low surface temperatures, an amorphous Si adlayer will be formed. SHG in a normal incidence measurement, which is expected only for ordered structures, should reflect the difference between these two cases. As shown in fig. 9 (Heinz et al. 1987), the SH signal allows one to distinguish between the two chemically

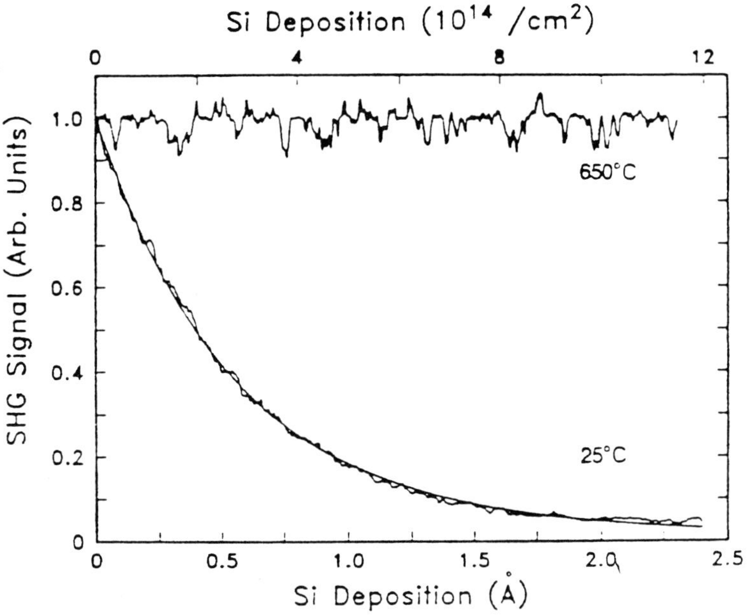

Fig. 9. Second-harmonic intensity from a Si(111)-7 × 7 surface during deposition of atomic Si at substrate temperatures of 25°C and 650°C. The heavy line for the room-temperature data corresponds to an exponential decay in the surface nonlinear susceptibility as a function of the density of adatoms (Heinz et al. 1987).

identical surfaces (all Si atoms) depending on the order present in the first few Ångstroms.

Up to this point, we have treated the anisotropic terms in the surface and bulk separately. In many experiments, they are both present simultaneously. Then the electric fields from each of the sources must be added together coherently. Some general comments are perhaps in order concerning the separation of the surface and bulk contributions. Unlike the case prevailing for the longitudinal bulk polarization (the γ term), it is always possible in *principle* to separate the surface and bulk anisotropic contributions on the basis of the dependence of the contributions on polarization and angle of incidence. There is, for example, one class of excitation geometries in which the bulk quadrupole terms can never contribute, namely, a mixing experiment in which the SH beam is produced by two counter-propagating waves of the same polarization. This can be verified directly from the form of the bulk quadrupole contributions, $\chi_q^{(2)}:\boldsymbol{E}\nabla\boldsymbol{E}$. In this configuration (most) anisotropic surface terms may still be measured. For the cubic materials of m3m- or 432-symmetry, the bulk anisotropic response is determined entirely by the single coefficient ζ, making it relatively easy to determine the bulk response. For high-symmetry surfaces, as we have illustrated above for the oxidized Si(001) surface, it may be possible to isolate the bulk anisotropic term. Once its value has been established, its influence may be taken into account in any measurement of the anisotropic components of $\chi_s^{(2)}$. This procedure was employed by Tom et al. (1983) in studies of oxidized Si(111) surfaces, where the bulk anisotropic term was isolated in measurements of the Si(001) face.

Finally, we make some more specific remarks about the SH radiation expected from (001), (110), and (111) faces of cubic materials when the surfaces (or interfaces) exhibit the same symmetry as the underlying bulk material. If no external anisotropic perturbation is introduced in forming the surface and the surface includes domains formed of differing atomic planes of the crystal, then the macroscopic symmetry of the surface (averaged over the different domains) should correspond to the symmetry operations in the point group of the bulk crystal that do not rotate the surface in question*. In this case (001), (111), and (110) surfaces of crystals with m3m-symmetry, should exhibit 4mm-, 3m-, and

*The symmetry of the surface may be lower than that of the bulk when either of the conditions mentioned is not satisfied. The cleaved Si(111)-2 × 1 surface discussed above illustrates how an external perturbation (enhanced by the surface reconstruction) can lower the surface symmetry. If the surface consists of a single atomic plane over a macroscopic distance, it is also not appropriate to consider operations in the point group of the bulk crystal. This result can be understood by realizing that the bulk symmetry operations included in the point group for describing macroscopic properties permits translations on an atomic scale. These operations may no longer be allowed for a surface comprised of a single atomic plane. An ideal bulk-terminated Si(001) surface, for example, will have only mm2, not the 4mm- (macroscopic) symmetry exhibited by a surface consisting of regions in different atomic planes.

mm2-symmetry, respectively. As discussed above, (001) faces with 4mm-symmetry have only the three isotropic elements of $\boldsymbol{\chi}_s^{(2)}$, namely, $\chi_{s,xzx}^{(2)} = \chi_{s,yzy}^{(2)}$, $\chi_{s,zxx}^{(2)} = \chi_{s,zyy}^{(2)}$, and $\chi_{s,zzz}^{(2)}$. (111) faces with 3m-symmetry have these three isotropic tensor elements and the single additional anisotropic element $\chi_{s,xxx}^{(2)} = -\chi_{s,xyy}^{(2)} = -\chi_{s,yxy}^{(2)}$. Referring to table 1, we see that (110) surfaces with mm2-symmetry have five independent tensor elements: $\chi_{s,xzx}^{(2)}$, $\chi_{s,yzy}^{(2)}$, $\chi_{s,zxx}^{(2)}$, $\chi_{s,zyy}^{(2)}$, and $\chi_{s,zzz}^{(2)}$. Explicit formulas for the SHG have been derived for these cases by Aktsipetrov et al. (1986), Tom (1984) and Sipe et al. (1987b). It is found that the variation in SH intensity as the sample is rotated about the surface normal can still be expressed in the form of eqs (53) and (54). The contribution of $\boldsymbol{\chi}_s^{(2)}$ enters generally into all of the terms other than those involving angular dependences of $\cos(4\varphi)$ or $\sin(4\varphi)$, as is to be expected for a *third*-rank tensor. Even when the variation with rotation about the surface normal of the sample assumes the same form, dependence on polarization and angle of incidence of the surface and bulk anisotropic terms will differ.

4. Microscopic models

4.1. General considerations

A complete description of second-order nonlinear optical effects at surfaces must certainly go beyond the macroscopic properties and explain the origin of the surface nonlinear susceptibility and the bulk quadrupole response in terms of microscopic quantities. In principle, the second-order nonlinear response of both the surface and the bulk can be computed by application of perturbation theory. In this approach the relevant generalized polarization (i.e., current) is calculated by treating the interaction Hamiltonian between the electromagnetic field and the material system to second-order (see, e.g., Shen 1984, ch. 2). In the surface or interface region, an accurate calculation would avoid any expansion in powers into multipole orders since the fields are known to change over a very small length scale. Such expressions have been given (in the random-phase approximation) by Bower (1976), generalizing the earlier work for the bulk nonlinear response (Jha 1966, Jha and Warke 1967, Rustagi 1968). While these relations are useful for displaying the sensitivity of the processes to surface states and resonances, they are complex and difficult to apply in practice. An accurate description of the linear fields at a surface or interface is already a daunting problem (see, e.g., Feibelman 1982). Only in the approximation of jellium (electrons with a homogeneous positive background charge) has it been possible to perform such calculations from first principles, as we discuss below. Nonetheless, these expressions show the suitability of resonant SHG or SFG for spectroscopic studies, with resonances appearing when either the pump or emitted photon energies coincides with that of a transition between empty and occupied states. (The resonances between pairs of unoccupied states may also exist under some circumstances (Dick 1984).) The direct application of such first-principles calculations to bulk properties is, perhaps, somewhat less difficult, both because of our better knowledge of the bulk properties and because a multipole expansion will be valid to high accuracy. Nonetheless the problem is a very challenging one, as an accurate prediction even of the bulk *linear* properties is difficult to obtain. Once again, only for the case of free electrons have such calculations been carried out in detail. The theory has, however, been applied to obtain approximate values for the nonlinear bulk response under well-defined approximations (Bloembergen et al. 1968).

In the remainder of this section, we outline some of the simplified models that have been applied to prediction of the surface and bulk nonlinear response. It is not within the scope of this chapter to review all aspects of the theory of surface and bulk second-order nonlinear optical effects. The primary emphasis shall be to relate the different approaches used in microscopic treatments to the phenomenological description developed above. Before discussing more detailed models, however, we consider the effects of the linear optical properties of the surface or interface in somewhat more detail. Here we show, following Guyot-Sionnest et al. (1986b), how the linear and nonlinear optical properties can be separated from one another in our formalism. Rather than considering the variation of the electric field to be described by abruptly changing local dielectric constants, we can introduce a function to characterize the variation of the fields more realistically. To this end, consider a wave of frequency Ω incident on the surface or interface with a field strength $\boldsymbol{E}^{\varepsilon'}(\Omega)$ within the idealized interfacial layer of dielectric constant ε' in the phenomenological model. In reality, the linear optical properties of the surface or interface lead to a complex variation in the field strength. This can be described by the dimensionless functions

$$S_i(z, \Omega) = E_i^{\varepsilon'}(\Omega)/E_i(z, \Omega), \tag{58}$$

where $\boldsymbol{E}(z, \Omega)$ is the actual electric field in a plane at a distance z. Here it has been implicitly assumed that the variation in the fields can be characterized solely in terms of the distance z. It is further assumed that normal components of the electric field are continuous across the interface, so that $S_x = S_y = 1$. In this case, Guyot-Sionnest et al. (1986b) have shown that the (generalized) surface nonlinear source polarization in the phenomenological model is given by

$$P_{s,i}(\Omega) = \int_{0^-}^{0^+} S_i(z, \Omega)\, P_i(z, \Omega), \tag{59}$$

with $\boldsymbol{P}(z, \Omega)$ denoting the true (volume) nonlinear source polarization at frequency Ω at a distance z. This relation shows that the linear optical properties do not directly influence the radiation by the nonlinear polarization (or current) parallel to the surface. For the component of the source polarization normal to the surface, however, the radiation is screened in just the same manner as the normal component of an electric field at the same frequency under the influence of the linear optical response of the medium. While this formal treatment does not predict what form $S_i(z, \Omega)$ might take, it does demonstrate that the linear optical properties enter into the computation of the radiated fields in the same manner as in the determination of the detailed variation of the pump fields at the surface or interface.

4.2. Electric-dipole response of surface

The notion of attributing the nonlinear response of a surface or interface of a centrosymmetric material to an electric-dipole response by the perturbed and, hence, no longer centrosymmetric interfacial region was already suggested in some of the early discussion of second-order nonlinear optical effects in centrosymmetric media. For example, Bloembergen and Shen (1966) compared the nonlinearity observed from the surfaces of centrosymmetric solids with the electric-dipole response expected from a layer of a non-centrosymmetric semiconductor with bulk properties but only a few Ångstroms in thickness. Most of the more quantitative analysis in the early work, however, was based on the approximation that the nonlinear optical response in the surface or interface region resembled that of the bulk. Strong surface contributions were then attributed to the response of the nonlocal (quadrupole) terms under the influence of the rapidly varying pump fields at the interface, as we shall discuss in section 4.3. In the more recent analysis of surface SHG and SFG, however, the electric-dipole approximation has been widely applied in the context of the nonlinear response of molecular layers. In this section, we review some of the key points in this treatment.

For a nonlinear response attributable to a molecular layer, we consider the surface nonlinear susceptibility (in the electric-dipole approximation) to be given as a sum of nonlinear polarizabilities of molecular species present at the surface*. Neglecting local-field effects, we can write

$$\boldsymbol{\chi}_s^{(2)} = \sum_{\text{unit area}} \boldsymbol{\alpha}^{(2)}. \tag{60}$$

Here the summation can be considered to be carried out over all of the molecules with a nonisotropic distribution. Assuming that the bulk materials are centrosymmetric, they will not contribute to this sum and only a localized region (typically on the order of a monolayer in thickness) will contribute to $\boldsymbol{\chi}_s^{(2)}$. Let us suppose for simplicity that the molecules all have the same chemical identity. We can define a surface density N_s corresponding to the total number of

*A similar approach has been developed to describe the nonlinear susceptibility of a semiconductor surface in terms of the nonlinear polarizability of the asymmetrically perturbed surface atoms. See Zadkov et al. (1987).

aligned molecules per unit area at the surface or interface. Then eq. (60) can be cast in the alternate form

$$\boldsymbol{\chi}_s^{(2)} = N_s \langle \boldsymbol{\alpha}^{(2)} \rangle, \tag{61}$$

where the angular brackets represent an average over the orientational distribution of the molecules in the surface layer. (This relation can be generalized in the obvious manner to include different molecular species or groups of molecules characterized by distinct orientational distributions such as might be present, e.g., in successive molecular layers.) Within this simple model, the surface nonlinear susceptibility tensor $\boldsymbol{\chi}_s^{(2)}$ is determined by the adsorbate density, nonlinear polarizability tensor of the adsorbed molecules $\boldsymbol{\alpha}^{(2)}$, and the molecular orientational distribution. In fact, all observable quantities also depend on the effective dielectric constant of the surface layer ε', which should also be considered as a part of the model. A few remarks can be made about each of these factors.

First, the surface density enters only as an overall scale factor and is not important for determining the polarization and angular dependences. Within the approximation of this model, however, we take the SH (or SF) electric field to be proportional to the adsorbate density. From the point of view of applying the technique to studies of surfaces and interfaces, this means that the strength of the SH electric field can be used to estimate the relative surface density. It should be noted that a simple application of this procedure is valid only if the orientational distribution of the adsorbates is not changing as a function of their surface density, which need not be the case.

The question of how to represent the nonlinear polarizability of the adsorbed species has not been treated in detail in the literature. If available, a good starting point would be the nonlinear polarizability of an isolated molecule. A more accurate representation would include the influence of perturbations on the electronic or vibrational structure of the molecules at the surface. In the language developed for linear surface vibrational spectroscopy, these corrections might be of 'chemical' or 'physical' origin. The former refer to changes in the properties of the adsorbed molecule itself; the latter includes long-range electromagnetic interactions with other molecules, which can also be described as local-field corrections. The chemical effects, while possibly of major importance, depend very much on the system under study and will not be discussed further, other than to note that shifts in the strengths and energy separations of electronic and vibrational levels in adsorbed molecules can be significant. The issue of local-field effects will be mentioned briefly below. When the molecules themselves are non-centrosymmetric, they are expected to have a non-zero nonlinear polarizability $\boldsymbol{\alpha}^{(2)}$ in the gas phase; for centrosymmetric molecules, $\boldsymbol{\alpha}^{(2)}$ will vanish, implying that the model under discussion will predict a finite nonlinear response only if the perturbations induced in $\boldsymbol{\alpha}^{(2)}$ by the surface are taken into account.

For many purposes, exact numerical values of the elements of $\boldsymbol{\alpha}^{(2)}$ may not be important, provided symmetry considerations or specific knowledge of the origin of the nonlinearity allow us to conclude that only a few elements are significant. This approach has generally been utilized in analysis of molecular orientation from the polarization dependence of the SHG or SFG process (i.e., from the relative values of the elements of $\boldsymbol{\chi}_s^{(2)}$, as will be elaborated below). From the opposite perspective, as Shen and co-workers (Rasing et al. 1986, Berkovic et al. 1987) have demonstrated, it may be possible to use well-defined molecular layers of known orientation and surface density to infer elements of $\boldsymbol{\alpha}^{(2)}$. In contrast to measurements of electric-field-induced SHG or SFG, a much higher degree of alignment may be attained at the surface, and the technique may be applicable for molecules difficult to study in gas phase. On the other hand, the interpretation of the measurement depends upon having a knowledge of the surface density and molecular orientation, as well as on the accuracy of eq. (61) relating the macroscopic and microscopic quantities.

The orientational averaging process of the molecular nonlinear polarizability is critical in predicting $\boldsymbol{\chi}_s^{(2)}$. If no orientational preference is exhibited, then $\boldsymbol{\chi}_s^{(2)}$ will vanish (except for SFG by chiral molecules, as mentioned in section 3.1). Conversely, a knowledge of the microscopic $\boldsymbol{\alpha}^{(2)}$ and macroscopic $\boldsymbol{\chi}_s^{(2)}$ responses may permit one to make inferences about the molecular orientation. To be more explicit, we can introduce the coordinate transformation tensor $\boldsymbol{T}$ between the molecular and laboratory systems. We can express eq. (61) as (Heinz et al. 1983, Mazely and Hetherington 1987)

$$\chi_{s,ijk}^{(2)} = N_s \langle T_{ii'} T_{jj'} T_{kk'} \rangle \alpha_{i'j'k'}^{(2)}, \tag{62}$$

where $\alpha_{i'j'k'}^{(2)}$ refers to the molecular nonlinear polarizability measured in the coordinate system of the molecule. Thus complete knowledge of $\boldsymbol{\chi}_s^{(2)}$, N_s, and $\boldsymbol{\alpha}^{(2)}$, may permit us to infer the moments of the orientational distribution of the molecules given by $\langle T_{jj'} T_{kk'} T_{jj'} \rangle$*. In our discussion, we examine the relation between $\boldsymbol{\chi}_s^{(2)}$ and $\boldsymbol{\alpha}^{(2)}$ in terms of cartesian coordinates. A powerful alternative approach is the use of irreducible spherical tensor components, as has been developed in detail by Dick (1985).

In regard to the inference of molecular orientation, the notion is to apply knowledge of $\boldsymbol{\chi}_s^{(2)}$, N_s, and $\boldsymbol{\alpha}^{(2)}$ to infer the moments of the orientational distribution of the molecules given by $\langle T_{jj'} T_{kk'} T_{jj'} \rangle$. In the absence of any simplification in the form of $\boldsymbol{\alpha}^{(2)}$, eq. (62) provides a complex set of equations, which cannot generally be inverted to obtain averages of orientational param-

*Both the magnitude and the *sign* of $\langle T_{jj'} T_{kk'} T_{jj'} \rangle$ carry information. For an isotropic system in the surface plane, the non-vanishing moments $\langle T_{jj'} T_{kk'} T_{jj'} \rangle$ are expected to change sign if the orientation of the molecules is reflected through the surface plane. Thus, the sign of the orientational moments gives information about whether the molecules are pointing into or away from the bulk. Experimentally, this requires measurement of the absolute phase of elements of $\boldsymbol{\chi}_s^{(2)}$ (together with knowledge of $\boldsymbol{\alpha}^{(2)}$). See Guyot-Sionnest et al. (1986a), Kemnitz et al. (1986) and Goh et al. (1988).

eters. However, if symmetry or resonant enhancements permit $\boldsymbol{\alpha}^{(2)}$ to be approximated by a small number of independent non-vanishing elements, the orientational averages can be determined more directly. To take the simplest example, suppose only one element $\alpha_{IJK}^{(2)}$ of the nonlinear polarizability is appreciable. Equation (62) can then be expressed as

$$[\chi_{s,ijk}^{(2)}]/[N_s\alpha_{IJK}^{(2)}] = \langle T_{iI}T_{jJ}T_{kK}\rangle. \tag{63}$$

If the numerical values of $\alpha_{IJK}^{(2)}$ and N_s are known, then different independent orientational moments $\langle T_{iI}T_{jJ}T_{kK}\rangle$ can be determined. When these quantities are not available, as is usually the case, ratios of different independent elements of $\boldsymbol{\chi}_s^{(2)}$ will still yield ratios of orientational moments.

A somewhat more general situation is considered now. Suppose we have molecules exhibiting a strong nonlinear response only in the $x'z'$-plane with a plane of mirror symmetry perpendicular to the x'-axis, where the primed coordinates refer to molecular axes. (This model is a good approximation for many planar molecules with conjugated bonds at frequencies below electronic resonances.) For SHG, the independent, non-vanishing elements of $\boldsymbol{\alpha}^{(2)}$, are $\alpha_{z'z'z'}^{(2)}$, $\alpha_{z'x'x'}^{(2)}$ and $\alpha_{x'z'x'}^{(2)} = \alpha_{x'x'z'}^{(2)}$. If we assume an orientational distribution that is isotropic in the plane of the surface, then eq. (62) yields for the three independent elements of $\boldsymbol{\chi}_s^{(2)}$:

$$\chi_{s,\perp\perp\perp}^{(2)} = N_s[\langle\cos^3\theta\rangle\alpha_{z'z'z'}^{(2)} + \langle\cos\theta\sin^2\theta\sin^2\psi\rangle(\alpha_{z'x'x'}^{(2)} + 2\alpha_{x'z'x'}^{(2)})], \tag{64a}$$

$$\chi_{s,\perp\parallel\parallel}^{(2)} = \tfrac{1}{2}N_s[\langle\cos\theta\sin^2\theta\rangle\alpha_{z'z'z'}^{(2)} - \langle\cos\theta\sin^2\theta\sin^2\psi\rangle \times (\alpha_{z'x'x'}^{(2)} + 2\alpha_{x'z'x'}^{(2)}) + \langle\cos\theta\rangle\alpha_{z'x'x'}^{(2)}], \tag{64b}$$

$$\chi_{s,\parallel\perp\parallel}^{(2)} = \tfrac{1}{2}N_s[\langle\cos\theta\sin^2\theta\rangle\alpha_{z'z'z'}^{(2)} - \langle\cos\theta\sin^2\theta\sin^2\psi\rangle \times (\alpha_{z'x'x'}^{(2)} + 2\alpha_{x'z'x'}^{(2)}) + \langle\cos\theta\rangle\alpha_{x'z'x'}^{(2)}]. \tag{64c}$$

The angles θ and ψ are the usual Euler angles (Goldstein 1950), corresponding, respectively, to the angle between the molecular z'-axis and the surface normal and the angle of rotation of the molecule about its z'-axis. An average over the Euler angle ϕ has been performed, in accordance with the assumption of isotropy in the surface plane. Since there are only three independent orientational averages in these relations it is possible to determine these quantities provided $\boldsymbol{\chi}_s^{(2)}$, N_s and $\boldsymbol{\alpha}^{(2)}$ are known. A useful relation may be obtained by combining the relations in eq. (64):

$$\chi_{s,\perp\perp\perp}^{(2)} + \tfrac{2}{3}\chi_{s,\perp\parallel\parallel}^{(2)} + \tfrac{4}{3}\chi_{s,\parallel\perp\parallel}^{(2)} = N_s\alpha_v^{(2)}\langle\cos\theta\rangle. \tag{65}$$

In this equation, we have introduced $\alpha_v^{(2)} = \alpha_{z'z'z'}^{(2)} + \frac{1}{3}\alpha_{z'x'x'}^{(2)} + \frac{2}{3}\alpha_{x'z'x'}^{(2)}$, the vector part of $\boldsymbol{\alpha}^{(2)}$ with respect to the molecular z'-axis. This quantity is measurable in a dc-induced SHG measurement for a molecule with its static dipole moment lying along $\hat{z}'$. In the literature, this framework has recently been used to analyze molecular orientation at a water surface (Goh et al. 1988).

The most common treatment assumes that only $\alpha^{(2)}_{z'z'z'}$ is appreciable (Heinz et al. 1983, Rasing et al. 1985a, b, 1986, Berkovic et al. 1986, 1987, Guyot-Sionnest et al. 1986a, Grubb et al. 1988b, Hicks et al. 1986, Kemnitz et al. 1986, Marowsky and Steinhoff 1988, Girling et al. 1985, 1987, Hsiung et al. 1989, Mullin et al. 1989) (or, in some cases, that $\alpha^{(2)}_{z'x'x'}$ dominates (Heinz et al. 1982, Peterson and Harris 1989)). Only two of the relations in eq. (64) will then be independent, but the ratio of these two gives information on the orientational distribution, without knowledge of the magnitude of the molecular nonlinear polarizability. For example, with $\boldsymbol{\alpha}^{(2)} = \alpha^{(2)}_{z'z'z'}\, \hat{z}'\hat{z}'\hat{z}'$, we find

$$\begin{aligned} \langle \cos\theta \rangle / \langle \cos^3\theta \rangle &= (2\chi^{(2)}_{s,\parallel\perp\parallel} + \chi^{(2)}_{s,\perp\perp\perp})/\chi^{(2)}_{s,\perp\perp\perp} \\ &= (2\chi^{(2)}_{s,\perp\parallel\parallel} + \chi^{(2)}_{s,\perp\perp\perp})/\chi^{(2)}_{s,\perp\perp\perp}. \end{aligned} \tag{66}$$

This ratio of orientational averages can be obtained without any knowledge of adsorbate density or the molecular nonlinear polarizabilities. Moreover, for a narrow distribution of molecular orientations, $\langle \cos\theta \rangle / \langle \cos^3\theta \rangle \approx \sec^2\theta_0$ yields the angle between the molecular z'-axis and the surface normal directly.

We return now to the question of the influence of the linear optical properties of the surface or interface. The linear optical response enters in this molecular model both through the choice of the dielectric screening ε' and through local-field corrections. From a fundamental point of view, one should be able to calculate both the effective linear dielectric constant and the local-field corrections from the linear polarizability of the molecules and knowledge of their geometry. This procedure, however, is difficult to apply because of the analytic complexities, as well as the lack of the necessary information. These difficult but significant issues have not been addressed extensively in the literature. Typically, the effective linear dielectric constant is taken to be that present in the absence of the aligned molecular layer, which is frequently a reasonable approximation in the low-density limit. In analysis of experimental data, local-field effects have generally not been accounted for. The importance of local fields has been treated theoretically by Ye and Shen (1983). As they point out, local fields arise both from the interaction of the molecules in the surface layer with the bulk materials (i.e., with their image dipoles) and with one another. Ye and Shen find that the former may be significant for molecules located less than 2.5 Å above a metal surface or 1.5 Å above an insulating surface, while the latter should be considered for molecules with separations of less than 10 Å. It should be noted that both of these sources of local-field corrections will enter into the analysis of molecular orientation, but just the term relating to molecule-molecule interactions will affect the linear dependence of $\boldsymbol{\alpha}^{(2)}$ on N_s for molecules with an unchanging orientational distribution. From the experimental perspective, only in the recent work of Hayden (1988) has the question of local-field effects in surface nonlinear optical processes been addressed directly. In this study, the density of a dye molecule with a large nonlinear polarizability was embedded with varying density in a Langmuir–Blodgett monolayer. For this system, it was inferred that local-field corrections between the molecules became significant at a separation of <10 Å.

In concluding this section, we address the general question of the accuracy of the electric-dipole approximation at a surface or interface, i.e., the extent to which the strong variation in electric fields in the vicinity of the surface influences the nonlinear optical response. In certain circumstances, the dipole approximation is likely to be quite good. Consider, for example, a dilute monolayer of well-oriented molecules, each exhibiting large nonlinear polarizability. Such a layer would be expected to give rise to a strong dipole response. If, moreover, the molecules are embedded in one of the bulk media (or vacuum), then variations in the electric field near the nonlinear molecules should be weak. Under these conditions, the electric-dipole terms should dominate. The situation is, of course, less favorable if either the expected electric-dipole contribution is weak (through poor molecular alignment, for example) or the nonlinear response is present in a region where the fields are changing rapidly. As such a typical case, let us return to the data in fig. 6, corresponding to the (effective) nonlinear susceptibility of the water surface in equilibrium with its vapor. Goh et al. (1988) take the strong temperature dependence of the (effective) surface nonlinear susceptibility as evidence that the surface electric-dipole terms are significant at least at the lower temperature range. This conclusion is based on the notion that higher-order terms, unlike the dipole contributions, will not be strongly affected by diminished surface ordering associated with increasing temperatures. On the other hand, the measurements were performed using laser radiation for which Kleinman symmetry ($\alpha^{(2)}_{ijk} = \alpha^{(2)}_{jik}$) is expected to hold to $\sim 10\%$ for the elements of the nonlinear polarizability $\boldsymbol{\alpha}^{(2)}$ of the water molecules. Within the electric-dipole approximation, we then expect, for example, $\chi^{(2)}_{s,\perp\parallel\parallel} \approx \chi^{(2)}_{s,\parallel\perp\parallel}$*. As the data in fig. 6 show, the disparity between these two elements of $\boldsymbol{\chi}^{(2)}_s$ is appreciable, indicating that the response apparently cannot be represented to high accuracy strictly within the electric-dipole approximation.

Only in one study in the literature has an attempt been made to evaluate directly the accuracy of the electric-dipole approximation at the surface. In the work of Guyot-Sionnest and Shen (1987), a stable adsorbed molecular monolayer of a solid substrate was immersed in different solvents. Assuming that the properties of the molecular layer remain unchanged in the different solvents, the same value for $\boldsymbol{\chi}^{(2)}_s$ should be deduced in each case provided that the electric-dipole response dominates. If, however, higher-order terms are important, they should lead to different measured values for $\boldsymbol{\chi}^{(2)}_s$, corresponding to the altered discontinuity in the dielectric constants of the bulk media. The authors found that the electric-dipole approximation worked well for aligned molecular layers

*The experimental data include a contribution from the bulk quadrupole (γ) term in $\chi^{(2)}_{s,zxx}$. The temperature dependence suggests, however, that this contribution is relatively minor.

with near resonant response, even in the presence of appreciable discontinuities in the electric field. In the opposite extreme of the non-resonant response from a solid substrate without any highly polarizable adsorbates, however, it was inferred that the electric-dipole contribution was relatively small. These results highlight the importance of justifying the absence of higher-order terms before attributing the surface nonlinear susceptibility solely to electric-dipole contributions. Further experimental and theoretical work is certainly warranted to elucidate these issues more fully.

Finally, it is important to emphasize the limitations inherent in ascribing the nonlinear response to the nonlinear polarizability of adsorbed species in real physical systems. The issue of the behavior of the substrate must be considered. In many cases, the nonlinearity of the substrate may be far greater than that of the adsorbed layer. This generally applies, for example, in the case of small molecules adsorbed on clean metal surfaces. If the nonlinear response of the substrate is not affected by the adsorbate, its contribution can, in principle, be accounted for in the analysis (Heinz et al. 1983). This may be a good approximation for insulating or oxidized surfaces. As was already appreciated by Brown and Matsuoka (1969) however, the observed change in the nonlinearity of the surface may be far too large to reflect the contribution of molecular adsorbates and a more detailed model of the interaction of the adsorbate with the substrate is required. To appreciate the subtlety of these interactions, the reader may wish to examine the recent results for the deposition of alkali atoms on clean metal surfaces (Tom et al. 1986a, Song et al. 1987, 1988).

4.3. Leading-order nonlocal response

As we have discussed in section 2.5, the quadrupole terms are responsible for the nonlinear response in the bulk of a centrosymmetric medium. By applying such a multipole expansion at the surface or interface, one may also estimate the surface contribution by taking into account the discontinuity in the electric fields between the two bulk media. Before entering into a more detailed treatment of these issues, we consider presently some general issues concerning the leading-order nonlinear response in a centrosymmetric medium. The discussion follows that of the recent work of Guyot-Sionnest and Shen (1988) and is restricted to the case of SHG.

In the formalism of a multipole expansion, the generalized polarization at a frequency Ω can be expressed as (Pershan 1963)

$$\boldsymbol{P}(\Omega) = \boldsymbol{P}_{\mathrm{d}}(\Omega) + \nabla \cdot \boldsymbol{Q}(\Omega) + (\mathrm{i}c/\Omega)\nabla \times \boldsymbol{M}(\Omega) + \dots, \tag{67}$$

where $\boldsymbol{P}_{\mathrm{d}}$, $\boldsymbol{Q}$, and $\boldsymbol{M}$ refer, respectively, to the electric-dipole moment, the electric-quadrupole moment, and the magnetic-dipole moment per unit volume. Only the terms with one or fewer spatial derivatives have been retained in this expansion. For the case of SHG in a centrosymmetric medium, each of these quantities is related to a quadratic function of the electric field at the fundamental frequency $\omega = \frac{1}{2}\Omega$ by

$$\boldsymbol{P}_{\mathrm{d}}(\Omega) = \chi_{\mathrm{q}1}^{(2)}:\boldsymbol{E}(\omega)\nabla\boldsymbol{E}(\omega), \tag{68a}$$

$$\boldsymbol{Q}(\Omega) = \chi_{\mathrm{q}2}^{(2)}:\boldsymbol{E}(\omega)\boldsymbol{E}(\omega), \tag{68b}$$

$$\boldsymbol{M}(\Omega) = \chi_{\mathrm{m}}^{(2)}:\boldsymbol{E}(\omega)\boldsymbol{E}(\omega). \tag{68c}$$

The usual electric-dipole susceptibility in eq. (68a) has been omitted since it vanishes in a centrosymmetric medium; only terms leading to a generalized nonlinear polarization proportional to a single spatial derivative of the pump field $\boldsymbol{E}(\omega)$ have been included. For SHG, no contribution is possible from $\chi_{\mathrm{m}}^{(2)}$, which is proportional to the fully antisymmetric third rank tensor (Guyot-Sionnest and Shen 1988). It follows that the generalized nonlinear polarizability can be written as

$$\boldsymbol{P}(\Omega) = \chi_{\mathrm{q}1}^{(2)}:\boldsymbol{E}(\omega)\nabla\boldsymbol{E}(\omega) - \nabla\cdot[\chi_{\mathrm{q}2}^{(2)}:\boldsymbol{E}(\omega)\boldsymbol{E}(\omega)]. \tag{69}$$

This expression can be related to the form of the bulk nonlinear polarization given in section 2.5 by combining the two tensors $\chi_{\mathrm{q}1}^{(2)}$ and $\chi_{\mathrm{q}2}^{(2)}$ in the following manner:

$$\chi_{\mathrm{q},ijkl}^{(2)} \equiv \chi_{\mathrm{q}1,ijkl}^{(2)} - (\chi_{\mathrm{q}2,ikjl}^{(2)} + \chi_{\mathrm{q}2,iklj}^{(2)}). \tag{70}$$

We then find that in a homogeneous medium eq. (35) holds, while generally we have the relation

$$\boldsymbol{P}(\Omega) = \chi_{\mathrm{q}}^{(2)}:\boldsymbol{E}(\omega)\nabla\boldsymbol{E}(\omega) - [\nabla\cdot\chi_{\mathrm{q}}^{(2)}]:\boldsymbol{E}(\omega)\boldsymbol{E}(\omega). \tag{71}$$

For a description of the bulk nonlinear response, the second term in eq. (71) can be omitted, but as Guyot-Sionnest and Shen have pointed out in their recent work (Guyot-Sionnest and Shen 1988), it should be included in a self-consistent description of the surface response in terms of leading-order nonlocal terms.

The influence of these nonlocal terms on the bulk nonlinear response has already been examined in section 2.5. An approach for describing the surface nonlinear response dating back to the early studies of SHG in centrosymmetric media (see, for example, Bloembergen et al. 1968), has been to apply these same expressions to the region of the surface. This can be accomplished by integrating the source terms in eq. (71) in the manner prescribed by eq. (59) to find the surface nonlinear polarization. In the early work on this subject, only the first term in eq. (71) was considered. This is equivalent to the approximation that the nonlinear response of the media described by $\chi_{\mathrm{q}2}^{(2)}$ is small or unchanging at the interface. The integration still requires, in general, a knowledge of the detailed variation of the electric fields across the interface, as well as the spatial dependence of the elements in $\chi_{\mathrm{q}}^{(2)}$. No attempt has been made in the literature to perform these calculations with a realistic model. The correct order of magnitude can be found by integrating, for example, under the assumption that the normal components of the electric fields vary linearly and that the elements of $\chi_q^{(2)}$ retain their bulk values up to the interface, where they change abruptly. [Some mathematical aspects of integrating the surface term have been discussed by Maystre et al. (1986).] For isotropic media, this procedure leads to two independent non-vanishing elements of the surface nonlinear susceptibility tensor $\chi_s^{(2)}$ (Guyot-Sionnest et al. 1986b):

$$\chi^{(2)}_{s,\perp\perp\perp} = \frac{[\varepsilon'(\Omega)\varepsilon'^2(\omega)][\varepsilon_1^2(\omega) - \varepsilon_2^2(\omega)]}{4\varepsilon_1(\omega)\varepsilon_2(\omega)}\left[\frac{\delta_1}{\varepsilon_1(\omega)} + \frac{\delta_2}{\varepsilon_2(\omega)}\right], \tag{72a}$$

$$\chi^{(2)}_{s,\parallel\perp\parallel} = \frac{[\varepsilon'(\omega)][\varepsilon_1(\omega) - \varepsilon_2(\omega)]}{2\varepsilon_1(\omega)\varepsilon_2(\omega)}[\beta_1 - \beta_2]. \tag{72b}$$

Here we have written the results in terms of the conventional parameters β, γ, and δ characterizing the elements of $\chi_q^{(2)}$ [eq. (37)]. It should be noted that no surface contribution for $\chi^{(2)}_{s,\perp\parallel\parallel}$ is predicted within these approximations, since the pump field does not exhibit any discontinuity at the surface. In the work of Guyot-Sionnest and Shen (1988), the effect of the second term in eq. (71) is considered. In this case, finite values are expected for all three of the independent elements of $\chi_q^{(2)}$, viz., $\chi^{(2)}_{s,\perp\perp\perp}$, $\chi^{(2)}_{s,\parallel\perp\parallel}$, and $\chi^{(2)}_{s,\perp\parallel\parallel}$. Indeed, by a suitable integration by parts, it can be shown that a contribution to each of these elements can be ascribed to the change in the bulk quadrupole moment between the two media (Guyot-Sionnest and Shen 1988).

We now briefly mention microscopic calculations of the quadrupole response embodied in $\chi_{q1}^{(2)}$ and $\chi_{q2}^{(2)}$. Formal expressions for these quantities can be derived by applying second-order perturbation theory with both the interaction Hamiltonian between the field and material system and the current operator expanded in a multipole series. This was already accomplished in the early work of Adler (1964), for example. Although direct evaluation of these expressions is difficult, they have served as a basis of a low-frequency approximation for insulating materials. For an isotropic centrosymmetric medium, Bloembergen et al. (1968) find in a closure approximation

$$\beta = -2\gamma = 3(4eN_v)^{-1}[\chi^{(1)}(\omega)]^2, \tag{73a}$$

$$\delta = 0, \tag{73b}$$

where N_v represents the density of valence electrons, $-e$ is the electronic charge, and $\chi^{(1)}$ denotes the linear susceptibility of the medium.

In a different approach, Guyot-Sionnest and Shen (1988) have developed a model suitable for liquids and other materials that can be treated with molecule-like units. It is assumed that the nonlinear response can be attributed entirely to the (electric-dipole) nonlinear polarizability of each molecule. In this case, an electric-dipole term appears at the surface, associated with aligned molecules. An effective nonlinear quadrupole moment is present in the bulk due to correlations between the orientation of molecules in proximity to one another and their concomitant imperfect cancellation. In this case, $\delta = \beta = \gamma = 0$, but the effect of the bulk quadrupole response is still felt by virtue of the second term in eq. (71). Such a model has been applied, for example, to describe the surfaces of liquid crystals (Guyot-Sionnest et al. 1986a).

Finally, we mention the nonlinear response expected from a non-interacting electron gas. As we shall discuss the surface effects in the next section, only the result for the response in the homogeneous bulk is considered here. In this case the polarization is longitudinal with the coefficient γ of eq. (36) given by

$$\gamma = \frac{N_0 e^3}{8m^2\omega^4}. \qquad (74)$$

Here N_0 is the density of electrons of charge $-e$, and mass m under excitation by a field of frequency ω. In a strictly classical description, this relation can be derived from the Lorentz-force equation by treating a single electron, or alternatively within the context of a hydrodynamic model (Sipe and Stegeman 1982). In fact, the same result also follows from a quantum mechanical calculation, as was first shown by Cheng and Miller (1964).

4.4. Complete nonlocal response

While the expansion to leading-order in a multipole expansion can be justified in the bulk of a centrosymmetric material, such an approximation may be quite inaccurate in the surface region, since the electric fields are changing on the scale of an atomic dimension. As argued above, this problem may be less acute for strongly nonlinear adsorbed molecules located away from regions of abrupt changes in dielectric properties, as on a solid surface in vacuum or in a liquid. For the elementary case of the second-order nonlinear response from a clean solid surface, however, the effect of the rapidly varying fields is likely to be important, as was emphasized by Rudnick and Stern (1971, 1974).

Although formal expressions for the surface nonlinear response can be derived from perturbation theory without any approximation concerning the structure of the optical fields at the surface (e.g., in the work of Bower (1976)), evaluation of such relations has not been attempted. The only system in which rigorous theoretical studies have been carried out is that of electrons in the presence of a smooth background charge (jellium). Unfortunately space does not permit us to enter into this important topic in any detail. Our aim here is only to relate the notation common to these studies to our phenomenological description and to make the reader aware of activity in this area.

Models of the free-electron response of surfaces date back to the period of the first measurements of SHG from metal surfaces (Brown et al. 1965, Brown and Parks 1966, Bloembergen et al. 1966, Sonnenberg and Heffner 1968, Krivoshchekov and Stroganov 1968, 1969). The early treatment of Jha (1965a, b) and

Bloembergen et al. (1968) examined a free-electron model with a step potential at the surface. In these studies, the influence of the breaking of inversion symmetry at the boundary was not taken into account and a long-wavelength limit was assumed. In the work of Rudnick and Stern (1971, 1974), a full calculation in the random-phase approximation was performed for a model of the surface consisting of a current sheet terminating a homogenous bulk medium. It was found that the component of the nonlinear current (or, in our language, generalized polarization) normal to the surface required a complete non-local calculation. The component of the current lying parallel to the surface could, however, be treated in a local approximation. For a more realistic model of a metal, Rudnick and Stern also included the influence of the breaking of inversion symmetry on the normal surface current and the effect of surface scattering on the parallel surface current. This leads to the introduction of two dimensionless parameters, a and b, which, it was suggested, should be on the order of unity.

The a and b parameters can be recast in terms of a surface nonlinear susceptibility within the phenomenological model developed in this chapter. We find

$$\chi^{(2)}_{s,\perp\perp\perp} = \frac{ae}{4m\omega^2}\left[\frac{\varepsilon(\omega)-1}{4\pi}\right]\varepsilon'(\Omega)\left[\frac{\varepsilon'(\omega)}{\varepsilon(\omega)}\right]^2, \tag{75a}$$

$$\chi^{(2)}_{s,\parallel\perp\parallel} = \frac{be}{4m\omega^2}\left[\frac{\varepsilon(\omega)-1}{4\pi}\right]\left[\frac{\varepsilon'(\omega)}{\varepsilon(\omega)}\right]. \tag{75b}$$

Once again, the electronic charge and mass are denoted by $-e$ and m, respectively. For consistency, we have included the interface dielectric constant ε' in the phenomenological model; to return to the assumptions generally applied modeling these systems, $\varepsilon'(\omega)$ should be taken to be that of the metal, $\varepsilon(\omega)$ and $\varepsilon'(\Omega)$ should be replaced by 1. In rewriting the expressions of Rudnick and Stern, we have represented the dielectric constant of the metal ε by $\varepsilon(\omega) = 1-\omega_p^2/\omega^2 = 1-4\pi N_0 e^2/m\omega^2$, with N_0 denoting the density of electrons in the bulk of the metal. Absent from eq. (75) is a surface contribution to $\chi^{(2)}_{s,\perp\parallel\parallel}$. In the theory of Rudnick and Stern (and all subsequent treatments of a surface with complete translational invariance), this surface contribution does not appear. From the standpoint of measurements, the apparent value $\chi^{(2)}_{s,\perp\parallel\parallel}$ is, however, nonzero because of the influence of the longitudinal bulk polarization γ, specified by eq. (74) and expressed as an effective surface term by eq. (47). In comparing theory with experiment, it should be borne in mind that $\chi^{(2)}_{s,\perp\parallel\parallel}$ will, in reality, include at least some true surface contribution and should not be assumed to be given solely by the bulk contribution. In this regard, it should be noted that Heskett et al. (1986, 1988) have observed significant changes in SHG efficiency from Ag(110) under s-polarized 532 nm excitation depending on the condition of the surface. By the same token, the possible existence of anisotropic response in single-crystal metal samples should not be neglected. Crystal faces of Ag with excitation at 1.06 μm, for example, exhibit appreciable anisotropy (Shannon et al. 1987).

A hydrodynamic model was developed by Sipe and co-workers (Sipe and Stegeman 1982, Sipe et al. 1980a, b) which included a pressure term appropriate for a quantum electron gas. Although the full equations were not solved in detail, it was concluded that $b = -1$ [in accordance with the idealized result of Rudnick and Stern (1971, 1974)] and a formal expression could be given for the a parameter: $a = -2(\omega_p^2 - \Omega^2)/(\omega_0^2 - \Omega^2)$. In this equation, ω_0 is an effective plasma frequency (defined by a complex integral equation) for the so-called 'selvedge' region between the surface and the bulk. The relation for the a parameter displays the sensitivity of normal current to the detailed nature of the electron distribution at the surface and suggests the possibility of an interesting resonant frequency response.

Prompted by the heightened interest in SHG as a surface probe, considerable effort has been devoted to obtaining a more complete understanding of the nonlinear response of a jellium surface. Keller has examined the problem on the basis of the semiclassical infinite barrier model (Keller 1985a, b, 1986). Corvi and Schaich (1986) refined the hydrodynamic model by treating the surface region as a series of several steps with the inclusion of an ohmic damping in the equation of motion. Among the interesting results of this analysis was the strong frequency dependence predicted for the SH response.

Recently realistic quantum mechanical calculations have been performed for jellium surfaces by Liebsch and co-workers. These calculations employed the density-functional theory within a local density approximation for the exchange and correlation forces. Calculations performed in the low-frequency limit on clean (Weber and Liebsch 1987a) surfaces and surfaces with model metallic overlayer of a different density (Weber and Liebsch 1987b) revealed values of $|a|$ well in excess of unity. Moreover, the strong nonlinear response could be associated with the behavior of the tail of the electron density and, as one would expect, was strongly influenced by metallic overlayers. Schaich and Liebsch (1988) proved within the context of the hydrodynamic model that finite frequencies could still be treated without the inclusion of retardation effects, i.e., formally taking the speed of light $c \to \infty$. The same approach was then applied to a time-dependent density functional theory (Liebsch 1988, 1989, Liebsch and Schaich 1989). This provided a realistic calculation of the nonlinear response of the jellium surface even at frequencies approaching the bulk plasma frequency. In addition to confirming the large value for the $|a|$ ($\sim$10) at low frequencies, a resonant enhancement was observed for frequencies in the range where the second-harmonic photon energy lies near the threshold for electron emission. Liebsch also examined other simpler theoretical approaches and concluded that the key property needed to obtain accurate results was a satisfactory description of the ground-state properties of the electrons at the surface. This finding was confirmed by Chizmeshya and Zaremba (1988), who were able to obtain results similar to those of Liebsch using a Thomas–Fermi–von Weizsacker theory. In this scheme, the total energy of the system can be written explicitly in terms of suitable integrals of the electron density, and the perturbation of an applied electric field can be readily determined.

With the great progress in theory of SHG from metal surfaces, there is clearly new impetus for additional experimental measurements on simple metal surfaces (such as aluminum) in ultrahigh vacuum. Although several experiments recently have investigated the nonlinear response of evaporated metal films with the aid of surface plasmons (Quail and Simon 1985, Simon et al. 1977, Tseng and Lue 1987), these studies involved either metals at solid interfaces or metals under ambient conditions and cannot be compared with the behavior predicted for a clean metal surface.*

4.5. External perturbations

The fact that second-order nonlinear optical effects are forbidden within the electric-dipole approximation in centrosymmetric media makes them very sensitive to a variety of external perturbations. In this section, we indicate how externally applied electric and magnetic fields and stress may influence the second-order response.

The case of static electric fields has attracted considerable attention over the years. Static electric fields can perturb the inversion symmetry of the medium and can, consequently, strongly influence the second-order nonlinear optical response. This process was already recognized in the early measurements of SHG from calcite (Terhune 1962), in which it was found that the SH signal could be increased dramatically in the presence of a static electric field. A few years later, Lee et al. (1967) observed similar effects for the surfaces of metals and semiconductors in electrolytic solutions. Recently, additional investigations have been performed on metal surfaces (Corn et al. 1984b) in a controlled field and on semiconductor surfaces in electrolytes (Aktsipetrov and Mishina 1984). From the point of view of theory, two cases can be distinguished. One is appropriate for metals where the applied electric field only penetrates into the surface region and consequently influences only the surface properties. The other, more complex case, arises when both the surface and bulk properties of the material may be altered.

If only the surface is perturbed by the applied electric field, we should be able to describe the effect as a modification of the surface nonlinear susceptibility tensor $\chi_s^{(2)} = \chi_s^{(2)}(\boldsymbol{E}_{dc})$. If we restrict ourselves to the usual situation where the dc electric field $\boldsymbol{E}$ is normal to the surface, we can conclude that the presence of the

*Measurements from bare aluminum surfaces under ultrahigh vacuum conditions have recently been reported by Murphy et al. (1989). The results confirm a value for $|a|$ considerably greater than unity.

electric field will not alter the form of the surface nonlinear susceptibility tensor, but can only change the values of its independent elements. In the literature, following Lee et al. (1967), the dependence of the surface nonlinear susceptibility is generally taken to be linear in $\boldsymbol{E}_{dc}$. Then the influence of the dc electric field can be considered as contributing an additional polarization to the surface given by

$$\begin{aligned} P_s(\Omega) &= \chi_s^{(2)}(\boldsymbol{E}_{dc}):\boldsymbol{E}(\omega)\boldsymbol{E}(\omega) \\ &\approx \chi_s^{(2)}(\boldsymbol{E}_{dc}=0):\boldsymbol{E}(\omega)\boldsymbol{E}(\omega) + \chi_s^{(3)}:\boldsymbol{E}_{dc}\boldsymbol{E}(\omega)\boldsymbol{E}(\omega), \end{aligned} \tag{76}$$

for SHG. The new coefficient $\chi_s^{(3)}$ can be viewed as a third-order nonlinear susceptibility for mixing a dc electric field with two optical fields*. While the simplified linear dependence of the surface nonlinear polarizability on $\boldsymbol{E}_{dc}$ is difficult to justify *a priori* for the strong fields produced at metal surfaces in an electrochemical cell, such a relation has been verified for a silver surface on a mica substrate under an external potential by Corn et al. (1984). Richmond (1986) and Richmond et al. (1987, 1988) have also seen qualitative agreement for the electric-field dependence of SHG implicit in eq. (76) in certain electrochemical environments. In these studies, the dc electric field was related to the measured charging of the surface by ions that do not interact strongly with the substrate. In a different study of physisorbed ions on a silver electrode, Furtak et al. (1987), however, failed to observe this simple behavior.

In general, the applied static electric field may extend into the bulk of the material. In this event, the field strength is expected to be appreciably lower than that experienced in the surface region and use of a third-order nonlinear susceptibility $\chi^{(3)}$ to describe the induced polarization should be quite accurate. In the limit where the applied field is relatively uniform over the relevant volume for the nonlinear interaction, eq. (40) can be applied to calculate the radiation from this bulk source.

In contrast to an electric field, the presence of a magnetic field will *not* break the inversion symmetry of the material. Thus no new dipole-allowed terms need to be considered in the bulk of the material.** An applied magnetic field can, however, alter the form and strength of the nonlocal bulk response and of the surface nonlinear susceptibility. In a recent publication, Pan et al. (1989) have examined theoretically the influence of static magnetic fields (i.e., of a static magnetization) on the surface nonlinear susceptibility tensor $\chi_s^{(2)}$ for SHG. Considering crystal surfaces with different symmetries, the authors identified tensor elements which could appear only in the presence of a finite magnetization.

*In the early work on surface SHG, it was suggested that the *entire* surface nonlinear susceptibility might be estimated by considering intrinsic static electric fields present at any surface or interface even in the absence of an applied field. See, for example, Brown and Matsuoka (1969).

**Here we assume that both the crystal structure and magnetic ordering are centrosymmetric. The interesting case of systems with magnetic ordering of lowered symmetry has been considered by Akhmediev et al. (1985).

Moreover, for a ferromagnetic material, the new tensor elements are predicted to have a magnitude of several percent of the normally allowed tensor elements. This analysis suggests that SHG may provide a viable probe of surface magnetism.

The effect of introducing homogeneous strain in a sample is somewhat analogous to the influence of an applied magnetic field, namely, it will not break the inversion symmetry of a centrosymmetric material. Once again, it may alter the magnitude and symmetry of the bulk quadrupolar response and the surface nonlinear susceptibility. This leads to the interesting possibility of examining deformations of a surface or interface layer, particularly if they result in a lowered symmetry. These issues have not been explored directly in the literature. What has been examined, however, is the interesting case of an *inhomogeneously* strained sample. In the presence of an inhomogeneous deformation, the inversion of the bulk material will be broken. This effect has been demonstrated by a Soviet group using oxidized Si samples (Abdullaev et al. 1987), as well as ion-implanted samples (Govorkov et al. 1989a, b). In these materials, the gradient in the strain (appearing along the direction of the surface normal) results from the changing structure as a function of depth into the sample. Since this mechanism leads to dipole-allowed bulk SHG, it may significantly increase the SH response. Indeed significant enhancements are reported in appropriately prepared Si samples. The bulk electric-dipole response has been modeled theoretically in a bond polarizability picture that takes into account the slight dissimilarity between bonds associated with the inhomogeneous deformation of the sample (Emel'yanov et al. 1987, Govorkov et al. 1989a, b).

5. *Concluding remarks*

In this chapter, we have developed a macroscopic description for second-order nonlinear optical effects in centrosymmetric media and examined various schemes employed to model the nonlinear susceptibility on a microscopic level in different physical systems. Such a division between a macroscopic description, founded in general considerations of symmetry and length scale, and a detailed microscopic description is a useful one. It establishes what information is experimentally attainable and allows us to decide on the most convenient method of obtaining it. It also prevents us from implicitly relying on specific conclusions of a microscopic model that may apply only with limited accuracy in reality. This point might be illustrated by the treatment of metal surfaces that has frequently appeared in the literature in which the SHG for s-polarized pump radiation is assumed to originate entirely in the bulk of the medium. Such a procedure is equivalent to taking $\chi^{(2)}_{s,\perp\parallel\parallel}$ as vanishing, which need not be the case on general grounds and, as mentioned in section 4.4, has been seen to be strongly violated in direct experimental measurements. It should be observed that the complete determination of $\chi_s^{(2)}$ can frequently be accomplished without undue effort, as outlined in section 3.1.

The macroscopic description of the second-order response also has flexibility. For example, the recently observed process of surface SH diffraction from a periodically modulated surface structure (Reider et al. 1988, Zhu et al. 1988, Zhu and Shen 1989, Heinz and Suzuki 1989, Suzuki and Heinz 1989) can be readily described by an extension of the phenomenological model presented in this chapter. The macroscopic characterization in terms of a surface nonlinear susceptibility and a bulk nonlocal response should, as argued in section 2, provide a suitable description for the polarization and angular dependence of second-order effects in centrosymmetric media. But both from the fundamental point of view and from the perspective of applications to the study of surfaces and interfaces, this level of understanding is insufficient.

A complete analysis based on first principles for a real system is undoubtedly challenging, since it involves a detailed knowledge of the surface structure in the ground state, the nature of surface excitations, and an understanding of electric fields at a surface. Such a level of sophistication appears to be attainable for the model system of jellium. While it will be difficult to perform such calculations when the ion cores of the atoms in a solid are considered, the results may apply reasonably well to certain materials and, more importantly perhaps, insight gained into the qualitative behavior will carry over to other materials. Despite the impressive success seen with first-principles calculations for jellium, there is clearly a need for further work based on more approximate treatments. For covalently bonded solids, for instance, bond polarizability models (Flytzanis 1975), originally developed to describe the electric-dipole response of non-centrosymmetric bulk materials, might be extended to treat both surface properties and the nonlocal bulk response. For molecular systems, a simple lowest-order description in terms of the nonlinear polarizability of aligned molecules is available. Much still remains to be accomplished in developing a correct treatment of the subtle but frequently important corrections. The role played by local-field effects and quadrupole terms, while investigated to some extent, merits further scrutiny.

With the current rapid advance in the application of second-order nonlinear processes as probes of surfaces and interfaces, we can confidently predict that our understanding of the fundamental underpinnings of the technique will advance rapidly as well. To this end, further theoretical studies will be required. Equally important, however, are additional experimental investigations on well-characterized and well-controlled systems. Only in this manner can the theoretical developments be guided and critically evaluated.

Acknowledgment

The author would like to thank Dr. D.P. DiVincenzo, Dr. P. Guyot-Sionnest, Dr. R. Hollering, Dr. G.A. Reider, and Prof. Y.R. Shen for valuable discussions and to acknowledge support by the Office of Naval Research.

References

Abdullaev, A.Yu., S.V. Govorkov, P.K. Kashkarov, N.I. Koroteev, G.I. Petrov and I.L. Shumai, 1987, Fiz. Tverd. Tela **29**, 1898 [Sov. Phys.-Solid State **29**, 1094].

Adler, E., 1964, Phys. Rev. **134**, A728.

Akhmanov, S.A., V.I. Emel'yanov, N.I. Koroteev and V.N. Seminogov, 1985, Usp. Fiz. Nauk **147**, 675 [Sov. Phys. Usp. **28**, 1084].

Akhmanov, S.A., S.V. Govorkov, N.I. Konoteev, G.I. Petrov, I.L. Shumai and A.V. Petukhov, 1989, Pis'ma Zh. Eksp. & Teor. Fiz. **49**, 257 [JETP Lett. **49**, 610].

Akhmediev, N.N., S.B. Borisov, A.K. Zvezdin, I.L. Lyubchanskii and Yu.V. Melikhov, 1985, Fiz. Tverd. Tela **27**, 1075 [Sov. Phys.-Solid State **27**, 650].

Aktsipetrov, O.A., and E.D. Mishina, 1984, Dokl. Akad. Nauk SSSR **274**, 62 [Sov. Phys.-Dokl. **29**, 37].

Aktsipetrov, O.A., N.N. Akhmediev, E.D. Mishina and V.R. Novak, 1983, Pis'ma Zh. Eksp. & Teor. Fiz. **37**, 175 [JETP Lett. **37**, 207].

Aktsipetrov, O.A., N.N. Akhmediev, I.M. Baranova, E.D. Mishina and V.R. Novak, 1985, Zh. Eksp. & Teor. Fiz. **89**, 911 [Sov. Phys.-JETP **62**, 524].

Aktsipetrov, O.A., I.M. Baranova and Yu.A. Il'inskii, 1986, Zh. Eksp. & Teor. Fiz. **91**, 287 [Sov. Phys.-JETP **64**, 167].

Aktsipetrov, O.A., I.M. Baranova, D.A. Esikov, L.L. Kulyuk, E.D. Mishina, E.E. Strumban, V.I. Tsytsanu and R.A. Ratseev, 1987, Dokl. Akad. Nauk SSSR **294**, 579 [Sov. Phys. Dokl. **32**, 396].

Aktsipetrov, O.A., I.M. Baranova, E.D. Mishina and A.V. Petukhov, 1987a, Pis'ma Zh. Tekh. Fiz. **13**, 156 [Sov. Tech. Phys. Lett. **13**, 64].

Aktsipetrov, O.A., I.M. Baranova and E.D. Mishina, 1987b, Dokl. Akad. Nauk SSSR **296**, 1348 [Sov. Phys. Dokl. **32**, 839].

Anderson, J.M., and J.C. Hamilton, 1988, Phys. Rev. B **38**, 8451.

Arjavalingam, G., T.F. Heinz and J.H. Glownia, 1986, in: Ultrafast Phenomena V, eds G.R. Fleming and A.E. Siegman (Springer, Berlin) p. 370.

Berkovic, G., Th. Rasing and Y.R. Shen, 1986, J. Chem. Phys. **85**, 7374.

Berkovic, G., Th. Rasing and Y.R. Shen, 1987, J. Opt. Soc. Am. B **4**, 945.

Berkovic, G., Y.R. Shen, G. Marowsky and R. Steinhoff, 1989, J. Opt. Soc. Am. B **6**, 205.

Bhattacharyya, K., E.V. Sitzmann and K.B. Eisenthal, 1987, J. Chem. Phys. **87**, 1442.

Bhattacharyya, K., A. Castro, E.V. Sitzmann and K.B. Eisenthal, 1988, J. Chem. Phys. **89**, 3376.

Biwer, B.M., M.J. Pellin, M.W. Schauer and D.M. Gruen, 1986, Surf. Sci. **176**, 337.

Biwer, B.M., M.J. Pellin, M.W. Schauer and D.M. Gruen, 1988, Langmuir **4**, 121.

Bjorkholm, J.E., and A.E. Siegman, 1967, Phys. Rev. **154**, 851.

Bloembergen, N., and R.K. Chang, 1966, in: Physics of Quantum Electronics, eds P.L. Kelley, B. Lax and P.E. Tannenwald (McGraw-Hill, New York) p. 80.

Bloembergen, N., and P.S. Pershan, 1962, Phys. Rev. **128**, 606.

Bloembergen, N., and Y.R. Shen, 1966, Phys. Rev. **141**, 298.

Bloembergen, N., R.K. Chang and C.H. Lee, 1966, Phys. Rev. Lett. **16**, 986.

Bloembergen, N., R.K. Chang, S.S. Jha and C.H. Lee, 1968, Phys. Rev. **174**, 813, Erratum **178**, 1528 (1969).
Bower, J.R., 1976, Phys. Rev. **14**, 2427.
Brown, F., and M. Matsuoka, 1969, Phys. Rev. **185**, 985.
Brown, F., and R.E. Parks, 1966, Phys. Rev. Lett. **16**, 507.
Brown, F., R.E. Parks and A.M. Sleeper, 1965, Phys. Rev. Lett. **14**, 1029.
Campbell, D.J., and R.M. Corn, 1987, J. Phys. Chem. **91**, 5668.
Campbell, D.J., and R.M. Corn, 1988, J. Phys. Chem. **92**, 5796.
Castro, A., S. Ong and K.B. Eisenthal, 1989, Chem. Phys. Lett. **163**, 415.
Chang, R.K., and T.E. Furtak, 1982, Surface Enhanced Raman Scattering (Plenum Press, New York).
Chang, R.K., J. Ducuing and N. Bloembergen, 1965, Phys. Rev. Lett. **15**, 6.
Chen, C.K., A.R.B. de Castro and Y.R. Shen, 1981a, Phys. Rev. Lett. **46**, 145.
Chen, C.K., T.F. Heinz, D. Ricard and Y.R. Shen, 1981b, Phys. Rev. Lett. **46**, 1010.
Chen, C.K., T.F. Heinz, D. Ricard and Y.R. Shen, 1981c, Chem. Phys. Lett. **83**, 455.
Chen, J.M., J.R. Bower, C.S. Wang and C.H. Lee, 1973, Opt. Commun. **9**, 132.
Chen, J.M., J.R. Bower and C.S. Wang, 1974, Jpn. J. Appl. Phys. Suppl., Vol. 2, Pt. 2, p. 711.
Chen, T.T., K.U. von Raben, D.V. Murphy and R.K. Chang, 1984, Surf. Sci. **142**, 369.
Chen, W., M.B. Feller and Y.R. Shen, 1989, Phys. Rev. Lett. **63**, 2665.
Chen, Z., W. Chen, J. Zheng, W. Wang and Z. Zhang, 1985, Opt. Commun. **54**, 305.
Cheng, H., and P.B. Miller, 1964, Phys. Rev. **134**, A683.
Chizmeshya, A., and E. Zaremba, 1988, Phys. Rev. B **37**, 2805.
Corn, R.M., M. Romagnoli, M.D. Levenson and M.R. Philpott, 1984a, Chem. Phys. Lett. **106**, 30.
Corn, R.M., M. Romagnoli, M.D. Levenson and M.R. Philpott, 1984b, J. Chem. Phys. **81**, 4127.
Corvi, M., and W.L. Schaich, 1986, Phys. Rev. B **33**, 3688.
di Lazzaro, P., P. Mataloni and F. de Martini, 1985, Chem. Phys. Lett. **114**, 103.
Dick, B., 1984, Chem. Phys. **91**, 1.
Dick, B., 1985, Chem. Phys. **96**, 199.
Dick, B., A. Gierulski, G. Marowsky and G.A. Reider, 1985, Appl. Phys. B **38**, 107.
Dinev, S.G., S.M. Saltiel, K.V. Stamenov, K.A. Stankov and V.G. Tunkin, 1978, Opt. Commun. **24**, 225.
DiVincenzo, D.P., 1987, Phys. Rev. B **37**, 3434.
Driscoll, T.A., and D. Guidotti, 1983, Phys. Rev. B **28**, 1171.
Edmonds, A.R., 1960, Angular Momentum in Quantum Mechanics (Princeton University Press, Princeton, NJ).
Emel'yanov, V.I., N.I. Koroteev and V.V. Yakovlev, 1987, Opt. Spektrosk. **62**, 1187 [Opt. Spectrosc. (USSR), **62**, 701].
Epperlein, E., B. Dick, G. Marowsky and G.A. Reider, 1987, Appl. Phys. B **44**, 5.
Feibelman, 1982, Prog. Surf. Sci. **12**, 287.
Felderhof, B.U., and G. Marowsky, 1987a, Appl. Phys. B **44**, 11.
Felderhof, B.U., and G. Marowsky, 1987b, Appl. Phys. B **43**, 161.
Flytzanis, C., 1975, in: Quantum Electronics, eds H. Rabin and C.L. Tang (Academic Press, New York).
Franken, P.A., A.E. Hill, C.W. Peters and G. Weinreich, 1961, Phys. Rev. Lett. **7**, 118.
Friedrich, A., B. Pettinger, D.M. Kolb, G. Lüpke, R. Steinhoff and G. Marowsky, 1989, Chem. Phys. Lett. **163**, 123.
Fukui, M., and G.I. Stegeman, 1982, in: Electromagnetic Surface Modes, ed. A.D. Boardman (Wiley, New York) p. 725.
Furtak, T.E., J. Miragliotta and G.M. Korenowski, 1987, Phys. Rev. B **35**, 2569.
Giordmaine, J.A., 1965, Phys. Rev. **138**, A1599.

Girling, I.R., N.A. Cade, P.V. Kolinsky, J.D. Earls, G.H. Cross and I.R. Peterson, 1985, Thin Solid Films **132**, 101.
Girling, I.R., N.A. Cade, P.V. Kolinsky, J.D. Earls, G.H. Cross and I.R. Peterson, 1987, J. Opt. Soc. Am. B **4**, 950.
Goh, M.C., and K.B. Eisenthal, 1989, Chem. Phys. Lett. **157**, 102.
Goh, M.C., J.M. Hicks, K. Kemnitz, G.R. Pinto, K. Bhattacharyya, K.B. Eisenthal and T.F. Heinz, 1988, J. Phys. Chem. **92**, 5074.
Goldstein, H., 1950, Classical Mechanics (Addison-Wesley, New York).
Govorkov, S.V., V.I. Emel'yanov, N.I. Koroteev, G.I. Petrov, I.L. Shumai and V.V. Yakovlev, 1989a, J. Opt. Soc. Am. B **6**, 1117.
Govorkov, S.V., V.I. Emel'yanov, N.I. Koroteev, G.I. Petrov, I.L. Shumai and V.V. Yakovlev, 1989b, Zh. Tekh. Fiz. **59**, 98 [Sov. Phys. Tech. Phys. **34**, 56].
Grubb, S.G., A.M. DeSantolo and R.B. Hall, 1988a, J. Phys. Chem. **92**, 1419.
Grubb, S.G., M.W. Kim, Th. Rasing and Y.R. Shen, 1988b, Langmuir **4**, 452.
Guidotti, D., T.A. Driscoll and H.J. Gerritsen, 1983, Solid State Commun. **46**, 337.
Guyot-Sionnest, P., and Y.R. Shen, 1987, Phys. Rev. B **35**, 4420.
Guyot-Sionnest, P., and Y.R. Shen, 1988, Phys. Rev. B **38**, 7985.
Guyot-Sionnest, P., H. Hsiung and Y.R. Shen, 1986a, Phys. Rev. Lett. **57**, 2936.
Guyot-Sionnest, P., W.Chen and Y.R. Shen, 1986b, Phys. Rev. B **33**, 8254.
Guyot-Sionnest, P., J.H. Hunt and Y.R. Shen, 1987a, Phys. Rev. Lett. **59**, 1597.
Guyot-Sionnest, P., Y.R. Shen and T.F. Heinz, 1987b, Appl. Phys. B **42**, 237.
Guyot-Sionnest, P., R. Superfine, J.H. Hunt and Y.R. Shen, 1988, Chem. Phys. Lett. **144**, 1.
Hamilton, J.C., and R.J. Anderson, 1989, J. Vac. Sci. Technol. A **7**, 100.
Hamilton, J.C., and R.J.M. Anderson, 1988, Chem. Phys. Lett. **151**, 455.
Haneman, D., 1987, Rep. Prog. Phys. **50**, 1045.
Harris, A.L., and N.J. Levinos, 1989, J. Chem. Phys. **90**, 3878.
Harris, A.L., C.E.D. Chidsey, N.J. Levinos and D.N. Loiacono, 1987, Chem. Phys. Lett. **141**, 350.
Hayden, L.M., 1988, Phys. Rev. B **38**, 3718.
Heinz, T.F., 1982, Ph.D. Dissertation (University of California, Berkeley, CA).
Heinz, T.F., and M.M.T. Loy, 1988, in: Advances in Laser Science III, eds A.C. Tam, J.L. Gole and W.C. Stwalley (American Institute of Physics, New York) p. 452.
Heinz, T.F., and G.A. Reider, 1989, Trends Anal. Chem. **8**, 235.
Heinz, T.F., and T. Suzuki, 1989, Proc. SPIE **1056**, 73.
Heinz, T.F., C.K. Chen, D. Ricard and Y.R. Shen, 1981, Chem. Phys. Lett. **83**, 180.
Heinz, T.F., C.K. Chen, D. Ricard and Y.R. Shen, 1982, Phys. Rev. Lett. **48**, 478.
Heinz, T.F., H.W.K. Tom and Y.R. Shen, 1983, Phys. Rev. A **28**, 1883.
Heinz, T.F., M.M.T. Loy and W.A. Thompson, 1985a, Phys. Rev. Lett. **54**, 63.
Heinz, T.F., M.M.T. Loy and W.A. Thompson, 1985b, J. Vac. Sci. & Technol. B **3**, 1467.
Heinz, T.F., M.M.T. Loy and S.S. Iyer, 1987, Mater. Res. Soc. Symp. Proc. **55**, 697.
Heinz, T.F., F.J. Himpsel, E. Palange and E. Burstein, 1989, Phys. Rev. Lett. **63**, 644.
Heskett, D., K.J. Song, A. Burns, E.W. Plummer and H.L. Dai, 1986, J. Chem. Phys. **85**, 7490.
Heskett, D., L.E. Urbach, K.J. Song, E.W. Plummer and H.L. Dai, 1988, Surf. Sci. **197**, 225.
Heuer, W., L. Schroeter and H. Zacharias, 1987, Chem. Phys. Lett. **135**, 299.
Hicks, J.M., K. Kemnitz, K.B. Eisenthal and T.F. Heinz, 1986, J. Phys. Chem. **90**, 560.
Hicks, J.M., L.E. Urbach, E.W. Plummer and H.L. Dai, 1988, Phys. Rev. Lett. **61**, 2588.
Hollering, R.W.J., 1990, J. Opt. Soc. Am. B, to be published.
Hollering, R.W.J., and W.J.O.V. Teeselink, 1990, Opt. Commun., to be published.
Hollering, R.W.J., D. Dijkkamp, H.W.L. Lindelauf, P.A.M. van der Heide and M.P.C.M. Krijn, 1990a, J. Vac. Sci. Technol., to be published.
Hollering, R.W.J., A.J. Hoeve and J.M. Lenssinck, 1990b, J. Vac. Sci. Technol., to be published.
Hsiung, H., G.R. Meredith, H. Vanherzeele, R. Popovitz-Biro, E. Shavit and M. L'ahav, 1989, Chem. Phys. Lett. **164**, 539.

Hunt, J.H., P. Guyot-Sionnest and Y.R. Shen, 1987, Chem. Phys. Lett. **133**, 189.
Iyer, S.S., T.F. Heinz and M.M.T. Loy, 1987, J. Vac. Sci. & Technol. **5**, 709.
Jha, S.S., 1965a, Phys. Rev. **140**, A2020.
Jha, S.S., 1965b, Phys. Rev. Lett. **15**, 412.
Jha, S.S., 1966, Phys. Rev. **145**, 500.
Jha, S.S., and C.S. Warke, 1967, Phys. Rev. **153**, 751.
Keller, O., 1985a, Phys. Rev. B **31**, 5028.
Keller, O., 1985b, J. Opt. Soc. Am. **2**, 367.
Keller, O., 1986, Phys. Rev. B **33**, 990.
Kemnitz, K., K. Bhattacharyya, J.M. Hicks, G.R. Pinto, K.B. Eisenthal and T.F. Heinz, 1986, Chem. Phys. Lett. **131**, 285.
Koos, D.A., 1989, J. Electrochem. Soc. **136**, 218 C.
Krivoshchekov, G.V., and V.I. Stroganov, 1968, Sov. Phys.-Solid State **9**, 2856.
Krivoshchekov, G.V., and V.I. Stroganov, 1969, Sov. Phys.-Solid State **11**, 89.
Lee, C.H., R.K. Chang and N. Bloembergen, 1967, Phys. Rev. Lett. **18**, 167.
Li, L., G. Yu, S. Dong, G. Wang and Z. Zhang, 1988, Appl. Phys. B **47**, 283.
Li, L., Y. Liu, G. Yu, W. Wang and Z. Zhang, 1989, Phys. Rev. B **39**, 8728.
Liebsch, A., 1988, Phys. Rev. Lett. **61**, 1233.
Liebsch, A., 1989, Appl. Phys. A **49**, 677.
Liebsch, A., and W.L. Schaich, 1989, Phys. Rev. B **40**, 5401.
Litwin, J.A., J.E. Sipe and H.M. van Driel, 1983, Phys. Rev. **31**, 5543.
Maker, P.D., 1970, Phys. Rev. A **3**, 923.
Marowsky, G., and R. Steinhoff, 1988, Opt. Lett. **13**, 707.
Marowsky, G., A. Gierulski and B. Dick, 1985, Opt. Commun. **52**, 339.
Marowsky, G., R. Steinhoff, G.A. Reider, D. Erdmann and D. Dorsch, 1987, Opt. Commun. **63**, 109.
Marowsky, G., L.F. Chi, D. Moebius, R. Steinhoff, Y.R. Shen, D. Dorsch and B. Rieger, 1988a, Chem. Phys. Lett. **147**, 420.
Marowsky, G., R. Steinhoff, L.F. Chi, J. Hutter and G. Wagnière, 1988b, Phys. Rev. B **38**, 6274.
Marshall, C.D., and G.M. Korenowski, 1986, J. Chem. Phys. **85**, 4172.
Marshall, C.D., and G.M. Korenowski, 1987, J. Chem. Phys. **91**, 1289.
Mate, C.M., G.A. Somorjai, H.W.K. Tom, X.D. Zhu and Y.R. Shen, 1988, J. Chem. Phys. **88**, 441.
Maystre, D., M. Neviere and R. Reinisch, 1986, Appl. Phys. A **39**, 115.
Mazely, T.L., and W.M. Hetherington III, 1987, J. Chem. Phys. **86**, 3640.
McGilp, J.F., 1987, J. Vac. Sci. & Technol. A **5**, 1442.
McGilp, J.F., 1989, J. Phys.: Condens. Matter **1**, SB85.
McGilp, J.F., and Y. Yeh, 1986, Solid State Commun. **59**, 91.
Miragliotta, J., and T.E. Furtak, 1988, Phys. Rev. B **37**, 1028.
Mizrahi, V., and J.E. Sipe, 1988, J. Opt. Soc. Am. B **5**, 660.
Mizrahi, V., F. Suits, J.E. Sipe, U.J. Gibson and G.I. Stegeman, 1987, Appl. Phys. Lett. **51**, 427.
Mizrahi, V., G.I. Stegeman and W. Knoll, 1989, Phys. Rev. A **39**, 3555.
Muenchausen, R.E., R.A. Keller and N.S. Nogar, 1987, J. Opt. Soc. Am. B **4**, 237.
Mullin, C.S., P. Guyot-Sionnest and Y.R. Shen, 1989, Phys. Rev. A **39**, 3745.
Murphy, D.V., K.U. von Raben, T.T. Chen, J.F. Owen, R.K. Chang and B.L. Laube, 1983, Surf. Sci. **124**, 529.
Murphy, R., M. Yeganeh, K.J. Song and E.W. Plummer, 1989, Phys. Rev. Lett. **63**, 318.
Pan, R.-P., H.D. Wei and Y.R. Shen, 1989, Phys. Rev. **39**, 1229.
Pershan, P.S., 1963, Phys. Rev. **130**, 919.
Peterson, E.S., and C.B. Harris, 1989, J. Chem. Phys. **91**, 2683.
Quail, J.C., and H.J. Simon, 1985, Phys. Rev. B **31**, 4900.

Rasing, Th., Y.R. Shen, M.W. Kim, P. Valint Jr and J. Block, 1985a, Phys. Rev. A **31**, 537.
Rasing, Th., Y.R. Shen, M.W. Kim and S.G. Grubb, 1985b, Phys. Rev. Lett. **55**, 2903.
Rasing, Th., G. Berkovic, Y.R. Shen, S.G. Grubb and M.W. Kim, 1986, Chem. Phys. Lett. **130**, 1.
Rasing, Th., T. Stehlin, Y.R. Shen, M.W. Kin and P. Valint Jr, 1988, J. Chem. Phys. **89**, 3386.
Rasing, Th., J. Huang, A. Lewis, T. Stehlin and Y.R. Shen, 1989, Phys. Rev. A **40**, 1684.
Reider, G.A., A.J. Schmidt and G. Marowsky, 1983, Opt. Commun. **47**, 223.
Reider, G.A., M. Huemer and A.J. Schmidt, 1988, Opt. Commun. **68**, 149.
Reif, J., P. Tepper, E. Mattias, E. Westin and A. Rosén, 1988, Appl. Phys. B **46**, 131.
Richmond, G.L., 1986, Langmuir **2**, 132.
Richmond, G.L., H.M. Rojhantalab, J.M. Robinson and V.L. Shannon, 1987, J. Opt. Soc. Am. B **4**, 228.
Richmond, G.L., J.M. Robinson and V.L. Shannon, 1988, Prog. Surf. Sci. **28**, 1.
Robinson, J.M., and G.L. Richmond, 1989, Electrochem. Acta **34**, 1639.
Rosenzweig, Z., and M. Asscher, 1988, Surface Sci. **204**, L732.
Rudnick, J., and E.A. Stern, 1971, Phys. Rev. B **4**, 4274.
Rudnick, J., and E.A. Stern, 1974, in: Polaritons, eds E. Burstein and F. de Martini (Pergamon Press, New York) p. 329.
Rustagi, K.C., 1968, Il Nuovo Cimento **53**, 346.
Schaich, W.L., and A. Liebsch, 1988, Phys. Rev. B **37**, 6187.
Shank, C.V., R. Yen and C. Hirlimann, 1983, Phys. Rev. Lett. **51**, 900.
Shannon, V.L., D.A. Koos, J.M. Robinson and G.L. Richmond, 1987a, Chem. Phys. Lett. **142**, 323.
Shannon, V.L., D.A. Koos and G.L. Richmond, 1987b, J. Chem. Phys. **87**, 1440.
Shannon, V.L., D.A. Koos and G.L. Richmond, 1987c, J. Chem. Phys. **91**, 5548.
Shannon, V.L., D.A. Koos, S.A. Keller, P. Huifang and G.L. Richmond, 1989, J. Phys. Chem. **93**, 6434.
Shen, Y.R., 1984a, Principles of Nonlinear Optics (Wiley, New York) ch. 25.
Shen, Y.R., 1986a, Ann. Rev. Mater. Sci. **16**, 69.
Shen, Y.R., 1986b, in: New Laser and Optical Investigation of Chemistry and Structure of Interfaces, eds R.B. Hall and A.B. Ellis (Verlag-Chemie, Weinheim) p. 151.
Shen, Y.R., 1989a, Nature **337**, 519.
Shen, Y.R., 1989b, Ann. Rev. Phys. Chem. **40**, 327.
Shen, Y.R., and F. de Martini, 1982, in: Surface Polaritons, eds V.M. Agranovich and D.L. Mills (North-Holland, Amsterdam) p. 629.
Simon, H.J., D.E. Mitchell and J.G. Watson, 1974, Phys. Rev. Lett. **33**, 1531.
Simon, H.J., R.E. Benner and J.G. Rako, 1977, Opt. Commun. **23**, 245.
Sipe, J.E., 1987, J. Opt. Soc. Am. B **4**, 481.
Sipe, J.E., and G.I. Stegeman, 1982, in: Surface Polaritons, eds V.M. Agranovich and D.L. Mills (North-Holland, Amsterdam) p. 661.
Sipe, J.E., V.C.Y. So, M. Fukui and G.I. Stegeman, 1980a, Solid State Commun. **34**, 523.
Sipe, J.E., V.C.Y. So, M. Fukui and G.I. Stegeman, 1980b, Phys. Rev. B **21**, 4389.
Sipe, J.E., V. Mizrahi and G.I. Stegeman, 1987a, Phys. Rev. B **35**, 9091.
Sipe, J.E., D.J. Moss and H.M. van Driel, 1987b, Phys. Rev. B **35**, 1129.
Sitzmann, E.V., and K.B. Eisenthal, 1988, J. Phys. Chem. **92**, 4579.
Sitzmann, E.V., and K.B. Eisenthal, 1989, J. Chem. Phys. **90**, 2831.
Song, K.J., D. Heskett, L. Urbach, H.L. Dai and E.W. Plummer, 1987, Mater. Res. Soc. Symp. Proc. **83**, 259.
Song, K.J., D. Heskett, H.L. Dai, A. Liebsch and E.W. Plummer, 1988, Phys. Rev. Lett. **61**, 1380.
Sonnenberg, H., and H. Heffner, 1968, J. Opt. Soc. Am. **58**, 209.
Stehlin, T., M. Feller, P. Guyot-Sionnest and Y.R. Shen, 1988, Opt. Lett. **13**, 389.
Superfine, R., P. Guyot-Sionnest, J.H. Hunt, C.T. Kao and Y.R. Shen, 1988, Surf. Sci. **200**, L445.

Suzuki, T., and T.F. Heinz, 1989, Opt. Lett. **14**, 1201.
Terhune, R.W., P.D. Maker and C.M. Savage, 1962, Phys. Rev. Lett. **8**, 404.
Tom, H.W.K., 1984, Ph.D. Dissertation (University of California, Berkeley, CA).
Tom, H.W.K., and G.D. Aumiller, 1986, Phys. Rev. B **33**, 8818.
Tom, H.W.K., T.F. Heinz and Y.R. Shen, 1983, Phys. Rev. B **51**, 1983.
Tom, H.W.K., C.M. Mate, X.D. Zhu, J.E. Crowell, T.F. Heinz, G.A. Somorjai and Y.R. Shen, 1984, Phys. Rev. Lett. **52**, 348.
Tom, H.W.K., C.M. Mate, X.D. Zhu, J.E. Crowell, Y.R. Shen and G.A. Somorjai, 1986a, Surf. Sci. **172**, 466.
Tom, H.W.K., X.D. Zhu, Y.R. Shen and G.A. Somorjai, 1986b, Surf. Sci. **167**, 167.
Tom, H.W.K., G.D. Aumiller and C.H. Brito-Cruz, 1988, Phys. Rev. Lett. **60**, 1438.
Tseng, C.-C., and J.T. Lue, 1987, Surf. Sci. **192**, 491.
van Wyck, N.E., E.W. Koenig, J.D. Byers and W.M. Hetherington III, 1985, Chem. Phys. Lett. **122**, 153.
Voss, D.F., M. Nagumo, L.S. Goldberg and K.A. Bunding, 1986, J. Phys. Chem. **90**, 1834.
Wang, C.C., 1969, Phys. Rev. **178**, 1457.
Wang, C.C., and A.N. Duminski, 1968, Phys. Rev. Lett. **20**, 668.
Weber, M.G., and A. Liebsch, 1987a, Phys. Rev. B **35**, 7411.
Weber, M.G., and A. Liebsch, 1987b, Phys. Rev. B **36**, 6411.
Wynne, J.J., and N. Bloembergen, 1969, Phys. Rev. **188**, 1211.
Ye, P., and Y.R. Shen, 1983, Phys. Rev. **B** 28, **4288.**
Zadkov, V.N., L.B. Meisner and T. Shreder, 1987, Opt. Spektrosk. **63**, 695 [Opt. & Spektrosk. (USSR) **63**, 411].
Zhu, X.D., and Y.R. Shen, 1989, Opt. Lett. **14**, 503.
Zhu, X.D., Y.R. Shen and R. Carr, 1985, Surf. Sci. **163**, 114.
Zhu, X.D., H. Suhr and Y.R. Shen, 1987, Phys. Rev. B **35**, 3047.
Zhu, X.D., Th. Rasing and Y.R. Shen, 1988, Phys. Rev. Lett. **61**, 2883.
Zhu, X.D., Th. Rasing and Y.R. Shen, 1989, Chem. Phys. Lett. **155**, 459.

CHAPTER 6

Nonlinear Surface Magnetoplasma Polaritons in Semiconductors

V. AMBRAZEVIČIENE, R. BRAZIS and A. KUNIGĖLIS

Institute of Semiconductor Physics
Lithuanian Academy of Sciences
Vilnius 232600

Nonlinear Surface Electromagnetic Phenomena
Edited by
H.-E. Ponath and G.I. Stegeman

© *Elsevier Science Publishers B.V., 1991*

Contents

1. Introduction

Electromagnetic surface excitations (Wood 1935, Fano 1941, Lifshits and Rosenzweig 1948, Ritchie 1957, 1973, Ferrell 1958, Gintsburg 1958, Landau and Lifshits 1960, Kheifets 1965, Kliewer and Fuchs 1966a, b, Parkhomov and Stepanov 1967, Khankina and Yakovenko 1967, Otto 1968, Ginzburg 1981) turned out to be one of the most interesting and important subjects in modern solid state physics. Surfaces and interfaces provide the possibility of coupling between the longitudinal oscillations of a *polarizable* medium (e.g., plasmas or ionic-crystal sublattices), and transverse electromagnetic waves (i.e., *photons*). As a result of the coupling, mixed states are created (Tolpygo 1950, Pekar 1957), including both electromagnetic and mechanic contributions to their total energy. These mixed excitations, referred to as light-excitons in pioneering works (Pekar 1957), obtained more recently the name *polaritons* (Barker 1964, Hopfield 1966), which is now generally accepted. The results of polariton investigations discussed in many original papers and reviews (*Polaritons* 1974, Bryksin et al. 1974, Otto 1974, Mills and Burstein 1974, *Surface Polaritons* 1982, Zhizhin et al. 1982, 1983, Martin et al. 1978, Halevi and Guerra-Vela 1978, Dmitruk et al. 1979, Chaplik and Krasheninnikov 1978, Abelès and Lopez-Rios 1982, Brazis 1981, Altshuler et al. 1983) are far from completely revealing all the variety of linear effects. Moreover, they brought about the potential abundance of nonlinear effects related to polariton excitation. However, nonlinear problems are predominantly confined either to theoretical studies (Bhat and Tripathi 1975, Mills 1977, Chen and Burstein 1977, Bonsall and Maradudin 1978, Sipe et al. 1980, 1986, Nurmikko et al. 1980, Agranovich et al. 1981, Gorobei et al. 1981, Reinisch and Neviere 1981, Agranovich and Chernyak 1982, Deck and Sarid 1982, Quail and Simon 1985, Leung 1985, Weber 1986, Innes and Sambles 1989, Pishreckji et al. 1989) or experiments in the optical region (Simon et al. 1974, 1977, Chen and Carter 1982, Akcipetrov et al. 1988).

The present paper deals with the low-frequency nonlinear polariton experimental studies initiated in 1978 (Ambrazevičiene and Brazis 1978, Ambrazevičiene et al. 1979). The low-frequency region provides many interesting possibilities, such as:

(i) a scaling approach to the experimental study of some complicated problems of thin-film optics;

(ii) surface mode investigations in nondegenerate free-carrier plasmas in semiconductors exhibiting nonlinear behavior at rather low pump power, as compared to metal plasmas, and; finally,

(iii) investigation of an almost unexplored direction which is nonlinear surface polaritons in gyrotropic media.

It is the low-frequency range which provides the possibility to obtain strong gyrotropy of the free-carrier system by applying a DC magnetic field.

A magnetic field is known to cause essential changes in the surface-mode dispersion (Chiu and Quinn 1972, Tyler et al. 1973, Kaplan et al. 1974, Wallis et al. 1974, Obunai and Sekiguchi 1974, Tarkhanian 1975, Ambrazevičiene and Brazis 1976, 1978, Davydov et al. 1977, Ambrazevičiene et al. 1980, Brazis 1981, Popov et al. 1983). The majority of works on the linear regime refer to the Voigt configuration in which the magnetic-field and propagation vectors are mutually perpendicular. This configuration is mostly suitable for investigations in the nonlinear regime as well, due to the possibility of separate TE- and TM-mode excitations.

Primarily we consider here the principles of experimental methods, and the peculiarities of the experimental set-up (section 2). The results, including harmonic radiation and electromagnetic-wave amplification, are discussed in section 3. In conclusion, the trends and prospects of the magnetoplasma-type surface-polariton investigation are discussed (section 4).

2. *Basic considerations and experimental set-up*

Let the plane $x=0$ define the interface between two isotropic nonmagnetic media, and the wave vector $\boldsymbol{k}$ lie in the xy-plane (fig. 1a). In the linear case, electromagnetic-wave propagation in each medium is described by the homogeneous differential equation

$$\nabla^2 \boldsymbol{E}_{1,2} - \nabla(\nabla \cdot \boldsymbol{E}_{1,2}) - \mu_0 \varepsilon_0 \hat{\varepsilon}_{1,2} \frac{\partial^2 \boldsymbol{E}_{1,2}}{\partial t^2} = 0, \tag{1}$$

where $\boldsymbol{E}_{1,2}$ is the electric field in medium 1 or 2, $\hat{\varepsilon}_{1,2}$ is the dielectric permittivity, μ_0 is the magnetic permeability and ε_0 is the electric permittivity. Solutions of eq. (1) for fields homogeneous along the z-axis

$$\{\boldsymbol{E}, \boldsymbol{H}\} = \{E, H\} \exp(-\mathrm{i}k_x x) \exp \mathrm{i}(\omega t - k_y y), \tag{2}$$

together with the conditions of tangential field component continuity on the interface lead to the relations (Landau and Lifshits 1960, *Polaritons* 1974, *Surface Polaritons* 1982)

$$k_y^2 = k_0^2 \varepsilon_1 \varepsilon_2 / (\varepsilon_1 + \varepsilon_2), \tag{3}$$

$$k_{x1,2}^2 = k_0^2 \varepsilon_{1,2} - k_y^2 = k_0^2 \varepsilon_{1,2}^2 / (\varepsilon_1 + \varepsilon_2), \tag{4}$$

where $k_0^2 = \omega^2 \varepsilon_0 \mu_0$. The magnitudes E and H are, in general, complex vectors representing elliptic polarization. Equations (3) and (4) determine the electromagnetic-mode dispersion and field decay constants across the boundary. Pure surface modes, i.e., real values of k_y and imaginary k_x, arise when the conditions

$$\varepsilon_1 \varepsilon_2 < 0 \quad \text{and} \quad \varepsilon_1 + \varepsilon_2 < 0 \tag{5}$$

are satisfied (Landau and Lifshits 1960, *Polaritons* 1974, *Surface Polaritons*

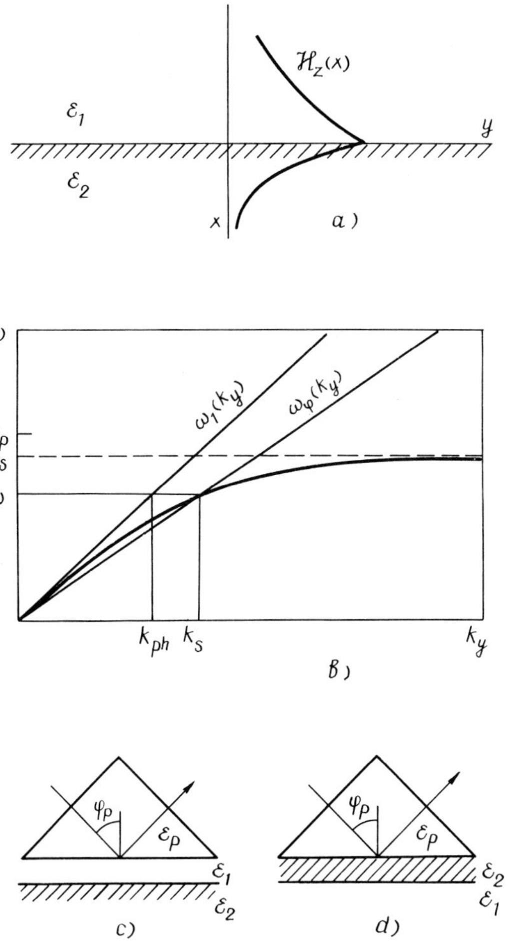

Fig. 1. The interface between the dielectric ($\varepsilon_1 = \text{const.}$) and semiconductor ($\varepsilon_2 = \varepsilon_2(\omega)$) half-spaces (a), the surface plasma-type polariton dispersion curve (b), and the FTR-configuration for the surface-mode excitation after Otto (1968, 1974) (c), and Kretschmann (1971) (d), where $\omega_1(k_y) = k_y/(\varepsilon_0\varepsilon_1\mu_0)^{1/2}$, $\omega_\phi(k_y) = k_y/(\varepsilon_0\varepsilon_p\mu_0)^{1/2} \sin \phi_p$. For other notations see text.

1982). Negative dielectric constants ε_1 or ε_2 required by these conditions are characteristic of metals and semiconductors in the frequency range below the free-carrier plasma frequency $\omega_p = (n_0 e^2/\varepsilon_0 m^* \varepsilon_L)^{1/2}$, where n_0 is the steady-state carrier density, m^* is the carrier effective mass, and ε_L is the lattice dielectric constant. The electromagnetic-wave dispersion relation in this case contains the surface-polariton branch which starts at $k_y = 0$, being initially linear $\omega_1(k_y)$ (fig. 1b). It approximates the surface-plasma oscillation frequency $\omega_s = \omega_p \times (1 + \varepsilon_1/\varepsilon_L)^{1/2}$ with the rise of k_y. The surface-polariton momentum $\hbar k_s$ exceeds that of photon $\hbar k_{ph}$, while the quasiparticle energy $\hbar\omega$ is the same (fig. 1b). Therefore, the surface modes of a given frequency cannot be excited by a plane wave of the same frequency propagating in dielectric medium 1. Momentum conservation, i.e., the matching of the bulk-mode and the surface-mode phase velocities, is achieved by using various techniques, e.g., the slot illumination (Zhizhin et al. 1979), the grating conversion (Fischer et al. 1973), or the frustrated total reflection – FTR (figs 1c, d). In the last case the tangential component of the photon momentum increases up to the value of $\hbar k_0(\varepsilon_p)^{1/2} \sin\phi_p$, owing to the prism, where ε_p is the prism's dielectric constant and ϕ_p is the angle of incidence. As a result, the resonance excitation of the surface polaritons takes place at energy $\hbar\omega_s$, determined by the surface-mode dispersion-branch intersection with the straight line $\omega_\phi(k_y)$ (fig. 1b). Energy transfer from the bulk incident electromagnetic mode to the surface mode results in a minimum reflectivity at this frequency. In the present work the FTR technique is chosen, primarily due to its advantage as a nondestructive tool of surface probing. On the other hand, the characteristics measured by this technique are quite easily simulated mathematically. In fact, all experimental structures include many media: the prism and the prism-sample spacer, the sample on a substrate, etc. Therefore, we shall consider the p-polarized wave reflection from a multilayer structure.

The structure (fig. 2) consists of media with flat parallel interfaces. These

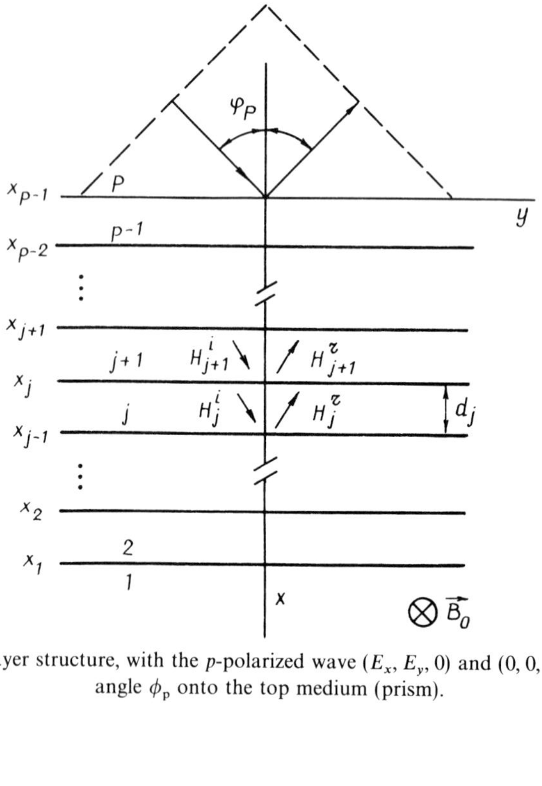

Fig. 2. The multilayer structure, with the p-polarized wave $(E_x, E_y, 0)$ and $(0, 0, H_z)$ incident at an angle ϕ_p onto the top medium (prism).

media are to be isotropic in initial state, and exhibit gyrotropy under the influence of the DC magnetic field $\boldsymbol{B}_0 = (0, 0, B_0)$ with the effective dielectric tensor (Ambrazevičiene et al. 1979, Tarkhanian 1975).

$$\varepsilon_j = \begin{bmatrix} \varepsilon_{xxj} & \varepsilon_{xyj} & 0 \\ \varepsilon_{yxj} & \varepsilon_{yyj} & 0 \\ 0 & 0 & \varepsilon_{zzj} \end{bmatrix}, \tag{6}$$

where, according to the Drude model,

$$\varepsilon_{xxj} = \varepsilon_{yyj} = \varepsilon_{\mathrm{L}j}\left\{1 - \sum \frac{\omega_{\mathrm{p}j}^2(\omega - \mathrm{i}\nu_j)}{\omega[(\omega - \mathrm{i}\nu_j)^2 - \omega_{\mathrm{c}j}^2]}\right\},$$

$$\varepsilon_{xyj} = -\varepsilon_{yxj} = -\mathrm{i}\varepsilon_{\mathrm{L}j}\sum \frac{\omega_{\mathrm{p}j}^2\omega_{\mathrm{c}j}}{\omega[(\omega - \mathrm{i}\nu_j)^2 - \omega_{\mathrm{c}j}^2]},$$

$$\varepsilon_{zzj} = \varepsilon_{\mathrm{L}j}\left\{1 - \sum \frac{\omega_{\mathrm{p}j}^2}{\omega(\omega - \mathrm{i}\nu_j)}\right\}, \tag{7}$$

and $\omega_{\mathrm{c}j} = q_j B_0/m^*$ is the cyclotron frequency, $q_j = +e$ or $-e$ is the carrier charge and ν_j is the carrier collision frequency. Summation in eq. (7) is carried out over all free-carrier types. In each layer of thickness $d_j = x_{j-1} - x_j$, as well as in the top medium, the electromagnetic waves are considered to be the superposition of the 'incident' ($k_{xj} > 0$) and 'reflected' ($k_{xj} < 0$) waves. In the bottom medium 1 (fig. 2) there is only one wave, either really transmitted or exponentially decaying with x depending on whether k_{x1} is real or imaginary.

Tangential-component continuity conditions for the electric and magnetic fields at the jth interface lead to recurrent formula relating the coefficients of reflection from the jth and $(j+1)$th boundaries. This relation was given in a previous work (Ambrazevičiene and Brazis 1978) for the general case of all gyrotropic media. We write down here the appropriate expression in compact coordinate-free form

$$r_{j+1} = \frac{H_{j+1}^{\mathrm{r}}}{H_{j+1}^{\mathrm{i}}} = \frac{(\eta_j^- - \eta_{j+1}^+)\exp 2\mathrm{i}k_{xj}d_j + (\eta_j^+ - \eta_{j+1}^-)r_j}{(\eta_{j+1}^+ - \eta_j^-)\exp 2\mathrm{i}k_{xj}d_j + (\eta_{j+1}^+ - \eta_j^+)r_j}, \tag{8}$$

where

$$\eta_j^{\pm} = \frac{k_y}{\varepsilon_{\mathrm{V}j}}\frac{\varepsilon_{xyj}}{\varepsilon_{xxj}} \pm \frac{k_{xj}}{\varepsilon_{\mathrm{V}j}}, \tag{9}$$

$$k_{xj} = k_0(\varepsilon_{\mathrm{V}j} - \varepsilon_{\mathrm{p}}\sin^2\phi_{\mathrm{p}})^{1/2}, \tag{10}$$

$$\varepsilon_{\mathrm{V}j} = (\varepsilon_{xxj}^2 + \varepsilon_{xyj}^2)/\varepsilon_{xxj}. \tag{11}$$

Using eqs. (8)–(11) and the Maxwell equations, we express the TM-mode fields as functions of coordinate x in the jth layer,

$$H_{zj}(x) = H^{i}_{zp} t_j \{\exp 2ik_{xj}(d_j - x) + r_j \exp 2ik_{xj}x\}, \tag{12}$$

$$E_{yj}(x) = H^{i}_{zp} t_j \{\eta_j^- \exp 2ik_{xj}(d_j - x) + r_j \eta_j^- \exp 2ik_{xj}x\}(\mu_0/\varepsilon_0)^{1/2}, \tag{13}$$

$$E_{xj}(x) = H^{i}_{zp} t_j \{(\eta_j^-/\beta_j^-) \exp 2ik_{xj}(d_j - x) + r_j(\eta_j^+/\beta_j^+) \exp 2ik_{xj}x\}(\mu_0/\varepsilon_0)^{1/2}, \tag{14}$$

where $x = 0$ is taken at the boundary x_j, and the coefficient

$$\beta_j^\pm = \frac{k_y \varepsilon_{xyj} \pm k_{xj}\varepsilon_{xxj}}{k_y \varepsilon_{xxj} \mp k_{xj}\varepsilon_{xyj}} \tag{15}$$

is the ratio of electric field components $E_x^\pm / E_y^\pm$ for the incident $(-)$ and reflected $(+)$ waves. The complex coefficient t_j determines the 'incident' *magnetic*-field amplitude at the boundary x_{j+1} in the jth layer divided by the incident-field amplitude H_{zp} at the boundary x_{p-1} in the pth medium (i.e., in the prism). This coefficient is

$$t_j = \prod \frac{1 + r_{s+1}}{\exp ik_{xs}d_s + r_k \exp(-ik_{xs}d_s)}, \tag{16}$$

where Π stands for the product of the relevant terms with the index s running from j to $p-1$.

Expressions (6)–(16) enable one to find the field structure and the measurable values of the power reflection coefficient $R = |r_\mathrm{p}|^2$ at various frequencies, magnetic fields and structure parameters. Typically, in the long-wave range experiments the frequency ω and the angle of incidence on the prism are fixed, and the value of B_0 is tuned. As a result, the normal-mode spectrum is changed due to the cyclotron-frequency dependence of the magnetoplasma effective dielectric constant [eq. (7)]. At any value of B_0 where the bulk modes in the jth layer or the surface modes at the jth boundary access the wave-vector value $k_0(\varepsilon_\mathrm{p})^{1/2} \sin \phi_\mathrm{p}$ given by the prism set-up, the total reflectivity is known to frustrate. The FTR-minimum position depends, first, on the semiconductor parameters such as the lattice dielectric constant, the carrier density, and the carrier effective mass (Brazis 1981, Ambrazevičiene et al. 1979, Ambrazevičiene and Brazis 1978). The minimum position on the B_0-scale is estimated from the normal-mode dispersion equation. The latter is known to be determined by zeroing the denominator of r_j with the simultaneous assumption that there is no source of incident wave ($H_{zp} = 0$) (Cardona 1971). At high values of $k_y = k_0(\varepsilon_\mathrm{p})^{1/2} \sin \phi_\mathrm{p}$, accessible by choosing a prism of high dielectric constant, the radicals in eq. (10) may be expanded in series of the small parameter $\varepsilon_\mathrm{V}/\varepsilon_\mathrm{p} \sin^2 \phi_\mathrm{p}$, except the values of ω_{cj} where the function $\varepsilon_{\mathrm{V}j}(\omega_{cj})$ has poles. Retaining the linear terms of the expansion, one can get a quite simple condition of the surface magnetoplasma-type polariton excitation in the Otto configuration (fig. 1c) (Brazis 1981),

$$B_0 = \frac{n_0 e}{\omega\varepsilon_0[\varepsilon_L + \varepsilon_d/(1-\delta)]} - \frac{m^*\omega}{e}, \tag{17}$$

where ε_d is the dielectric constant of the medium contacting with the semiconductor, and δ stands for $\varepsilon_d/\varepsilon_p \sin^2 \phi_p$. Relation (17) is valid at the condition of negligible absorption, i.e., $\omega_{cj} \gg \nu_j$.

When the dielectric layer between the prism and the sample is very thin, $d_{p-1} \ll k_{x(p-1)}$, the surface mode decays due to the leakage of its energy into the prism. This radiation loss together with the absorption result in the FTR-line broadening and shifting toward low magnetic fields (Ambrazevičiene et al. 1980). A similar shift arises for thin samples, due to the coupling of surface-mode fields at opposite interfaces (Ambrazevičiene and Brazis 1978). Equations (8)–(17) are obtained for layers which are unbounded in the *yz*-plane. Therefore, the applicability of the above method for experimental data-modelling is restricted by diffraction and is subject to testing by detailed comparison with experiments. The main conclusion of such a comparison is that the diffraction effects make the absolute measurements of reflectivity difficult. Nevertheless, the FTR-minima positions predicted by the theory for various parameters of the semiconductor and the dielectric layer are in fairly good agreement with the experimental data (Brazis 1981, Ambrazevičiene et al. 1979, Ambrazevičiene and Brazis 1976, 1978). Moreover, the linear regime of experiments provides the possibility of obtaining kinetic parameters of semiconductor samples by fitting calculated FTR lines to the experimental ones. Thus, the FTR method is

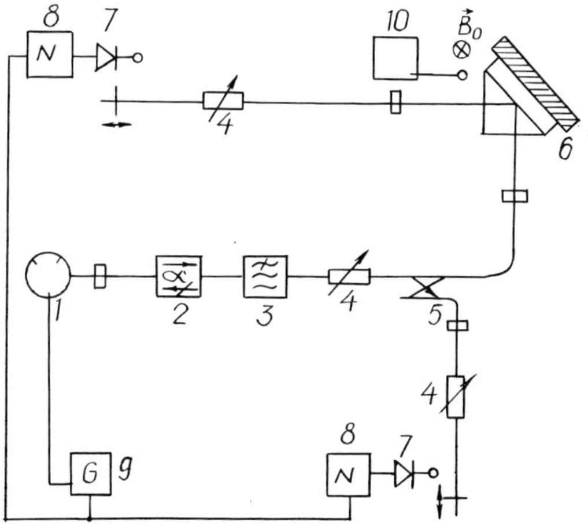

Fig. 3. Schematic diagram of the spectrometer including magnetron (1), isolator (2), low-pass filter (3), variable attenuators (4), directional coupler (5), FTR-unit (6), detector sections (7), sampling oscilloscope (8), pulse generator (9) and magnetic-field sensor (10).

concluded to give reliable data on the low-frequency surface- and waveguide-mode dispersion. This method enables one to distinguish between the wave processes in the bulk and that on different boundaries of semiconductors.

Figure 3 shows the experimental set-up which is used for the nonlinear surface magnetoplasma-type polariton investigation since the first works (Ambrazevičiene and Brazis 1978, Ambrazevičiene et al. 1979). Electromagnetic waves outgoing from the magnetron passed through the isolator and the low-pass filter to the FTR unit containing the semiconductor sample. A pulse duration of 0.35 μs with repetition rate below 40 Hz was used to avoid sample heating. Calibrated directional couplers and precision attenuators terminated by point detectors were used to measure the incident and reflected pulses. Detected signals were sampled and integrated, and, finally, plotted as a function of magnetic field.

Harmonic radiation was studied using high-pass waveguide filters and selective radiometers. Experiments were carried out in the temperature range from 77 to 300 K. Samples in the form of single-crystal plates or epitaxial semiconductor structures were used.

3. Nonlinear effects

Excitation of the surface magnetoplasma mode resonance enhances the amplitude of the electromagnetic field near the surface, giving rise to nonlinear effects (Bhat and Tripathi 1975, Mills 1977, Chen and Burstein 1977, Bonsall and Maradudin 1978, Sipe et al. 1980, Nurmikko et al. 1980, Agranovich et al. 1981, Gorobei et al. 1981, Reinisch and Neviere 1981, Agranovich and Chernyak 1982). These are modelled in conducting media, as a rule, by expanding the current density $\boldsymbol{j}$ in a power series of the electric field. Higher-order contributions give rise, first, to the change of effective dielectric permittivity ε_{ij} on the left side of the wave eq. (1), resulting in the change of the principal-frequency surface wave vector. This effect is referred to as the surface-mode self-action. Secondly, the right side of the equation turns out to be a nonzero quantity

$$\boldsymbol{S}^{\mathrm{NL}} = \mu_0(\partial \boldsymbol{j}_2/\partial t + \partial \boldsymbol{j}_3/\partial t + \cdots), \tag{18}$$

which is the source of the even and odd harmonics corresponding to the parity of the power in the expansion of $\boldsymbol{j}$ against $\boldsymbol{E}$.

Note that until the late 70s there were no data on nonlinear effects related to surface polaritons in semiconductors. Filling this gap was the aim of the experimental investigations discussed.

3.1. Surface magnetoplasma polariton self-action

The intensity-induced change of the effective dielectric constant resulting in the bulk light beam self-focusing or defocusing is known to be peculiar to the odd-order nonlinearity (Bloembergen 1965, Akhamanov and Khokhlov 1964). The surface-mode counterpart of this nonlinear effect is expected to manifest itself in shifting the reflection minimum with the rise of pumping-wave intensity (Ambrazevičiene et al. 1979, Ambrazevičiene and Brazis 1981). Figure 4 illustrates the experimental confirmation of the expected effect in indium-antimonide crystals. The FTR minimum shifts to higher magnetic fields with the rise of pumping power. Experiments on wave reflection without the prism have shown high reflectivity, which is characteristic for the region of negative dielectric constant, in agreement with calculations (fig. 5). Thus, the FTR minimum is concluded to arise due to the surface-mode excitation.

The FTR-minimum shift to higher magnetic fields according to eq. (18), is ascribed to the intensity-induced increase of carrier density. Fitting the calculated FTR line to the experimental one, with the carrier density n_0 as an adjustable parameter, gives the possibility to find the values of n_0 and ε_V at the minima positions. The obtained values of ε_V are negative at various pumping-power levels: $\varepsilon_V = -(13 \pm 1)$, as marked by crosses in fig. 5.

Following formulas (6)–(16) one can calculate the field amplitudes in the sample. The result (see inset in fig. 4) confirms the surface nature of the excited

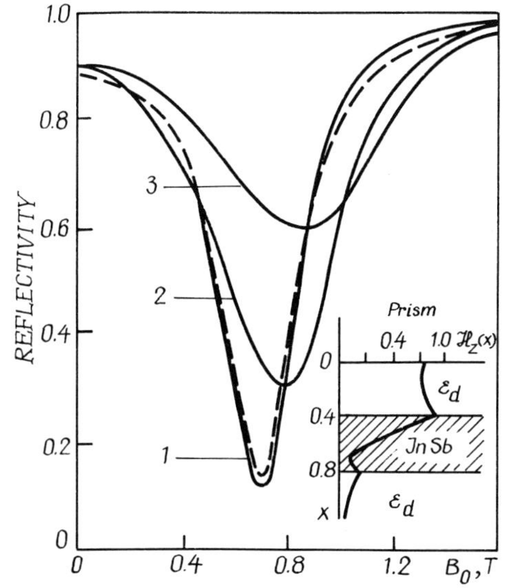

Fig. 4. FTR curves measured at various pumping-power levels P_1 (solid lines) and calculated for the lowest pumping power (dashed line) in the n-type InSb crystal. The values of P_1 (kW) are: (1) 0.099, (2) 3.0, (3) 7.5. Dielectric constants are: $\varepsilon_p = 12.34$ (the prism), $\varepsilon_d = 2.07$ (the prism-sample spacer), $\varepsilon_L = 16.8$ (the InSb-lattice contribution). The carrier density $n_0 = 2.15 \times 10^{20}\ \mathrm{m}^{-3}$ and the mobility $\mu = 8\ \mathrm{m}^2\ \mathrm{V}^{-1}\ \mathrm{s}^{-1}$ are obtained by fitting curve 1 to the experimental data. $T = 77$ K, $\omega_1 = 2.3 \times 10^{11}$ Hz. The inset shows the calculated amplitude dependence on the coordinate x (mm) for $B_0 = 0.7$ T.

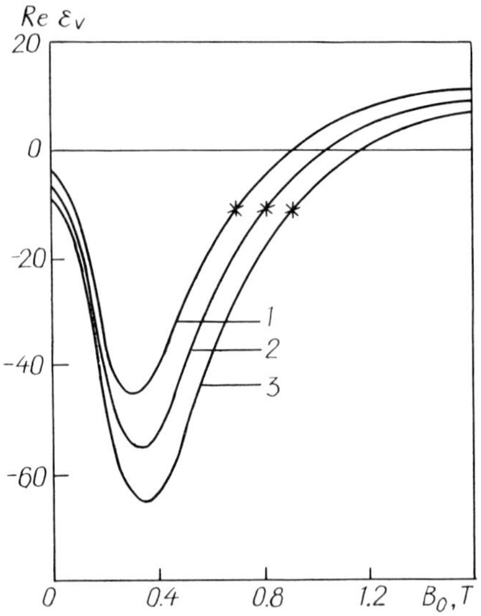

Fig. 5. Calculated real-part values of the Voigt dielectric function for the InSb crystal with electron densities corresponding to the minima positions in fig. 4, n_0 (10^{20} m^{-3}): (1) 2.15, (2) 2.45, (3) 2.85; other parameters are the same as in fig. 4. Crosses mark the values of Re(ε_V) and B_0 corresponding to the observed FTR minima.

modes. With the known values of the wave amplitude E_s outgoing from the source, as well as the relevant parameters of the structure, the amplitude of the electric field in the sample is evaluated. This gives the value of $E = 0.3\,E_s$ at the surface of InSb in the absence of the magnetic field B_0. At the magnetic field corresponding to the FTR minimum the amplitude is found to be 3 times higher, due to the surface-mode formation. This approach enables us to determine the threshold amplitude for the avalanche breakdown in InSb from the self-action experiments. The FTR minimum, positioned initially at $B_0 = 0.7$ T for InSb, starts to shift at pumping-power levels exceeding 1.9 kW. Using this value, a threshold field $E_{th} > 1$ kV/cm is obtained, in agreement with data of independent experiments (Brazis and Mironas 1977, Brazis et al. 1980).

With the rise of pumping power the FTR line broadens due to the increase of electron–phonon scattering probability. In some (more compensated) samples of InSb the broadening is preceded by a slight narrowing at the pumping-wave power below 1.9 kW. This is caused by the decreasing rate of electron scattering on ionized impurities with the growth of the mean electron energy. At higher power levels there is no simple way of determining the carrier density, collision frequency and other parameters from the line shape, because all the quantities change from one point to point, due to competing ionization–recombination and dissipative processes.

In the n-type GaAs samples, the increasing pumping-wave power shifts the FTR minimum toward low magnetic fields (fig. 6). This shift, evaluated according to eq. (17), may be caused either by the increase of the carrier effective mass m^* or the decrease of their density n_0. However, the wide-gap crystals like GaAs are known to have only slightly nonparabolic conduction bands. As a matter of fact, the electron effective mass increases significantly only at quite high energies, corresponding to the onset of intense electron transfer from the principal Γ-valley to the X- and L-valleys. Along with this transfer the carrier density n_0 in the Γ-valley would decrease. However, these effects are known to arise at electric fields $E > 3.5$ kV/cm (Ridley and Watkins 1961, Hilsum 1962, Gunn 1963, Rees 1969). In our experiments the field amplitude did not exceed 2 kV/cm even at highest pumping power of 20 kW. Therefore, the intervalley transfer cannot be responsible for the self-action.

The cause of the FTR-minimum shift to low magnetic fields is to be sought in the change of the electron distribution. At low temperatures, pure GaAs crystals subject to an electric field are known to exhibit inelastic electron scattering on optic phonons. In crossed electric (E_0) and magnetic (B_0) fields, the emission of optic phonons results in the build-up of a two-group electron distribution in momentum space (fig. 7). One group occupies the spindle-like region containing those electron trajectories centered on the $k_c = -m^* E_0/\hbar B_0$ which

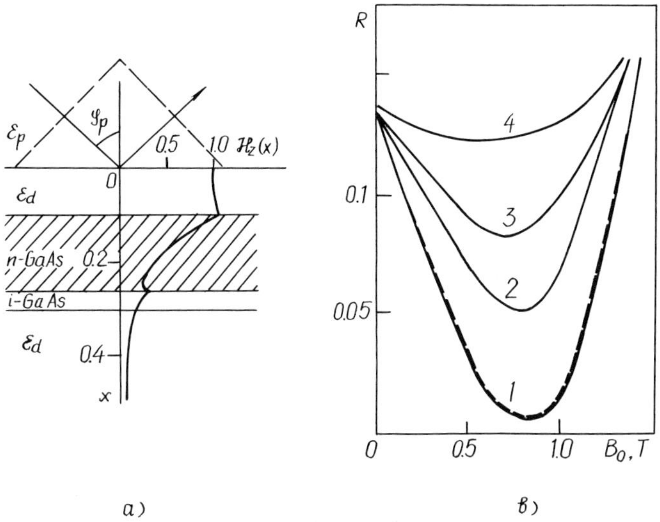

Fig. 6. Experimental FTR lay-out with the n-type GaAs epitaxial layer (a), and the corresponding FTR curves (b) at various pumping-power values (kW): (1) 0.032, (2) 1.8, (3) 5.6, (4) 10.0. The dashed line (b) represents the best-fit calculations, and the $H_z(x)$ line (a) is the calculated amplitude at the same fitting parameters. The scale of the x-axis is given in millimeters, $\omega_1 = 2.3 \times 10^{11}$ Hz, $\varepsilon_p = 12.34$, $\varepsilon_d = 2.07$, $\phi_p = 45°$, $T = 77$ K.

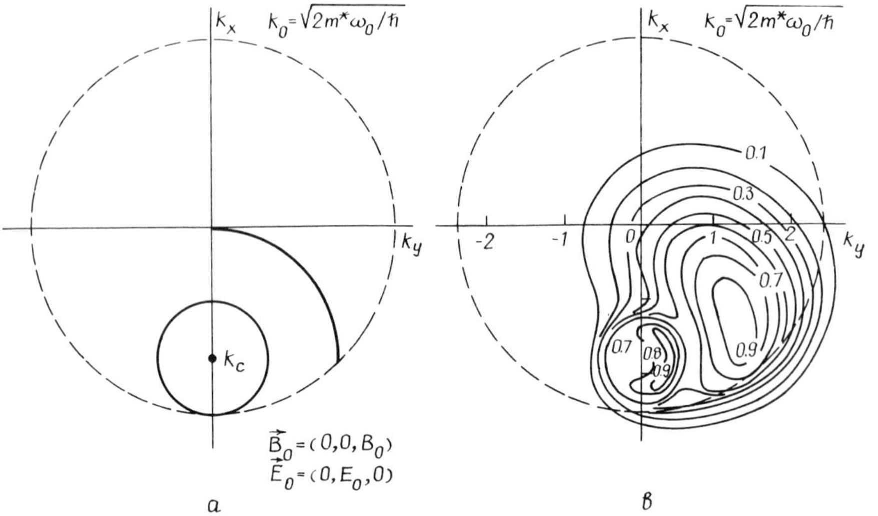

Fig. 7. Electron distribution in momentum space in crossed steady-state electric and magnetic fields when there are no other scattering mechanisms except the instantaneous emission of the optic phonons (a) (Levinson 1983), and when the quasielastic scattering is included together with the finite lifetime of electrons at energies $\varepsilon > \hbar^2 k_0^2/2m^*$ (b) (Brazis et al. 1983). The solid contour lines are the integrated values of the distribution function $f(k_x, k_y) = \int f(\boldsymbol{k})\, dk_z$ normalized to its maximum value (b) for the n-type GaAs at $T = 77$ K, $E_0 = 2.25 \times 10^5$ V/m, $B_0 = 0.25$ T, and $\hbar\omega_0 = 36$ meV. Two maxima representing two groups of electrons are clearly seen.

do not cross the sphere of radius k_0. The latter is determined by equalling the electron kenetic energy $\hbar k_0^2/2m^*$ to the optic-phonon energy $\hbar\omega_0$ (Levinson 1983). It is this group of electrons which performs, in average, many cycles of rotation in a magnetic field between successive acoustic-phonon collisions, and brings about the magnetoplasma-type modes. With the rise of the E_0/B_0-ratio, the electron density n_c in the region of closed trajectories decreases, resulting in the FTR-line shift to low magnetic fields. Along with this, the increased number of electrons on truncated trajectories (see arc line in fig. 7a) create an absorbing background which manifest itself in the FTR-line broadening. Penetration of electrons to energies $\varepsilon > \hbar\omega_0$ and their scattering on acoustic phonons in the region of $\varepsilon < \hbar\omega_0$ smears off the distribution (fig. 7b), though the main feature of two-electron groups is preserved.

The above consideration is valid for either stationary or circularly-polarized wave electric fields. Therefore, it provides only the guiding idea of what one would expect in arbitrarily polarized fields. Formation of different electron groups in a linearly polarized electric field $\boldsymbol{E}_1 \perp \boldsymbol{B}_0$ is, in fact, more complicated (Brazis and Shiktorov 1980, Kurosawa 1980) and gives rise to highly nonlinear drift response, resulting in radiation frequency conversion.

3.2. The third- and fifth-harmonic generation

Consider now relations between the surface magnetoplasma-type waves and the odd-harmonic radiation primarily observed in early works (Ambrazevičiene et al. 1980, 1982). The radiation was received in the direction of pumping-wave reflection. This is a natural approach to harmonics observations as far as the prism and the prism-sample spacer are free of dispersion (Bloembergen 1965, Akhmanov and Khokhlov 1964).

The third- and the fifth-harmonic radiation intensities exhibit maxima as functions of the magnetic field (fig. 8). The maxima are shifted to low magnetic fields as compared to the pumping-wave FTR-minimum position (Ambrazevičiene and Brazis 1981). This means that the maxima of radiation intensity have no evident relation with the pumping-wave resonance.

The intensity maxima may be explained as the result of surface-polariton excitation by the Nth-harmonic current in the sample. The pumping wave of frequency ω_1 incident on the prism at angle ϕ_p (fig. 8a), excites in the lower-

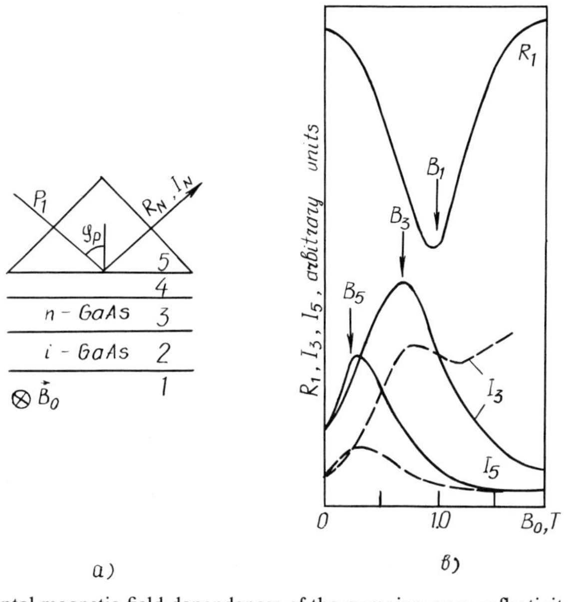

Fig. 8. Experimental magnetic-field dependences of the pumping-wave reflectivity R_1 and the odd-harmonic intensities I_N (a) for the n-type GaAs epitaxial structure (b). The dashed lines are calculated assuming homogeneous current distribution over the epitaxial layer thickness. The arrows show the calculated magnetic-field values for the surface-polariton excitation at frequencies ω_1, $3\omega_1$, and $5\omega_1$. The structure consists of teflon (1), an insulating substrate (2) of thickness $d_2 = 310$ μm, the n-type GaAs layer (3) of thickness $d_3 = 70$ μm, a teflon layer (4) of thickness $d_4 = 100$ μm, and the silicon prism (5). $\omega_1 = 2.3 \times 10^{11}$ Hz, $n_0 = 7.7 \times 10^{20}$ m^{-3}, $\phi_p = 45°$, $T = 77$ K (Ambrazevičiene et al. 1982).

lying, optically less dense layer the electromagnetic field which propagates along the layer with phase velocity $v_y = \omega_1/(\mu_0\varepsilon_0\varepsilon_p)^{1/2} \sin\phi_p$. This field drives the GaAs-sample current which may be expressed as the sum of harmonics,

$$j = \sum [j_{Nx}(x), j_{Ny}(x), 0] \exp iN\omega_1(t - y/v_y). \tag{19}$$

At a certain value of the magnetic field the Nth harmonic of the current is brought into phase synchronism with the normal mode of the same frequency, i.e., $v_{ph}(N\omega_1) = v_y$. As a result, the amplitude of the normal-mode field is enhanced resonantly, and the radiation intensity accesses a maximum.

The magnetic-field dependences of the intensities may be obtained by solving the magnetoactive-slab radiation problem (Ivanov and Ryutov 1965). The current amplitudes depend on the x-coordinate, therefore, the intensity calculations accounting for the self-consistent field- and current amplitudes are computer-time consuming. Reasonable estimations may be obtained by supposing a homogeneous distribution of the Nth-harmonic current over the layer thickness. With this assumption, and taking the dielectric tensor for the n-type GaAs in the form of eq. (7), the calculated radiation intensity is found to exhibit the expected resonances, as is seen in fig. 8 (dashed lines).

Calculations of the reflection coefficient [eq. (8)] give the FTR-minimum position B_1 for the fundamental frequency ω_1 in agreement with the experiment (fig. 8b). Similarly, if we suppose the incident-wave frequency to be $3\omega_1$ or $5\omega_1$, with the same angle of incidence ϕ_p, we obtain the calculated FTR-minima positions B_3 or B_5, respectively. The latter agree fairly well with the positions of the experimental maxima of radiation intensity. The field calculations show the amplitude distribution corresponding to the surface mode with the maximum value of the tangential component E_y at the GaAs-sample boundary facing the prism.

Summing up the above consideration, one can state that the observed effect of resonant enhancement of the third- or the fifth-harmonic intensity is caused by the build-up of the surface magnetoplasma-type polaritons.

The radiation intensity in GaAs approximated $1\ \mathrm{W/cm^2}$, the pumping intensity being of the order of $1\ \mathrm{kW/cm^2}$.

The odd harmonics as well as observed self-action are caused by electron heating. With the rise of energy, electrons are less subject to ionized-impurity scattering. At energies $\varepsilon > \hbar\omega_0$, the optic-phonon emission limits the free-carrier drift velocity. Further increase of electron energy gives rise to intervalley transfer. All these processes result in highly nonsinusoidal drift response to the sine-law time-dependent electric field (Brazis and Shiktorov 1980, Kurosawa 1980, Pozhela and Reklaitis 1979). The nonlinearities originating from the change of scattering rate are referred to as *resistive*.

Besides the scattering, the essential role in nonlinearities belongs to the nonparabolic character of electron energy bands. This may be considered as the relativistic analog of the energy-dependent effective mass. The theoretical accounts for the nonparabolic bands in the third-harmonic generation in magnetic fields have been proposed in the 60s (Lax et al. 1967) with the assumptions that the electron motion is collisionless and the effective-mass deviation is small compared to the equilibrium value. Considerations of electron motion in crossed DC fields $E_0 \perp B_0$, have shown the electron trajectories in momentum space to be noncircular even in the case of an isotropic spectrum. The trajectories are either elliptic, or parabolic and hyperbolic, depending on the E_0/B_0-ratio (Brazis et al. 1983). The effective-mass change on the trajectory turns out to be not small compared to its equilibrium value. In the alternating sine-law electric field the trajectories are more complicated, giving rise to strongly nonsinusoidal drift response. The nonlinearities caused by the changing effective mass of carriers are referred to as *reactive*.

All these phenomena defy analytical description. As an alternative, the numerical Monte Carlo methods turned out to be quite efficient in determining the nonlinear drift response of electrons in crossed fields (Brazis and Shiktorov 1980, Kurosawa 1980). It is instructive, however, to get some idea about the peculiarities of the resistive and inductive mechanisms in magnetic fields by considering a simple model of the equation of motion. Taking into account the velocity dependences of the effective mass m^* (Lax et al. 1967, Belyantsev et al. 1971) and the collision frequency v (Brazis and Mironas 1983) in the form $\{m^*, v\} = \{m_0^*, v_0\}(1 + \gamma_{m,v} v^2)$, where $\gamma_{m,v}$ is the nonlinearity parameter, one can find the approximate solutions of the equation of motion for the odd harmonics. With these solutions, the third-harmonic source [eq. (18)] is evaluated as $S_3 = -A\mu_0 e n_0 v_1^2 \boldsymbol{v}_1$, where $A = (2i\omega_1 + v_0)\gamma_m + v_0\gamma_v$ and v_1 is the fundamental harmonic drift velocity amplitude. In the fields $\boldsymbol{E}_1 = (E_{1x}, E_{1y}, 0)$ and $\boldsymbol{B}_0 = (0, 0, B_0)$ this source is expressed as

$$S_{3i} = iA\mu_0\varepsilon_0^3\omega_1^3(en_0)^{-2}(\chi_{xx}^2 + \chi_{xy}^2)(E_{1x}^2 + E_{1y}^2)\chi_{ij}E_{1j}, \tag{20}$$

where the free-electron susceptibility χ_{ij} is determined from eq. (7) as $\chi_{ij} = \varepsilon_{ij} - \varepsilon_L \delta_{ij}$, and δ_{ij} is the Kronecker-delta. The nondiagonal components of susceptibility are seen to make an important contribution to eq. (20), manifesting the significant role of the magnetic-field-induced gyrotropy in the odd-order nonlinearity. The resistive and reactive sources are not separable, as one can judge from the expression for the coefficient A. The experiments at $\omega_1 = 2.3 \times 10^{11}$ Hz show a much more intense third harmonic in GaAs compared to that in InSb, just opposite to what should be expected for the nonparabolic-band nonlinearity. The resistive nonlinearity seems to be dominating in this frequency range. The reactive source is expected to dominate in the far-infrared range where the condition $2\omega_1 > v_0\, \gamma_v/\gamma_m$, following from the $A(\omega_1)$-dependence, is fulfilled. The same trends are expected in the fifth-harmonic frequency dependence.

3.3. Second-harmonic radiation

Besides the odd harmonics, the second-harmonic radiation was observed in magnetoactive semiconductors in the absence of the DC electric field (Ambrazevičiene et al. 1987). This result is rather striking because electron heating is known to bring about the second harmonic in the presence of a DC electric field only (Nurmikko et al. 1980, Seeger 1973, Dienys and Dargys 1981).

Let us consider in detail the results obtained on indium antimonide. The second-harmonic radiation intensity exhibits a maximum at certain magnetic fields (fig. 9), as it was the case for the odd harmonics as well. The studies of samples having various carrier densities n_0 revealed both the harmonic intensity maxima and the fundamental-frequency reflection minimum shift to high magnetic field with the rise of n_0 (fig. 10). This is the evidence of free-electron plasma contribution to the observed resonances. For the InSb samples of thickness $d > 0.15$ mm, the resonance values of the magnetic field obey the density- and frequency dependences predicted by eq. (17) for the surface-magnetoplasma modes. Using eqs (7) and (8), we obtain corrections for the resonance positions accounting for the finite thicknesses of the sample d_p, and the prism-sample spacer d_d. This results in a fairly good agreement between the calculated and measured data (fig. 9 (arrows), fig. 10).

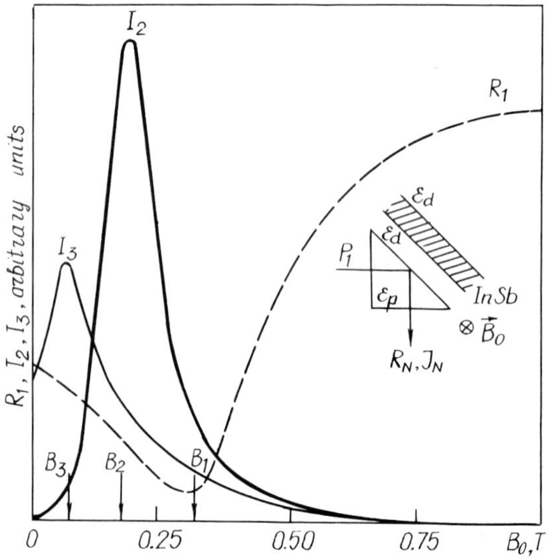

Fig. 9. Experimental pumping-wave reflectivity R_1 and the second- and the third-harmonic radiation intensities as functions of the DC magnetic field at $T = 77$ K. The InSb-sample parameters are $n_0 = 9 \times 10^{19}\ \mathrm{m}^{-3}$ and $d = 260\ \mu$m, and those of the sample-prism spacer are $\varepsilon_d = 2.07$ and $d_d = 200\ \mu$m. The fundamental frequency is $\omega_1 = 2.3 \times 10^{11}$ Hz and the angle of incidence on the prism ($\varepsilon_p = 12.34$) is $\phi_p = 45°$. The calculated surface-mode resonant magnetic-field values are shown by arrows (Ambrazevičiene et al. 1987).

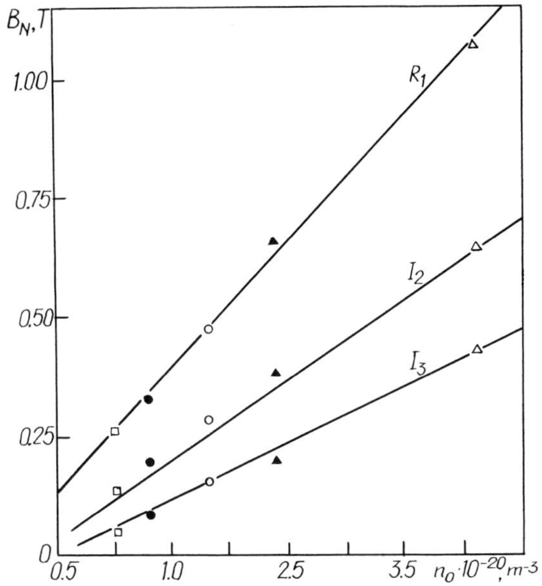

Fig. 10. Carrier-density dependences of the resonant magnetic fields corresponding to the minimum pumping-wave reflectivity R_1 and the maxima of harmonic intensity $I_{2,3}$ for the n-type indium antimonide. Solid lines – theory, points – experiment (Ambrazevičiene et al. 1987).

The second-harmonic power P_2 increases as the square of the pumping power P_1, provided that $P_1 < 600$ W (fig. 11). Deviation from the linear P_2 against P_1^2-dependence at high pumping levels is caused by both the increase of the electron–phonon collision rate and the onset of impact ionization in high electric fields. The observed square-law nonlinearity is not the exclusive property of the surface polaritons because it is observed in experiments on normal incidence as well (Ambrazevičiene et al. 1986, 1987a, b). However, the surface-polariton-enhanced second harmonic is almost ten times more intense as that observed in normal-transmission experiments (fig. 11). It is worth noting that the second harmonic was not found at $B_0 = 0$, even at the highest pumping-power level $P_1 = 1$ kW.

In the above experiments, as well as in analogous optical experiments on the second-harmonic radiation in metals, the condition $(2\omega_1)^2 \ll \omega_p^2$ typical of the surface plasmon-polariton excitation, was fulfilled. However, the measured value of the transformation coefficient $\rho_2 = I_2/I_1^2$ in optical experiments did not exceed 2×10^{-18} cm^2/W (Quail and Simon 1985). The coefficient is predicted to be a squared function of the pumping-wave frequency (Mills 1977), therefore the second-harmonic intensity is expected to drop by many orders of magnitude, going from the optic range to the millimeter waves. The experimental result is just opposite: the resonance value of the transformation coefficient in the n-type InSb $\rho_2 = 1.5 \times 10^{-10}$ cm^2/W exceeds the optical value by 8 orders of magnitude. We have, seemingly, a completely different nonlinearity source.

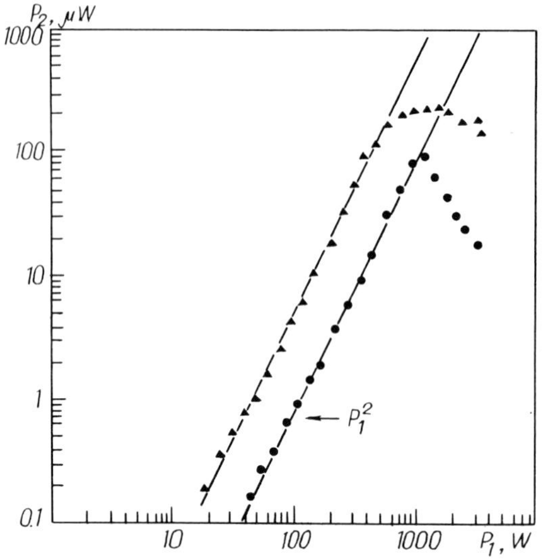

Fig. 11. The second-harmonic radiation power at normal transmission (points) and in the case of frustrated total reflection (triangles) for the same InSb sample (Ambrazevičiene et al. 1987).

Let us evaluate the second-harmonic source on the basis of the hydrodynamic approximation. Retaining the second-order terms in the hydrodynamic equation, we obtain several different contributions,

$$\begin{aligned} \boldsymbol{S}_2 = {} & \mu_0\varepsilon_0(e/m^*)(\nabla\cdot\hat{\varepsilon}\boldsymbol{E}_1)(\boldsymbol{E}_1+\boldsymbol{v}_1\times\boldsymbol{B}_0-\nu m^*\boldsymbol{v}_1/e) \\ & +\mu_0(n_0e^2/m^*)\boldsymbol{v}_1\times\boldsymbol{B}_1-\mu_0n_0e(\boldsymbol{v}_1\nabla)\boldsymbol{v}_1 \\ & -\mu_0\varepsilon_0(e/m^*)D\{\boldsymbol{B}_1\times\nabla(\nabla\cdot\hat{\varepsilon}\boldsymbol{E}_1)\}-\mu_0n_0(e/m^*)k_B\nabla T_e, \end{aligned} \tag{21}$$

where the tensor ε_{ij} is determined by eqs. (6) and (7) at frequency ω_1, and the parameters m^* and ν defined in preceding sections, as well as the diffusion coefficient D, are considered to be scalar. The spatial and temporal variations of these parameters give the third- and the higher odd-order harmonics. The diffusion term turns out to be small, both for the third- and the second harmonics. The same is said for the last term in eq. (21), which originates from the gradient of kinetic pressure and contains the electron temperature T_e. The contributions of different mechanisms were evaluated using typical experimental field values and parameters of the n-type InSb: $E_1 = 10^4$ V/m, $v_1 = 3\times10^5$ m/s, $B_1 = 10^{-4}$ T, $n_0 = 10^{20}$ m^{-3}, $\nu = 3\times10^{11}$ s^{-1}, $\omega_1 = 2.3\times10^{11}$ Hz, $\nabla \simeq \boldsymbol{k} = 2\times10^3$ m^{-1}, $\varepsilon_0 = -50$ (at $B_0 \simeq 0.25$ T). The main contribution of the order of 1.3×10^{10} V/m^2 comes from the terms containing the products $(\nabla\cdot\hat{\varepsilon}\boldsymbol{E}_1)\boldsymbol{v}_1\times\boldsymbol{B}_0$ and $\boldsymbol{E}_1\times\boldsymbol{B}_1$, as well as the convective term $(\boldsymbol{v}_1\nabla)\boldsymbol{v}_1$. The gradient term $(\nabla\cdot\hat{\varepsilon}\boldsymbol{E}_1)\boldsymbol{E}_1$ and the resistive one containing the collision frequency are lower by an order of magnitude. Experiments show dramatic magnetic-field influence on the second-harmonic intensity drawing our attention to the field-dependent sources like that containing the product $\boldsymbol{v}_1 \times \boldsymbol{B}_0$. It may be considered as a specific source governed by the magnetic field. On the other hand, the DC magnetic field acts indirectly on all the nonlinear sources through the field dependence of $\boldsymbol{v}_1$ and $\hat{\varepsilon}$. Generally, these are the effects of the magnetic-field-induced gyrotropy.

Considering the second-harmonic sources we did not take into account the energy dependence of the carrier effective mass. The role of this source of nonlinearity may be evaluated quite easily using the model of susceptibility modulation in the presence of the pumping-wave magnetic field (Buchsbaum 1964, Bate and Wisseman 1969). The second-harmonic source, defined by the second derivative of the polarization, $\partial^2 \boldsymbol{P}/\partial t^2$, may be evaluated for the case of parallel fields $\boldsymbol{B}_0 \parallel \boldsymbol{B}_1$ as

$$S_{2i} = -3\mu_0\varepsilon_0\omega_1\left\{(e/m_0^*) + eB_0\left(\frac{\partial\langle 1/m\rangle}{\partial B_0}\right)\right\}\frac{\partial\chi_{ij}}{\partial\omega_c E_{1j}}, \tag{22}$$

where $\langle 1/m\rangle$ is the mean (over an electron ensemble) inverse cyclotron effective mass. It is expected to depend on the magnetic field when an electric field $\boldsymbol{E}_1 \perp \boldsymbol{B}_0$ is present.

The nonparabolic-band contribution to the second-harmonic source described by the second term in eq. (22), to the authors' knowledge, have not been considered in literature. The second harmonic observed experimentally in narrow-gap semiconductors (InSb, $Hg_{1-x}Cd_xTe$) was much more intense than that in wide-gap material (GaAs). Therefore, the above specific source related to the nonparabolic bands seems to deserve theoretical studies.

Summing up the experimental results on harmonic radiation in nondegenerate semiconductors, it is instructive to list some general features:

(i) the harmonics are caused by the free-carrier plasma;

(ii) the second harmonic arises in DC magnetic fields without a DC electric field;

(iii) the surface magnetoplasma-type polariton excitation by the current harmonics results in the enhancement of radiation intensity by many orders of magnitude.

The observed electromagnetic-wave generation brings about the possibility of another interesting phenomenon, which is wave amplification in a similar experimental set-up. Let us consider the proper experimental results.

3.4. *Electromagnetic-wave amplification*

Consider the small signal of frequency ω_s entering the prism at the same angle as the strong pumping wave of lower frequency $\omega_1 < \omega_s$. Let both waves be

p-polarized. As a result of the wave interaction in the nonlinear semiconductor layer placed below the ground face of the prism, the reflected small-signal wave is expected to be amplified. The amplification has been observed for a signal frequency $\omega_s = 2\omega_1$ and for incommensurate frequencies (Ambrazevičiene et al. 1984a, b).

In order to get the double-frequency signal, a part of the pumping-wave power (50–70 mW) was directed to the side shoulder of the waveguide set-up (fig. 12, inset). The frequency-doubling Gunn diode fabricated from the epitaxial GaAs was placed in the transmitting cavity belonging to this shoulder.

In the absence of the pumping wave, the resonant attenuation of the small signal was observed (fig. 12), the resonance position being in agreement with the calculated one for the surface magnetoplasma polaritons. The FTR line was found to transform with the rise of the pumping-wave power, bringing about the FTR *maximum*.

The amplification was observed for the pumping-wave power exceeding a certain threshold value. For InSb crystals this was $P_1 > 20$ W. A moderate threshold overflow ($P_1 \simeq 50$ W) resulted in the amplification confined to a narrow range of magnetic fields corresponding to the fundamental-frequency surface magnetoplasma mode resonance. At higher pumping-power levels the

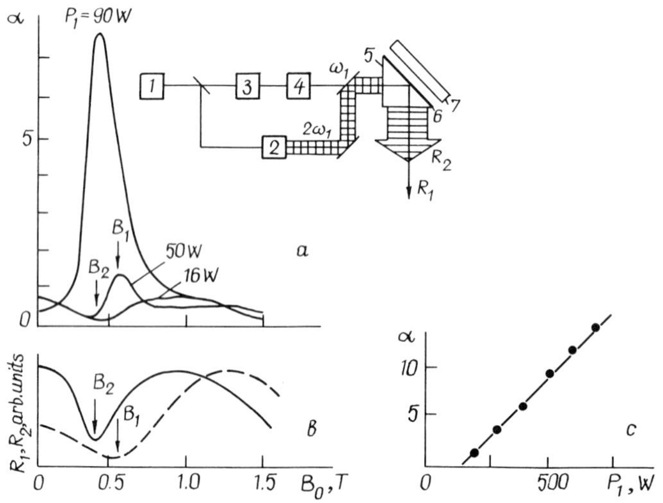

Fig. 12. Magnetic-field dependences of the small double-frequency signal amplification coefficient at various pumping power levels (a), the FTR line for the same signal in the absence of pumping (b, solid line), the FTR line for the pumping wave at $P_1 = 160$ W (b, dashed line), and the pumping-power dependence of the amplification coefficient at the point B_2 (c). The calculated positions of FTR minima are shown by the arrows. The inset shows the experimental lay-out including: magnetron (1) ($\omega_1 = 2.3 \times 10^{11}$ Hz); second-harmonic generator (2); phase shifter (3); attenuator (4); prism (5) ($\varepsilon_5 = 12.34$); dielectric (6) ($\varepsilon_6 = 2.07$, $d_6 = 50$ μm); and the InSb crystal (7) ($n_0 = 3.5 \times 10^{20}$ m^{-3}, $d_7 = 500$ μm), and $T = 77$ K.

maximum of the amplification coefficient was found to grow and shift toward the magnetic-field value B_2 corresponding to the signal-frequency surface-mode excitation. After the maximum positioning at B_2, the amplification coefficient was found to increase linearly with the rise of the pumping power (fig. 12).

The amplification coefficient α turned out to be dependent on the pumping-wave phase (fig. 13). The dependence of α on B_0 at a given phase shift $\Delta\phi$ repeated those obtained at $\Delta\phi + N\pi$, where N is an integer. Changing the phase shift $\Delta\phi$, the absolute maximum of amplification coefficient was found to exist at the magnetic field corresponding to the signal-frequency surface-mode excitation (fig. 13, curve 1).

In the case of the n-type GaAs samples, the amplification was observed at a higher threshold ($P_1 > 300$ W), as compared to the case of indium antimonide. The pumping-wave amplitude evaluated with the use of expressions (12)–(14) was found to be insufficient for the intervalley transfer of electrons.

The amplification in the case of the n-type GaAs exhibited the same main features as in the case of indium antimonide, retaining the resonant character of the dependence of α on B_0, and the phase-shift dependence of the type $\alpha(\Delta\phi) = \alpha(\Delta\phi + N\pi)$ (fig. 14).

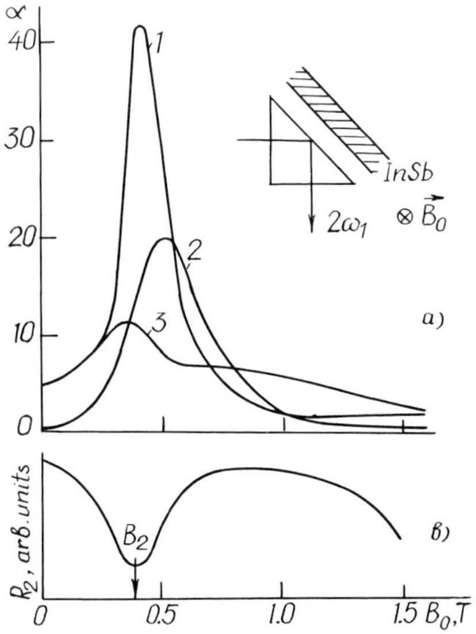

Fig. 13. Experimental magnetic-field dependences of the amplification coefficient in the FTR configuration at various pumping-wave phase shifts (a) and the FTR line for the same signal in the absence of pumping (b). The calculated minimum position is shown by the arrow B_2. (1) $\Delta\phi = 0°$, (2) 70°, (3) 140°. $P_1 = 2$ kW, $\omega_1 = 2.3 \times 10^{11}$ Hz, $\omega_s = 2\omega_1$, $\varepsilon_p = 12.34$ (prism), $\varepsilon_d = 2.07$, $d_d = 50$ μm (prism-sample spacer), $n_0 = 3.5 \times 10^{20}$ m^{-3}, $d = 500$ μm (InSb crystal), and $T = 77$ K (Ambraževičiene et al. 1984b).

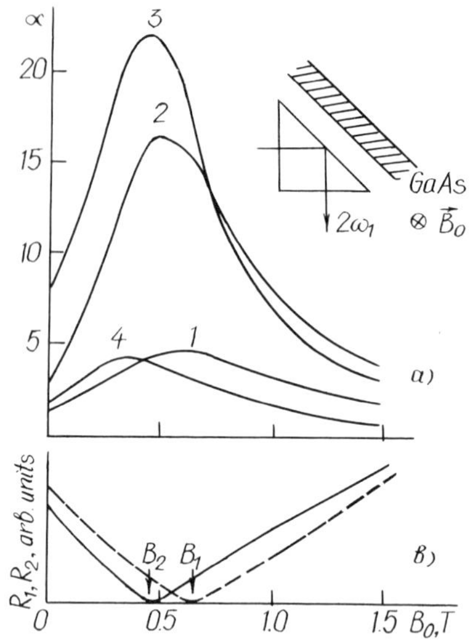

Fig. 14. Experimental magnetic-field dependences of the amplification coefficient for the n-type GaAs at various pumping-wave phase shifts (a), the FTR line for the same signal in the absence of pumping (b), the FTR line for the pumping wave (dashed line). The pumping-wave frequency $\omega_1 = 2.3 \times 10^{11}$ Hz and the signal frequency $\omega_s = 2\omega_1$. Calculated FTR-minima positions are shown by the arrows $B_{1,2}$. 169 μm-thick GaAs layer with electron density $n_0 = 3.2 \times 10^{20}\ \mathrm{m}^{-3}$ on 350 μm-thick insulating GaAs substrate is used. The values of $\Delta\phi$ are: (1) 0; (2) 22°; (3) 45°; (4) 140° (Ambrazevičiene et al. 1984a).

Consider now briefly the results on the amplification of the incommensurate frequencies. The signal radiated by the magnetron contained the high-intensity fundamental component ($\lambda = 8$ mm) and the complementary waves of weak intensity in the range of $\lambda = 6.6$ mm which were usually filtered. However, if the complementary waves were directed to the prism along with the fundamental wave, their power at the prism output was found to increase as a result of the wave interaction in the nonlinear sample. The maximum amplification coefficient reached 25 dB for the case of n-type GaAs. The maximum gain was observed at the magnetic field corresponding to the signal-frequency surface magnetoplasma polariton excitation.

Summing up the experimental results on the wave amplification, the main peculiarities of the phenomenon are worth noting:

(i) the small-amplitude high-frequency signal amplification arises on account of the lower frequency high-power wave in FTR experiments with the nondegenerate semiconductor as a nonlinear element;

(ii) the amplification coefficient depends on the phase shift $\Delta\phi$ between the pump- and the signal waves obeying the rule $\alpha(\Delta\phi) = \alpha(\Delta\phi + N\pi)$, with the maximum of α at a certain phase-shift value;

(iii) the absolute maximum of $\alpha(B_0)$ arises when the condition of the signal-frequency surface polariton excitation is fulfilled.

4. Conclusion

The principal result of the present work is the experimental observation of the electromagnetic-wave self-action and their radiation and amplification related to the surface magnetoplasma polaritons. The role of the external magnetic field is crucial both in the resonance-system tuning and in controlling the dominating nonlinearity source. In the most general meaning the magnetic-field action results in the *gyrotropy* of the semiconducting medium, therefore, one can expect the above nonlinear effects to arise in other gyrotropic media, such as optically active solids or liquid crystals.

Returning to the nonlinear surface plasmon-polaritons in solids, we would like to note that the optical experiments on metals and the present millimeter-wave experiments on semiconductors mark two limiting points of the spectral region which is completely unexplored in the discussed subject. The far-infrared laser experiments carried out recently on bulk nonlinearities (Mayer and Keilmann 1986) await for the extension to surface modes. The experiments, widely spanned over the frequency range, would bring about the decisive selection between the dominating nonlinearities. Such a selection requires also theoretical studies devoted, particularly, to the revealing of the nonparabolic-band effects in the nonlinearities. Concerning the aspects of practical applications one can conclude that the achieved transformation- and amplification coefficients are insufficient to be applied in the secondary sources of radiation for the communication systems. Nevertheless, the resonance maxima of the harmonic radiation intensity closely related to the magnetoactive plasma parameters may be used just now in the nonequilibrium electron system diagnostics.

References

Abeles, F., and T. Lopez-Rios, 1982, in: Surface Polaritons (North-Holland, Amsterdam) p. 241.

Agranovich, V.M., and V.Ya. Chernyak, 1982, Phys. Lett. A **88**, 423.

Agranovich, V.M., V.I. Rupasov and V.Ya. Chernyak, 1981, Pis'ma Zh. Eksp. & Teor. Fiz. **33**, 196.

Akcipetrov, O.A., E.N. Dubinina, S.S. Elovikov, E.D. Mishina, A.A. Nikulin, N.N. Novikova and M.S. Strebkov, 1988, Pis'ma Zh. Eksp. & Teor. Fiz. **48**, 92.

Akhmanov, S.A., and R.V. Khokhlov, 1964, Problemy Nelineinoi Optiki, Moskva, VINITI Akad. Nauk SSSR, p. 295.

Altshuler, E.Yu., L.I. Kats and V.V. Popov, 1983, Obzory po elektronnoi tekhnike, Ser. 1 Elektronika SVCh **7/940**, 60.

Ambrazevičienė, V., and R. Brazis, 1976, Solid State Commun. **18**, 415.

Ambrazevičienė, V., and R. Brazis, 1978a, in: Fizika soedinenii A^3B^5 (Leningrad) p. 71.

Ambrazevičienė, V., and R. Brazis, 1978b, Fiz. Tekh. Poluprovodn. **12**, 1114.

Ambrazevičienė, V., and R. Brazis, 1981, in: Tekhnologiya, Polucheniye i Elektricheskiye Svoistva Soedinenii A^3B^5, Leningrad: LPI, p. 68.
Ambrazevičienė, V., R. Brazis and A. Kunigėlis, 1979, LFR **19**, 401.
Ambrazevičienė, V., R. Brazis, A. Kunigėlis and P. Shiktorov, 1980, in: Plasma i Tokovye Neustoichivosti v Poluprovodnikakh, Vilnius, p. 19.
Ambrazevičienė, V., R. Brazis, A. Kunigėlis and P. Shiktorov, 1982, Fiz. Tekh. Poluprovodn. **16**, 1502.
Ambrazevičienė, V., R. Brazis and J. Paršeliūnas, 1984a, Fiz. Tekh. Poluprovodn. **18**, 741.
Ambrazevičienė, V., R. Brazis and A. Kunigėlis, 1984b, in: 10th Int. Conf. on Infrared and MM Waves, Takarazuka, Japan, p. 532.
Ambrazevičienė, V., R. Brazis and A. Kunigėlis, 1986, Fiz. Tekh. Poluprovodn. **20**, 152.
Ambrazevičienė, V., R. Brazis and A. Kunigėlis, 1987a, Pis'ma Zh. Tekhn. Fiz. **13**, 816.
Ambrazevičienė, V., R. Brazis and A. Kunigėlis, 1987b, Int. J. Infrared & MM Waves **8**, 373.
Barker Jr, A.S., 1964, Phys. Rev. **136**, A1290.
Bate, R.T., and W.R. Wisseman, 1969, Phys. Rev. **181**, 763.
Belyantsev, A.M., V.A. Kozlov and B.A. Trifonov, 1971, Phys. Status Solidi b **48**, 581.
Bhat, B., and V.K. Tripathi, 1975, J. Appl. Phys. **46**, 1141.
Bloembergen, N., 1965, Nonlinear optics (Benjamin, New York).
Bonsall, L., and A.A. Maradudin, 1978, J. Appl. Phys. **49**, 253.
Brazis, R., 1981a, Litov. Fiz. Sb. **21**, 73.
Brazis, R., 1981b, Fiz. Tekh. Poluprovodn. **15**, 1192.
Brazis, R., and A. Mironas, 1977, Fiz. Tekh. Poluprovodn. **11**, 795.
Brazis, R., and A. Mironas, 1983, Fiz. Tekh. Poluprovodn. **17**, 720.
Brazis, R., and P. Shiktorov, 1980, in: 10 soveshchanye po teorii poluprovodnikov (abstracts), Novosibirsk: IFP SO AN SSSR **1**, 88.
Brazis, R., A. Mironas and A. Filipavichyus, 1980, Fiz. Tekh. Poluprovodn. **14**, 1420.
Brazis, R., E. Starikov and P. Shiktorov, 1983a, Fiz. Tekh. Poluprovodn. **17**, 40.
Brazis, R., E. Starikov and P. Shiktorov, 1983b, Fiz. Tekh. Poluprovodn. **17**, 13.
Bryksin, V.V., D.N. Mirlin and Yu.A. Firsov, 1974, Usp. Fiz. Nauk **113**, 29.
Buchsbaum, S.J., 1965, in: Proc. Symp. Plasma Effects in Solids, 1964, Paris, Dunod, p. 3.
Cardona, M., 1971, Am. J. Phys. **39**, 1277.
Chaplik, A.V., and M.V. Krasheninnikov, 1978, Solid State Commun. **27**, 1516.
Chen, Y.J., and E. Burstein, 1977, Nuovo Cimento B **39**, 807.
Chen, Y.J., and G.M. Carter, 1982, Appl. Phys. Lett. **41**, 307.
Chiu, K.W., and J.J. Quinn, 1972, Nuovo Cimento B **10**, 1.
Davydov, A.B., V.A. Zakharov and I.E. Podchinyonov, 1977, Fiz. Tverd. Tela **19**, 1676.
Deck, R.T., and D. Sarid, 1982, J. Opt. Soc. Am. **72**, 1613.
Dienys, V., and A. Dargys, 1981, J. Phys. Colloq. (France) **42**(10), C7-33.
Dmitruk, N.L., V.G. Litovchenko and T.N. Gorbunova, 1979, Ukr. Fiz. Zh. **24**, 1516.
Fano, U., 1941, Opt. Soc. Am. **31**, 213.
Ferrell, R.A., 1958, Phys. Rev. **111**, 1214.
Fischer, B., N. Marschall and H.J. Queisser, 1973, Surf. Sci. **34**, 50.
Gintsburg, M.A., 1958, Zh. Eksp. & Teor. Fiz. **34**, 1635.
Ginzburg, V.L., 1981, Usp. Fiz. Nauk **134**, 469.
Gorobei, N.N., I.P. Ipatova and A.V. Subashiev, 1981, Pis'ma Zh. Eksp. & Teor. Fiz. **34**, 157.
Gunn, J.B., 1963, Solid State Commun. **1**, 88.
Halevi, P., and C. Guerra-Vela, 1978, Phys. Rev. B **18**, 5248.
Hilsum, C., 1962, Proc. IRE **50**, 185.
Hopfield, J.J., 1966, J. Phys. Soc. Jpn. **21**, Suppl. 77.
Innes, R.A., and J.R. Sambles, 1989, J. Phys. Condens. Matter **1**, 6231.
Ivanov, A.A., and D.D. Ryutov, 1965, Zh. Eksp. & Teor. Fiz. **48**, 1366.
Kaplan, H., E.D. Palik, R. Kaplan and R.W. Gammon, 1974, J. Opt. Soc. Am. **64**, 1551.

Khankina, S.I., and V.M. Yakovenko, 1967, Fiz. Tverd. Tela **9**, 2943.
Kheifets, M.I., 1965, Fiz. Tverd. Tela **7**, 3485.
Kliewer, K.L., and R. Fuchs, 1966a, Phys. Rev. **144**, 495.
Kliewer, K.L., and R. Fuchs, 1966b, Phys. Rev. **150**, 573.
Kretschmann, E., 1971, Zt. Phys. **241**, 313.
Kurosawa, T., 1980, in: Proc. 15th Int. Conf. Physics of Semiconductors, Kyoto, p. 345.
Landau, L.D., and E.M. Lifshitz, 1960, Electrodynamics of Continuous Media (Addison-Wesley, Reading, Mass.).
Lax, B., W. Zawadzki and M.H. Weiler, 1967, Phys. Rev. Lett. **18**, 462.
Leung, K.M., 1985, Phys. Rev. B **32**, 5093.
Levinson, I.B., 1983, Usp. Fiz. Nauk **139**, 347.
Lifshitz, I.M., and L.N. Rosenzweig, 1948, Zh. Eksp. & Teor. Fiz. **18**, 1012.
Martin, B.G., A.A. Maradudin and R.F. Wallis, 1978, Surf. Sci. **77**, 416.
Mayer, A., and F. Keilmann, 1986, Phys. Rev. B **33**, 6954.
Mills, D.L., 1977, Solid State Commun. **24**, 669.
Mills, D.L., and E. Burstein, 1974, Rep. Prog. Phys. **37**, 817.
Nurmikko, A.V., D.M. Bolle and S. Talisa, 1980, Int. J. Infrar. & MM Waves **1**, 3.
Obunai, T., and T. Sekiguchi, 1974, Jpn. J. Appl. Phys. **13**, 93.
Otto, A., 1968, Z. Phys. **216**, 398.
Otto, A., 1974, Festkörperprobleme **14**, 1.
Pakhomov, V.I., and K.N. Stepanov, 1967, Zh. Eksp. & Teor. Fiz. **37**, 1393.
Pekar, S.I., 1957, Zh. Eksp. & Teor. Fiz. **33**, 1022.
Pishneckij, N.F., A.N. Sudarkin and V.V. Jakimenko, 1989, Poverkhn. Fiz. Khim. Mekh. **1**, 88.
Polaritons, 1974, Proc. 1st Taormina Conf. on Structural Matter, eds E. Burstein and F. de Martini (Pergamon Press, New York) p. 389.
Popov, V.V., R.M. Revzin and M.A. Safonova, 1983, Radiotekh. & Elektron. **28**, 2008.
Pozhela, Yu.K., 1981, Plasma and Current Instabilities in Semiconductors (Pergamon Press, Oxford) p. 301.
Pozhela, Yu.K., and A. Reklaitis, 1979, Fiz. Tekh. Poluprovodn. **13**, 1127.
Quail, J.C., and H.J. Simon, 1985, Phys. Rev. B **31**, 4900.
Rees, H.D., 1969, IBM J. Res. Dev. **13**, 537.
Reinisch, R., and M. Neviere, 1981, Phys. Rev. B **24**, 4392.
Ridley, B.K., and T.B. Watkins, 1961, Proc. R. Soc. (London) **78**, 293.
Ritchie, R.H., 1957, Phys. Rev. **106**, 874.
Ritchie, R.H., 1973, Surf. Sci. **34**, 1.
Seeger, K., 1973, Semiconductor Physics (Springer, New York).
Simon, H.J., D.E. Mitchell and J.G. Watson, 1974, Phys. Rev. Lett. **33**, 1531.
Simon, H.J., R.E. Benner and J.G. Rako, 1977, Opt. Commun. **23**, 245.
Sipe, J.E., V.C.Y. Fukui and G.I. Stegeman, 1980, Solid State Commun. **34**, 523.
Sipe, J.E., V.C.Y. So, M. Fukui and G.I. Stegeman, 1986, Phys. Rev. B **21**, 4389.
Surface Polaritons, 1982, (North-Holland, Amsterdam).
Tarkhanian, R.G., 1975, Phys. Status Solidi b **72**, 111.
Tolpygo, K.B., 1950, Zh. Eksp. & Teor. Fiz. **20**, 497.
Tyler, I.L., B. Fisher and R.J. Bell, 1973, Opt. Commun. **8**, 145.
Wallis, R.F., J.J. Brion, E. Burstein and A. Hartstein, 1974, Phys. Rev. B **9**, 3424.
Weber, M.G., 1986, Phys. Rev. B **33**, 6775.
Wood, R.W., 1935, Phys. Rev. **48**, 488.
Zhizhin, G.N., M.A. Moskalyova, E.V. Shomina and V.A. Yakovlev, 1979, Pis'ma Zh. Eksp. & Teor. Fiz. **29**, 533.
Zhizhin, G.N., E.A. Vinogradov, M.A. Moskalyova and V.A. Yakovlev, 1982/1983, Appl. Spectr. Rev. **18**, 171.

CHAPTER 7

Light Scattering from Phonons and Spin Waves on Surfaces

F. NIZZOLI

Department of Mathematics and Physics
and Consorzio Interuniversitario Fisica della Materia
University of Camerino
62032 Camerino, Italy

and

D.L. MILLS

Department of Physics
University of California
Irvine, CA 92717, USA

Nonlinear Surface Electromagnetic Phenomena
Edited by
H.-E. Ponath and G.I. Stegeman

© *Elsevier Science Publishers B.V., 1991*

Contents

1. Introductory remarks

A central issue in the physics of solids is the elucidation and understanding of the diverse array of collective excitations encountered in various materials. An important and powerful probe of these modes is light scattering spectroscopy, which has as its basis the nonlinear interaction between a laser photon and the collective excitation of interest.

Consider, for example, a thermally excited phonon described by a normal coordinate we denote schematically by $Q(\boldsymbol{x}, t) = Q_0 \exp(\mathrm{i}\boldsymbol{q}\cdot\boldsymbol{x} - \mathrm{i}\omega t)$, with $\boldsymbol{q}$ and ω the wave vector and frequency of the mode. As the wave propagates through the solid, it modulates the dielectric tensor $\varepsilon_{\alpha\beta}$ so in its presence we have

$$\varepsilon_{\alpha\beta}(\boldsymbol{x}, t) = \varepsilon_{\alpha\beta} + \left(\frac{\partial \varepsilon_{\alpha\beta}}{\partial Q}\right) Q(\boldsymbol{x}, t). \tag{1.1}$$

The dielectric tensor thus fluctuates in space and time, with the result that the light is scattered inelastically from the dynamic diffraction grating created by the phonon. If the incident photon has wave vector $\boldsymbol{k}_{\mathrm{I}}$ and frequency Ω_{I}, we obtain scattered light with the frequencies $\Omega_{\mathrm{S}} = \Omega_{\mathrm{I}} \pm \omega$, and wave vector $\boldsymbol{k}_{\mathrm{S}} = \boldsymbol{k}_{\mathrm{I}} \pm \boldsymbol{q}$. From the frequency spectrum and propagation direction of the scattered light, we obtain both the frequency and wave vector of the phonon. The scattering process is a nonlinear mixing between the incident photon and the thermally excited phonon, which produces the scattered photon as the output.

Any collective excitation which modulates the dielectric tensor of a solid in the manner just described leads to inelastic light scattering, and as a consequence a wide variety of elementary excitations can be studied with the method. These include both optical and acoustical phonons in metals, semiconductors or insulators, spin waves in magnetically ordered materials, and plasmons in semiconductors. An excellent introduction to the field is found in the proceedings of a conference held a number of years ago (Wright 1969) on the light scattering spectra of solids.

There is a restriction on the range of wave vectors that may be probed by the method. Note that the magnitude of the wave vector $\boldsymbol{q}$ of the mode that participates in the scattering process is $|\boldsymbol{k}_{\mathrm{S}} - \boldsymbol{k}_{\mathrm{I}}|$. Since both the incident and scattered photons lie in the visible frequency range in a typical experiment, $\boldsymbol{k}_{\mathrm{S}}$ and $\boldsymbol{k}_{\mathrm{I}}$ each has a magnitude the order of $10^5\ \mathrm{cm}^{-1}$, and so does $\boldsymbol{q}$. Thus, the method probes modes with wavelength very long compared to a lattice constant, or in the language of solid state physics it probes only the near vicinity of the center of the Brillouin zone.

If we return to the example of phonons, at such small wave vectors, there are two distinct classes of phonons in a typical crystal. One has acoustical phonons whose frequency is proportional to $|\boldsymbol{q}|$; the frequency of such an acoustical phonon lies in the microwave regime. The very small frequency shift of the light is measured through use of a Fabry–Pérot interferometer, and light scattering spectroscopy carried out by this means is referred to as Brillouin spectroscopy. One also has optical phonons whose frequency is insensitive to $|\boldsymbol{q}|$ (but dependent on the direction of $\boldsymbol{q}$ in anisotropic crystals), when $|\boldsymbol{q}| \sim 10^5\ \mathrm{cm}^{-1}$. These modes have frequencies in the range of several hundred wave numbers, and a grating spectrometer is used to measure the frequency of the scattered light. This is the method referred to as Raman spectroscopy. In ferromagnetic crystals, the spin-wave frequencies explored in the light scattering lie in the Brillouin regime, while in typical antiferromagnets or ferrimagnets, there are modes in the Raman regime.

In recent years, the light scattering method has been used to study both phonons and spin waves in very thin films, and in multilayer structures such as superlattices. This chapter is devoted to a discussion of this exciting new application of light scattering spectroscopy.

There are several reasons why the method is particularly useful. First of all, in very thin films and multilayer structures, as we shall see one encounters new classes of collective modes not present in the infinitely extended, homogeneous crystal. We know very well that the response of a solid to a variety of external probes is controlled by the nature of its phonon spectrum, or its spin-wave spectrum if it is a magnetic crystal. In films and multilayer structures, the dispersion relations and properties of the collective modes are influenced critically by the dimensions of the constituents, and are thus subject to design and control. Light scattering spectroscopy offers us a means of directly measuring the dispersion relations of the various modes; useful information is contained also in the relative intensities of the various features in the spectrum. A number of the modes probed by this technique are accessible by no other experimental method at present. Quantitative comparison between measured mode properties and calculations based on idealized theoretical models is a means of assessing how close an actual multilayer structure is to the theoretical models with their perfectly sharp interfaces.

It is also the case that the physical constants which characterize a very thin film or multilayer structure are difficult to measure. For example, it is a straightforward matter to measure the elastic constants of a macroscopic crystal or solid material. This is quite impossible to do by conventional methods for a very thin film grown on a substrate. As we shall see, Brillouin spectroscopy allows one to measure phonons that are standing-wave resonances of the thin film. From the frequencies of such modes, accurate data on elastic constants may be obtained. On superlattices, one may probe Rayleigh waves (surface acoustic phonons) whose displacement field penetrates many layers of the structure. The frequencies of such modes provide average elastic constants for the superlattice. In ultra-thin films, spin waves are studied by the method. Here one may extract values for the magnetization in the film, and on anisotropy energies operative in the film that are absent in the bulk crystal.

Thus, light scattering spectroscopy has proved a most powerful means of elucidating the properties of the collective modes of thin films and multilayer structures, and it also provides us with quantitative data on important parameters which describe these new materials. We turn next to a discussion of general aspects of the technique, as applied to such structures, and then to examples of both experiments and theoretical calculations.

2. General comments on light scattering from surfaces

This section is devoted to general comments on light scattering from finite samples, where the nonlinear interaction responsible for the scattering event takes place near a surface or interface, or within a thin film. A discussion of the qualitative features present in the spectrum requires considerations somewhat more complex than the case touched upon in section 1, where for light scattering in infinitely extended, transparent media we have both conservation of energy ($\Omega_S = \Omega_I \pm \omega$) and all components of wave vector ($\boldsymbol{q} = \pm(\boldsymbol{k}_S - \boldsymbol{k}_I)$). Also, the nature of the excitations themselves are more complex, near surfaces or interfaces and within films.

Consider first the scattering of light from a semi-infinite sample. The experiment is illustrated schematically in fig. 1. We have a plane wave incident on the sample from the outside. If the surface is perfectly smooth, and the material quiescent, we have specularly reflected light whose frequency equals that of the incident light. Thermally excited phonons or other excitations modulate the dielectric tensor as in eq. (1.1), and as a consequence we have light scattered inelastically from the specular direction.

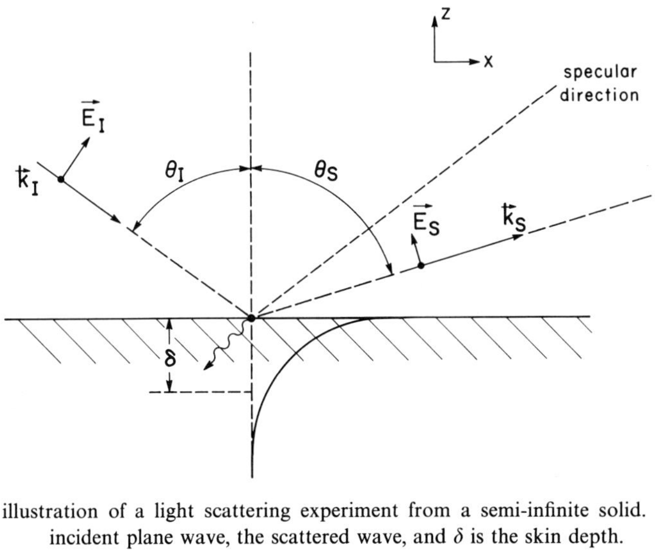

Fig. 1. An illustration of a light scattering experiment from a semi-infinite solid. We have the incident plane wave, the scattered wave, and δ is the skin depth.

In many of the experiments, the material absorbs the incident light strongly, resulting in a skin depth δ, as indicated in the figure. In a typical metal, in the visible range of frequencies, $\delta \cong 150$ Å, a length much shorter than the wavelength of light, or the wavelength of the thermal excitations discussed in section 1.

The theory of light scattering from opaque media was developed some years ago (Mills et al. 1970). A key element in the analysis is the matrix element $M(\mathrm{IS})$ between the incident and scattered photon states. In the medium, each light wave in the interaction is described by a wave vector whose component parallel to the surface $\boldsymbol{k}_{\|}$ necessarily equals that of the wave outside the crystal and whose component normal to the surface $k_{\perp}$ is complex, with a real part $k_{\perp}^{1}$ and imaginary part $k_{\perp}^{2}$ equal to $\frac{1}{2}\delta$, with δ the skin depth. An essential ingredient in $M(\mathrm{IS})$ is then the integral

$$I = \frac{1}{(2\pi)^3} \int \mathrm{d}^2 x_{\|}\ \mathrm{d}z\ \mathrm{e}^{\mathrm{i}(\boldsymbol{k}_{\|\mathrm{I}} - \boldsymbol{k}_{\|\mathrm{S}}) \cdot \boldsymbol{x}_{\|}}\ \mathrm{e}^{\mathrm{i}(k_{\perp\mathrm{I}}^{1} - k_{\perp\mathrm{S}}^{1})z} \times \mathrm{e}^{-\frac{1}{2}\left(\frac{1}{\delta_{\mathrm{I}}} + \frac{1}{\delta_{\mathrm{S}}}\right)z} Q(\boldsymbol{x}, t), \tag{2.1}$$

where z is normal to the surface, and the superscript $\|$ designates components of vectors parallel to the plane of the surface. If $Q(\boldsymbol{x}, t)$ has the spatial variation $\exp(\mathrm{i}\boldsymbol{q} \cdot \boldsymbol{x})$ of a plane wave bulk excitation, the absorption lengths δ_{I} and δ_{S} of the incident and scattered wave are assumed infinite, and the integration on z extends over all space, eq. (2.1) imposes wave vector conservation $\boldsymbol{q} = \boldsymbol{k}_{\mathrm{S}} - \boldsymbol{k}_{\mathrm{I}}$ on the scattering process.

In the situation of present interest, we need to reexamine the structure of the contribution to the matrix element in eq. (2.1).

First consider scattering by a surface mode, such as a Rayleigh surface phonon, or a surface spin wave. For such modes, $Q(\boldsymbol{x}, t)$ exhibits a wavelike variation in the two directions parallel to the surface, so $Q(\boldsymbol{x}, t)$ is proportional to $\exp(\mathrm{i}\boldsymbol{q}_{\|} \cdot \boldsymbol{x}_{\|})$, with $\boldsymbol{q}_{\|}$ the wave vector of the surface wave. In the direction normal to the surface, the amplitude of the excitation decays to zero exponentially. We then have, in general (Landau and Lifshitz 1959, Wallis 1973):

$$Q(\boldsymbol{x}, t) = \exp(\mathrm{i}\boldsymbol{q}_{\|} \cdot \boldsymbol{x}_{\|}) \sum_i Q_i \exp(-\alpha_i z) \mathrm{e}^{-\mathrm{i}\omega_{\mathrm{S}} t}, \tag{2.2}$$

where the one or more decay constants α_i has a magnitude comparable to $q_{\|}$. Then we find, with $\Delta k_{\perp} = k_{\perp\mathrm{I}}^{1} - k_{\perp\mathrm{S}}^{1}$

$$I = \frac{1}{2\pi} \delta(\boldsymbol{k}_{\|\mathrm{S}} - \boldsymbol{k}_{\|\mathrm{I}} - \boldsymbol{q}_{\|}) \sum_i \frac{Q_i}{\alpha_i + \frac{1}{2}\left(\frac{1}{\delta_{\mathrm{I}}} + \frac{1}{\delta_{\mathrm{S}}}\right) + \mathrm{i}\Delta k_{\perp}}, \tag{2.3}$$

which requires wave vector components parallel to the surface are conserved in the scattering process. When this is combined with energy conservation, $\Omega_{\mathrm{S}} = \Omega_{\mathrm{I}} + \omega_{\mathrm{S}}$, with ω_{S} the frequency of the surface wave, the direction of the scattered photon is uniquely determined. In almost all cases of interest to us, the frequency of the phonon or spin wave can be ignored compared to Ω_S and Ω_I when the scattering angle θ_S in fig. 1 is determined, so the direction of the scattered photon is controlled only by the magnitude of the wave vector transfer in the interaction.

The light scattering spectrum contains contributions not only from surface waves, but also from bulk excitations that propagate up to the surface, and modulate the dielectric constant within the skin depth. If, for simplicity, we suppose we can view the bulk excitation as a simple plane wave, then eq. (2.2) becomes

$$Q(\boldsymbol{x}, t) = \exp(\mathrm{i}\boldsymbol{q}_{\parallel} \cdot \boldsymbol{x}_{\parallel}) Q_B \exp(\mathrm{i}q_z z) \mathrm{e}^{-\mathrm{i}\omega_B t}, \tag{2.4}$$

and eq. (2.3) is replaced by

$$I = \frac{1}{2\pi} \delta(\boldsymbol{k}_{\parallel S} - \boldsymbol{k}_{\parallel I} - \boldsymbol{q}_{\parallel}) \frac{Q_B}{\frac{1}{2}\left(\frac{1}{\delta_I} + \frac{1}{\delta_S}\right) + \mathrm{i}(\Delta k_{\perp} - q_z)}. \tag{2.5}$$

We again have conservation of wave vector components parallel to the surface, and it follows just as in the case of the surface waves, that the direction of the scattered photon is controlled by $\boldsymbol{q}_{\parallel}$, to excellent approximation. Hence, when one examines the spectrum of scattered light that emerges from the surface in a given direction, one sees *lines* produced by scattering from surface waves of the appropriate wave vector $\boldsymbol{q}_{\parallel}$, and in addition there are structures produced by scattering from those bulk and evanescent waves whose wave vector $\boldsymbol{q} = \boldsymbol{q}_{\parallel} + \hat{z} q_z$ has the same wave vector projection $\boldsymbol{q}_{\parallel}$ onto the plane of the surface. In principle, bulk modes with all possible values of q_z contribute to the spectrum, so the bulk modes contribute a feature that is a *continuous band*, referred to also as the 'mixed-modes' band in the literature on Brillouin scattering by phonons.

The shape of the bulk band is influenced importantly by the value of the skin depth. The cross-section is proportional to $|I|^2$, and thus contains the factor

$$f(q_z) = \frac{|Q_B|^2}{\frac{1}{4}\left(\frac{1}{\delta_I} + \frac{1}{\delta_S}\right)^2 + (\Delta k_{\perp} - q_z)^2}. \tag{2.6}$$

If the skin depth is very long compared to a wavelength, so $\delta_{I,s}|\Delta k_{\perp}| \gg 1$, this function is peaked sharply about the value $q_z = \Delta k_{\perp}$, the value of the z-component of wave vector that enters the wave vector conservation condition appropriate to the infinitely extended, transparent solid. In this limit, the bulk waves contribute a line to the loss spectrum, peaked about the frequency appropriate to the transparent solid. The line is broadened by the presence of absorption, which partially breaks down conservation of wave vector components normal to the surface; one refers to such a structure as an opacity-broadened line.

For metals, one is in the opposite limit $\delta_{\mathrm{I,S}}|\Delta k_\perp| \ll 1$, and one may ignore the factor $\Delta k_\perp$ in eq. (2.6) to obtain

$$f(q_z) \cong \frac{|Q_{\mathrm{B}}|^2}{\frac{1}{4}\left(\frac{1}{\delta_{\mathrm{I}}} + \frac{1}{\delta_{\mathrm{S}}}\right)^2 + q_z^2}. \tag{2.7}$$

The bulk mode feature is now a broad band, which receives contributions from modes with q_z in the range from 0 to $\delta_{\mathrm{I,S}}^{-1}$. When $q_z = 0$, we have bulk modes which propagate parallel to the surface, and when $q_z \sim \delta_{\mathrm{I,S}}^{-1}$ we have bulk modes which are propagating very nearly perpendicular to the surface.

Our discussion of the bulk mode contribution to the spectrum is highly schematic, of course. The modes cannot be viewed as simple plane waves, but in fact are to be regarded as linear combinations of a wave approaching the surface and all of the various reflected waves. The above conclusions remain valid, when the proper eigenvectors are employed, though a full discussion leads one into rather involved algebra.

If we have a thin film, rather than the semi-infinite solid discussed above, then all values of q_z are not allowed. Boundary conditions select out particular, discrete values of $q_z \cong n\pi/d$, where d is the film thickness and n is an integer. The bulk spectrum then becomes a sequence of lines rather than a continuum, with a contribution from the modes with $q_z \cong \pi/d$, $2\pi/d$, etc. This simple picture, valid for a single mode, can be generalized to more complicated situations where modes of different polarization are coupled: although the existence condition for q_z becomes complicated the discrete nature of the spectrum is preserved. Similar features are also exhibited by supported films, provided that the shear velocity of the film is smaller than the shear velocity of the substrate. A single light scattering spectrum thus provides one with information on the surface waves, and all of the bulk standing-wave resonances whose matrix element is sufficiently large. By varying the scattering angle θ_{S} and thus $\boldsymbol{q}_\parallel$, one may study the dependence of the frequency of each mode on $\boldsymbol{q}_\parallel$. These remarks enable one to appreciate the power of the light scattering technique.

In a superlattice, one encounters a rich spectrum of modes that are collective modes of the structure as a whole. Various possibilities are illustrated in fig. 2, in a schematic manner. In the subsequent section, we shall explore explicit cases in detail. In all cases illustrated, the wave field exhibits a variation with $\boldsymbol{x}_\parallel$ in the form $\exp(\mathrm{i}\boldsymbol{q}_\parallel \cdot \boldsymbol{x}_\parallel)$, and fig. 2 then illustrates possible variations in the direction normal to the interfaces.

We have a wave localized at the surface of the structure in fig. 2a. This would be the Rayleigh wave of the structure, if we have acoustical phonons in mind. Its propagation velocity, which can be measured by the light scattering method, provides average elastic constants for the structure, in the limit the Rayleigh wave amplitude extends over a large number of films. For superlattices with films whose thickness is tens of Angstroms, or a small number of hundreds of

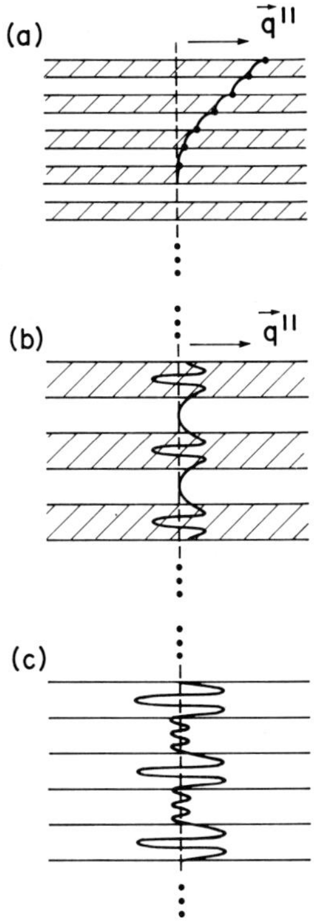

Fig. 2. Various classes of collective modes in a superlattice structure. We have (a) a surface mode, (b) a mode localized largely in one of the two constituents, and (c) a mode with standing-wave character in both constituents.

Angstroms, this condition is well satisfied by the modes probed by Brillouin spectroscopy.

In fig. 2b, we show a mode whose field has oscillatory character in one of the constituents, and which decays exponentially as one moves away from the interface into the second. When a large fraction of the energy density is stored in the constituent within which the wave field is oscillatory, then one readily obtains data on the elastic constants of these films from the frequencies of such modes, if we have acoustical phonons in mind once again. Finally, fig. 2c shows a mode in which the wave field is oscillatory in both constituents.

In this discussion, we have considered light scattering from surfaces, films and multilayers within the framework of a picture which assumes the light senses the excitation responsible for the scattering event through its modulation of the dielectric tensor of the structure, as in eq. (1.1). This is the same coupling mechanism which controls light scattering in the bulk of large crystals. Near surfaces and interfaces, entirely new coupling mechanisms may be encountered.

Consider, for example, the scattering of light by a Rayleigh surface wave on a crystal surface. As the wave propagates, the surface itself is distorted in a wavelike manner, to create a sinusoidal diffraction grating which moves with the speed of the Rayleigh wave. Incoming light is scattered inelastically by this surface ripple, even in the absence of modulation of the bulk dielectric constant by the strain field of the Rayleigh wave. A bulk phonon will also create a surface ripple, unless the amplitude of the displacement normal to the surface is zero by symmetry (as in a shear-horizontal wave), or if it vanishes right at the surface as a consequence of a 180° phase shift upon reflection. In metals, where the skin depth is very small, the surface ripple contribution provides the dominant contribution to the matrix element for Brillouin scattering, while in semiconductors or insulators, where the penetration depth is much larger, its role is lessened but in many instances it remains comparable in influence to the contribution from the bulk of the material. Of course, in a multilayer structure, one has a ripple contribution from each interface.

In this section, we have given a qualitative discussion of light scattering from excitations near surfaces and interfaces, with emphasis on aspects common to different cases. The presentation has been schematic and over-simplified, to place focus on the central features of the process. We now turn to explicit examples of light scattering studies of phonons and spin waves near surfaces and interfaces, along with the theory required to interpret the spectra.

3. *Brillouin scattering from acoustical phonons in thin films and multilayer structures*

The increasing large number of experimental groups using multi-pass tandem Fabry–Pérot interferometers (Sandercock 1982) in light scattering from opaque materials has made possible the collection of a vast amount of data concerning Brillouin scattering in thin films and multilayer structures. Combinations of transparent or opaque films supported by transparent or opaque substrates have been studied. Also free-standing films have been investigated with this technique. We will not review here Brillouin scattering from liquid films and interfaces, which have been considered by Loudon (1984) and by Albuquerque et al. (1985).

3.1. Main features of Brillouin spectra from semi-infinite media

Before we examine the results obtained in the case of films we must briefly review the main features of Brillouin spectra taken in opaque, semi-infinite materials. After the early work of Mills et al. (1970), the complete theory has been formulated independently by several research groups at the end of the 70s (Loudon 1978a, b, Subbaswamy and Maradudin 1978, Rowell and Stegeman

1978b, Marvin et al. 1980a, b, Velasco and Garcia-Moliner 1980). Both the elasto-optic coupling and the ripple scattering were considered, so that the theory accounts for the interaction of light with bulk and surface waves at the same time. Although the theoretical work published so far is restricted to the simple case of cubic materials, the general expressions for the cross-section are quite complicated. In addition they also depend on the polarization of the incoming and scattered light. Therefore, we prefer to give here approximate but simpler theoretical results which are valid for particular but very common cases. Unless otherwise stated, we will restrict our attention to both incident and scattered light with the electric field in the scattering plane (*p*-polarized, or TM light) and to phonons polarized in the same plane. According to the reference frame in fig. 1 the longitudinal modes are *x*-polarized and the shear-vertical modes are *z*-polarized.

The spectrum shown in fig. 3 refers to the (110) surface of GaAs and it is rich enough in information to exemplify all the scattering mechanisms from bulk and surface phonons mentioned in the previous section. The most prominent feature of this spectrum is the sharp peak (at about 10 GHz) due to scattering from the surface Rayleigh wave (RW). Neglecting the finite phonon lifetime and the collection angle of the experimental apparatus, which produces some uncertainty in the value of $\boldsymbol{q}_{\parallel}$, the RW would be just a line (Dirac δ-function), as we have seen in section 2. The properties of Rayleigh surface waves in isotropic and anisotropic media are reviewed, for example, by Auld (1973) and Farnell (1970).

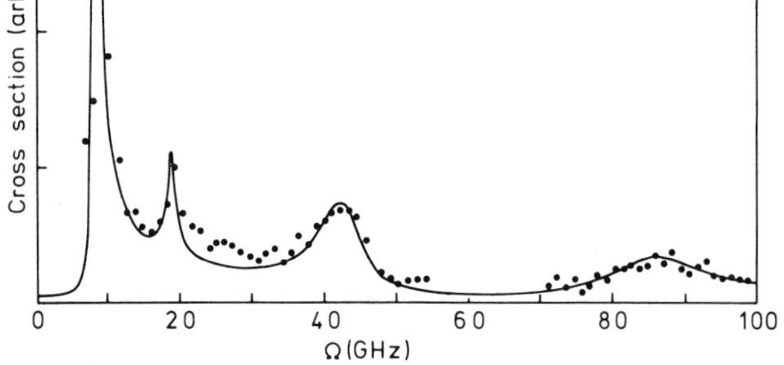

Fig. 3. Experimental (dots) and calculated (solid line) Brillouin scattering intensities for the (110) surface of GaAs with $\boldsymbol{q}_{\parallel}$ along $[1\bar{1}0]$. Backscattering geometry p → p. (After Marvin et al. 1980b; the original experimental data are after Loudon and Sandercock 1980.)

All the other features of fig. 3 belong to a continuous spectrum produced by bulk phonons, i.e., have an intrinsic linewidth. In semiconductors, like GaAs with a wavelength $\lambda = 5145$ Å, the penetration depth of the light is comparable to the phonon wavelength, so that both the elasto-optic and the ripple mechanisms contribute to the scattering strength of the RW. The relative importance of the two coupling mechanisms can be estimated from the following approximate formula for the cross-section, valid for opaque materials and backscattering geometry ($\theta_\mathrm{I} = \theta_\mathrm{S}$) with p-polarized light (Marvin et al. 1980b):

$$I^{\mathrm{p}\to\mathrm{p}} = \frac{k_\mathrm{B} T}{2\pi^2 \rho \omega_\mathrm{S}^2} (\Omega_\mathrm{I}/c)^4 \frac{\cos^3 \theta_\mathrm{I}}{(\cos \theta_\mathrm{I} + \varepsilon^{-1/2})^4} \delta(\boldsymbol{k}_{\parallel \mathrm{S}} - \boldsymbol{k}_{\parallel \mathrm{I}} \pm \boldsymbol{q}_\parallel)$$
$$\times |\varepsilon^{-3/2}(k_{11} - k_{13}c_{13}/c_{11}) \sin \theta_\mathrm{I} w_x + (1 + \sin^2 \theta_\mathrm{I}) w_z|^2, \qquad (3.1)$$

where ρ is the mass density, k_B the Boltzmann constant, T the temperature, ε the dielectric constant, k_{ij} and c_{ij} the Voigt components of the elasto-optic and of the elastic stiffness tensors respectively. Then w_z and w_x are the phonon displacement components evaluated at the surface for a given value of $\boldsymbol{q}_\parallel$ and the corresponding frequency ω_S of the surface wave. Equation (3.1) is valid for the (001) surface of cubic crystals and $\boldsymbol{q}_\parallel$ along the [100] direction, but it can be also applied to other cases after a proper rotation of the reference system (Marvin et al. 1980b). The displacement ellipse has a vertical-to-longitudinal axis ratio $|w_z/w_x|$ usually in the range 1–2, giving the RW a dominant shear-vertical character. In opaque materials $\varepsilon^{-3/2}(k_{11} - k_{13}c_{13}/c_{11})$ is usually smaller than 1, so that the ripple term in eq. (3.1), proportional to w_z, is the dominant one; however, in many cases the elasto-optic contribution (proportional to the coefficients k_{ij} and to the longitudinal displacements w_x) is not negligible and can give rise to interference effects.

The two high-frequency broad structures of fig. 3 (at about 42 and 87 GHz) are clearly due to interaction of light with transverse and longitudinal bulk phonons, respectively. The large intrinsic linewidth of these peaks is due to the skin depth in GaAs, which is comparable to the wavelength of the light. These are the opacity-broadened lines, already mentioned in the discussion of eqs (2.5) and (2.6). It is interesting to consider the broadening effect as an application of the Heisenberg principle: the spatial uncertainty in the occurrence of the phonon creation (annihilation) in opaque materials is strictly related to the uncertainty in the z-component of the phonon wave vector. The opacity-broadened bulk lines are obviously detected in Brillouin scattering through the elasto-optic mechanism only: the asymmetry of the spectral lines has been explained by Dervisch and Loudon (1976) as due to reflection of bulk waves at the boundary.

Finally, we have to explain the continuum of modes between the RW and the transverse bulk wave in fig. 3, especially the peak at about 19 GHz. This is the continuous band already mentioned in the previous section, coming from those bulk and evanescent waves whose total wave vector $\boldsymbol{q} = \boldsymbol{q}_\parallel + \hat{z} q_z$ has the same surface wave vector $\boldsymbol{q}_{\parallel}$ (mixed modes). Note that, although this continuous band originates mainly from bulk waves, it exhibits a real surface character: in other words, this part of the spectrum has dispersion relations of the kind $\omega = q_{\parallel} v$, where v is the propagation velocity, in complete analogy to the RW and unlike the opacity-broadened bulk lines, whose frequency scales as the total wave vector q. The lineshape of the mixed-mode spectrum is today well understood in terms of density of phonons states projected onto the plane of the surface. Figure 4 shows the behavior of the surface spectral density of longitudinal and shear-vertical phonons in an isotropic medium for different values of the Poisson ratio σ. Analytic expressions can be found in the literature (Loudon 1978a, b, Camley and Nizzoli 1985). The shear-vertical phonons contribute to the scattering with a sort of flat background: because this component is responsible for the ripple mechanism we expect that strongly absorbing materials have a Brillouin spectrum resembling very much the shear-vertical spectral density of fig. 4. This actually happens in metals (Sandercock

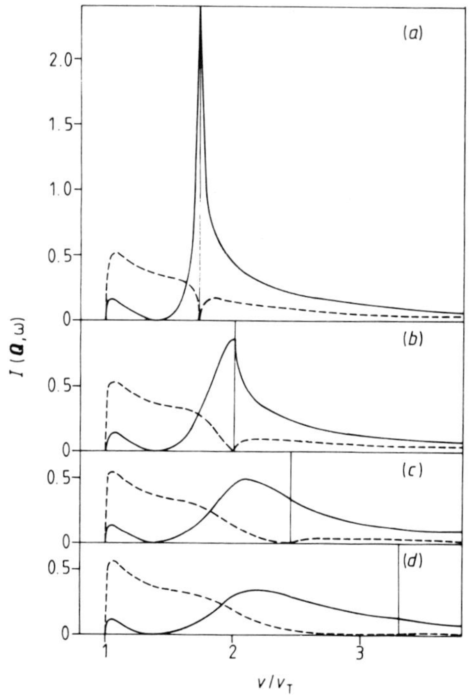

Fig. 4. Calculated surface spectral density of longitudinal (solid lines) and shear-vertical (dashed lines) phonons in an isotropic elastic medium, versus the adimensional velocity v/v_T at fixed $\boldsymbol{q}_{\parallel}$. v_T is the shear velocity. The vertical lines mark the longitudinal velocity. The Poisson ratio σ is equal to (a) 0.25, (b) 0.336, (c) 0.40, and (d) 0.45. The Rayleigh wave has been omitted. (After Camley and Nizzoli 1985.)

1978, Loudon 1978a). The longitudinal component is more structured, especially for small values ($\sigma \leqslant 0.33$) of the Poisson ratio, when a sharp peak appears for a velocity almost equal to the longitudinal bulk velocity. The peak of fig. 3 at about 19 GHz has been explained by Bortolani et al. (1978) in terms of the high density of the longitudinal phonon modes at that frequency and of the elasto-optic coupling. It has been found by Glass and Maradudin (1983) that the high phonon density actually corresponds to the propagation, parallel to the surface, of a so-called 'leaky' mode, sometimes also referred to as 'longitudinal resonance'. It has been also observed in polycrystalline gold (Bassoli et al. 1986), where, however, it has a short lifetime and, therefore, a large frequency spread, due to the value of the Poisson ratio $\sigma = 0.43$, as can be verified in fig. 4.

The results obtained in Brillouin scattering from semi-infinite opaque crystals can be summarized as follows. The theory developed so far has proved to be fully successful and explains all the features of the measured spectra. The elastic constants, evaluated from the position of the spectral lines, agree with those measured previously with other techniques, once experimental problems arising from the finite collection angle of the scattered light are properly taken into account (Loudon and Sandercock 1980). In several cases, when the elasto-optic constants were unknown, it has been possible to fit the amplitude of the spectral lines, calculated from the theory, using the elasto-optic constants as free parameters and therefore determining their values. We remark that this is possible because the spectra contain features due to both the elasto-optic and the ripple mechanisms, the last one being not dependent on any free parameter. In this way the elasto-optic constants of silicon and GaAs (Marvin et al. 1980b) and of gold (Bassoli et al. 1986) were determined for $\lambda = 5145$ Å.

We are now in a position to review the more complex field of light scattering from phonons in films, either homogeneous or multilayered. In the case of films there is a strong interest in studying their elastic behavior, which sometimes is unknown or not necessarily the same as that of the bulk constituent(s).

3.2. Scattering geometries

Different scattering geometries have been used so far to study surface acoustic waves in films by Brillouin scattering. We briefly describe four of them.

(i) Scattering on reflection from a surface (backscattering geometry). This works well with an opaque film on any substrate or with a transparent film on an opaque substrate. In all these versions, the light is confined to a limited spatial region and intense Brillouin scattering from the bulk volume does not occur, due to the small optical skin depth. The Brillouin signals are very small, typically 7 orders of magnitude smaller than the quasi-elastically scattered light. This is the scattering geometry most commonly used, and several systems have been studied so far. The results will be reviewed in the following sections 3.5, 3.6 and 3.7.

(ii) To increase the signal intensity, Moretti et al. (1985) used an attenuated total reflection technique (Kretschmann) for exciting surface plasmon polaritons in a metal film (silver) deposited on a glass hemisphere. In this way the electric field in the film is strongly enhanced. Then the surface acoustic wave reconverts the surface plasmon into an electromagnetic radiating wave, on both sides of the metal film. A theoretical prediction of the feasibility and magnitude of surface enhanced Brillouin scattering was presented by Fukui et al. (1980, 1981). In a transmission geometry Moretti et al. (1985) found an enhancement factor of about 25. Recently, Robertson et al. (1987) measured Brillouin enhanced signals off the RW in reflection geometry, obtaining an enhancement of about 750. Both transmission and reflection measurements were quantitatively analyzed with the theory of Marvin et al. (1982), which includes ripple scattering from the dynamical corrugations induced by the RW at both the film interfaces (film–vacuum and film–glass). The agreement between the measured and calculated Brillouin intensities versus the scattering angle θ is very good. In particular, strong maxima are found when θ equals the attenuated total reflection angle θ_{ATR}. Although this scattering geometry is interesting from the physical point of view, because it relates different kinds of excitations, and because it allows fast measurements, it seems unlikely that it can become a routine technique. In fact, it is restricted to the rather specialized case of metal films of thickness 400–800 Å deposited on a transparent substrate.

(iii) In this geometry both the film and the substrate are transparent. The incident light does penetrate into the substrate and intense Brillouin scattering off bulk phonons of the substrate occurs. It is intense relative to the thin film case because the scattering volume is so much larger. The scattered light can be collected either in backscattering or in transmission. This kind of experiment is possible if the acoustical properties of the film and substrate are very different, so that the Brillouin spectra are well separated. Dil et al. (1981) used this geometry to measure guided acoustic waves and bulk waves in polycarbonate films on a Pyrex glass substrate.

(iv) The fourth geometry relies on guided optical waves, i.e., electromagnetic waves which are guided parallel to the surface in an 'integrated optics' waveguide. A dielectric film of higher refractive index than the two surrounding media is required. In the experimental realization of Stegeman and co-workers (Rowell and Stegeman 1978a, Rowell et al. 1982) a prism has been used to couple the light into the waveguide. The scattered light can be observed as a guided electromagnetic mode or in a direction radiated out of the waveguide, with the aid of another prism. The light propagates in the waveguide by multiple total internal reflection at the two film interfaces. In this way the scattering intensity is enhanced because the same laser beam continuously resamples the phonon modes. Since the optical field decays exponentially into both bounding media, Brillouin scattering can be studied in the surface region of the bounding media as well as in the film. The theory was formulated by Rowell and Stegeman

(1982a, b). This geometry takes advantage of the limited scattering volumes, so that intense scattering off the bulk modes of the substrate is avoided. However, its application is limited to particular systems. As an example, we show in fig. 5 a Brillouin spectrum in a thin glass film on a glass substrate exhibiting peaks due to bulk longitudinal phonons of the two media and to surface and guided acoustical waves in the film (Rayleigh and Sezawa). One of the first reports of Brillouin scattering off Rayleigh and Sezawa waves, as well as the first measurement of surface phonon lifetimes via Brillouin scattering, was published by Rowell and Stegeman (1978a). The technique has been mainly used to detect the changes of the elastic constants of glasses obtained with different spattering processes.

All the experimental scattering geometries discussed above need high-contrast interferometers to resolve the small Brillouin signals with respect to the strong-elastically scattered light. This is achieved with the use of multi-pass Fabry–Pérot filters. The spectra taken in films contain several spectral lines of surface, guided and sometimes bulk modes. It is therefore necessary to use an instrumentation with a large free spectral range, in order to avoid the

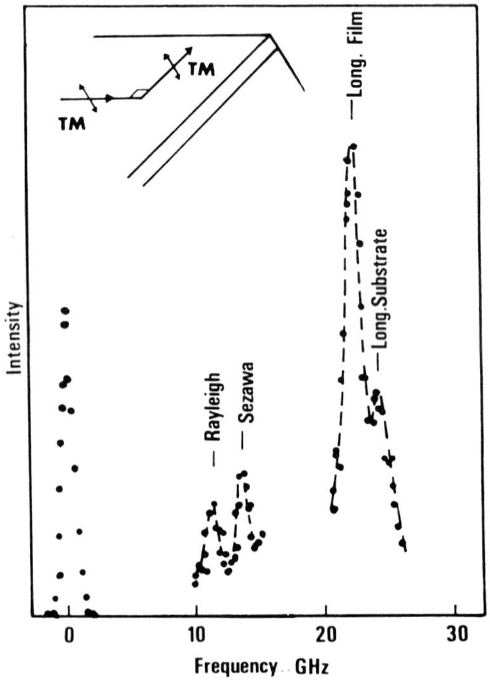

Fig. 5. The Brillouin spectrum obtained from a thin glass film. As indicated in the inset, both the incident and scattered light were propagated as TM-polarized guided waves. The various peaks observed are labeled. (After Rowell et al. 1982.)

overlapping of the spectral lines, typical of Fabry–Pérot interferometers. Therefore, tandem interferometers have been designed, which suppress signals from near order spectra. Two systems have been used so far to synchronously scan two tandem Fabry–Pérots: a mechanical one (Sandercock 1982) and an electronic driven one (Dil et al. 1981). The reader can find more details concerning the instruments for Brillouin scattering in these references and in the references therein.

3.3. Brief overview of acoustic modes in free-standing and supported films

Before presenting the results obtained in Brillouin scattering from films, it is convenient to make the reader acquainted with the various families of acoustical modes of films and interfaces. More details can be found in the exhaustive presentations of Auld (1973) and Farnell and Adler (1972). The basic idea is that the lack of translational invariance perpendicular to the surface allows one to write the phonon field in the film (initially considered as unsupported) as a superposition of partial waves of different perpendicular wave vector q_z:

$$\boldsymbol{w}^{(\mathrm{f})}(z) = \sum_{\mu} a_\mu \boldsymbol{u}_\mu \mathrm{e}^{\mathrm{i}q_z^\mu z}, \tag{3.2}$$

where μ labels the complex normal wave vectors at a given frequency ω and surface wave vector $\boldsymbol{q}_\parallel$, $\boldsymbol{u}_\mu$ is the polarization vector of the modes of the infinite medium and a_μ are coefficients determined by the mechanical boundary conditions (stress-free surfaces). In the simplest case (considered in most practical applications) the phonon modes factor into shear-horizontal modes (y) and sagittal modes (x, z). For given $\boldsymbol{q}_\parallel$ and film thickness h, the boundary conditions will be satisfied for a discrete set of frequencies ω_j, which corresponds to propagation of guided modes in the film. We distinguish between Love modes (y-polarized) and Lamb modes (x-, z-polarized). In the limit $q_\parallel h \to \infty$ the two lower-order Lamb modes merge as the RW of the semi-infinite crystal.

When the film is supported, the situation is complicated by the presence of the substrate, whose phonon field can be developed in analogy to eq. (3.2), as a superposition of partial waves of the substrate:

$$\boldsymbol{w}^{(\mathrm{s})}(z) = \sum_{\lambda} b_\lambda \boldsymbol{v}_\lambda \mathrm{e}^{\mathrm{i}q_z^\lambda z} \tag{3.3}$$

The phonon field of the composite system can now be found by imposing the proper mechanical boundary conditions: stress-free surface and continuity of displacements and stresses at the interface. Under particular conditions (shear velocity of the film smaller than the shear velocity of the substrate) in the supported film there are guided modes resembling the modes of the free plate, but with an evanescent tail extending into the substrate. These modes are called generalized Love or Lamb modes, according to their polarization. The number of the guided modes increases with $q_{\|}h$. Among the generalized Lamb modes, the two lower-order ones are of particular interest. The first one becomes the RW of the substrate for $q_{\|}h \to 0$, and the RW of the film for $q_{\|}h \to \infty$. The second one (called the Sezawa mode), in the limit $q_{\|}h \to \infty$, can be trapped at the interface between the two media, provided their acoustical proprieties are favorable. In the latter case we have a true interface mode, named after Stoneley. In the Brillouin scattering literature the generalized Lamb waves are often called Sezawa waves. In the following we adopt this convention.

Sezawa modes present an upper cut-off frequency ω_T^S corresponding to propagation parallel to the surface of shear bulk waves in the substrate. For $\omega > \omega_T^S$, there is the continuous band of mixed-modes already mentioned. The Sezawa modes actually extend also into this frequency range, where they become resonant or 'leaky' modes. They are degenerate with bulk modes of the substrate, which reflect at the interface and take energy off the film. The 'leaky' Sezawa modes have an intrinsic linewidth and usually appear in Brillouin spectra as broad structures. However, when the localization in the film is strong, they can be as sharp as the RW itself. Examples will be presented later.

3.4. Theory of Brillouin cross-section in thin films

The theory of Brillouin scattering in supported films has been given by Bortolani et al. (1983). It applies to the scattering geometries (i) and (iii) of section 3.2 and to crystals of cubic symmetry. They consider both ripple and elasto-optic mechanisms and deal with a film of any thickness. Scattering off Lamb and Love modes of unsupported films can be also obtained as limiting cases of their general results, by letting (for the substrate) $c_{ij} \to 0$, $\rho \to 0$, $\varepsilon \to 1$. The cross-sections of Love and Lamb modes in free-standing isotropic films have been explicitly studied by Albuquerque et al. (1980) and by Albuquerque et al. (1984), respectively: in these cases analytical expressions can be derived. Within the same framework Albuquerque (1980) has published a paper on Brillouin scattering from interface waves.

The theory of Bortolani et al. (1983) requires numerical calculations for the phonon modes, otherwise analytical expressions for the scattered electromagnetic field are given. The final expressions are very lengthy. The Brillouin intensity for p→p scattering has the following schematic form:

$$I^{\mathrm{p}\to\mathrm{p}} \propto \Big| \sum_{\mu} a_{\mu}[A_{\mu,x}u_{\mu,x} + A_{\mu,z}u_{\mu,z}] + Aw_z^{(\mathrm{f})}(h) + \sum_{\lambda} b_{\lambda}[B_{\lambda,x}v_{\lambda,x} + B_{\lambda,z}v_{\lambda,z}] + Bw_z^{(\mathrm{s})}(0)\Big|^2 . \tag{3.4}$$

The first term in the squared modulus refers to the elasto-optic coupling in the film and the third one to the elasto-optic coupling in the substrate. The quantities a, u, b and v are related to the phonon displacement field in the two media, according to eqs (3.2) and (3.3), and require numerical calculations in anisotropic crystals. The weighting factors $A_{i,j}$ and $B_{i,j}$ are linear combinations of the elasto-optic constants and are determined by the electromagnetic boundary conditions. Note that the elasto-optic terms depend on both the transverse and the longitudinal components of the partial waves of the film $u_{\mu,z}$, $u_{\mu,x}$ and of the substrate $v_{\lambda,z}$, $v_{\lambda,x}$. The term $Aw_z^{(f)}(h)$ represents ripple scattering from the free surface of the film and is proportional to the normal component of the displacement field (3.2) calculated at the free surface ($z = h$). A is a function of the scattering angles, of the film thickness and of the dielectric functions of the media. The term $Bw_z^{(s)}(0)$ represents ripple scattering from the film–substrate interface and is therefore proportional to the normal displacements at the interface ($z = 0$). The form of eq. (3.4) leads to an immediate important conclusion: the four contributions to the cross-section coherently interfere and, whenever their amplitudes are comparable, modulate the cross-section in a complicated way with strong interference effects. The next sections are devoted to the illustration of experimental results for films, and many applications of the theory will be also presented.

3.5. Brillouin scattering in opaque films

This field, as well as Brillouin scattering from semi-infinite opaque media was pioneered by John Sandercock. In the first paper showing the possibility of detecting guided modes in supported metal films (Bortolani et al. 1979) a spectrum taken by Sandercock for a 400 Å film of aluminum on silicon was satisfactory explained in terms of ripple scattering at the film surface. The spectrum gave indications of the RW, of the first-order Sezawa wave and of the continuum of mixed-modes due to resonant Sezawa waves. Measurements and calculations for Al films of various thicknesses allowed Sandercock et al. (1980) to reproduce the dispersion relations and the scattering intensities of the guided modes using the values of elastic constants in polycrystalline aluminum found in the literature. The surface waves of aluminum films on transparent substrates have been studied by Sussner et al. (1980).

Gold films on semiconductors or insulators have been also investigated. Bortolani et al. (1980, 1984) found that the scattering intensity of guided modes in gold films on silicon were reproduced only by including in the cross-section (3.4) the elasto-optic coupling in gold. Hillebrands et al. (1986) reported on investigations of elastic surface-wave excitations near the cluster-to-layer transition of gold in NaCl. In the cluster regime ($h \leqslant 150$ Å) a new type of surface-localized mode was observed in the mixed-modes spectrum, with an anomalous dispersion relation. To the best of our knowledge, no convincing explanation of this experimental fact has been found so far.

Gold has been produced in the form of thin unsupported films by Grimsditch et al. (1987) and Brillouin spectra were recorded. A simple but clever technique was devised to obtain films with thicknesses as small as 200 Å. The quadratic dispersion of the first-order, low-frequency, Lamb bending mode was measured. They were also able to extract information on a number of elastic constants of the film.

The problem of determining the elastic constants of supported films of hexagonal symmetry was considered in detail by Nizzoli et al. (1988) for aluminum deposited on the (100) surface of NaCl. These authors were unable to fit uniquely the experimental spectra to obtain the four c_{ij} independently, since different combinations of the c_{ij} lead to similar spectra and dispersion curves. Further problems were found in this system: the calculated linewidths of the guided modes (belonging in this case to the mixed-mode spectrum) turned out to be much greater than the experimental ones. The calculations assumed perfect adhesion between the film and the substrate (ideal boundary conditions). Only by assuming that the Al films were not completely bounded to the substrate, the authors were able to reconcile calculations and experiments. This paper showed for the first time the capability of Brillouin scattering to be a convenient and nondestructive technique to provide information about adhesion.

In films of molybdenum with thicknesses 4000 and 8000 Å deposited onto a glass substrate by sputtering, Bell et al. (1988) measured for the first time by Brillouin scattering the Stoneley wave of a solid interface. Figure 6 shows two spectra taken at the Mo free-surface and at the glass-Mo interface. The interface was accessible thanks to the transparent substrate. For this film thickness the

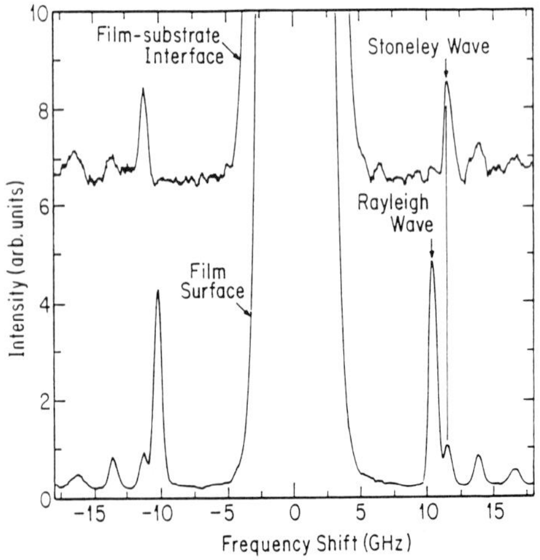

Fig. 6. Detection of the Stoneley wave. Measured Brillouin spectra (p→p scattering) for a 4000 Å film of Molybdenum on glass substrate. Scattering angles $\theta_I = 59°$ and $\theta_S = 70°$. The spectra were taken at the Mo-glass interface (upper curve) and at the Mo free-surface (lower curve). (After Bell et al. 1988.)

RW was observed only at the film surface, the Stoneley wave mainly at the interface, as expected from their different localizations. Mo films in the thickness range 300–8000 Å turned out to be homogeneous and with the same elastic properties. The Stoneley wave has been also measured by Jorna et al. (1989) at the nickel-glass interface: in this case, however, thick Ni films ($h > 5000$ Å) were found to be not homogeneous.

Finally, we mention that Brillouin scattering has been also used to measure the elastic properties of surface layers obtained by modifying semiconductor surfaces with various techniques. Grimsditch et al. (1978) studied hydrogenated amorphous silicon, Vacher et al. (1980) amorphous silicon obtained by bombardment, Sapriel et al. (1985) ion implanted GaAs.

3.6. Brillouin scattering in transparent films

Transparent films on opaque materials give serious problems of interpretation of the spectra, because the complete penetration of light in the film makes at least three scattering mechanisms active in the cross-section: the ripple scattering from both interfaces and the elasto-optic coupling in the film. In other words, a full understanding of the spectra cannot be achieved by shortcuts in the evaluation of eq. (3.4). We consider in this section only scattering experiments taken in the geometry (i). Brillouin measurements in transparent films using the geometries (iii) and (iv) were already considered in the previous section 3.2.

In order to test the theory of Brillouin scattering in supported films, Bortolani et al. (1982) faced the difficult task of explaining the surface spectra of isotropic SiO_2 films on crystalline silicon. For this system, there is no free parameter in the theory, because elastic constants, elasto-optic coefficients and dielectric functions are known from the literature. Figure 7 shows a comparison between an experimental spectrum and the calculated intensities (full theory and ripple approximation). The full theory, including the elasto-optic contributions, is needed to explain the experimental data. This spectrum shows peaks due to the RW, two guided localized modes (Sezawa or generalized Lamb modes) and three 'leaky' Sezawa modes belonging to the continuum. The elasto-optic coupling cannot be neglected and is the major contribution to the cross-section for the high-frequency leaky modes. The interference effects are very large.

Transparent Langmuir–Blodgett films of Cd-arachidate deposited on molybdenum were studied by Zanoni et al. (1986) by varying the number of monolayers. Three elastic constants (c_{11}, c_{33}, c_{13}) of these strongly anisotropic hexagonal films were determined by Brillouin scattering and upper limits were found for the other elastic constants. The determination of the elastic constants of transparent films of unknown properties is a very promising field of application of a Brillouin scattering. Recently a refinement of the technique has allowed the determination of the whole set of elastic constants of polymeric Langmuir–Blodgett films (Nizzoli et al. 1989).

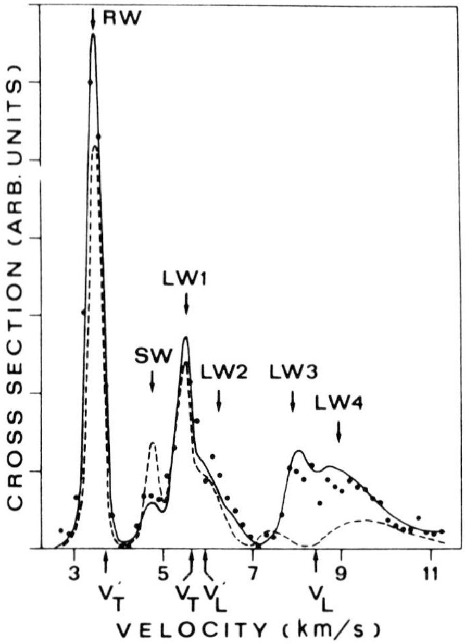

Fig. 7. Measured (dots) and calculated cross-section (p → p) for a 2250 Å film of amorphous SiO_2 on (001) silicon. Solid curve: full theory. Dashed curve: calculated cross-section with ripple terms only. RW = Rayleigh wave, SW = Sezawa wave, LWn = generalized Lamb (Sezawa) mode of n-th order. v'_T and v'_L are the transverse and longitudinal velocities of the film. v_T and v_L are the same for the substrate. (After Bortolani et al. 1982.)

Recently, Hillebrands et al. (1988b) reported theoretical as well as experimental evidence for a new class of long-wavelength acoustical excitations, called 'longitudinal guided modes' (LGM). In ZnSe films grown on (001) GaAs they observed, for film thicknesses in the range 0.5–2 μm, a pronounced peak of intensity comparable to the Rayleigh wave line. Its linewidth is only slightly larger than that of the RW peak. The spectra are shown in fig. 8. With decreasing the film thickness h, the intensity of the LGM decreases, accompanied by a linewidth broadening and a slight frequency increase. The frequency of the LGM is slightly above that of the longitudinal bulk wave. Using data taken from the literature and the elastic constants of ZnSe determined from the dispersion relations of the guided modes themselves (Lee et al. 1988), Hillebrands et al. (1988b) very accurately reproduced the scattering intensity of the measured spectra. It has been found that the LGM is strongly localized in the film and couples with light solely through the elasto-optic mechanism. The authors conclude that the classification of acoustical sagittal surface and film excitations now exhibits complete symmetry with respect to shear and longitudinal waves and includes surface modes (Rayleigh wave, longitudinal reso-

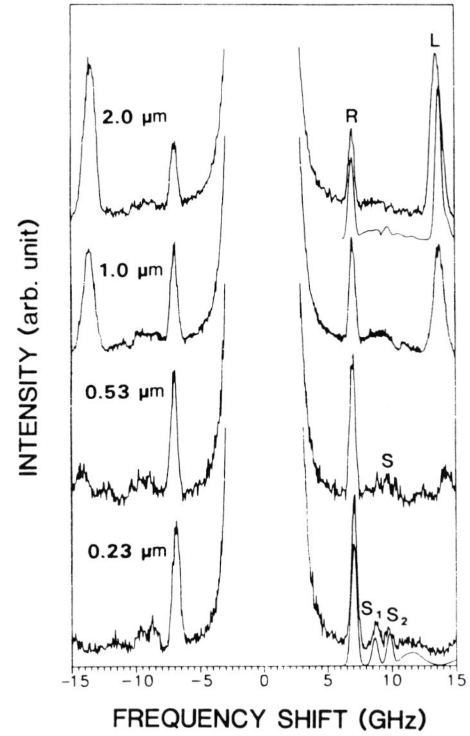

Fig. 8. Brillouin spectra of (001) ZnSe films on a (001) GaAs substrate. $\boldsymbol{q}_{\parallel}$ is in the [100] direction. Angle of incidence $\theta_\mathrm{I} = 60°$. The film thicknesses h are as indicated. In two cases ($h = 2$ and 0.23 μm) a comparison is made with the calculated intensities (thin lines). R: Rayleigh wave, L: longitudinal guided mode, S; Sezawa modes. (After Hillebrands et al. 1988b).

nance) and film modes (Sezawa modes, longitudinal guided modes). The new LGM offers, as major application, the determination of the longitudinal sound velocity in transparent films, a parameter which is otherwise difficult to evaluate.

3.7. Brillouin scattering in superlattices

The elastic properties of a stratified medium composed of two alternate layers of thicknesses d_1 and d_2 have been studied about 50 years ago by Bruggeman (1937). Since then and for a long time the problem has attracted interest mainly in seismology. More recently, Rytov (1956), in the context of physical acoustics, calculated explicitly the elastic constants of a stratified medium, made of two isotropic materials, in the long-wavelength limit, when the excitation wavelength is larger than the modulation wavelength of the superlattice $\Lambda = d_1 + d_2$, viz. $q\Lambda < 1$. As already remarked in section 2, in practical Brillouin experiments this condition is satisfied for a modulation wavelength of a small number of hundreds of Angstroms. The Rytov approximation leads to the concept of 'effective elastic constants', that is, in this limit, the superlattice behaves as a homogeneous medium with a well-defined elastic-constant tensor. The symmetry of the effective medium depends on the two constituents: for example, two isotropic media give rise to an effective crystal of hexagonal character.

Explicit and general expressions for the effective elastic constants of stratified media in terms of the $c_{ij}(n)$ and of the d_n of the two components ($n = 1, 2$) are given by Grimsditch (1985) and Grimsditch and Nizzoli (1986). The same method has been applied by Djafari-Rouhani and Sapriel (1986) to the calculation of the effective dielectric and photoelastic tensors of superlattices. An important feature of the method is that the effective elastic constants do not depend on the modulation wavelength Λ, rather on the relative thicknesses d_n/Λ. For example, the effective c_{44} of a stratified medium made of two orthorhombic crystals with a principal axis along the superlattice normal is given by:

$$c_{44} = \left[\frac{d_1}{\Lambda c_{44}(1)} + \frac{d_2}{\Lambda c_{44}(2)} \right]^{-1}. \tag{3.5}$$

Many recent papers deal with the theory of bulk and surface acoustical waves beyond the effective medium approximation, by retaining the discrete nature of the superlattice. In this case one has to write the wave equations in each layer in terms of the elastic constants $c_{ij}(n)$ of the layer and then to suppose that the assumed superposition of partial waves obeys the boundary conditions at each interface. Obviously the solutions must satisfy Bloch's theorem, reflecting the translational symmetry of the system. Then the power spectrum of fluctuations of the acoustical field can be calculated.

The theory of bulk acoustical modes in superlattices has been given by Camley et al. (1983a, b), Djafari-Rouhani et al. (1983), Babiker et al. (1985a, b). The striking features of the theory, indeed already contained in the work of Rytov (1956), are the folding of the Brillouin zone due to the macroscopic periodicity of the superlattice and the opening of gaps at the zone boundaries. However, these effects are outside the reach of Brillouin scattering: Raman scattering has been used instead (Colvard et al. 1980).

Semi-infinite media have been studied in order to obtain the localized surface waves of a superlattice. Superlattices with the surface parallel to the layers were investigated by Auld et al. (1977), Djafari-Rouhani et al. (1983), Camley et al. (1983a, b), Sapriel et al. (1983a), Dobrzynski et al. (1984). They found gaps in the continuum of bulk modes, with the possibility of surface-localized phonons corresponding to optical frequencies: the existence of these phonons mainly depends on the thickness and type of the outermost layer. Also the possibility of acoustical 'Love-type' waves has been demonstrated.

Djafari-Rouhani et al. (1984) studied the interesting case of the RW of a superlattice stratified normal to the surface: near the Brillouin boundary they found two different Rayleigh branches, which result from the folding of the dispersion curves of the RW of the homogeneous medium. Finally, piezoelectric superlattices have been investigated by Fernandez and Velasco (1987) and by Nougaoui and Djafari-Rouhani (1987).

The experimental measurements of the RW in superlattices have been initiated by Kueny et al. (1982) in metal systems (Nb/Cu) and by Sapriel et al. (1983b) in semiconductor systems ($GaAs/Ga_{1-x}Al_xAs$). The result obtained by Kueny et al. (1982) was surprising: the RW velocity as a function of the modulation wavelength Λ showed a marked anomaly (a dip for $20 \leqslant \Lambda \leqslant 40$), instead of being independent of Λ, as predicted by eq. (3.5). On the other hand, there is no reason why eq. (3.5) is not valid in this range of Λ. This has been also numerically verified by Kueny and Grimsditch (1982). The explanation may be found in a change of the elastic properties of at least one material, when produced in form of a thin film in the superlattice.

Other metal superlattices show the same behavior of Nb/Cu: Mo/Ni (Khan et al. 1983), V/Ni (Danner et al. 1986), Mo/Ta (Bell et al. 1987a). The RW velocity of Au/Cr, instead, is enhanced for modulation wavelengths in the range $\Lambda =$ 50–100 Å (Bisanti et al. 1987). It seems that by now only the Fe/Pd superlattice has not shown any anomaly (Baumgart et al. 1986).

Schuller and Rahman (1983) performed a molecular dynamics calculation and were able to relate the elastic anomaly in Mo/Ni to the lattice expansion of Ni in the direction perpendicular to the surface, actually observed by X-ray in coincidence with the softening of the RW. Also the other superlattices exhibiting RW anomalies show lattice expansion (or contraction). Schuller and Grimsditch (1986) used a simple expression (derived from the Murnaghan equation of state) to express the RW velocity as a function of the lattice constant responsible for the lattice expansion, and obtained good agreement between the experimental data for Nb/Cu (fig. 9) and the calculated RW velocity. The same simple model

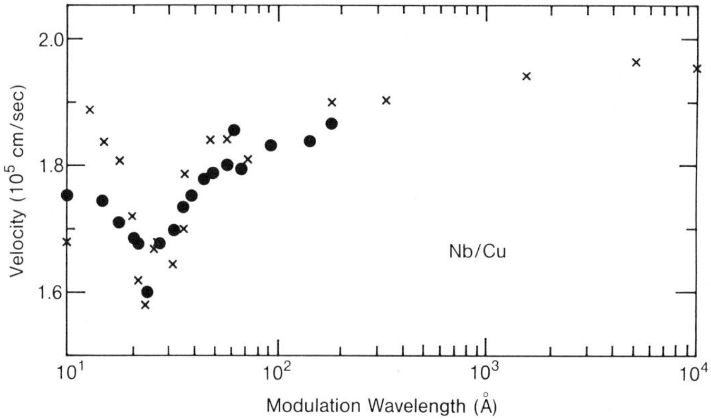

Fig. 9. Surface-wave velocity measured by Brillouin scattering in Nb/Cu superlattices versus the bilayer thickness (modulation wavelength). Crosses are experimental results from Kueny et al. (1982) and dots are the calculated values. (After Schuller and Grimsditch 1986.)

explains the anomalous behavior in all the systems for which the lattice-constant data are available. One has now to explain the physical origin of the changes in the lattice constants. Different mechanisms have been suggested, but the phenomenon is still not well understood. A discussion of the possible band-structure effects is given by Cammarata (1986). Wolf and Lutsko (1988) suggest that the presence of structurally disordered solid interfaces induces the expansion (or contraction). Huberman and Grimsditch (1989) have recently proposed an explanation based on an electronically driven lattice deformation extending uniformly throughout each layer.

Lets go back now to the problem of determining the unknown elastic constants of a supported film, in this case a metal superlattice. Bell et al. (1987a, b) measured several (up to nine) generalized Lamb (Sezawa) modes in Mo/Ta and Nb/Cu superlattices. As an example, in fig. 10 we show the experimental dispersion relations of the guided modes in a 4480 Å-thick Mo/Ta superlattice on sapphire. The calculated best-fit curves are also shown. This figure offers a nice example of dispersion relations of Sezawa modes in a supported film because it illustrates the behavior of the guided modes both in the discrete spectrum and in the continuum ($v > v_T$), where the modes are actually 'leaky'. Figure 11 shows a very clean experimental spectrum for the Mo/Ta superlattice, compared with the calculated ripple cross-section in the effective medium approximation: the agreement is excellent. Bell et al. (1987a) tried to fit the four independent elastic constants of the hexagonal Mo/Ta superlattice. They found that the dependences of c_{11}, c_{13} and c_{33} on the mode velocities are nearly inseparable and thus a fit to the measured velocities is very

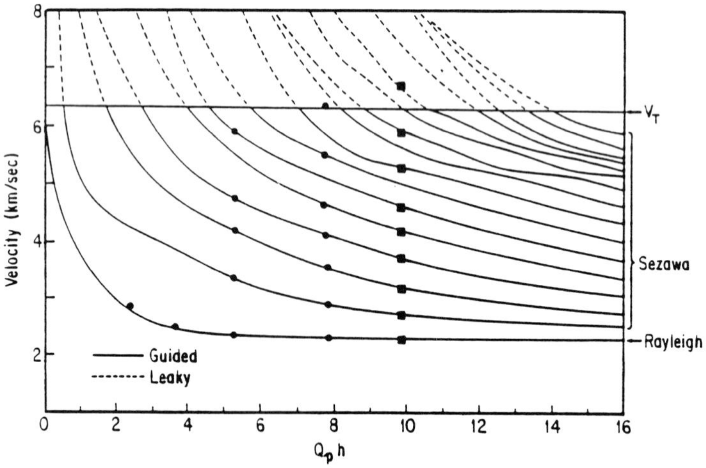

Fig. 10. Experimental velocities obtained at five different scattering angles for a 4480 Å-thick Mo/Ta superlattice on sapphire with modulation wavelength $\Lambda = 160$ Å. The solid curves are the best fit to the data (squares). (After Bell et al. 1987a.)

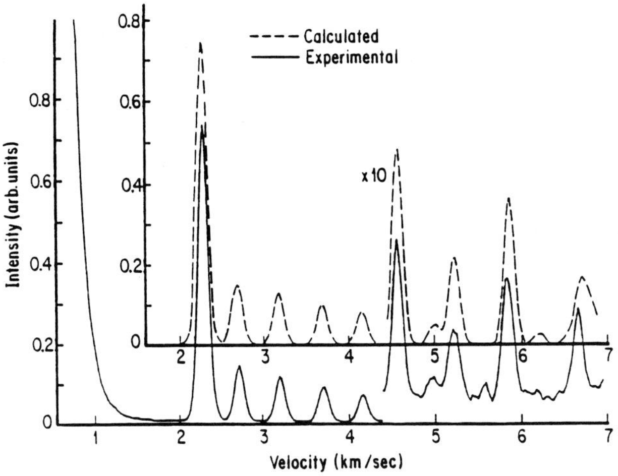

Fig. 11. Comparison of the light scattering data for the Mo/Ta superlattice of fig. 10 with the (ripple) calculated intensity based on the elastic constants obtained from least-squares fitting the velocities of the surface waves. (After Bell et al. 1987a.)

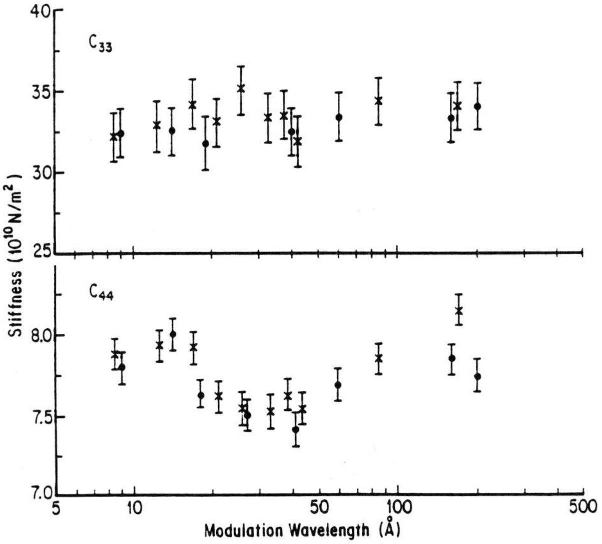

Fig. 12. Mo/Ta superlattices. Superlattice modulation-wavelength dependence of the elastic constants c_{33} and c_{44} determined by fitting the measured phase velocities to an acoustic model. Samples with 50% Mo are shown as crosses and 54% Mo as dots. (After Bell et al. 1987a.)

sensitive to small errors. By fixing c_{11} and c_{33} to their respective averages over separate sets of data and allowing only c_{33} and c_{44} to freely vary during the least-square fitting, variations of c_{33} were much reduced. The best-fit results obtained for c_{33} and c_{44} versus the modulation wavelength are shown in fig. 12: c_{44} exhibits a softening, while c_{33} is nearly constant. The high correlation of c_{11}, c_{13} and c_{33} in fitting the dispersion relations of the Sezawa modes in supported films requires a strong effort by the researchers in the field in order to find other ways to evaluate independently a couple of these elastic constants.

Finally, we have to mention that generalized Love waves have been recently measured by Brillouin scattering in Nb/Cu superlattices on sapphire. Bell et al. (1987c) were able to detect these shear-horizontal guided modes in depolarized p→s scattering, as required by the selection rules (Sandercock 1982). The measurements were performed at the superlattice-sapphire interface in order to couple, through the elasto-optic mechanism, with the exponential tails of the Love modes extending into sapphire. In fact, the ripple mechanism is not active for shear-horizontal modes. A measured spectrum showing excitation of two Love modes in Nb/Cu is shown in fig. 13, and is compared with an ordinary p→p spectrum for sagittal guided modes. The measurement of the Love modes allowed the authors to determine the effective elastic constant c_{12} of the superlattice.

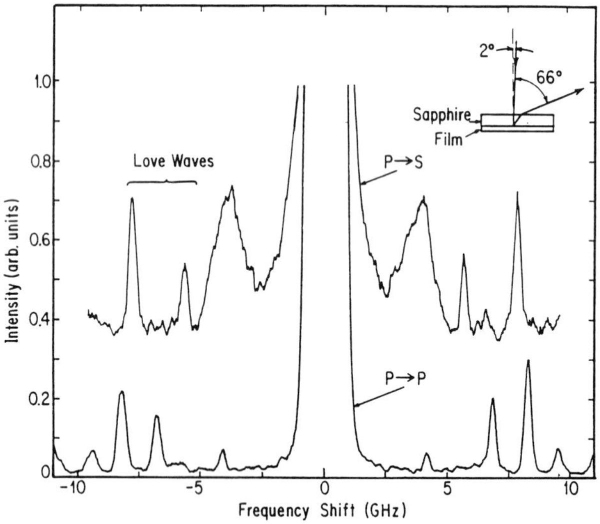

Fig. 13. Detection of generalized Love modes. Brillouin spectra obtained from the substrate-film interface for two polarization conditions for the scattered light. Film is a Cu/Nb superlattice of thickness 3180 Å and modulation wavelength $\Lambda = 91$ Å. The scattering and sample geometry is shown in the inset. (After Bell et al. 1987c.)

4. *Raman scattering from optical phonons and polaritons in thin films and multilayer structures*

In section 3, we discussed the scattering of light by surface acoustic phonons (Rayleigh waves), and by thermally excited bulk and guided phonons in films, and near the surface of opaque media. From the discussion of section 2, it is evident that any elementary excitation that modulates the dielectric tensor of the medium within which it is excited will scatter light inelastically. In this section, we turn our attention to the scattering of light by surface polaritons in films and multilayers, and by companion excitations, the standing-wave bulk optical phonons. The experiments here are Raman scattering experiments, since, as we shall see, the frequency shift experienced by the scattered light lies in the range of a few hundred wave numbers.

Before we turn to a discussion of the experiments, it will prove useful to introduce the reader to the basic properties of these modes.

4.1. *The basic properties of optical phonons and surface polaritons in ionic insulators*

A phenomenological treatment of the long-wavelength normal modes of insulators was provided many years ago (Born and Huang 1954). We begin with a brief review of this material, cast in a more modern notation. For simplicity, we confine our attention to a cubic crystal, though it is a simple matter to extend the treatment to anisotropic media (Mills and Burstein 1974). Furthermore, we suppose the crystal has two atoms per unit cell, so we have only three long-wavelength optical modes. The discussion here, while restricted in the interest of simplicity, none the less can be applied to quite a number of common solids, such as the alkali halides, and semiconductors of the zinc blende crystal structure.

We let $\boldsymbol{u}$ be the normal coordinate associated with the optical vibration. The motions of the ions within the unit cell in general generates a macroscopic electric field $\boldsymbol{E}$; associated with it is also a dynamic dipole moment $\boldsymbol{p} = e^*\boldsymbol{u}$, where e^* is the Born effective charge. The equation of motion for $\boldsymbol{u}$ thus reads

$$\ddot{\boldsymbol{u}} + \omega_{\mathrm{TO}}^2 \boldsymbol{u} = \frac{e^*}{M}\boldsymbol{E}, \tag{4.1}$$

where M is the reduced mass of the unit cell and ω_{TO} is the frequency of the long-wavelength transverse optical phonon (see below). For the magnetic fields, we have $\boldsymbol{B} = \boldsymbol{H}$, so we supplement eq. (4.1) by the Maxwell equations

$$\boldsymbol{\nabla} \cdot \boldsymbol{D} = \varepsilon_\infty \boldsymbol{\nabla} \cdot \boldsymbol{E} + 4\pi n e^* \boldsymbol{\nabla} \cdot \boldsymbol{u} = 0, \tag{4.2a}$$

$$\boldsymbol{\nabla} \times \boldsymbol{E} = -\frac{1}{c}\dot{\boldsymbol{B}}, \tag{4.2b}$$

$$\nabla \times \boldsymbol{B} = \frac{\varepsilon_\infty}{c} \dot{\boldsymbol{E}} + 4\pi n e^* \dot{\boldsymbol{u}}. \tag{4.2c}$$

The dipole moment/unit volume associated with the lattice motion is $\boldsymbol{P} = ne^*\boldsymbol{u}$, and ε_∞ describes the contribution to the dielectric response of the medium from polarization of the valence electrons.

For oscillations with frequency ω, eq. (4.1) gives

$$\boldsymbol{u} = \frac{e^*}{M} \frac{1}{\omega_{\mathrm{TO}}^2 - \omega^2} \boldsymbol{E}, \tag{4.3a}$$

and then eqs (4.2) may be reduced to the two statements

$$\varepsilon(\omega) \nabla \cdot \boldsymbol{E} = 0, \tag{4.3b}$$

and

$$\nabla^2 \boldsymbol{E} - \nabla(\nabla \cdot \boldsymbol{E}) - \varepsilon(\omega) \frac{\omega^2}{c^2} \boldsymbol{E} = 0, \tag{4.3c}$$

where

$$\varepsilon(\omega) = \varepsilon_\infty + \frac{4\pi n e^{*2}}{M} \frac{1}{\omega_{\mathrm{TO}}^2 - \omega^2}. \tag{4.4}$$

Equations (4.3) describe the electric fields in a dielectric medium, whose dielectric constant is given by eq. (4.4). The striking feature is the resonance at ω_{TO} introduced by the lattice motion.

There are two classes of wavelike solution to eqs (4.3). If $\boldsymbol{q}$ is the wave vector of the mode, then we have:

(i) *Longitudinal solutions* ($\boldsymbol{q} \parallel \boldsymbol{E}$, and also $\boldsymbol{q} \parallel \boldsymbol{u}$):
For this case eq. (4.3b) requires the frequency to be such that

$$\varepsilon(\omega) = 0, \tag{4.5}$$

which gives

$$\omega^2 \equiv \omega_{\mathrm{LO}}^2 = \omega_{\mathrm{TO}}^2 + \frac{4\pi n e^*}{M\varepsilon_\infty}, \tag{4.6}$$

which can be recognized as the longitudinal optical phonon frequency of the crystal. Note that for this case eq. (4.3c) is satisfied identically, and also that $\boldsymbol{B} = 0$.

This is then the bulk longitudinal optical phonon, described within dielectric theory.

(ii) *Transverse solutions* ($\boldsymbol{q} \perp \boldsymbol{E}$, and also $\boldsymbol{q} \perp \boldsymbol{u}$):
Here we have the dispersion relation

$$\frac{c^2 q^2}{\omega^2} = \varepsilon(\omega). \tag{4.7}$$

These transverse waves, which may be viewed as coupled modes of the electromagnetic field and lattice motion, are referred to as polaritons (Mills and Burstein 1974). The reader will recognize that eq. (4.7) is nothing more than the dispersion relation of a transverse electromagnetic wave in a dispersive dielectric. The description we have just used, based on eqs (4.1) and (4.2), allows one to break down the description of the electromagnetic wave into components consisting of the lattice motion responsible for a portion of the dipole moment per unit volume in the material, and a 'bare' electromagnetic field. The term polariton is used when this decomposition is either useful or essential. We shall appreciate its usefulness below, when we discuss the coupling responsible for the scattering of light from polaritons.

For any value of q, we have two frequencies which emerge from eq. (4.7). Thus, there are two branches of the polariton dispersion curves. As $q \to \infty$ (the limit $cq \gg \omega$), the frequency of one branch approaches ω_{TO}, the frequency which emerges from eq. (4.1) when $\boldsymbol{E} \to 0$. Thus, for this branch, we can infer $\boldsymbol{E} \to 0$ in the limit $cq \gg \omega$; the mode is the transverse optical phonon which emerges from conventional lattice dynamics, which ignores the influence of retardation on the electric fields generated by the lattice motion. In such a description, the transverse optical phonon generates no macroscopic electric field. However, at sufficiently long wavelengths ($cq \lesssim \omega$), retardation effects lead to the generation of a macroscopic field, and the properties of the mode are altered profoundly.

There is a frequency regime where the dielectric constant $\varepsilon(\omega)$ in eq. (4.7) is negative. One sees easily this is the frequency band $\omega_{\mathrm{TO}} \leqslant \omega \leqslant \omega_{\mathrm{LO}}$, often called the Restrahl band; the reflectivity of our mode crystal is unity in this region.

Consider now a semi-infinite crystal made from our model dielectric. A new class of modes, surface polaritons, now make their appearance. These are, for the present problem, the analogue of the Rayleigh surface acoustic waves that played such a prominent role in the discussion of section 2.

Suppose we have a crystal in the lower half-space $z < 0$, with surface coincident with the xy-plane. We can obtain a description of surface polaritons by seeking solutions to the basic wave equations for which the electromagnetic field decays exponentially to zero, as one moves in either direction away from the surface. The modes are also p-polarized, and when this is combined by the requirement $\boldsymbol{\nabla} \cdot \boldsymbol{E} = 0$, then in the vacuum above the crystal one has

$$\boldsymbol{E} = E^{>}\left(\hat{q}_{\|} + \mathrm{i}\frac{q_{\|}}{\alpha_0}\hat{z}\right)\mathrm{e}^{\mathrm{i}\boldsymbol{q}_{\|}\cdot\boldsymbol{x}_{\|}}\mathrm{e}^{-\alpha_0 z}, \tag{4.8}$$

where the subscript $\|$ describes vectors parallel to the surface plane.

The wave equation requires

$$\alpha_0 = \left(q_{\|}^2 - \frac{\omega^2}{c^2}\right)^{1/2}, \tag{4.9}$$

so we are limited to the domain $q_{\|} > \omega/c$ for the field to be 'bound' to the surface. In the substrate, assuming that the frequency of the wave is such that $\varepsilon(\omega) \neq 0$, so again $\nabla \cdot \boldsymbol{E} = 0$,

$$\boldsymbol{E} = E^{<}\left(\hat{q}_{\|} - \mathrm{i}\frac{q_{\|}}{\alpha}\hat{z}\right)\mathrm{e}^{\mathrm{i}\boldsymbol{q}_{\|}\cdot\boldsymbol{x}_{\|}}\mathrm{e}^{+\alpha z}, \tag{4.10}$$

where the wave equation gives

$$\alpha = \left(q_{\|}^2 + \varepsilon(\omega)\frac{\omega^2}{c^2}\right)^{1/2}. \tag{4.11}$$

There is a further constraint on the solution. The boundary conditions of electromagnetic theory must be obeyed at the surface. Conservation of tangential $\boldsymbol{E}$ leads to the requirement $E^{>} = E^{<}$, while conservation of normal $\boldsymbol{D}$ gives

$$\varepsilon(\omega) = -\frac{\alpha}{\alpha_0}. \tag{4.12}$$

The result in eq. (4.12) is a key result. Since α and α_0 are both positive, surface polaritons exist only when $\varepsilon(\omega) < 0$, i.e., only in the Restrahl band. It is also the case that eq. (4.12) may be rearranged to give the dispersion relation of the waves:

$$\frac{c^2 q_{\|}^2}{\omega^2} = \frac{\varepsilon(\omega)}{\varepsilon(\omega) + 1}. \tag{4.13}$$

The frequencies given by eq. (4.13) all lie within the restrahl band. A sketch of the dispersion curves of the bulk and surface polaritons is given in fig. 14. Clearly the interesting region is where $q_{\|} \cong \omega_{\mathrm{TO}}/c$, since it is here that the lattice motion and electromagnetic field 'mix' strongly. For parameters typical of either alkali halides or zinc blende semiconductors, one has $q_{\|} \cong 10^3\ \mathrm{cm}^{-1}$, a wave vector quite small compared to that of visible light. We discuss the implications of this fact below.

A film may be treated by an extension of the above discussion. We refer the reader elsewhere for a detailed discussion of surface polaritons in a film grown on a substrate (Mills and Maradudin 1973). For a film of material described by a dielectric constant such as displayed in eq. (4.4), surrounded by material (possibly vacuum) with positive, frequency-independent dielectric constant, one finds two branches to the surface polariton spectrum. Each surface supports a surface wave, and the waves on each surface interact, to produce the two branches.

A medium within which the dielectric constant may be negative is referred to often as a 'surface active' medium, since its surface supports propagation of surface polaritons. One may form superlattices out of such films. The most common example is the well-known GaAs/AlAs system. In frequency regimes

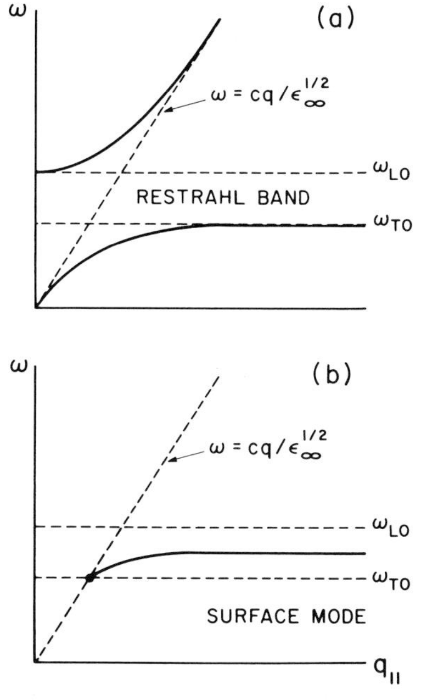

Fig. 14. We show (a) the dispersion relation of bulk polaritons and (b) the dispersion relation of surface polaritons, for the cubic ionic crystal with two atoms per unit cell, as discussed in section 4.1.

where the dielectric constant of GaAs is negative, that of AlAs is positive and conversely. Then in the appropriate frequency regime, each interface in the superlattice may support a surface polariton. These link together, to form collective excitations of the structure as a whole. A simple description of the superlattice collective modes may be given, in the absence of retardation (Camley and Mills 1978). The influence of retardation on these modes has been explored subsequently (Szenics et al. 1986). It is interesting to note that if the superlattice is terminated, one encounters collective modes of the structure localized to its surface. These may be viewed as coherent superpositions of interface modes, linked together by an exponentially decaying envelope.

4.2. The nonlinear interaction between surface polaritons and light

Before we turn to a description of experimental studies of surface polaritons by Raman scattering, we discuss the nature of the nonlinear interaction that leads to the light scattering process.

In section 1, we saw that it is the modulation of the dielectric tensor by an excitation such as a bulk or surface polariton that leads to the inelastic scattering of light. This is described schematically in eq. (1.1).

The dielectric tensor $\varepsilon_{\alpha\beta}(\boldsymbol{x}, t)$ may be written as $\varepsilon_{\alpha\beta}(\boldsymbol{x}, t) = \delta_{\alpha\beta} + 4\pi\chi_{\alpha\beta}(\boldsymbol{x}, t)$, where $\chi_{\alpha\beta}(\boldsymbol{x}, t)$ is the (fluctuating) electric susceptibility tensor. The discussion is normally phrased in terms of the latter quantity, in the literature on polaritons (see, for example, Benson and Mills 1970).

We have seen that associated with a polariton is a displacement of optical character $\boldsymbol{u}$, and in addition the macroscopic field. Each can modulate the electrical susceptibility in an independent manner. That is, if we move the ions, so $\boldsymbol{u} \neq 0$ in a wave that does not generate an electric field, $\chi_{\alpha\beta}$ is modulated, and if we apply a field with nuclei clamped in place, we once again modulate $\chi_{\alpha\beta}$. The latter is just the electro-optic effect. Thus, when a polariton (surface or bulk) is excited, the fluctuating part of the electric susceptibility tensor is

$$\delta\chi_{\alpha\beta}(\boldsymbol{x}, t) = \sum_{\lambda} \left(\frac{\partial\chi_{\alpha\beta}}{\partial u_{\lambda}}\right)_{E} u_{\lambda} + \sum_{\lambda} \left(\frac{\partial\chi_{\alpha\beta}}{\partial E_{\lambda}}\right)_{u_{\lambda}} E_{\lambda} \tag{4.14a}$$

$$\equiv \sum_{\lambda} \{a_{\alpha\beta\lambda} u_{\lambda} + b_{\alpha\beta\lambda} E_{\lambda}\}, \tag{4.14b}$$

where the coefficients $a_{\alpha\beta\lambda}$ and $b_{\alpha\beta\lambda}$ are introduced.

In eq. (14.14b), eq. (4.1) may be used to eliminate E_{λ}, so we have

$$\delta\chi_{\alpha\beta}(\boldsymbol{x}, t) = \sum_{\lambda} \tilde{a}_{\alpha\beta\lambda}(\omega) u_{\lambda}, \tag{4.15}$$

where

$$\tilde{a}_{\alpha\beta\lambda}(\omega) = a_{\alpha\beta\lambda} + \frac{M}{e^{*}}(\omega_{\mathrm{TO}}^{2} - \omega^{2}) b_{\alpha\beta\lambda}. \tag{4.16}$$

The atomic displacement and electro-optic contributions to $\delta\chi_{\alpha\beta}$ thus interfere, and the renormalized coupling constant $\tilde{a}_{\alpha\beta\lambda}(\omega)$ acquires a strong dependence on the frequency ω of the surface or bulk polariton as a consequence (Faust and Henry 1966).

The tensors $a_{\alpha\beta\lambda}$ and $b_{\alpha\beta\lambda}$ are third-rank tensors. They thus vanish in any crystal which contains an inversion center. Thus, while our description of polaritons applies nicely to alkali halide crystals, in fact, $\tilde{a}_{\alpha\beta\lambda}$ vanishes identically for these materials. This is not the case for semiconductors of the zinc blende structure, such as GaAs and GaP, and for other more complex materials of lower symmetry, which also lack an inversion center.

It is possible to break down the Raman selection rule just described, if one utilizes a laser whose frequency lies very close to the adsorption edge. In this case, $\delta\chi_{\alpha\beta}$ is influenced by terms which depend on the wave vector of the incident or scattered photon (and the polariton), and the ordinary Raman selection rules fail (Martin and Falicov 1983).

There is no equivalent to the surface ripple mechanism in the case currently under discussion, so eq. (4.15) may be used as the basis for theoretical discussions of light scattering from both surface and bulk polaritons. We shall discuss various theoretical analyses in the next subsection.

4.3. Light scattering from surface and bulk polaritons

We first begin with a discussion of light scattering from bulk polaritons in bulk crystals. Then we turn to films and multilayers, where surface and bulk polaritons make their appearance.

In the discussion above, we have seen that the interesting region of the bulk polariton dispersion curve is the region where the wave vector $q \sim 10^3\ \mathrm{cm}^{-1}$. Our earlier discussions of light scattering experiments showed that with visible light, the wave vector transfer is of the order of $10^5\ \mathrm{cm}^{-1}$, a value much larger than $10^3\ \mathrm{cm}^{-1}$. In such an experiment, we are well into the region where the polariton on the lower branch is indistinguishable from a simple transverse optical phonon. The spectrum in the Restrahl frequency region will consist of a loss at the TO phonon frequency, and also at that of the LO phonon, with relative intensity controlled by the considerations of the previous subsection.

It is possible to carry out a Raman experiment, and explore the interesting region where $q = 10^3\ \mathrm{cm}^{-1}$. This requires use of a scattering geometry where the deflection angle of the photon is very small, of the order of 10^{-2} radians or $1°$. This is then near forward scattering. The geometry presents a formidable experimental challenge, because one must look for the scattered light at angles very close to the transmitted beam (the sample must be nominally transparent, to carry out such experiments). Early experimental studies of bulk polaritons were carried out in the middle sixties (Henry and Hopfield 1965, Porto et al. 1966), and there is now an extensive literature on the subject.

It is also the case that for surface polaritons on the surface of a thick sample, or in a film, wave vectors $q_{\parallel}$ in the range of $10^3\ \mathrm{cm}^{-1}$ are of interest, as we see from fig. 14b. Now for surface polaritons, as we see from the general discussion presented in section 2, only wave vector components *parallel* to the surface are conserved. Then the regime of small $q_{\parallel}$ may be accessed in two scattering geometries, as illustrated in fig. 15. We have the backscattering geometry shown in fig. 15a, where $q_{\parallel}$ is controlled by $\Delta\theta$, and also the forward scattering geometry in fig. 15b. Any particular value of $q_{\parallel}$ may be probed in either geometry. Of course, in an experiment on a very thick or an opaque sample, the backscattering geometry must be used.

One should expect to observe surface polaritons in the Raman spectrum of light backscattered from typical semiconductors, if the skin depth is made small by using a laser frequency beyond the absorption edge. A crude estimate of the surface polariton contribution to $\langle E^2 \rangle_T$, the thermal fluctuations in the electric field within the skin depth, shows this to be comparable to the contribution from the LO phonons, which are seen clearly in such experiments. However, early attempts to detect surface polaritons in backscattering from bulk crystals were quite unsuccessful.

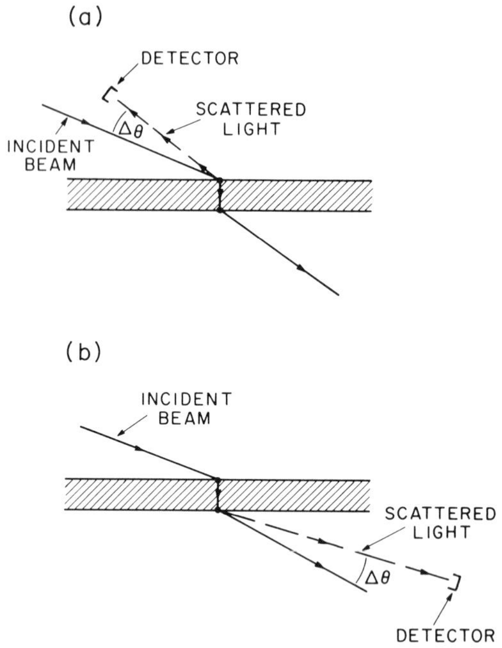

Fig. 15. A schematic illustration of (a) the backscattering and (b) the forward-scattering geometry.

The first experimental study of surface polaritons by the light scattering method was carried out in the forward scattering geometry (Evans et al. 1973). The sample was a thin (2500 Å) GaAs film on a sapphire substrate; the substrate is transparent to the laser radiation used in the experiment. Deflection angles $\Delta\theta$ (fig. 15) as small as 1° were used, and the dispersion relation of the lower of the two surface polariton branches are measured.

Theoretical studies of the scattering from surface polaritons, based on a formalism erected for this problem similar in philosophy to that discussed in section 3, provided an explanation for the failure to observe the surface polariton signal in backscattering. There is a large forward/backward asymmetry in the surface polariton cross-section (Chen et al. 1975, Mills et al. 1976), with the backscattering signal weaker than the forward scattering signal by two to three orders of magnitude. The culprit is the factor $\Delta k_{\perp}$ in the denominator of eq. (2.3), which is much larger in backscattering than it is in forward scattering. This kinematical factor greatly reduces the backscattering cross-section. The theory provides an excellent quantitative account of relative intensity of the surface polariton, bulk polariton, and LO phonon features in the forward scattering Raman spectrum.

Further detailed measurements by Prieur and Ushioda (1975) allowed the branch of the surface polariton spectrum to be resolved. This was done by a careful study of the shape of the LO phonon peak in the forward and back-scattering spectra, realizing the upper branch of the surface polariton is present as a shoulder in forward scattering, but not backscattering (Prieur and Ushioda 1975).

The experiments just described explore films so thin that all the standing-wave bulk polaritons (wave vector $\approx n\pi/D$, with D the film thickness) have wave vectors very large compared to 10^3 cm^{-1}. Thus, the frequency of these modes lie very close to ω_{TO}, and they manifest themselves in the Raman spectrum as a line at the TO phonon frequencies. Ushioda and colleagues carried out a series of experiments on rather thick, free standing GaP films (thickness 5 μm–125 μm) (Valdez et al. 1978). A number of features of these spectra are in accord with theoretical predictions (Subbaswamy and Mills 1978), but at the same time there are discrepancies which suggest that roughness or inhomogeneities may influence the spectra.

The theoretical aspects of these and other early studies of light scattering from surfaces and films are discussed in a review article that has appeared elsewhere (Mills and Subbaswamy 1981) and a companion review discusses in detail the surface polariton scattering experiments just described (Ushioda 1981).

Recent developments in materials technology, most particularly the method of molecular beam epitaxy (MBE) allow the preparation of multilayer structures of extraordinarily high quality. One makes superlattices which incorporate GaAs films, in combination with other semiconducting materials such as AlAs, or $Ga_xAl_{1-x}As$. Free carriers may be present in one or the other constituent. In such MBE synthesized multilayers, the interfaces are nearly perfect on the atomic scale.

As remarked earlier in this section, various interfaces in such structures may support surface polaritons. These link together, very much as the waves on the two surfaces of a film, to form collective excitations of the structure as a whole. In the superlattices, one has a remarkably rich spectrum of such modes, whose dispersion relation is controlled by the relative dimensions and physical properties of the films in the structure, the nature and spatial distribution of free carriers, and so on. We have entered an era where the spectrum of collective excitations of the structure can be tailored in advance, by design. Since a review of the theoretical description of such modes has been prepared quite recently (Mills 1989) we omit detailed discussion of their properties here.

Raman scattering proves a powerful and detailed probe of the collective excitations in superlattice structure. The first study of superlattice collective modes by the light scattering method (for superlattices fabricated from semiconducting materials) is the work of Cardona and his colleagues (Sood et al. 1985a, b), who studied superlattices fabricated from GaAs and AlAs. In the volume cited in the previous paragraph, the articles by Jusserand and Cardona (1989) and by Pinczuk and Abstreiter (1989) provide detailed reviews of recent experimental activity in this area. The concepts required to understand the new experiments and the underlying theory are very similar to those outlined above.

5. Light scattering by spin waves in thin films and multilayer structures

The collective excitations of spins in an ordered magnetic material are spin waves. These excitations can be studied by light scattering; the frequency of spin waves can vary over a very considerable range, so both the Brillouin and Raman technique are used in their study. In this chapter, our primary attention will be directed toward spin waves in ferromagnets. In the long-wavelength regime probed by light scattering, the frequency of these waves lies in the microwave regime. As a consequence, they are studied by Brillouin scattering.

The introduction of multi-pass Fabry–Pérot spectrometers into Brillouin spectroscopy allows one to study spin waves in ultra-thin films, as thin as a few monolayers in thickness. The Brillouin scattering technique has evolved into a primary means of deducing the magnetic parameters of such ultra-thin films, and of thin magnetic films which have been incorporated into superlattice structures. We have seen in section 3 that the Brillouin study of phonons in very thin films offers a means of deducing the elastic constants of such materials. In ferromagnetic films, often spin-wave and phonon features appear in the same light scattering spectrum. Thus, the method offers one the possibility of obtaining information on both the elastic constants and the magnetic parameters in a single experimental study.

We provide an introduction to the properties of bulk and surface spin waves in section 5.1, we discuss the coupling between light and spin motions in section 5.2, and then in section 5.3 we discuss light scattering from surface and bulk spin excitations in semi-infinite ferromagnets, in thin ferromagnetic films, and superlattices which incorporate ferromagnetic films.

At the time of this writing, there has been very little experimental attention devoted to experimental studies of surface spin waves on antiferromagnets, or of the excitations in thin films or multilayer structures which incorporate antiferromagnets, or materials whose ground-state spin arrangement has symmetry lower than ferromagnetic. There has been theoretical activity which suggests this area will prove fascinating, and recent advances in materials technology has led to the appearance of high-quality samples possibly suitable for such studies. Thus, we comment on possible areas of experimental study, and the predictions of theory below, though space considerations do not allow a full exposition of background material in this area. Magnetism is a rich but complex field, and it is difficult to describe the basic physics of the diverse materials encountered in a brief review.

5.1. Surface and bulk spin waves in ferromagnets; comments on antiferromagnets

There is an analogy between the long-wavelength spin waves in ferromagnets, and optical phonons in ionic crystals with a preferred axis that is useful, in our view. While there are clear and distinct differences that become dramatic when we discuss surface spin waves, nonetheless the comparison is useful.

The unique direction in the ferromagnet, analogous to the optic axis of an ionic dielectric of tetragonal symmetry, is defined by the direction of the magnetization $\boldsymbol{M}$. We assume $\boldsymbol{M}$ is aligned parallel to the z-axis by an external magnetic field $\boldsymbol{H}_0$.

A set of noninteracting spins placed in an external magnetic field $\boldsymbol{H}_0$ processes at the frequency γH_0, where γ is the gyromagnetic ratio, equal to 1.8×10^7 rad/gauss sec for the magnetic moments in bulk Fe. Each spin in such an ensemble would process independently of the others, so if we construct a picture of the elementary excitation spectrum of such a spin array, we have a set of Einstein oscillators, each with frequency γH_0. This frequency is the analogue of the TO optical phonon frequency in the theory of lattice vibrations of long wavelength. Of course, γH_0 lies in the microwave frequency range, while ω_{TO} lies in the infrared.

Each spin in the ensemble has a magnetic moment, and as the spin array processes coherently, a macroscopic magnetic field is generated, just as the out-of-phase optical motions of atoms within the unit cell of an ionic crystal generate a macroscopic electric field (see section 4.1). In ferromagnets, and at microwave frequencies, the influence of retardation is quite negligible, so it is an excellent approximation to ignore its influence.

When the influence of the macroscopic magnetic field generated by the spin motion is incorporated into the theory of spin waves, the frequency of the wave is found to depend on the angle $\theta_{\boldsymbol{q}}$ between the wave vector of the spin wave, and the direction of $\boldsymbol{M}$, which we assume parallel to $\boldsymbol{H}_0$. It is also the case that in an ionic crystal with optic axis, the frequencies of optical phonons depend on the angle between the wave vector and the preferred axis (see Mills and Burstein 1974). The frequency $\omega_{\boldsymbol{q}}$ of the spin wave (there is only one branch) is given by (Kittel 1963):

$$\omega_{\boldsymbol{q}} = \gamma[H_0(H_0 + 4\pi M \sin^2 \theta_{\boldsymbol{q}})]^{1/2}. \tag{5.1}$$

For propagation along the magnetization ($\theta_{\boldsymbol{q}} = 0$), no macroscopic field is generated. For bulk Fe, $4\pi M \cong 18$ kG, so the contribution from the dipole field and the external field is quite comparable, under usual laboratory conditions.

We have one remaining interaction. Two nearby spins are coupled by strong exchange interactions of quantum-mechanical origin, which have the form $-J_{12}\boldsymbol{S}_1 \cdot \boldsymbol{S}_2$. It is these strong interactions (typically two or three orders of magnitude larger than the Zeeman and magnetic dipole energies) that lock the spins into their ferromagnetic alignment. When a spin wave with wave vector $\boldsymbol{q}$ is excited, two nearby spins are no longer aligned perfectly, and there is a contribution to the excitation energy from the exchange interactions. In the limit $q \to 0$, when all spins remain strictly parallel, quite clearly exchange couplings will fail to contribute to the excitation energy. Thus, the exchange interactions 'stiffen' the response of the spin system at finite wave vector very much as the elastic restoring forces do in a lattice; the latter enter only when nearby atoms are displaced by differing amounts. In a ferromagnet, the influence of exchange on the excitation energy of a spin wave is described by an exchange stiffness constant D. Equation (5.1) then becomes

$$\omega_{\boldsymbol{q}} = \gamma[(H_0 + Dq^2)(H_0 + Dq^2 + 4\pi M \sin^2 \theta_{\boldsymbol{q}})]^{1/2}. \tag{5.2}$$

For bulk Fe, once again, $D = 2.5 \times 10^{-9}$ gauss cm^2. Thus, for the spin waves studied in light scattering, where we have q in the range 10^5 cm^{-1} to 10^6 cm^{-1} (we have in mind standing spin waves in thin films, or bulk spin waves excited within the optical skin depth, where wave vector conservation breaks down as described in section 2), the value of Dq^2 is typically small compared to H_0 or $4\pi M$, but the influence of exchange is nonetheless significant, as we shall see.

If the ferromagnet is terminated, with magnetization $\boldsymbol{M}$ in the plane of the surface, one finds surface spin waves propagate on it. Their modes have been known for many years, and are referred to as Damon–Eshbach modes (Damon and Eshbach 1961). For the case where the influence of exchange is ignored, and for propagation in the direction normal to the surface, the frequency of the Damon–Eshbach wave is given by a simple expression,

$$\omega_{\boldsymbol{q}_\parallel} = \gamma(H_0 + 2\pi M). \tag{5.3}$$

In the discussion of surface polaritons given in section 4, when retardation is ignored, the frequency of the surface wave is independent of $\boldsymbol{q}_\parallel$. One may see this by examining fig. 14, in the regime $q_\parallel \gg \omega/c$. As we shall see below, in the surface spin wave case, the frequency depends on the direction of propagation. The frequency in eq. (5.3) lies above that of the long-wavelength bulk spin waves, as described by eq. (5.1).

There is one dramatic difference between surface spin-wave propagation, and that of surface polaritons on the simple ionic crystal. The surface spin wave propagates in only one direction across the magnetization as one looks down on the crystal; there is a solution which propagates from right to left, but not from left to right! If one considers a finite film, as opposed to a semi-infinite material, then on the upper surface, there will be a wave which propagates from right to left, and on the lower surface a wave which propagates from left to right.

The asymmetric propagation characteristics are described as nonreciprocal propagation in the literature. There are other examples of nonreciprocal propagation, where application of a magnetic field parallel to the surface renders the dispersion relation (and eigenvector) of the wave different when the wave crosses the magnetic field from the left, as compared to from the right. This is the case for Rayleigh surface phonons on the surface of a ferromagnet (Scott and Mills 1977), and for surface polaritons on doped semiconductors in a magnetic field which induces cyclotron motion in the free carriers (Brion et al. 1972).

In all the cases studied so far, the nonreciprocal propagation characteristics are exhibited by surface waves, but never by the corresponding bulk waves. This rules out what might be one's first guess about the origin of this phenomena, the breakdown of time-reversal symmetry by application of the magnetic field. Scott and Mills argue that the breakdown of reflection symmetry in the presence of the surface, combined with the pseudo-vector character of the magnetic field under reflection, explains the presence of nonreciprocity in only surface wave propagation (Scott and Mills 1977). We shall see that the influence of the nonreciprocity has a dramatic influence on the surface spin wave contributions from surface spin waves.

As remarked earlier, the spin wave spectrum of antiferromagnets is very different from that of ferromagnets. In practice, it is influenced importantly by crystalline anisotropy of spin orbit origin, or with origin in internal dipole fields. The two classic and much studied antiferromagnets, MnF_2 and FeF_2, each have a unique axis of crystallographic origin. The two antiparallel sublattices are each aligned (antiparallel to each other) along the *c*-axis, and 'pinned' to it by an effective anisotropy field of strength H_A. Long-wavelength spin waves are then influenced by exchange, even at $\boldsymbol{q} = 0$. The exchange effects are described by an effective field $\boldsymbol{H}_E$ exerted by one sublattice on the second; in both materials, H_E is very large, with magnitude in the range 500 kG.

In the limit $\boldsymbol{q} = 0$, one has two spin wave branches in these materials, with frequency (Kittel 1963):

$$\omega_{\pm} = \gamma\{[2H_E H_A + H_A^2]^{1/2} \pm H_0\}, \tag{5.4}$$

with H_0 an external magnetic field applied parallel to the unique axis. The spin wave is twofold degenerate in zero magnetic field, and application of a field splits the above degeneracy.

The above discussion ignores the dipolar fields which played such a critical role in the discussion of spin waves in the ferromagnet. Dipolar contributions are indeed present in the antiferromagnet, and when they are included in the treatment once again the frequency of the spin waves depends on the direction of the wave vector relative to the unique axis. Quite frequently, the term $(2H_E H_A + H_A^2)^{1/2}$ is sufficiently large that in the antiferromagnet the dipolar fields generated by the spin motion are a modest (but at times important) correction to eq. (5.4). The frequency in eq. (5.4) will generally lie in the far infrared, in zero external field, in the frequency regime from a few wave numbers to a few tens of wave numbers.

The frequency of the lower branch drops with field, and eventually vanishes. For fields H_0 above $(2H_E H_A + H_A^2)^{1/2}$, the frequency of the lower branch becomes negative. The low-field state, with both magnetizations parallel to $\boldsymbol{H}_0$ and the optic axis, becomes unstable. There is a first-order phase transition with increasing field to a new state. The low-field and high-field states are illustrated in fig. 16. In the literature, this field-induced phase transition is referred to as the 'spin flop' transition.

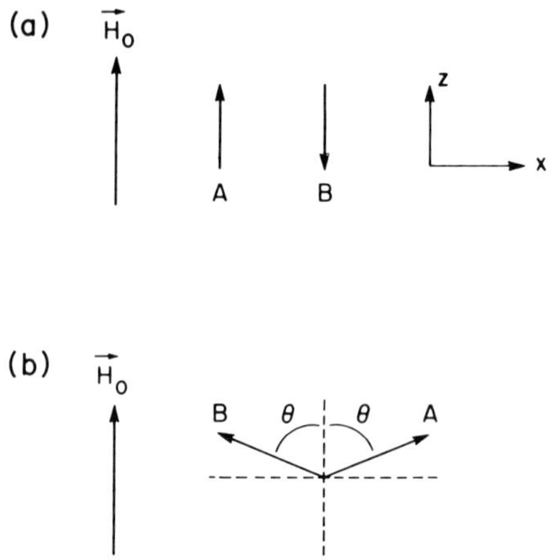

Fig. 16. We illustrate schematically the arrangement of spins in the two sublattices of a uniaxial antiferromagnet in an external magnetic field (a) below that required to induce a spin flop transition and (b) above that required to induce the spin flop transition.

Various authors have explored the theory of surface spin waves on antiferromagnets. Camley provided an early treatment of surface modes in the classic two-sublattice antiferromagnet, within the framework of a calculation that ignored retardation effects (Camley 1980). Other authors have elaborated on this basic calculation (Lüthi et al. 1983), and recently a theory of surface spin waves with magnetic field normal to the easy axis has appeared (Almeida and Mills 1988a). These various studies show that one realizes nonreciprocal propagation when the external magnetic field $\boldsymbol{H}_0 \neq 0$, and that such surfaces admit a rich spectrum of surface waves.

As remarked earlier, at the time of this writing, there are no experimental studies of surface magnons on antiferromagnets. New materials developments offer promise for the near future, however. Thin films of FeF_2 of extraordinary quality have been fabricated, and within these bulk polaritons formed by the hybridization of the electromagnetic field with the antiferromagnetic spin waves have been observed (Touissant et al. 1984). Such films, with suitable geometry, should support surface magnons.

There also are new classes of materials which, when viewed from a macroscopic perspective, may be labeled artificial antiferromagnets. For example, by MBE methods, one may synthesize very-high-quality superlattices of (nonmagnetic) Y and (ferromagnetic) Gd. In each Gd film, the magnetization is parallel to the interfaces. It is established that for a selected range of Y film thicknesses, an antiferromagnetic coupling occurs between neighboring Gd films, evidently with origin in conduction-electron polarization induced in the Y by its proximity to the ferromagnetic Gd (Majkrzak et al. 1986). The magnetic ground state then has the character of an antiferromagnet, with the direction alternating as one moves down the superlattice. Recently, an effective medium theory of the dynamic response of such a structure has been developed (Almeida and Mills 1988b), and the superlattice admits a rich spectrum of 'bulk' and 'surface' magnons.

It has been shown also that very thin films of Fe separated by a few atomic layers of Cr couple antiferromagnetically (Zhang and Zinn 1987). Preparation of superlattices from such layers would offer another means of realizing an artificial antiferromagnet.

It is the view of the present authors that artificial antiferromagnets such as those just discussed offer intriguing possibilities. In true antiferromagnets, exchange couplings between the two sublattices are very strong, and the dipole interactions which enter the theory of surface spin waves are rather weak in comparison to the other strong couplings. It is also the case that very strong magnetic fields are required to induce the sublattice canting which leads to intriguing new surface modes (Almeida and Mills 1988a). In the artificial antiferromagnets, the interlayer exchange couplings are quite weak, and modest external fields can induce very substantial cantings of the magnetic-moment array. Light scattering studies of such structures will prove intriguing.

5.2. *The coupling between light and spins in magnetic media*

In this section, we confine our attention to ferromagnetic materials, for which light scattering studies have been carried out of surface magnons on the surface of thick crystals, in films, and also superlattice structures.

In a ferromagnetic medium, the dielectric tensor is influenced by the direction of the magnetization. Thus, we may expand the dielectric tensor in powers of the magnetization, since this dependence is typically weak. Proceeding in the spirit of eq. (1.1), we have

$$\varepsilon_{\alpha\beta}(\boldsymbol{x}, t) = \varepsilon_{\alpha\beta} + \delta\varepsilon^{(1)}_{\alpha\beta}(\boldsymbol{x}, t) + \delta\varepsilon^{(2)}_{\alpha\beta}(\boldsymbol{x}, t) + \ldots, \tag{5.5}$$

where

$$\delta\varepsilon^{(1)}_{\alpha\beta}(\boldsymbol{x}, t) = \sum_{\lambda} K_{\alpha\beta\lambda} M_{\lambda}(\boldsymbol{x}, t), \tag{5.6a}$$

and

$$\delta\varepsilon^{(2)}_{\alpha\beta}(\boldsymbol{x}, t) = \sum_{\lambda\mu} G_{\alpha\beta\lambda\mu} M_{\lambda}(\boldsymbol{x}, t) M_{\mu}(\boldsymbol{x}, t). \tag{5.6b}$$

When the system is in thermal equilibrium, the time average of the magnetization $\langle \boldsymbol{M}(\boldsymbol{x}, t)\rangle = M\hat{z}$ is independent of position. Thermal excitation of spin waves leads to fluctuations in the direction of the magnetization. Thus, we write

$$\boldsymbol{M}(\boldsymbol{x}, t) = \hat{z}M + \boldsymbol{m}(\boldsymbol{x}, t), \tag{5.6c}$$

where, for small amplitude fluctuations in direction, $\boldsymbol{m}(\boldsymbol{x}, t)$ lies in the xy-plane.

At first glance, it would seem that $\delta\varepsilon^{(1)}_{\alpha\beta}(\boldsymbol{x}, t)$ should vanish, since it is linear in the magnetization, and thus changes sign under time reversal. However, if $K_{\alpha\beta\lambda}$ is complex, this term need not vanish. In a perfectly transparent crystal, in fact, $K_{\alpha\beta\lambda}$ is pure imaginary, and is also antisymmetric in the indices α and β (Landau and Lifschitz 1960). For a cubic crystal, one has

$$K_{\alpha\beta} = \mathrm{i}\lambda\varepsilon_{\alpha\beta\lambda} \tag{5.7}$$

where $\varepsilon_{\alpha\beta\lambda}$ is the Levi–Civita symbol, equal to $+1$ when $\alpha\beta\lambda$ is an even permutation of xyz, -1 when it is an odd permutation, and zero when any two indices are equal.

Both terms in eq. (5.6a) enter not only in the description of light scattering from thermal fluctuations in the spin system, but are well known in magneto optics. Setting $\boldsymbol{M}(\boldsymbol{x}, t) = \hat{z}M$ in $\delta\varepsilon^{(1)}_{\alpha\beta}$ generates the terms that control the phenomenon of Faraday rotation, and the same procedure in $\delta\varepsilon^{(2)}_{\alpha\beta}$ generate the terms that control the Cotton–Mouton effect.

The contribution to the Brillouin spectrum from scattering off single spin waves (the one-magnon spectrum) is controlled by those terms in eq. (5.6) linear in the amplitude of the spin fluctuation $\boldsymbol{m}(\boldsymbol{x}, t)$. The symmetry of the tensor $G_{\alpha\beta\lambda\mu}$ is the same as that of the elastic constants. Thus, in a cubic material, the only nonzero terms in the fluctuating part of the dielectric tensor are

$$\delta\varepsilon_{xz} = \mathrm{i}\lambda m_y + G_{44}Mm_x, \tag{5.8a}$$

$$\delta\varepsilon_{yz} = -\mathrm{i}\lambda m_x + G_{44}Mm_y. \tag{5.8b}$$

In general, a spin wave is elliptically polarized, and thus there is coherent motion of both m_x and m_y when the wave is excited. We see that there are terms contributed to $\delta\varepsilon_{xz}$ and $\delta\varepsilon_{yz}$ by both the linear and the quadratic terms in eq. (5.16). Comparisons between the theoretical calculations summarized in the next subsection and the data discussed there suggest that in the ferromagnetic metals, the contribution from the terms in G_{44} are quite small.

From the fact that only $\delta\varepsilon_{xz}$ and $\delta\varepsilon_{yz}$ are nonzero, we obtain a selection rule for scattering from spin waves in ferromagnetic media: if the incident photon has its electric vector parallel to the magnetization (which is parallel to $\hat{z}$), then the electric vector of the scattered photon must lie in the xy-plane, or conversely. The scattering event rotates the plane of polarization.

In a typical experiment performed on a semi-infinite sample, a film or a superlattice, the magnetization is parallel to $\hat{z}$ which then lies parallel to the surface. We may suppose the surface to be parallel to the xz-plane, and the y-axis normal to it. The surface spin waves discussed in the previous section propagate perpendicular to the magnetization, so $\boldsymbol{q}_{\parallel}$ is directed along $\hat{x}$. One may then choose the xy-plane to contain the wave vector of the incident and scattered photon. In this geometry, if the incident photon is p-polarized, the scattered photon will be s-polarized, and conversely. The Rayleigh surface acoustic wave (or any phonon with displacement sagittal plane) has vanishing cross-section for this geometry in a cubic crystal, so one supresses the dominant contributions to the Brillouin spectrum from phonons by using this geometry.

By using polarizers, in the same measurement one may study the phonon spectrum or the spin wave spectrum simply by rotating the polarizer for the scattered beam by ninety degrees. As we have seen, from the phonon spectrum of a thin film or multilayer sample, one may deduce the elastic constants, and we shall see that the spin wave spectra allow detailed study of the key magnetic parameters. The technique is thus a most powerful probe of the new materials emerging from thin film and multilayer synthesis studies.

5.3. *Light scattering studies of spin waves at surfaces, in thin films and in superlattices*

The first observation of a surface spin wave by the method of Brillouin spectroscopy was reported by Grünberg and Metawe (1977), in their study of ferromagnetic EuS. As we shall see in more detail below, the nonreciprocal propagation characteristics of the Damon–Eshbach surface spin waves have a dramatic effect on their Brillouin spectra, as these authors first demonstrated. Sometime later, Sandercock and Wettling reported the study of surface and bulk spin waves in Brillouin studies of thick crystals of Fe (Sandercock and Wettling 1979).

In fig. 17, we show the data of Sandercock and Wettling. The solid line is a theoretical calculation discussed below. The skin depth of this material is roughly 150 Å, and a scattering geometry similar to that described in section 5.2 was used in the experiment.

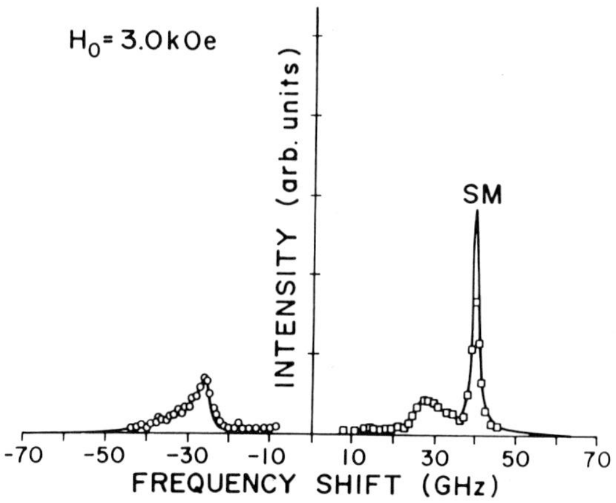

Fig. 17. The Brillouin spectrum for scattering of light by spin waves on the surface of Fe (Sandercock and Wettling 1979). The data is shown as squares, and the solid line is a theoretical fit (Camley et al. 1981).

The prominent peak labeled SM is the feature produced by the Damon–Eshbach surface spin wave. The most striking aspect of the spectrum in fig. 17 is that the surface spin wave peak appears on the Stokes side of the spectrum, but is absent from the anti-Stokes side. This is dramatic confirmation of the 'one way' nature of the propagation of the Damon–Eshbach mode. If the angle of incidence of both the incident and scattered photon is fixed, then one sees easily that if the spin wave responsible for the Stokes feature has wave vector $+\boldsymbol{q}_{\parallel}$, that responsible for the anti Stokes spectrum must have the wave vector $-\boldsymbol{q}_{\parallel}$. Thus, if the Damon–Eschbach wave appears on the Stokes side, it must be absent from the anti-Stokes side, and conversely.

The weaker bands which peak to the low-frequency side of the surface spin wave are produced by scattering from bulk spin waves, which penetrate the skin depth and scatter back off the surface. This is an opacity broadened structure, in the language of our section 2. Careful examination of shape of the bulk spin wave band on the Stokes side of the spectrum shows that it differs distinctly from that on the anti-Stokes side. This is a more subtle consequence of the symmetry breakdown that leads to the dramatic nonreciprocity in the propagation characteristics of the Damon–Eshbach wave. While the frequency of the bulk spin waves is necessarily an even function of wave vector, when a bulk wave approaches the surface, assuming its wave vector is not strictly normal to the surface, the amplitude and phase of the reflected wave differ for the two senses of propagation across the magnetization. A consequence is that the shape of the bulk spin wave contribution to the Stokes side of the laser line differs from that of the same feature on the anti-Stokes side.

The theory of the Brillouin scattering of light from spin waves within the skin depth of a semi infinite ferromagnet was developed by Camley and Mills (1978), and extended to scattering from a thin film a short time later (Camley et al. 1981). The theory is equivalent in content to that mentioned in section 3 except, of course, a description of thermal fluctuations in the spin system must be developed. These authors used a description of the incident and scattered electromagnetic field (generated by a Green's function method) valid for an arbitrary complex dielectric constant. Certain correlation functions which describe the thermal fluctuations in a spin system are required, and these are generated within a Green's function framework also. The theory of the spin wave contribution to the spin fluctuations incorporates the influence of the external magnetic field, the dipolar fields generated by the fluctuations in the spin system, and the influence of exchange, described by introduction of the exchange stiffness D. We refer the reader to the original papers for the details.

The solid line in fig. 17 is a theoretical spectrum calculated within the above framework. Since the optical constants in Fe are very well known, as are the magnetic parameters that characterize the material, no adjustable parameters have been introduced. The fit is excellent. Note that the theory reproduces very nicely the Stokes/anti-Stokes asymmetries found in the shape of the opacity-broadened bulk spin wave bands.

The first study of a thin film by the Brillouin technique was reported by Grimsditch and co-workers, who used the technique to study the spin excitations in an amorphous FeB film (Grimsditch et al. 1979). These authors detected the Damon–Eshbach mode on the side exposed to the laser beam. In addition, these authors detected several 'standing-wave resonances' of bulk spin waves in the films. The spectra have also been calculated theoretically (Camley et al. 1981) and are in impressive agreement with the data. There is a Stokes/anti-Stokes asymmetry in the intensities of the various standing-wave resonances nicely reproduced by the theory. A substantial volume of new data, and further comparisons between theory and experiment, are contained in a paper by Camley and Grimsditch (1980).

The combination of experiment and theory described above establish Brillouin spectroscopy as a viable and quantitative study of spin waves at ferromagnetic surfaces and in thin films. The spin waves of thin films have been studied traditionally by the method of ferromagnetic resonance. It is perhaps useful to compare the two techniques.

In a ferromagnetic resonance experiment, the sample is placed in a microwave cavity with one or perhaps a small number of resonance frequencies. The wavelength of the microwaves is very long compared to the typical size of a sample, so only modes with $\boldsymbol{q}_{\|} = 0$ are excited. (Actually, for a sample of dimensions $L \times L \times D$ with D the thickness, one can excite modes with $\boldsymbol{q}_{\|} = (n\hat{x} + m\hat{y})\pi/L$, where the surface is in the xy-plane. Interpretation of such spectra then requires knowledge of the boundary condition at the film edges. These are often poorly understood). One then typically sees standing-wave resonances of the film.

An external magnetic field is used to sweep the spin wave frequencies through the cavity resonance frequencies. One can study the spin waves at one, or at best, a small number of particular frequencies.

The Brillouin scattering method allows one to measure the frequency of each mode as a *function* of the applied magnetic field, over a rather wide range of fields. Data of this sort appears in the work of Grimsditch, Malozemoff and Brunsch cited above. In addition, for a sample of thickness d, and transverse dimensions L, the inequality $q_{\|} L \gg 1$ is very well satisfied. The spectra are never influenced by the boundary conditions at the film edges.

There is one more important advantage of Brillouin spectroscopy. Very-high-quality samples are required for ferromagnetic resonance studies. The film must be very homogeneous in thickness; this requirement places a limit on the range of materials that may be studied by the method. The requirements are far less severe for light scattering studies. The sample only need be homogeneous over a spatial region a bit larger than the laser spot.

Of course, the resolution achieved by Brillouin spectroscopy is far worse than achieved in ferromagnetic resonance. The line-widths observed by the former method are instrumental, while ferromagnetic resonance often allows resolution sufficient to study linewidths is considerable detail. Such linewidth analyses have proved very valuable. Thus, each method has its strengths; it is clear that Brillouin spectroscopy is a powerful new method in the study of the magnetic response of thin films.

More recently, high-quality superlattices which incorporate ferromagnetic films have been fabricated (see Schuller 1986). An example is provided by the Mo–Ni system, where one has alternating layers of ferromagnetic Ni and nonmagnetic Mo; the individual layers can be as thin as 10 Å, or as thick as several hundred Angstroms.

Theoretical analyses show that such superlattices admit a rich spectrum of spin wave modes (Camley et al. 1983b, Grünberg and Mika 1983). One may view them as follows. Each interface between a magnetic and nonmagnetic film may support Damon–Eshbach modes. The modes on the various interfaces link together to form collective excitations of the whole structure. One has propagating modes, that transport energy *normal* to the interfaces, and in addition there are surface modes that consist of interface Damon–Eshbach modes 'linked' together by an envelope that decays exponentially as one moves away from the surface into the sample. The theory makes an interesting prediction. First, if d_1 is the thickness of the ferromagnetic constituent of the superlattice, and d_2 that of the nonmagnetic constituent, the surface mode exists only in the parameter range $d_1 > d_2$. Furthermore, and quite remarkably, its frequency is given by eq. (5.3) for $\boldsymbol{q}_\parallel$ perpendicular to the magnetization, where M is now the magnetization of one of the ferromagnetic films.

Brillouin scattering studies have subsequently verified the above predictions in considerable detail, for the Mo–Ni system fabricated from films in the 100 Å–300 Å range (Grimsditch et al. 1983, Kueny et al. 1984). In the end, successful contact between theory and experiment allows one to infer the value of the magnetization within the ferromagnetic films.

More recent experimental work has focused on the study of ultra-thin films, a few tens of Angstroms in thickness or less. A key issue here is that in such ultra-thin films, the spins are subject to anisotropy fields very much larger than those characteristic of bulk magnetic matter. While the microscopic origin of such fields is not clear at the time of this writing, simple arguments based on the low symmetry of the atomic environment in such films lead one to expect such large anisotropy fields. Recent Brillouin scattering work shows that these anisotropy fields strongly influence the frequency of spin waves in superlattices fabricated from ultra-thin films. A very detailed and quantitative analysis of such effects is provided by recent studies of Fe–Pd superlattices (Hillebrands et al. 1988a).

It will be very interesting to use light scattering to study the dynamical response of the new MBE-grown rare earth superlattices, such as the Y–Gd system mentioned at the end of section 5.1. When the Gd film moments are aligned antiparallel, the spin wave frequencies lie above the microwave range used in magnetic resonance studies. Light scattering spectroscopy thus emerges as the most appropriate means of probing their dynamic response.

Another intriguing structure that has been synthesized recently is the Y-Dy superlattice (Salamon et al. 1986). Dysprosium is a hexagonal crystal; in the bulk the spins order into an incommensurate spiral. Within each plane perpendicular to the *c*-axis, one realizes ferromagnetic alignment of spins, and as one progresses along the *c* direction, the moments in each sheet rotate, to produce a spiral spin state. In the superlattice, neutron studies detect no magnetic moment within the Y films, but the spiral state is realized within the Dy films. The ground state is coherent, in that the spiral in a given Dy film is coupled to that in neighboring films by magnetic couplings mediated by the intervening Y layers. These materials are a fascinating new form of magnetic matter, and the study of their dynamical response will prove fascinating.

References

Albuquerque, E.L., 1980, J. Phys. C **13**, 2623.
Albuquerque, E.L., R. Loudon and D.R. Tilley, 1980, J. Phys. C **13**, 1775.
Albuquerque, E.L., M.C. Oliveros and D.R. Tilley, 1984, J. Phys. C **17**, 1451.
Albuquerque, E.L., N.S. Almeida and M.C. Oliveros, 1985, Z. Phys. B **59**, 311.
Almeida, N.S., and D.L. Mills, 1988a, Phys. Rev. B **37**, 3400.
Almeida, N.S., and D.L. Mills, 1988b, Phys. Rev. B **38**, 6698.
Auld, B.A., 1973, in: Acoustic Fields and Waves in Solids, Vol. 2 (Wiley, New York).
Auld, B.A., G.S. Beaupre and G. Hermann, 1977, Electron. Lett. **13**, 525.
Babiker, M., D.R. Tilley, E.L. Albuquerque and C.E.T. Gongalves da Silva, 1985a, J. Phys. C **18**, 1269.
Babiker, M., D.R. Tilley and E.L. Albuquerque, 1985b, J. Phys. C **18**, 1285.
Bassoli, L., F. Nizzoli and J.R. Sandercock, 1986, Phys. Rev. B **34**, 1296.
Baumgart, P., B. Hillebrands, R. Mock, G. Guntherodt, A. Boufelfel and C.M. Falco, 1986, Phys. Rev. **34**, 9004.
Bell, J.A., W.R. Bennett, R. Zanoni, G.I. Stegeman, C.M. Falco and F. Nizzoli, 1987a, Phys. Rev. B **35**, 4127.
Bell, J.A., W.R. Bennett, R. Zanoni, G.I. Stegeman, C.M. Falco and C.T. Seaton, 1987b, Solid State Commun. **64**, 1339.
Bell, J.A., R. Zanoni, C.T. Seaton, G.I. Stegeman, W.R. Bennett and C.M. Falco, 1987c, Appl. Phys. Lett. **51**, 652.
Bell, J.A., R. Zanoni, C.T. Seaton, G.I. Stegeman, J. Mokous and C.M. Falco, 1988, Appl. Phys. Lett. **52**, 610.
Benson, H., and D.L. Mills, 1970, Phys. Rev. B **1**, 4835.
Bisanti, P., M.B. Brodsky, G.P. Felcher, M. Grimsditch and L.R. Sill, 1987, Phys. Rev. B **35**, 7813.
Born, M., and K. Huang, 1954, in: Dynamical Theory of Crystal Lattices (Oxford University Press, Oxford).
Bortolani, V., F. Nizzoli and G. Santoro, 1978, Phys. Rev. Lett. **41**, 39.
Bortolani, V., F. Nizzoli, G. Santoro, A.M. Marvin and J.R. Sandercock, 1979, Phys. Rev. Lett. **43**, 224.
Bortolani, V., F. Nizzoli, G. Santoro, J.R. Sandercock and A.M. Marvin, 1980, in: Proc. 7th Int. Conf. on Raman Spectroscopy, ed. W.F. Murphy (North-Holland, Amsterdam) p. 442.
Bortolani, V., F. Nizzoli, G. Santoro and J.R. Sandercock, 1982, Phys. Rev. B **25**, 3442.
Bortolani, V., A.M. Marvin, F. Nizzoli and G. Santoro, 1983, J. Phys. C **16**, 219.

Bortolani, V., F. Nizzoli and G. Santoro, 1984, J. Phys. Colloq. (France) **45**, C5-45.
Brion, J.J., R.F. Wallis, A. Hartstein and E. Burstein, 1972, Phys. Rev. Lett. **28**, 1455.
Bruggeman, D.A.G., 1937, Ann. Phys. (Leipzig) **29**, 160.
Camley, R.E., 1980, Phys. Rev. Lett. **45**, 283.
Camley, R.E., and M. Grimsditch, 1980, Phys. Rev. B **22**, 5420.
Camley, R.E., and D.L. Mills, 1978, Phys. Rev. B **18**, 4821.
Camley, R.E., and F. Nizzoli, 1985, J. Phys. C **18**, 4795.
Camley, R.E., T.S. Rahman and D.L. Mills, 1981, Phys. Rev. B **23**, 1226.
Camley, R.E., B. Djafari-Rouhani, L. Dobrzynski and A.A. Maradudin, 1983a, Phys. Rev. B **27**, 7318.
Camley, R.E., T.S. Rahman and D.L. Mills, 1983b, Phys. Rev. B **27**, 261.
Cammarata, R.C., 1986, Scr. Metall. **20**, 479.
Chen, Y., E. Burstein and D.L. Mills, 1975, Phys. Rev. Lett. **34**, 1516.
Colvard, C., M. Merlin, M.V. Klein and A.C. Gossard, 1980, Phys. Rev. Lett. **45**, 298.
Damon, R., and J. Eshbach, 1961, J. Phys. & Chem. Solids **19**, 308.
Danner, R., R.P. Huebener, C.S.L. Chun, M. Grimsditch and I.K. Schuller, 1986, B **33**, 3696.
Dervisch, A., and R. Loudon, 1976, J. Phys. C **9**, 1669.
Dil, J.G., N.C.J.A. van Hijningen, F. van Dorst and R.M. Aarst, 1981, Appl. Opt. **20**, 1374.
Djafari-Rouhani, B., and J. Sapriel, 1986, Phys. Rev. B **34**, 7114.
Djafari-Rouhani, B., L. Dobrzynski, O. Hardhouin Duparc, R.E. Camley and A.A. Maradudin, 1983, Phys. Rev. B **28**, 1711.
Djafari-Rouhani, B., A.A. Maradudin and R.F. Wallis, 1984, Phys. Rev. B **29**, 6465.
Dobrzynski, L., B. Djafari-Rouhani and O. Hardhouin Duparc, 1984, Phys. Rev. B **29**, 3138.
Evans, D.J., S. Ushioda and J.D. McMullen, 1973, Phys. Rev. Lett. **31**, 372.
Farnell, G.W., 1970, Properties of Elastic Surface Waves, in: Physical Acoustics, Vol. 6, eds W.P. Mason and R.N. Thurston (Academic Press, New York) p. 109.
Farnell, G.W., and E.L. Adler, 1972, Elastic Wave Propagation in Thin Layers, in: Physical Acoustics, Principles and Methods, Vol. 9, eds W.P. Mason and R.N. Thurston (Academic Press, New York) p. 35.
Faust, W.L., and C.H. Henry, 1966, Phys. Rev. Lett. **17**, 1265.
Fernandez, L., and V.R. Velasco, 1987, Surf. Sci. **185**, 175.
Fukui, M., O. Toda, V.C.-Y. So and G.I. Stegeman, 1980, Solid State Commun. **36**, 995.
Fukui, M., O. Toda, V.C.-Y. So and G.I. Stegeman, 1981, J. Phys. C **14**, 5591.
Glass, N.E., and A.A. Maradudin, 1983, J. Appl. Phys. **54**, 79.
Grimsditch, M., 1985, Phys. Rev. B **31**, 6818.
Grimsditch, M., and F. Nizzoli, 1986, Phys. Rev. B **33**, 5891.
Grimsditch, M., W. Senn, G. Winterling and M.H. Brodsky, 1978, Solid State Commun. **26**, 229.
Grimsditch, M., A. Malozemoff and A. Brunsch, 1979, Phys. Rev. Lett. **43**, 711.
Grimsditch, M., M.R. Khan and I.K. Schuller, 1983, Phys. Rev. Lett. **51**, 498.
Grimsditch, M., R. Bhadra and I.K. Schuller, 1987, Phys. Rev. Lett. **58**, 1216.
Grünberg, P., and F. Metawe, 1977, Phys. Rev. Lett. **39**, 1561.
Grünberg, P., and K. Mika, 1983, Phys. Rev. B **27**, 2955.
Henry, C.H., and J.J. Hopfield, 1965, Phys. Rev. Lett. **15**, 964.
Hillebrands, B., R. Mock, G. Güntherodt, P.S. Bechthold and N. Herres, 1986, Solid State Commun. **60**, 649.
Hillebrands, B., A. Boufelfel, C.M. Falco, P. Baumgart, G. Güntherodt, E. Zirngiebl and J.D. Thompson, 1988a, J. Appl. Phys. **63**, 3880.
Hillebrands, B., S. Lee, G.I. Stegeman, H. Cheng, J.E. Potts and F. Nizzoli, 1988b, Phys. Rev. Lett. **60**, 832.
Huberman, M.L., and M. Grimsditch, 1989, Phys. Rev. Lett. **62**, 1403.
Jorna, R., D. Visser, V. Bortolani and F. Nizzoli, 1989, J. Appl. Phys. **65**, 718.

Jusserand, B., and M. Cardona, 1989, in: Light Scattering in Solids, Vol. 5, eds M. Cardona and G. Güntherodt (Springer, Berlin) ch. 3, p. 49.
Khan, M.R., C.S.L. Chun, G.P. Felcher, M. Grimsditch, A. Kueny, C.M. Falco and I.K. Schuller, 1983, Phys. Rev. **27**, 7186.
Kittel, C., 1963, Quantum Theory of Solids (Wiley, New York) ch. 4.
Kueny, A., and M. Grimsditch, 1982, Phys. Rev. B **26**, 4699.
Kueny, A., M. Grimsditch, K. Miyano, I. Banerjee, C.M. Falco and I.K. Schuller, 1982, Phys. Rev. Lett. **48**, 166.
Kueny, A., M.R. Khan, I.K. Schuller and M. Grimsditch, 1984, Phys. Rev. B **29**, 2879.
Landau, L.D., and E.M. Lifshitz, 1959, Theory of Elasticity (Pergamon Press, London) pp. 109–113.
Landau, L.D., and E.M. Lifshitz, 1960, Electrodynamics of Continuous Media (Pergamon Press, Oxford) pp. 331ff.
Lee, S., B. Hillebrands, G.I. Stegeman, H. Cheng, J.E. Potts and F. Nizzoli, 1988, J. Appl. Phys. **63**, 1914.
Loudon, R., 1978a, Phys. Rev. Lett. **40**, 581.
Loudon, R., 1978b, J. Phys. C **11**, 403.
Loudon, R., 1984, Ripples on liquid interfaces, in: Surface Excitations, eds V.M. Agranovich and R. Loudon (North-Holland, Amsterdam) p. 589.
Loudon, R., and J.R. Sandercock, 1980, J. Phys. C **13**, 2609.
Lüthi, B., R.E. Camley and D.L. Mills, 1983, Phys. Rev. B **28**, 1475.
Majkrzak, C.F., J.W. Cable, J. Kwo, M. Hong, D.B. McWhan, Y. Yafet, J.V. Wasszczak and C. Vettier, 1986, Phys. Rev. Lett. **56**, 2700.
Martin, R.M., and L.M. Falicov, 1983, in: Light Scattering in Solids, Vol. 1, ed. M. Cardona (Springer, Heidelberg) p. 79.
Marvin, A.M., V. Bortolani and F. Nizzoli, 1980a, J. Phys. C **13**, 299.
Marvin, A.M., V. Bortolani, F. Nizzoli and G. Santoro, 1980b, J. Phys. C **13**, 1607.
Marvin, A.M., V. Bortolani, F. Nizzoli, G. Santoro and V. Celli, 1982, J. Phys. C **15**, 3273.
Mills, D.L., 1989, in: Light Scattering in Solids, Vol. 5, eds M. Cardona and G. Güntherodt (Springer, Berlin) ch. 2, p. 13.
Mills, D.L., and E. Burstein, 1974, Rep. Prog. Phys. **37**, 417.
Mills, D.L., and A.A. Maradudin, 1973, Phys. Rev. Lett. **31**, 372.
Mills, D.L., and K.R. Subbaswamy, 1981, Progress in Optics, Vol. 19, ed E. Wolf (North-Holland, Amsterdam) p. 47.
Mills, D.L., A.A. Maradudin and E. Burstein, 1970, Ann. Phys. **56**, 504.
Mills, D.L., Y. Chen and E. Burstein, 1976, Phys. Rev. B **13**, 4419.
Moretti, A.L., W.M. Robertson, B. Fisher and R. Bray, 1985, Phys. Rev. B **31**, 3361.
Nizzoli, F., R. Bhadra, O.F. deLima, M.B. Brodsky and M. Grimsditch, 1988, Phys. Rev. B **37**, 1007.
Nizzoli, F., B. Hillebrands, S. Lee, G.I. Stegeman, G. Duda, G. Wegner and W. Knoll, 1989, Phys. Rev. B **40**, 3323.
Nougaoui, A., and B. Djafari-Rouhani, 1987, Surf. Sci. **185**, 154.
Pinczuk, A., and G. Abstreiter, 1989, in: Light Scattering in Solids, Vol. 5, eds M. Cardona and G. Güntherodt (Springer, Berlin) ch. 4, p. 153.
Porto, S.P.S., B. Tell and T.C. Damon, 1966, Phys. Rev. Lett. **16**, 450.
Prieur, J.Y., and S. Ushioda, 1975, Phys. Rev. Lett. **34**, 1012.
Robertson, W.M., A.L. Moretti and R. Bray, 1987, Phys. Rev. B **35**, 8919.
Rowell, N.L., and G.I. Stegeman, 1978a, Phys. Rev. Lett. **41**, 970.
Rowell, N.L., and G.I. Stegeman, 1978b, Phys. Rev. B **18**, 2598.
Rowell, N.L., and G.I. Stegeman, 1982a, Can. J. Phys. **60**, 1788.
Rowell, N.L., and G.I. Stegeman, 1982b, Can. J. Phys. **60**, 1804.

Rowell, N.L., V.C.-Y. So, R. Normandin and G.I. Stegeman, 1982, Can. J. Phys. **60**, 1821.
Rytov, S.M., 1956, Akust. Zh. **2**, 71 [Sov. Phys.-Acoust. **2**, 681].
Salamon, M.B., S. Sinha, J.J. Rhyne, J.E. Cunningham, R.W. Erwin, J. Borchers and C.P. Flynn, 1986, Phys. Rev. Lett. **56**, 259.
Sandercock, J.R., 1978, Solid State Commun. **26**, 547.
Sandercock, J.R., 1982, Trends in Brillouin Scattering, in: Light Scattering in Solids, Vol. 3, eds M. Cardona and G. Güntherodt (Springer, Berlin) p. 173.
Sandercock, J.R., and W. Wettling, 1979, J. Appl. Phys. **50**, 7784.
Sandercock, J.R., F. Nizzoli, V. Bortolani, G. Santoro and A.M. Marvin, 1980, Suppl. to Le Vide, Les Couches Minces **201**, 754.
Sapriel, J., B. Djafari-Rouhani and L. Dobrzynski, 1983a, Surf. Sci. **126**, 197.
Sapriel, J., J.C. Michel, J.C. Toledano, R. Vacher, J. Kervarec and A. Regreny, 1983b, Phys. Rev. B **28**, 2007.
Sapriel, J., J.C. Michel, P. Henoc and Y.I. Nissim, 1985, Inst. Phys. Conf. Ser. No. 74, ch. 2, p. 115.
Schuller, I.K., 1986, in: Frontiers in Electronic Materials and Processing, ed. L.J. Brillson, Am. Inst. Phys. Conf. Proc. **138**, 93.
Schuller, I.K., and M. Grimsditch, 1986, J. Vac. Sci. & Technol. B **4**, 1444.
Schuller, I.K., and A. Rahman, 1983, Phys. Rev. Lett. **50**, 1377.
Scott, R.Q., and D.L. Mills, 1977, Phys. Rev. B **15**, 3545.
Sood, A.K., J. Menendez, M. Cardona and K. Ploog, 1985a, Phys. Rev. B **32**, 1412.
Sood, A.K., J. Menendez, M. Cardona amd K. Ploog, 1985b, Phys. Rev. Lett. **54**, 2115.
Subbaswamy, K.R., and A.A. Maradudin, 1978, Phys. Rev. B **18**, 4181.
Subbaswamy, K.R., and D.L. Mills, 1978, Solid State Commun. **27**, 1085.
Sussner, H., J. Pelous, M. Schmidt and R. Vacher, 1980, Solid State Commun. **36**, 123.
Szenics, R., R.F. Wallis, G.F. Guiliani and J.J. Quinn, 1986, Surf. Sci. **166**, 45.
Toussaint, R.M., D.W. Hone, V. Jaccarino and S.M. Rezende, 1984, Phys. Rev. B **30**, 3859.
Ushioda, S., 1981, Progress in Optics, Vol. 19, ed. E. Wolf (North-Holland, Amsterdam) p. 141.
Vacher, R., H. Sussner and M. Schmidt, 1980, Solid State Commun. **34**, 279.
Valdez, J.B., G. Mattei and S. Ushioda, 1978, Solid State Commun. **27**, 1085.
Velasco, V.R., and F. Garcia-Moliner, 1980, Solid State Commun. **33**, 1.
Wallis, R.F., 1973, Prog. Surf. Sci. **4**, 233.
Wolf, D., and J.F. Lutsko, 1988, Phys. Rev. Lett. **60**, 1170.
Wright, G.B., 1969, Light Scattering Spectra of Solids (Springer, New York).
Zanoni, R., C. Naselli, J.A. Bell, G.I. Stegeman and C.T. Seaton, 1986, Phys. Rev. Lett. **57**, 2838.
Zhang, P.X., and W. Zinn, 1987, J. Vac. Sci. & Technol. A **5**, 1910.

CHAPTER 8

Surface Magnetic and Structural Phase Transitions

M.I. KAGANOV, A.P. LEVANYUK, S.A. MINYUKOV
and A.V. CHUBUKOV

Institute of Physical Problems
Academy of Sciences of the USSR
Uliza Kosigna 2
117334 Moscow, USSR

Nonlinear Surface Electromagnetic Phenomena
Edited by
H.-E. Ponath and G.I. Stegeman

© *Elsevier Science Publishers B.V., 1991*

Contents

1. Introduction

Phase transitions occurring on the surface of crystals have repeatedly attracted the attention of researchers in recent years. This was largely caused by the fact that in the neighborhood of the point of surface transition (of the second order) the system is nearly as 'soft' as it is in the neighborhood of a transition point in an unbounded specimen. As a result, the role of nonlinear affects is considerably enhanced, as it happens in bulk transitions.

The experimental investigation of surface transitions is still at a very early stage. Only a few communications appeared on continuous reconstruction of surface structures (Compuzano et al. 1985, Weller et al. 1985, Francis and Richardson 1986). A much greater number of papers were devoted to a theoretical analysis of surface transitions. A number of different theoretical approaches to study the role played by the surface are known by now. In fact, all techniques which were earlier employed to analyze the phase transitions in unbounded specimens found their use in the new field: high-temperature and low-temperature expansions, Monte-Carlo computations, ε-expansion, scaling analysis, and search for exact solutions for specific models. We shall not cite all the relevant publications (they are too numerous). The reader can find the necessary information in the review by Binder (1983).

This paper is aimed at acquainting the readers with the work carried out in the framework of the phenomenological approach traceable to the Ginzburg–Landau method in the theory of superconductivity (Ginzburg and Landau 1950). Presumably, this approach is best suited to a qualitative description of the effects and to the evaluation of the feasibility of observing the predicted anomalous properties of the system. We will begin with a phenomenological description of the phase transition, assuming sign reversal in one of the coefficients in the expansion of the free-energy density in powers of the order parameter. A difference between the surface and bulk forces results in an inhomogeneous order parameter and, therefore, affects the phase transition. The peculiarities in the behavior of the system are described by introducing the surface energy, which is assumed to be proportional to the squared order parameter. The proportionality factor (denoted by q) is an additional phenomenological parameter of the theory (Kaganov and Omel'yanchuk 1971, Kaganov 1972, Mills 1971). Higher-order terms in the expansion of surface energy in powers of the order parameter are not needed because the parameter q can hardly be expected to vanish in the temperature range which is of interest here. Note that in some cases the phenomenological functional can be obtained by a straightforward averaging of the microscopic Hamiltonian. Thus the relation between q and Δ for magnetics with the exchange interaction between nearest neighbors equal to $J(1 + \Delta)$ for the spins on the surface and to J for all other spins is (Mills 1971, Binder and Hohenberg 1972, 1974, Lubensky and

Rubin 1975)

$$q = \frac{1}{a}(1 - 2(D-1)\Delta), \tag{1}$$

where a is the interatomic spacing and D is the dimensionality of space.

The first part of the review treats the surface magnetic phase transitions of the order–disorder type (at the Curie temperature T_C) and of the orientational type. In the former case the order parameter is the magnetic-moment density $M(\boldsymbol{r})$ and in the latter, the angle $\theta(\boldsymbol{r})$ between the magnetic moment and one of the axes of the crystal. The cases of a specimen occupying the half-space $z \geqslant 0$ and of a finite-thickness plate $(-d \leqslant z \leqslant d)$ are treated separately.

Obviously, the phenomenological approach to the problem of phase transition is valid only outside the fluctuation region. This region is known to be fairly wide for order–disorder transitions: the Ginzburg–Levanyuk criterion (Levanyuk 1959, Ginzburg 1960) is violated even for $|T - T_C|/T_C \leqslant 1$, and hence, the phenomenological approach can at best give a qualitative description of a phase transition; furthermore, the values of the critical exponents deviate from the predictions of the Landau theory practically everywhere in the region where the expansion of the thermodynamic potential in powers of the order parameter is valid.* In contrast to this, in the case of orientational transitions the condition of validity of the Landau theory is violated only in the immediate vicinity of the transition point, because the inhomogeneous term in the free energy is determined by relatively large exchange forces, while reorientation is determined by the dependence of small (relativistic) anisotropy constants on temperature. We will show that the fluctuation region for half-space is wider than in an unbounded specimen, and even more so for a finite-thickness plate, but that the Ginzburg–Levanyuk parameter remains much less than unity at reasonable values of surface-energy constant and plate thickness.

Another important feature is the two-dimensional character of the system in the vicinity of the phase transition point; this is a corollary of the boundedness of the specimen in one dimension. It is well known (see, e.g., Patashinsky and Pokrovsky 1982) that magnetic ordering is impossible in 2D systems with a continuous symmetry group. As we want to avoid discussing the fluctuations in 2D systems (it would disagree with the spirit of this review), we always treat systems of Ising type, with one-component order parameter, for which the 2D space is not critical for fluctuations. This restriction to Ising systems not only facilitates the justification of the phenomenological treatment of surface transitions but, in addition, serves to avoid unwieldy formulas. Finally, note that the analysis is restricted to relatively simple situations in which the magnetic moment on the surface and in the bulk is parallel to the surface of the specimen

*Note that the dependence of thermodynamic characteristics on plate thickness also changes (Kaganov and Karpinskaya 1979).

and, therefore, there are no demagnetizing fields. Otherwise, the surface results in the formation of a striped domain structure (Bespyatnikh et al. 1980, Chubukov 1982).

It is fairly obvious that the main ideas of the theory of magnetic surface transitions can be extended to structural phase transitions, including ferroelectric ones. It is found, however, that these two cases are essentially different both regarding the feasibility of the phase transition (Levanyuk and Minyukov 1983a, b) and the possibility of observation of different anomalies.

For structural transitions in 2D systems with short-range interactions, the Landau theory proves to be inapplicable in all temperature regions, both for order–disorder and displacive systems. However, some formulae of the Landau theory can be easily modified for the fluctuation region. Moreover, for ferroelectrics with a one-component order parameter, the region of applicability of Landau theory exists even in the two-dimensional case, due to long-range field effects. That is why we shall use below the language of phenomenological theory.

The review is organised as follows: sections 2 and 3 treat the order–disorder and orientational transitions, respectively. Section 4 discusses the role of fluctuations. The structural phase transitions are analysed in section 5. Sections 2–4 are written by Kaganov and Chubukov and section 5 by Levanyuk and Minyukov.

2. *Order–disorder transitions*

Our starting point is the expansion of the energy density in integral (even) powers of the magnetic moment and its gradient

$$f\{M(r)\} = AM^2 + BM^4 + C(\nabla M)^2, \tag{2}$$

where

$$A = A(\tau) = \beta\left(\frac{T_C}{\mu M_0}\right)\tau, \qquad B = \beta'\left(\frac{T_C}{\mu M_0}\right)\frac{1}{M_0^2},$$

$$C = \beta''\left(\frac{T_C}{\mu M_0}\right)a^2, \qquad \tau = \frac{T - T_C}{T_C}, \quad \beta, \beta', \beta'' > 0,$$

a is the interatomic spacing, M_0 is the magnetic moment of a unit volume at saturation, μ is the Bohr magneton, and the coefficients β, β' and β'' are of order unity. In the framework of the phenomenological description, the introduction of a surface adds to the bulk component of total free energy of a plate (the integral of $f\{M(r)\}$) another term, $2F_S$, where

$$F_S = qCM^2(d), \quad [q] = 1/\text{cm}, \tag{3}$$

is the surface component of the free energy (the factor 2 appears because of the two interfaces, assumed to be identical, of the plate). At equilibrium, the magnetic-moment density can depend only on the coordinate z along the perpendicular to the plate surface. The distribution of magnetization being symmetrical, it is convenient to look for $M(z)$ in the interval $0 \leqslant z \leqslant d$, adding a boundary condition at $z = 0$:

$$M'(0) = 0, \tag{4}$$

the prime symbolizing the derivative with respect to z.

Minimizing the free energy, we arrive at the equation

$$CM'' = AM + 2BM^3. \tag{5}$$

The boundary condition for eq. (5) is obtained by demanding the variation in $M(d)$ to vanish:

$$M'(d) + qM(d) = 0. \tag{6}$$

Finding the first integral of eq. (5) is elementary:

$$C(M')^2 = \tilde{f}\{M(z)\} - \tilde{f}\{M(0)\}, \tag{7}$$

where $\tilde{f}\{...\}$ is given by eq. (2) without the last term.

Obviously, the sign of M' is determined by the sign of the coefficient in the expression for surface energy:

$$M'(z) = -\frac{\operatorname{sign} q}{\sqrt{C}}[\tilde{f}\{M(z)\} - \tilde{f}\{M(0)\}]^{1/2}. \tag{8}$$

In principle, eqs. (6) and (7) allow the determination of the magnetic moment as a function of the coordinate z. In fact, it is found that only the half-space solution can be obtained as a combination of elementary functions. On the other hand, if one wishes to construct the phase diagram and determine the thermodynamic functions, it is convenient to use eq. (8) and express $\mathrm{d}z$ via $\mathrm{d}M$, converting the functional $F\{M(z)\}$ into a function in two variables, $M(0)$ and $M(d)$. Obviously, the condition of the minimum of the function in two variables coincides with boundary condition (6) and eq. (8) integrated over the plate thickness. The approach based on substituting a function in two variables for the functional is especially convenient for the case of finite-thickness plates, since a straightforward analysis of eq. (8) is then difficult.

2.1. *Half-space*

In this section it is convenient to place the origin of coordinates at the boundary, i.e., to assume that the magnet occupies the space $z \geqslant 0$. Condition (4) is 'extended' to infinity and condition (6) to zero, so that the sign of M' is reversed:

$$M'(0) = qM(0). \tag{9}$$

Deep into the specimen the gradient plays a negligible role, so that

$$M_\infty = \begin{cases} 0, & \tau \geqslant 0, \\ (-A/2B)^{1/2}, & \tau < 0. \end{cases} \tag{10}$$

In this case the solution of eq. (8) is found readily, and has identical form for positive and negative q when $\tau < 0$ (i.e., if there is long-range ordering in the bulk):

$$M(z) = M_\infty \frac{(M_S + M_\infty) + (M_S - M_\infty)\exp(-z/\delta_V)}{(M_S + M_\infty) - (M_S - M_\infty)\exp(-z/\delta_V)}, \tag{11}$$

where $\delta_V = (-C/2A)^{1/2}$ and $M_S \equiv M(0)$ is the value taken by the magnetic moment at the boundary:

$$M_S = \left(\frac{q^2 C}{4B} + M_\infty^2\right)^{1/2} - \frac{q}{2}\left(\frac{C}{B}\right)^{1/2}. \tag{12}$$

If $q > 0$, the surface 'resists' the phase transition, so that $M_S < M_\infty$, and when M_∞ vanishes, M_S vanishes as well. In fact, M_S tends to zero faster than M_∞. According to eq. (12),

$$M_S|_{T \lessapprox T_c} \approx \frac{M_\infty^2}{q}\left(\frac{B}{C}\right)^{1/2} \sim M_0|\tau|/q. \tag{13}$$

Therefore, if $q > 0$, the surface produces a rather insignificant effect. Namely, the transition takes place simultaneously throughout the specimen's bulk, and the only difference is a change in the temperature dependence of the surface magnetization (eq. (13)) in comparison with that of the bulk. According to the classification suggested by Lubensky and Rubin (1975) and Bray and Moore (1977), this transition is said to be ordinary.

If the sign of surface energy is reversed (i.e., the surface is predisposed to magnetic ordering), a qualitatively different situation is produced. Namely, eq. (12) implies that the surface moment does not vanish at $\tau = 0$:

$$M_S|_{T=T_C} = |q|\left(\frac{C}{B}\right)^{1/2}. \tag{14}$$

This means that before magnetization develops in the bulk, certain ordering (magnetization) appears in the subsurface layer and decays into the bulk. This phenomenon was called surface magnetism. According to eqs. (8) and (9), at $\tau > 0$.

$$M(z) = M \sinh^{-1}(\kappa + z/\delta_S), \tag{15}$$

where

$$\kappa = \tfrac{1}{2}\ln\frac{(\tilde{M}^2 + M_S^2)^{1/2} + \tilde{M}}{(\tilde{M}^2 + M_S^2)^{1/2} - \tilde{M}}, \quad \tilde{M} = \left(\frac{A}{B}\right)^{1/2}, \quad \delta_S = \left(\frac{C}{A}\right)^{1/2}$$

(the subscript S emphasizes that surface magnetism is meant), and M_S is found, as always, from the boundary condition, eq. (9):

$$BM_S^2 = Cq^2 - A. \tag{16}$$

The threshold for the existence of surface magnetism, T_S, is found by setting M_S to zero:

$$\tau_S = \beta''(qa)^2/\beta. \tag{17}$$

As we see from eq. (15), deep into the specimen, as $z \to \infty$, the magnetic moment tends to zero exponentially. As $\tau \to 0$, the depth of penetration δ_S tends to infinity and ultimately the bulk magnetization appears at $\tau = 0$. At this point $M(z)$ is a power function:

$$M(z) = M_S/(1 + |q|z), \quad \tau = 0. \tag{18}$$

At $q < 0$ the transition from the paramagnetic to the ferromagnetic state thus occurs in the following way: first (at $T = T_S > T_C$) a magnetization appears close to the surface and propagates into the specimen as temperature is reduced; then, at $T = T_C$, the bulk magnetic moment sets in. In accepted terminology (Lubensky and Rubin 1975, Bray and Moore 1977), the transition at $T = T_S$ is referred to as the surface transition, and that at $T = T_C$ as the extraordinary transition. This last transition is nontrivial in that it is 'felt' by surface characteristics: at the transition point the second derivative (with respect to temperature) of surface magnetization undergoes a finite jump.* The phase diagram close to T_C is schematically shown in fig. 1. Note a special position of the point P at which three curves of phase transition converge. Each of the curves is characterized by specific surface characteristics. If we move strictly along the τ-axis ($q = 0$), the features of surface characteristics in the neighborhood of P will differ from those in the neighborhoods of each of the three phase-transition curves converging to P. For this reason the transition at $q = 0$ can be subsumed under the fourth, special type of phase transition connected with the surface. In the terminology suggested by Bray and Moore (1977), this transition is referred to as 'special'.** The information on the surface characteristics in the vicinity of each of the mentioned transition types is given in table 1 [compiled mostly on the basis of the work of Bray and Moore (1977)]. The equations of state are written for the formal case of a magnetic field acting only on the surface of a specimen. Note that the values of the critical exponents of the ordinary and extraordinary transitions are found from the equation of state

*Recall that so far the description stays within the framework of the Landau theory.

**The statement that a 'special' transition occurs at $q = 0$ is not quite correct, because as any other parameter in the Landau functional, q is obtained by averaging the microscopic spin Hamiltonian and therefore depends on the chosen cut-off parameter. One can only state that the three phase transition lines join at certain values of the parameters of the microscopic Hamiltonian.

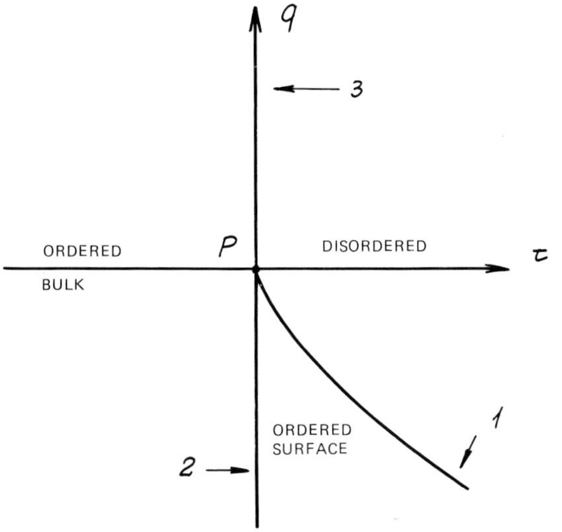

Fig. 1. Phase diagram of a semibounded magnet close to the Curie point T_C, coordinates: $\tau = (T - T_C)/T_C$ vs. q (Bray and Moore 1977). Phase transition lines for: (1) surface transition, (2) extraordinary transition, (3) ordinary transition. 'Special' transition takes place at the point P as we move along the τ-axis ($q = 0$).

$M_S = M_S(H, \tau)$ after the regular contributions (those without singularities) have been subtracted.

2.2. Finite-thickness plate

The main difference between the cases of finite-thickness plate and half-space is that the shift of the transition point in the plate exists independently of the sign of q: if $q > 0$ (the surface is not predisposed to the transition) then the critical point shifts toward lower temperatures ($T_{Cd} < T_C$), but if $q > 0$, then magnetization appears at temperatures higher than T_C. Furthermore, the spontaneous magnetic moment sets in a plate simultaneously throughout its thickness (although the distribution of the magnetic moment is far from homogeneous). As a result, only two of the four transition types occurring in half-space are retained: the surface (if $q \neq 0$, the sign of q is irrelevant) and the 'special' ones. We have already mentioned that the shift of the transition point and the magnetic characteristics close to the critical point are conveniently calculated if the free-energy functional is first converted to a function of $M(0)$ and $M(d) = M_S$. Calculations show (Kaganov and Omel'yanchuk 1971, Kaganov 1972, Kaganov and Karpinskaya 1979, Karpinskaya 1979) that the dependence of the Curie temperature $T_{Cd} = T_C(1 + \tau_d)$ on plate thickness is of the form

$$A(\tau_d) = \left(\frac{C\Delta^2}{d^2}\right) \text{sign}\, q \tag{19}$$

Table 1

Surface equation of state, temperature dependence of surface magnetization for four types of transition in a semibounded magnet. The following notation is used for the surface, ordinary and extraordinary transitions:

$$h = \frac{H_S}{2q^2 C}\left(\frac{B}{C}\right)^{1/2}; \qquad m_S = \frac{M_S}{|q|}\left(\frac{B}{C}\right)^{1/2}; \qquad \hat{\tau} = A/cq^2,$$

where H_S is the magnetic field applied only at the surface. The values of the critical exponents for the ordinary and extraordinary transitions are found after the regular (having no singularity) contributions to M_S and χ_S have been subtracted:

$$M_S - M_S^{\text{reg}} \sim (\tau_C - \tau)^{\beta_S}; \qquad \chi_S - \chi_S^{\text{reg}} \sim |\tau_C - \tau|^{\gamma_S}.$$

Type of transition	Critical point	Surface equation of state	Temperature dependence of surface magnetization ($h=0$)	Critical exponents	
				β_S	γ_S
Ordinary ($q>0$)	$\tilde{\tau}_C = 0$	$\begin{cases} h = m_S[1+(\tilde{\tau}+m_S^2)^{1/2}], & \tilde{\tau}>0 \\ h = m_S + m_S^2 - \frac{1}{2}\|\tilde{\tau}\|, & \tilde{\tau}<0 \end{cases}$	$m_S = \begin{cases} 0, & \tilde{\tau}>0, \\ \frac{1}{2}\|\tilde{\tau}\|, & \tilde{\tau}<0. \end{cases}$	1	$-\frac{1}{2}$
Surface ($q<0$)	$\tilde{\tau}_C = 1$	$\begin{cases} h = m_S[(\tilde{\tau}+m_S^2)^{1/2}-1], & \tilde{\tau}>0 \\ h = m_S^2 - m_S - \frac{1}{2}\|\tilde{\tau}\|, & \tilde{\tau}<0 \end{cases}$	$m_S = \begin{cases} 0, & \tilde{\tau}>1, \\ (1-\tilde{\tau})^{1/2}, & 0<\tilde{\tau}<1, \\ \frac{1}{2}+\frac{1}{2}(1+2\|\tilde{\tau}\|)^{1/2}, & \tilde{\tau}<0. \end{cases}$	$\frac{1}{2}$	1
Extraordinary ($q<0$)	$\tilde{\tau}_C = 0$	$\begin{cases} (\tilde{\tau}+m_S^2)^{1/2} = 1 + h/m_S, & \tilde{\tau}>0 \\ m_S = \frac{1}{2}+\frac{1}{2}(1+2\|\tilde{\tau}\|+4h)^{1/2}, & \tilde{\tau}<0 \end{cases}$	$m_S = \begin{cases} 1-\frac{1}{2}\tilde{\tau}-\frac{1}{2}\tilde{\tau}^2+O(\tilde{\tau}^3), & \tau>0, \\ 1-\frac{1}{2}\tilde{\tau}-\frac{1}{4}\tilde{\tau}^2+O(\tilde{\tau}^3), & \tilde{\tau}<0. \end{cases}$	2	-2
Special ($q=0$)	$\tilde{\tau}_C = 0$	$\begin{cases} \frac{H_S}{2(AC)^{1/2}} = M_S\left(1+\frac{M_S^2 B}{A}\right)^{1/2}, & A>0 \\ \frac{H_S}{2(BC)^{1/2}} = M_S^2\left(1-\frac{\|A\|}{2BM_S^2}\right), & A<0 \end{cases}$	$M_S = \begin{cases} 0, & A>0, \\ \left(\frac{\|A\|}{2B}\right)^{1/2}, & A<0. \end{cases}$	$\frac{1}{2}$	$\frac{1}{2}$

where

$$\varDelta \tan \varDelta = qd, \qquad q>0,$$
$$\varDelta \tanh \varDelta = |q|d, \quad q<0. \tag{20}$$

The expressions for the spontaneous magnetic moment, specific heat, and other characteristics close to phase-transition curves $T \approx T_{\mathrm{Cd}}$ are listed in table 2 for the specific case of $q<0$ (Kaganov and Karpinskaya 1979). Similar formulas for $q>0$ were given by Karpinskaya (1979). Note that if $q<0$, the limiting case $d \to \infty$ corresponds to a transition to a state with nonzero surface moment: the jump in the specific heat, ΔC, becomes independent of d, the mean magnetic moment and magnetic susceptibility tend to zero, and the surface moment tends to a constant value.

3. *Orientational transitions*

The spin-reorientation transitions in magnets, which are vigorously investigated in recent years (Belov et al. 1979), can be described by introducing an effective

Table 2

Heat-capacity jump ΔC and the behavior of magnetic characteristics close to the phase transition point T_{Cd} in a plate of thickness $2d$. χ^+ and χ^- are the magnetic susceptibilities above and below the transition point, respectively.

Jump in specific heat	$\varDelta C = \dfrac{T_{\mathrm{Cd}}}{T_{\mathrm{C}}} \dfrac{\beta^2}{\beta'\beta''} \dfrac{M_0}{\mu} d \times \begin{cases} 1-\frac{1}{3}\lvert q\rvert d, & \lvert q\rvert d \ll 1, \\ \dfrac{1}{\lvert q\rvert d}, & \lvert q\rvert d \gg 1. \end{cases}$
Magnetization in the middle of the specimen	$M(0) = M_0 \left(\dfrac{\beta}{2\beta'}\right)^{1/2} (T_{\mathrm{Cd}} - T)^{1/2} \times \begin{cases} 1-\frac{1}{6}\lvert q\rvert d, & \lvert q\rvert d \ll 1, \\ 2\sqrt{2}\exp(-\lvert q\rvert d), & \lvert q\rvert d \gg 1. \end{cases}$
Surface magnetization	$M(d) = M_0 \left(\dfrac{\beta}{2\beta'}\right)^{1/2} (T_{\mathrm{Cd}} - T)^{1/2} \times \begin{cases} 1+\frac{1}{3}\lvert q\rvert d, & \lvert q\rvert d \ll 1, \\ \sqrt{2}, & \lvert q\rvert d \gg 1. \end{cases}$
Mean magnetization	$\bar{M} = M_0 \left(\dfrac{\beta}{2\beta'}\right)^{1/2} (T_{\mathrm{Cd}} - T)^{1/2} \times \begin{cases} 1+\frac{1}{6}\lvert q\rvert d, & \lvert q\rvert d \ll 1, \\ \sqrt{2}/\lvert q\rvert d, & \lvert q\rvert d \gg 1. \end{cases}$
Susceptibility	$\chi^+ = 2\chi^- = \dfrac{2\mu M_0 \beta''}{\beta} \dfrac{1}{\lvert T_{\mathrm{Cd}} - T\rvert} \times \begin{cases} \frac{1}{4}(1-\frac{1}{2}(qd)^2), & \lvert q\rvert d \ll 1, \\ \dfrac{1}{\lvert q\rvert d}, & \lvert q\rvert d \gg 1. \end{cases}$

anisotropy constant k_1 which reverses sign when temperature changes. As a rule, the analysis can be reduced to studying the anisotropy energy

$$f\{\theta(r)\} = -k_1 \sin^2\theta - k_2 \sin^4\theta + C(\nabla\theta)^2, \tag{21}$$

where

$$k_1 = k_1(\tau) = \beta M_0^2 \tau; \qquad \tau = \frac{T - T_0}{T_0}, \quad T_0 \sim \mu M_0$$

and k_2 is the second anisotropy constant ($|k_2| \ll |k_1|$, far from $T = T_0$). The rest of the notation is the same as in section 2.

The bulk transition type is determined by the sign of k_2: if $k_2 > 0$, the transition from the state $\theta = 0$ to the state $\theta = \pi/2$ occurs as a first-order transition (thermodynamic potentials equalize at $k_1 = -k_2$, the phase $\theta = 0$ ceases to be stable at $k_1 = 0$, and the phase $\theta = \pi/2$ at $k_1 = -2k_2$), and if $k_2 < 0$, the reorientation goes through an intermediate phase via two second-order phase transitions at the points $k_1 = 0$ and $k_1 = -2k_2$. Note that the validity of expansion (21) in the case of a first-order phase transition is justified because the expansion in eq. (21) is in fact an expansion in powers of v_a/c (v_a is the velocity of atomic electrons, and c is the speed of light).

When writing the expression for the surface energy, we take into account that not only the value of the anisotropy constant in the subsurface layer may be changed in this case, but also the directions of the easy and hard magnetization. Assuming (as in section 2) that the magnetic moment both at the surface and in the bulk of the specimen is parallel to the surface, we can present the surface energy in the form $2F_S$, where

$$F_S = k_S \sin^2(\theta_S - \varphi), \tag{22}$$

k_S is the surface anisotropy constant (having the dimension of length), θ_S is the value that θ assumes on the surface of the specimen, and the angle φ prescribes, for $k_S > 0$, the direction of the easy-magnetization axis on the surface. The dimensionless combination

$$\sigma = \frac{k_S}{(C|k_2|)^{1/2}} \tag{23}$$

is a measure of surface energy. Since $k_S = \tilde{k}_S a$, and $\tilde{k}_S$ is a dimensionless quantity of the same nature as k_1 (besides $\tilde{k}_S \gg k_1$ in the neighborhood of T_0), $\sigma \sim \tilde{k}_S(\mu M_0/|k_2| T_C)^{1/2}$ may be both greater and less than unity. The effect of the surface is calculated in the same way we did it for the order–disorder transition, which allows us to pass directly to discussing the results. Clearly, in this case the surface leads to much more diverse and profound consequences because, first, the transitions allowed in the bulk can be of first-order as well, and, second, a new parameter φ has appeared. Let us also note that even if second-order transitions occur in the bulk, and $\varphi = 0$, the effect of the surface will nevertheless be more profound than in order–disorder transitions, since if the surface is not predisposed to one of the transitions, it is definitely predisposed to another, so that even in the half-space case one of the surface transitions is bound to occur earlier than the corresponding transition in the unbounded specimen. Now we can give a consistent description of the effects of the surface. To be specific, we assume $k_S > 0$. We begin with half-space.

3.1. Half-space

Let us first consider the case of a first-order transition in the bulk ($k_2 > 0$). If $\varphi \neq 0, \pi/2$, then the surface energy has a term linear in the order parameter; hence, its effect is to some degree similar to the effect of an external field. Accordingly, only a first-order transition is possible on the surface. An analysis shows (Kaganov 1980) that, firstly, the value of the angle θ_S on the specimen surface is always distinct both from 0 and from $\pi/2$, and secondly, the hysteresis loop for θ_S is narrower than within the specimen (fig. 2). The critical values of k_{1c} and $\tilde{k}_{1c}$ at which the surface counterparts of the bulk states $\theta = 0$ and $\theta = \pi/2$ lose their stability, take the following forms in the two limiting cases $\sigma \gg 1$ and $\sigma \ll 1$:

$$
\begin{aligned}
k_{1c} &\approx -k_2(2\sigma^2)^{1/3}(\sin 2\varphi)^{2/3}, \quad && \sigma \ll 1,\\
&\approx -k_2 \sin^2 \varphi, && \sigma \gg 1.\\
\tilde{k}_{1c} &\approx -2k_2[1 - \tfrac{1}{2}(2\sigma^2)^{1/3}(\sin 2\varphi)^{2/3}], && \sigma \ll 1,\\
&\approx -2k_2(1 - \tfrac{1}{2}\cos^2 \varphi), && \sigma \gg 1.
\end{aligned}
\qquad (24)
$$

Note that if $\sigma \to \infty$, the width of the hysteresis loop remains finite.

Let us consider a situation in which the surface and bulk anisotropy axes are perpendicular to each other ($\varphi = \pi/2$).* In this case the surface favors the state of

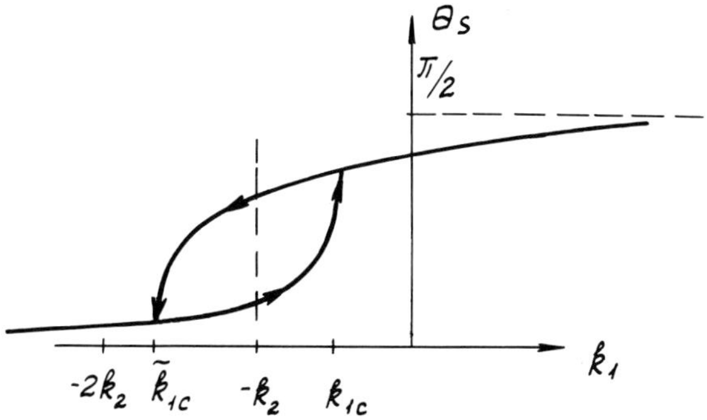

Fig. 2. Surface hysteresis loop in the case of first-order bulk transition and $\varphi \neq 0, \pi/2$. The values of k_{1c} and $\tilde{k}_{1c}$ are given in the text [eq. (24)].

*The results obtained below can immediately be extended to the case of $\varphi = 0$, because the total free energy is invariant under the substitution $\pi/2 - \varphi \to \varphi$; $\pi/2 - \theta \to \theta$; $-k_1 - 2k_2 \to k_1$.

$\theta_S = \pi/2$ and there are no terms linear in the order parameter in the surface energy. This last factor is quite significant because it cancels the 'ban' on a large shift of phase liability points. Calculations show (Kaganov 1980) that if $\varphi = \pi/2$, the introduction of a boundary results in a qualitatively new effect: for $\sigma > 1$ the state $\theta_S = 0$, which is not favorable as regards the surface energy, becomes unstable earlier than the potentials of the phases $\theta_S = 0$ and $\theta_S = \pi/2$ are equalized (fig. 3). The surface reorientation then starts with a second-order phase transition (at the point $k_{1c} = -k_2\sigma^2$), whereby an intermediate surface state is nucleated. The transition from the intermediate surface state to the phase $\theta_S = \pi/2$ is always a first-order transition, but θ_S changes jumpwise if $1 < \sigma < \sqrt{2}$ and continuously if σ is still greater (fig. 3). Reverse reorientation (from the phase $\theta = \pi/2$) is always accompanied by a jump in θ_S at a point $k_1 = -2k_2$ since the introduction of a surface at $\varphi = \pi/2$ obviously cannot shift the liability threshold of the state $\theta = \pi/2$.

Let us now turn to the case of $k_2 < 0$, in which reorientation proceeds via two close to each other second-order transitions. If $\varphi \neq 0, \pi/2$, then the term linear in $\sin\theta_S$ in the surface energy completely blurs the transitions on the surface, so

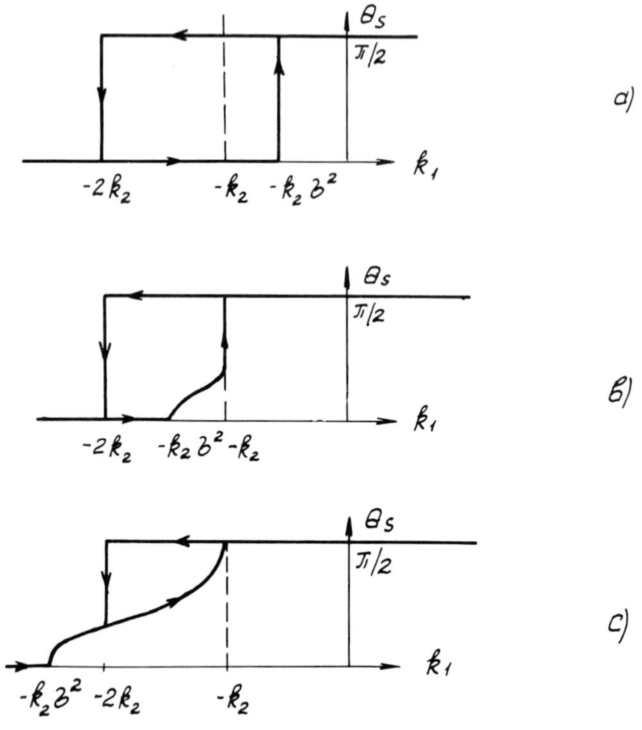

Fig. 3. Surface hysteresis loop in the case of first-order bulk transition and $\varphi = \pi/2$: (a) $\sigma < 1$, (b) $1 \leqslant \sigma < \sqrt{2}$, (c) $\sigma \geqslant \sqrt{2}$.

that the angle θ_S becomes a smooth function of k_1, without singularities, at any nonzero value of σ. If $\varphi = 0$ or $\pi/2$, the situation is different. Namely, both second-order transitions still exist as surface ones, but one of the surface transition points (at $\varphi = \pi/2$ this point is that of the transition from the state $\theta_S = 0$ to the intermediate phase) shifts by $|k_2|\sigma^2$ as compared with the corresponding transition point in the bulk of the specimen (fig. 4). Note the classification, according to the accepted terminology, of the transitions at $\varphi = \pi/2$: the surface transition at $k_1 = -|k_2|\sigma^2$, the extraordinary transition at $k_1 = 0$, and the ordinary transition at $k_1 = 2|k_2|$. The singular features of each transition are clearly the same as in the vicinity of order–disorder transitions. We will also give θ_S as an explicit function of σ for $\varphi = \pi/2$ ($b = k_1/2|k_2|$):

$$\sin^2\theta_S = \begin{cases} 0, & k_1 < -\sigma^2|k_2|, \\ 2b + \sigma^2/(1+\sigma^2), & -\sigma^2|k_2| \leqslant k_1 < 0, \\ \dfrac{2b + \sigma^2 + \sigma[\sigma^2 + 4b(1-b)]^{1/2}}{2(1+\sigma^2)}, & 0 \leqslant k_1 < 2|k_2|, \\ 1, & k_1 \geqslant 2|k_2|. \end{cases} \tag{25}$$

As could be expected (see section 2), $\theta_S \sim (k_1 + \sigma^2|k_2|)^{1/2}$ close to a surface transition and $\pi/2 - \theta_S \sim (2|k_2| - k_1)$ close to an ordinary one. Completing the discussion of orientational transitions in half-space, let us point to an interesting corollary of taking the surface energy into consideration in the case when the reorientation is induced by an external magnetic field $\boldsymbol{H}$ (Ivanov and Kaganov 1984). To be specific, we direct the field along the anisotropy axis. Neglecting the exchange interaction term, we write the bulk component of the free energy in the form

$$F(\theta) = \int \mathrm{d}z\,(-k_1 \sin^2\theta - h\sin\theta), \quad h = \frac{H}{M_0}, \quad \theta = \theta(z) \tag{26}$$

We are interested in the case of $k_1 < 0$. Then the reorientation from the state $\theta = -\pi/2$ to the state $\theta = \pi/2$ occurs by two second-order transitions (at $h = \pm 2k_1$)

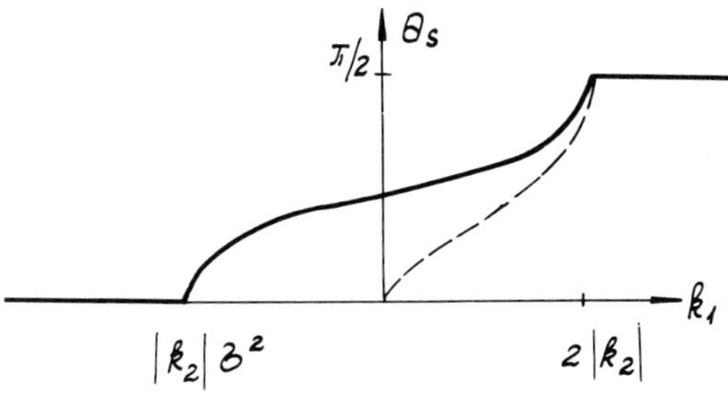

Fig. 4. The function $\theta_S = \theta_S(k_1)$ for $k_2 < 0$ and $\varphi = \pi/2$. The function $\theta(k_1)$ deep into the specimen is given by the dashed curve.

via an intermediate phase. Let $\varphi = \pi/2$. The state 'favored' by the surface is that with $\cos\theta_S = 0$, the surface energy being identical for the states with $\theta = \pi/2$ and $\theta = -\pi/2$. Calculations show (Ivanov and Kaganov 1984) that this 'indifference' leads to an effect which in a certain sense is the reverse of those discussed above: when $\tilde{\sigma} = \tilde{k}_S/(C|k_1|)^{1/2} > 1$, the continuity of θ_S in the intermediate phase breaks down and a hysteresis loop appears on the surface (fig. 5). As σ grows, the loop widens but stays narrower than $4|k_1|$ so that the reorientation (regardless of direction) starts with a second-order transition.

3.2. Finite-thickness plate

With the plate thickness taken into account, the effects of the surface energy become even more pronounced. Without citing all the relevant results [they are given in the paper of Kaganov and Chubukov (1982)], we shall give the essential

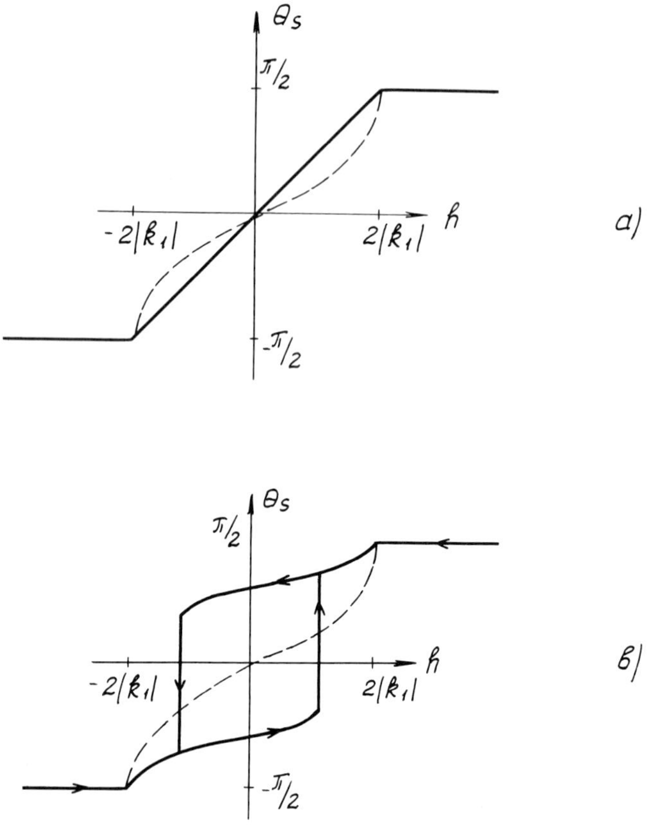

Fig. 5. θ_S as a function of the magnetic field h for $\varphi = \pi/2$: (a) $\sigma < 1$, (b) $\sigma > 1$. The dashed curve gives $\theta(h)$ deep into the specimen.

ones. If a first-order phase transition occurs in the bulk, then for $\varphi \neq 0, \pi/2$ the hysteresis loop collapses at a certain critical thickness of the plate, d_{cr}, so that the reorientation involves no phase transition if $d < d_{cr}$. The value of d_{cr} is easily found if $\varphi = \pi/4$:

$$d_{cr} = \left(\frac{C}{k_2}\right)^{1/2} \arctan \sigma. \tag{27}$$

If $\varphi = \pi/2$, then if $\sigma > \sqrt{3}$ and $d < \tilde{d}_{cr}$, where

$$\tilde{d}_{cr} = \left(\frac{C}{|k_1^*| - 2k_2}\right)^{1/2} \arctan \frac{k_S}{(C(|k_1^*| - 2k_2))^{1/2}},$$

$$k_1^* = \begin{cases} \dfrac{-12k_S^2}{5C(\sigma^2 - 3)}, & \sigma \gtrsim \sqrt{3}, \\ -2k_2 + O\left(\dfrac{1}{\sigma}\right), & \sigma \gg 1, \end{cases} \tag{28}$$

the reorientation proceeds through two phase transitions of second order, in complete contrast to the situation in an unbounded specimen (fig. 6). If phase transitions in the bulk are of second order, the fact that d is finite is essential if $\varphi \neq 0, \pi/2$; indeed, in this case the reorientation occurs in the entire sample at any finite d without a phase transition (as was the case for the angle θ_S in half-space). And finally, if the reorientation is induced by a magnetic field and the geometry

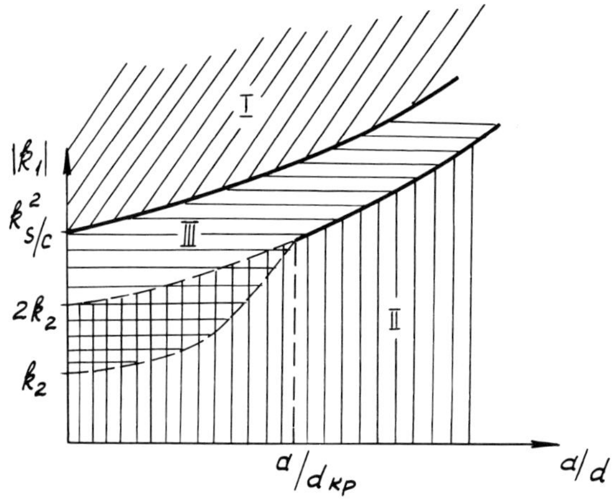

Fig. 6. Phase diagram of a finite-thickness plate in the case of $\sigma > \sqrt{3}$. Solid curves trace second-order transition lines. Dashed lines show the phase liability boundaries for $\theta \equiv 0$ (phase 1), $\theta \equiv \pi/2$ (phase 2), and for the intermediate phase (phase 3). The tilted, vertical and horizontal hatching mark the regions of stability of the phases 1, 2, and 3, respectively. The value of d_{cr} is given in the text [eq. (28)].

is the same as in the half-space, then, firstly, a hysteresis loop will be observed in sufficiently thin plates within the intermediate phase at any finite σ, and secondly, if $\sigma > \sqrt{6}$, the reorientation in plates thinner than a certain critical value proceeds as one first-order transition, that is, the situation is again quite opposite to that in the bulk.

4. The role of fluctuations

The analysis, given in the preceding sections, of the role played by surface energy operates with the Landau theory; it assumes that the fluctuations of the order parameter are small. We have mentioned in section 1 a quite familiar observation that this is not the case for order–disorder transitions: the fluctuation region begins very nearly when $|(T - T_C)/T_C| = 1$, so that it is important to take fluctuations into consideration in the entire relevant temperature range.

The effect of fluctuations on surface characteristics was discussed in a large number of papers [see bibliography in the review by Binder (1983)]. Without attempting to present a complete survey of these papers, we emphasize the factor of major importance: the values of the exponents which determine the behavior of surface characteristics in the surface and extraordinary transitions can be expressed in terms of the bulk transition exponents and those in the ordinary transition in terms of two bulk exponents and a single surface exponent. This is fairly natural for surface transitions because the two-dimensional (2D) nature of fluctuations indicates that the exponents' values must be the same as in the unbounded 2D Ising model.* The relation between the surface and bulk exponents of the ordinary and extraordinary transitions is less obvious. Nevertheless, the analysis of Bray and Moore (1977) shows that the singularity of surface magnetization in the extraordinary transition is identical to that in the bulk free-energy, i.e., after the regular contribution has been subtracted.

$$M_S \sim |\tau|^{\beta_1^e}, \quad \text{where } \beta_1^e = 2 - \alpha. \tag{29}$$

The surface magnetization in the ordinary transition (again after the subtraction of the regular contribution) behaves as follows (Diehl and Dietrich 1981):

$$M_S \sim |\tau|^{\beta_1^0}, \quad \text{where } \beta_1^0 = \tfrac{1}{2}\nu(D - \eta + \eta_1). \tag{30}$$

The symbols α, ν and η in formulas (29) and (30) are the bulk critical exponents of specific heat, correlation length and correlation function at the transition point, respectively, and D is the dimensionality of space (in our analysis, $D = 3$), η_1 is the single-surface critical exponent, accepting the value $\eta_1 = -5/9$ for

*Recall that our discussion is restricted to systems with a one-component order parameter.

$D = 3$ (according to the extrapolation of the first two terms in ε-expansion) (Diehl and Dietrich 1981).

Let us turn now to orientational transitions. We have already mentioned in the introduction that the fluctuation region in an unbounded specimen is very narrow, so that the Landau theory fails only in the immediate vicinity of T_0 (when $|(T - T_0)/T_0| \sim 10^{-8}$) (Belov et al. 1979). The determination of the width of the fluctuation region in a bounded specimen (half-space or plate) implies the calculation of the fluctuation correction to the order parameter averaged over a scale of the order of the correlation length and the comparison of this correction with the value found in the Landau theory (Landau and Lifshitz 1979). Such calculations were carried out by Kaganov and co-workers for surface transitions (Kaganov and Karpinskaya 1979, Kaganov and Chubukov 1982). The difference from the case of an unbounded specimen is that the fluctuation correction $\delta\theta_{\mathrm{fl}}(\boldsymbol{r})$ must be expanded not in plane waves but in a complete system of functions given by the equation

$$\Delta\Psi + \frac{|k_1|}{C}\Psi = 0, \tag{31}$$

with the boundary condition

$$C\left.\frac{\mathrm{d}\Psi}{\mathrm{d}z}\right|_{z=d} = k_{\mathrm{S}}\Psi|_{z=d} \tag{32}$$

(to be specific, we chose the case of $\varphi = \pi/2$ and a surface analog of the bulk phase transition at $k_1 = 0$).

Condition (32) ensures the discreteness of the eigenvalues μ_i which are the factors with z in the eigenfunctions $\phi_i(\boldsymbol{\rho}, z)$ ($\boldsymbol{\rho}$ is a two-dimensional vector in the plane of the specimen's surface). An analysis shows that in accordance with the prediction, the only 'dangerous' fluctuation is the one with a specific eigenvalue μ_1; in other words, fluctuations become two-dimensional in the vicinity of the transition point. A comparison of the value $(\delta\theta_{\mathrm{fl}})^2$, averaged over the correlation area, with the equilibrium value $\theta^2(z)$ demonstrates that the fluctuation region is much wider than in an unbounded specimen even for half-space; namely, fluctuations are dangerous if

$$|\Delta k_1| \lessapprox \frac{\tilde{T}_0}{T_{\mathrm{C}}}\frac{\mu M_0}{T_{\mathrm{C}}}k_2^2\sigma^2(\sigma^2 + 1) \tag{33}$$

($\tilde{T}_0$ is the surface-transition temperature), while in the unbounded specimen they are dangerous if

$$|\Delta k_1| \lessapprox \left(\frac{T_0}{T_{\mathrm{C}}}\right)^2\frac{\mu M_0}{T_{\mathrm{C}}}k_2^2. \tag{34}$$

The fluctuation region is wider in finite-thickness plates than in half-space and increases as the plate thickness is increased. Nevertheless, the fluctuation region is extremely narrow for all thicknesses d up to $d \sim a$ at any reasonable value of σ because $T_0 \sim \mu M_0 \ll T_C$, and the Landau theory remains valid practically up to the transition point.

5. Structural phase transitions

The background of the phenomenological theory is the same both for magnetic and structural phase transitions. So it is natural that the basic idea about surface magnetic phase transitions can be applied to the problem of surface structural phase transitions. The latter, however, differ from magnetic transitions in some important aspects. This is why a special section is devoted to the treatment of surface structural phase transitions. In this section we attract the reader's attention just to these differences.

5.1. Symmetry arguments

We shall show first of all that for a structural surface phase transition more restrictive conditions are to be fulfilled than for a magnetic one (Levanyuk and Minyukov 1983a). It is well understood now that ferroelectric and similar phase transitions are symmetry-breaking phenomena, i.e., spontaneous violation of symmetry takes place at the transitions. Thus the necessary condition for observation of a surface phase transition (as well as a bulk one) is the possibility of a high-symmetry state of the surface. For a magnetic phase transition this possibility is beyond a shadow of doubt because the appearance of magnetic ordering means the violation of time-reflection symmetry and any crystal structure may have this symmetry. But for structural phase transitions this symmetry is irrelevant and the question has to be considered more closely.

The symmetry of a crystal facet is the symmetry of the semi-infinite crystal. As a result only mirror planes and rotation axes perpendicular to the facet can be symmetry elements of the facet with the exclusion for screw axes and glide mirror planes with the glide directions which are not parallel to the facet. Consequently, most of the crystal facets have no planes or axes of symmetry and no phase transitions with the lowering of point symmetry are possible at the surface. One may say also that most of the facets are locally ferroelectric and only a very restricted number of surfaces can be in a nonpolar state. It should be emphasized that we mean here the spontaneous polarization parallel to the facet because any facet is polar as to the axis perpendicular to the facet owing to nonequivalence of two directions of this axis.

In the above discussion the facet was considered implicitly as a plane one. Virtually it is the case only for close-packed facets, while other facets consist of terraces of close-packed ones. Nevertheless, the above conclusions remain to be valid for such facets also with the only restriction that symmetry-lowering distortions inherent for a nonclose-packed facet are inhomogeneous and are concentrated near the ends of the terraces (provided, of course, that the close-packed facet is a high-symmetry one).

5.2. Long-range field effects

Let us consider specific features of surface structural phase transitions on high-symmetry facets. If macroscopic long-range fields do not arise due to space inhomogeneities of the order parameter, the phenomenological theory is just the same as that for order–disorder magnetic phase transitions (section 2). One has to take into account that the difference between T_C and T_S is normally very great (Levanyuk and Minyukov 1983a) so the bulk correlation radius of the order parameter is small at $T \sim T_S$. As a result the surface structural phase transition will influence the crystal structure only very closely to the surface at the distance of the order of magnitude of an atomic distance a.

Let us consider the cases when macroscopic long-range fields may be important. It is seemingly the case of a proper ferroelectric (the order parameter can be identified with components of the polarization vector). Indeed, in the three-dimensional case the macroscopic (depolarization) field is known to lead to the formation of a domain structure in a nonshortcircuited crystal and also to influence greatly the critical fluctuations (see, e.g., Bruce and Cowley 1981, Strukov and Levanyuk 1983). For surface ferroelectrics the role of this field is less important than for the bulk ones. Indeed, in the case of a homogeneously polarized facet this field is created by bound charges localized at the edges of the facet, that is, by linear charges, whereas in the bulk ferroelectric it is the plane charges that create the field. The same argument is valid also for critical fluctuations and space inhomogeneities of the order parameter due to defects.

Long-range elastic fields are known to be important for proper ferroelastic phase transitions in three-dimensional systems. The order parameter at such a transition can be considered as a component (or a set of components) of the strain tensor. The corresponding component(s) of the elastic stiffness tensor go(es) to zero as $T \to T_C$ as well as the velocity of certain acoustic waves ('acoustic soft mode').

To begin the discussion of the possibility of a surface ferroelastic phase transition let us assume that the stiffness (λ) corresponding to the U_{xz}-component becomes negative within a surface layer, the surface being perpendicular to the z-axis. Having in mind the results of the first part of the chapter one seemingly can expect that when the absolute value of the stiffness becomes great enough, the local strain arises in the surface layer (fig. 7). It is hardly meaningful, however, to regard this phase transition as a ferroelastic one. Let us mention first of all that the symmetry transformation properties of U_{xz} coincide with those of P_x because P_z is invariant. It means that the phase transition under consideration is the proper ferroelastic one. Moreover, there is no soft acoustic mode associated with the transition. To make sure of it let us consider the system with a *macroscopic* 'soft' layer. Two surface acoustic waves, in this case the Rayleigh wave and the Love wave, penetrate into the bulk to the distance of a wavelength. As a result the velocities of the waves are governed by the bulk stiffness if the wave vectors are small and both the waves are not 'soft'. The deformation displayed at fig. 7, is, in fact, an optic rather than an acoustic one, because it consists in the shift of the surface atoms with respect to the neighboring ones. This result is quite transparent also from the point of view of mathematics. Indeed, the contribution of the surface layer to the elastic energy is given by the expression

$$\int \lambda_{\mathrm{S}}(z)\, U_{xz}^2 \,\mathrm{d}V \simeq a \int \lambda_{\mathrm{S}} \delta(z)\, U_{xz}^2 \,\mathrm{d}V, \tag{35}$$

where a is of the order of magnitude of the interatomic distance, and $\int \delta(z)\,\mathrm{d}z = 1$. Using Fourier representation one sees that the main contribution to the energy is due to great wavevectors and the difference between optic and acoustic vibration is known to disappear in this region of the $\boldsymbol{k}$-space.

The same argument is valid also for the case when the stiffness with respect to U_{xy} becomes locally negative (the surface being again perpendicular to the z-axis). It is clear that the homogeneous surface U_{xy} deformation is impossible because the soft surface is clamped by the rigid bulk. Let us remember that a ferroelastic phase transition in the clamped sample is possible only at the expense of the domain formation. Using routine estimation of the domain width, one can easily see that the width is of the order of magnitude of the lattice constant and may be great under special conditions only (Andreev 1980). This result illustrates once more that in this case the softness is displayed in the region of large wave vectors.

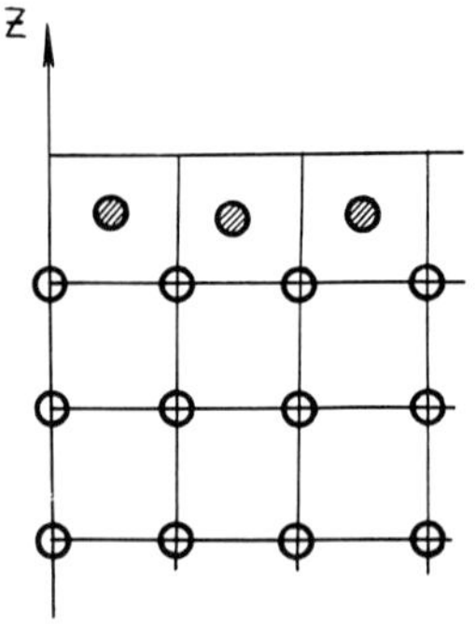

Fig. 7. Local deformation of the surface layer (surface atoms are marked by hatching).

5.3. Fluctuations

The less the space dimensionality the more important are the critical fluctuations. This is why the deviations from mean-field (Landau) theory are more significant for two-dimensional systems than for three-dimensional ones and fluctuation-induced anomalies, in particular scattering ones, are more pronounced.

In the case when space inhomogeneities in the order parameter do not lead to the appearance of long-range fields, the Fourier-transform of the correlation function of η-fluctuations has the same form as for three-dimensional systems (see, e.g., Landau and Lifshitz 1979):

$$\langle \eta_{k_\parallel} \eta_{-k_\parallel} \rangle = \frac{k_B T}{S(2A_S + 2C_S k_\parallel^2)}, \quad T > T_S, \tag{36}$$

where $\boldsymbol{k}_\parallel$ is a two-dimensional wavevector, A_S and C_S are the coefficients of the surface free energy renormalized by the bulk distortions accompanying those in the surface layer, $A_S \to 0$ as $\tau_S = (T - T_S)/T_S \to 0$.

For a three-dimensional system, the macroscopic electric fields are known to suppress strongly the critical fluctuations in proper ferroelectrics with one component of the order parameter (uniaxial ferroelastics). It has already been mentioned that for two-dimensional systems this suppression is somewhat weaker. Nevertheless, it proves to be strong enough to be responsible for some important effects (Maleev 1976). Thus it is worthwhile to discuss this question in some detail. The conventional method to take into account the macroscopic electric field accompanying the polarization fluctuations is to add to the expression for thermodynamic potential [see eqs. (2) and (3)] the term

$$-\int_V \tfrac{1}{2}\boldsymbol{P} \cdot \boldsymbol{E}\, \mathrm{d}v = -a \int_S \tfrac{1}{2}\boldsymbol{P}_\parallel \cdot \boldsymbol{E}_\parallel\, \mathrm{d}S, \tag{37}$$

where $\boldsymbol{P}_\parallel$ and $\boldsymbol{E}_\parallel$ are components parallel to the surface, a is a distance of the order of the atomic magnitude. As mentioned above, the thickness of the surface layer influenced by the surface transition (the bulk correlation radius) is of the order of the atomic magnitude. That is why eq. (37) is valid. Furthermore, we neglect for simplicity the difference between dielectric constants of the medium surrounding the surface layer. In our case we may put

$$\boldsymbol{P}(\boldsymbol{r}) = \boldsymbol{P}_\parallel(\boldsymbol{\rho}) a \delta(z), \tag{38}$$

where $\boldsymbol{\rho}$ is the two-dimensional radius vector parallel to the surface and z is the axis perpendicular to the surface. For the Fourier components one has

$$\boldsymbol{P}(\boldsymbol{k}) = \frac{1}{V}\int_V \boldsymbol{P}(\boldsymbol{r})\, \mathrm{e}^{-\mathrm{i}\boldsymbol{k}\cdot\boldsymbol{r}}\, \mathrm{d}\boldsymbol{r} = \frac{a}{V}\int_S \boldsymbol{P}_\parallel(\boldsymbol{\rho})\, \mathrm{e}^{-\mathrm{i}\boldsymbol{k}_\parallel\cdot\boldsymbol{\rho}}\, \mathrm{d}\boldsymbol{\rho} = \frac{a}{L}\boldsymbol{P}_\parallel(\boldsymbol{k}_\parallel), \tag{39}$$

where $V = SL$. From the Maxwell equations one obtains

$$\boldsymbol{E}_{\|}(\boldsymbol{k}) = -\frac{4\pi(\boldsymbol{P}_{\|}(\boldsymbol{k}) \cdot \boldsymbol{k}_{\|})\boldsymbol{k}_{\|}}{k^2}, \tag{40}$$

and

$$E_{\|}(\boldsymbol{k}_{\|}, z=0) = \frac{L}{2\pi}\int \boldsymbol{E}(\boldsymbol{k})\,\mathrm{d}k_z = -\frac{2\pi(\boldsymbol{P}_{\|}(\boldsymbol{k}_{\|}) \cdot \boldsymbol{k}_{\|})\boldsymbol{k}_{\|}\,a}{k_{\|}}. \tag{41}$$

Just as for three-dimensional systems, the fluctuating electric field changes the critical behavior for an uniaxial ferroelastic only. For this case, putting $\eta \equiv \mathrm{P}_x$, $P_y \equiv 0$, one has (within the framework of the Landau theory):

$$\langle \eta_{k_{\|}} \eta_{-k_{\|}} \rangle = \frac{k_{\mathrm{B}} T}{S\left(2A_{\mathrm{S}} + 2C_{\mathrm{S}} k_{\|}^2 + \dfrac{2\pi k_x^2 a^2}{k_{\|}}\right)}. \tag{42}$$

Mention that taking into account the noncritical component of the polarization is fairly useless for the two-dimensional case because it leads to no more than renormalization of the coefficient C_{S}.

Let us discuss the applicability of the Landau theory for surface structural phase transitions. One may use, for example, the condition that the first fluctuation correction to the coefficient A_{S} is small (see, e.g., Strukov and Levanyuk 1983). For displacive transitions one can estimate the value of coefficient A_{S} as $(T_{\mathrm{at}}/T_{\mathrm{S}})\tau_{\mathrm{S}}$ (where $T_{\mathrm{at}} \sim 10^4$–$10^5$ K), but the values of the other coefficients have 'atomic' orders of magnitude. As a result, using eqs. (36) and (42), we obtain the estimate of the region of validity of the Landau theory:

$$\tau_{\mathrm{S}} > \left(\frac{T_{\mathrm{S}}}{T_{\mathrm{at}}}\right)^{(D-2)/(4-D)}, \tag{43}$$

where D is the 'effective' dimension of space, which differs from 2 and equals 2.5 for a surface uniaxial ferroelectric. We see that for $D = 2$ the Landau theory is applicable up to $\tau_{\mathrm{S}} \sim 1$. For order–disorder phase transitions one has to substitute T_{S} for T_{at} in eq. (43). In the two-dimensional case this substitution has no effect and the condition of applicability of the Landau theory in two-dimensional systems is the same both for order–disorder and displacive nonferroelectric phase transitions. The same is valid for two-axial proper ferroelectrics. For uniaxial ferroelectrics it follows from eq. (43) that the Landau theory is applicable if $\tau_{\mathrm{S}} > (T_{\mathrm{S}}/T_{\mathrm{at}})^{1/3}$.

5.4. *Anomalies*

The bulk phase transitions are known to manifest themselves in anomalies of various physical quantities. For surface phase transitions the anomalies are, of course, more difficult to observe. Let us inspect, however, if some of them are observable. As tentative estimates one can put $P \sim P_{at}|\tau_S|$ for an order–disorder phase transition and $P \sim P_{at}|\tau_S|^{\beta}(T_C/T_{at})^{1/2}$ for a displacive one ($P_{at} \sim E_{at} \sim 10^8$–$10^9$ V/cm). The voltage between the facets perpendicular to the facet in question is estimated as Pa ($a = 10^{-8}$ cm). For $|\tau_S| \sim 1$, the order of magnitude of this voltage is 1–10 V for order–disorder phase transitions and 10^{-1}–1 V for displacive ones. We see that the polarization anomaly due to surface ferroelectric transition is quite observable. The same is valid also for the anomalies in the dielectric constant and the macroscopic electric quadrupole moment (Levanyuk and Minyukov 1983b).

One of the anomalies being widely studied now is the anomaly of light scattering (see, for example, Cummins and Levanyuk 1983). Let us consider the light scattering near a surface structural phase transition ($T > T_S$). As in the case of a bulk phase transition one should distinguish between the cases of linear and of bilinear coupling between the dielectric tensor ε_{ij} (in the optic region) and the order parameter η. For a proper ferroelectric surface phase transition the first case is realized. In reality, since P_z is invariant for a facet which is perpendicular to the z-axis, one finds $\varepsilon_{xz} = \varepsilon_{xz_0} + bP_x$. The scattered light intensity is known to be determined by the correlation function of ε_{ij} fluctuations. Since the surface phase transition affects, as has been stated above, only a surface layer with thickness a, having an atomic order of magnitude, we are interested in light scattering from this surface layer. Therefore, one can put $\Delta\varepsilon(\boldsymbol{r}) = \Delta\varepsilon(\boldsymbol{\rho})a\delta(z)$ and find for the intensity of the scattered light:

$$I_S(\boldsymbol{q}) \sim \left\langle \int \Delta\varepsilon(\boldsymbol{r})\,\Delta\varepsilon^*(\boldsymbol{r}')\,e^{i\boldsymbol{q}\cdot(\boldsymbol{r}'-\boldsymbol{r})}\mathrm{d}\boldsymbol{r}\,\mathrm{d}\boldsymbol{r}' \right\rangle$$

$$\simeq a^2 S \left\langle \int \Delta\varepsilon(0)\,\Delta\varepsilon^*(\boldsymbol{\rho})\,e^{i\boldsymbol{q}\cdot\boldsymbol{\rho}}\,\mathrm{d}\boldsymbol{\rho} \right\rangle, \tag{44}$$

where $\boldsymbol{q}$ is the scattering wave vector. In the case under consideration the ε_{xz} fluctuations are proportional to the order parameter fluctuations. As a result one obtains for the scattering intensity:

$$I_S(\boldsymbol{q} \simeq 0) \sim a^2 S^2 b^2 \langle \eta_{\boldsymbol{q}_\parallel} \eta_{-\boldsymbol{q}_\parallel} \rangle_{\boldsymbol{q}_\parallel \simeq 0} \simeq aSb^2 \tau_S^{-\gamma}, \tag{45a}$$

for displacive phase transitions and

$$I_S(\boldsymbol{q} \simeq 0) \sim aSb^2 \tau_S^{-\gamma} \left(\frac{T_C}{T_{at}}\right) \tag{45b}$$

for order–disorder ones.

The ratio of the intensity I_S to the intensity of light scattering from the bulk (the volume $V = LS$) can be estimated as $(T_{at}/T_S)(a/L)\tau_S^{-\gamma}$ for displacive phase transition (if no bulk displacive phase transition is realized) and as $(a/L)\tau_S^{-\gamma}$ for order–disorder ones. For $L = 10^5 a \sim 10^{-3}$ cm this ratio is equal to unity by the order of magnitude for $\tau_S \sim 10^{-1}$–10^{-2} and $\tau_S \sim 10^{-2}$–10^{-3} respectively. If one takes into account the deviations from the Landau theory, then the scattering from the surface becomes important even for larger values of τ_S.

Thus, in carrying out measurements on thin films or for a totally reflected incident beam it is quite possible to find anomalies of light scattering near ferroelectric surface phase transitions.

For nonferroelectric phase transitions only quadratic coupling of ε_{ij} and η is permissible and the light scattering anomaly is due to the increase of second-order scattering only. The estimations show that the intensity of this scattering is at least by two orders of magnitude smaller than that for the first-order scattering discussed above. Thus, the observation of these anomalies is a more complicated experimental problem.

Of course, the mentioned difference between ferroelectric and nonferroelectric surface phase transitions disappears in investigations of order parameter fluctuations by diffuse scattering of low-energy electrons. However, this method is still too complicated to be widely used.

Finally, let us emphasize once more that the anomalies near ferroelectric surface phase transitions are quite observable. Unfortunately, as far as we know these phase transitions have not been observed up to now. The only reason is, probably, difficulties in obtaining crystals with perfect surfaces. Indeed, from the general point of view one sees no reason why surface ferroelectric phase transitions should be rarer than the bulk ones.

6. Conclusion

We hope that our considerations will lead to the conclusion that the observations and investigations of magnetic and structural surface phase transitions is a surmountable problem of fundamental interest for the physics of phase transitions. However, experimental works on this problem were not sufficiently intensive till now. The hope to stimulate these investigations was the main motivation to write this chapter.

References

Andreev, A.F., 1980, Pis'ma Zh. Eksp. & Teor. Fiz. **32**, 654.

Belov, K.P., A.K. Zvezdin, A.M. Kadomtseva and R.Z. Levitin, 1979, in: Orientational Transitions in Rare Earth Magnets (Nauka, Moscow) in Russian.

Bespyatnikh, V.I., I.E. Dikshtein and V.V. Tarasenko, 1980, Fiz. Tverd. Tela **22**, 3335.

Binder, K., 1983, in: Phase Transitions and Critical Phenomena, eds C. Domb and J.L. Lebowitz, Vol. 8 (Academic Press, New York).

Binder, K., and P.C. Hohenberg, 1972, Phys. Rev. B **6**, 3461.

Binder, K., and P.C. Hohenberg, 1974, Phys. Rev. B **9**, 2194.

Bray, A.J., and M.A. Moore, 1977, J. Phys. A **10**, 1927.
Bruce, A.D., and R.A. Cowley, 1981, Structural Phase Transitions (Taylor & Francis Ltd., London).
Campuzano, J.C., A.M. Lahee and G. Jennings, 1985, Surf. Sci. **152 & 153**, 68.
Chubukov, A.V., 1982, Fiz. Tverd. Tela **24**, 2465.
Cummins, H.Z., and A.P. Levanyuk, 1983, Light Scattering Near Phase Transitions (North-Holland, Amsterdam).
Diehl, H.W., and S. Dietrich, 1981, Z. Phys. B **42**, 65.
Francis, S.M., and N.V. Richardson, 1986, Phys. Rev. B **33**, 662.
Ginzburg, V.L., 1960, Fiz. Tverd. Tela **2**, 2031.
Ginzburg, V.L., and L.D. Landau, 1950, Zh. Eksp. & Teor. Fiz. **20**, 1064.
Ivanov, N.B., and M.I. Kaganov, 1984, Fiz. Tverd. Tela **26**, 1101.
Kaganov, M.I., 1972, Zh. Eksp. & Teor. Fiz. **62**, 1196.
Kaganov, M.I., 1980, Zh. Eksp. & Teor. Fiz. **79**, 1544.
Kaganov, M.I., and A.V. Chubukov, 1982, Zh. Eksp. & Teor. Fiz. **82**, 1617.
Kaganov, M.I., and N.S. Karpinskaya, 1979, Zh. Eksp. & Teor. Fiz. **76**, 2143.
Kaganov, M.I., and A.M. Omel'yanchuk, 1971, Zh. Eksp. & Teor. Fiz. **61**, 1679.
Karpinskaya, N.S., 1979, Fiz. Tverd. Tela **21**, 1160.
Landau, L.D., and E.M. Lifshitz, 1979, in: Statistical Physics, Course of Theoretical Physics, Vol. 5 (Pergamon Press, Oxford).
Levanyuk, A.P., 1959, Zh. Eksp. & Teor. Fiz. **36**, 810.
Levanyuk, A.P., and S.A. Minyukov, 1983a, Fiz. Tverd. Tela **25**, 2617.
Levanyuk, A.P., and S.A. Minyukov, 1983b, Fiz. Tverd. Tela **25**, 3485.
Lubensky, T.C., and H.H. Rubin, 1975, Phys. Rev. B **12**, 3885.
Maleev, S.V., 1976, Zh. Eksp. & Teor. Fiz. **70**, 2375.
Mills, D.L., 1971, Phys. Rev. B **3**, 3887.
Patashinsky, A.Z., and V.L. Pokrovsky, 1982, in: Fluctuation Theory of Phase Transitions (Nauka, Moscow) in Russian.
Strukov, B.A., and A.P. Levanyuk, 1983, in: Physical Background of Ferroelectricity (Nauka, Moscow) in Russian (English Edition to be published by Gordon & Breach, New York).
Weller, D., S.F. Alvarado, W. Gudot, K. Schröder and M. Campagna, 1985, Phys. Rev. Lett. **54**, 1555.

CHAPTER 9

Laser-induced Surface Gratings

A.M. PROKHOROV, I.A. AVRUTSKY, P.V. BAZAKUTSA,
V.A. SYCHUGOV and A.V. TISCHENKO

General Physics Institute
USSR Academy of Sciences
117942 Moscow, USSR

Nonlinear Surface Electromagnetic Phenomena
Edited by
H.-E. Ponath and G.I. Stegeman

© *Elsevier Science Publishers B.V., 1991*

Contents

Introduction

Since the advent of lasers, the research of laser radiation action upon condensed media has been developing. Even at an early stage of this research, it has been revealed that at radiation-power densities close to the material damage threshold the surfaces of the irradiated media develop periodic structures (PSS) (Birnbaum 1985). Further experiments have shown that this phenomenon is of a universal character, i.e., the formation of PSS is possible on metals, semiconductors, as well as on dielectrics (Keilmann and Bai 1982, van Driel et al. 1982). As an example we present in fig. 1 a photo of periodic structures emerging on a germanium surface. PSS feature a spontaneous development on an originally smooth surface under only one laser beam with a uniform intensity distribution, the period and orientation of PSS being determined in this case by wavelength, polarization, and angle of incident radiation. A spatially inhomogeneous heating of the surface needed to fix PSS on the irradiated surface, is a result of interference of an incident wave with that appearing due to laser radiation scattering on the surface roughness or on the material density inhomogeneities in the subsurface region. The 'surviving' and growth of one or several quite specific spatial harmonics from all the ensemble of inhomogeneities on the medium surface are due to the appearance of feedback between the forming microrelief and acting radiation (Bunkin et al. 1969). Thus, the development of PSS is in essence similar to the phenomena of induced light scattering and presents a typical example of a nonlinear optical effect (Akhmanov et al. 1985).

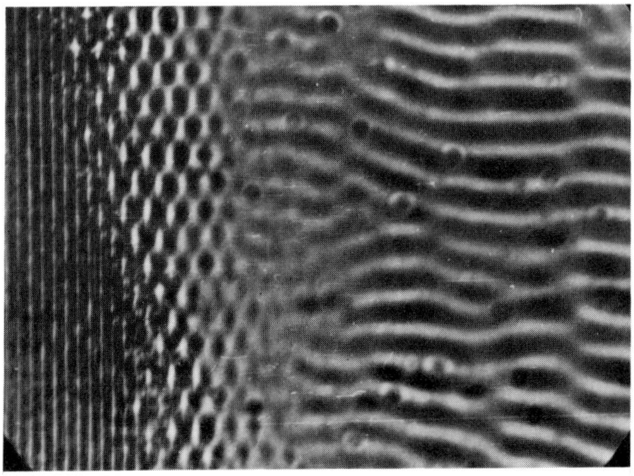

Fig. 1. Periodic structure on germanium.

1. Electrodynamic model of periodic microrelief formation

There are two major items in the problem of laser-induced development of periodic surface structures (PSS). Firstly, it is important to know how the action of a homogeneous light beam upon the absorbing medium surface gives rise to a periodic energy release on it, and secondly, it is needed to find out which thermodynamic mechanism makes this inhomogeneous energy release fixed on the surface in each particular case. The first part of the problem is typically electrodynamic and it has been solved earlier, but for nonabsorbing media, while considering laser-induced generation of capillary waves on the liquid surface (Gavrikov et al. 1969). Since in case of PSS formation on the surface of an absorbing condensed medium, light action produces periodic structures, when solving the electrodynamic part of the problem, it is natural to consider the diffraction of the acting radiation on a periodic structure and to assume that it is interference of the diffraction orders in the absorbing medium that leads to a periodic energy release on its surface.

1.1. Light diffraction on a corrugated interface

Let us consider the processes of light scattering and absorption on a sinusoidally corrugated boundary surface of two media (see fig. 1) (Sychugov and Tischenko 1982):

$$z = \sigma \sin Ky. \tag{1}$$

Assume that the light is incident on the surface of an absorbing condensed medium with $\varepsilon_1 = n_1^2 - \kappa_1^2 + \mathrm{i}(2n\kappa)$ from the medium with $\varepsilon_2 = n_2^2$ $(\kappa_2 = 0)$. Here n_i is the refraction index and κ_i is the absorption coefficient of the corresponding medium, σ is the corrugation amplitude, and K is the grating vector with the grating period: $\Lambda = 2\pi/K$.

Let $\boldsymbol{K} = (K_x, K_y, K_z)$ be the wave vector of radiation incident on a ripple from the medium with permeability ε_2:

$$K_z = -kG_2^0 = -k\sqrt{\varepsilon_2 - \frac{K_x^2 + K_y^2}{k^2}}, \tag{2}$$

where $k = 2\pi/\lambda$ is the wave vector of light in vacuum. Due to modulation of boundary conditions over the y-coordinate, there appear diffracted waves in both media. The y components of wave vectors of different-order diffractional waves differ by a whole number of vectors of grating K:

$$K_p = K_y + pK, \quad p = 0, \pm 1, \pm 2, \ldots, \tag{3}$$

and the electromagnetic field has the form:

$$
\boldsymbol{H}=\begin{cases} \boldsymbol{c}_0 \exp[\mathrm{i}(K_x x+K_y y-kG_2^0 z)]+\sum\limits_{p=-\infty}^{\infty} \boldsymbol{b}_p \exp[\mathrm{i}(K_x x+K_p y+kG_2^p z)], \\ \qquad\qquad\qquad\qquad z \geqslant \sigma \sin Ky, \\ \sum\limits_{p=-\infty}^{\infty} \boldsymbol{a}_p \exp[\mathrm{i}(K_x x+K_p y-kG_1^p z)], \quad z \leqslant \sigma \sin Ky, \end{cases} \tag{3'}
$$

where $\boldsymbol{c}_0$ is the amplitude of the incident wave, $\boldsymbol{b}_p$ are the amplitudes of the waves scattered into medium 2, $\boldsymbol{a}_p$ are the amplitudes of the waves scattered into medium 1,

$$
\begin{aligned} &G_2^p=\sqrt{\varepsilon_2-k_p^2/k^2}, \\ &G_1^p=\sqrt{\varepsilon_1-k_p^2/k^2}, \quad k_p^2=K_x^2+K_p^2. \end{aligned} \tag{4}
$$

In eqs. (2) and (4) roots with positive real and imaginary parts are taken. In the approximation of a small corrugation

$$
k\sigma G_1^p \ll 1, \qquad k\sigma G_2^p \ll 1, \quad p=0, \pm 1, \pm 2, \ldots, \tag{5}
$$

and small amplitude of nonzero-order diffractional waves

$$
|\boldsymbol{b}_p|, |\boldsymbol{a}_p| \ll |\boldsymbol{b}_0|, |\boldsymbol{a}_0|, \tag{6}
$$

in the first approximation over $k\sigma$ the solution of the diffraction problem has the form:

$$
\tilde{a}_p=\tilde{b}_p=pk\sigma\frac{\varepsilon_1-\varepsilon_2}{G_2^p+G_1^p}G_2^0\left\{\frac{G_1^0 \sin(\varphi_p-\varphi_0)}{\varepsilon_1 G_2^0+\varepsilon_2 G_1^0}c_0+\frac{\cos(\varphi_p-\varphi_0)}{G_2^0+G_1^0}\tilde{c}_0\right\}, \tag{7}
$$

$$
\begin{aligned} a_p=pk\sigma&\frac{(\varepsilon_1-\varepsilon_2)G_2^0\varepsilon_1}{\varepsilon_1 G_2^p+\varepsilon_2 G_1^p} \\ &\times\left\{\frac{G_2^p \sin(\varphi_p-\varphi_0)}{G_2^0+G_1^0}\tilde{c}_0+\left[G_1^0 G_2^p \cos(\varphi_p-\varphi_0)+\frac{k_0 k_p}{k^2}\right]\right. \\ &\qquad\left.\times c_0(\varepsilon_1 G_2^0+\varepsilon_2 G_1^0)^{-1}\right\}, \end{aligned} \tag{8}
$$

$$
\begin{aligned} b_p=pk\sigma&\frac{(\varepsilon_1-\varepsilon_2)G_2^0\varepsilon_2}{\varepsilon_1 G_2^p+\varepsilon_2 G_1^p} \\ &\times\left\{\frac{G_1^p \sin(\varphi_p-\varphi_0)}{G_2^0+G_1^0}\tilde{c}_0-[G_1^0 G_1^p \cos(\varphi_p-\varphi_0)-\varepsilon_1 k_0 k_p/\varepsilon_2 k^2]\right. \\ &\qquad\left.\times c_0(\varepsilon_1 G_2^0+\varepsilon_2 G_1^0)^{-1}\right\}, \quad p=\pm 1 \end{aligned} \tag{9}
$$

$$a_0 = \frac{2\varepsilon_1 G_2^0 c_0}{\varepsilon_1 G_2^0 + \varepsilon_2 G_1^0} + \sum_{p=\pm 1} p\frac{k\sigma}{2}\Bigg\{[-(\varepsilon_2 G_1^p + \varepsilon_1 G_2^0)\tilde{a}_p + \varepsilon_2(G_2^0 - G_1^p)\tilde{b}_p]\sin(\varphi_p - \varphi_0) - \left[(\varepsilon_2 + G_2^0 G_1^p)\cos(\varphi_p - \varphi_0) - \frac{\varepsilon_2}{\varepsilon_1}\frac{k_0 k_p}{k^2}\right]a_p + \left[(\varepsilon_2 - G_2^0 G_2^p)\cos(\varphi_p - \varphi_0) - \frac{k_0 k_p}{k^2}\right]b_p\Bigg\} \times \frac{\varepsilon_1}{\varepsilon_1 G_2^0 + \varepsilon_2 G_1^0}, \tag{10}$$

$$b_0 = \frac{\varepsilon_1 G_2^0 - \varepsilon_2 G_1^0}{\varepsilon_1 G_2^0 + \varepsilon_2 G_1^0} c_0 + \sum_{p=\pm 1} p\frac{k\sigma}{2}\Bigg\{[\varepsilon_1(G_1^0 - G_1^p)\tilde{a}_p - (\varepsilon_1 G_2^p + \varepsilon_2 G_1^0)\tilde{b}_p]\sin(\varphi_p - \varphi_0) + \left[(\varepsilon_1 + G_1^0 G_2^p)\cos(\varphi_p - \varphi_0) - \frac{\varepsilon_1 k_0 k_p}{\varepsilon_2 k^2}\right]b_p - \left[(\varepsilon_1 - G_1^0 G_1^p)\varepsilon_2\cos(\varphi_p - \varphi_0) - \frac{k_0 k_p}{k^2}\right]a_p\Bigg\} \times \frac{\varepsilon_2}{\varepsilon_1 G_2^0 + \varepsilon_2 G_1^0}, \tag{11}$$

$$\tilde{a}_0 = \frac{2G_2^0 \tilde{c}_0}{G_2^0 + G_1^0} + \sum_{p=\pm 1} p\frac{k\sigma}{2}\{[(G_2^0 + G_1^p)a_p + (G_2^p - G_2^0)b_p]\sin(\varphi_p - \varphi_0) - [(\varepsilon_1 + G_2^0 G_1^p)\cos(\varphi_p - \varphi_0) - k_0 k_p/k^2]\tilde{a}_p + [(\varepsilon_2 - G_2^0 G_2^p)\cos(\varphi_p - \varphi_0) - k_0 k_p/k^2]\tilde{b}_p\} \times (G_2^0 + G_1^0)^{-1}, \tag{12}$$

$$\tilde{b}_0 = \frac{G_2^0 - G_1^0}{G_2^0 + G_1^0}\tilde{c}_0 + \sum_{p=\pm 1} p\frac{k\sigma}{2}\{[(G_1^p - G_1^0)a_p + (G_2^p + G_1^0)b_p]\sin(\varphi_p - \varphi_0) - [(\varepsilon_1 - G_1^0 G_1^p)\cos(\varphi_p - \varphi_0) - k_0 k_p/k^2]\tilde{a}_p + [(\varepsilon_2 + G_1^0 G_2^p)\cos(\varphi_p - \varphi_0) - k_0 k_p/k^2]\tilde{b}_p\} \times (G_2^0 + G_1^0)^{-1}, \tag{13}$$

where a_p, b_p are the wave components of the magnetic field parallel to the surface $z=0$ and perpendicular to the propagation vector of the pth order; and $\tilde{a}_p$, $\tilde{b}_p$ are those of the electric field; φ_p is the angle between the fringes and the propagation plane of the pth diffraction order.

1.2. Light absorption on the corrugated boundary of an absorbing medium

The volume density of the light energy released in the absorbing medium is

$$U = \frac{\omega}{8\pi} \boldsymbol{E}\boldsymbol{E}^* \operatorname{Im}(\varepsilon_1). \tag{14}$$

In a small corrugation approximation we confine ourselves to considering the zeroth- and first-order diffraction:

$$\boldsymbol{E}\boldsymbol{E}^* = \boldsymbol{E}_0\boldsymbol{E}_0^* + 2\sum_{p=\pm 1} \operatorname{Re}(\boldsymbol{E}_p\boldsymbol{E}_0^*). \tag{15}$$

Energy absorption in a thin surface layer near the corrugated boundary of an absorbing medium will be determined by the following formula

$$Q = \Bigg\{(\boldsymbol{a}_0\boldsymbol{a}_0^*) + 2k\sigma \sin Ky \operatorname{Im}(G_1^0)(\boldsymbol{a}_0\boldsymbol{a}_0^*) + 2\sum_{p=\pm 1} [\operatorname{Re}(\boldsymbol{a}_p\boldsymbol{a}_0^*) \cos Ky - p \operatorname{Im}(\boldsymbol{a}_p\boldsymbol{a}_0^*) \sin Ky]\Bigg\} \frac{\omega}{8\pi} \operatorname{Im}(\varepsilon_1), \tag{16}$$

where

$$(\boldsymbol{a}_0\boldsymbol{a}_0^*) = |\tilde{a}_0|^2 + \frac{|G_1^0|^2 + k_0^2/k^2}{|\varepsilon_1|^2}|a_0|^2, \tag{17}$$

$$(\boldsymbol{a}_p\boldsymbol{a}_0^*) = \tilde{a}_p\left[\tilde{a}_0^* \cos(\varphi_0 - \varphi_p) - \frac{(G_1^0)^*}{(\varepsilon_1)^*}\sin(\varphi_0 - \varphi_p) a_0^*\right] + a_p\left\{\frac{G_1^p \tilde{a}_0^*}{\varepsilon_1}\sin(\varphi_0 - \varphi_p) + a_0^*\left[\frac{G_1^p (G_1^0)^*}{|\varepsilon_1|^2}\cos(\varphi_0 - \varphi_p) + \frac{k_0 k_p}{k^2|\varepsilon_1|^2}\right]\right\}. \tag{18}$$

It is evident that formula (16) has the form

$$Q = \Big\{F_0 + F_1 \sin Ky + F_2 \cos Ky\Big\} \frac{\omega}{8\pi} \operatorname{Im}(\varepsilon_1), \tag{19}$$

from which it follows that the density of energy release, Q, is modulated over the y-axis with a period equal to that of initial grating [eq. (1)]. This means that if the affecting radiation varies the shape of the surface as a result of melting, evaporation, thermal expansion etc., it may also facilitate the deepening and

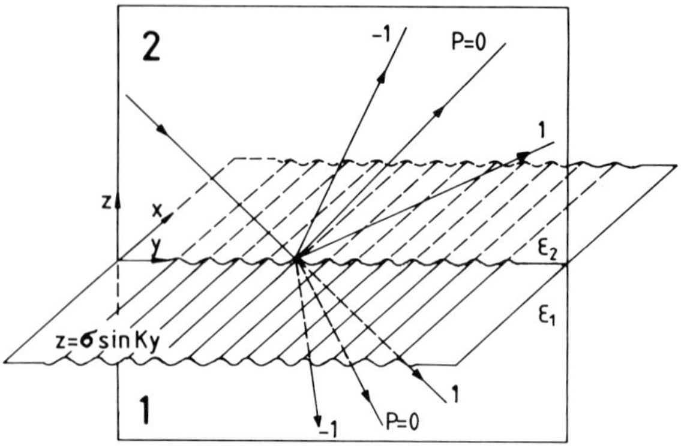

Fig. 2. A schematic representation of light diffraction on a corrugated boundary of two media.

smoothing of ripples. In formula (19) the term F_0 determines the average heating of the surface, the term F_1 is equal to half of the difference in energy release on the 'hills' and 'valleys' of the ripples, i.e., this term determines the light action upon the ripple depth; F_2 stands for periodic energy release shifted in phase by $\pi/2$ in respect to a ripple. Thus, besides a general heating of the corrugated surface, the light action can cause variation in the ripple depth and its shift along the surface (Maslennikov et al. 1983).

1.3. Phase relationships between diffractional waves as a selector of PSS growth

According to the experience (Tsucada et al. 1983), the properties of periodic structures developing under laser radiation action upon the surface of solids, i.e., their period and orientation, do not mainly depend on the original state of the surface. This allows, when considering laser-induced periodic structures, not to take into account the peculiarities of the initial spectrum of inhomogeneities on the surface of a condensed medium and to assume that all the harmonics make approximately the same contribution. Therefore, the value proportional to the difference of energy release at the valleys and hills of the ripples and normalized over the ripple depth, $\Delta Q = -2F_1/k\sigma$, may be used as a parameter determining the rate of the ripple growth. Let us consider periodic structures whose ripples are directed perpendicular to the vector $\boldsymbol{E}$ of radiation. The calculations by formulae (16)–(18) have shown that for such structures the dependence $\Delta Q(\lambda/\Lambda)$ on the fringes separation Λ has two extrema near $\lambda/\Lambda \sim 1$ (fig. 3). This means that under irradiation, from all the spectrum of surface roughness, only those harmonics must be deepened which correspond either to the minimum of $\Delta Q(\lambda/\Lambda)$, or to the maximum, depending on the thermodynamic mechanism. Closeness of λ/Λ to unity attests that the ripple self-enhancement may occur, for instance, in case of weakly absorbing media only when one of the diffractional

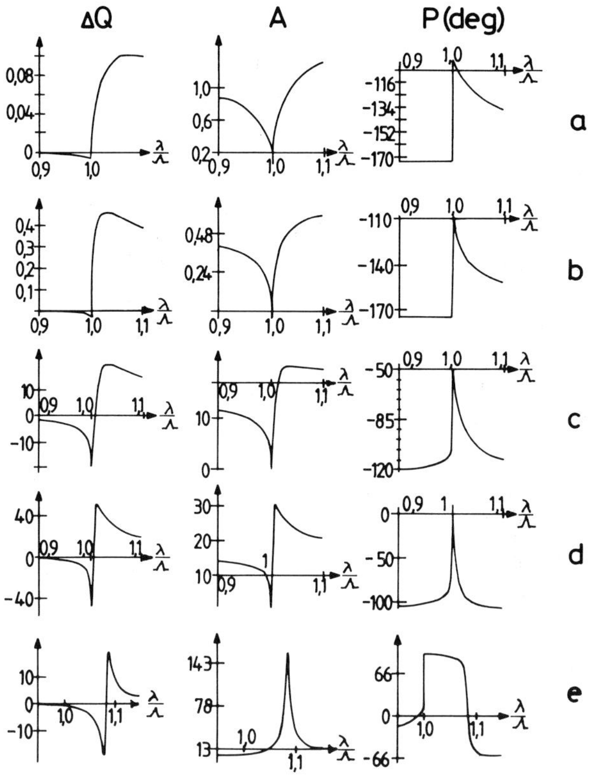

Fig. 3. Calculated dependences of ΔQ, amplitude A, phase P of the diffractional wave propagating in the absorbing medium upon λ/Λ for the case of normal light incidence from the air to the media surfaces with the following values of optical constants: (a) $n = 1.9$, $\kappa = 0.005$, (b) $n = 4$, $\kappa = 0.1$, (c) $n = 4.2$, $\kappa = 3.4$, (d) $n = 4.8$, $\kappa = 7.4$, (e) $n = 0.05$, $\kappa = 2.6$.

waves generated by it propagates in this medium in the direction close to the critical one, that is almost at the angle of total internal reflection. The same is also valid for strongly absorbing media despite the fact that in this case it is doubtful to speak of light propagation.

With a view to elucidate those peculiarities of the diffraction, which determine the form of the function ΔQ, we have calculated the dependences on (λ/Λ) of the value ΔQ, of the amplitude of the first-order diffractional wave and of its phase in respect to the transmitted wave (zeroth-order diffraction). (Formulae (8)–(11) and (19) were used in the calculations.) Figure 3 presents these dependences for the media with various values of optical constants n and κ in case of $\theta = 0°$. First of all, it needs to be noted that the amplitudes of the first-order diffractional waves in the absorbing medium are equal to zero at $\Lambda = \lambda$, but already at a small difference of λ/Λ from unit, they rapidly increase and achieve a maximum at some ratio $\lambda/\Lambda > 1$ depending on n and κ. With increasing κ and decreasing n the amplitude of the diffractional wave maximum grows, its width decreases, and the position of this maximum approaches the value $\lambda/\Lambda = 1$; particularly, at $n \to 0$ and $\kappa > 1$ the maximum of the diffractional wave is a narrow high peak localized at $\lambda/\Lambda = \sqrt{\kappa^2/\kappa^2 - 1)}$. But in the general case, a spike of the value ΔQ in the vicinity of $\lambda/\Lambda \approx 1$ is determined not so much by the growth of amplitude and localization of its maximum, as by a sharp change of the diffractional wave phase in respect to the transmitted zeroth-order wave, as just this phase shift provides an optimum spacing of a thermal grating (periodic distribution of the absorbed energy) in respect to the initial grating. As discerned in fig. 3, near $\lambda/\Lambda = 1$ the phase P of the diffractional wave first increases with increasing λ/Λ, reaches its maximum at $\lambda/\Lambda = 1$, and than decreases. The value of the phase change depends on optical constants: for instance, if for weakly absorbing substances ($\kappa \to 0$) the phase jump makes up 90°, for substances with $n \ll \kappa$ it is as much as 180°. The character of the function $P(\lambda/\Lambda)$ also depends on n and κ: for $n \gtrsim \kappa$ it has the form of a sharp peak, and for $n \ll \kappa$ there appears a 'plateau' in the dependence $P(\lambda/\Lambda)$ between $\lambda/\Lambda = 1$ and $\lambda/\Lambda = \sqrt{\kappa^2/(\kappa^2 - 1)}$. Typical is that the positive extrema of ΔQ are associated with that part of the dependence $P(\lambda/\Lambda)$, where P decreases, i.e., these extremum values of ΔQ are within $\lambda/\Lambda > 1$. At increasing κ the positions of the maximum of amplitude A and its phase jump P draw together, and as a result, the extremum values of ΔQ increase.

At the values of κ exceeding n, to be more exact, at $n^2 - \kappa^2 < -1$, a surface electromagnetic wave (SEW) may be excited at the absorbing medium–air interface. According to the calculations at $\kappa \gtrsim n$ the maximum of the diffractional wave amplitude in the absorbing medium does not coincide with the maximum of the diffractional wave in the air. If the first maximum, for example molten Ge ($\lambda/\Lambda = 1.009$), is close to the maximum positive value $\Delta Q(\lambda/\Lambda = 1.012)$, the second corresponds to the value λ/Λ equal to $n^*_{SEW} = \mathrm{Re}\sqrt{\varepsilon_1\varepsilon_2/(\varepsilon_1 + \varepsilon_2)} = 1.003$, i.e., it corresponds to the excitation of SEW and is in the vicinity of the negative maximum $\Delta Q(\lambda/\Lambda = 1.0006)$ (Bazakutsa et al. 1984).

The picture essentially changes of $n \ll \kappa$, for example, for silver (Ag) having $n = 0.06$ and $\kappa = 2.6$ at $\lambda = 0.63$ μm. In this case the maximum values of the amplitudes of diffractional waves coincide and correspond to the value $\lambda/\Lambda = n^*_{SEW}$. It follows that for Ag-like materials ($n \ll \kappa$) the formation of periodic microrelief under laser irradiation is facilitated by SEW, whereas for materials with $\kappa \gtrsim n$, such as molten germanium, this formation is due to either the diffractional mechanism, or SEW excitation, depending on the thermal mechanism of microrelief fixing.

In the cases where $\kappa < n$, periodic microrelief evolves on the surfaces of the irradiated media only due to the diffractional mechanism.

Thus, as a result of calculations, it is established that the formation of periodic surface structures is only one of the peculiarities of light diffraction on a corrugated interface, a decisive point in this process being essential changes in the phase of the diffractional wave.

1.4. Periodic surface structures at normal and oblique light incidence

Among the variety of the observed periodic structures most frequent are those whose fringes run perpendicular to the vector of the incident electric field and, therefore, they are most studied. Less studied are the periodic structures with the fringes oriented along the electric field vector. The former are frequently referred to as 'transverse', and the latter as 'longitudinal'.

Transverse PSS formed both by *P*-polarized (vector $\boldsymbol{E}$ lies in the plane of light incidence), and by *S*-polarized (vector $\boldsymbol{E}$ is parallel to the sample surface) light, feature 'resonance' dependence $\Delta Q(\lambda/\Lambda)$. It has two extrema in the vicinity of $\lambda/\Lambda=1$ ($\theta=0°$), i.e., a negative minimum and a positive maximum. Those harmonics of the surface inhomogeneities which correspond to the extremum values of ΔQ are most subjected to light influence as a result of its absorption in the subsurface layer. Whether this light action would lead to the ripple deepening or suppression depends on a particular thermal mechanism of microrelief fixing on the surface. If the thermal mechanism is such that the ripple deepens at larger energy release in the valleys of the ripples (evaporation (Tribel'sky and Goldberg 1982), light pressure (Gavrikov et al. 1969), etching (Prokhorov et al. 1984), etc.), then those harmonics of the surface roughness will deepen, which correspond to the maximum of the function $\Delta Q(\lambda/\Lambda)$. If the thermal mechanism is such that more energy is released in the ripple hills than in their valleys (thermal expansion (Gousheng et al. 1982), inhomogeneous melting (Young et al. 1983)), then enhanced will be the harmonics corresponding to the negative minimum of the function $\Delta Q(\lambda/\Lambda)$.

At variation of the incident light angle, the positions of the extremum values of ΔQ vary (fig. 4) so that for *S*-polarized radiation the angular dependence of the position of extremum values of ΔQ has the form:

$$\frac{\lambda}{\Lambda}=\sqrt{n^{*2}-\sin^2\theta}, \tag{20}$$

whereas for *p*-polarization of light:

$$\frac{\lambda}{\Lambda}=n^{*}\pm\sin\theta. \tag{21}$$

The value of n^* and the values of ΔQ corresponding to the maximum and minimum of the function $\Delta Q(\lambda/\Lambda)$, depend on optical constants of the absorbing medium (n and κ).

Figure 5a shows how the values of n^* corresponding to the maximum of ΔQ change, as well as the change of ΔQ itself, normalized over $k\sigma$, at the angle of light incidence $\theta=0°$. In this fig. 5a, region I represents the materials on whose

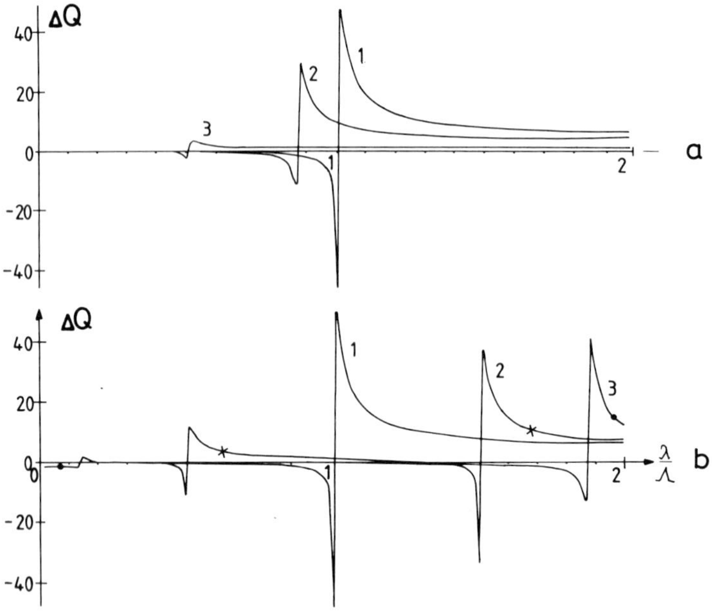

Fig. 4. The function ΔQ for S-polarization (a) and P-polarization (b) of radiation and various angles of incidence: (1) 0°, (2) 30°, (3) 60°; the case of light incidence from the air to the material surface with $n = 4.8$, $\kappa = 7.4$.

surfaces transverse ripples may deepen under irradiation. Region II corresponds to those values of the constants at which the maximum of ΔQ is absent.

It is discerned from fig. 5b how the values of ΔQ corresponding to the minimum of function $\Delta Q(\lambda/\Lambda)$ vary. It is typical that for the minimum of ΔQ $n^* > 1$, but always less than for the maximum. For the values of n and κ lying above curve I in fig. 5b, the diffence of n^* from unity does not exceed 10^{-4}. In region II the minimum of the function ΔQ disappears. In the vicinity of the n-axis at $n < 1$ and $\kappa \lesssim 10^{-2}$, the minimum of ΔQ may lead to the development of transverse structures with $n^* \simeq n$ in the presence of a corresponding thermodynamic mechanism.

Note that the functions presented in fig. 5 may noticeably transform with variation of the angle θ, nevertheless, according to the numerous experiments, formulae (20) and (21) describe well variation in the periods of transverse structures with angle θ.

Calculation of the dependence of ΔQ on λ/Λ for both 'longitudinal' and 'transverse' structures reveals two extrema: a small positive maximum at $\lambda/\Lambda = 1$ and a large negative minimum at $\lambda/\Lambda \sim n(\theta = 0°)$ (see fig. 6). Transition from maximum to minimum in the function ΔQ is due to essential variations in the

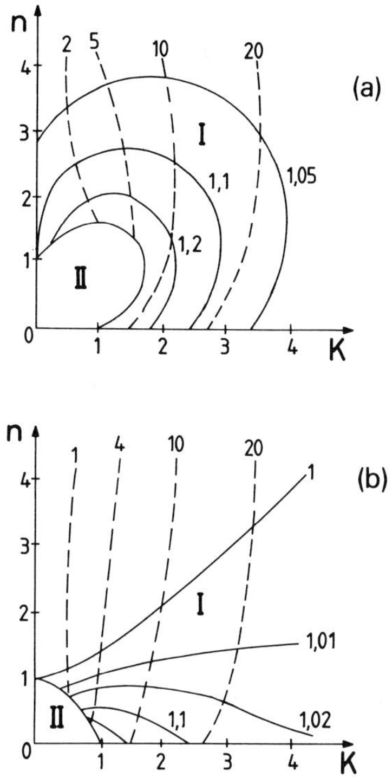

Fig. 5. The regions of variation of the material optical constants where transverse periodic structures may develop at $\theta = 0°$, solid lines correspond to constant n^*, dotted lines to constant values of ΔQ: (a) for spatial periods corresponding to maximum of ΔQ; (b) for spatial spacing corresponding to minimum of ΔQ.

phase of diffractional waves at variation of the ratio λ/Λ from 1 to n. Note that, unlike the case of transverse structures, the value n^*, corresponding to the maximum of ΔQ is always, i.e., at all n and κ, equal to 1. With variation of optical constants only the value of ΔQ changes, which however remains always essentially smaller than that of ΔQ for 'transverse' structures.

Since the function ΔQ has a large magnitude of minimum in the vicinity of $\lambda/\Lambda = n$, then at a corresponding thermodynamic mechanism of ripple deepening, a small-scale longitudinal relief with a period $\Lambda \sim \lambda/n$ may occur on the surface of the irradiated materials. With varying absorption coefficient, the position and shape of the considered ΔQ extremum vary: the value $n^* = (\lambda/\Lambda)_{ext}$ first decreases and then grows, its shape becomes smoother with the increasing κ, and at large κ the extremum disappears. Figure 7 presents the region of variation of constants n and κ, in which the minimum of $\Delta Q(\lambda/\Lambda)$ exists (Bazakutsa et al. 1985). In this figure curve I is described by the equation

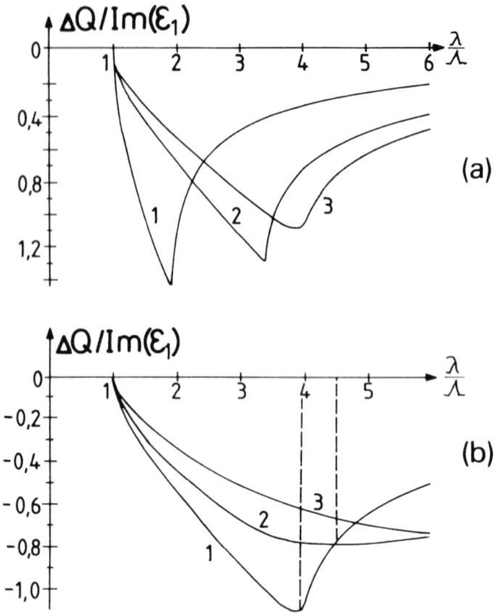

Fig. 6. The function ΔQ for 'longitudinal' ripples in case of (a) $\kappa = 0.1$, $\theta = 0°$, (1) $n = 1.9$, (2) $n = 3.4$, (3) $n = 4$; (b) $n = 4$, $\theta = 0°$, (1) $\kappa = 0.1$, (2) $\kappa = 1.5$, (3) $\kappa = 3$.

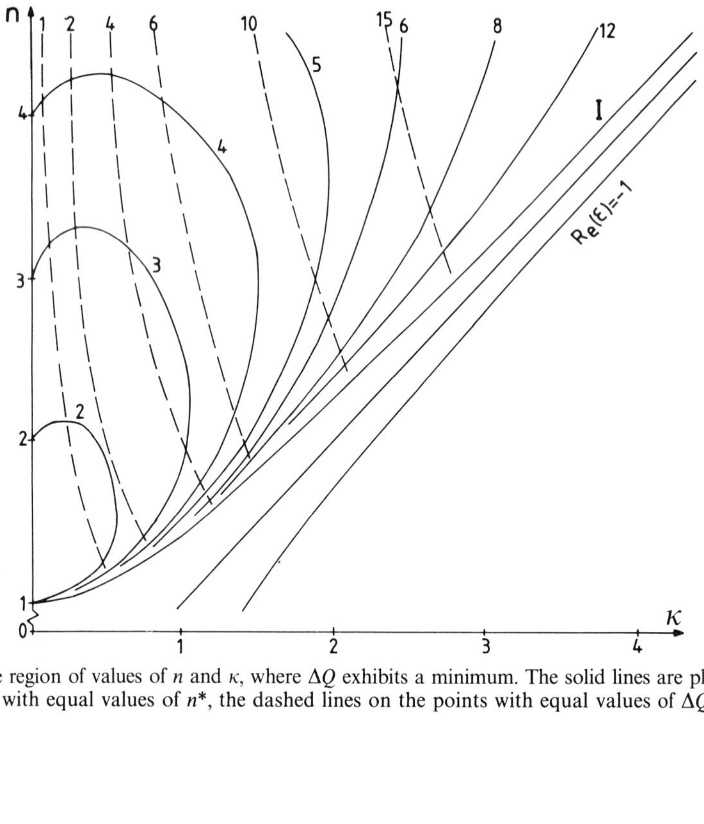

Fig. 7. The region of values of n and κ, where ΔQ exhibits a minimum. The solid lines are plotted on points with equal values of n^*, the dashed lines on the points with equal values of ΔQ_{min}.

$\mathrm{Re}(\varepsilon) = n^2 - \kappa^2 = 1$. Thus, longitudinal structures with a period $\Lambda \lesssim \lambda/n$ may develop on the surfaces of semiconductors and dielectrics, and not on the surfaces of metals for which $\mathrm{Re}(\varepsilon) < -1$.

In case of oblique light incidence, the position of the minimum is transient for both P- and S-polarizations, in the first case the position of maximum being well described by formula (20), and in the second case the minimum bifurcates and the corresponding values of $(\lambda/\Lambda)_{\min}$ are determined by formula (21).

1.5. Transfer of periodic structures under pulsed periodic irradiation

As was already mentioned, under laser irradiation periodic surface structures may not only vary in depth, but also transfer along the surface. This is due to the fact that at a given surface profile, for instance $z(y) = \sigma \sin Ky$, the coordinate dependence of the variable component of energy deposition on a ripple has a sinusoidal form, in general cases shifted in phase by an angle ψ with respect to the initial ripple:

$$Q(y) = Q \sin(Ky + \psi). \tag{22}$$

If inhomogeneous energy deposition deepens ripples due to some of the thermodynamic mechanisms, the shift of phase, ψ, leads to an asymmetric deepening of periodic structure, i.e., to its transfer along the surface (see fig. 8). The phase shift can be easily calculated, if the values F_1 and F_2 are known. Figure 9 presents the dependence of phase ψ on the angle, θ, of incidence on the germanium surface for the case of a ripple development whose spacing period corresponds to the minimum of ΔQ. It is evident from fig. 9, that with varying angle of incidence, the direction of the ripple transfer should vary, and for some angle θ_0, the ripple should develop without shifting. The value of the angle is determined by the optical constants of the material.

The experiments on observation of nonsteady-state patterns at irradiation of the germanium surface with pulsed periodic YAG-Nd^{3+}-laser light have been described by Avrutsky et al. (1985a). The laser-pulse lifetime was $\tau \sim$ 20 ns, $\lambda = 1.06$ μm and the light beam was incident at $\theta > 40°$ and P-polarized.

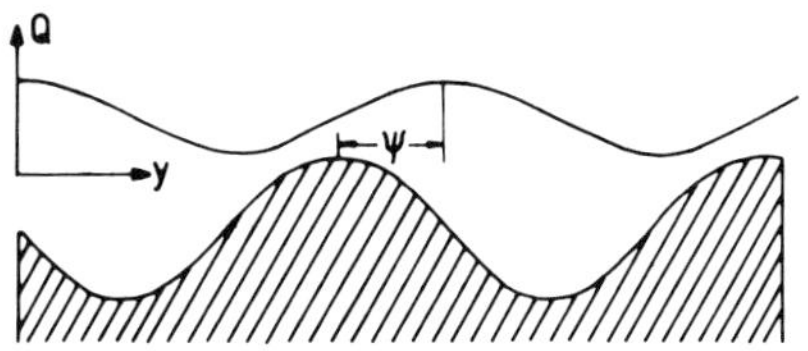

Fig. 8. Phase shift ψ of energy deposition on a periodic surface of the medium absorbing light. The ripple 'transition' proceeds to the right, if its growth is determined by such a thermal mechanism at which more energy is deposited in the 'hills' than in the 'valleys' of the ripple. In the other situation the transition proceeds to the left.

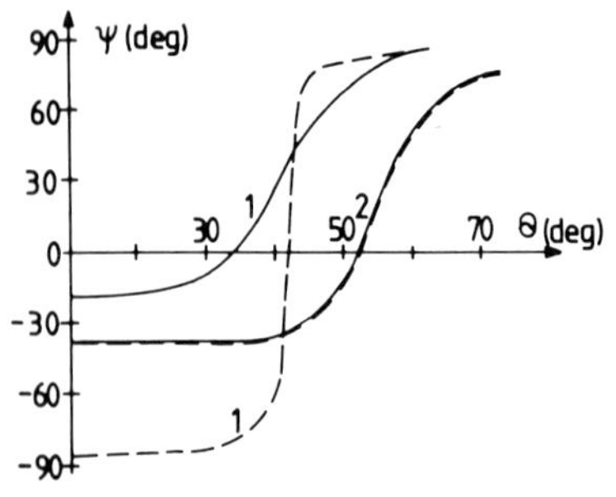

Fig. 9. The dependence of phase ψ on the angle θ of radiation incidence on the surface of solid germanium ($\varepsilon = 16 + i0.8$) (curves 1) and liquid germanium ($\varepsilon = 32 + i72$) (curve 2) at $\lambda = 1.06$ μm; for the maximum of the function ΔQ (solid lines) and the minimum of ΔQ (dashed lines).

Large angles of incidence were needed to make it possible to observe periodic structures via an optical microscope in the process of their evolution (fig. 10). The development of 'transverse' ripples with a period $\Lambda = \lambda/(1 - \sin\theta)$ was observed. It was established by the shape of the intensity distribution of the probe beam scattered on the ripple ($\lambda = 0.63$ μm) in the vicinity of its first-order diffraction, that the development of relief was due to such a mechanism at which more energy is deposited in the 'hills' than in the 'valleys' of the ripples (Avrutsky et al. 1984a). With the sample irradiation, besides the deepening of the relief and growth of its area, one could observe the shift of each strip with respect to fixed marks of the microscope eyepiece (fig. 11). It was revealed that while the ripple develops, the rate of its transition increases and then drops almost to zero after the beginning of the stage of its suppression. This points to a close association of the processes of ripple deepening and 'movement'. During its movement from the initial position to a stop, a ripple shifts by not more than ten periods. With increasing laser fluence the rate of the ripple 'movement', as well as the rate of its growth (Prokhorov et al. 1982) become higher, also attesting to the association of the processes of deepening and movement of the ripple. At the pulse repetition rate 12.5 Hz, $\theta = 60°$, the energy density of the laser pulse is 264 mJ/cm^2, the ripple shift per minute was as high as 4 μm, and at energy density of the pulse ~ 300 mJ/cm^2, the rate of the shift was already ~ 10 μm/s.

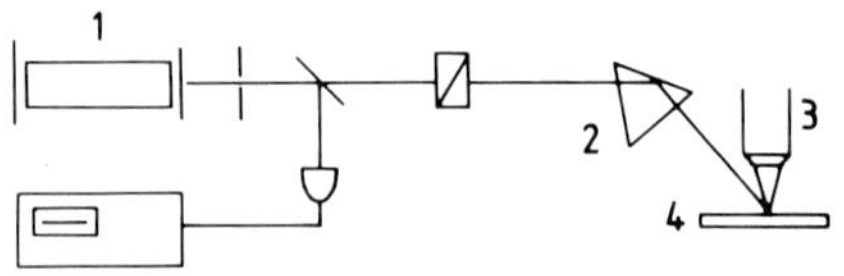

Fig. 10. A schematic diagram of observation of the periodic structure 'movement': (1) YAG-laser, (2) prism, (3) microscope, (4) germanium sample.

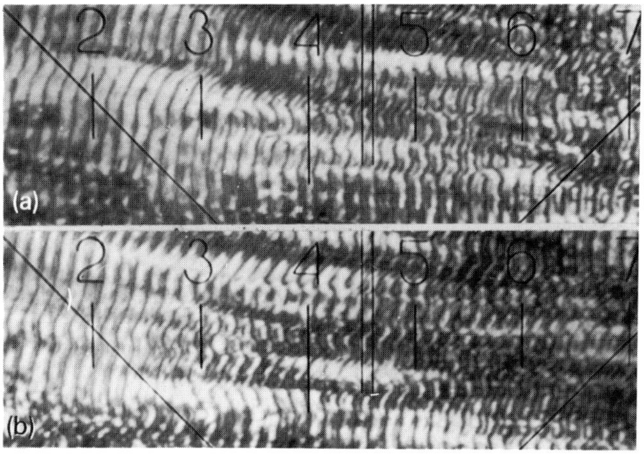

Fig. 11. A view of the developed periodic structure (a); the same view after ten additional radiation pulses (b): after additional irradiation the fringe between two marks in picture (a) shifted by half a period to the left (according to visual observation).

At energy density less than 260 mJ/cm^2, the ripple 'movement' was not noticeable. The dependence of the rate of ripple transition on the intensity of the irradiation manifests itself in the fact that in the center of the radiation spot the rate is higher than at the edges, and this results in the fringes deformation.

It should be noted that at radiation intensity close to the threshold of the beginning of ripple movement ($\sim$260 mJ/cm^2), it was revealed that the strips move first to one side (at a distance approximately equal to the ripple period), and then to the opposite side which coincides with the direction of ripple movement at higher laser fluences. According to the calculations based on the electrodynamic model, the observed direction of the ripple movement at larger intensity of radiation corresponds to its development due to the minimum of $\Delta Q(\lambda/\Lambda)$, whereas the direction of the ripple formed due to the maximum of ΔQ would be opposite. This permits us to assume that the change of the direction is associated with the change of the thermodynamic mechanism of ripple deepening, which is observed in the process of germanium irradiation in some range of radiation power densities (Avrutsky et al. 1984a).

1.6. Longitudinal periodic structures

There are far less reports on the formation of periodic structures of the longitudinal type, than of the transverse type. Nevertheless, it was noted in (Willis and Emmony 1975, Anisimov et al. 1983) that on semiconductor materials, such as Si, Ge, under CO_2-laser irradiation ($\lambda \approx 10$ μm) the structures develop with the spacings lying in the range of 0.8–3 μm, whose fringes run parallel to the vector $\boldsymbol{E}$ of the radiation. It was also reported in Avrutsky et al. (1984b) on observation of longitudinal structures with period $\Lambda = 1.5$ μm ($\lambda = 9.6$ μm) (see fig. 12), and an attempt was made to account for their appearance. When considering the electrodynamic model of PSS it was noted that in case of longitudinal fringes the function ΔQ has a deep negative minimum at $\lambda/\Lambda \gtrsim n$. In our opinion, it is this extremum of $\Delta Q(\lambda/\Lambda)$, that determines the development of a small-scale longitudinal periodic surface on Ge and Si. Figure 6 shows how the shape and position of the minimum of ΔQ change with increasing absorption coefficient of the material. A more complete idea of its position one may get from the diagram in fig. 7. Ge and Si are transparent at 10 μm, and if periodic structures develop on their surfaces under CO_2-laser irradiation, this means that their absorption coefficients strongly vary at this wavelength. As a result, the periods of the longitudinal structures will be less than λ/n and the definiteness of the periods will be less marked, since the minimum of ΔQ is very smoothed at large values of κ. But in the cases where one may speak of some period (at least average) of the longitudinal structure, the absorption coefficient of an irradiated material can be estimated, provided it changes weakly. For instance, we can conclude from fig. 12 that the development of fringes on germanium with an average period of 1.5 μm is accompanied by the growth of the absorption coefficient from 0 to 2.5.

In order to attain a longitudinal structure with a period λ/n on the surface of some material, it is necessary to have such conditions of irradiation under which fringes would evolve, whereas the optical constants of the material remain steady. These conditions turned out to be attainable if a thin absorbing film on the surface of an irradiated material is deposited, on which the structures are

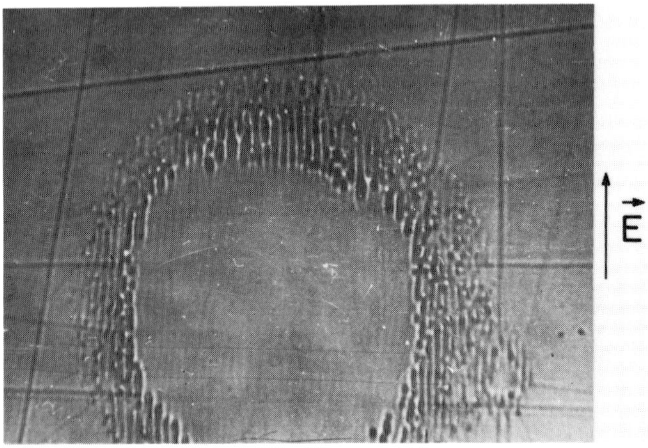

Fig. 12. CO_2-laser-induced small-scale longitudinal relief on germanium surface.

recorded at small radiation powers. Very appropriate for the case considered is a thin titanium film ($h \sim 100$ Å), on which the structures are recorded due to its oxidation. Figure 13 shows a reflexogram of the surface of a glass sample ($n = 1.51$) with a titanium film irradiated with a YAG-ND^{3+}-laser beam ($\lambda = 1.06$ μm). The angle of incident radiation is $\theta = 0°$. The diffraction pattern in fig. 13 is obtained by means of the He–Cd laser ($\lambda = 0.44$ μm). Marked reflexes on it correspond to the structures with a period $\Lambda = 0.6$ μm. Identical longitudinal structures were obtained on the samples of Al_2O_3, ZnS, etc.

The longitudinal structures with a period $\Lambda \simeq \lambda/\cos\theta$ are experimentally observed in the cases, where the angle of incidence of P-polarized radiation is greater than 35–40° (Young et al. 1982). These structures on the Ge surface are shown in figs. 1 and 15. The electromagnetic model presented above predicts an opportunity of forming such structures, but it cannot account for why the angle of incidence cannot be small in this case. It also fails to account for the

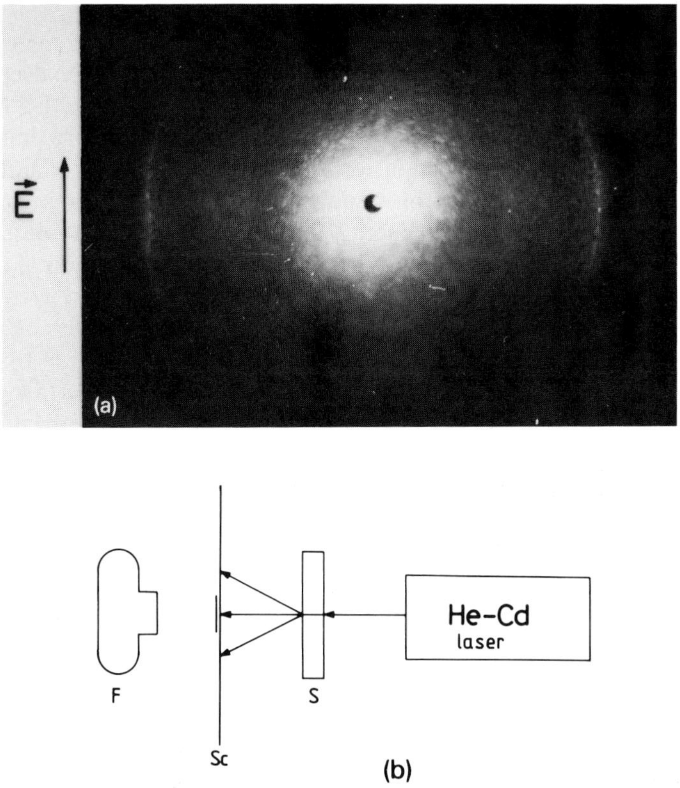

Fig. 13. Reflexogram of the surface of a glass sample ($n = 1.51$) with a titanium film irradiated with YAG-laser light. (a) Angle $\theta = 0°$, (b) schematic diagram for obtaining a reflexogram.

experimental fact that at repeated radiation the longitudinal structure on, for instance, a germanium surface occurs after decomposition of the transverse structure with a small period ($\Lambda = \lambda/(1 + \sin\theta)$), and then in the place of the longitudinal structure forms the transverse one with a period $\Lambda = \lambda/(1 - \sin\theta)$. According to the above model, the longitudinal structure should form after the transverse ones, as the corresponding maximum of the function ΔQ is far less in magnitude than the extremum for the transverse structures. All this stimulated us to turn to another electrodynamic model, in which diffraction of incident light occurs on inhomogeneities of the medium dielectric constant ε. As in the case of surface corrugation, in this model self-enhancement of periodic inhomogeneities in ε under incident light may lead, after all, to the formation of a microrelief due to local melting (evaporation, etching and so on) of the sample surface.

When solving the problem of light diffraction on a periodic inhomogeneity ε, it was assumed that this inhomogeneity exponentially weakens with the medium depth (Gousheng et al. 1982)

$$\varepsilon(y, z) = \varepsilon_1 + \Delta\varepsilon \exp(\alpha z)\cos(K_y y + K_z z), \tag{23}$$

where ε_1 is the complex bulk dielectric constant of the absorbing material, $\Delta\varepsilon$ is the amplitude of modulation and α is the attenuation factor of $\Delta\varepsilon$ deep into the medium. In an approximation of small modulation depth $\varepsilon(|\Delta\varepsilon/\varepsilon_1| \ll 1)$, we can obtain an expression for the energy absorbed in the nearest proximity to the surface, which turns out to have a form similar to that of formula (19). The efficiency of self-enhancement of some periodic inhomogeneity, $\Delta\varepsilon$, can be estimated by the difference of energy deposition ΔQ, which must be here rigorously positive. We shall not dwell on the results of application of the model to transverse and small-scale longitudinal structures, and proceed to considering what this model provides in case of longitudinal structures with $\Lambda \simeq \lambda/\cos\theta$. Figure 14 presents the function $\Delta Q(\lambda/\Lambda)$ for solid germanium ($\varepsilon = 16 + \mathrm{i}0.8$) at different angles, θ, of incidence of P-polarized radiation ($\lambda = 1.06$ μm). The calculations were done for two types of dielectric constant variation, $\Delta\varepsilon = \pm 1 + \mathrm{i}$, describing the heating of the sample or the increasing concentration of free carriers of the charge in it under irradiation ($nv > E_g$). It is seen from the figure that with increasing angle of incidence the shape of the function $\Delta Q(\lambda/\Lambda)$ strongly varies, exhibiting extrema in the vicinity of the values $\lambda/\Lambda = \sqrt{n^{*2} - \sin^2\theta}$, where $n^* \simeq 1$. With growing angle θ, the extrema become more marked, which is consistent with the experimental observation of longitudinal structure formation only at large angles of incidence in case of P-polarized radiation. As it follows from fig. 14 the change of the real part of $\Delta\varepsilon$ weakly influences the value of ΔQ, though noticeably influences the shape of curve $\Delta Q(\lambda/\Lambda)$. The change of the imaginary part of ε plays a decisive role in the formation of longitudinal periodic structures. In case of semiconductors this variation always occurs under high-power laser radiation with $nv > E_g$. For instance, on heating the silicon surface up to $T = 600$°C, a local-temperature

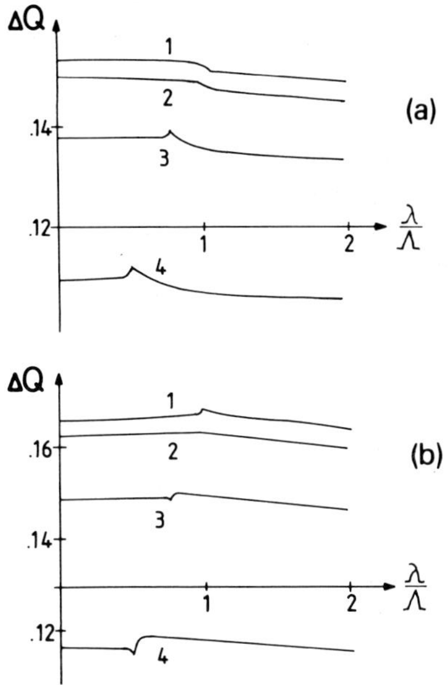

Fig. 14. The function ΔQ for solid germanium at P-polarized incident radiation in case (a) $\Delta\varepsilon = 1 + i1$, (b) $\Delta\varepsilon = -1 + i1$; (1) $\theta = 0°$, (2) $\theta = 20°$, (3) $\theta = 40°$, (4) $\theta = 60°$.

change only by 25°C entails the growth of $\mathrm{Im}(\Delta\varepsilon)$ up to 0.1. With the increasing temperature of the semiconductor surface (up to the melting temperature) the corresponding ($\mathrm{Im}(\Delta\varepsilon) = 0.1$) local temperature difference decrease, that may, in principle, provide a predominant growth of the longitudinal structure over the transverse one.

2. *Thermodynamic mechanisms of developing periodic structures on germanium*

An unambiguous theoretical determination of a particular thermal mechanism of formation of various periodic structures is formidable because of the lack of reliable evidence on optical, mechanical and other properties, especially for semiconductors, under conditions of high-power laser irradiation. Therefore, information gained from experimental observations is most important for developing theoretical models. In particular, it is important to know whether various periodic structures develop on a uniformly molten surface, or under conditions of inhomogeneous melting. This information can be obtained by observing how the surface defects, evolved before irradiation, are distorted by the developing periodic structures. The scratches and roughness on the surfaces remaining after mechanic polishing may serve as these defects.

Figure 12 presents a picture of a germanium surface with a periodic structure (period $\Lambda = 1.5$ μm, the fringes run parallel to the vector $\boldsymbol{E}$ of radiation), evolving after irradiation with several laser pulses ($\lambda = 9.6$ μm, $\theta = 0°$, energy density in the pulse $I \simeq 15$ J/cm^2, pulse lifetime $\tau = 100$ ns). A bright region is distinguished in the picture, where the radiation intensity is maximal and the surface is molten, which is confirmed by the absence of scratches and roughness remaining after mechanic polishing and well discerned in the part of the surface not subjected to radiation. The periodic structure occurs near the melt perimeter with the scratches and roughness remaining between the fringes, i.e., the material has not been molten, whereas there are no fringes on a completely molten surface. It should be noted that there is a sharp boundary between the completely molten portion of the surface and the laser-spot periphery subjected only to local melting. This boundary points to a qualitative difference between the two regions, and the fact that the scratches were conserved in one of them and absent in the other unambiguously attests that the spot periphery has been locally molten, and the center has been completely molten. Thus, it is evident that a microrelief of the type presented in fig. 12 evolves due to the processes developing on a solid surface and at its local melting.

Another type of the structures is presented in fig. 15. These structures occur at irradiation of germanium with P-polarized light at $\lambda = 1.06$ μm at angles of incidence $\theta > 40°$. It is also well seen here that the scratches and roughness remained between the grooves. The fringes have flat 'hills' and V-shaped 'grooves' (fig. 16). The most shallow fringe constitutes an inhomogeneous, in

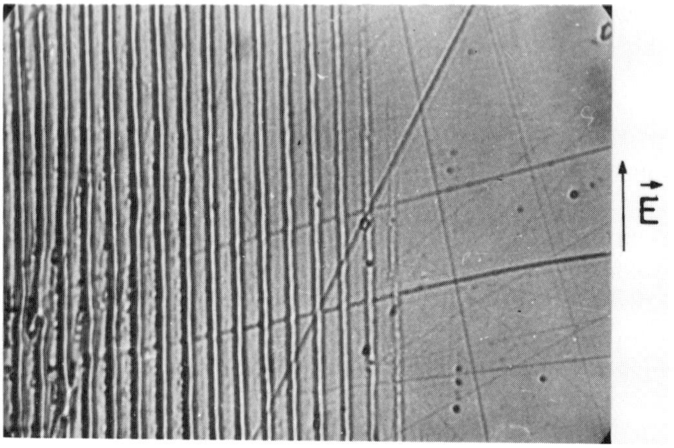

Fig. 15. Microrelief induced by P-polarized, 1.06 μm radiation incident at $\theta = 80°$ on a germanium surface; fringes are parallel to vector $\boldsymbol{E}$ of radiation.

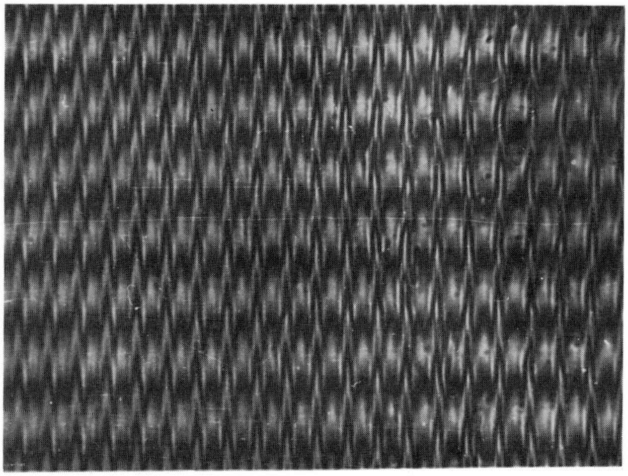

Fig. 16. Interferogram of the surface with a relief of the type presented in fig. 15.

depth, riffle broken interspersed with non-molten areas. With decreasing depth these riffles transform into points on a smooth surface almost not connected with each other, in which must have been the melting centers. All these peculiarities attest that a relief of the type presented in fig. 15 develops as a result of the solid surface melting.

Similar pictures of reliefs on a germanium surface have been obtained by S-polarized radiation ($\lambda = 1.06\ \mu\text{m}$, $\theta = 30°$, $\Lambda = 1.2\ \mu\text{m}$). This points out that inhomogeneous melting of the surface is, in this case, a thermodynamic mechanism of structure formation.

Periodic modulation of the density of energy absorbed by the material selvedge leads to modulation of the melt bulk. Since the density of germanium increases at melting (the density of solid germanium is $\rho = 5.46\ \text{g/cm}^3$, and that of liquid germanium is $\rho = 5.57\ \text{g/cm}^3$ (Ehrlich et al. 1982)), the air–liquid-germanium interface becomes corrugated due to depth modulation of the melt. Such a model does not, however, account for the appearance of a residual, i.e., preserved after laser irradiation, grating, which is most frequently considered in the experiments. To elucidate the appearance of a residual grating, the above mechanism of forming periodic structures has to be supplemented with a natural notion that the properties of a germanium selvedge essentially differ from the properties in its bulk. First of all, such a property is the material density, which may be less in the selvedge than in the bulk. Variation of the density in the selvedge, compared to the bulk one, is due to the fact that mechanic processing of the surface gives rise to microcracks breaking the structure of the material and decreasing its density. After the material's melting these microcracks disappear and, therefore, the density near the surface becomes equal to that of the bulk.

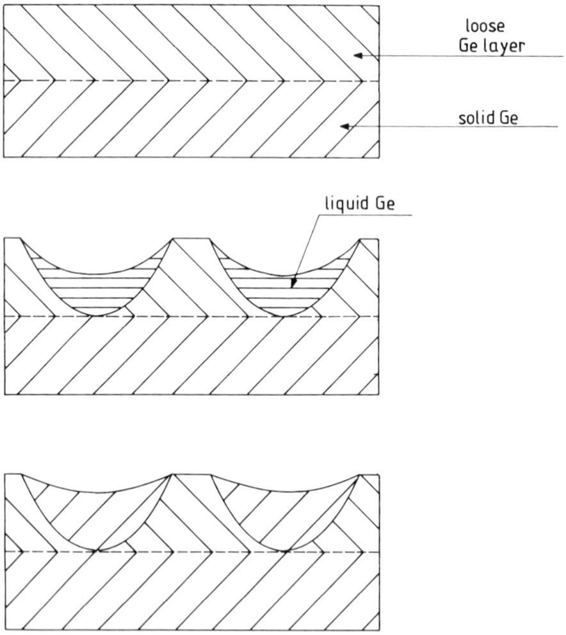

Fig. 17. A scheme of fringe formation on a loose germanium surface.

Due to this, after resolidification of the surface, a residual profile may appear on a solid germanium surface (see fig. 17). Thus, if before resolidification of the surface its corrugation is a result of the difference between the densities of solid and liquid phases, then after resolidification it is a result of the difference between the densities of the solid phase in the bulk and on the surface. To verify this suggestion, the dynamics of PSS formation on germanium was studied in Bazakutsa et al. (1983) at irradiation with periodically pulsed YAG-Nd^{3+}-laser pulses ($\lambda = 1.06$ μm, *S*-polarization), whose radiation density on a target was 100 mJ/cm^2. The fringe depth was estimated by the diffraction intensity of a weak (10 mW) probe beam from a He–Ne laser ($\lambda = 0.63$ μm). In the experiments were compared the dependences of the diffractional efficiencies of the occurring gratings on the time of irradiation (number of pulses) of the surface for two samples polished in a single wafer, one of which was bombarded by argon ions with an energy of 120 keV (dose 10^{15} cm^{-2}). These dependences are shown in fig. 18a. The rate of the grating formation and the rate of its disappearance on a sample processed by argon ions is higher than those on a polished sample. In our opinion, this is due to the fact that the germanium selvedge is loosened by argon ions. This notion is verified by the results of experiments on irradiation of germanium films deposited on germanium substrates and being more loose. It was revealed in the experiments that the appearance, growth and decomposi-

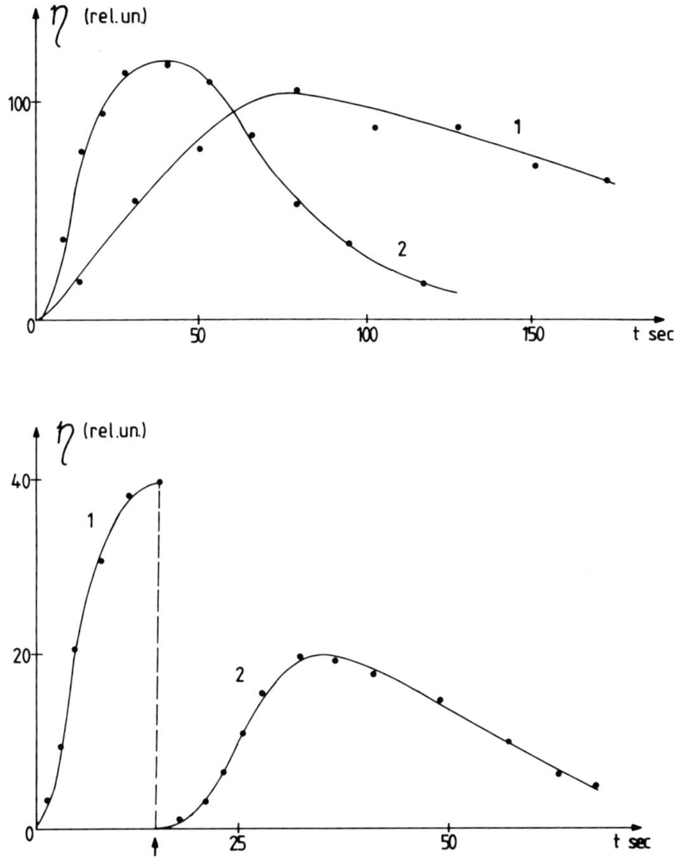

Fig. 18. (a) The dependence of the grating diffraction efficiency η on the exposure time t for a sample: (1) polished, (2) treated with Ar ions; (b) The dependence of η on t for a twice-irradiated sample: (1) for the first grating formed on the nominally smooth surface, (2) for the second grating perpendicular to the first one and formed on the same surface portion.

tion of periodic structures on germanium films proceed during the first five pulses. A convincing confirmation of the action of the considered mechanism is the results of the experiments on multiple irradiation of one and the same portion of Ge surface. For instance, if a grating is formed on a sample surface by irradiation (curve 1, fig. 18b) and then, turning the sample in the plane of its surface by 90°, continue irradiation, another grating will develop on the same surface portion, perpendicular to the first one and with a maximum depth less than that of the first one (curve 2, fig. 18b). If the first grating is damaged before the repeated irradiation, the second grating will not appear at the same radiation power. Thus, if the radiation intensity is close to the threshold of periodic-structure formation, the way of surface treatment essentially influences the dynamics of fringe formation on the surface irradiated with high-power radiation.

If there is no damaged layer in the vicinity of the sample surface, the above mechanism cannot provide the appearance and growth of a residual periodic relief. It was shown in Emmony et al. (1973) that a locally molten portion of a germanium surface assumes a specific shape on resolidification: a cone-like hill appears on the spot of maximum melt depth. Let us make an attempt to use this effect for developing a thermodynamic model of PSS formation under conditions of little lifetime (compared to the period of capillary waves) of local periodic portions of the melt.

Consider heat spreading in a half-space $z > 0$ with uniform properties in the presence of heat sources with a density:

$$F(x, z, t) = Q_0(1 + \alpha \sin Kx)\delta(z, t), \tag{24}$$

and initial condition

$$T(x, z, 0) = T_0. \tag{25}$$

In formula (24), $\delta(z, t)$ is a delta-function, $K = 2\pi/\Lambda$ and α is the modulation amplitude of the sources, T_0 is the sample temperature at an initial instant.

The solution of the Cauchy problem for the heat-transfer equation is given by the Poisson formula and represents a convolution of the fundamental solution of the heat-transfer equation with the boundary conditions (Vladimirov 1981). For boundary conditions (24) and (25), the solution of the heat-transfer equation is:

$$T(x, z, t) = T_0 + Q_0 \frac{1}{a\sqrt{\pi t}} \exp\left(-\frac{z^2}{4a^2 t}\right) \{1 + \alpha \exp(-a^2 t K^2) \sin Kx\}, \tag{26}$$

where T_0 is the initial temperature of the material, $a^2 = \lambda_T / C\rho$, λ_T is the coefficient of thermal conductivity, C is the thermal capacity, and ρ is the material density. It follows from eq. (26) that the isotherm corresponding to the melting temperature T_m has the form:

$$z_{T_m}(x, t) = \sqrt{4a^2 t \ln\left(\frac{Q_0(1 + \alpha \exp(-a^2 tK^2) \sin Kx}{(T_m - T_0) a\sqrt{\pi t}}\right)}. \tag{27}$$

Let us calculate the positions of these isotherms as a function of time. In fig. 19 they are shown for solid germanium over specific time intervals, by condition; $\rho = 5.46$ g/cm^3, $a = 0.6059$ cm c$^{-1/2}$, $T_m - T_0 = 916$°C, $K = 5.9 \times 10^4$ cm^{-1}, $\alpha = 0.1$, and $Q_0 = 0.3$ grad cm.

Since liquid germanium has a density less than that of solid germanium, with the transition of the front of melting (crystallization), the matter is also

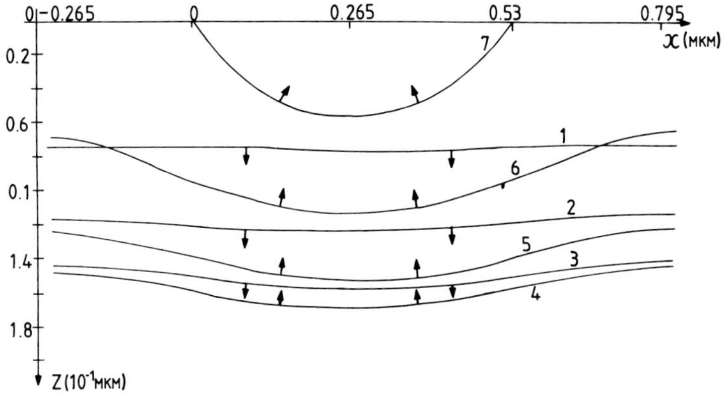

Fig. 19. Isotherms corresponding to the melting temperature at different instances: (1) $t = 1.7 \times 10^{-11}$ s, (2) $t = 7.5 \times 10^{-11}$ s, (3) $t = 17 \times 10^{-11}$ s, (4) $t = 34 \times 10^{-11}$ s, (5) $t = 68 \times 10^{-11}$ s, (6) $t = 75 \times 10^{-11}$ s, (7) $t = 95 \times 10^{-11}$ s. The arrows on the isotherms indicate the direction of transfer of the solid/liquid interface of germanium.

transferred in the same direction. For the case presented in fig. 19, the regions corresponding to a smaller value of $F(x)$ cool down faster than those with larger $F(x)$; therefore, in each point (x, z) the crystallization front is inclined more than that of melting. In this connection, on germanium resolidification, at melting more matter is brought to the region with a higher temperature than taken away.

Thus, asymmetry of the processes of melting and crystallization leads to the development of a residual relief. This conclusion is verified by the model calculation of the matter transfer and variation of the surface shape at melting–crystallization. For simplicity, it was assumed in the calculations that the flow of germanium mass along the surface, $\mathrm{d}M_x$, starting with the transfer of the front of melting (crystallization), and passing via each element of the volume $\mathrm{d}x\,\mathrm{d}z$ with the coordinate $x = x_1$, is uniformly distributed between the points x_1, $\Lambda - x_1$ and makes the height of the surface element $[x_1, \Lambda - x_1]$ to grow by

$$\mathrm{d}h(x, z, t) = \frac{\mathrm{d}M_x(x_1, z, t)}{\Lambda - 2x_1}, \tag{28}$$

where $\mathrm{d}M_x = \Delta\rho \cos\xi \,\mathrm{d}x\,\mathrm{d}z$, ξ is the angle of inclination of the melting–crystallization front, $\Delta\rho$ is the value of variation of material density (taken negative at melting, and positive at crystallization) (see fig. 20).

A complete variation of the height $h(x_1)$ of each surface element $(x_1, x_1 + \mathrm{d}x)$ after crystallization is

$$h(x_1) = \sum_{t=0}^{t'} \sum_{x<x_1} \sum_{z_{T_m}(t)} 2\,\mathrm{d}h - \sum_{t=0}^{t'} \sum_{z_{T_m}(t)} \mathrm{d}M_x(x_1, z, t)\frac{1}{\mathrm{d}x}, \tag{29}$$

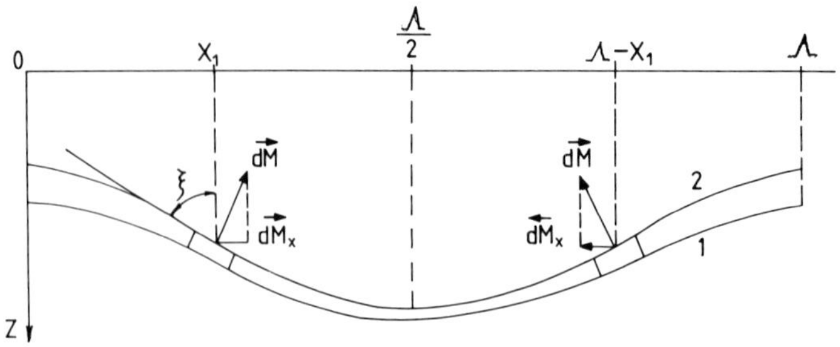

Fig. 20. Movement of the crystallization front and matter transfer along the x-axis. Curves 1 and 2 are the fronts of crystallization at instances t and $t + \Delta t$.

where t' is the time of complete resolidification of the surface. The second term in eq. (29) stands for the mass carried away from each element of the bulk under the surface element $(x_1, x_1 + dx)$. Figure 21 presents the results of the calculation of the shape variation of the initially smooth surface at its melting and resolidification. As it is seen, a fringe hill forms at the maximum of energy deposition. Near the fringe hill forms a valley lying below $z = 0$. According to the calculations (fig. 22), the increasing depth α of modulation of energy-deposition density $F(x)$ entails the growth of the forming fringe amplitude. The dependence $\sigma(\alpha)$ (fig. 22a) illustrates, in fact, the process of the fringe-amplitude growth at its pulsed periodic irradiation, since with the growth of the fringe amplitude, the density modulation of the deposited energy generated by it also increases (Avrutsky et al. 1985).

Besides the fringe deepening under high-power radiation, its phase may also vary, i.e., transfer along the surface. This is due to the fact that in the general case

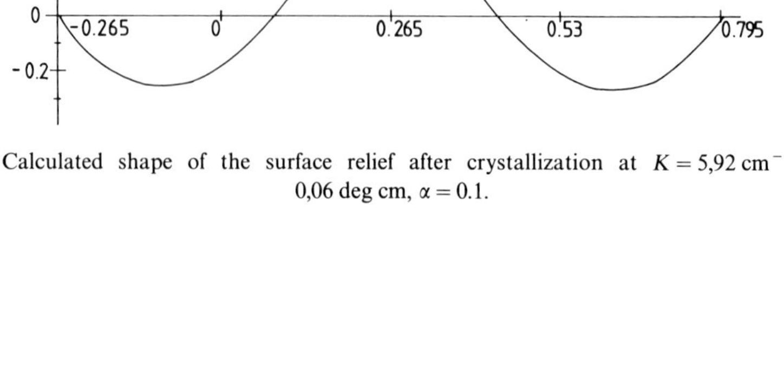

Fig. 21. Calculated shape of the surface relief after crystallization at $K = 5{,}92\ \mathrm{cm}^{-1}$, $Q_0 =$ 0,06 deg cm, $\alpha = 0.1$.

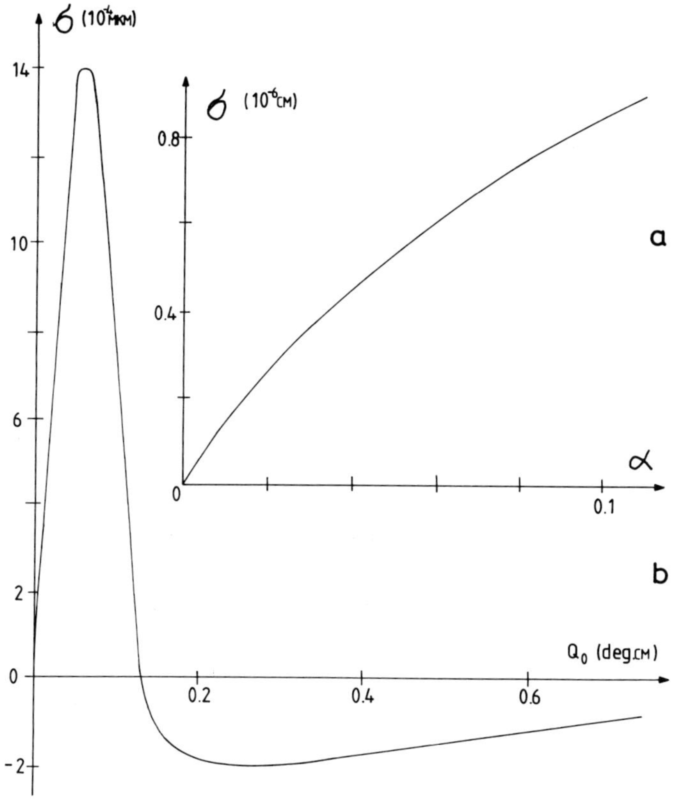

Fig. 22. The amplitude of σ of the fringes versus the depth α of density modulation of the deposited energy (a); the fringe depth versus the density of the deposited energy for $\alpha = 0.01$, $\Lambda = 1.06$ μm (b).

of an arbitrary angle of radiation incidence, the maximum of energy deposition is shifted with respect to the 'hills' and 'valleys' of the fringes. For instance, for the parameters of liquid germanium ($\varepsilon = -32 + i72$) at $\lambda = 1.06$ μm and $\theta = 60°$, the phase shift between the sinusoids of the grating and energy deposition is equal to 37.3°. The estimates based on the above model show that at the density of the absorbed energy ~100 mJ/cm^2, $\theta = 60°$, fringe depth ~100 Å, after one pulse the hill shifts by 0.16 μm ($\Lambda = 8$ μm), i.e., 24 pulses are needed to shift it by half a period. This result is in qualitative agreement with the experimental data presented earlier.

The process of melting and solidification considered results in fringes whose 'hills' are in maxima of energy deposition. But for the same values of α and K but large Q_0 (>0.13), the picture is quite opposite – the 'valleys' form in the maxima of energy deposition. The matter is that at large values of Q_0 the lifetime of the melt becomes comparable with that of heat transfer over distances $\sim\frac{1}{2}\Lambda$. As a result, the crystallization front moves faster in the maxima of energy deposition than in the minima. This leads to the formation of 'valleys' in the maxima of energy deposition. Therefore, a general dependence of the groove depth σ on the energy density may be both above and below zero at a given α. The $\sigma(Q_0)$ function has two extrema: at $Q_0 = 0.06$, for the case of 'hills' in the maxima of energy deposition, and at $Q_0 = 0.3$, for the case of 'valleys' in the maxima of energy deposition (fig. 22b).

It should be noted here that the problem of heat transfer (eqs. (24)–(29)) has been considered for the case where the laser-pulse lifetime is far less than that of the melting–crystallization process, and the depth of the absorbing layer is far less than that of the melt, the variation of thermodynamic constants at phase transition being not taken into account. An account of the light action duration, of the depth of the absorbing layer and of the variation of thermophysical constants at phase transition may lead to some change of quantitative characteristics of the process, such as the time of melting and resolidification, shape of isotherms, etc. Invariable is the fact that the fronts of melting and crystallization have different shapes, leading to surface swelling in the region of maximum energy deposition.

To conclude this section we would like to note that we have considered two lowest-energy thermodynamic mechanisms of forming residual fringes on the surface of an absorbing, condensed medium and, if the first of them fixes periodic structures corresponding to the maximum of ΔQ, the second of them corresponds to the minimum of ΔQ under conditions of inhomogeneous melting.

3. *The development of a periodic microrelief on the surface of a laminated material at laser irradiation*

The action of laser radiation frequently influences the layers near the material surface (Bazhenov et al. 1982), whose optical constants noticeably differ from the constants of the material beneath. Sometimes, such layers are purposely developed at the material surfaces (Ageev and Miloslavsky 1984). In this case a microrelief induced by laser radiation may contain information on a laminated medium. Besides, the investigation of PSS evolving on the surfaces of laminated materials with waveguide properties may be of special importance, since here the role of the guided waves in the process of PSS formation can be unambiguously revealed, provided that the waveguide is not damaged by laser radiation.

Earlier it was noted that for observation of induced PSS formation, it is appropriate to use a thin absorbing film deposited on the surface of a condensed medium. PSS develop on this film at essentially smaller powers of laser radiation, thus making it possible to save a sample from damage. In case of laser irradiation of a laminated medium surface, this technique is also very suitable.

In our experiments on studying the role of waveguide modes in the development of a periodic microrelief we have used a waveguide structure consisting of a 1.3 μm-thick ZnO film deposited on a glass sublayer. A thin titanium layer ($\sim$200 Å) was deposited on the ZnO film and played the role of an indicator film on which PSS were formed. The YAG-Nd^{3+}-laser, with $\lambda = 1.06$ μm and a pulse duration of 20 ns, was used in the experiments. The diameter of the laser-beam spot on the target was $d = 1.5$ mm, and the radiation power was slightly above the threshold of PSS formation ($I = 0.1$ MW) on titanium film.

Periodic structures developed on the waveguide surface after a train of pulses, the titanium film became oxidized and transparent and, as a result, PSS appeared on it. Aluminizing of the irradiated surface of the waveguide structure has shown that a microrelief formed on its surface.

Figure 23b presents the reflexogram of microrelief obtained at normal YAG-laser radiation incidence on the waveguide surface. It follows from the reflexogram that both 'longitudinal' and 'transverse' structures develop on the waveguide, i.e., the grooves are parallel and perpendicular to the vector $\boldsymbol{E}$ of the incident radiation. According to the intensity of the reflexes, longitudinal-type gratings are more efficient. The efficiency of the development of a periodic structure on the sample surface is estimated by the value ΔQ, equal to the difference of the energy densities absorbed in the 'valleys' and 'hills' of the fringes. The calculation of this value performed for a waveguide structure consisting of a thin ($kh = 8.5$) absorbing layer ($\kappa = 0.005$) with refractive index $n_0 = 1.98$ (ZnO), deposited on a transparent substrate with refractive index $n_1 = 1.51$ (glass), has shown that for longitudinal gratings this value is indeed greater than for transverse ones (see fig. 23a). Sharp spikes on the curve $\Delta Q(\lambda/\Lambda)$, correspond, depending on the ratio λ/Λ, to the periods of gratings where excitation of waveguide modes occurs. In particular, it follows from the reflexogram and the calculated curve in fig. 23, that at laser action, the gratings corresponding to excitation of higher-order modes are formed more efficiently.

The measurement of the periods of the longitudinal gratings formed on the waveguide surface has shown that the efficient refractive indices n_m^* of the modes have the following values: $n_1^* = 1.92$, $n_2^* = 1.84$, $n_3^* = 1.73$ and $n_4^* = 1.54$. The pointed spectrum n_m^* turned out not to correspond to the initial waveguide (ZnO film on glass), and to obtain this spectrum n_m^*, one has to take into account the titaniumoxide layer developing under irradiation. Our estimates made on the basis of the dispersion relationship for a two-layer waveguide have shown that the thickness of the oxide layer is as high as $\sim$400 Å.

It is worth mentioning that the bright reflex nearest to the center of the pattern in fig. 23b corresponds to the longitudinal structure with a period equal to λ/n, where $n = 1.4$ is the value close to that of the substrate's refractive index. Proceeding from the very fact of the development of longitudinal grating with a period λ/n and from the form of the light-intensity distribution in reflexes, one

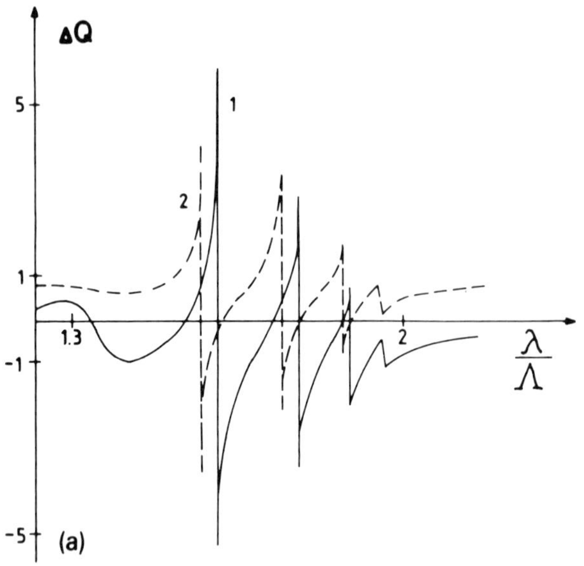

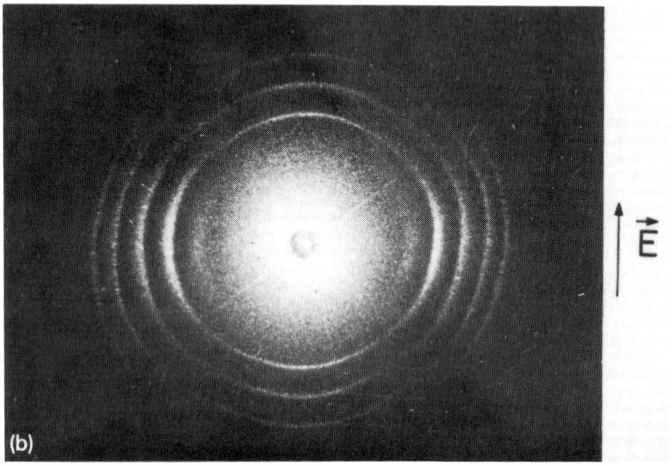

Fig. 23. The dependence of the value ΔQ on the ratio λ/Λ for a waveguide film ($n_0 = 1.98$) on a glass substrate ($n_1 = 1.51$). The film thickness is $kh = 8.5$. Curve 1 is for TE-polarized incident light, curve 2 for TM-polarization (a). Reflexogram of the waveguide surface after YAG-Nd^{3+}-laser irradiation. Normal incidence (b).

can conclude that the mechanism of periodic microrelief fixing on the considered waveguide structure is such, that more energy is deposited in the hills than in the valleys of the fringes, i.e., a negative extremum of the function $\Delta Q(\lambda/\Lambda)$ is involved here (see fig. 23a) (Avrutsky et al. 1987a).

As noted above, a microrelief occurs on the waveguide structure surface due to titanium oxidation. In this case, oxidation develops faster in the places where more energy is deposited and, therefore, thicker oxides form the hills of the fringes. It is this effect that makes it possible to elucidate interesting peculiarities of PSS formation and, in particular, of longitudinal periodic structures with a period $\Lambda = \lambda/n$, as well as the peculiarities of PSS forming on waveguides with leaky modes. In these waveguides the refraction index of the waveguide layer is less than that of the substrate, and the propagating wave undergoes total internal reflection only on one side of the interface, whereas on the other side the light is only partially reflected causing light losses due to its radiation into the substrate (Ding and Garmire 1983). To solve the question which periodic structures might develop on the waveguide surface with the leaky modes, we have calculated the values $\Delta Q(\lambda/\Lambda)$. The results of this calculation for weakly absorbing (at $\lambda = 1.06$ μm and $\kappa = 5 \times 10^{-4}$) SiO_2 film ($n_0 = 1.46$) on a sapphire sublayer ($n_1 = 1.76$) are presented in fig. 24a. In case of normal light incidence the extrema of the value ΔQ occur if the ratio λ/Λ approximately coincides with the value of efficient refractive index of the modes of the considered waveguide. The number of extrema for longitudinal periodic structures coincides with or is by a unit greater than the number m of TE-modes in the waveguide, and the number of extrema for transverse structures at $m > 3$ is a little less than the number of TM-modes. The latter fact is due to essential losses for higher-order TM-modes in the waveguides with leaky modes.

To observe experimentally the PSS development we have used the above waveguide structure, i.e., SiO_2 films on sapphire with the thickness h equal to 1.15 μm. A 100 Å-thick titanium layer was deposited over the SiO_2 film. The titanium layer served as an indicator film on which periodic structures formed. The YAG-Nd^{3+} laser was used as a source of radiation. Figure 24b shows the reflexogram of the microrelief obtained at normal radiation incidence on the waveguide surface. It follows from the picture that periodic structures develop on the waveguide with leaky modes due to excitation of TE-modes in it. Note that both modes become excited, even that of the higher order for which the calculated losses are 2200 cm^{-1}. An estimation of the values n^* on the basis of the obtained reflexogram has shown that these values noticeably exceed those obtained from the dispersion relationship:

$$kh\sqrt{n_0^2 - n_m^{*2}} = \pi(m - \tfrac{1}{2}) + \operatorname{arc\,tg}\sqrt{\frac{n_m^{*2} - 1}{n_0^2 - n_m^{*2}}}. \tag{30}$$

Comparison of the values of n^*, with n^* obtained from the calculation of ΔQ, has shown that the latter surpass the former due to radiational losses χ in the substrate and are as high as $\Delta n^* \simeq \chi/k$ (Avrutsky et al. 1987b).

Thus, according to the experimentally verified calculations, analysis of the laser-induced surface structures at the surface of a laminated material might,

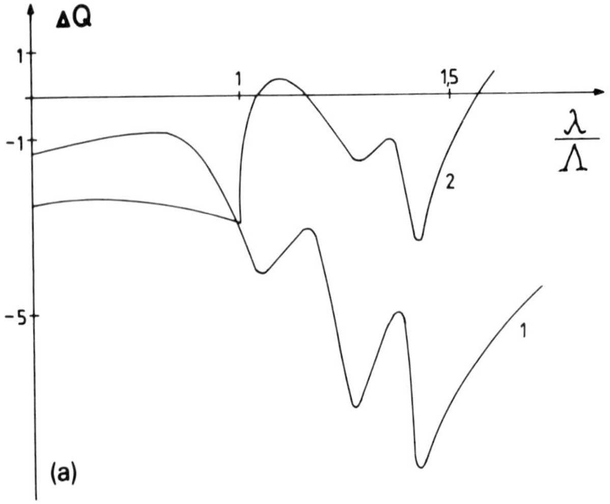

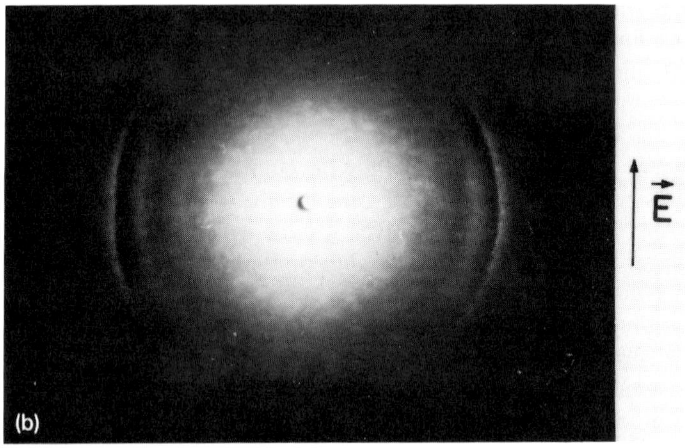

Fig. 24. (a) The dependence of ΔQ on the ratio λ/Λ for the waveguide with leaky modes: $n_0 = 1.46$, $n_1 = 1.76$, $n_2 = 1$, and $h = 1.15$ μm. Curve 1 – TE-polarized incident radiation, curve 2 – TM-polarized incident radiation. (b) Reflexogram of the waveguide surface.

indeed, provide interesting information on the parameters of the layers forming the material and of the substrate itself.

We note here, for example, that using nitrocellulose film (indicator film) deposited on a Ge–Al_2O_3 waveguide, it is possible to measure the refractive index of the germanium film in a 10 μm wavelength region analysing the surface structures. Its value has been found to be equal to $n_{Ge} = 3.59$, that is 10% lower than that of bulky n_{Ge}. This gives evidence of the looseness of evaporated germanium films, that was mentioned in the observation of thermodynamical mechanisms of developing PSS.

Acknowledgment

The authors are grateful to A.S. Svakhin, V.L. Maslennikov, G.A. Bufetova, N.A. Merkushina and L.V. Marchenko for their help in preparing the chapter.

References

Ageev, L.A., and V.K. Miloslavsky, 1984, TE-grating diffraction properties photoinduced in a AgCl–Ag thin film, Zh. Tekhn. Fiz. **54**, 888–895.

Akhmanov, S.A., V.I. Emel'yanov, N.I. Koroteev and V.N. Seminogov, 1985, Influence of the powerful laser radiation upon the metal and semiconductor surface: nonlinear-optical effects and nonlinear-optical diagnostics, Sov. Phys. Usp. **147**, 675–745.

Anisimov, V.N., V.Yu. Baranov, L.A. Bol'shov, A.M. Dykhne, D.D. Malyuta, V.D Pismenny, A.Yu. Sebrant and M.A. Stepanova, 1983, Formation of periodic structures on metal and semiconductor surface under action of pulsed CO_2-laser radiation, Surface, Physics, Chemistry, Mechanics, No. 7, p. 138–145.

Avrutsky, I.A., P.V. Bazakutsa, A.M. Prokhorov and V.A. Sychugov, 1984a, Periodical microrelief on Ge surface and thermophysical mechanism of its formation, Pis'ma Zh. Tekhn. Fiz. **10**, 1333–1339.

Avrutsky, I.A., P.V. Bazakutsa, A.M. Prokhorov, V.A. Sychugov and A.V. Tischenko, 1984b, Self-enhancement of modulation of dielectric constant on germanium surface affected by powerful light wave, Pis'ma Zh. Tekhn. Fiz. **10**, 1089–1094.

Avrutsky, I.A., P.V. Bazakutsa, A.M. Prokhorov and V.A. Sychugov, 1985a, A movement of the periodic surface relief under the action of powerful laser radiation, Sov. J. Quantum Electron. **12**, 650–652.

Avrutsky, I.A., P.V. Bazakutsa, A.M. Prokhorov and V.A. Sychugov, 1985b, Thermophysical mechanism of deepening and movement of periodic structures on germanium, Pis'ma Zh. Tekhn. Fiz. **11**, 621–625.

Avrutsky, I.A., G.A. Golubenko, A.S. Svakhin, V.A. Sychugov and A.V. Tischenko, 1987a, An analysis of structural changes on the surface of a layered medium exposed to the laser radiation, Sov. J. Quantum Electron. **14**(1), 67–70.

Avrutsky, I.A., G.A. Bufetova, A.S. Svakhin, V.A. Sychugov and A.V. Tischenko, 1987b, Planar waveguides with leaky modes and determination of their parameters, Sov. J. Quantum Electron. **14**(4).

Bazakutsa, P.V., A.M. Prokhorov, V.A. Sychugov and A.V. Tischenko, 1983, The state of the germanium surface and its reaction to powerful laser radiaton, Pis'ma Zh. Tekhn. Fiz. **9**, 541–545.

Bazakutsa, P.V., V.A. Sychugov and A.M. Prokhorov, 1984, Light diffraction on the surface roughness and its rule in formation of a periodic surface microrelief, Sov. J. Quantum Electron. **11**, 2127–2128.

Bazakutsa, P.V., V.L. Maslennikov, A.M. Prokhorov and V.A. Sychugov, 1985, On the possibility of investigation of a periodic surface micro-relief for determination of the optical constants of substances under the condition of high-power laser radiation, Surface, Physics, Chemistry, Mechanics, No. 6, pp. 82–85.

Bazhenov, V.V., A.M. Bonch-Bruevich, M.N. Libenson, V.S. Makin, S.D. Pudkov and V.V. Trubaev, 1982, Surface waveguide mode excitation by laser radiation and their effect upon a character of the surface condensed media damage, Izv. Akad. Nauk SSSR, Ser. Fiz. **46**, 1186–1193.

Birnbaum, M., 1985, Semiconductor surface damage produced by ruby laser, J. Appl. Phys. **36**, 3688–3689.

Bunkin, F.V., A.A. Samokhin and M.V. Fedorov, 1969, To a theory of stimulated light scattering on the liquid surface, Zh. Eksp. & Teor. Fiz. **56**, 1057–1064.

Ding, T.N., and E. Garmire, 1983, Measuring refractive index and thickness of thin films; a new technique, Appl. Opt. **22**, 3177–3181.

Ehrlich, D.J., S.R.J. Brueck and J.Y. Tsao, 1982, Time-resolved measurements of stimulated surface polariton wave scattering and grating formation in pulsed-laser-annealing germanium, Appl. Phys. Lett. **41**, 630–632.

Emmony, D.C., R.P. Howson and L.J. Willis, 1973, Laser mirror damage in germanium at 10.6 μm, Appl. Phys. Lett. **23**, 598–600.

Gavrikov, V.K., A.V. Kats and V.M. Kontorovich, 1969, Stimulated scattering on surface waves, Dokl. Akad. Nauk SSSR **186**, 1052–1054.

Gousheng, Z., P.M. Fouchet and A.E. Seigman, 1982, Growth of spontaneous periodic structures on solids during laser illumination, Phys. Rev. B **26**, 5366–5380.

Keilmann, F., and Y.H. Bai, 1982, Periodic surface structures frozen into CO_2-laser melted quartz, Appl. Phys. **29**, 9–18.

Maslennikov, V.L., A.M. Prokhorov, V.A. Sychugov and A.V. Tischenko, 1983, Diffraction of a high-power light wave on the corrugated boundary of two media, Pis'ma Zh. Tekhn. Fiz. **9**, 679–682.

Prokhorov, A.M., V.A. Sychugov, A.V. Tischenko and A.A. Khakimov, 1982, Kinetics of corrugation formation on germanium surface at irradiation with powerful laser radiation, JETP Lett. **8**, 1409–1413.

Prokhorov, A.M., V.A. Sychugov and T.V. Tulaikova, 1984, The effect of liquid waveguide upon the process of grating formation at semiconductor photo-etching, Phys. & Technol. Semicond. **18**, 2125–2129.

Sychugov, V.A., and A.V. Tischenko, 1982, Propagation and transformation of the light waves in corrugated structures, Sov. J. Quantum Electron. **9**, 1451–1458.

Tribel'sky, M.I., and S.M. Goldberg, 1982, On resonance excitation of the surface electromagnetic waves at laser evaporation of the condensed media, Pis'ma Zh. Tekhn. Fiz. **8**, 1227–1230.

Tsucada, N., S. Sugata and Y. Mita, 1983, New experimental evidence of surface ripples on gallium arsenide in laser annealing, Appl. Phys. Lett. **42**, 424–426.

van Driel, H.M., J.E. Sipe and J.F. Young, 1982, Laser-induced periodic surface structure on solids: A universal phenomenon, Phys. Rev. Lett. **49**, 1955–1958.

Vladimirov, V.S., 1981, Equations of Mathematical Physics (Nauka, Moscow) pp. 260–267.

Willis, L.J., and D.C. Emmony, 1975, Laser damage in germanium, Opt. & Laser Technol., October 1975, pp. 222–228.

Young, J.F., J.E. Sipe, J.S. Preston and H.M. van Driel, 1982, Laser-induced periodic surface damage and radiation remnants, Appl. Phys. Lett. **41**, 261–264.

Young, J.F., J.S. Preston, J.E. Sipe and H.M. van Driel, 1983, Time-resolved evolution of laser-induced periodic surface structure on germanium, Phys. Rev. B **27**, 1424–1427.

CHAPTER 10

Laser-Induced Surface Polaritons and Optical Breakdown

A.M. BONCH-BRUEVICH and M.N. LIBENSON

S.I. Vavilov
State Optics Institute
199304 Leningrad, USSR

Nonlinear Surface Electromagnetic Phenomena
Edited by
H.-E. Ponath and G.I. Stegeman

© *Elsevier Science Publishers B.V., 1991*

Contents

Introduction

In recent years one of the most interesting effects on the surface of a condensed medium, the excitation of the surface polaritons (SP), also known as light-induced surface electromagnetic waves (SEW), has been studied. At the boundary of the medium with air, or some other transparent dielectric, such waves are excited in a wide frequency range up to the surface plasma frequency band or the residual frequency band for semiconductors and dielectrics, i.e., under the condition that the real-valued part of the dielectric constant of the medium is negative. The SEW propagation length along the surface may be as long as several centimeters. This length appears to be highly sensitive to the surface state, i.e., to the presence of absorbed layers, thin films, roughness, etc. It opens good prespectives for SEW applications for the investigation of surface properties and characteristics (*The Surface Polaritons* (1975)).

One of the important features of SEW is its possibility to be directly excited by light incident on the rough surface. Thanks to the SEW field localization near the boundary, even at the insignificant energetic conversion of the incident radiation in it, there takes place a considerable redistribution of light-field intensity near the surface and its amplification as compared with the incident wave field. The ideas of electrodynamic amplification of the light field as a result of its partial transformation into surface polaritons and localized excitations, recently greatly broadened our notion about optical processes on the surface. Also this allowed to explain partially such phenomena as giant Raman scattering by the absorbed particles, second-harmonic generation at light reflection from metals and some others (cf. Cosobukin 1983).

The idea about the important role of partial conversion of laser radiation in SEW (and waveguide mode) was used also for the explication of some effects of laser-radiation interaction with matter. First of all, the effect of surface periodic structure (SPS) formations on damaging and heating of condensed media. Such structures are usually oriented in a certain way towards the projection of the light-wave electric vector on the surface and they have a period d of the order of the incident radiation wavelength λ (fig. 1). Along with them there sometimes appear structures with period $d \sim \lambda/n$, where n is the refractive index of the material. The SPS formation has a universal character, it is observed on metals, semiconductors and dielectrics in a wide range of wavelengths (from IR to UV spectral region), durations and intensities of the incident laser radiation (from a continuous one with flux density $q_0 \sim 100\ \mathrm{W/cm^2}$ to picosecond pulses $q_0 \sim \mathrm{GW/cm^2}$).

The first to observe SPS was Birnbaum (1965). During the next 15 years some regularities and features of SPS formation (the orientation towards the incident radiation polarization plane, the connection of the period with the angle of incidence of the radiation, thermal mechanisms of periodic relief formation) were established or discussed (Siegrist et al. 1973, Emmony et al. 1973, 1975,

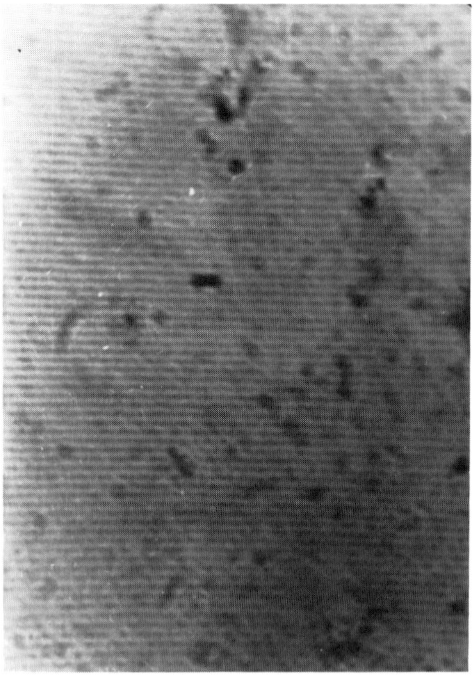

Fig. 1. The typical SPS picture. The structure is obtained on Ge under action of a series of pulses of a Nd laser with a duration at 30 ns and at a flux density of 50 MW/cm^2.

Willis and Emmony 1975, Isenor 1977, Leamy et al. 1978, Bahin et al. 1979). However, the most important problem about the nature of SPS formation remained open. The explanation of this phenomenon was given in 1981 (Kochengina et al. (1981); subsequent publication in Bonch-Bruevich et al. (1982) and also in Aksenov and Gurkin (1982), Keilmann and Bai (1982), Brueck and Ehrlich (1982) , Prochorov et al. (1982a), Anisimov et al. (1983), Tribelsky and Gol'berg (1982)). It is based on the interference of the incident radiation and the SEW generated by it on the rough boundary of the surface-active medium.* In some cases, the incident radiation interferes not with SEW but with a waveguide mode excited by it in the surface layer of the media (Kochengina et al. 1981, Bonch-Bruevich et al. 1982). In the appearing interfacial field there takes place a spatially modulated heating (and damping) of the surface, accompanied by SPS formation. It is important that the SPS period conforms to the condition of the resonance generation of the surface polariton (or the waveguide mode), the efficiency of which is determined by the SPS amplitude. Thus there takes place a positive feedback between modulation of temperature

*The medium which has a negative real part of dielectric constant ($\mathrm{Re}\,\varepsilon < 0$).

field depth and the interferential distribution of the light intensity. The suggested qualitative explanation of the SPS formation mechanism (it might be called 'polaritonic') greatly stimulated investigations in this field (Aksenov 1982, Prochorov et al. 1982b, Fauchet and Siegman 1983, Keilmann 1984, Konov et al. 1983, Bagenov et al. 1983, 1984, Gromov and Uphimzev 1983). The number of investigations is continuously increasing. Recently, there has been given a generalization of the SPS formation model (the so-called diffraction modes). With the help of this model not only a proper electromagnetic surface mode but also the radiation scattered on the periodical grating in the direction of the medium volume (at a certain angle) has been analyzed. The interference of the latter with the refracted radiation creates structures growing with the greatest increment at some spatial frequencies. The model is applicable to the materials with an arbitrary value of the dielectric constant.

In the process of SPS formation the most important result of SEW excitation is the interference of their fields with the incident radiation field, whereas in other surface phenomena a high surface-wave intensity is of importance. The low-threshold optical gas breakdown at the surface (Kochengina et al. 1981, Bonch-Bruevich et al. 1982, Kovalev et al. 1983), photosorbing processes, light-induced thermochemical reactions (Bonch-Bruevich et al. 1985), etc., are an indication for these phenomena.

Below, the basic physical notions about the polariton mechanism of SPS formation at heating and damaging (optical breakdown) of the surface, and its theory are discussed. It is preceded by a short characterization of basic properties of SEW of optical range and regularities of their excitation on a periodical grating. There has been studied the SEW generation influence on the effective absorption of the intensive coherent radiation in a surface-active medium and on the physical processes taking place in the optical field.

1. Basic properties of the optical-range surface polaritons

A surface polariton (SP) is a combined elementary excitation existing in the vicinity of two media with dielectric constants of different signs. Remember that in a dielectric an SP is of photon-like character and in a solid it is either a surface phonon (then we speak about phonon-polariton) or an exciton (exciton-polariton) or an excitation in an electronic plasma (surface plasmon-polaritons). In the optical frequency range the localization region of such an SP exceeds interatomic distances. In a classical description this allows us to use the formalism of dielectric permitivities and the Maxwell equations with standard boundary conditions and to treat the excitations themselves as waves propagating along the boundary of two media (with a wave vector $\boldsymbol{K}_s$) and being damped exponentially on both sides of boundary (see fig. 2). In the most simple case the media are considered as one of which is a dielectric without loss with

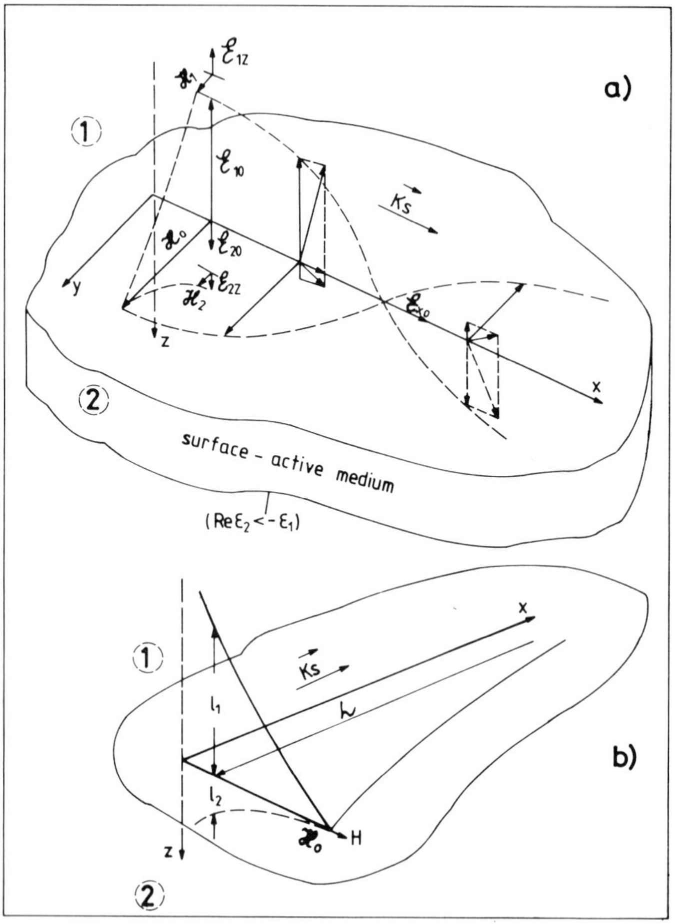

Fig. 2. Instantaneous distribution of the electric and magnetic fields in SEW of TM-type (a) and tangent and normal distribution of the field amplitude on the surface (b). The notation used is explained in the text.

permittivity $\varepsilon_1 = n_1^2$ and the other is a surface-active medium which has $\varepsilon_2 < 0$, $|\varepsilon_2| > \varepsilon_1$.

Let us examine the field structure in SEW running along the x-axis (fig. 2). The magnetic vector $\boldsymbol{H}$ lies in the boundary plane and has one component, $\mathscr{H}_y$, perpendicular to the wave vector. The electrical vector has two components. One of them, $\mathscr{E}_z$, changes in phase with $\mathscr{H}_y$ and together with it is responsible for the transfer of wave energy along the surface; the other, $\mathscr{E}_x$, shifted for media without a loss in phase on a quarter period, is responsible for the energy transfer from one medium to the other. The SEW damping in the media without losses is nondissipative and shows the fact that SEW of the TM-type is not a purely transverse wave. The scale of such damping on both sides of the boundary $(l_1; l_2)$ depends on the polarization characteristics of the bordering media,

$$l_1 = \frac{[-(\varepsilon_1 + \varepsilon_2)]^{1/2}}{K_0 \varepsilon_1}, \qquad l_2 = \frac{[-(\varepsilon_1 + \varepsilon_2)]^{1/2}}{K_0 \varepsilon_2}. \tag{1}$$

The SEW (or SP) frequency dispersion is determined by the well-known expression (cf., for example, *The surface polaritons* (1975), Agranovich (1975))

$$K_s = \frac{\omega}{c_0} \sqrt{\frac{\varepsilon_1 \varepsilon_2}{\varepsilon_1 + \varepsilon_2}} = K_0 \eta, \tag{2}$$

where $K_0 = \omega / c_0$ is the wave number of the incident light with frequency ω, c_0 is the light velocity in vacuum, and η is the refractive index of the boundary for SEW. The field distribution in SEW is described by an expression of the form

$$A_{1,2}(x, z, t) = A_0 \mathrm{e}^{\pm z/l_{1,2}} \mathrm{e}^{\mathrm{i}(K_s x - \omega t)}, \tag{3}$$

where A_0 is the field amplitude; the plus sign $(+)$ refers to the surface-nonactive medium I (dielectric), the minus sign $(-)$ refers to the surface-active medium II (cf. fig. 2). Assuming the amplitude of the magnetic vector on the surface, $\mathscr{H}_0$, to be given then the other amplitudes are connected with it by the following relations

$$\mathscr{E}_{1z} = \mathscr{E}_{10} = -\sqrt{\frac{\varepsilon_2}{\varepsilon_1(\varepsilon_1 + \varepsilon_2)}} \mathscr{H}_0, \qquad \mathscr{E}_{2z} = \mathscr{E}_{20} = \sqrt{\frac{\varepsilon_1}{\varepsilon_2(\varepsilon_1 + \varepsilon_2)}} \mathscr{H}_0,$$

$$\mathscr{E}_{x0} = \mathrm{i} \frac{\mathscr{H}_0}{\sqrt{\varepsilon_1 + \varepsilon_2}}. \tag{4}$$

For definiteness, assume that the surface-active medium is a solid electronic plasma with a dielectric constant corresponding to Drude's model:

$$\varepsilon_2(\omega) = \varepsilon = \varepsilon' + \mathrm{i}\varepsilon'' = n_\infty^2 - \frac{\omega_p^2}{\omega^2 + \gamma^2} + \mathrm{i} \frac{\omega_p^2 \gamma}{\omega(\omega^2 + \gamma^2)}, \tag{5}$$

where $\omega_p = (4\pi N e_0^2 / m_*)^{1/2}$ is the plasma frequency determined by the concentration N, the charge e_0 and the mass m_* of free electrons; γ is the frequency of electron collisions; $n_\infty^2 = \varepsilon_\infty$ is the high-frequency limit of dielectric permittivity. In such an approximation it is easy to describe optical properties of metals in the IR spectral region ($\gamma < \omega < \omega_p \sim 10^{16}\ \mathrm{s}^{-1}$) which interest us. It is also suitable to describe melted semiconductors which become surface-active in this frequency region only when the melt temperature is exceeded. From eq. (5) it follows that $|\varepsilon'| \gg \varepsilon'' > 1$, therefore, $\eta \approx 1$, i.e., the wave number of an SP is a little greater than K_0 in the frequency region where ω approaches the surface-plasmon frequency, $\omega_{ps} = \omega_p / (n_1^2 + n_\infty^2)^{1/2}$. At $\gamma < \omega < \omega_p$, the depth of the SEW penetration into the metal is equal to $l_2 = \delta = c_0/\omega_p$ and is 10–50 nm; in a dielectric it is damped in the layer $l_1 \approx \delta\omega_p^2/n_1^2\omega^2$, which exceeds the metal skinlayer by 10^2–10^4 times depending on the frequency. As a result it appears that at frequencies $\omega \ll \omega_p$ the greatest part of the SEW energy is transferred through a transparent dielectric, for example, air.

The energy dissipation does, principally, not change the described SEW behavior features in the optical frequency region. The length of the SEW path along the L surface up to its absorption becomes an additional characteristic:

$$L = \frac{c_0}{\gamma}\frac{\omega_p^2}{\omega^2}. \tag{6}$$

(Here the effects of SEW scattering are ignored.)

The evaluations show that at $\lambda = 10.6$ μm on noble metals (Ag, Au, Cu) the SEW path can reach 2–5 cm and the lifetime of the surface polariton can reach 10^{-10} s. In the visual spectrum range L is considerably smaller ($\sim 10^{-2}$–10^{-1} cm) but it considerably exceeds the SEW wavelength. It is possible to give the following clear physical explanation for the polariton path to be much greater than the length of the photon free-path in a metal (up to absorption, which is equal to $l_f = c_0/\gamma \sim 10^{-4}$–$10^{-3}$ cm). As has been mentioned before, the SEW is propagated basically on a nonabsorptive dielectric but, being partially a longitudinal wave, it penetrates the skinlayer of an absorbing medium. Due to the penetration in this medium per c/γ-length, a part of the wave energy, equal to the relation between the lengths of SEW localization in it and in a dielectric, is absorbed. Therefore, L is of order $l_f l_1/l_2$ that, taking into account eq. (1), reduces to eq. (6). It is evident that the SEW pulse is transferred along the surface together with the energy. It should result in such phenomena as carrier entrainement by surface polaritons and the appearance of a photocurrent.

2. *The transformation of coherent radiation into SEW*

The question about the mechanism and efficiency of SEW excitation by light incident on the medium surface is of great importance. On the plane surface the direct transformation of light into an SEW is impossible, as in such a case it is impossible to satisfy the law of conservation of momentum which, at the given frequency, for SEW is greater than for the exciting light. For a light to SEW transformation it is necessary to have either a spatial inhomogeneity of the dielectric permittivity or a spatial dispersion of the conductivity in the medium. Depending on the inhomogeneity parameters and the surface electromagnetic-mode type, the transformation efficiency is different. Two types of such modes are distinguished: localized ($L \ll \lambda$) and wave ones ($L \gg \lambda$). The localized plasmon-polariton excitation on separate roughnesses of a plane surface and on metal particles of different forms plays an important role in the field amplification, which reaches 10^3–10^4 near the plasma resonance point. It explains many features of gigantic combi-Raman scattering of molecules absorbed on the metal surface and also the generation of the second harmonic at light reflection (Cosobukin 1983, Achmanov et al. 1985). For the excitation of a wave, and which interests us, it is necessary, as a rule to have a extensive relief inhomogeneity or, in a more general sense, an inhomogeneity in the dielectric constant of medium. The typical modal inhomogeneity here is a harmonic one-dimensional grating $\xi(x) = h\exp(igx)$, where h is the amplitude and g is the spatial frequency.

The analysis of the electrodynamical problem for the plane-electromagnetic-wave diffraction grating (Jha et al. 1980) shows that in the scattered radiation there is a component propagating along the surface. Other conditions being equal, its amplitude is maximal when the wave number coincides with the wave number of the surface polariton at the given light frequency. For every angle of incidence, θ, of plane-polarized radiation, with wave vector $\boldsymbol{K}_0$ and projection $\boldsymbol{K}_t$, there exists such a harmonic grating, with wave vector $\boldsymbol{g}$ (which is numerically equal to $g = 2\pi/d$ and is perpendicularly oriented to its ripples), on which there is excited a SEW propagating at an angle β to $\boldsymbol{K}_t$ (fig. 3a). The vector equation

$$\boldsymbol{K}_t + \boldsymbol{g} = \boldsymbol{K}_s, \tag{7}$$

corresponds to the law of momentum conservation in the process of transformation of radiation into a surface wave without frequency changes. In the process of this transformation the reciprocal orientation of $\boldsymbol{K}_t$, $\boldsymbol{g}$ and $\boldsymbol{K}_s$ is easily explained by the circular vector diagram (Aksenov and Gurkin 1982, Keilmann and Bai 1982). The latter easily determines the vector $\boldsymbol{g}$ of that 'resonance' grating on which the SEW with the given $\boldsymbol{K}_s$ at arbitrary angle β is excited (fig. 3b). At the known values of K_0, θ and hence K_t ($K_t = K_0 \sin\theta$) and the observance of the law of SEW dispersion [eq. (2)], as follows from this diagram, the resonance grating has a spatial frequency

$$g = (K_t^2 + K_s^2 - 2K_tK_s\cos\beta)^{1/2} = K_0(\sin^2\theta + \eta^2 - 2\eta\sin\theta\cos\beta)^{1/2}. \tag{8}$$

Such a grating is always, to a greater or smaller degree, present in the spatial spectrum of unevennesses of the real rough surface.

The amplitude-grating linear-approximation theory of the transformation of a plane wave into an SEW on a harmonic grating (discussed by Achmanov et al. 1985), which holds true for $h \ll \lambda$, leads to an SEW and the following correlation between SEW electric fields and incident radiation:

$$\mathscr{E}_{x0} = \mu h(\cos\beta - \sin\,\theta)E_t^p; \qquad \mathscr{E}_{x0} = \mu h\sin\beta E_t^s. \tag{9}$$

They connect the longitudinal component of the SEW electric vector on the surface, $\mathscr{E}_{x0}$, with the resulting field of the light wave, E_t^p or E_t^s (depending on the

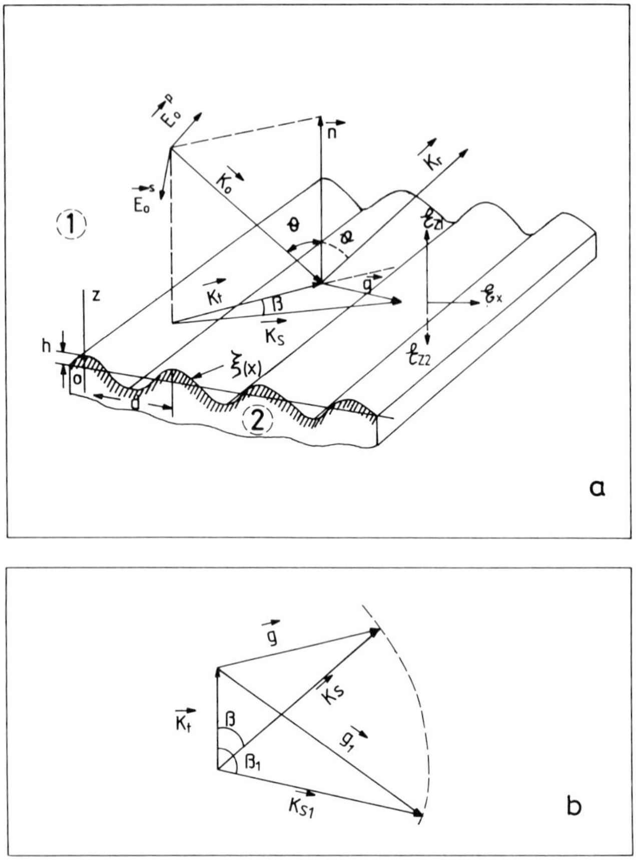

Fig. 3. (a) Reciprocal orientation of wave-vector projection of incident radiation ($\boldsymbol{K}_t$) and wave vector of SEW ($\boldsymbol{K}_S$) on the surface in the conversion process on a grating with wave vector $\boldsymbol{g}$ and period g, θ is the angle of light incidence, (b) Circular vector diagram illustrating $\boldsymbol{K}_t$, $\boldsymbol{g}$ and $\boldsymbol{K}_s$ vector module relation in the conversion process of incident radiation into SEW.

polarization). This field appears on the surface at the interference of incident and reflected waves. The choice of the mentioned field components is connected with the fact that they are responsible for thermoeffects in the medium. It follows from the formulas that the maximal field of the SEW is achieved at its excitation along the electric-vector projection of the light wave; at the same time the SEW is not excited in directions orthogonal to $\boldsymbol{E}_t^{p,s}$, and, in addition, for p-polarized radiation, not in the direction $\beta = \arccos(\sin\theta)$.

The theoretical analysis shows that in a first approximation, when $\theta = 0°$ and $n = 1$ (air), the transformation coefficient μ is equal to

$$\mu = \frac{\kappa_1 \kappa_2 (1 - \varepsilon)}{\kappa_2 + \varepsilon \kappa_1} = |\mu| e^{i\varphi}, \tag{10}$$

where

$$\kappa_1^2 = K_s^2 - K_0^2, \qquad \kappa_2^2 = K_s^2 - \varepsilon K_0^2, \tag{11}$$

and φ is the phase shift (see below).

The above expressions factually describe the line form of the surface-polariton excitation by light at normal incidence on the resonance grating ($g = K_s$). Without giving a detailed calculation, it should be noted that the position of the polariton line maximum K_{sm} coincides with the wave number of the SEW when there are no losses (see also fig. 4),

$$K_{s0} = K_0 \sqrt{\frac{\varepsilon'}{\varepsilon' + 1}}, \tag{12}$$

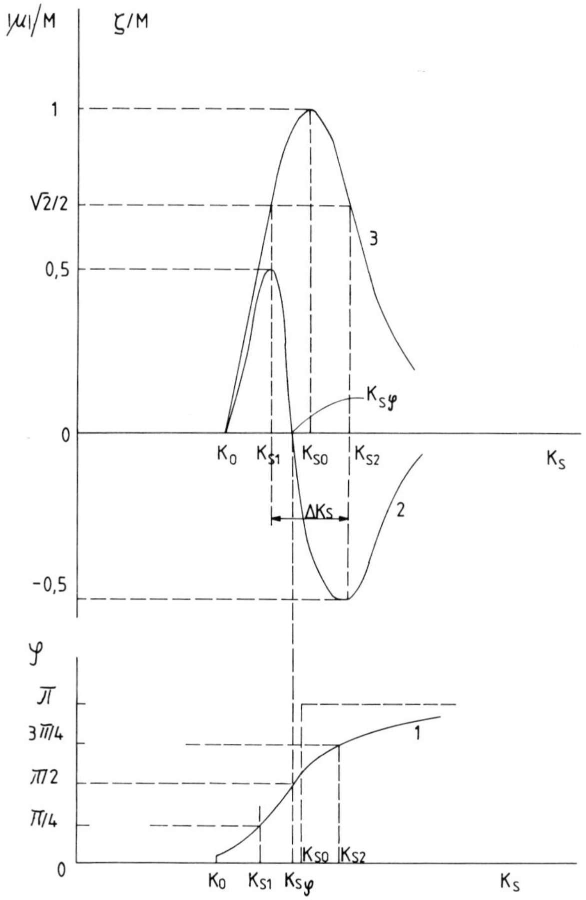

Fig. 4. Characteristics of the SEW excitation efficiency depending on K_s wave-vector modulus: phase shift φ (1); the real part $\zeta = \operatorname{Re} \mu = |\mu| \cos \varphi$ (2) and the modulus $|\mu|$ (3) of the conversion coefficient. Curves 2 and 3 are in relative units. The notation is given in the text.

and when they are present this position is different from it by an amount

$$\Delta K_{s0} \approx K_{s0}(\varepsilon'')^2/8(\varepsilon')^3 < 0, \tag{13}$$

because $\varepsilon' < 0$.

It is possible to show that at $|\varepsilon'| \gg 1$ the maximal transformation amplitude is equal to

$$|\mu|(K_{sm}) = M = \frac{K_0}{\varepsilon''}\{2[(\varepsilon')^2 + (\varepsilon'')^2][\sqrt{(\varepsilon')^2 + (\varepsilon'')^2} - \varepsilon']\}^{1/2}. \tag{14}$$

The expression given describes the frequency dependence of the efficiency of SEW excitation by light at normal incidence on the resonance grating. If we assume, in particular, that the electronic plasma is responsible for the optical properties of the medium, i.e. [see eq. (5)]

$$\varepsilon' \approx -\frac{\omega_p^2}{\omega^2 + \gamma^2}; \qquad \varepsilon'' \approx \frac{\gamma\omega_p^2}{\omega(\omega^2 + \gamma^2)},$$

from eq. (14) we get:

$$M = \frac{\omega}{\delta\gamma}\left\{2\left[1 + \left(1 + \frac{\gamma^2}{\omega^2}\right)^{1/2}\right]\right\}^{1/2}. \tag{15}$$

This dependence, expressed by the dimensionless units $M\delta$ and γ/ω, is given in fig. 5. The magnitude $M\delta$ characterizes the SEW excitation efficiency on the grating with height $\delta = c_0/\omega_p$ equal to the depth of the skinlayer of the plasma ($\sim 10^{-6}$ cm). It follows from eq. (15) that $M\delta = 1$ (or $\mathscr{E}_x = E_t$) when the radiation frequency $\omega = 0.35\gamma$, where γ is the electronic collision frequency. Such an equation usually corresponds to the middle-IR spectrum region.

The higher the light frequency the smaller is the required amplitude of the

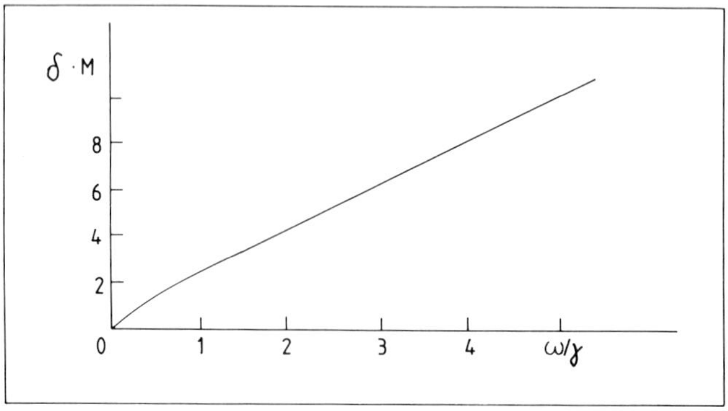

Fig. 5. Frequency dependence of SEW excitation efficiency (in dimensionless units) on the boundary of an electronic solid-state plasma and vacuum; $\delta = c_0/\omega_p$ is the skin-layer depth.

resonance grating (h_0) at which the equality is achieved of SEW fields and radiation acting on the surface ($\mathscr{E}_x = E_t$). This follows from the physical meaning of the μ parameter and eq. (9), in accordance with the latter, $\mu h_0 = 1$. As a result it is easy to obtain that at $\omega \ll \gamma$

$$M = \frac{1}{\delta}\sqrt{\frac{2\omega}{\gamma}}, \qquad h_0 = \delta\sqrt{\frac{\gamma}{2\omega}}, \tag{16}$$

and at $\omega \gg \gamma$

$$M = \frac{2\omega}{\delta\gamma}, \qquad h_0 = \frac{\gamma\delta}{2\omega} = \frac{\lambda}{4\pi}\frac{\gamma}{\omega_p}. \tag{17}$$

As $\gamma/\omega_p < 0.01$, the strength of the SEW electric field longitudinal component may become equal to the light field acting on the surface, even for $h_0 \sim 10^{-3}\lambda$.

It should also be emphasized that the spikewidth of SP excitation (the polariton linewidth) corresponded to the 0.5-level (see fig. 4), as it can be shown to equal

$$\Delta K_{s0} \approx \frac{K_0\varepsilon''}{\sqrt{2e_1^2}} = \frac{1}{\sqrt{2}L}, \qquad e_1 = -\varepsilon', \tag{18}$$

and if $K_0 \sim 6 \times 10^4\ \text{cm}^{-1}$ is ~ 10–$100\ \text{cm}^{-1}$.

In the SPS theory the central role is played by the real part of the transformation coefficient

$$\zeta = \text{Re}\ \mu = |\mu| \cos\varphi, \tag{19}$$

which, as is shown below, determines the structure increase increments at different mechanisms of surface relief changes and allows to separate the most effective SPS. The angle φ from eq. (19) is a phase shift between the field distribution $\mathscr{E}_x$ in the running surface wave and the grating. It also characterizes the spatial phase difference between the interferential distribution of the acting light intensity and the grating.

The dependence $\varphi(K_s)$ is shown in fig. 4 (curve 1). Without losses in the surface-active medium, the angle φ abruptly changes from 0 to π when going through the resonance point $K_s = K_{s0}$ of the SP excitation (it is shown in fig. 4 by the dotted line). The presence of the SEW absorption in the medium results in 'spreading' of the φ-jump near K_{s0}. The value $K_s = K_{s\varphi}$ for $\cos\varphi = 0$ and $\zeta = 0$ is equal to

$$K_{s\varphi} \approx K_{s0} - \tfrac{5}{8}K_0\frac{(\varepsilon'')^2}{e_1^3}. \tag{20}$$

In contrast to the magnitude $|\mu|$, the parameter ζ, in the limit of polariton linewidth, experiences two extrema thanks to $\cos\varphi$ sign-change at transfer through the point $K_s = K_{s\varphi}$. The function $\zeta(K_s)$ and its basic features are shown in fig. 4 (curve 2). In fig. 4 is also shown, for comparison, the dependence $|\mu| = f(K_s)$ (curve 3) and, schematically, the reciprocal localization of typical points on the K_s-axis. When $e_1 \gg 1$,

$$\zeta = \zeta_{max} \approx \frac{K_0 e_1^{3/2}}{\varepsilon''} \quad \text{at} \quad K_s = K_{s1} \approx K_{s0} - \frac{K_0 \varepsilon''}{2e_1^2}, \tag{21}$$

$$\zeta = \zeta_{min} \approx -\zeta_{max} \quad \text{at} \quad K_s = K_{s2} \approx K_{s0} + \frac{K_0 \varepsilon''}{2e_1^2}. \tag{22}$$

The magnitude $\zeta = 0$ at $K_s = K_{s\varphi}$. One can see that in the given approximation

$$\zeta_{max} = |\zeta_{min}| = \tfrac{1}{2}M, \qquad |\mu|(K_{s1}, K_{s2}) = M/\sqrt{2};$$
$$\varphi(K_{s1}) = \tfrac{1}{4}\pi, \qquad \varphi(K_{s2}) = \tfrac{3}{4}\pi, \qquad \Delta K_s = K_{s2} - K_{s1} = 1/2L. \tag{23}$$

The frequency dependence of the ζ_{max} and ζ_{min} magnitudes as follow from eq. (23) is determined by the frequency dependence of the M parameter [eq. (15)].

Thus near the SP excitation-line center there exist two extrema, of equal magnitude, of the function Re μ. They differ from each other only in phase φ by $\pi/2$. From the presented relations it follows directly that the field gain on the surface connected with partial light transformation into a coherent plane SEW on the grating of unlimited size may be quite great. So, at $\omega \gg \gamma$,

$$M = 2\zeta_{max} \approx 2K_0 e_1^{3/2}/\varepsilon'' \approx 2\omega/\gamma\delta, \tag{24}$$

from which follows that the $\mathscr{E}_{x0}$ field achieves the outer one on the surface $E_t^{p,s}$ when the grating height is $h \approx \delta\gamma/\omega \ll \delta$, constituting only 10–100 Å for radiation with $\lambda = 1.06$ μm. The possibility of considerable amplification of the light field on the surface as a result of SEW excitation is the consequence of the coherent pumping of space-wave energy into the SEW at its propagation along the resonance grating and also is the result of the high localization of the energy distribution in an SEW. The influence of the general amplification and field redistribution near the surface at SEW generation on the character of laser action on the surface-active media will be specially considered below.

3. *The interference of SEW and the incident radiation*

The generation and amplification of an SEW at its path along the resonance grating under coherent laser radiation leads to a complicated picture of the resulting interference light field on the surface. The additional complication is connected with the fact that on a random surface there should not be excited one but a whole set of SEW with different directions. The spatial spectrum of geometrical roughness or phase in heterogeneity contains resonance harmonics for different β angles between $\boldsymbol{K}_t$ projection of the $\boldsymbol{K}_0$ wave vector and the direction of SEW propagation (see fig. 3). The $\boldsymbol{g}$ orientation and spatial frequency g of the corresponding grating are determined by vector equation (7) and expression (8), respectively. The efficiency of SEW excitation depends on β and θ angles (the angle of light incidence) and on the light polarization. In the linear approximation for the grating amplitude this is described by formula (9) in which $E_t^{p,s}$ is the strength of the resulting light field on the surface, noticeably different for p- and s-radiation polarizations, at given

$$E_t^p = E_0 \frac{2\sqrt{\varepsilon}\cos\theta}{\sqrt{\varepsilon - \sin^2\theta} + \varepsilon\cos\theta} \approx \frac{2E_0}{\sqrt{\varepsilon}},$$

$$E_t^s = E_0 \frac{2\cos\theta}{\cos\theta + \sqrt{\varepsilon - \sin^2\theta}} \approx \frac{2E_0}{\sqrt{\varepsilon}}\cos\theta, \tag{25}$$

where E_0 is an incident vector. Let us analyze the interference-field structure for electrical vector components parallel to the surface plane and responsible for the energy transfer into the medium and also for its heating. This field ($\boldsymbol{E}_{\|}$) may be represented in the following way

$$\boldsymbol{E}_{\|}(\boldsymbol{r}, z) = \left[\boldsymbol{E}_t^{p,s} e^{i(\boldsymbol{K}_t \boldsymbol{r})} + \int_{-\pi}^{\pi} \mathscr{E}_x(\beta) \hat{k}_\beta e^{i(\boldsymbol{K}_{s\beta}\boldsymbol{r} + \alpha_\beta)} \, d\beta \right] \times \exp(-i\omega t) \exp(-z/\delta), \tag{26}$$

where $\hat{k}_\beta$ is the unit vector in an arbitrary direction β (see fig. 6); $\boldsymbol{r}$ is the radius vector in the surface plane; $z > 0$ is the coordinate, perpendicular to the surface, into the medium; α_β is the spatial phase of the resonance grating for a given angle β, in principle dependent on the position on the surface ($\boldsymbol{r}$). In eq. (26) is also taken into account the fact that a refracted wave and an SEW are damped

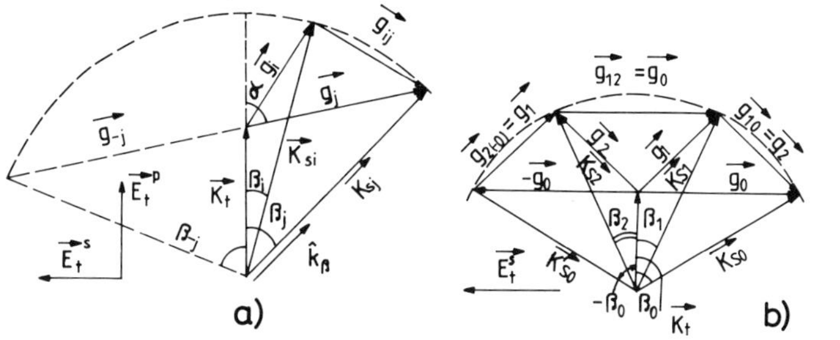

Fig. 6. Vector diagrams explaining the interference of SEW and incident radiation and also the reciprocal interference of SEW; (a) the reciprocal interference with the formation of the 'intensity' grating $\boldsymbol{g}_{ij}$ which does not coincide with the basic one, (b) the reciprocal interference with the formation of 'intensity' gratings which coincide with the basic ones. The notation is given in text.

in the surface-active medium in a layer as thick as $\delta \ll \lambda$. The intensity distribution in the interference field may be found with the help of the well-known relation $q = \frac{1}{2}\sigma E E^*$, where σ is conductivity; it can be presented as:

$$
\begin{aligned}
q(\boldsymbol{r}, z) = q_0 \cos\theta A^{\mathrm{p,s}}(\theta) \\
\times \Bigg\{ 1 + 2|\mu| \int_{-\pi}^{\pi} h_{\mathrm{g}}(\beta) \begin{Bmatrix} (\cos\beta - \sin\theta)\cos\beta \\ \sin^2\beta \end{Bmatrix} \\
\times \cos[(\boldsymbol{K}_{\mathrm{s}\beta} - \boldsymbol{K}_{\mathrm{t}})\boldsymbol{r} + \varphi + \alpha_\beta]\,\mathrm{d}\beta \\
+ |\mu|^2 \int_{-\pi}^{\pi} h_{\mathrm{g}}(\beta) h_{\mathrm{g}}(\beta') \begin{Bmatrix} (\cos\beta - \sin\theta)(\cos\beta' - \sin\theta) \\ \sin^2\beta - \sin\beta' \end{Bmatrix} \\
\times \cos(\beta - \beta') \cos[(\boldsymbol{K}_{\mathrm{s}\beta} - \boldsymbol{K}_{\mathrm{s}\beta'})\boldsymbol{r} + \alpha_\beta - \alpha_{\beta'}]\,\mathrm{d}\beta\,\mathrm{d}\beta' \Bigg\} \\
\times [\exp(-2z/\delta)], \quad \begin{cases} \text{for p-polarized radiation,} \\ \text{for s-polarized radiation,} \end{cases}
\end{aligned}
\tag{27}
$$

where q_0 is the intensity of incident radiation; $A^{\mathrm{p,s}}(\theta)$ is the absorbing metal ability which depends on the angle of incidence, θ (if it is not very large) as follows:

$$
A^{\mathrm{p}}(\theta) = \frac{A_0}{\cos\theta}, \qquad A^{\mathrm{s}}(\theta) = A_0 \cos\theta, \qquad A_0 = A|_{\theta=0}. \tag{28}
$$

In expression (27) the upper co-factors in the inner brace refer to p-polarization, the lower ones refer to s-polarization. It is possible to distinguish three components in the structure of this expression:

(1) the average absorbing flux formed by the incident radiation and excited SEW (to which $\beta = \beta'$ corresponds in the double integral);

(2) the flux connected with the interference of SEW and incident radiation, which is proportional to the resonance grating height h_{g};

(3) the flux due to reciprocal SEW interference of different directions and proportional to the product of two grating heights.

It is important to note that the SEW and incident radiation interference form a periodical distribution of intensity in the medium having the same period and orientation as the grating $\boldsymbol{g}_i$ (fig. 6a) on which an SEW of given number i is excited. However, this distribution is shifted in phase relatively to this grating by an angle φ [see eq. (27)]. This ensures the condition for the appearance of feedbacks, which influence the formation of periodical surface structures out of initial 'seed' grating. Below, such gratings will be called 'basic'. As follows from eq. (27) and the vector diagram (fig. 6a) for an unspecified direction along the surface (which is characterized, e.g., by the angle γ between the $\boldsymbol{K}_{\mathrm{t}}$ and $\boldsymbol{g}$ vectors) it is possible to indicate two basic gratings with vectors $\boldsymbol{g}_j$ and $\boldsymbol{g}_{-j}$. Their modules are determined by relations of the type of eq. (8),

$$g_{j,-j}^2 = K_t^2 - 2K_t K_s \cos\beta_{j,-j},$$

and are connected to one another by additional conditions,

$$g_j(\gamma) g_{-j}(\gamma) = K_s^2 - K_t^2, \qquad \beta_j + \beta_{-j} = \pi + 2\gamma,$$
$$\cos\beta_j = \sin\theta \sin^2\gamma + \cos\gamma\sqrt{1 - \sin^2\gamma \sin^2\theta}.$$

The mutual interference of SEW, as follows from eq. (27), makes a contribution to the general distribution of light intensity compared with the discussed interference type for the case that the height of the grating $h_g \gtrsim 2\,\mathrm{Re}\,\mu/|\mu|^2$. If we assume that SEW are excited at the values K_s, corresponding to the parameter $\zeta = \mathrm{Re}\,\mu$ extrema, then the following relation should be satisfied

$$h_g > h_{g*} = 4\zeta_{\max}/M^2 = 2/M = \delta\gamma/\omega.$$

When the wavelength is 1.06 μm then the magnitude h_{g*} is 10–50 Å.

In eq. (27) a number of characteristic properties of the obtained angular distribution of the intensity at different spatial frequencies are of interest. One of them is the presence of distinguished directions determined by the orientation of the vector $\boldsymbol{E}_t$ for which the efficiency of SEW introduction is maximal ($\boldsymbol{K}_s \parallel \boldsymbol{E}_t$; $\boldsymbol{K}_s \parallel -\boldsymbol{E}_t$) or minimal ($\perp \boldsymbol{E}_t$ and the direction $\cos\beta = \sin\theta$ for p-polarized radiation). A second one is the absence of an intensity modulation at superposition of two SEW with orthogonal wave vectors ($\beta - \beta' = \pm\frac{1}{2}\pi$). The third one is the sign change of the interferential components of the intensity distribution at some values of the angle β, e.g., at $\beta = \arccos(\sin\theta)$ and $\beta = \frac{1}{2}\pi$ for p-polarized radiation.

The reciprocal SEW interference causes essential changes in the spatial spectrum of the appearing interferential picture. As is seen from eq. (27) it leads to a light intensity modulation at spatial frequencies corresponding to the vector equation $\boldsymbol{g}_{ij} = \boldsymbol{K}_{si} - \boldsymbol{K}_{sj}$ (see fig. 6a) and changing from 0, at $\boldsymbol{K}_{si} \parallel \boldsymbol{K}_{sj}$, up to $2K_s$, at $\boldsymbol{K}_{si} \parallel -\boldsymbol{K}_{sj}$ (the period varies from ∞ up to $\frac{1}{2}\lambda$). When the corresponding grating $\boldsymbol{g}_{ij}$ does not coincide with any ot the basic ones, $\boldsymbol{g}_i = \boldsymbol{K}_{si} - \boldsymbol{K}_t$, then SEW are not excited on it and this grating should be called an additional one. If $\boldsymbol{g}_{ij}$ coincides with one of the vectors $\boldsymbol{g}_i$ then both types of interference play a role in the evolution of the basic gratings. A visual evidence of such a situation is the closed combination of coinciding basic and additional gratings shown in fig. 6b. The vector diagram shows that the coinciding basic gratings $\boldsymbol{g}_0$ and $-\boldsymbol{g}_0$, which are formed as a result of the interference between incident radiation and SEW propagating at an angle $\pm\beta_0 = \arcsin(\cos\theta)$, coincide also with an additional grating $\boldsymbol{g}_{1,2}$. It appears in the distribution of the light intensity as a result of the interference of the SEW with $\boldsymbol{K}_{s1}$ and $\boldsymbol{K}_{s2}$ propagating along the direction $\beta_1 = -\beta_2 = \arcsin(\cos\frac{1}{2}\theta)$. In turn, the basic gratings $\boldsymbol{g}_1$ and $\boldsymbol{g}_2$ on which corresponding SEW are excited,

$$g_1^2 = g_2^2 = K_0^2(1 + \sin^2\theta - \sin\theta\sqrt{3 + \sin^2\theta}),$$

coincide with additional ones $\boldsymbol{g}_{2(-0)}$ and $\boldsymbol{g}_{10}$ formed at reciprocal interference of SEW with the wave vectors $\boldsymbol{K}_{s2}$, $\boldsymbol{K}_{s0}$ and $\boldsymbol{K}_{s1}$, $\boldsymbol{K}_{s0}$. It should be noted that in the general case two pairs of interfering SEW contribute to the change of height of every basic grating formed with the help of several SEW; therefore the dynamics of all basic gratings is interdependent. This considerably complicates not only the analysis of SEW formation-process but also the qualitative analysis of the character of the resulting feedbacks between the relief height on different g and the efficiency of SEW excitation. However, the necessity of the analysis of such a general situation, as a rule, does not exist. Investigation of the topology of experimentally observed SPS shows that they form a macrostructure of the block-type on the surface (fig. 7). In the limits of each block with dimensions $\sim$10–100 periods of SPS, there is formed either only one or several basic gratings out of their full set. It is not difficult to understand, taking into account that as a result of the nonuniformity of the relief of the real surface in every

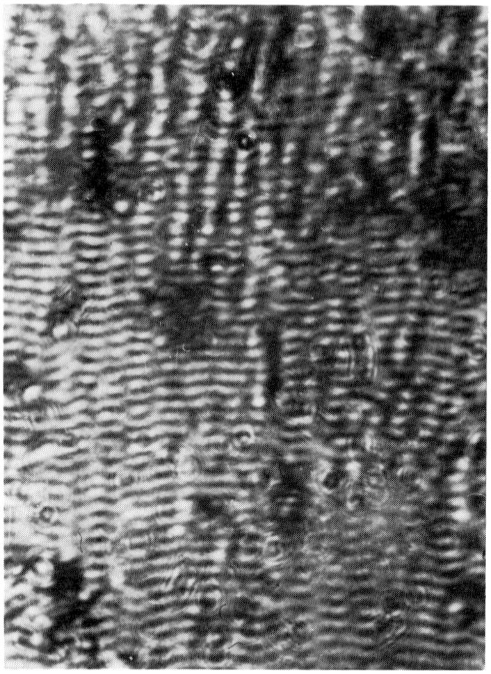

Fig. 7. The periodical structures on the germanium surface forming the typical block mosaic. The radiation conditions: a series of pulses of plane-polarized laser radiation with $\lambda = 1.06$ μm, of 20 ns pulse duration, angle of incidence is close to 0, $q_0 = 1.2q_m$. The minute structure step ≈ 1 μm.

concrete place, there will be developing that grating which has the maximal initial amplitude and increase increment. It is much easier to trace the influence of reciprocal SEW interference on SPS dynamics under such conditions. This influence is revealed in the interaction of SEW and the structures only near the block boundary in the range of the SEW path length. The characteristics mentioned are taken into account when developing the SPS theory in the plane-surface-wave approximation in which the interference of the SEW set with incident radiation is considered (the plane surface waves are considered as the chief components of the interference field).

Remember that the plane-wave approximation probably is not universal. The analysis of cylindrical SEW which, theoretically, can be excited not on a grating but on an isolated sharp relief nonuniformity (protrusion, cavity) should play a no less important role. The investigation of such modes, having a field structure of the type (Libenson and Rumjanzev 1986)

$$E, H(r, \varphi, z, t) \sim H_n^{(1)}(ar) \cos(n\varphi) \exp(\pm \kappa z - \mathrm{i}\omega t), \quad a \approx K_0,$$

where $H_n^{(1)}(u) = \mathscr{I}_n(u) + \mathrm{i}N_n(u)$, the cylindrical function satisfying the radiation principle (Vaganov and Kazenelenbaum 1982), is of great interest. Consistent analysis of these excitations and their interaction on the statistical surfaces evidently will allow to create an adequate theory of the origin of the periodical structure from the initially accidental relief and to evaluate correctly the effects of electrodynamic field amplification and their contribution to the optical surface breakdown. At the same time, the fundamental regularities of SPS evaluation and also their parameter connections with the characteristics of the surface-active medium and radiation may be obtained on the basis of the theory given below. This theory is based on the assumption that there exists a quasi-continuous set of initial 'resonance' gratings on the surface.

4. The SPS formation mechanisms on the surface-active medium boundaries

Let us examine the features of the heating of a homogeneous semi-infinite medium by one of the interference components obtained above:

$$q(x, z) = q[1 + \rho \cos(gx + \varphi)] \exp(-\alpha z),$$

$$q = q_0 \cos\theta\, A^{\mathrm{p,s}}(\theta), \qquad g = g(\beta, \theta), \qquad \alpha = 2/\delta, \tag{29}$$

$$\rho = \rho_{\mathrm{p,s}} = 2|\mu|h \begin{cases} (\cos\beta - \sin\theta)\cos\beta, & \text{for p-polarized light,} \\ \sin^2\beta, & \text{for s-polarized light,} \end{cases} \tag{30}$$

Here, for definiteness, the distribution of the flux density of the surface-active medium connected with SEW and incident radiation interference is described.

The x-axis is directed along the wave vector of the resonance grating. The magnitude of ρ, which depends on the light polarization (p, s) and characterizing the angle dependence of the coefficient of light transformation into SEW [see eq. (30)], is considered to be small enough to disregard the contribution of the average flux density to SEW in comparison with $q_0 A \cos\theta$ in eq. (29). The power of heat emission in the medium, f, as usual, is connected with $q(x, z)$ by the relation $f = -\mathrm{d}q/\mathrm{d}z = \alpha q(x, z)$. The solution of the corresponding linear problem of the heat conduction is not given here. Let us inspect the results obtained. In accordance with eq. (29), in the heat influence of radiation it is possible to distinguish two components: the middle-heat flux q and the spatially modulated dope $q\rho\cos(gx + \varphi)$. The first component creates the temperature field $T_1(z, t)$ which does not depend on x. The second component transfers the spatial intensity modulation to the temperature distribution so that

$$T_2(x, z, t) = U_2(z, t)\cos(gx + \varphi). \tag{31}$$

The total temperature field of the medium is equal to

$$T(x, z, t) = T_1(z, t) + U_2(z, t)\cos(gx + \varphi). \tag{32}$$

It is of great importance for the further discussion that in distributions (31) and (32) the spatial phase shift φ is retained without any change. The functional dependence $T_1(z, t)$, allowing for eq. (29), is determined by a well-known expression (see, e.g., Veiko et al. 1967),

$$T_1(z, t) = \frac{q}{2K\alpha}\left\{e^{\alpha^2 at}\left[e^{-\alpha z}\Phi^*\left(\alpha\sqrt{at} - \frac{z}{2\sqrt{at}}\right) + e^{\alpha z}\Phi^*\left(\alpha\sqrt{at} + \frac{z}{2\sqrt{at}}\right)\right]\right.$$
$$\left. + 4\alpha\sqrt{at}\,\mathrm{i}\Phi^*\left(\frac{z}{2\sqrt{at}}\right) - 2e^{-\alpha z}\right\}, \tag{33}$$

where a, K are the temperature conduction and the heat conductance of the materials; $\Phi^*(u)$, $\mathrm{i}\Phi^*(u)$ are an additional error function and its integral, respectively.

The function $U_2(z, t)$ is the temperature field of systems of thermal dipoles separated by a distance $d = 2\pi/g$. The general characteristics of such fields are the thermal saturation appearing at $t \gg d^2/a$ when $U_2(z, t) \to U_2(z)$ and also the rapid damping of temperature field (at $\sim d$ distance from dipole planes).

In this case it can be shown that

$$U_2(z, t) = \frac{q\rho\alpha}{2K(\alpha^2 - g^2)}$$
$$\times\left\{e^{(\alpha^2 - g^2)at}\left[e^{-\alpha z}\Phi^*\left(\alpha\sqrt{at} - \frac{z}{2\sqrt{at}}\right) + e^{\alpha z}\Phi^*\left(\alpha\sqrt{at} + \frac{z}{2\sqrt{at}}\right)\right]\right.$$
$$\left. + \frac{\alpha}{g}\left[e^{-gz}\Phi^*\left(\frac{z}{2\sqrt{at}} - g\sqrt{at}\right) - e^{gz}\Phi^*\left(\frac{z}{2\sqrt{at}} + g\sqrt{at}\right)\right] - 2e^{-\alpha z}\right\}. \tag{34}$$

As in the optical range of frequencies for surface plasmon-polaritons and incident radiation $\alpha \sim 10^5$–10^6 cm^{-1} at action times $t \gtrsim 10^{-10}$ s, the skinlayer of the medium may be considered thermally infinitely thin ($\delta \ll \sqrt{at}$). In this case, it is possible to exclude α from the expression for $T_1(z, t)$. In order to exclude α from the expression for $U_2(z, t)$ it is necessary, in addition to carry out the condition $\alpha \gg g$. Then

$$T_1(z, t) = 2\frac{q\sqrt{at}}{K}\mathrm{i}\Phi^*\left(\frac{z}{2\sqrt{at}}\right),$$

$$U_2(z, t) = \frac{q\rho}{2Kg}\left[\mathrm{e}^{-gz}\Phi^*\left(\frac{z}{2\sqrt{at}} - g\sqrt{at}\right) - \mathrm{e}^{gz}\Phi^*\left(\frac{z}{2\sqrt{at}} + g\sqrt{at}\right)\right]. \quad (35)$$

Without considering any longer the changes of the temperature $T_1(z, t)$ along x, we may state that the characteristic time for the distribution $U_2(z, t)$ is $t_s = 1/ag^2 = d^2/4\pi^2 a$ and the spatial scales are $\sqrt{at}$ and $1/g$. At $t \gg t_s$,

$$U_2(z, t) \to \tilde{U}_2(z) = \frac{q\rho}{Kg}\exp(-gz), \quad (36)$$

or in the more general case from eq. (34)

$$\tilde{U}_2(z) = \frac{q\rho\alpha}{K(\alpha^2 - g^2)}\left[\frac{\alpha}{g}\exp(-gz) - \exp(-\alpha z)\right]. \quad (37)$$

The tilde (~) here indicates that $U_2(z)$ is adjunct to the stationary distribution of the temperature. It follows that for the temperature field of thermal dipole systems the layers with dimensions $b > 1/g$ are thermophysically thick. These layers can be uniformly heated even at $t \gg b^2 a$. The relation $g/\alpha = 2\pi\delta/d \sim 2\pi\delta/\lambda$ is here an indication that the skinlayer is thin.

The temperature variation of the surface $T_2(x, 0, t) = T_{20}(x, t)$, according to eq. (34), appears to equal

$$T_{20}(x, t) = \frac{q\rho\alpha}{K(\alpha^2 - q^2)}\left\{\frac{\alpha}{g}\Phi(g\sqrt{at}) + \mathrm{e}^{(\alpha^2 - g^2)t}\Phi^*(\alpha\sqrt{at}) - 1\right\}\cos(gx + \varphi). \quad (38)$$

Hence, at $t \gg t_s$ it follows that:

$$\tilde{T}_{20}(x) = \frac{q\rho\alpha}{Kg(\alpha + g)}\cos(gx + \varphi), \qquad t_s = \frac{1}{g^2 a}. \quad (39)$$

This quasi-stationary distribution of the temperature of a dipole thermal source at $g \sim 5 \times 10^4$ cm^{-1} ($\lambda \sim 1$ μm) is settled after a characteristic time $t_s \sim 10^{-9}$–10^{-8} s. So, the action of laser radiation of duration over 10 ns on the metal and melted semiconductors can be described in the given approximation. The only limitation, as is shown below, is the limitation of the maximal light intensity. If $\alpha \gg g$, one can get from eq. (38) the most simple formula given by Bonch-Bruevich et al. (1982):

$$T_{20}(x, t) = \frac{q\rho}{Kg} \Phi(g\sqrt{at}) \cos(gx + \varphi), \qquad \tilde{U}_{\max} = \frac{q\rho}{Kg}. \tag{40}$$

It should be noted that when $\alpha \gg g$, formulae of the type of eq. (39) describe the heating by a dipole source of the thermically thin layer which is on the surface of a massive body, e.g., on the melted surface of the semiconductor. In this case the role of the parameter α is played by the magnitude $1/b$, and the final expression is

$$\tilde{T}_{20}(x) = \frac{q\rho}{Kg(1 + bg)} \cos(gx + \varphi). \tag{41}$$

The evaluation from the given formulas show that the temperature drops between the maximum and minimum of the thermal fields on the surface at $q_0 = 10^7$ W/cm^2, $g \sim 10^4$–10^5 cm^{-1} and $\rho \sim 0.1$ may be 10–100° and result in the appearance of the spatially modulated surface-relief changes, thanks to some effective thermal mechanisms which begin to act.

It is possible to give several physico-chemical processes on the surface influencing its relief at moderate, average and high temperatures: thermal deformation, intensive sound excitation, the matter density jump at melting, the melt displacement, evaporation, and different thermo-chemical reactions. The most effective mechanisms appear to be those which are connected with activation-type processes, the velocity of which depends exponentially on the temperature: $V = V_0 \exp(-T_a/T)$. Among them we should note (see Bonch-Bruevich et al. 1982, 1984a, Tribelsky and Gol'berg 1982, Konov et al. 1983) sublimation and evaporation; the hydrodynamic mechanism when the surface melts and evaporates and the relief is formed under the influence of excessive vapour and also under the influence of ponderomotor and thermocapillar forces (Bonch-Bruevich et al. 1982, Anisimov et al. 1983, Keilmann 1984, Achmanov et al. 1985, Emelyanov and Seminogov 1984a); the thermo-chemical mechanism when the relief is formed at the activation of some chemical reaction, e.g., oxidation or dissociation (Bonch-Bruevich et al. 1982, Gromov and Uphimzev 1983, Bagenov et al. 1985). For the above-mentioned mechanisms, SPS formations are typical excesses of the activation temperature T_a over the real T (of one order or more), because the relative difference of the relief-formation velocity between maximum and minimum of the light-intensity distribution is much higher than the relative temperature drop,

$$\frac{[V(T_{\max}) - V(T_{\min})]}{\bar{V}(T)} \Bigg/ \frac{(T_{\max} - T_{\min})}{\bar{T}} \approx \frac{T_a}{\bar{T}} \gg 1.$$

Under these conditions, even a weak temperature modulation may be heavily expressed in the forming relief, the equation given above playing the role of some kind of coefficient of modulation depth amplification ($K_f = T_a/\bar{T}$) which is, as a rule, 10–40. As, on the one hand, the SEW excitation efficiency and, consequently, the surface temperature is a function of relief height at the spatial resonance frequency [see, for example, eqs (30) and (39)], and, on the other hand, the relief formation velocity V is determined by the temperature, in the process of the periodical structure formation there appears a feedback.* It might appear that such a feedback is always positive: the higher the grating, the stronger the process of radiation transformation in SEW (in the frame of the discussed linear theory) and the deeper the modulation of the light intensity, temperature and, consequently, the relief formation velocity. All this immediately contributes to the grating height growth. However, it is not really quite that. The grating evolution depends not only on the amplitude but also on the phase φ of the laser radiation in SEW transformation (Bonch-Bruevich et al. 1984a). So, for the greatest majority of the above-mentioned mechanisms of structure formation at which the medium boundary moves against the exterior normal to the surface (see fig. 8a), the positive feedback appears only if the light intensity maxima fall at the concave surface parts. It corresponds to a change of the angle φ in the interval $\frac{1}{2}\pi < \varphi < \frac{3}{2}\pi$; the equation being $\varphi_{\text{opt}} = \pi$. If the heating is accompanied by the growth of the surface convexity (fig. 8b), as happens at the thermal deformation owing to linear material broadening or relief formation as a result of excitation and the amplification of surface acoustic waves synchronous with SEW (Emelyanov and Seminogov 1984b), then the positive feedback corresponds to the φ-diapason from $-\frac{1}{2}\pi$ to $\frac{1}{2}\pi$, the φ_{opt} being 0. The grating for

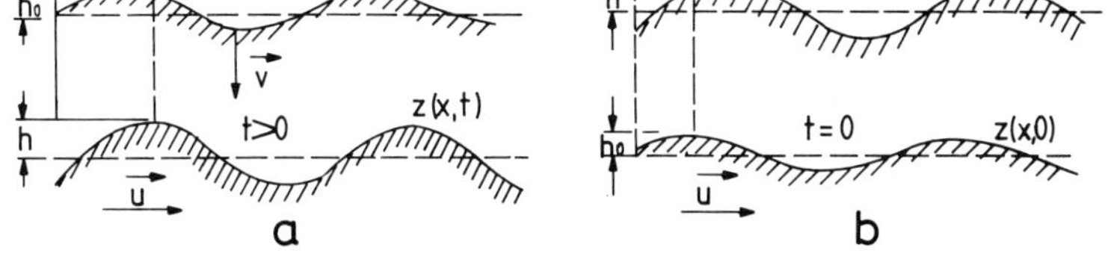

Fig. 8. The diagrams illustrating the spatial phase shift φ between the dynamical gratings $z(x, t)$ 'running' along the x-axis with velocity $\boldsymbol{u}$ and interferential distribution of radiation intensities $q(x)$ at $t = 0$ (1) and $t > 0$ (2): (a) for $\boldsymbol{V} \| -\boldsymbol{n}$ when $\varphi > \frac{1}{2}\pi$, (b) for $\boldsymbol{V} \| \boldsymbol{n}$ when $\varphi < \frac{1}{2}\pi$.

*This has been mentioned already in our first paper on the polariton mechanism, but in a more clear way it has been discussed in Tribelsky and Gol'berg 1982.

which the spatial phase-shift is situated in another interval should degradate as the intensive light affects.

It has been shown above (see section 2, fig. 4) that the angle φ changes its value from $\varphi \approx 0$ to $\varphi = \pi$ when a plasmon-polariton is excited in the limits of the polariton linewidth. As a result, there are no special limitations on the concrete mechanism of the SPS formation. At the same time, according to $\varphi(K_s)$ dependence form in fig. 4 (curve 1) for φ, it is possible to confirm that for deformational SPS, growth mechanisms are characteristic that result in structures with greater period rather than mechanisms which are accompanied by matter removal (all other conditions being equal). The reason is the fact that the corresponding K_s value in the first case is smaller than in the second. It follows from this that if in the process of laser action on a material, conditions are created favorable for the SPS formation with $K_s < K_{s0}$ $(\varphi < \frac{1}{2}\pi)$ and $K_s > K_{s0}$ $(\varphi > \frac{1}{2}\pi)$, then when the irradiation is finished both structures may be 'fixed'. In this case, the SPS with a near-period growth mechanism may be observed (for a more detailed consideration see below).

Passing on to the theoretical analysis of general regularities of SPS growth within the mentioned thermal mechanisms, let us divide them into two groups: solid-phase ones and liquid-phase ones. Notwithstanding the presence of many common features, there exists a difference between them which is important for the further discussion. For solid-phase mechanisms of relief formation the 'favorable' spatial phase shift φ between the grating and interference field should be strictly kept. At liquid-phase ones, due to the melt vibrations, such phase limitations are nullified. It should also be noted that later we are going to discuss the linear theory of the periodical structure growth. It is based on the linear connection between the SPS formation velocity and a small deviation (ΔK) of a local parameter (for example, temperature) from the average $(\bar{K})$. The periodical structures themselves are approximated by harmonical gratings of the type

$$z(x, t) = h(t) \cos[gx - \psi(t)], \tag{42}$$

where $h(t)$ is the amplitude and $\psi(t)$ is the phase, which are in the general case the time functions which should be determined.

Characteristic for the enumerated solid-phase SPS growth mechanisms (sublimation, deformation, thermo-chemical) is, in a wide temperature range, the monotony of the local velocity change $V_n(x, t)$ of the surface movement $z(x, t)$ along the exterior normal ($\boldsymbol{n}$) to the undisturbed plane boundary (fig. 8),

$$V_n(x, t) = \frac{\mathrm{d}z}{\mathrm{d}t}(x, t). \tag{43}$$

In many cases this allows to describe local relief distortions in the linear in $T_{20}(x, t)$ approximation. If the medium surface sublimates (or some other Arrhenius-type process is activated on it), then at $T_{20} \ll T_{10}$:

$$V(x,t) = V_0 \exp\left[-\frac{T_a}{T_0(x,t)}\right]$$
$$\approx V_0 \exp\left(-\frac{T_a}{T_{10}(t)}\right)\left[1 + \frac{T_a}{T_{10}^2(t)} T_{20}(x,t)\right]$$
$$= \bar{V}\left[1 + \frac{T_a}{\bar{T}_0^2(t)} T_{20}(x,t)\right]. \tag{44}$$

Here and below, $\bar{T}_0(t) = T_{10}(t)$ is the average temperature of the surface; the magnitude $T_{20}(x,t)$, as a matter of fact, is the local deviation from the average; $\bar{V} = V(\bar{T})$. The simplest dynamic regularities for the resonance gratings may be deducted from the assumption that the magnitude $T_{20}(x,t)$ is equal to its quasi-stationary value $\tilde{T}_{20}(x)$ which corresponds, e.g., to eq. (39) or eq. (41). Then, omitting the first component ($\bar{V}$) in the function decomposition of $V(x,t)$ in eq. (44) as the one which does not play any role in the local relief change, we get

$$\frac{\mathrm{d}z}{\mathrm{d}t}(x,t) = V_n(x,t) = \pm f_1(t)\,\tilde{T}_{20}(x), \tag{45}$$

where $f_1(t)$ is the function (or constant) connecting the SPS growth velocity with the temperature [e.g., from eq. (44) $f_1(t) = \bar{V}(t)\,T_a/\bar{T}_0^2(t)$]. The plus sign (+) corresponds to the case where with the temperature growth the boundary-movement velocity coincides with the exterior normal direction to the plane surface (fig. 8b); the minus sign (−) corresponds to the case where these directions are opposite (fig. 8a). Substituting into eq. (45) expression (42) instead of $z(x,t)$ and taking into account that the distribution $\tilde{T}_{20}(x)$, as has been shown above, is phase-shifted relative to the forming grating $z(x,t)$ we finally get

$$\frac{\mathrm{d}}{\mathrm{d}t}\{h(t)\cos[gx - \psi(t)]\} = \pm f(t)\,h(t)\cos[gx - \psi(t) + \varphi].$$

From this it is possible to deduce two equations for the harmonic-grating dynamics for the functions h and ψ,

$$\frac{\mathrm{d}h}{\mathrm{d}t} = \pm f(t)\,h\cos\varphi, \tag{46}$$

$$\frac{\mathrm{d}\psi}{\mathrm{d}t} = \mp f(t)\sin\varphi, \tag{47}$$

which have the following solutions (Bonch-Bruevich et al. 1984a),

$$h(t) = h_0 \exp\left(\pm\cos\varphi \int_0^t f(t)\,\mathrm{d}t\right), \tag{48}$$

$$\psi(t) = \mp\sin\varphi \int_0^t f(t)\,\mathrm{d}t + \psi_0. \tag{49}$$

Here, h_0 is the initial height, ψ_0 is the initial phase of the resonant harmonical grating with spatial frequency $g = g(\beta, \theta)$,

$$z(x, 0) = h_0 \cos(gx - \psi_0),$$

and $f(t)$ is some reverse time of the grating amplitude change which is characteristic for the given SPS formation mechanism and, generally speaking, is a function of time because of the change of average surface temperature. Introducing $\Gamma(t) = \cos\varphi \int_0^t f(t)\,\mathrm{d}t$ and assuming $\psi_0 = 0$ we finally get,

$$z(x, t) = h_0 \exp[\pm\Gamma(t)] \cos[gx \pm \Gamma(t) \tan\varphi]. \tag{50}$$

The function $\Gamma(t)$ actually plays the role of a dimensionless increment of instability (damping decrement) of SPS. According to the above sign-law, when $V_n > 0$ these structures grow for $\varphi < \frac{1}{2}\pi$; if $V_n < 0$, then SPS grow for $\frac{1}{2}\pi < \varphi < \pi$. Below, we shall analyze, for concreteness, the latter case, in which $\tan\varphi < 0$ and

$$z(x, t) = h_0 \exp[-\Gamma(t)] \cos[gx - \Gamma(t) \tan\varphi]. \tag{51}$$

The function $\Gamma(t)$ is determined by the law of the surface-movement average velocity change. If $V(t) = \text{const.}$, then $\Gamma(t) = y\bar{V}t$. If $\bar{V}$ depends on the temperature according to a law of an activational exponent, then

$$\Gamma(t) = y\int_0^t \bar{V}(t)\frac{T_\mathrm{a}}{\bar{T}_0^2(t)}\,\mathrm{d}t \approx y\bar{V}(t)\frac{T_\mathrm{a}}{\dfrac{\mathrm{d}\bar{T}_0(t)}{\mathrm{d}t}}.$$

The coefficient y, which is determined by the stationary temperature $(\tilde{T}_{20})$ dependence on the optical and thermophysical characteristics of the medium, on the coefficient of radiation conversion into SEW, and also on the polarization, is the combination of many parameters. In the special case that $\alpha\sqrt{at} \gg 1$ and the average surface temperature increases in time as

$$\bar{T}_0(t) = \frac{2q_0 \cos\theta\, A^{\mathrm{p,s}}(\theta)\sqrt{at}}{\sqrt{\pi}K},$$

and $\bar{T}_{20}$ is described by eq. (39) together with eq. (30), for activation-type mechanisms we finally get:

$$\Gamma_\mathrm{p}(t) = 2V(\bar{T}_0)\sqrt{\frac{t}{a}} \cdot \frac{\cos\beta(\cos\beta - \sin\theta)}{g(1 + g/\alpha)}|\mu| \cos\varphi, \tag{52}$$

$$\Gamma_\mathrm{s}(t) = 2V(\bar{T}_0)\sqrt{\frac{t}{a}} \cdot \frac{\sin^2\beta}{g(1 + g/\alpha)}|\mu| \cos\varphi. \tag{53}$$

The magnitude $\Gamma_{\mathrm{p,s}}$ depends in a complex way on the angles β and θ, whereas the resonance spatial frequencies g and $\bar{T}_0(t)$ also depend on them. It is also evident that the SPS growth increment is determined only by the real part of the complex coefficient of coherent radiation transformation into SEW,

$$\zeta = \operatorname{Re}\,\mu = |\mu| \cos \varphi .$$

This result has a general character and holds for any other SPS formation mechanisms. As the magnitude ζ has within the polariton linewidth, two extrema (see fig. 4, curve 2) for the analyzed case $\frac{1}{2}\pi < \varphi < \pi$, the increment is maximal at $\varphi = \frac{3}{4}\pi$, $\zeta = -K_0 e_1^{3/2}/\varepsilon''$ and $K_s \approx K_{s0} + K_0 \varepsilon''/2e_1^2 > K_{s0}$. In particular, at normal radiation incidence ($\theta = 0$) and $\beta = 0$.

$$g \approx K_0, \quad \Gamma_p = \Gamma_s = \Gamma = -2V(\bar{T}_0)\sqrt{\frac{t}{a}\frac{e_1^{3/2}}{\varepsilon''}} \approx -2V(\bar{T}_0)\sqrt{\frac{t}{a}\frac{\omega_p}{\gamma}}. \tag{54}$$

The evaluations according to the obtained formula show that for $V(\bar{T}_0) \sim 1$–10 cm/s and $\omega_p/\gamma \sim 10$–100, the magnitude $-\Gamma$ becomes equal to 1 at $t = t_* \sim 10^{-7}$–10^{-3} s, i.e., SPS may develop during times corresponding to the action as a series of, as well as isolated, light pulses.

The performed analysis concerned the mechanisms which lead to irreversible surface relief changes and its 'fixing' in form of SPS at a quick medium cooling after laser action. Taking into account the activation effects gives an exponential dependence of the value Γ on the radiation intensity and, correspondingly, a weak (logarithmic) dependence of the average temperature $\bar{T}_0$, which is necessary for SPS growth in a given time, on the basic process parameters,

$$\bar{T}_0 = \frac{T_a}{\ln\left(\dfrac{2V_0\omega_p}{\gamma}\sqrt{\dfrac{t_*}{a}}\right)}, \qquad \Gamma(t_*) = -1. \tag{55}$$

The relations obtained can be applied both to sublimational and evaporational (for viscous liquids) SPS growth mechanisms and to the thermochemical one, in which the corresponding reaction (oxidation or partial decomposition) is activated in the solid phase. The oxidation mechanism of periodical structure formation in air is, at present, well-known for steels, molybdenum, nickel and some other metals and alloys (Bagenov et al. 1985). The validity condition of the theory developed here is the smallness of the average thickness of the oxidized film (or some other reaction product) in connection with its influence on the shift of the center of the plasmon-polariton surface excitation line and by this on the spatial resonance frequency g and also on the magnitude of the angle φ. This influence may become one of the basic causes of the structure degeneration at the passing of the angle φ through the value $\frac{1}{2}\pi$.

It should also be noted that the dynamic grating becomes 'running' along the x-axis. For $V_n < 0$, its movement direction coincides with the direction of the vector $\boldsymbol{K}_S$, and the movement velocity in the general form is determined from eq. (47). During the time $t = t_*$ (when $\Gamma = -1$) at $\varphi = \frac{3}{4}\pi$, the grating shifts over a distance $1/g = d/2\pi$.

Special analysis is required for the hydrodynamic mechanism of the SPS growth which, evidently, is basic for semiconductor materials which change into surface-active states at melting. Analyzing this mechanism, we shall assume that on the surface of the medium there is a thin layer of melt which does not change by the process of the subsequent laser heating. When applied to semiconductors, the constancy of the melt thickness proves its value by the fact that, because of the increase of reflection at melting, the movement of the phase-transition boundary into depth is delayed for a certain time (see, e.g., Libenson and Shandybina 1984). Let us use for the description of the appearing relief $z(x, t)$ in the melt the method given in Anisimov et al. (1983) which is based on the equations of the non-viscous liquid thin-film-layer hydrodynamics. If the melt depth is smaller than the expected SPS period (see fig. 9), these equations are reduced to one equation for $z(x, t)$,

$$\frac{\partial^2 z}{\partial t^2}(x, t) = -\frac{b}{s}\frac{\partial^2 p(z)}{\partial x^2}, \tag{56}$$

where b is the melt thickness, s is the melt density and p is the external pressure on its boundary. As has been mentioned before, this pressure is distributed nonuniformly along the liquid surface and is modulated by the interference distribution of radiation intensity and, consequently, the temperature. Among the components determining the magnitude p, let us take into account the evaporation pressure p_v and the capillary pressure p_σ, respectively, equal to

$$p_v = p(\bar{T})\left[1 + \frac{T_a}{\bar{T}^2}\tilde{T}_{20}(x)\right], \qquad p_\sigma = \sigma g^2 z(x, t),$$

where $p(\bar{T})$ is the average vapour pressure, which can be assumed equal to the atmospheric pressure (then $\bar{T} = T_b$ is the boiling temperature); σ is the surface tension of the melt.

Introducing in the same way as above, the two unknown functions $h(t)$ and $\psi(t)$, let us describe as before the appearing dynamic grating by eq. (42), only

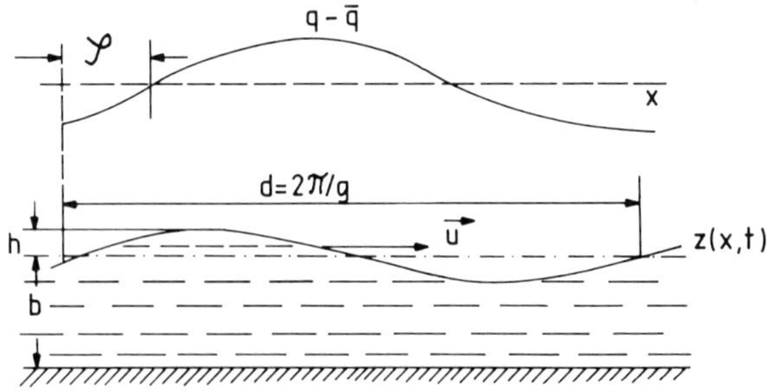

Fig. 9. The capillary wave in a thin layer of melt ($b \ll d$) under the influence of an inhomogeneous light flux q. Qualitative diagram. The notation is given in the text.

with the simplification that the velocity $d\psi/dt$ of the grating is assumed to be constant. Then,

$$z(x, t) = h(t) \cos[g(x + ut)], \tag{57}$$

where u is the parameter under consideration. Taking into account that the temperature $\tilde{T}_{20}(x)$ is proportional to the height of the grating out of phase by φ relative to $z(x, t)$ in eq. (57), and designating for short the proportionality factor κ we get

$$\frac{\partial^2 p_v}{\partial x^2} = -\kappa g^2 h(t) \cos[g(x + ut) + \varphi].$$

Then eq. (56) takes the form

$$\frac{\partial^2}{\partial t^2}\{h(t) \cos[g(x + ut)]\} = -\frac{\sigma b g^4}{s} h(t) \cos[g(x + ut)] - \kappa g^2 \frac{b}{s} \cos[g(x + ut) + \varphi] h(t),$$

from which follows that

$$\frac{d^2 h}{dt^2} - h(\omega_u^2 - \omega_0^2 - \Gamma_q^2 \cos\varphi) = 0,$$

$$\omega_u \frac{dh}{dt} = -\frac{\Gamma_q^2 \sin\varphi}{2} h. \tag{58}$$

The following designations were introduced in this equation,

$$\omega_u = ug, \qquad \omega_0 = g^2 \sqrt{\frac{\sigma b}{s}}, \qquad \Gamma_q = g \sqrt{\frac{\kappa b}{s}}. \tag{59}$$

Here ω_0 is the frequency of the free capillary oscillations in the liquid film which have spatial period $d = 2\pi/g$, Γ_q is the effective instability increment, connected with the evaporation pressure by the way of a parameter κ which depends on the flux density q; ω_u is the frequency of the capillary wave with $\lambda = 2\pi/g$ and propagating with velocity u. At $\omega_u = \omega_0$, the velocity u is equal to $u_0 = g\sqrt{\sigma b/s}$. At $\sigma \sim 10^3$ e/cm^2, $b = 10^{-5}$ cm, $s = 10$ g/cm^3, $g \approx 6 \times 10^1$ cm^{-4} ($d = \lambda \sim 1$ μm), ω_0 is $\sim 10^8$ s^{-1} and the velocity $u_0 \sim 2 \times 10^3$ cm/s. Such waves can appear spontaneously on the melt surface (at $gb \ll 1$) in the form of a system of standing waves in the absence of the external influences. Therefore, as an initial condition for the equations system obtained for the function $h(t)$ and the magnitude ω_u it is appropriate to add the condition

$$z(x, 0) = h_0 \cos gx \quad \text{or} \quad h|_{t=0} = h_0, \quad \left.\frac{dh}{dt}\right|_{t=0} = 0.$$

It should also be noted that as the initial relief $z(x, 0)$ at high spatial frequencies (when the lifetime of the capillary waves sharply fall, allowing for viscosity) the solid-phase relief may appear.

It is not difficult to show that the equation system of eq. (58), satisfying the given initial condition, has the following solution

$$z(x, t) = \tfrac{1}{2}h_0\{e^{\Gamma t}\cos[g(x-ut)] + e^{-\Gamma t}\cos[g(x+ut)]\}, \tag{60}$$

where

$$\omega_u = \frac{1}{\sqrt{2}}[(\omega_0^4 + \Gamma_q^4 + 2\omega_0^2\Gamma_q^2\cos\varphi)^{1/2} + (\omega_0^2 + \Gamma_q^2\cos\varphi)]^{1/2}, \tag{61}$$

$$\Gamma = \frac{1}{\sqrt{2}}[(\omega_0^4 + \Gamma_q^4 + 2\omega_0^2\Gamma_q^2\cos\varphi)^{1/2} - (\omega_0^2 + \Gamma_q^2\cos\varphi)]^{1/2}, \tag{62}$$

$$\Gamma_q^2\cos\varphi = 2\frac{q_0\cos\theta\, A_0\zeta gb}{Ks(1+gb)}\bar{p}\frac{T_a}{\bar{T}^2}f_{p,s},$$

$$f_p = \cos\beta(\cos\beta - \sin\theta)/\cos\theta, \qquad f_s = \cos\theta\sin^2\beta. \tag{63}$$

According to the expressions obtained, the dynamic relief at the resonance spatial frequency is the superposition of two capillary waves running into each other with velocity u, one of which ($\boldsymbol{u} \parallel \boldsymbol{K}_s$) increases with time, the other ($\boldsymbol{u} \parallel -\boldsymbol{K}_s$) being damped. Figure 10 shows that the time dependence of the relief height h is the cross-section $x = 0$:

$$z(0, t) = h_0\cosh(\Gamma t)\cos(\omega_u t). \tag{64}$$

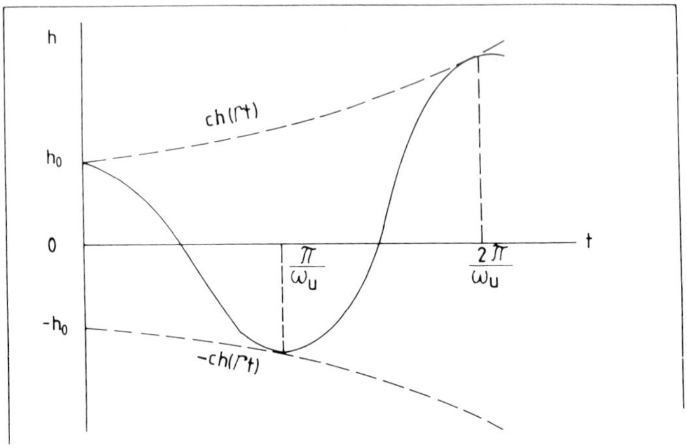

Fig. 10. The dependence of resonance relief height h on the surface of the thin melt layer on the action time t of laser radiation. The notation is given in the text.

In contrast to the solid-phase mechanism of SPS growth, here the resonance grating height oscillates in time. The oscillation frequency ω_u and the instability increment Γ depend on the radiation intensity ω_0 and on the magnitude $\zeta = |\mu| \cos\varphi$ (see expressions (61)–(63)). The functions $\omega_u(\varphi)$ and $\Gamma(\varphi)$ shown in fig. 11 are of special interest. When $\varphi = 0$, then $\Gamma = 0$ and $\omega_u = \sqrt{\omega_0^2 + \Gamma_q^2}$. This sufficiently evident result is the consequence of the phase coincidence of capillary and evaporation pressures. At $\varphi > 0$, the instability increment magnitude is different from zero in the whole range of phase shift from 0 up to π (fig. 11, curves 1–5). Such a result is principally different from the results obtained above. It can be explained by the fact that in the analyzed conditions the melted film curvature in every point periodically changes sign (with frequency ω_u) and surface, independently of the angle φ $(0 < \varphi \leqslant \pi)$, and is always, for a certain period of time, in 'favour' of the relief formation thanks to the excess vapour pressure. The angle $\frac{1}{2}\pi$ becomes in such cases the optimal shift angle. However, as fig. 11 shows and as follows from expressions (62)–(63), the Γ magnitude maximum is shifted in the direction of angles $\varphi > \frac{1}{2}\pi$,

$$\varphi_{\text{opt}} = \arccos(-\Gamma_q^2/2\omega_0^2), \qquad \frac{\Gamma_{\max}}{\omega_0} = \frac{1}{2}\left(\frac{\Gamma_q}{\omega_0}\right)^2.$$

This shift is due to the influence of the evaporative pressure on the oscillation frequency ω_u, which decreases monotonically as φ increases (fig. 11, dash-dotted curves) and becomes minimal at $\varphi = \pi$,

$$\omega_{u_{\min}} = \sqrt{\omega_0^2 - \Gamma_q^2}, \qquad u_{\min} = \frac{\omega_{u_{\min}}}{g},$$

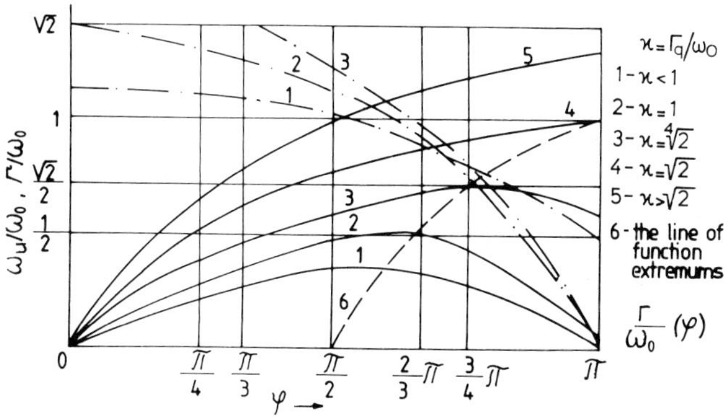

Fig. 11. The dependences of the dimensionless instability increment Γ/ω_0 (full lines 1–5) and dimensionless frequencies ω/ω_0 (dash-dotted lines) on the phase-shift angle φ for capillary waves appearing on the melt surface when SEW and incident radiation interfere. The notation is given in the text.

when capillary and evaporative pressure have opposite phases. That is why the phase shifts $\varphi = 0$ and $\varphi = \pi$ are not equivalent. At $\varphi = 0$, the melt surface may oscillate only harmonically. At $\varphi = \pi$ the oscillatory character of the movement

$$z(x, t) = h_0 \cos(\omega_{u_{\min}} t) \cos gx, \quad \Gamma = 0,$$

is kept up to the value $\Gamma_q = \omega_0$. At $\Gamma_q > \omega_0$ the relief of the melted film changes in time as

$$z(x, t) = h_0 \cosh(\Gamma t) \cos(gx), \quad \Gamma = \sqrt{\Gamma_q^2 - \omega_0^2}.$$

The solution $\Gamma_q = \omega_0$ also separates from each other the two types of functions $\Gamma(\varphi)$. At $\Gamma_q \leqslant \omega_0$, $\Gamma(\varphi)$ becomes zero at both ends of the interval of change of φ $(0 \leqslant \varphi \leqslant \pi)$, when $\Gamma_q > \omega_0$, then $\Gamma(\pi) > 0$ and at $\Gamma_q > \sqrt{2}\, \omega_0$ (the full curve 4 in fig. 11) the function $\Gamma(\varphi)$ becomes monotonous and the instability increment maximum corresponds to the phase shift $\varphi = \pi$ (see curve 5 in fig. 11).

From the above consideration it also follows that when the hydrodynamic SPS formation mechanism is realized two structures grow simultaneously: one corresponding to $K_s = K_{s1}$, $\varphi = \frac{1}{4}\pi$ and the other corresponding to $K_s = K_{s2}$, $\varphi = \frac{3}{4}\pi$ (see expressions (21) and (22)) corresponding to two extremal parameter values $\zeta = |\mu| \cos \varphi$ (fig. 4) within the polariton linewidth. In passing let us mention that the magnitude ζ corresponding to eqs. (62) and (63) determines the relief dynamics also for this mechanism. From eq. (62) it is possible to find the relation between the instability increments Γ in both structures. If we ascribe to the parameters of the K_{s1} structure the index '1' and of the K_{s2} structure the index '2' then, taking into consideration the condition $\cos \varphi_1 = -\cos \varphi_2 = 1/\sqrt{2}$, we get

$$\left(\frac{\Gamma_2}{\Gamma_1}\right)^2 = \frac{\sqrt{2(\omega_0^4 + \Gamma_q^4 - \sqrt{2}\omega_0^2\Gamma_q^2)} + \Gamma_q^2 - \sqrt{2}\omega_0^2}{\sqrt{2(\omega_0^4 + \Gamma_q^4 + \sqrt{2}\omega_0^2\Gamma_q^2)} - \Gamma_q^2 - \sqrt{2}\omega_0^2}.$$

The increase increments are close to each other only at relatively low radiation intensities when $\Gamma_q \ll \omega_0$, at $\Gamma_q \gg \omega_0$ their ratio tends to a constant value $(\Gamma_2/\Gamma_1)_{\max} = \sqrt{2} + 1 \approx 2.41$. Nevertheless, it is not so great that it should be impossible to observe the formation of both structures in the compared time intervals. In connection with this, let us evaluate the relative difference in SPS periods for the case when the incident light is close to the normal and $g_1 \approx K_{s1}$, $g_2 \approx K_{s2}$. Taking into account eqs. (21) and (22), we have

$$\frac{d_2 - d_1}{d} \approx \frac{K_{s2} - K_{s1}}{K_{s0}} \approx \frac{\varepsilon''}{e_1^2} = \frac{\lambda}{2\pi L} \sim 5(10^{-3}\text{–}10^{-2}). \qquad (65)$$

Such an explanation of the 'double' SPS formation with close periods observed experimentally appears to be the most probable.

As has already being mentioned at the realization of the hydrodynamic relief formation mechanism, its height generally changes according to the harmonic oscillation law with the amplitude growing in time. It is evident that the relief oscillation should continue even after light-pulse action up to the time when either the melt film becomes solid or the oscillations are damped as a result of the viscous losses. In the latter case, the characteristic damping time is equal to

$$t_v \approx \frac{3\nu}{\omega_0^2 b^2} + \frac{b^2}{\nu},$$

where ν is the kinematic viscosity ($\sim 10^{-3}\ \mathrm{cm^2\ s^{-1}}$) and at $b \sim 10^{-5}$ cm holds true at $t_v \sim 10^{-7}$ s. The solidification time is, on average, of the order of $t_s \sim b^2/a \ll t_v$. Therefore, at the realization of the hydrodynamic mechanism of structure formation under surface cooling, the relief has no time to smooth out and is crystallized in the form of SPS. If $t_s \gg t_i$ then after the pulse termination during the time $t_i < t < t_s$ the relief dynamics represents the harmonic oscillations at frequency ω_0, the amplitude of which is determined by the mechanical pulse obtained during the light pulse with duration t_i. It is easy to ensure that in the approximation mentioned at $t_i \ll 1/\omega_0,\ 1/\Gamma$,

$$z(x, t) = h_0 \left[\cos(\omega_0 t) \cos(gx) - \frac{\int_0^{t_i} \Gamma_q^2(t)\, \mathrm{d}t}{\omega_0} \sin(\omega_0 t) \cos(gx + \varphi) \right],$$

where the magnitude $h_0 \Gamma_q(t)/\omega_0$ has the meaning of the 'initial' velocity V_{in} of oscillations. It should be noted here that we succeeded in a general way to take into account the function $\Gamma_q^2(t)$ connected with the growth of the average vapour pressure during the melted-layer heating. If $V_{in}/\omega_0 > h_0$, the maximal height of the resonance relief h_i formed by the action of the short laser pulse is equal to

$$h_i = h_0 \frac{2\sqrt{2\pi t_i}}{\omega_0 \sqrt{a}} \frac{gb}{(1 + gb)} \frac{\omega}{\gamma\delta} \frac{p_*}{s} \times \begin{cases} \cos\beta(\cos\beta - \sin\theta), & \text{p-polarization,} \\ \sin^2\beta, & \text{s-polarization.} \end{cases}$$

At $t_i \sim 10^{-8}$ s, $p_* \sim 1$ atm (on average), $h_i \sim (1\text{–}100) h_0$, depending on the ratio ω/γ.

When describing the dynamic gratings, it was assumed that the local surface temperature deviations from the average one always correspond to their quasi-stationary values. The applicability condition of such an approximation was the smallness of the time of the quasi-stationary distribution determination $\tilde{T}_{20}(x)$ in comparison with the magnitude $1/\Gamma$,

$$\Gamma/ag^2 \ll 1.$$

Here, taking as an example a simple model situation, we shall consider what peculiarities introduces temperature lag account into the grating dynamics. It is evident that this effect should influence the SPS growth when the light intensities are relatively high. Let us assume that the periodical relief is formed in the solid phase and is described by an equation of the type of eq. (46). Let us also assume that the achievement of some average temperature T_* is the condition for its development. The local deviations from this temperature, $U(z, t)\cos(gx+\varphi)$, determine the relief change velocity V,

$$\frac{\mathrm{d}h}{\mathrm{d}t} = -V\cos\varphi = -\chi U(0,t)\cos\varphi, \tag{66}$$

where χ is a proportionality factor and in the law of the grating amplitude growth [eq. (46)] the minus sign ($-$) is chosen for definiteness. Assuming also that the average temperature $\bar{T}$ is stabilized on the level T_* (or changes slowly) let us use the heat conduction equation for finding the function $U(z, t)$,

$$\frac{\partial U}{\partial t} + ag^2U - a\frac{\partial^2 U}{\partial z^2} = 0, \qquad (67)^*$$

with the following boundary and initial conditions,

$$-K\left.\frac{\partial U}{\partial z}\right|_{z=0} = q_0\kappa h, \qquad U|_{z\to\infty}\to 0, \qquad U_{\mathrm{in}}|_{t=0} = 0. \tag{68}$$

In eq. (68), the grating-height linear approximation and its influence on the SPS generation efficiency and on the modulation depth of the interference field is taken into consideration. Thus, feedbacks in the problem are characterized by the factors χU in eq. (66) and κh in eq. (68). To complete the system of equations let us add the initial condition of the grating height $h|_{t=0} = h_0$. The problem solution can be represented in the following form

$$\begin{aligned}\frac{h(t)}{h_0} = {} & \frac{m}{2m+n}\mathrm{e}^{nt}\Phi^*(-\sqrt{mt}) + \frac{\mathrm{e}^{-mt/2}}{2\sqrt{m-4n}(2m+n)} \\ & \times \Big\{[(m+n)\sqrt{m-4n}+(m-n)\sqrt{m}]\mathrm{e}^{\sqrt{m(m-4n)}t/2} \\ & \times \Phi^*\left(\frac{\sqrt{m}+\sqrt{m-4n}}{2}\sqrt{t}\right) \\ & + [(m+n)\sqrt{m-4n}-(m-n)\sqrt{m}]\mathrm{e}^{-\sqrt{m(m-4n)}t/2} \\ & \times \Phi^*\left(\frac{\sqrt{m}-\sqrt{m-4n}}{2}\sqrt{t}\right)\Big\},\end{aligned} \tag{69}$$

*The term ag^2U corresponds to the component $-a\,\partial^2U/\partial x^2$.

where m and n are solutions of the system of equations

$$\begin{cases} m - n = v = ag^2, \\ n\sqrt{m} = f = -\dfrac{q_0 \kappa \chi \sqrt{a} \cos \varphi}{K}, \quad \cos \varphi < 0. \end{cases} \tag{70}$$

The asymptotic behaviour of $h(t)$ in the limit of large times ($mt > 1$) is of significant interest. When $m > 0$ and $n > 0$, it is determined only by the asymptotes of the first term in eq. (69),

$$h(t) \approx \frac{2h_0 m}{2m + n} \exp(nt), \quad \Gamma = n. \tag{71}$$

The connection of the m and n parameters with the v and f magnitudes depends on the dimensionless parameter

$$\xi = \frac{\varphi}{v^{3/2}} = \frac{q_0 \kappa \chi |\cos \varphi|}{aKg^2}.$$

One can make sure that only at low intensities, when $\xi \ll 1$, the heat conduction delay can be ignored and the temperature dipole field can be considered in the quasi-stationary approximation. In reality, at $m \gg n$ when $m \approx v$ and $n \approx f/\sqrt{v}$, we get from eq. (71)

$$h(t) \approx h_0 \exp\left(\frac{ft}{\sqrt{v}}\right) = h_0 \exp\left(-\frac{q_0 \kappa \chi \cos \varphi}{Kg} t\right). \tag{72}$$

With increasing ξ, there must be observed differences from the simplest temporal law [eq. (72)]. The cause is a weaker than predicted (by the quasi-stationary approximation) local temperature deviation from the average one. Therefore, for obtaining the same value of instability increment, greater light intensities are required. The results of exact calculations on the basis of the solutions of the system of equations given by eq. (70) show that at relatively high light intensities the instability occurs under the condition that the temperature field of thermal dipoles is far from saturation and the instability increment depends nonlinearly on q_0,

$$\Gamma = f^{2/3} = \left(\frac{q_0 \kappa \chi \sqrt{a} |\cos \varphi|}{K}\right)^{2/3}.$$

The contribution of heat-conducting lag to grating dynamics should be observed at q_0 fluxes answering the condition $\xi(q_0) \gtrsim 1$ which as a result gives

$$q_0 \gtrsim \frac{Ka\bar{T}^2}{4\bar{V}T_a} \left(\frac{\omega}{c_0}\right)^2.$$

At $K_0 \approx 6 \times 10^4\ \text{cm}^{-1}$, $\bar{V} \sim 10^2$ cm/s, $\bar{T} \sim 10^3$ K, $T_a \sim 4 \times 10^4$ K and $q_0 \gtrsim 10^8\ \text{W/cm}^2$.

It is evident that the analyzed effect greatly influences the SPS dynamics whatever the thermal mechanisms of their formation might be.

The theory developed allows to trace the instability increment Γ dependence (or increase time $t_* = \Gamma^{-1}$ dependence) of the structure on the integral parameters: the angle of light incidence θ, the angle β between the directions of $\boldsymbol{K}_t$ and $\boldsymbol{K}_s$ (see fig. 3), the flux density q_0 and some others, and also to indicate the structure developing with greatest velocity.

For the p-polarized radiation for which the interference component of the flux density [eq. (30)] depends on the angles as $\cos\beta(\cos\beta - \sin\theta)$, the structures with wave vectors $\boldsymbol{g}_1$ and $\boldsymbol{g}_2$ oriented along the vector $\boldsymbol{E}_t^p \parallel \boldsymbol{K}_t$ are the most quickly developing (fig. 12a),

$$\boldsymbol{g}_1 = \boldsymbol{K}_s^{(1)} - \boldsymbol{K}_t, \quad \beta = 0,$$

$$\boldsymbol{g}_2 = \boldsymbol{K}_s^{(2)} - \boldsymbol{K}_t, \quad \beta = \pi.$$

Their periods as a function of the angle of incidence are determined from the vector diagram

$$d_1 = \frac{\lambda}{\eta - \sin\theta} \approx \frac{\lambda}{1 - \sin\theta}; \qquad d_2 \approx \frac{\lambda}{1 + \sin\theta}. \tag{73}$$

These dependences are shown in fig. 12b (curves 1 and 2); they are in accordance with corresponding experimental curves given by Aksenov and Gurkin (1982), Keilmann and Bai (1982), Prochorov et al. (1982a).

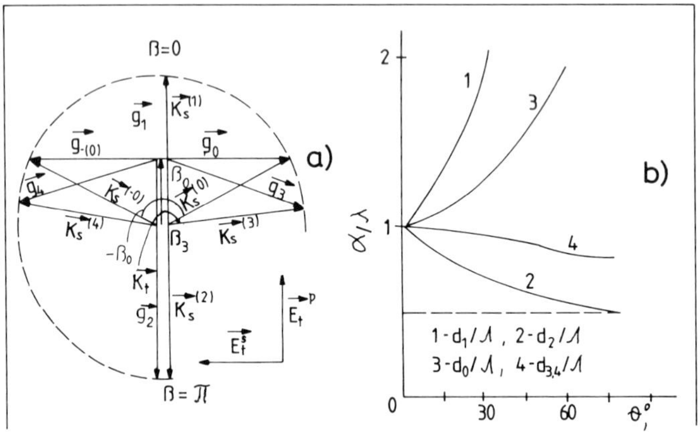

Fig. 12. The vector diagram (a) illustrating the orientation and the period of improved-quality SPS formed under the influence of p- and s-polarized laser radiation, and the dependence of SPS period d, referred to wavelength λ on the angle of incidence θ (b).

Let us consider the angular dependence of the instability increments of periodical structures formed by p-polarized radiation using eq. (52)

$$\Gamma_p = \Gamma_0 \frac{\cos\beta(\cos\beta - \sin\theta)K_0}{g(\beta,\theta)[1 + g(\beta,\theta)/\alpha]}. \tag{74}$$

Here Γ_0 is unconnected with the angular characteristic factor. Its structure and typical value have been discussed before; $g(\beta, \theta)$ is the modulus of the wave vector $\boldsymbol{g}$ of the resonance grating determined by formula (8) at $\eta = 1$,

$$g \approx K_0(1 + \sin^2\theta - 2\sin\theta\cos\beta)^{1/2}.$$

The typical form of the dependence of the relative instability increment $\gamma_p = \Gamma_p/\Gamma_0$ on the angle β is shown in fig. 13 (curve 1). The magnitude γ_p is maximal at $\beta = 0$ and $\beta = \pi$, when it is equal to

$$\gamma_p|_{\beta=0} = \frac{\alpha}{\alpha + K_0(1 - \sin\theta)}, \qquad \gamma_p|_{\beta=\pi} = \frac{\alpha}{\alpha + K_0(1 + \sin\theta)}. \tag{75}$$

It tends to zero at $\beta_0 = \arccos(\sin\theta)$ and $\beta = \frac{1}{2}\pi$. Extrema γ_p in the interval between the angles β_0 and $\frac{1}{2}\pi$ correspond to the value

$$\beta = \arccos\frac{2\sin^2\theta + 1 - \sqrt{1 + 4\sin^2\theta}}{4\sin\theta},$$

and its modulus is much smaller than 1. The change of sign of γ_p means that in the interval between the angles $\beta_0 \leqslant \beta \leqslant \frac{1}{2}\pi$ the structures may develop only by the hydrodynamic mechanism, when $\Gamma > 0$ at any value of $\cos\varphi$ within the polariton linewidth. It is clear that this discussion concerns the structure with

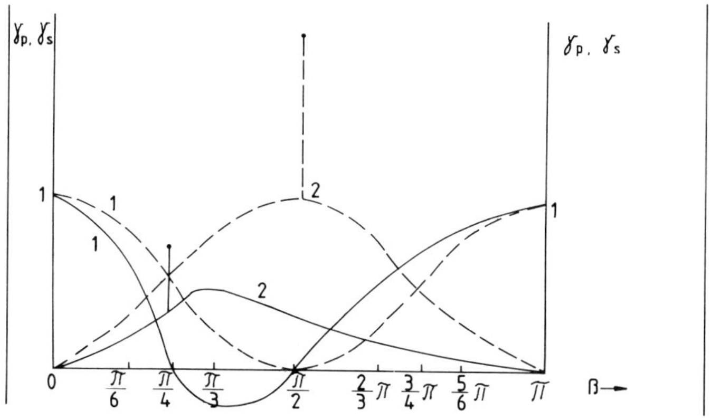

Fig. 13. The angular dependences of relative increments of SPS instability forming under the influence of p- and s-polarized radiation (curves 1 and 2) at an angle of incidence $\theta = 0$ (dotted line) and $\theta = 45°$ (full line).

close but different values of SEW wave numbers K_{s1} and K_{s2}. In connection with this let us mention that for a hydromechanic mechanism of relief formation there takes place a slightly different dependence than the angular dependence of γ_p in eq. (74). As follows from a careful consideration of expressions (62) and (63), these differences are of quantitative character as a result of the monotonous change of Γ (with the growth Γ_q^2) and are determined by the angular dependence of ω_0 and Γ_q. So,

$$\Gamma \approx \Gamma_q^2(\sin\varphi)/2\omega_0 \quad \text{at} \quad \Gamma_q^2 \cos\varphi \ll \omega_0^2,$$

$$\Gamma \approx \Gamma_q \sin\tfrac{1}{2}\varphi \quad \text{at} \quad \Gamma_q^2 \cos\varphi \gg \omega_0^2.$$

For example, at low intensities and $|\cos\varphi| = \sin\varphi = 1/\sqrt{2}$, when it is possible to assume

$$\Gamma_0 = \frac{2\sqrt{2}q_0}{K}\bar{p}\frac{T_a}{\bar{T}^2}\sqrt{\frac{b}{\sigma s}},^*$$

the angular dependence of the instability increment is given by

$$\gamma_p = \frac{\Gamma_p}{\Gamma_0} = \frac{K_0\cos\beta(\cos\beta - \sin\theta)}{g(\beta,\theta)[1 + bg(\beta,\theta)]}, \tag{76}$$

and has the same characteristics as the curves 1 in fig. 13. At the same time, at $\beta = 0$ and $\beta = \pi$, the values of γ_p according to eq. (76) are somewhat different from eq. (75). If in one case the magnitude

$$\frac{\Delta\gamma_p}{\bar{\gamma}_p} = 2\frac{\gamma_p|_{\beta=0} - \gamma_p|_{\beta=\pi}}{\gamma_p|_{\beta=0} + \gamma_p|_{\beta=\pi}},$$

is equal to $2K_0\sin\theta/(\alpha + K_0)$, in the other case it is equal to $2K_0 b\sin\theta/(1 + K_0 b)$. From this it is evident that, firstly, the instability increment of the $\boldsymbol{g}_1$ structure is bigger and, secondly, the difference in increments depends on incident angle of the light and on the thickness of the surface-active medium layer in which energy emission takes place. The former of the characteristics mentioned is well confirmed by experiment. As for the other one, it cannot be used for the refinement of the SPS formation mechanism according to the experimental results of $\Delta\gamma_p/\bar{\gamma}_p$ or for the evaluation of the absorbing layer thickness.

There is one more isolated structure ($\boldsymbol{g}_0$) which differs in the fact that two SEW (see the vector diagram in fig. 12a), with $\boldsymbol{K}_s^{(0)}$ and $\boldsymbol{K}_s^{(-0)}$ vectors, contribute simultaneously to its formation. The gratings ($\boldsymbol{g}_0$ and $\boldsymbol{g}_{(-0)}$) formed by SEW interference with incident radiation have a similar orientation and period,

$$\boldsymbol{g}_0 = -\boldsymbol{g}_{(-0)}, \qquad d_0 = \frac{\lambda}{\sqrt{\eta^2 - \sin^2\theta}} \approx \frac{\lambda}{\cos\theta}, \qquad \cos\theta = \sin\beta_0. \tag{77}$$

*In particular at $\bar{p} = 1$ atm, $\bar{T}_a = 4 \times 10^4$ K, $\bar{T} = 2 \times 10^3$ K, $\sigma = 10^3$ e/cm^2, $b = 10^{-5}$ cm, $q_0 = 10^7$ W/cm^2 and the magnitude Γ_0 is equal to 10^7 s^{-1}.

This important characteristic has been discussed by Bagenov et al. (1983). For p-polarized radiation the lines of the structure with wave vector $\boldsymbol{g}_0$ are parallel with the $\boldsymbol{E}_t^p$ vector, so far as $\boldsymbol{g}_0 \perp \boldsymbol{E}_t^p$. However, the SEW excitation efficiency in this direction goes, formally, to zero (in the linear approximation; see e.g., expression (9), such SPS do not develop at all or grow very slowly). At the same time, the structure with $\boldsymbol{g} = \boldsymbol{g}_0$ is most qualitative under the excitation of s-polarized laser radiation. The examination of the angular dependence of instability increments of SPS for radiation polarized perpendicularly to the incident plane shows that during solid-state mechanisms of relief formation

$$\gamma_s = \frac{\Gamma_s}{\Gamma_0} = \frac{\sin^2\beta \cos^2\theta\, K_0}{g(\beta,\theta)[1 + g(\beta,\theta)/\alpha]}. \qquad (78)^*$$

For the analyzed structure $\boldsymbol{g}_0$ ($g_0 = K_0 \cos\theta$, $\sin\beta = \sin\beta_0 = \cos\theta$), which is doubly degenerated (because it is created by physically coinciding gratings $\boldsymbol{g}_0$ and $\boldsymbol{g}_{(-0)}$), the instability increments of the $\boldsymbol{g}_0$ and $\boldsymbol{g}_{(-0)}$ gratings are added, and, therefore,

$$\gamma_s = \gamma_{s0} = \frac{2(\cos^3\theta)\alpha}{\alpha + K_0\cos\theta}. \qquad (79)$$

The appearance of the coefficient 2 in eq. (79) may be explained by a more rigorous derivation based on the contribution of the two SEW ($\boldsymbol{K}_s^{(0)}$ and $\boldsymbol{K}_s^{(-0)}$) to the interference field structure. Instead of eq. (29) here it may be assumed that

$$\begin{aligned} q(x,z) &= q\{1 + \rho[\cos(g_0x + \varphi) + \cos(-g_0x + \varphi)]\}\exp(-\alpha z) \\ &= q[1 + 2\rho\cos\varphi\cos(g_0x)]\exp(-\alpha z). \end{aligned} \qquad (80)$$

In this case, the factor $\cos\varphi$ transfers directly into the modulated term amplitude and the phase shift disappear in the intensity distribution. The equations of the harmonic grating dynamics [eqs. (46) and (47)] take the form

$$\frac{dh}{dt} = \pm 2f(t)h\cos\varphi, \qquad \frac{d\psi}{dt} = 0,$$

from which formula (79) follows.

An interesting detail in the dynamics of the generated SPS is revealed by the analysis of the hydrodynamic mechanism of its formation when it is described by expressions (60)–(63). Subject to eq. (80) we have,

$$\begin{aligned} &z(x,t) = h_0\cos(gx)\cos(\omega_u t), \quad \omega_u = \sqrt{\omega_0^2 - \Gamma_q^2}, \quad \Gamma = 0 \quad \text{at} \quad \Gamma_q^2 < \omega_0^2, \\ &z(x,t) = h_0\cosh(\Gamma t)\cos(gx), \quad \omega_0 = 0 \qquad \text{at} \quad \Gamma_q^2 > \omega_0^2. \end{aligned}$$

*In principle, Γ_0 differs for s- and p-polarized radiation because of the difference in the values of absorption capacity A_s and A_p at incident angles of the light $\theta > 0$, and therefore differs for $\bar{V}$.

Here

$$\Gamma = \sqrt{\Gamma_{q0}^2 - \omega_0^2},$$

$$\Gamma_{q0}^2 = \frac{4q_0 A_0 \xi}{Ks} \bar{p} \frac{T_a}{\bar{T}^2} \frac{bK_0 \cos^5\theta}{(1 + bK_0 \cos\theta)},$$

$$\omega_0^2 = g_0^4 \frac{\sigma b}{s} = K_0^4 \frac{\sigma b}{s} \cos^4\theta. \tag{81}$$

From eq. (81) it follows that there exists an energetic threshold for the degenerated SPS instability (in contrast to nondegenerated structures, when formally, $\Gamma > 0$ for all $q_0 > 0$ and factually for $q_0 > q_m$, where q_m is the threshold of medium melting). Such a feature is connected with a worse spatial phasing of the light intensity distribution [eq. (80)], for which the phase shift is equal to π as compared with $\varphi = \frac{3}{4}\pi$. The threshold flux corresponding to this condition follows from eq. (81),

$$q_0 > q_{0*} = \frac{\sigma K_0^2 K \bar{T}^2 (1 + bK_0 \cos\theta)}{8\bar{p} T_a \cos\theta}.$$

For the above mentioned typical parameter values and with $\cos\theta = 1$, $bK_0 \ll 1$, $- q_{0*} \approx 5 \times 10^7$ W/cm^2. It can be shown that the instability increment of the degenerated structure will exceed the value for the analogous SPS if it were formed only by one SEW with optimal shift angle $\varphi = \frac{3}{4}\pi$ already at $q_0 > \frac{8}{7} q_{0*}$.

Returning, again, to the general angular dependence $\gamma_s(\beta)$ for the solid-phase mechanisms of SPS formation let us emphasize that along with the isolated peak at $\cos\beta_0 = \sin\theta$ this dependence contains a smooth maximum at*

$$\cos\beta = \cos\beta_3 = \frac{1 + \sin\theta}{3 \sin\theta} \left[1 - \sqrt{1 - \frac{3\sin^2\theta}{(1 + \sin^2\theta)^2}} \right],$$

which, at $\theta < 60°$, can be sufficiently rigorously described by the simplified formulas

$$\cos\beta_3 = \frac{\sin\theta}{2(1 + \sin^2\theta)}, \qquad \gamma_{s_{max}} = \tfrac{1}{2}\cos^2\theta \sqrt{\frac{4 + 7\sin^2\theta + 4\sin^4\theta}{(1 + \sin^2\theta)^3}}.$$

Typical dependences $\gamma_3(\beta)$ for two light incidence angles containing the above-mentioned characteristics are shown in fig. 13 (curve 2). Figure 12a shows the vector diagrams illustrating the orientation of the most qualitative structures under the action of s-polarized radiation ($\boldsymbol{g}_0$, $\boldsymbol{g}_3$ and $\boldsymbol{g}_4$) and fig. 12b shows the corresponding dependences of the period of these structures on the angle θ. The

*This extremum has been found experimentally by Aksenov and Gurkin (1982) and has been explained by Keilmann (1984), Emelyanov and Seminogov (1984b).

dependence $d_0/\lambda = f(\theta)$ described by the simple law given by eq. (77) is in good accordance with experimental data.

The described physical notions and the linear theory of periodical structures on the surface-active medium boundary allows to explain the conditions of their appearance and to describe the growth regularities at different mechanisms of relief formation. Although the theory is based on the polariton model of the appearance of the spatially inhomogeneous light field on the medium surface, many theoretical conclusions and results can be used for the explanation of SPS formation under different conditions, for example, for the incident light interference and the waveguide mode generated by it (such a mechanism has been suggested by Kochengina et al. (1981) for the explanation of SPS with period $\sim \lambda/n$ where n is the diffractive index of a weakly absorbing medium). Emelyanov et al. (1987a) give an example of a consistent application of the general approach to the SPS analysis. In this paper, on the basis of the diffraction model of the spatially modulated intensity distribution in the medium, an explanation is given for the SPS formation under CO_2-laser action on quartz. At the same time, the level of theoretical development is not yet high enough for the explanation of the quantitative regularities of SPS formation on the majority of concrete materials. One of the overlooked factors, which should be taken into account is the nonlinearity of the radiation transformation into SEW on real gratings. The first step in this direction has already been taken (Gandelman and Kondratenko 1985). The other factor is the influence of the change of state and the properties of the surface layer of materials under laser action on dispersion, damping and excitation efficiency of the surface polariton. In particular, it is of great importance for semiconductors which become surface-active during melting. As the melt thickness under pulse-laser radiation can be small enough ($\lesssim 1$ μm) and can change with time, the necessity of SEW characteristic consideration in a complicated system becomes evident enough. The analysis, which is not given here, shows that in the optical region of the frequencies the dispersion and the SEW damping in the metal-like layer does not differ from that for a semi-infinite metal, for thicknesses with

$$b > b_* = \frac{n_3^2}{\sqrt{n_3^2 - n_1^2}} \delta \frac{\omega}{\omega_p},$$

where n_1 and n_3 are the refraction index of the air and the semiconductor, respectively, δ is the skinlayer depth in the metalloid semiconductor melt. Usually, $b_* \lesssim 100$ A. This explains well the practical stability of the SPS period on a semiconductors, notwithstanding the possible spread of the melt thickness on its surface under laser action.

5. The contribution of powerful surface electromagnetic waves generation in the process of intensive light action on a condensed medium

The partial transformation of incident coherent radiation into SEW and other surface electromagnetic excitations lead to a general redistribution of the resulting light field near the condensed medium surfaces. Under intensive radiation it is accompanied by a set of effects connected with a great power of the generated surface waves. For performing the corresponding evaluations it is convenient to use the above given general expression for the fields of SEW excited by coherent light diffraction on the periodical grating. As follows from eq. (23), for normal light incidence the expressions given below relate the field strengths ($\mathscr{E}_x$, $\mathscr{E}_{1z}$) and the densities of the flux transported by SEW along ($q_{s\parallel}$) and normal ($q_{s\perp}$) to the surface, on the one hand, and the incident field strength (E_0) and the incident radiation density (q_0), on the other hand,

$$\frac{\mathscr{E}_x}{E_0} \approx 4\frac{\omega^2}{c_0\gamma}h, \qquad \frac{\mathscr{E}_{1z}}{E_0} \approx 4\frac{\omega}{\gamma}\frac{h}{\delta}, \tag{82}$$

$$\frac{q_{s\parallel}}{q_0} = \chi = 16\left(\frac{\omega\omega_p}{\gamma c_0}\right)^2 h^2, \qquad \frac{q_{s\perp}}{q_0} = \xi = 8\left(\frac{\omega}{c_0}h\right)^2 \frac{\omega_p}{\gamma}. \tag{83}$$

Here, $\mathscr{E}_x$ and $\mathscr{E}_{1z}$ are the parallel and perpendicular components of the SEW electrical vector on the division boundary ($z = 0$). $\mathscr{E}_{1z}$ characterizes the field at dielectrics with $\varepsilon = 1$. By Jha et al. (1980) the process of transformation of light in SEW is considered in a linear approximation according to the resonance grating height, neglecting the SEW scattering on the periodical relief. A more strict analysis, taking into consideration the SEW scattering processes in recent times, has been carried out by Gandelman and Kondratenko (1983) and Emelyanov et al. (1987b). From eqs. (82) and (83) it follows that at coherent 'pumping' of incident energy into SEW propagating along the resonance grating, the general field amplification may be very significant even for $h < \delta \sim 10^{-6}$ cm, where δ is the skin-layer depth. Thus, the magnitude $q_{s\parallel} = q_0$ at $h = \delta\gamma/4\omega_p$, which corresponds to $h \approx 0.1\delta$ for the radiation wavelength $\lambda = 1.06$ μm.

One of the most demonstrative effects of the powerful SEW influence on laser action consists in the appearance of the specific dimensional–orientational dependence of the surface-active medium absorption capability (AC), its absolute value simultaneously growing considerably (Bonch-Bruevich et al. 1984b, 1985). The cause of the effect is connected with dissipative SEW damping during propagation along the surface, as a result of which the radiation energy transferred into SEW converts into heat but not within the limits of the immediate radiation region and partially outside of it at the distance of the track length $L = c_0\omega_p^2/\gamma\omega^2$. To analyze the basic effect characteristics and its influence on the regularities of thermal action of the laser radiation let us use the following model. Let the rectangle region of dimension $2d \times s$ be radiated on the surface when $s \gg d$ (s is assumed to be infinite, see fig. 14a) and the direction of the projection of the electric vector of the light wave at its normal incidence on the surface $\boldsymbol{E}_t$ makes an angle ψ with the symmetry axis of the action zone. Let us also assume that within the radiation zone there appeared (or there was from the very beginning) a resonance grating of height h on which the coherent incident radiation with flux density q_0 converted into two SEW with wave vectors $\boldsymbol{K}_{s1}$ and $\boldsymbol{K}_{s2}$ propagating in opposite directions along the x-axis parallel to $\boldsymbol{E}_t$ (fig. 14a). The SEW excitations in other directions are neglected for the sake of simplicity.

The strict definition of the field strengths in SEW is very difficult in this case because the grating has finite dimensions ($2l = 2d/\sin \psi$) and it is necessary to

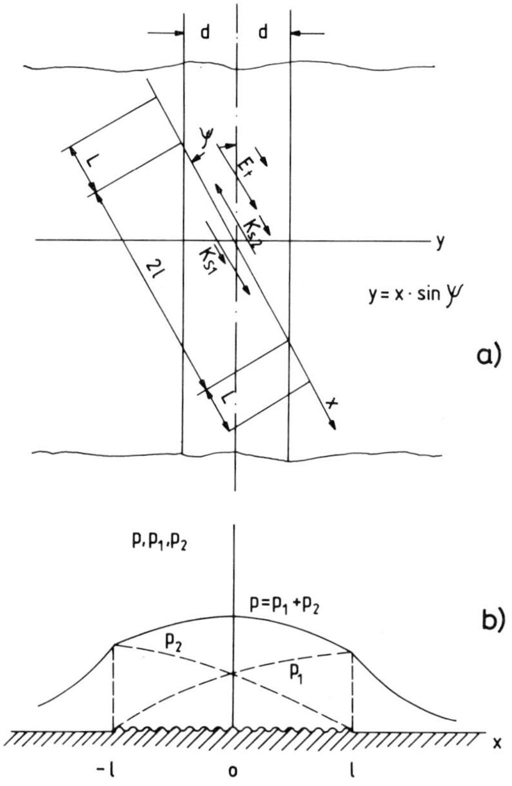

Fig. 14. Dimensional dependences of absorption capability of the medium in the process of intensive SEW generation on the surface: (a) auxiliary diagram illustrating the SEW generation at an angle ψ with the symmetry axis of the action zone; (b) spatial distribution of the resulting SEW power. The notation is given in text.

consider diffractional effects. At the same time, when describing the appearing specific effects in the simplest theoretical way, it is possible to make use of 'the flux' approximation, the essence of which resides in the following. If the change of both SEW powers (P_1 and P_2) takes place as a result of 'pumping' on the resonance grating and of damping, then the power balance may be described by

$$\frac{\mathrm{d}P_1}{\mathrm{d}x} = \xi q_0 - \frac{P_1}{L}, \qquad \frac{\mathrm{d}P_2}{\mathrm{d}x} = -\xi q_0 + \frac{P_2}{L}, \tag{84}$$

where ξ is the energetic conversion coefficient depending, strictly speaking, on the x position. The basic assumption is connected with the fact that the value of ξ is taken from formula (83)

$$\xi = 8\left(h\frac{\omega}{c_0}\right)^2 \frac{\omega_\mathrm{p}}{\gamma}.$$

Physically, this assumption means that on a resonance grating with a finite length $2l$ the coherent radiation is transformed into SEW in the same way as on an infinite grating. The maximal value of SEW power and, connected with its dissipation, the value of the flux $q_{\mathrm{s}\perp}$ being defined by the exact expression of eq. (83). When solving eq. (84) with boundary conditions: $P_1|_{x=-l} = 0$ and $P_2|_{x=l} = 0$ (see fig. 14b) it is possible to find the total SEW power $P = P_1 + P_2$ within the region of their light pumping, $-d \leqslant y \leqslant d$,

$$P = 2\xi q_0 L\left[1 - \exp\left(-\frac{d}{L\sin\psi}\right)\cosh\left(\frac{y}{L\sin\psi}\right)\right], \tag{85}$$

where y is an axis perpendicular to the symmetry axis of the action zone (fig. 14a). Outside this region both waves are damped on a characteristic length L along the direction $\pm x$, and then

$$P = 2\xi q_0 L \sinh\left(\frac{d}{L\sin\psi}\right) \times \begin{cases} \exp\left(-\dfrac{y}{L\sin\psi}\right), & y \geqslant d, \\ \exp\left(\dfrac{y}{L\sin\psi}\right), & y \leqslant -d. \end{cases} \tag{86}$$

From eqs. (85) and (86) it is easy to find an effective absorption capability (AC) of the surface-active medium A_eff, defined as $A_\mathrm{eff} = q_{\mathrm{s}\perp}/q_0 = P/q_0 L$,

$$A_\mathrm{eff} = \begin{cases} 2\xi \sinh\left(\dfrac{d}{L\sin\psi}\right)\exp\left(\dfrac{y}{L\sin\psi}\right), & y \leqslant -d, \\ A + 2\xi\left[1 - \exp\left(-\dfrac{d}{L\sin\psi}\right)\cosh\left(\dfrac{y}{L\sin\psi}\right)\right], & -d \leqslant y \leqslant d, \\ 2\xi \sinh\left(\dfrac{d}{L\sin\psi}\right)\exp\left(-\dfrac{y}{L\sin\psi}\right), & y \geqslant d. \end{cases} \tag{87}$$

Here A is the medium AC. It is evident that the parameter A_{eff} depends not only on the efficiency of the SEW excitation but also on the dimension of radiation zone d^*, on the SEW running length (L) and on the angle ψ, i.e., on the orientation of the vector $\boldsymbol{E}_t$ relative to the symmetry axis of this zone. It leads to interesting macroscopic peculiarities in the kinetics of general regularities of the heating (and the processes accompanying it) of the medium by the polarized coherent radiation.

Omitting intermediate computations we shall give only the final expression for nonstationary heating of a thermally thin plate (of thickness b) by the heat source [eqs. (85), (86)] under the condition that heat removal by heat conduction takes place only along the y direction. On the axis $y = 0$, where the heating is maximal, the temperature is equal to (Bonch-Bruevich et al. 1985)

$$T_0(t) = q_0 \frac{(A + 2\xi)t}{cb}\left[1 - 4\mathrm{i}^2\Phi^*\left(\frac{d}{2\sqrt{at}}\right)\right] + \frac{q_0 \xi L^2 \sin^2\psi}{Kb}$$
$$\times \left\{\exp\left(\frac{at}{L^2 \sin^2\psi}\right)\left[\exp\left(\frac{d}{L \sin\psi}\right)\Phi^*\left(\frac{\sqrt{at}}{L \sin\psi} + \frac{d}{2\sqrt{at}}\right)\right.\right.$$
$$\left. - \exp\left(-\frac{d}{L \sin\psi}\right)\Phi^*\left(\frac{\sqrt{at}}{L \sin\psi} - \frac{d}{2\sqrt{at}}\right)\right] - 2\Phi^*\left(\frac{d}{2\sqrt{at}}\right)$$
$$\left. + 2\exp\left(-\frac{d}{L \sin\psi}\right)\right\}. \tag{88}$$

It is evident that the heating rate of the material together with the energetic parameters of the acting radiation is also influenced by the efficiency of its transformation into SEW (ξ) and by the relation of the three spatial lengths d, $l = L \sin\psi$ and $\sqrt{at}$. The magnitude $l = L \sin\psi$ entering eq. (88) is specific for the situation analyzed and shows that the temperature depends on the length and the direction of the SEW path. Other conditions being equal, T_0 is maximal if $\psi = 0$ (i.e., the SEW is propagating along the irradiated region, see fig. 14a) and is minimal if $\psi = \frac{1}{2}\pi$. The efficiency of this influence is the greatest under short action times ($t \ll d^2/a$, l^2/a) and for a thermally thin plate ($t > b^2/a$),

$$T_0(t) = \frac{q_0 t}{cb}\left\{A + 2\xi\left[1 - \exp\left(-\frac{d}{L \sin\psi}\right)\right]\right\}. \tag{89}$$

Hence it follows that the intensive SEW generation under laser action leads to the appearance of a new dimensional dependence of the heating velocity (and, consequently, threshold intensities) which is saturated when the spot size is of the order l – the projection of the SEW path length L on the direction of the

*The dimensional AC dependence connected with the SEW excitation by light, was found both by Bonch-Bruevich et al. (1984b) and, within a different approximation, by Konov and Tokarev (1985).

temperature gradient. Depending on the wavelength of the light, the kind of material and also the angle ψ, the magnitude l can vary in a wide range and reach ~ 1 cm.

At $t \gg d^2/a$, l^2/a, the dimensional dependence of the thin-plate heating rate gradually smoothes. However, if for great action time the heat exchange is important then it remains unchanged. One can make sure that for the linear convective heat exchange of the surface (with coefficient β_c) and the above-mentioned action geometry, the stationary temperature on the $y=0$ axis is equal to

$$T_0 = \frac{q_0(A+2\xi)}{\beta_c}\left[1-\exp\left(-\sqrt{\frac{\beta_c}{Kh}}d\right)\right] + \frac{2q_0\xi L^2\sin^2\psi}{Kb-\beta_c L^2\sin^2\psi} \times\left[\exp\left(-\frac{d}{L\sin\psi}\right)-\exp\left(-\sqrt{\frac{\beta_c}{Kb}}d\right)\right]. \tag{90}$$

If the magnitude T_0, as function of the angle ψ, is maximal at $\psi=0$, then the heating of the points outside the radiation zone ($|y|>d$), with the other conditions being equal, is maximal at $\psi=\frac{1}{2}\pi$, when the SEW is propagating along the y-axis.

The above-mentioned peculiarities of laser heating should be taken into consideration when analyzing different physico-chemical processes on a surface activated by thermal light action: emission, sorption–desorption, evaporation, thermochemical reactions and some others. These characteristics are also of practical interest. Thus, using polarized radiation it is possible to control the action characteristics changing, e.g., the angle between the polarization direction and the laser beam transport direction, according to Bagenov et al. (1986). Conform the developed notions, it influences the heat rate and the dimension of the resulting zone of thermal action of laser radiation.

If SEW are excited under the influence of intensive visible light then, as a result of the amplification and the redistribution of the field near the medium surface, it is possible to wait for the efficiency augmentation of different photo-processes: resonant photoexcitation and photodesactivation of atoms in the gas layer adjoining to the surface-active medium boundary; photochemical reactions; photoemission and some others. As an illustration let us examine the influence of SEW excitation on the photochemical reaction velocity of, e.g., photodissociation of chemically active gas accompanied by the condensation of one of the reaction products on the surface. Such a reaction obtained at the resonant absorption of visible and ultraviolet laser radiation (see, e.g., Brueck and Ehrlich 1982, Ehrlich et al. 1980) may be used for metal deposition from the gas phase.

For evaluation of the photochemical deposition velocity V let us assume that the absorption cross-section (σ) of photons and plasmon-polaritons in gas are equal and V is determined by the diffusional flux of the dissociated gas molecules on the surface,

$$V \sim j, \qquad j = D\frac{\partial n}{\partial z}\bigg|_{z=0} = \tfrac{1}{4} n|_{z=0}\bar{u}.$$

Here n is the concentration of the photo-chemical reaction product, D is the coefficient of its diffusion in gas and $\bar{u}$ is average thermal velocity of particles in it. The distribution $n(z)$, where z is the coordinate measured from the surface along the outer normal to it, is, in the simplest case, described by the one-dimensional stationary equation of diffusion,

$$D\frac{d^2 n}{dz^2} - \frac{n}{\tau_0} = -\sigma n_0 I(z),$$

where τ_0 is the effective lifetime of photoexcited and photodissociated molecules, n_0 is the concentration of the initial matter in the gas phase ($n \ll n_0$) and $I(z)$ is the distribution of the light intensity.

The difference in the magnitude and in the distribution $I(z)$ in the absence or presence of SEW determines the difference in average reaction rate. If on the medium surface the radiation does not transfer into SEW and is strongly reflected, in the first approximation for the normal incident light,

$$I(z) = I_1(z) \approx q_0(1 - \cos(4\pi z/\lambda)), \qquad I_1(0) = I_{1_{\min}} = 0. \tag{91}$$

If the surface polaritons are effectively excited on the boundary of the surface-active medium and their field greatly exceeds the light field on the surface, then

$$I(z) = I_2 \approx q_0 \chi \exp(-2z/l_1), \qquad I_2(0) = I_{2_{\max}} = q_0\chi, \tag{92}$$

where χ is determined by eq. (83) and l_1 is a characteristic length of the polariton field localization in the gas phase. In the approximation considered $l_1 = \delta(\omega_p/\omega)^2$.

From the solution of both diffusional problems, which are not given here, it follows that the ratio of the deposition process velocities in the situations analyzed is equal to

$$\frac{V(I_2)}{V(I_1)} = \chi\frac{1 + \lambda^2/4\pi^2 D\tau_0}{1 + 2\sqrt{D\tau_0}/l_1},$$

and in the typical case of $\lambda \lessapprox \sqrt{D\tau_0} < l_1$ is of the same order as χ, reaching 5–50. Such a strong effect is completely connected with the field amplification at the expense of the normal component of the electric vector of SEW in a chemically active gas ($\mathscr{E}_z$) and the flux transmitted along the surface ($q_{s\parallel}$).

Thus, the SEW generation effect for intensive laser action on surface-active media may show up in different physical processes. It adds to the physical notions about the interaction of laser radiation with matter a great class of phenomena in which polarization and light coherence play the leading role. In a purely optical sense, the transformation of the radiation into SEW on the rough boundary of the two-medium division may be as important an phenomenon as the reflection and refraction of light. In conclusion, let us pay special attention to the situation when as a result of the interaction of the particles in a gas near the surface or immediately on it there occurs a stimulated emission (or an absorption) of surface polaritons (and also of localized plasmons) which may be the initiating reaction of a series of catalytic reactions on the surface of a metallic catalyst (Bonch-Bruevich et al. 1984b, 1985).

Acknowledgments

The authors wish to express their thanks to V.S. Makin and S.D. Pudkov for fruitful discussion and T.V. Gagarina for her great help in the preparation of this manuscript.

Note added in proof

Since the article was completed some new results, published in works of Bonch-Bruevich and Libenson (1988) and Bonch-Bruevich et al. (1989), have been obtained by the authors. The authors consider it necessary to outline two earlier works in which similar ideas on the nature of SPS formation were suggested (Guosheng Zhow et al. 1982, Young et al. 1983).

References

Achmanov, S.A., V.I. Emelyanov, N.I. Koroteev and V.N. Seminogov, 1985, Usp. Fiz. Nauk **147**, 675–745.

Agranovich, V.M., 1975, Usp. Fiz. Nauk **115**(2), 199–237.

Aksenov, V.P., 1982, Preprint FIAN USSR (194), 17 pages.

Aksenov, V.P., and B.G. Gurkin, 1982, Dokl. Akad. Nauk SSSR **256**(6), 1365–1366.

Anisimov, V.N., V.I. Baranov, L.A. Bolshov, A.M. Dychne, D.D. Malyuta, V.D. Pismennyi, A.Yu. Sebrant and M.A. Stepanova, 1983, Povekchn. (7), 138–144.

Bagenov, V.V., A.M. Bonch-Bruevich, M.N. Libenson, V.S. Makin, S.D. Pudkov and V.V. Trubaev, 1983, Pis'ma Zh. Tekhn. Fiz. **9**(15), 932–937.

Bagenov, V.V., A.M. Bonch-Bruevich, M.N. Libenson and V.S. Makin, 1984, Pis'ma Zh. Tekhn. Fiz. **10**(24), 1520–1526.

Bagenov, V.V., V.S. Makin, G.M. Rubanova and V.V. Trubaev, 1985, Izv. Akad. Nauk SSSR Ser. Phys. **49**(6), 1229–1232.

Bagenov, V.V., A.M. Bonch-Bruevich, R.V. Kanapenas, M.N. Libenson, V.S. Makin, R.I. Petrushkevichus and I.I. Rexnis, 1986, Pis'ma Zh. Tekhn. Fiz. **12**(3) 151–156.

Bahin, G., T. Bernstein and R. Kalish, 1979, Appl. Phys. Lett. **34**(8), 486–488.

Birnbaum, M., 1965, J. Appl. Phys. **36**(11) 3688–3689.

Bonch-Bruevich, A.M., and M.N. Libenson, 1988, Optiko-Mechan. Promisl. (12), 35–47.

Bonch-Bruevich, A.M., M.K. Kochengina, M.N. Libenson, V.S. Makin, S.D. Pudkov and V.V. Trubaev, 1982, Izv. Akad. Nauk USSR, Ser. Phys. **46**(6), 1186–1183.
Bonch-Bruevich, A.M., M.N. Libenson and V.S. Makin, 1984a, Pis'ma Zh. Tekhn. Fiz. **10**(1), 3–8.
Bonch-Bruevich, A.M., M.N. Libenson and V.S. Makin, 1984b, The Materials of the 6th Conf. Non-resonant Int. Opt. Rad. with Matter (Vilnius) p. 4–5.
Bonch-Bruevich, A.M., M.N. Libenson and V.S. Makin, 1985, Opt. Spektrosk. **59**(6), 1350–1354.
Bonch-Bruevich, A.M., M.N. Libenson, V.S. Makin and A.G. Rumjanzev, 1989, Izv. Akad. Nauk USSR, Phys. **53**(4), 769–776.
Brueck, S.R.J., and D.J. Ehrlich, 1982, Phys. Rev. Lett. **48**(24), 1678–1681.
Cosobukin, V.A., 1983, Poverkhn. (12) 3–18.
Ehrlich, D.J., R.M. Osgood and T.F. Deutsch, 1980, IEEE J. Quantum Electron. **QE-16**(11), 1233–1243.
Emelyanov, V.I., and V.N. Seminogov, 1984a, Kvantovaya Elektron. **11**(5), 871–873.
Emelyanov, V.I., and V.N. Seminogov, 1984b, Zh. Eksp. & Teor. Fiz. **86**(3) 1026–1036.
Emelyanov, V.I., V.I. Konov, V.N. Seminogov and V.N. Tokarev, 1987a, Preprint Inst. Ob. Fiz. Akad. Nauk USSR (89), Moscow, 37pp.
Emelyanov, V.I., V.N. Seminogov and V.I. Sokolov, 1987b, Kvantovaya Elektron. **14**(1), 33–46.
Emmony, D.C., R.P. Howson and L.J. Willis, 1973, Appl. Phys. Lett. **23**(11), 598–600.
Emmony, D.C., N.J. Phyllips, J.H. Toger and L.J. Willis, 1975, J. Phys. D **8**(4), 1472–1475.
Fauchet, P.M., and A.E. Siegman, 1983, Appl. Phys. A **32**(3), 135–140.
Gandelman, G.M., and P.S. Kondratenko, 1983, Pis'ma Zh. Eksp. & Teor. Fiz. **38**(5), 246–248.
Gandelman, G.M., and P.S. Kondratenko, 1985, Zh. Eksp. & Teor. Fiz. **88**, 1470–1475.
Gromov, G.G., and V.B. Uphimzev, 1983, Pis'ma Zh. Tekhn. Fiz. **9**(10), 580–582.
Isenor, N.R., 1977, Appl. Phys. Lett. **31**(3), 148–150.
Jha, S.S., J.R. Kirtley and J.C. Tsang, 1980, Phys. Rev. B **22**(8), 3973–3982.
Keilmann, F., 1984, Phys. Rev. Lett. **51**(23), 2097–2100.
Keilmann, F., and Y.H. Bai, 1982, Appl. Phys. A **29**(I), 9–18.
Kochengina, M.K., M.N. Libenson, V.S. Makin and S.D. Pudkov, 1981, The Materials of 5th Conference on Nonresonant Interaction of Optical Radiation with Matter, Leningrad, pp. 381–382.
Konov, V.I., and V.N. Tokarev, 1985, Preprint Inst. Ob. Fiz. Akad. Nauk USSR (70), Moscow, 43pp.
Konov, V.I., A.M. Prochorov, V.A. Sychugov, V.N. Tokarev and N.I. Chapliev, 1983, Preprint Fiz. Inst. Akad. Nauk USSR (122), Moscow, 34pp.
Kovalev, A.S., A.M. Popov and O.B. Pjatigorskaja, 1983, Zh. Tekhn. Fiz. **53**(5), 939–941.
Leamy, H.J., C.A. Rozgonyi, T.T. Sheng and G.K. Celler, 1978, Appl. Phys. Lett. **32**(10), 535–537.
Libenson, M.N., and A.G. Rumjanzev, 1986, Opt. Spektrosk. **60**(4), 675–677.
Libenson, M.N., and G.D. Shandybina, 1984, Pis'ma Zh. Tekhn. Fiz. **10**(20), 1238–1243.
Prochorov, A.M., V.A. Sychugov, A.V. Tischenko and A.A. Chakimov, 1982a, Pis'ma Zh. Tekhn. Fiz. **8**(16), 961–966.
Prochorov, A.M., V.A. Sychugov, A.V. Tischenko and A.A. Chakimov, 1982b, Pis'ma Zh. Tekhn. Fiz. **8**(23), 1409–1413.
Siegrist, M., G. Kaech and F.K. Kneubühl, 1973, Appl. Phys. **2**(1), 45–48.
The Surface Polaritons, 1975, Nauka, Moscow, 525pp.
Tribelsky, M.I., and S.M. Gol'berg, 1982, Pis'ma Zh. Tekhn. Fiz. **8**(20), 1227–1230.
Vaganov, R.B., and B.Z. Kazenelenbaum, 1982, in: Foundations of Diffraction Theory (Nauka, Moscow) 272pp.

AUTHOR INDEX

SUBJECT INDEX

CUMULATIVE INDEX, VOLUMES 1–29

Monographs

Contributed Volumes